Springer Series in
Surface Sciences

10

Editor: Robert Gomer

Springer Series in Surface Sciences

Editors: G. Ertl and R. Gomer   Managing Editor: H. K. V. Lotsch

R. Vanselow   R. F. Howe (Eds.)

Chemistry and Physics of Solid Surfaces VII

With 315 Figures

Springer-Verlag Berlin Heidelberg New York
London Paris Tokyo

Professor Ralf Vanselow

Department of Chemistry and Laboratory for Surface Studies,
The University of Wisconsin-Milwaukee,
Milwaukee, WI 53201, USA

Dr. Russell Howe

Department of Chemistry, University of Auckland,
Private Bag. Auckland, New Zealand

Series Editors

Professor Dr. Gerhard Ertl

Fritz-Haber-Institut der Max-Planck-Gesellschaft, Faradayweg 4–6
D-1000 Berlin 33

Professor Robert Gomer

The James Franck Institute, The University of Chicago, 5640 Ellis Avenue,
Chicago, IL 60637, USA

Managing Editor

Dr. Helmut K. V. Lotsch

Springer-Verlag, Tiergartenstrasse 17,
D-6900 Heidelberg, Fed. Rep. of Germany

ISBN-13: 978-3-642-73904-0   e-ISBN-13: 978-3-642-73902-6
DOI:10.1007/ 978-3-642-73902-6

Softcover reprint of the hardcover 1st edition 1988

Offsetprinting: Druckhaus Beltz, 6944 Hemsbach/Bergstr.

2154/3150-543210   Printed on acid free paper

Preface

This volume contains review articles written by the invited speakers at the eighth International Summer Institute in Surface Science (ISISS 1987), held at the University of Wisconsin-Milwaukee in August of 1987.

During the course of ISISS, invited speakers, all internationally recognized experts in the various fields of surface science, present tutorial review lectures. In addition, these experts are asked to write review articles on their lecture topic. Former ISISS speakers serve as advisors concerning the selection of speakers and lecture topics. Emphasis is given to those areas which have not been covered in depth by recent Summer Institutes, as well as to areas which have recently gained in significance and in which important progress has been made.

Because of space limitations, no individual volume of Chemistry and Physics of Solid Surfaces can possibly cover the whole area of modern surface science, or even give a complete survey of recent progress in the field. However, an attempt is made to present a balanced overview in the series as a whole. With its comprehensive literature references and extensive subject indices, this series has become a valuable resource for experts and students alike. The collected articles, which stress particularly the gas-solid interface, have been published under the following titles:

Surface Science: Recent Progress and Perspectives, Crit. Rev. Solid State Sci. **4**, 125-559 (1974)
Chemistry and Physics of Solid Surfaces, Vols. I, II, and III (CRC Press Boca Raton, FL 1976, 1979, and 1982); Vols. IV, V, and VI: Springer Ser. Chem. Phys., Vols. 20 and 35, and Springer Ser. Surf. Sci., Vol. 5 (Springer, Berlin, Heidelberg 1982, 1984, and 1986)

The present volume begins with the themes of adsorption and surface reaction. The first group of authors is led by *Ehrlich* who extensively reviews activated chemisorption, dealing with the mechanisms involved, with corresponding sticking coefficients and experimental methods. *Kern* and *Comsa*, authors of the second chapter, focus on physisorption. They describe the structure and dynamics of noble gas layers (Xe, Kr, Ar/Pt{111}) as primarily investigated by high resolution thermal helium scattering. Adsorption on semiconductors, investigated by infrared spectroscopy, is covered by *Chabal*, while *Maradudin* deals with the more fundamental aspects of surface phonons on clean as well as adsorbate covered surfaces.

A group of three reviews is based upon near edge X-ray absorption fine structure (NEXAFS): *Horsley* theoretically investigates spectra of adsorbates using multiple scattering calculations; *Outka* and *Stöhr* report on measurements of adsorbate molecules, starting with simple molecules and finishing with polymers; *Gland* describes how surface kinetics can be monitored with NEXAFS, reporting, for example, the interesting finding that chemisorbed CO is displaced from Ni{100} by more weakly adsorbed hydrogen at $\simeq 10^{-3}$ Torr.

The current status of strong metal-support interaction, a widely studied phenomenon in the area of heterogeneous catalysis, is reviewed by *Baker*, with particular emphasis on the use of high-resolution TEM. *Kreuzer* presents a systematic approach to adsorption-desorption kinetics of two-phase adsorbates based on the Onsager approach to non-equilibrium thermodynamics. He also includes the Becker-Döring theory of nucleation and droplet growth. *Pfeifer* applies the concept of fractals to various surface effects, and theoretical aspects of critical phenomena at surfaces are reviewed by *Einstein*.

A feature of every Summer Institute is the presentation of a historical review lecture by a prominent member of the surface science community. In 1987, the review lecture was presented by *Hagstrum*. His chapter is devoted to surface electron interactions underlying ion neutralization and metastable deexcitation spectroscopies (INS, MDS).

Another area on which a number of reviewers focus their attention involves phenomena from the field of surface crystallography: *Wortis* thoroughly reviews the equilibrium shape of crystals. His treatise serves as a stepping stone for the following authors. *Engel* deals with a phenomenon which was first proposed by Burton, Cabrera, and Frank in 1951 - "surface roughening", while *van der Veen* et al. review an effect the latter should not be confused with - "surface melting". *Bonzel* and *Dückers* investigate the anisotropy of the specific surface free energy using surface self-diffusion and surface core-level shifts and relate it to surface reconstruction. Clean and CO covered Pt{110} serve as examples. Some of the phenomena reviewed within this group of articles are highlighted in an amazingly ideal system - the helium liquid-solid interface. *Maris* describes how this system provides unique opportunities to study these phenomena, but he also points out some properties which are a result of the highly quantum nature of helium. *Van Hove* shows how the familiar technique of LEED can be used to solve complex and disordered surface structures. The emphasis is on theoretical approaches to the calculation of the multiple scattering of electrons through these structures.

In volume VI of this series *Behm* and *Hösler* presented a very thorough review of a new and powerful technique for surface analysis - scanning tunneling microscopy. Because of the exciting progress in this area, we asked *Tromp* to provide us with a short update.

The final article by *Kasper* and *Jorke* deals with a very applied subject, the growth kinetics of Si molecular beam epitaxy. It emphasizes the important role of equilibrium adatom properties and of monatomic steps on the growth kinetics, and shows that adsorption, desorption and segregation govern the incorporation of dopant atoms from neutral beams.

As in previous volumes, a thorough subject index is provided, together with extensive lists of references. We would like to thank the sponsors of ISISS: the Office of Naval Research (Grant No. N00014-87-G0195) as well as the Graduate School, the College of Letters and Science and the Laboratory for Surface Studies at the University of Wisconsin-Milwaukee. Their support made both the conference and the publication of this volume possible. The cooperation of the authors and the publisher is gratefully acknowledged.

Milwaukee, Auckland
January 1988

Ralf Vanselow
Russell Howe

Contents

1. Activated Chemisorption

Gert Ehrlich

Coordinated Science Laboratory and
Department of Materials Science and Engineering
University of Illinois at Urbana – Champaign, Urbana, IL 61801, USA

Activated chemisorption, that is adsorption in which the gas must pass over an activation barrier in order to chemisorb to the surface, is a subject dating back to the dark ages of surface studies. It is also of great current interest, and the rush of recent papers is most impressive. Nevertheless, activated chemisorption is generally considered an isolated curiosity, rather than the widespread phenomenon it actually is.

Before plunging into the maelstrom of current activities it is therefore useful to examine the early history of activated chemisorption, which establishes the scope of the important phenomena. Based on this survey we offer in Sect.1.2 a tentative classification of the different types of systems in which activation is likely to be necessary for chemisorption to occur. Of course, on many clean metal surfaces chemisorption is rapid and does not require activation. Expectations about chemisorption in general have been largely built upon the substantial body of information concerned with fast chemisorption. In Sect.1.3 some of the important formal distinctions between the kinetics of fast and activated chemisorption are therefore outlined, in order to isolate the unusual features expected if chemisorption requires activation. Only thereafter do we examine detailed studies in a few of the important systems for which chemisorption is thought to be activated. The primary aim of these case studies is to evaluate the experimental evidence indicating that chemisorption is activated, but we will also seek to define more clearly the important kinetic events contributing to activation and to illustrate the many different categories of activated chemisorption. At the end of this survey it should be more evident that activated chemisorption is a quite pervasive and important process ripe for greater exploration.

1.1 A Brief History

Langmuir [1.17], in his monumental 1916 paper presenting his vision of chemical processes in general, also offered his remarkably modern concept of how to understand the reactivity of surfaces. According to him, the work necessary to create a surface goes into breaking chemical bonds in the solid; the surface thus formed can be viewed as a giant array of free radicals, with the high chemical reactivity expected of such species. Dissociative chemisorption on a surface should therefore occur with a high probability. However, these expectations were not always borne out by early experimental studies of chemisorption, often done on powders. A good (if rather late) example of careful work of this type are the studies on tungsten by *Frankenburg* and *Hodler* [1.2]. They observed that on raising the temperature of their specially prepared tungsten powder, above 90°C for hydrogen and above 20°C for nitrogen, the amount adsorbed actually increased.

Earlier findings of this type led *Taylor* [1.3] to enunciate the idea that chemisorption is activated. A quotation from his 1931 paper gives a clear account of his position: "...the adsorption process per se is not necessarily a rapid process, may indeed be a very slow process, too slow to be measurable, and...for each adsorption process there is a characteristic velocity which is determined by the same factors which determine the velocity of chemical reactions.

It seems necessary,...to abandon entirely the assumption implicit in older theories that the processes of adsorption are rapid. The assumption that all processes of adsorption possess their own characteristic activation energies permits a single general treatment of adsorption."

Taylor's ideas on chemisorption had an immediate impact. They led *Lennard-Jones* [1.4] to formulate a quantum mechanical description for such phenomena. According to *Lennard-Jones*, a molecule A_2 is attracted to the surface by dispersion forces, described by the potential curve $2M+A_2$ in Fig.1.1. Atoms of the dissociated molecular gas can interact much more strongly with the surface, forming a chemisorption bond, as suggested by the curve $2M+2A$, with a potential minimum in close proximity to the surface. The intersection of the atomic and molecular potential energy curves sets the value of the activation energy. This important concept, which is still the start for most discussions of activated chemisorption even today, was presented at the General Discussion on the Adsorption of Gases by Solids, organized by the Faraday Society in 1932.

It was not a time of triumph for *Taylor*, however. His ideas came under attack by *Allmand* and *Chaplin* [1.5], who suggested that the experimental evidence for activated adsorption was misleading in that "...the process requiring activation may be the displacement of residual gas from the adsorbing surface by the adsorbed molecules...", a stand we now know is often correct. In his General Introduction to the Discussion, *Taylor* [1.6] felt compelled to note about his theory that "This extraordinarily simple generalization of actual experimental observations seems to have aroused a quite unreasonable amount of opposition." Despite his vigorous defense, a seed of doubt had been planted.

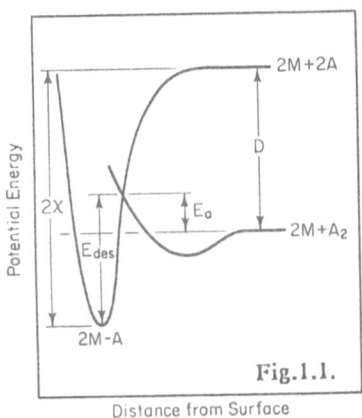

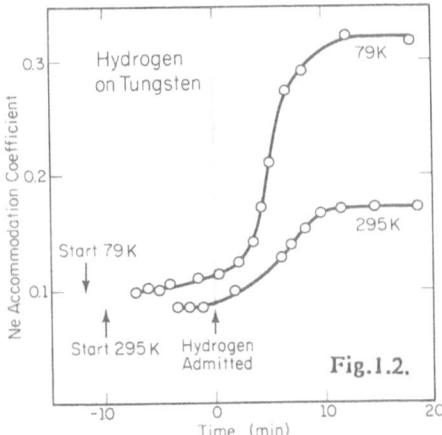

Fig.1.1. Schematic potential diagram for activated chemisorption, after *Lennard-Jones* [1.4]. $2M+A_2$ designates the potential between the diatomic molecule A_2 and a metal surface arising from van der Waals interactions; $2M+2A$ gives the potential from chemical interactions between atom A and the metal. (χ: energy for atom desorption from the surface; E_a: activation energy for dissociative chemisorption of molecule from the gas phase; D: dissociation energy for molecule in the gas phase)

Fig.1.2. Measurements of hydrogen chemisorption on tungsten filament by *J.K. Roberts* [1.7]. Change in neon accommodation coefficient, which is proportional to hydrogen coverage, indicates that chemisorption is rapid even at 79 K

Not many years later, in 1935, *Roberts* [1.7] came out with his elegant experiments, in which he used measurements of the thermal accommodation coefficient of helium on a freshly cleaned tungsten wire to study the chemisorption of hydrogen. As indicated in Fig.1.2, these experiments revealed that hydrogen chemisorbed with a high probability on colliding with the surface; even at 79 K, adsorption was complete on a time scale of minutes, not hours, as expected from prior work on powders. That chemisorption of hydrogen was not activated was also established by extensive studies carried out in *Beeck's* group in the Emeryville laboratories of the Shell Development Corporation [1.8]. Even though much of this work was not published until after the war [1.9], it was already clear in the late thirties that on clean metal surfaces chemisorption of many simple gases occurred rapidly, without requiring activation.

The end of World War II brought with it the development of ultrahigh vacuum technology and equipment, which made it possible to routinely characterize chemisorption rates on well-defined specimens. By resorting to field emission experiments, *Gomer* [1.10-12] was able to demonstrate that chemisorption of hydrogen on clean tungsten and nickel, and of oxygen on tungsten, took place even with the surface cooled below liquid nitrogen temperatures. At the Bell Telephone [1.13] and the GE Research Laboratories [1.14-16], experiments on rates of nitrogen and carbon monoxide chemisorption on tungsten were done by thermal desorption methods, devised earlier by *Apker* [1.17]. These experiments revealed that the fraction of the impinging molecules chemisorbing (the sticking coefficient) *decreased* with increasing temperature, rather than increasing as expected if chemisorption is opposed by an activation barrier. This early work was followed at an accelerating rate by studies which demonstrated unequivocally that for a wide variety of gases, chemisorption on metals and also on semiconductors was rapid, and this has become the accepted norm of behavior [1.18].

However, concomitant with the rise of surface studies by modern techniques there appeared a series of important publications, drawing on the methodology of the thirties, which convincingly demonstrated the importance of activated chemisorption. First of these was the work of *Kemball* [1.19], who examined chemisorption of methane on evaporated films of nickel. The rate of chemisorption increased rapidly with temperature, as in Fig.1.3, and he deduced an activation energy of 11 kcal/mol for this process. Some years later, in 1957, *Eley* and *Rossington* [1.20] used the ortho-to-para hydrogen conversion to study chemisorption of hydrogen on copper films; they deduced an activation energy of 5 kcal/mol for chemisorption, and arrived at the potential diagram in Fig.1.4. At much the same time *Tamaru* [1.21], while at Princeton, examined the dissociative chemisorption of molecular hydrogen on germanium films produced by the decomposition of germane. His results, summarized in the diagram in Fig.1.5, pointed to

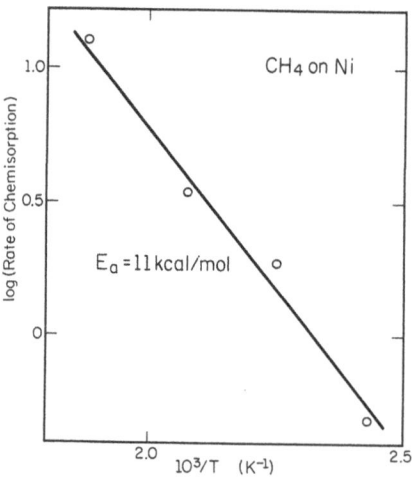

Fig.1.3. Temperature dependence of methane chemisorption rate on evaporated nickel films, after *Kemball* [1.19]. As the temperature is raised, the rate increases appreciably

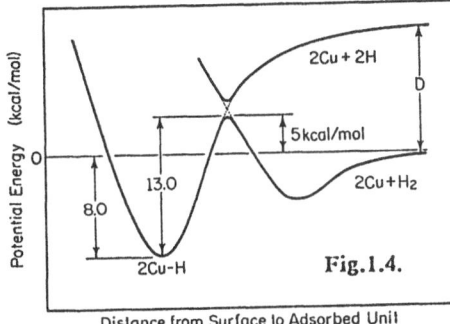
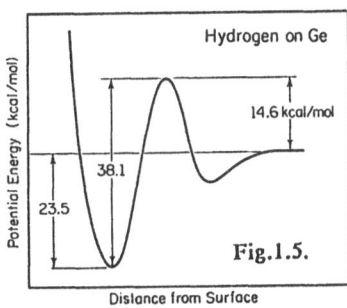

Fig.1.4. Potential diagram for hydrogen chemisorption deduced by *Eley* and *Rossington* [1.20] from ortho-to-para hydrogen conversion studies

Fig.1.5. Potential diagram for hydrogen chemisorption on germanium films, after *Tamaru* [1.21]

an activation energy of 14.6 kcal/mol for the adsorption process. These early experiments were subsequently confirmed and extended in various laboratories throughout the world, and now constitute classical examples of activated chemisorption, which stand in contrast to the rapid chemisorption found in so many other instances.

Apart from confirmatory activities following the trends set in these three seminal studies, there now intervened a hiatus in new research on activated chemisorption. This lasted for a decade and a half, a time of considerable activity devoted to examinations of various rapid chemisorption processes. However, one of the very early studies of the crystallographic dependence of such a fast process, the chemisorption of nitrogen on tungsten [1.22], led to a surprising observation: on the {110} plane, the most closely packed of the body-centered cubic lattice, no chemisorption was observed at room temperature upon moderate exposures to nitrogen; on other, rougher crystal planes of tungsten, such as {100} and {111}, much lower exposures gave full coverage. Subsequent work in other laboratories [1.23,24] demonstrated that nitrogen was in fact able to adsorb on the {110} plane, forming a state with the same desorption energy as on rougher surfaces, but at a rate orders of magnitude slower. The idea thus arose of a kinetic inhibition to chemisorption on low index planes of metals otherwise known for their ability to chemisorb rapidly.

In experiments on perfect W{110} planes formed by field evaporation at low temperatures, *Polizzotti* [1.25,26] was able to demonstrate that hydrogen does not chemisorb at temperatures in the vicinity of 40 K. Instead, as shown in Fig.1.6, a weakly held, presumably molecular state formed, which desorbed on warming to 70 K, leaving the surface essentially bare. Chemisorption occurred at temperatures of

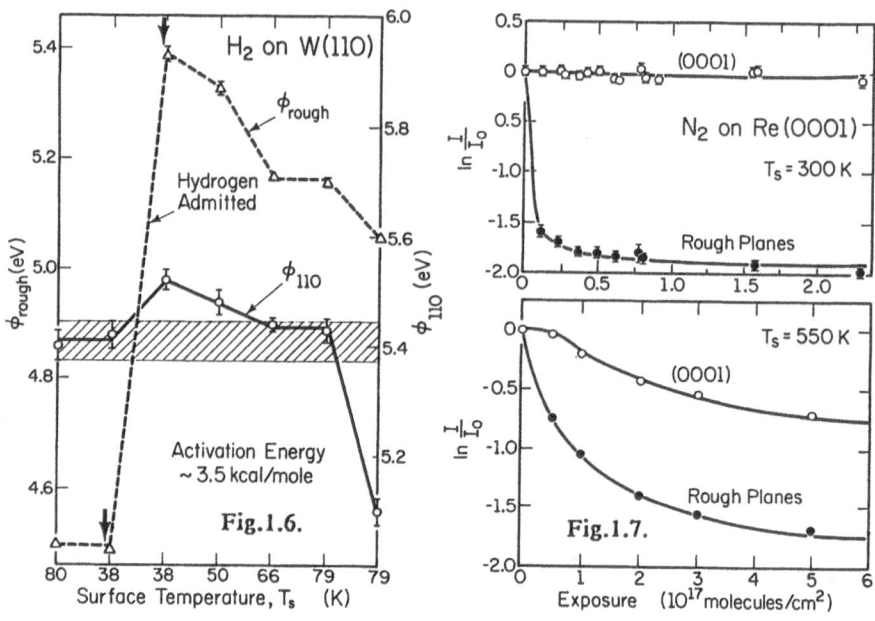

Fig.1.6. Hydrogen adsorption on W{110}, compared to adsorption on rough surfaces [1.25,26]. Change in work function ϕ due to hydrogen adsorbed at low temperatures is opposite in sign to that for chemisorbed hydrogen

Fig.1.7. Chemisorption of nitrogen on Re{0001}, compared to adsorption on rough surfaces [1.27,28], as indicated by field emission current I. At room temperature, Re{0001} remains clean on exposure to N_2, but on warming to $\simeq 500$ K it fills quickly

$\simeq 80$ K and above, but apparently by diffusion of hydrogen from the rougher areas surrounding the {110} plane. To account for these effects, chemisorption on the perfect {110} plane of tungsten had to be activated, involving a barrier >3.5 kcal/mol. Of course on the rougher planes of tungsten, chemisorption of hydrogen occurred rapidly, as had been found in many previous experiments. Later studies by *Liu* [1.27], relying on the same techniques and summarized in Fig.1.7, showed that on the close packed {0001} plane of rhenium, chemisorption of nitrogen required heating the surface to $T \simeq 500$ K. *Liu* [1.28,29] also examined chemisorption on the {110} plane of tungsten kept at different temperatures, and these experiments revealed an activation barrier to nitrogen chemisorption on this surface as well.

In the meantime there had been some interesting advances in understanding the classical examples of activated chemisorption. *Stewart* [1.30], using molecular beam techniques, showed that in the chemisorption of methane on rhodium, heating the gas alone while keeping the surface cold accelerated the rate of chemisorption. His measurements, in Fig.1.8, demonstrated that excitation of the gas is sufficient to overcome the activation barrier, and this immediately

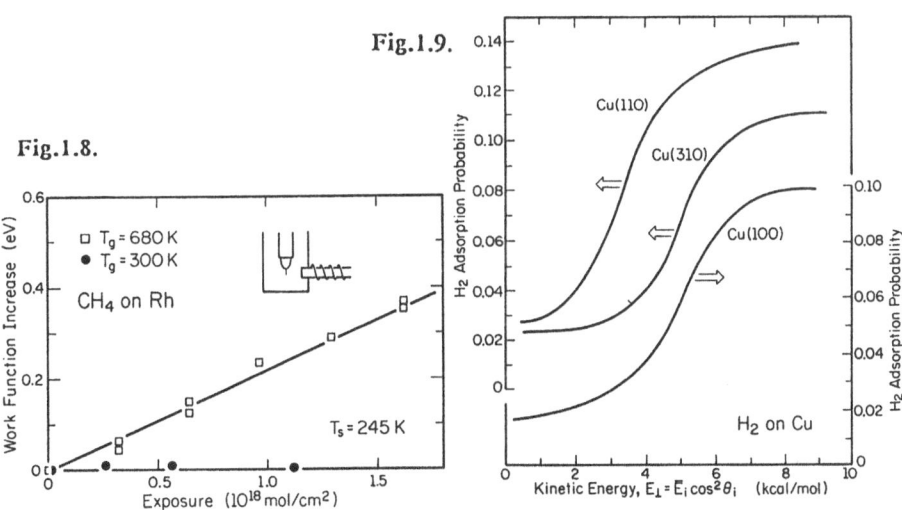

Fig.1.8.

Fig.1.9.

Fig.1.8. Effect of gas temperature T_g upon rate of methane chemisorption on rhodium [1.30]

Fig.1.9. Probability of hydrogen chemisorption on single crystal faces of copper, for different translational energies normal to the surface [1.31]. θ_i = angle of incidence

raises the question: What type of molecular excitation is most effective in promoting chemisorption? Trendsetting work dealing with this matter was done in *Stickney's* laboratory at MIT. In molecular beam studies of hydrogen chemisorption on different crystal planes of copper, *Balooch* et al. [1.31] demonstrated a significant difference in the rates of adsorption on the {110}, {310}, and {100} planes, apparent in Fig.1.9. Most important, however, was the finding that putting energy into the translational motion of the molecules brings about chemisorption. This opened up a field of investigation now intensively pursued in many laboratories.

A general rebirth of interest in activated chemisorption appears to have occurred in the late seventies. At this stage we note only a few of the important new theses that developed. One aspect of chemisorption known since the thirties is the limited range of elements which strongly chemisorb nitrogen. For example, chemisorption of nitrogen from the molecular gas has not been found on nickel nor on the metals of the platinum family [1.32], all of which are highly active in the chemisorption of other gases. This unusual selectivity was generally ascribed to thermodynamic effects: only very strong interactions between nitrogen atoms and a surface can compete effectively against the high dissociation energy of nitrogen molecules in the gas phase. However, in 1976 *Wilf* and *Dawson* [1.33] demonstrated that chemisorption of nitrogen did occur on platinum at room temperature, forming a strongly bound layer. Kinetic rather than thermodynamic factors therefore had to be responsible for the absence of nitrogen chemisorption, and later adsorption studies by *Grunze* et al. [1.34] extended this conclusion to nickel.

Support for this view was provided by the wide-ranging work with NO on various platinum family members, much of it in *Ertl's* laboratory in Munich, which established that nitrogen layers could be formed on these elements from molecules less strongly bonded than nitrogen [1.35]. Nitrogen desorption from such layers occurs only at elevated temperatures, again suggesting that slow chemisorption from the molecular gas rather than thermodynamics was accountable for the previous failure to form atomically-bound nitrogen layers. Work in Munich, shown in Fig.1.10, also cleared up another very important adsorption process involving nitrogen. *Bozso* et al. [1.36,37] were able to quantitatively study the interactions of nitrogen with the {111}, {100}, and {110} planes of iron and demonstrated that chemisorption was activated. In the early thirties, *Emmett* and *Brunauer* [1.38] had already surmised that nitrogen chemisorption was the slow step in ammonia synthesis over iron. The modern work verified this nicely.

More recently there have also been several very interesting studies of the activated chemisorption of hydrogen. In 1979, following upon earlier and sometimes conflicting studies of hydrogen chemisorption on platinum {111}, *Salmerón* et al. [1.39] in *Somorjai's* laboratory demonstrated that this chemisorption was in fact activated, a conclusion since verified in other laboratories. Here we have another striking example of activated chemisorption on a low index plane of a metal for which the atomically rougher surfaces are known to be highly active. Just as fascinating are the studies of the chemisorption of hydrogen on nickel{111} in *Ertl's* laboratory [1.40], and at Graz by *Rendulic* and *Winkler* [1.41]. For years this system had been examined intensively, but it is only recently that these two groups were able to show that hydrogen on Ni{111} constitutes another example of chemisorption requiring activation on a close packed plane.

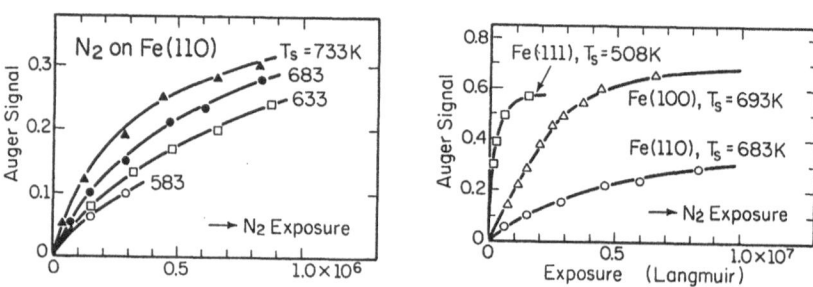

Fig.1.10. Amount of nitrogen chemisorbed on single crystal planes of iron [1.36,37] at different temperatures T_s of the surface

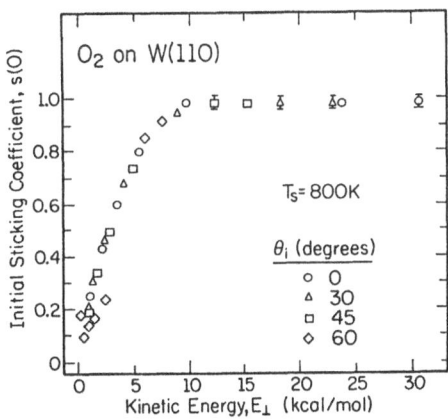

Fig.1.11. Initial sticking coefficient of oxygen on W{110} at 800 K, for different translational energies perpendicular to the surface [1.42]. (θ_i: angle of incidence)

Surface structure is one of the important factors affecting activated chemisorption to emerge from this survey, and in the next section we will attempt to outline others. As more work is being reported, however, some unexpected and even confusing results are also appearing, especially from work with supersonic molecular beams. Consider for example the chemisorption of oxygen on tungsten {110}, which has been known to occur with a very high sticking coefficient. In recent studies shown in Fig.1.11, *Rettner* et al. [1.42] have reported that chemisorption could be accelerated significantly by imparting translational energy to oxygen molecules striking a W{110} plane. Obviously there still are surprises in store for us.

1.2 Classification of Activated Chemisorption

The historical survey makes it clear that in spite of occasionally contradictory results, there are many combinations of gases and solid surfaces for which chemisorption is known to occur only slowly and over an activation barrier. For what sort of system can we expect chemisorption to be activated?

In dissociative chemisorption, bonding between the atoms in a stable molecule is disrupted and new chemical combinations are formed with the surface. For this reorganization of bonds to occur easily, interactions between the valence electrons in the molecule and in the solid are necessary. To make predictions of the circumstances under which this condition will *not* be met obviously requires a manageable and general theory of chemisorption, something not currently available. However, from the examples cited above, it is possible to discern some headings under which to systematize the discussion of activated chemisorption.

1.2.1 Molecular Structure

Methane is an excellent example of the role of molecular structure in imposing an activation energy to chemisorption. In the course of chemisorption, a C-H bond in the molecule is broken, and the H atom as well as the C-H fragment are attached to the surface. However, C-H bonds in methane are well shielded against interactions with the surface - a CH_4 molecule looks much like an inert sphere, the outer shell of which must be perturbed to make it accessible to chemical interactions. Other species in which the bond to be broken is similarly buried on the inside of the molecule should behave in chemisorption much as does CH_4.

1.2.2 Electronic Structure of the Gas

This is a catch-all designed to account for the lack of reactivity of molecular nitrogen with surfaces. Although this low reactivity is not really understood, it may be that nitrogen, with its π-bonding, is able to form on the surface stable molecular entities that compete successfully with even more strongly bound dissociated states.

1.2.3 Electronic Structure of the Solid

At semiconductor surfaces bonding is known to be more localized than on metals [1.43]. Inasmuch as interatomic spacings in simple molecules are considerably shorter than in solids (0.746 Å for H_2, 2.352 Å for Si), the preconditions for easy bond breaking, namely strong overlap between the wave functions of electrons in the solid and in the molecule, will not easily be satisfied on semiconductors or for that matter on insulators. Activated chemisorption will therefore be the rule.

1.2.4 Atomic Arrangement of the Solid

Slow chemisorption can be expected on the densely packed planes of otherwise active metals. For bcc solids, that means chemisorption is more likely to be activated on {110} planes; for fcc elements the {111} planes and for hexagonal crystals the {0001} planes may be unreactive. It is of course the distribution of valence electrons upon which bond breaking at the surface depends. The arrangement of atom cores affects the former, and it is convenient to correlate lack of reactivity with a high geometrical coordination of surface atoms.

1.2.5 Strength of Chemisorption Bonding

Activated chemisorption should be the rule when the energy change in chemisorption is only weakly negative, as for example in the interactions of molecular hydrogen with copper. Consider atom recombination

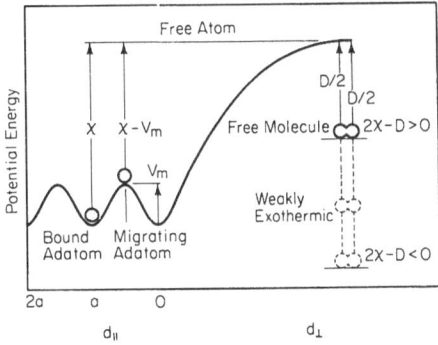

Fig.1.12. Schematic potential diagram for atom recombination at a surface [1.44]. (V_m: barrier to atom diffusion over the surface; a: crystal spacing; $d_{\parallel}$: distance on the surface; $d_{\perp}$: distance normal to surface; 2χ-D: heat of adsorption from molecular gas; D: dissociation energy of molecule)

[1.44], which is the inverse of dissociative adsorption. Even if chemisorption is endothermic, that is if the adsorbed state is higher in energy than the molecular state in the gas phase, the recombination of adatoms to form a molecule will still be activated: adatoms have to diffuse over the surface to encounter each other. The activation energy for recombination will therefore have to be at least as large as the barrier to diffusion of an adatom. The activated state in this example will consist of one adatom in its ground state, with another at the saddle position for surface diffusion, as suggested in Fig.1.12. If chemisorption is only weakly exothermic, this configuration may have an energy higher than the molecular gas, so that chemisorption would require an activation energy in order to occur.

1.2.6 Surprises

To this category we assign newly discovered examples of activated chemisorption, such as oxygen on W{110} [1.42], which are not yet understood and do not obviously fit into the previous categories.

Although these rough classifications may help to bring some order into the phenomenology of activated chemisorption, only the detailed examination of individual systems can reveal how activated chemisorption actually takes place. Before embarking upon such an examination, we will outline the formal kinetics of chemisorption occurring over an activation barrier in order to make clear how such systems are recognized and studied.

1.3 Formal Kinetics

The hallmark of activated chemisorption is of course its temperature dependence. Just like in any other activated process, the rate of activated chemisorption increases as the temperature is raised. In dealing with surface phenomena, it is a simple matter to control the temperature of the solid and the gas independently. This feature introduces special elements into the kinetics and we will consider how this affects the rate of chemisorption, both activated and otherwise.

The presence of an activation barrier separating the gas and the chemisorbed layer has a bearing not only on the temperature dependence of the rate - it affects other aspects of the kinetics as well. For example, the change in the rate upon increasing the surface concentration is rather different when chemisorption is fast than when it is activated. The effects of surface concentration are important in interpreting rate phenomena, and we will consider this topic first of all. The dependence of the rate of adsorption on the angle at which gas collides with the surface is significantly affected by an activation barrier to chemisorption; the same is true for the angular dependence of the rate of desorption, which has recently been emphasized as a probe for adsorption barriers. Both concentration and angular dependencies will be briefly examined to see how activated chemisorption differs from the more usual fast adsorption processes, and to describe approaches useful in exploring activated chemisorption.

1.3.1 Concentration Dependence of Chemisorption Rates

We will consider both activated and unactivitated chemisorption processes in which a molecular precursor bound on the surface can participate in the kinetics [1.45,46]. The potential diagrams appropriate to these reactions are shown in Fig.1.13. Molecules striking the surface from the gas phase can chemisorb directly, with a probability given by

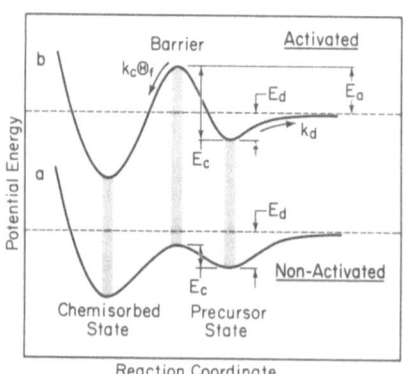

Fig.1.13. Schematic potential diagram for (a) nonactivated and (b) activated chemisorption [1.45]. E_d = barrier to precursor desorption; E_c = barrier to precursor conversion; $k_c \Theta_f$ = rate constant for precursor conversion

$s_a \Theta_f$, where s_a is the sticking coefficient on an empty site and Θ_f is the fraction of the surface sites free to chemisorb. Of the molecules that do not chemisorb, a fraction s_m may condense into the precursor state to form a reservoir, at a concentration of n molecules per unit area, from which subsequent conversion into the chemisorbed state may occur at the rate $nk_c \Theta_f$. If $p(2\pi mkT)^{-1/2}$, the impingement rate on unit area, is written as Imp, then in the usual formulation the total rate of chemisorption is given by

$$\text{Imp} s(\Theta_f) = \text{Imp} s_a \Theta_f + nk_c \Theta_f \ , \tag{1.1}$$

where $s(\Theta_f)$ is the overall sticking coefficient for chemisorption.

Under normal conditions the precursor concentration n will be so small that it can be assumed in a steay state, dictated by competition between condensation from the gas, conversion from the precursor into the chemisorbed state, and loss of precursor by evaporation back into the gas phase at the rate nk_d. The precursor concentration therefore becomes

$$n = \frac{s_m \text{Imp}(1 - s_a \Theta_f)}{k_c \Theta_f + k_d} \ , \tag{1.2}$$

and the overall sticking coefficient $s(\Theta_f)$ can be written as

$$s(\Theta_f) = s_a \Theta_f \left[1 - \frac{s_m}{1 + k_d/k_c \Theta_f} \right] + \frac{s_m}{1 + k_d/k_c \Theta_f} \ . \tag{1.3}$$

It is useful to write the rate constants k_d and k_c as the usual product of a frequency factor and a Boltzmann term in the barrier height, so that

$$k_d = \nu_d \exp(-E_d/kT_s) \ , \quad k_c = \nu_c \exp(-E_c/kT_s) \ , \tag{1.4}$$

where T_s is the temperature of the surface and k is just Boltzmann's constant. The sticking coefficient $s(\Theta_f)$ therefore appears as

$$s(\Theta_f) = s_a \Theta_f \left[1 - s_m / (1 + \frac{\nu_d}{\nu_c \Theta_f} \exp \frac{E_c - E_d}{kT_s}) \right]$$

$$+ s_m / (1 + \frac{\nu_d}{\nu_c \Theta_f} \exp \frac{E_c - E_d}{kT_s}) \ . \tag{1.5}$$

If chemisorption is not activated, the potential diagram in Fig. 1.13a is appropriate. The barrier to conversion of the precursor E_c will be smaller than that to desorption E_d, and the second term in the brackets, which gives the dependence upon the concentration of adsorbed gas, will therefore make only a small contribution to the rate as long as Θ_f is large. The behavior of the kinetics is more immediately obvious provided direct adsorption of the gas, given by the product $s_a \Theta_f$, is negligible and chemisorption occurs entirely from the precursor. The sticking coefficient can then be written in the more familiar form

$$\frac{s(\Theta_f)}{s_m} = 1/(1 + \frac{\nu_d}{\nu_c \Theta_f} \exp \frac{E_c - E_d}{kT_s}) . \tag{1.6}$$

During the initial stages of adsorption, that is if Θ_f is not too far from unity, adsorption will proceed under the condition

$$1 \gg \frac{\nu_d}{\nu_c \Theta_f} \exp \left[\frac{E_c - E_d}{kT_s} \right] ,$$

and the sticking coefficient will be independent of the amount adsorbed, that is, independent of Θ_f. As the amount adsorbed increases and Θ_f becomes small, the magnitude of the second term at the right of (1.6) increases and may become comparable to unity. At that point the sticking coefficient will begin to diminish with increasing concentration.

The initial lack of a concentration dependence of the sticking coefficient is a characteristic long associated with fast chemisorption processes. The physics responsible for this absence is simple. Provided conversion is rapid compared to evaporation of precursor, the concentration n of precursor is set by the competition between the supply from the gas phase and the rate of conversion. If the rate constant for conversion decreases as the number of sites available diminishes, the precursor concentration increases, so that the overall conversion rate, given by the product $nk_c \Theta_f$, stays exactly constant. This compensation can of course be maintained only as long as conversion, that is $k_c \Theta_f$, is much faster than the rate constant for evaporation k_d.

These considerations are not applicable to activated chemisorption described by the potential diagram in Fig. 1.13b, as for such a process the activation energy E_c for conversion of the precursor exceeds the activation energy E_d for desorption of precursor. For activated chemisorption we will have the condition

$$\frac{\nu_d}{\nu_c \Theta_f} \exp \left[\frac{E_c - E_d}{kT_s} \right] > 1 ,$$

and the overall sticking coefficient can be approximated by

$$s(\Theta_f) = s_a \Theta_f + s_m^0 \Theta_f \exp \left[- \frac{E_c - E_d}{kT_s} \right] \tag{1.7}$$

$$s_m^0 \equiv \frac{s_m \nu_c}{\nu_d} . \tag{1.8}$$

Even if direct chemisorption from the gas does not contribute significantly and precursor processes dominate, the rate of activated chemisorption will depend sensitively upon coverage. The rate of conversion of precursor into the chemisorbed state is now always small compared to evaporation and the precursor concentration is dictated by competition between condensation from the gas and reevaporation. If the rate constant for conversion $k_c \Theta_f$ drops, this does not bring about an increase in precursor concentration, and the rate of chemisorption therefore falls.

Sticking coefficient values which fall with increasing surface coverage starting from the initially clean surface therefore suggest either rapid chemisorption without intervention of a precursor, or else slow, activated adsorption. A distinction between the two should be possible based on the magnitude of the sticking coefficient. A low sticking coefficient strongly dependent upon coverage should be a good indicator that chemisorption is activated. A survey of some measured sticking coefficients [1.45], given in Fig.1.14, indicates that real systems do indeed behave in accord with the expectations outlined here. It should be noted parenthetically that if the sticking coefficient is small and chemisorption is activated, then the dependence upon coverage does not provide us with any information about the possible contribution of precursors to the chemisorption process.

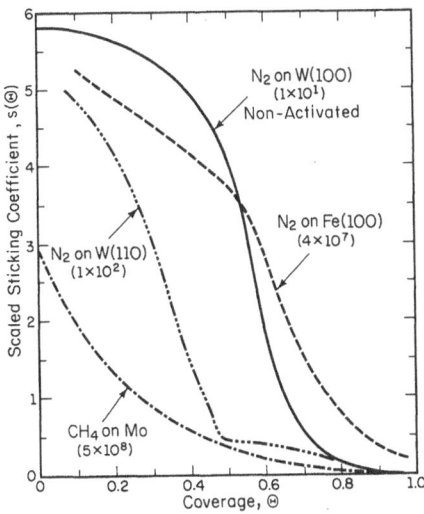

Fig.1.14. Concentration dependence for activated and nonactivated chemisorption [1.45]. Surface cover at $\Theta=1$: N_2 on W(100) at 300 K [1.18], 8×10^{14} atoms/cm²; N_2 on Fe(100) at 508 K [1.36], 1.22×10^{15} atoms/cm²; N_2 on W(110) at 300 K [1.24], 1.8×10^{14} atoms/cm²; CH_4 on Mo at 323 K [1.47], $2.1 \cdot 10^{14}$ mol/cm². Figure in parentheses gives factor by which plotted values are to be multiplied to yield the sticking coefficient

1.3.2 Temperature Dependencies

The effect of temperature upon the rate of chemisorption is apparent from (1.7). For fast chemisorption, in which the rate initially does not depend sensitively upon coverage, the first term, accounting for direct chemisorption from the gas phase, should not make a significant contribution. The sticking coefficient dominated by precursor processes can be represented [1.18] as

$$\frac{s(\Theta_f)}{s_m - s(\Theta_f)} = \frac{v_c \Theta_f}{v_d} \exp\left[-\frac{E_c - E_d}{kT_s}\right], \tag{1.9}$$

with the barrier to desorption E_d larger than E_c, the barrier to conversion of precursor. The ratio $s(\Theta_f)/[s_m - s(\Theta_f)]$ will therefore decrease with temperature. In the examination of limited experiments, the value of s_m, the condensation efficiency into the precursor, is not generally available, yet (1.9) depends sensitively upon this quantity. It is therefore more usual to examine the temperature dependence of the sticking coefficient itself, which is more complicated, and is given by

$$\frac{s(\Theta_f)}{s_m} = \left[1 + \frac{v_d}{v_c \Theta_f} \exp\left(\frac{E_c - E_d}{kT_s}\right)\right]^{-1}. \tag{1.10}$$

At low temperatures, where the second term in the denominator is negligibly small, that is where evaporation of precursor is not important, temperature has little effect upon the rate of chemisorption. At high temperatures also, where $kT_s > E_c - E_d$, that is in the temperature regime where essentially all precursor molecules have enough energy to go over either of the two barriers, the sticking coefficient will again be insensitive to temperature. Although it is true that for fast chemisorption from the precursor the sticking coefficient is expected to fall with rising surface temperatures, this fall constitutes a reliable test for unactivated chemisorption via a precursor only if measurements span a sufficiently large temperature range.

When chemisorption is activated, the sticking coefficient given by (1.7) behaves quite differently [1.48]. Provided the barrier to conversion of precursor is significantly higher than that to desorption, the rate expression becomes

$$\frac{s(\Theta_f)}{\Theta_f} = s_a^0 \exp\left[-\frac{E_a}{kT_g}\right] + s_m^0 \exp\left[-\frac{E_c - E_d}{kT_s}\right]. \tag{1.11}$$

Here the contribution from direct chemisorption out of the gas phase at temperature T_g is given by the first term on the right, which previously was just represented by $s_a \Theta_f$. To understand the dependence of the rate upon temperature, the activation energies E_a and $E_c - E_d$ must

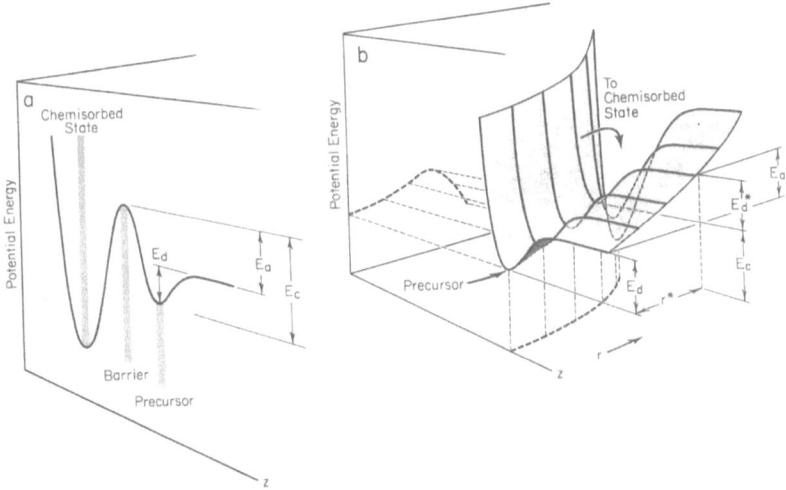

Fig.1.15. Schematic potential diagrams for activated chemisorption with participation of precursor [1.48]. z = distance from surface. (a) Chemisorption by translational activation; (b) Chemisorption by excitation of internal molecular coordinate r. Starred quantities refer to precursor with r at the critical value for going over the barrier

be related to each other. From the standard one-dimensional potential diagram in Fig.1.15a, it appears that

$$E_a = E_c - E_d . \qquad (1.12)$$

However, this relation will hold true only under special circumstances, namely when the kinetic energy gained by an incoming molecule as it is accelerated by the precursor potential can be effectively utilized in going over the barrier to chemisorption E_c. If translational energy of the incoming molecules is indeed used in overcoming the barrier to chemisorption, then the equality between activation energies in (1.12) should hold, and the temperature dependence of chemisorption both from the gas and out of the precursor should be the same.

Detailed trajectories in activated chemisorption proceeding through a precursor have recently been considered by *Gadzuk* and *Holloway* [1.49,50]. In the model examined here, the details of atomic encounters are ignored; molecules are just classified according to whether their energy distribution is best characterized by the temperature of the gas T_g or the temperature of the surface T_s. Even in this approximate scheme it is clear, from considering Fig.1.15b, that when internal degrees of molecular motion play an important part in chemisorption, differences can be expected in the activation energy for chemisorption from the gas and from the precursor. For precursor molecules converting into the chemisorbed state, the net activation energy is still given by $E_c - E_d$; it is as usual dictated by competition between conversion and evaporation of precursor molecules. However, in chemisorption

17

directly from the gas phase, the energy of translational motion may not be readily converted into the degree of freedom most effective in going over the barrier.

In the limit of no conversion of the translational energy gained by the gas phase molecules in rolling down the precursor potential into the appropriate internal degree of freedom, the activation energy from the gas phase becomes

$$E_a \simeq E_c \; . \tag{1.13}$$

That is, the activation energy for chemisorption from the gas phase obtained by looking at the dependence of the sticking coefficient upon T_g may be significantly larger than the activation energy for chemisorption out of the precursor, exceeding the latter by as much as E_d, the desorption energy of the precursor. Under these circumstances, measurements with gas and surface temperature varied in unison will be dominated by precursor processes and will yield an activation energy equal to $E_c - E_d$. Experiments in which only the gas temperature is changed while the surface is maintained cold will give the higher activation energy E_c. This difference can therefore serve as a useful indicator for the participation of internal molecular motions in overcoming the barrier to chemisorption.

The various comparisons have assumed throughout that precursors are important in activated chemisorption. If this assumption is not applicable, then changing the temperature of the surface may have only a minor effect, possibly helping to facilitate the direct interaction with the gas. The temperature of the gas striking the surface will of course be all important in dictating the rate of chemisorption.

1.3.3 Angular and Mass Dependence

From the preceding paragraphs it appears that separate measurements of the dependence of the sticking coefficient upon the temperature of the surface and of the gas are desirable in characterizing activated chemisorption. These measurements are most readily accomplished by molecular beam methods. Such methods also open up a new way of characterizing activated chemisorption, by examining the angular dependence of the sticking coefficient s or the rate of adsorption upon the angle of incidence of the impinging molecules.

For a Maxwellian gas of number density n_g, the number of molecules striking unit area of surface per unit time at an angle θ, with speed between v and v+dv and with directions in the element of solid angle $d\Omega$, is given by the Knudsen formula [1.51]:

$$d\dot{N} = n_g \left[\frac{m}{2\pi kT} \right]^{3/2} \exp\left[-\frac{mv^2}{2kT} \right] v^3 \cos\theta \; dv \; d\Omega \; . \tag{1.14}$$

18

The rate of activated chemisorption from such a beam when a strictly one-dimensional barrier of magnitude E_a has to be overcome was first examined by *van Willigen* [1.52]. Provided the potential energy depends only upon the distance from the surface, and only the translational energy of the incoming molecules matters, then

$$s = 0 \quad \text{if} \quad \tfrac{1}{2}m \ (v\cos\theta)^2 < E_a \ , \tag{1.15}$$

$$s = 1 \quad \text{if} \quad \tfrac{1}{2}m \ (v\cos\theta)^2 \geq E_a \ . \tag{1.16}$$

Equation (1.14) is now readily integrated over all values of v for which the sticking coefficient s is unity, to give s_θ, the sticking coefficient as a function of the angle θ:

$$\frac{s_\theta}{s_0} = \frac{E_a + kT \cos^2\theta}{(E_a + kT)\cos\theta} \exp\left(-\frac{E_a}{kT}\tan^2\theta\right) . \tag{1.17}$$

As illustrated in Fig.1.16, the spherical polar plot which describes the incidence from the gas, and also the rate if chemisorption occurs on collision with unit efficiency, is replaced by a much more highly peaked distribution if there is a one-dimensional activation barrier opposing chemisorption and this barrier can be overcome by molecules with a higher translational energy. Observations of the dependence of the rate upon the angle of incidence can therefore be used to examine activated chemisorption. More than that, if the principle of detailed balance can safely be applied, then desorption from a system of the type considered here will have exactly the same angular dependence. Such measurements may be more readily implemented than adsorption studies, but should provide the same information.

Studies of the angular dependence of either adsorption or desorption are at best tedious and difficult, and in his original work *van Willigen* actually examined the distribution of gas (hydrogen) after permeation through metal membranes. This of course has the advantage of affording a continuously replenished source of gas on which measurements can be made reasonably rapidly. However, it has been empha-

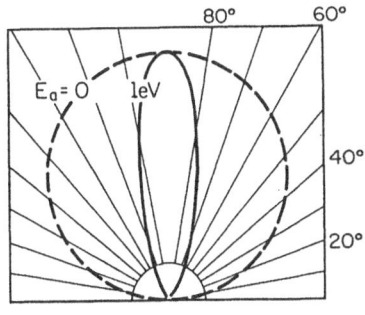

Fig.1.16. Angular distribution of molecular flux striking surface with translational energy perpendicular to the surface larger than the barrier height E_a [1.52]. Curves are arbitrarily normalized

sized by *Comsa* [1.51,53] that there is in principle no reason why evolution of gas subsequent to permeation must necessarily be equivalent to evolution after adsorption. Unless it can be demonstrated that the two processes are in fact equivalent, permeation experiments cannot be used to provide direct information about chemisorption phenomena. This proviso must be kept in mind in considering the more detailed descriptions of activated chemisorption which follow.

The conclusions about the angular dependence of adsorption or desorption outlined here are based upon important assumptions - the activation barrier is one-dimensional, and is overcome by the translational motion of the molecules. If internal molecular degrees of freedom predominate in promoting passage over the barrier, then the angular distribution may not differ at all from the $\cos\theta$ dependence characteristic of a Maxwellian gas striking a surface.

The role of translational motion in activated chemisorption can be explored by carrying out identical molecular beam experiments, but with isotopically substituted molecules. For a molecular beam source operating under Knudsen conditions, the fraction $f(\epsilon)d\epsilon$ of the molecules striking the surface per unit time with translational energy between ϵ and $\epsilon+d\epsilon$ is immediately found from (1.14) as

$$f(\epsilon) \, d\epsilon = \frac{\epsilon}{kT} \exp\left[-\frac{\epsilon}{kT}\right] \frac{d\epsilon}{kT} . \tag{1.18}$$

This distribution does not depend upon the mass of the incident molecules, and we therefore should not find any isotope effect if classical translation overcomes the activation barrier.

Knowledge of the extent to which translational or internal motions facilitate chemisorption provides us with at least qualitative information about the shape of the relevant potential diagram. In accord with *Polyani's* rules [1.54,55] we expect that if there is a barrier in the exit channel to dissociative chemisorption, as in Fig.1.17a, then vibrational

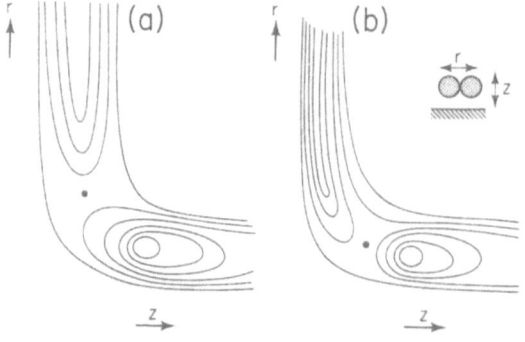

Fig.1.17. Schematic potential contours in activated chemisorption [1.40]. Position of saddle point marked by •. (a) Barrier in exit channel; (b) Barrier in entrance channel to chemisorption

20

motion will be required for chemisorption. On the contrary, if the barrier is in the entrance channel (Fig.1.17b), then the translational motion of the incoming molecule will be most helpful in expediting chemisorption. These various tests are put to good use in the survey of activated chemisorption presented in the following sections.

1.4 Case Studies in Activated Adsorption

The understanding of activated chemisorption processes is still fragmentary and quite uneven. In order to emphasize what is known, we will examine the state of the literature concerned with a few of the systems already cited in the introduction, largely in the sequence in which these systems were first studied. This examination will be concerned entirely with the kinetics of the chemisorption process. Where possible we will focus upon efforts to reveal the mechanism of activation, but a major aim will be to establish the experimental facts about activated chemisorption. In keeping with the definition adopted at the start, we will not consider molecules such as CO, CO_2, or NO, for which dissociation may be limited by an activation barrier separating differently bound chemisorbed entities.

1.4.1 Methane and Other Alkanes on Metal Surfaces

The interaction between simple hydrocarbons and metals is one of the most extensively studied examples of activated chemisorption. About a decade after *Kemball*'s [1.19] initial work, *Suhrmann* et al. [1.56] in 1965 examined photoelectron emission as well as resistivity of evaporated nickel films exposed to methane. These measurements are consistent with an activation energy of 7.1 kcal/mol, lower than that indicated by the original work of *Kemball*. A somewhat later study on single crystal planes of nickel [1.57] found rapid adsorption, but this appears to have been an artifact caused by fragmentation of the gas in the analytical instrumentation. Extensive investigations of methane chemisorption on evaporated metal films were carried out in *Frennet*'s laboratory in the late sixties [1.47,58]. These measurements, at pressures in the micron range, decisively demonstrated that chemisorption was activated on metals other than nickel, and established the general trends in reactivity.

The early history of adsorption studies with methane has been extensively reviewed [1.59,60]. It is worth noting, however, that the record of these investigations, even up to the present day, is confused by the occasional failure to explicitly recognize that chemisorption is activated and can occur only slowly compared to the rate at which diatomics adsorb on active transition metals, for example. The variety of results obtained is perhaps best illustrated by work on tungsten, all done using field emission of electrons to follow the interactions with methane. Field emission measurements on tungsten at room temperature, by *Yates* and *Madey* [1.61], did not reveal any chemisorption, presumably because observations were not carried to high enough exposures. In similar studies, also done with methane on tungsten, *Hellwig* [1.62] did observe rapid adsorption; his findings were, however, obscured by the possibility of background reactions and fragmentation processes. *Hopkins* and *Shah's* [1.63] field emission measurements on tungsten at 300 K gave indications of chemisorption, but only for exposures above 10^4 Langmuir. *Shigeishi* [1.64] carried out field emission studies of methane interacting with tungsten maintained at different temperatures. He not only observed chemisorption, but deduced a sticking coefficient of $\simeq 5 \times 10^{-4}$ at temperatures below 600 K; with rising surface temperatures the sticking coefficient increased.

These reports may be overtly confusing, but once artifacts are recognized and rejected the available work clearly indicates that chemisorption does occur at high exposures and is activated. Most important, these studies suggest the precautions required for a reliable examination of activated chemisorption. The surface under study must be exposed to high doses of methane, inasmuch as the sticking coefficient for activated chemisorption is likely to be much less than unity. Because of this low reactivity, more reactive impurities in the gas, as well as the possibility of side reactions induced by the measuring technique during the adsorption experiments, must be carefully eliminated in order to isolate the course of the activated chemisorption itself.

As was indicated earlier, the more recent efforts to explore the mechanism of methane chemisorption began in 1972 with the work of *Stewart* [1.30], which demonstrated that heating the gas alone was enough to bring about chemisorption on a cold rhodium surface. More quantitative studies [1.59,65] done soon thereafter indicated that increasing the temperature of the molecular beam impinging upon the rhodium field emitter used as a sample increased the rate of chemisorption, and yielded an activation energy of $\simeq 7$ kcal/mol. Measurements with CD_4 also were attempted; background reactions made reliable measurements impossible, but it appeared that the sticking coefficients for CD_4 were very much smaller than for CH_4. For a molecular beam source operating under Knudsen conditions, the fraction of the molecules striking the surface with translational energy in a specified range is given by (1.18) and does not depend upon the mass of the gas molecules. The large isotope effect apparent in these early experiments was interpretated as arising from a quantal effect. Indeed, as indicated by Fig.1.18, the experimental results for methane chemisorption on rhodium were in good agreement with the predictions of *Slater's* quantum mechanical rate theory [1.66]. In this theory, the limiting step in the dissociation of the molecule is assumed to the be extension of

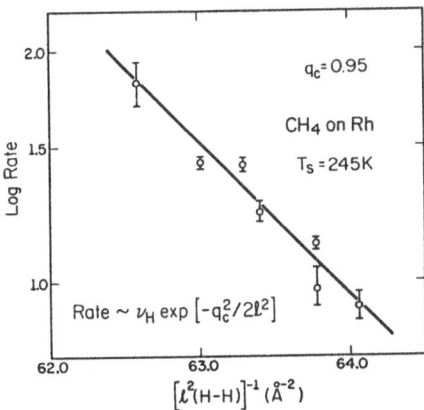

Fig.1.18. Chemisorption rate on rhodium for methane at different temperatures plotted in accord with *Slater*'s quantum mechanical rate theory [1.65,66]. (ℓ^2: mean square amplitude of H-H distance, q_c: critical value of breaking coordinate, ν_H: frequency factor)

some bond length beyond a critical value q_c, through superposition of normal vibrational amplitudes of the molecule.

Rather different experiments were done at the same time by *Winters* [1.67]. In these studies a tungsten wire was heated to various temperatures and the rate of dissociative chemisorption on the surface was monitored by following the pressure diminution in the system. Subsequent measurements [1.68] with methanes deuterated to different degrees revealed an activation energy of $\simeq 11$ kcal/mol at high surface temperatures, and a rather small dependence of the activation energy upon the mass of the molecules. *Winters* [1.68] accounted for this behavior by invoking a tunneling model; with four adjustable parameters he was able to adequately fit the measured temperature dependence. Although the experiments as well as the results of *Stewart* and of *Winters* were quite different, both studies were interpreted by invoking vibrational excitation of the molecules as a prerequisite to chemisorption, and both relied on quantum mechanical tunneling to account for the isotope effects.

These early measurements on rhodium, and the suggestion that vibrational excitation was involved in overcoming the barrier to chemisorption, had considerable appeal, and for good reason. The normal modes of CH_4, shown in Fig.1.19, all have energies below 9 kcal/mol, which is in the range of the activation energy reported for the chemisorption of methane. Furthermore, the asymmetric stretching mode of methane, ν_3, with a vibrational energy of 8.63 kcal/mol, can be readily excited with a helium-neon laser [1.70] operating in the infrared at 3.39 μ. Two independent experiments to test the efficacy of vibrational excitation were soon undertaken - one by *Yates* et al. [1.71], the other by *Brass* et al. [1.72]. In the former, adsorption was tested on a rhodium {111} plane by measuring the amount of hydrogen desorbed

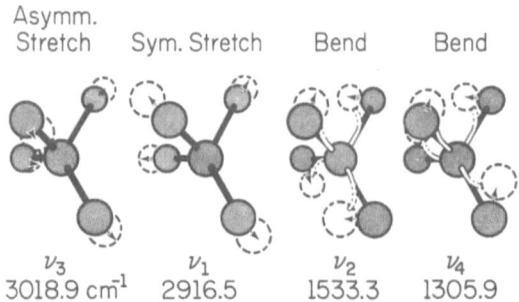

Asymm.
Stretch   Sym. Stretch   Bend   Bend

ν_3 ν_1 ν_2 ν_4
3018.9 cm^{-1} 2916.5   1533.3   1305.9

Fig.1.19. Normal modes and vibrational frequencies for CH_4 [1.69]

from the surface after exposure to methane with and without irradiation in the infrared. In the experiments by *Brass* et al. a rhodium film evaporated onto the walls of a cylindrical tube was the test surface; adsorption both with and without a laser beam down the tube axis was measured by recording the pressure decrease in the tube. Only the results of the latter experiments are shown in Fig.1.20, but the findings of both groups were essentially the same. There was no indication that exciting either the ν_3 asymmetric stretching mode or the $2\nu_4$ mode of CH_4 had any significant effect on its reactivity with rhodium. The mechanism by which methane chemisorbs was again an entirely open question.

Brass [1.73] at this stage undertook a series of adsorption measurements aimed at obtaining more quantitative data about the adsorption of methane on rhodium films. Again relying on standard pressure

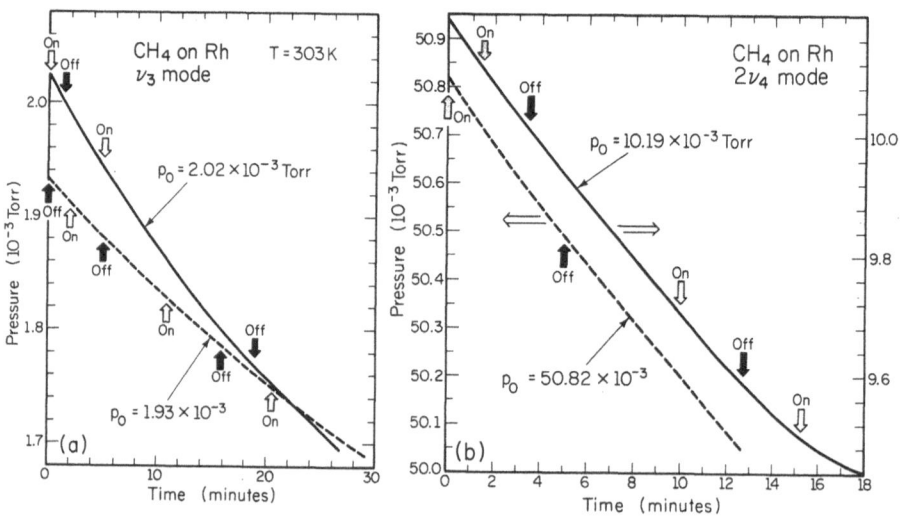

Fig.1.20. Effect of irradiation with He-Ne laser at 3.39 μ upon chemisorption of CH_4 on rhodium films [1.72]. (p_0: initial pressure). At low pressures (a) irradiation excites ν_3 mode of methane molecules colliding with the surface. At higher pressures (b) collision of an excited methane with other methane molecules yields methane in $2\nu_4$ state. Neither excitation appears to affect chemisorption

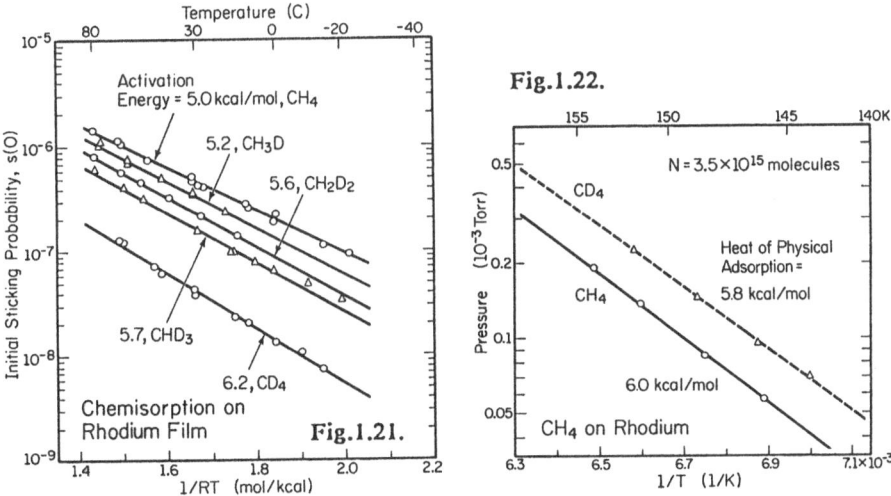

Fig.1.22.

Fig.1.21. Temperature dependence of chemisorption rate for differently deuterated methanes on rhodium films prepared as in Fig.1.20 [1.60,73]. Substrate and gas are always maintained at the same temperature

Fig.1.22. Temperature dependence of the pressure required to maintain a constant coverage of 3.5×10^{15} molecules of methane molecularly adsorbed on evaporated rhodium films [1.60,73]. Heat of adsorption derived from Clausius-Clapeyron relation

measurements with a high sensitivity Pirani gauge which did not significantly perturb the adsorbing system, *Brass* studied the rate of chemisorption of methane and of all the deuterated methanes, on rhodium films prepared much as in the previous infrared experiments. From measurements (shown in Fig.1.21) in which both the gas and film temperature were varied simultaneously over the range from 250 to 350 K, an activation energy of 5 kcal/mol was deduced for CH_4. Substitution of a deuterium for a hydrogen atom increased the barrier in roughly 0.3 kcal/mol increments, and for CD_4 the activation energy for chemisorption was found to be 6.2 kcal/mol. Molecular adsorption of both methane and deuteromethane was also explored (Fig.1.22), but at temperatures below 105 K, where chemisorption occurs too slowly to make a significant contribution. At low coverages, an adsorption energy of 6 kcal/mol was found for both isotopes.

At first sight these results appear to confirm the impression that vibrational excitation of methane molecules plays no important role in the chemisorption process. The vibrational energy imparted to methane in previous experiments with helium-neon lasers, 8.63 kcal/mol, is rather larger than the activation energy for chemisorption measured here. This impression is, however, incorrect. In Sect.1.3.2 it was noted that for experiments of the type done here, in which both the surface and the gas are kept at the same temperature, the activation energy deduced from the temperature dependence is given by $E_c - E_d$, the difference between the barrier to precursor conversion and to desorption. If vibrational excitation of the molecules were a requirement for chemisorption, then the energy that must be imparted to the incoming molecules should be comparable to $\simeq E_c$. From *Brass's* experiments, the barrier to conversion E_c is $\simeq 11$ kcal/mol, significantly larger than the energy supplied to CH_4 in the various unsuccessful attempts at laser excitation. The laser experiments therefore provide no information about the possible role of vibrational excitation in the chemisorption of methane.

From the considerations in Sect.1.3.2 it appears, however, that the contribution of vibrational excitation should be noticeable in measurements of the activation energy done under different conditions. If vibrational excitation is effective in accelerating the passage of molecules into the chemisorbed state, then the activation energy found in experiments in which only the gas is heated should amount to $\simeq E_c$, compared to a value of $E_c - E_d$ in experiments where both gas and surface are heated and precursor processes dominate. Measurements of chemisorption from hot methane gas on a cool rhodium film were done by *Brass* [1.60,74] in an apparatus (shown in Fig.1.23) specially devised to allow separate temperature control over gas and surface. The results in Fig.1.24 are quite striking: the activation energy for chemisorption obtained when only the gas is heated is 11 kcal/mol, 6 kcal/mol higher than in previous experiments with gas and surface at the same temperature. This difference is just equal to the molecular desorption energy E_d for methane. The conclusion is that vibrational excitation of the molecular gas plays a dominant role in overcoming the barrier to chemisorption.

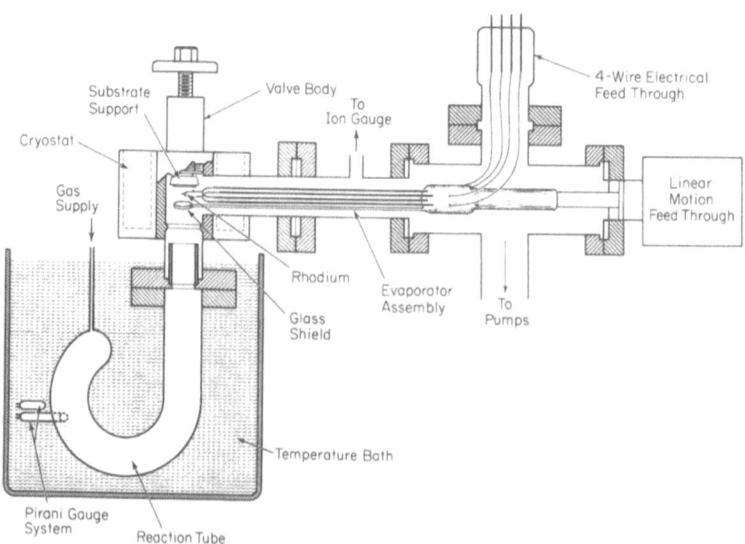

Fig.1.23. Schematic of reactor for chemisorption studies on evaporated film maintained at temperature different from impinging gas [1.74]

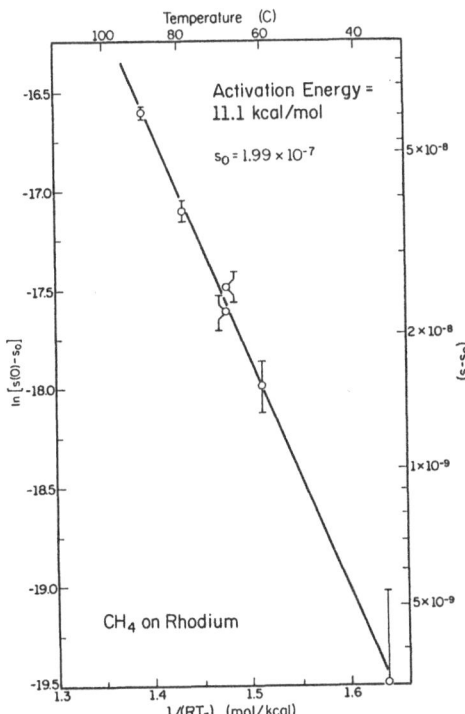

Temperature (C)

Activation Energy = 11.1 kcal/mol

$s_0 = 1.99 \times 10^{-7}$

CH$_4$ on Rhodium

Fig.1.24. Effect of gas temperature upon chemisorption rate of methane on rhodium film at 273 K [1.60,74]. ($s(0)$: initial sticking coefficient, s_0: contribution from gas accommodated to temperature of rhodium surface)

All these measurements were made on evaporated films of uncertain surface structure. In order to assess the extent to which surface orientation affects the chemisorption of methane, Lo [1.75] has carried out rate measurements on differently oriented planes of tungsten. A thermal beam of either CH$_4$ or CD$_4$ was allowed to impinge on a tungsten single crystal, and the change in surface coverage was determined by measuring changes in the Kelvin contact potential. Investigated were the {111}, {211}, {411}, {811}, and {100} planes. The activation energy, determined from the dependence of the rate of chemisorption upon the temperature of the molecular beam, was lowest on {111} at 7.2 kcal/mol, and highest on W{100} at 9.7 kcal/mol. As shown by the data for W{211} in Fig.1.25, the rate for CD$_4$ is always lower and the activation energy is higher than for CH$_4$. On almost all the planes this difference in activation energies amounts to ≃25%.

With the completion of these studies (but prior to their appearance in the standard literature) there intervened a brief lull, which has more recently been replaced by a period of great activity devoted to the examination of methane and other saturated hydrocarbons. Rettner et al. [1.76] have explored the rate of methane chemisorption on the W{110} plane, the densely packed and generally nonreactive face of tungsten. These studies were carried out in a modern molecular beam system, of the type illustrated in Fig.1.26, relying on a supersonic beam for information about the effect of translational energy upon the

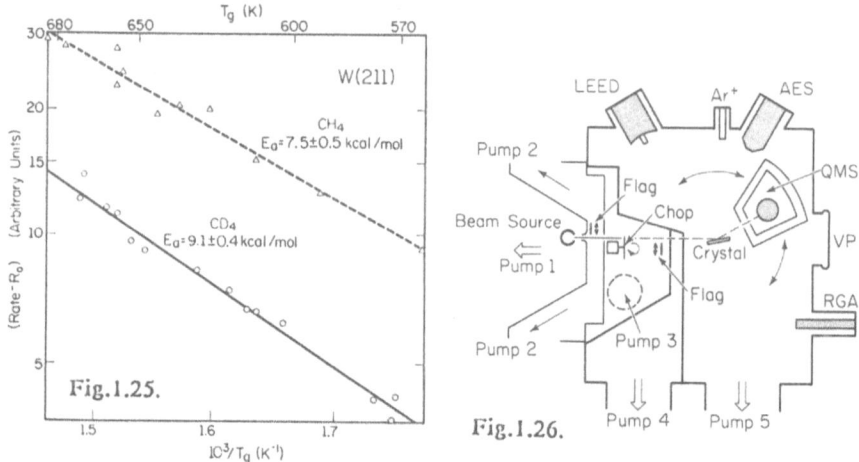

Fig.1.25. Chemisorption rate of methane at different temperatures T_g on W{211} at $T_s = 300$ K [1.75]

Fig.1.26. Schematic of ultrahigh vacuum system for adsorption studies using a supersonic molecular beam [1.42]

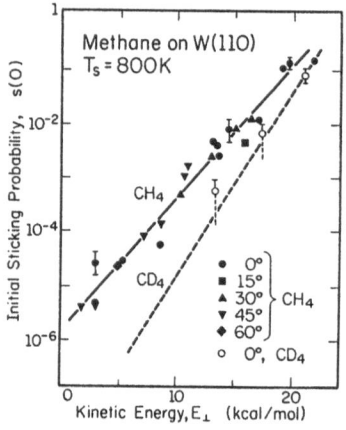

Fig.1.27. Chemisorption rate for methane at different translational energies on W{110} at 800 K [1.76]. $E_\perp$: component of translational energy perpendicular to surface). Symbols indicate different angles of incidence

rate of chemisorption. On a surface kept at 800 K, *Rettner* et al. found an impressive increase in the rate, shown in Fig.1.27, on increasing the kinetic energy of translation perpendicular to the surface. A one-dimensional barrier appears to do justice to these experiments, just as in the earlier work by *Balooch* et al. [1.31] on the chemisorption of hydrogen on copper. Limited measurements were also made with CD_4. The rate of dissociation was significantly smaller than for CH_4 and *Rettner* et al. obtained reasonable agreement with these experiments by assuming the reaction occurred by tunneling of H or D through a one-dimensional parabolic barrier. In a subsequent refinement in the interpretation of these measurements, *Rettner* et al. [1.77] isolated experiments done at the same translational energy, but with the beam at different temperatures. A roughly 400 K increase in beam temperature was found to raise the measured sticking coefficient by a factor of $\simeq 5$. *Rettner* et al. reject contamination caused by the higher beam temperature as a significant effect, and conclude that vibrational excitation contributes to the dissociative chemisorption of methane on W{110}. From their measurements it appears that energy put into vibrational excitation is at least as effective as in translation.

Most of the subsequent studies have been focused on testing the effect of translational excitation upon the rate of chemisorption. A note of caution is therefore in order concerning the interpretation of isotope effects on reactivity. If classical translational motion of the gas overcomes the barrier to chemisorption, then we should not expect any difference in reactivity with mass. The lower rate of CD_4 compared to CH_4 has therefore been routinely attributed to tunneling of some kind, even though there are alternative rationalizations available [1.75]. A demonstration that tunneling is indeed the correct explanation is not easy, but the expressions for the sticking coefficient are simple enough. If translational motion alone is important in the reaction, and the activation barrier is one-dimensional and can be represented by a parabola, then the sticking coefficient can be written as

$$ s = \int_0^\infty f(\epsilon)\, D(\epsilon)\, d\epsilon \ . \tag{1.19} $$

For a thermal beam, the fraction of the molecules $f(\epsilon)$ striking the surface with translational energy between ϵ and $\epsilon+d\epsilon$ is given by (1.18). In the WKB approximation [1.78], the transmission coefficient $D(\epsilon)$ for a parabolic barrier of height V_0 and full width at half maximum ω can be represented by

$$ 1/D(\epsilon) = 1 + \exp\left[\frac{\pi\omega}{\hbar} \sqrt{m/V_0}\ (V_0 - \epsilon)\right] . \tag{1.20} $$

Extensive comparisons between the reactivity of CH_4 and CD_4 at various gas temperatures on a single crystal plane of tungsten are available from the work of *Lo* [1.75]. These results are compared in Fig.1.28 with the predictions of this tunneling model [1.79] obtained by numerical integration [1.80] of (1.19). In Fig.1.28a the data for CH_4 have been used to derive the barrier constants in (1.20); the rate for CD_4 is then predicted by letting the mass m increase from 1 to 2 in (1.20). The calculated values come close to the experiments, but there are two significant problems: the slope of the predicted curve, that is the predicted activation energy, is not correct. What is even worse, the calculated rates are quite insensitive to the value of the effective mass; this can be chosen as 3 instead of 2 and the predictions are not much affected. Small deviations between the predictions and experiments

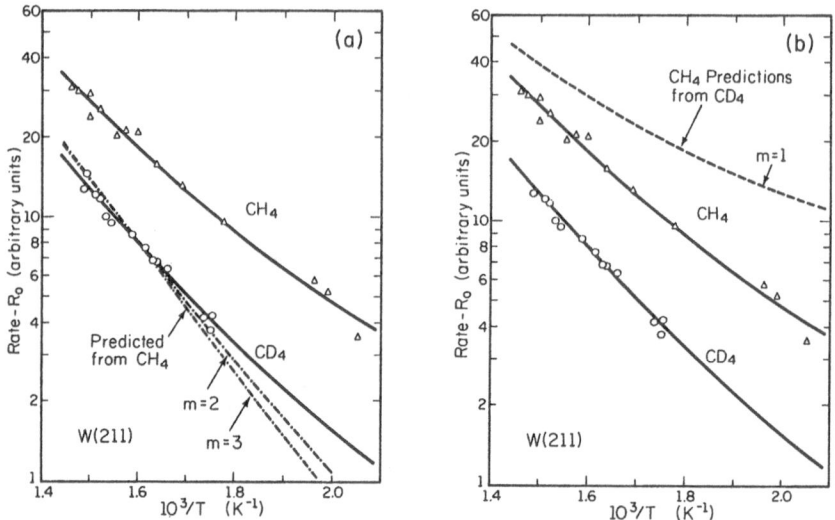

Fig.1.28. Chemisorption rates for CH_4 and CD_4 on W{211} [1.75] and predictions of simple tunneling model from (1.18-20) [1.79,80]. (a) Predictions for m=2 and m=3 from data for CH_4; (b) Predictions for CH_4 based on data for CD_4

must therefore be taken very seriously. A more sensitive test of the tunneling model is available by fitting the CD_4 data, and predicting the rate of CH_4. This is done in Fig.1.28b, and indicates more clearly the serious discrepancy with the experiments. Comparisons between model predictions and data will obviously have to be done more carefully in the future to properly evaluate the true role of tunneling.

In finishing this section it is worth noting that during the past year there has been an outpouring of interesting studies on the chemisorption of simple hydrocarbons. Most extensive has been the work in *Madix's* laboratory [1.81-84], devoted to exploring the effect of translational excitation upon the reactivity of hydrocarbons as a function of the length of the carbon backbone. In Fig.1.29 are shown the results obtained by *Hamza* et al. [1.84] on iridium {110}-(1x2) kept at $\simeq$1000 K. It seems that low kinetic energies do *not* enhance the rate of dissociation at the surface. However, as the kinetic energy imparted to the molecule is increased above 12 kcal/mol, the sticking coefficient on the initially clean surface rises. For methane and ethane it is the

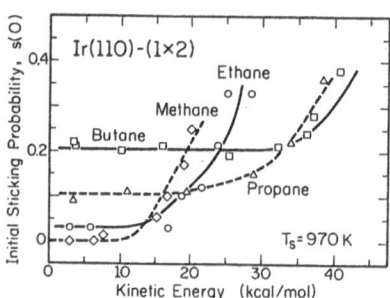

Fig.1.29. Effect of translational energy upon chemisorption rates of simple hydrocarbons on Ir(110) at $T_s = 970$ K [1.84]

normal component of the kinetic energy that promotes chemisorption - the barrier is in effect one-dimensional. However, with increasing chain length, that is on going from methane to butane, more translational energy must be pumped into the molecules to enhance their reactivity. *Hamza* et al. attribute this to transfer of energy during the collision from molecular translation to vibrations of the surface atoms. This process is expected to become more important for larger hydrocarbons, as is indeed found. From their studies on Ir{100}, *Hamza* et al. conclude that translational activation is effective because molecules with kinetic energy of translational motion above some critical value can penetrate close to the metal surface, where a reorganization of molecular bonding is possible.

It is interesting that for molecules other than methane the sticking coefficient at low translational energies is not zero. There appears to be a reaction path for molecules leading from the gas phase to chemisorption but requiring less than 3 kcal/mol activation. This is also apparent from the earlier studies by *Szuromi* and *Weinberg* [1.85], who report chemisorption of ethane, molecularly adsorbed on iridium{110} at 130 K, over a barrier of $\simeq 7$ kcal/mol. At surface temperatures below 1000 K, *Hamza* et al. [1.84] find that raising the translational energy of the incoming molecules leads to an initial *decrease* in sticking probability, which is shown for butane in Fig.1.30. This decrease has been interpreted as reaction after the molecules are trapped in a precursor state, where a reorganization of molecular bonding is possible.

Rather similar behavior was noted by *Hamza* and *Madix* [1.83] in the chemisorption of alkanes on Ni{100}. The initial sticking coefficients on the surface at 500 K are displayed in Fig.1.31. On nickel {100}, molecules that have not been given $\simeq 8$ kcal/mol translational energy do not react, and in this sense there is a quantitative difference from the behavior on iridium {110}. However, if the iridium data are

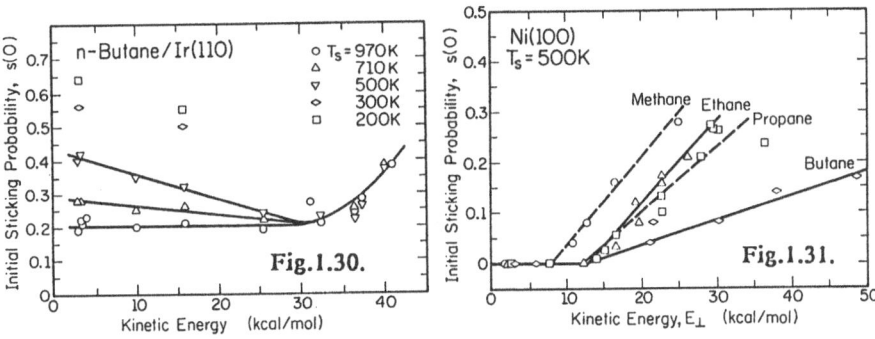

Fig.1.30. Effect of translational energy on chemisorption of n-butane on Ir{110} kept at different temperatures [1.84]

Fig.1.31. Chemisorption of simple hydrocarbons at different translational energies on Ni{100} at 500 K [1.83]. ($E_\perp$: kinetic energy perpendicular to {100})

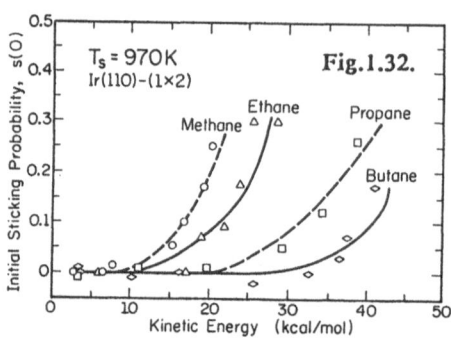
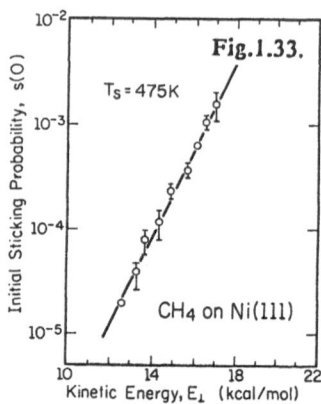

Fig.1.32. Effect of translational energy upon the rate of activated chemisorption of simple hydrocarbons on Ir{110} [1.84]. Data obtained from Fig.1.29 by subtracting contributions from nonactivated processes

Fig.1.33. Chemisorption of methane on Ni{111} at 475 K and its dependence upon the translational energy perpendicular to the surface [1.86]

corrected for this effect, as is done in Fig.1.32, then the trends on the two metals are quite comparable. The critical energy, above which translational activation raises the rate of chemisorption, again moves to higher values with increasing chain length. *Hamza* and *Madix* [1.83] also did scattering studies at incident energies below the critical value and find that as the molecular chain length increases, the energy losses out of the molecule into the surface increase as well. This of course is in keeping with their explanation for the increase in the translational energy required for chemisorption of longer molecules.

Studies on Ni{111} have been reported from *Cyer's* laboratory. In early work [1.86] with supersonic methane beams it was observed that the sticking coefficient scaled with the component of the translational energy perpendicular to the surface. It is apparent from Fig.1.33 that for translational energies below 12 kcal/mol dissociation is below the limit of detection; at 17 kcal/mol, however, the rate has increased by orders of magnitude. The adsorbed species were characterized by high resolution electron energy loss spectroscopy (HREELS); these measurements indicate the formation of methyl radicals in the chemisorption process. In more recent work, *Ceyer* et al. [1.87] have done measurements at different nozzle temperatures but constant translational energy. These studies again point to vibrational excitations as providing an effective way of overcoming the barrier to dissociative chemisorption.

After many alarums and excursions, some order is at last evident in the activated chemisorption of methane and other alkanes: vibrational as well as translational excitations both appear capable of accelerating the rate. However, there are clear indications in some of the systems studied that translational motion only becomes effective at higher energies, and that dissociation can occur by alternative, possibly unactivated, channels. It will be of special interest to explore these low energy paths, as they should dominate in ordinary thermal processes. Much still remains to be done. Quantitative measurements of the relative effectiveness of different excitations, and how this varies with the orientation as well as the chemical constitution of the surface, are obviously in order. The contributions from tunneling processes will also have to be more critically examined.

In turning now to other systems, a note of caution is in order. Simple hydrocarbons have many vibrational modes accessible at reasonably low energies, and in this and other ways they are rather different from some of the diatomic molecules that undergo activated chemisorption; the latter may therefore behave quite differently than the heavily studied alkanes.

1.4.2 Hydrogen on Copper

Although there had been early but uncertain indications of activated hydrogen adsorption on copper powders [1.88,89] the first work to clearly demonstrate the activated chemisorption of hydrogen from the molecular gas was that of *Eley* and *Rossington* [1.20]. The potential diagram they came up with in Fig.1.4, showing an activation energy of 5 kcal/mol, was deduced from measurements of the kinetics for the ortho-to-para hydrogen conversion on copper wires, films, and foils. Hydrogen adsorption itself was not measured. This was done later by *Pritchard* and *Tompkins* [1.90], who examined surface potential changes on copper films exposed to hydrogen atoms. From rough measurements of the temperature dependence of the desorption rate they arrived at an activation energy of 12 kcal/mol; the lack of adsorption from the molecular gas at 70 K was interpreted as arising from an activation energy to adsorption of 4 to 5 kcal/mol. *Pritchard* [1.91] made additional measurements of the desorption rate with an improved thermionic diode and found a somewhat higher desorption energy of 17 kcal/mol.

In all these more recent studies hydrogen was observed on the copper surface only in the presence of H atoms. *Alexander* and *Pritchard* [1.92], however, were able to observe hydrogen chemisorption from the molecular gas by resorting to a Kelvin probe which could be operated at pressures as high as 60 Torr of molecular hydrogen. In observations on various copper films they found isosteric heats of adsorption that varied from ≃9.5 to 12 kcal/mol from one specimen to the next, but were reasonably independent of coverage. That this variability stems from structural differences was established by *Pritchard* et al. [1.93] working with single crystal planes of copper. On Cu{100}, {111}, and {110} explosed to molecular hydrogen at room temperature, they found no indication of chemisorption at 1 Torr.

Changes in the contact potential were, however, observed on {211}, {311}, and {755}. On the {311} plane they measured an isosteric heat of adsorption of ≈9.5 kcal/mol. These more refined studies therefore confirm the earlier picture of *Eley* and *Rossington* [1.20], which was actually based on a partially inappropriate analysis [1.92]. Chemisorption of hydrogen on copper from the molecular gas occurs over an activation barrier of roughly 5 kcal/mol; the layer so formed is only weakly held, however, so that at room temperature and low pressures it is not thermodynamically stable.

The presence of an activation barrier to chemisorption was also inferred by *Balooch* and *Stickney* [1.94], from measurements of the angular distribution of hydrogen evolved from differently oriented planes of copper after permeation through the crystal. From the observed distribution, and using (1.17), they arrived at activation energies for adsorption of hydrogen amounting to 6 kcal/mol on {111}, 5 kcal/mol on {100}, and 2 kcal/mol on {110}, values which are in reasonable accord with the direct measurements of the activation energy cited above. The results from permeation studies were substantiated by *Balooch* et al. [1.31] in separate experiments measuring the angular distribution of HD desorbing from Cu{100}, {110}, and {310}. All of these distributions are highly peaked, as expected for adsorption in which translational energy perpendicular to the surface overcomes the activation barrier. The results in Fig.1.34 are in reasonable agreement with (1.17) if E_a = 5 kcal/mol on {100} and 3 kcal/mol on {110}.

Balooch et al. [1.31] also measured sticking coefficients as a function of the translational energy of the supersonic beam used in their work. The rates of adsorption, already shown in Fig.1.9, scale nicely with the component of the translational energy perpendicular to the surface. If the energy corresponding to the maximum in the derivative of the energy distribution curve is taken as the activation energy, values of 3 kcal/mol are obtained for copper {110} and 5 kcal/mol for {100} and {310}. Both the angular distributions and the energy depen-

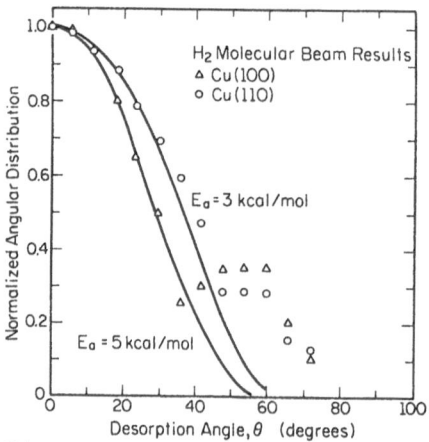

Fig.1.34. Angular distribution of HD desorbed from copper single crystal planes [1.31]. Solid curves give predictions from (1.17)

dence of the sticking coefficient in this trend-setting investigation indicate an adsorption process in which translational motion overcomes the barrier to dissociative adsorption. *Gelb* and *Cardillo* [1.95] have since carried out classical trajectory studies for different empirical potential surfaces. They conclude that vibrational excitation should be very effective in increasing the reaction rate of hydrogen with the copper surfaces, but there is no direct evidence from the molecular beam experiments to support this.

Recently, the nicely unified picture achieved from permeation, desorption, and adsorption experiments has been disturbed by more detailed permeation studies. *Comsa* and *David* [1.96] examined the velocity of D_2 evolved at different angles from the {111} and {100} planes of copper after permeation. It is evident from their data (Fig.1.35) that there is no significant difference in the behavior of the two crystal planes, nor is there a significant dependence of the energy of the molecules on the angle of emission. This of course is quite contrary to what is expected from *van Willigen*'s model and (1.17). *Comsa* and *David* point out that evolution of gas after permeation need not occur in the same way as desorption from an adsorbed layer - they view the properties of molecules evolved in permeation as dictated by bulk behavior.

Kubiak et al. [1.97] have examined the internal energy distribution of H_2 and D_2 evolved after permeation experiments from copper {110} and {111} kept at 850 K and higher. Their most interesting conclusion is that the number of molecules in the first excited vibrational level far exceeds the number expected in an equilibrium distribution at the temperature of the surface. For H_2, $P_{v''=1}/P_{v''=0} = 0.052 \pm 0.014$ from copper {110} and 0.084 ± 0.03 from {111}, whereas the equilibrium ratio at 850 K amounts to only 9×10^{-4}. In view of the previous find-

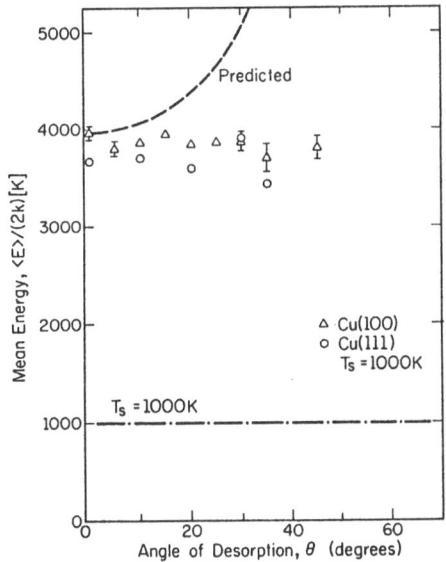

Fig.1.35. Energy of D_2 molecules desorbing from Cu{100} and Cu{110} at different polar angles [1.96]. Dashed curve gives predictions from (1.17), dot-dash curve indicates Maxwellian distribution

35

ings which suggest that permeation and desorption are not necessarily equivalent processes, these results cannot be immediately applied in interpreting the activated adsorption of hydrogen on copper. They suggest that vibrational excitation should assist the rate of permeation, and it would certainly be interesting to directly examine the effect of vibrational excitation upon the rate of hydrogen chemisorption on copper. There are other features of this system that also merit further study, the crystallographic effects being among the more important. It would be worthwhile to establish the thermodynamics of adsorption on crystal planes of different orientations, and to correlate these with the results of detailed measurements of the kinetics of chemisorption as well as of surface diffusion on the same plane.

1.4.3 Hydrogen on Elemental Semiconductors

The early studies of activated adsorption on elemental semiconductors present a rather confusing picture. *Law* [1.98] in 1955 reported hydrogen adsorption on germanium from the molecular gas at room temperature, but this may have been a spurious effect, produced by atomization in the ion gauge used. Somewhat later work by *Dell* [1.99] on germanium powder obtained by reducing the oxide did not reveal any hydrogen uptake. It remained for *Tamaru's* studies [1.21] in the same year to establish that chemisorption of hydrogen on germanium was activated. His work, on Ge films obtained by decomposition of GeH₄, provided the first quantitative data on the kinetics and energetics of hydrogen adsorption and desorption at a semiconductor.

Tamaru's results were largely verified by *Bennett* and *Tompkins* [1.100], who studied adsorption on evaporated germanium films. In the vicinity of room temperature, chemisorption from the molecular gas was found to increase with temperature, suggesting an activation energy of 16.6 kcal/mol, compared to *Tamaru's* [1.21] estimate of 14.6 kcal/mol. Desorption from hydrogen layers produced by exposing the germanium film to a mixture of atomic and molecular hydrogen occurred over a barrier of 38 kcal/mol, again in good agreement with *Tamaru's* results. Adsorption and desorption of hydrogen on Ge{111} was examined by *Belyakov* and *Kompaniets* [1.101]; they were unable to find any adsorption from the molecular gas, which is not surprising in view of their limited exposures. *Belyakov* et al. [1.102] give a sticking coefficient <10⁻⁸ for H₂ on Ge{111}. Adsorption did occur in the presence of an incandescent filament; desorption of the layers so formed was second order, and occurred over a barrier of 18 kcal/mol, considerably smaller than the height found in other studies. Except for this last result, the picture for adsorption of hydrogen on germanium is clear. The heat of adsorption from the molecular gas is quite appreciable (23 kcal/mol), but molecular hydrogen requires a large activation energy, amounting to ≃15 kcal/mol, for dissociation on the surface.

Rather less is known about the adsorption of hydrogen on silicon. Early work by *Law* and *Francois* [1.103] on silicon filaments gave a sticking coefficient of 10^{-5} for adsorption from the molecular gas. This was soon recognized as due to atomization of hydrogen on the ionization gauge used to do the desorption studies. Subsequent work by *Law* [1.104] on evaporated silicon films and on filaments, in which gauge effects were minimized, revealed only minor hydrogen adsorption from the molecular gas, amounting to 2×10^{13} atoms/cm^2, and a sticking coefficient lower than previously estimated. More detailed studies were made by *Datsiev* and *Belyakov* [1.105] as well as by *Belyakov* et al. [1.102] on Si{111} and {110}. In these studies the cathodes of their mass spectrometer were shielded, so that there was no direct path to the sample. Nevertheless, adsorption from molecular hydrogen was observed and they estimated an initial sticking probability of $1-2 \times 10^{-3}$. It is to be noted that in comparable studies of Ge{111} no adsorption from the molecular gas was found. The hydrogen layer on Si{111} desorbed by second-order kinetics with an activation energy of 42 kcal/mol.

This desorption energy is in good agreement with the work of *Kleint* et al. [1.106], also on Si{111}. They again found second-order desorption, suggesting evolution of H_2 by recombination of gas atomically adsorbed on the surface over a barrier of $\simeq 40$ kcal/mol. However, *Kleint* et al. report only negligible amounts of hydrogen adsorbed from the molecular gas at a pressure of 5×10^{-5} Torr. *Brzóska* and *Kleint* [1.107] reviewed earlier studies on single crystal silicon as well as on evaporated films, and conclude that the sticking probability for hydrogen chemisorbed from the molecular gas does not exceed 10^{-5} at room temperature. This conclusion is supported by the more recent work of *Schulze* and *Henzler* [1.108]; they find no adsorption from molecular hydrogen on Si{111} and give a figure of less than 10^{-6} for the sticking coefficient. In desorption of hydrogen chemisorbed from the atomized gas, *Schulze* and *Henzler* find several adsorption states. At low coverages, desorption occurs entirely in a second-order process with an activation energy of 57.7 kcal/mol. More recently *Froitzheim* et al. [1.102] have done HREELS studies of hydrogen layers on Si{111}. They conclude that adsorption does occur from the molecular gas, but that the hydrogen is held at or close to steps, whereas in adsorption from the atomized gas the hydrogen atoms are bound at terrace sites.

For silicon no studies are available of how chemisorption from the molecular gas depends upon the gas temperature. However, there have been trajectory studies by *Rice* et al. [1.110] of the dissociation of molecular hydrogen on Si{111} with potentials giving a dissociation barrier of 4.15 kcal/mol. In these simulations, energy transfer out of vibrational motions of the molecules during collision with the surface was not found significant: the energy released in dissociation of the gas at the surface goes primarily into vibrational excitation of Si-H. The Polanyi criteria (Fig.1.17) therefore imply that translational energy is most effective in overcoming the barrier to adsorption. At 1000 K a sticking coefficient of roughly $4.6 \cdot 10^{-3}$ was estimated. In separate Monte Carlo simulations of the desorption process, an activation energy of 57.7 kcal/mol was found by *Raff* et al. [1.111].

Although these studies are very interesting, they are no substitute for quantitative measurements of the rate of chemisorption from the molecular gas. Despite the great technological and scientific interest of hydrogen interactions with silicon, such measurements are not available. In their absence, can we be certain that chemisorption is in fact activated? Could the various observations cited here be compatible with endothermic chemisorption, in which desorption is limited by the barrier to surface diffusion, as described in Sect.1.2?

In this connection it is significant that the barrier to desorption of hydrogen layers produced by exposing silicon to atomic hydrogen is quite high, in excess of 40 kcal/mol. This value is higher than the activation energy for the diffusion of any monatomic gas on a metal. More than that, the energy to break a Si-H bond estimated on the assumption that diffusion is *not* the limiting step is comparable to the dissociation energy of 70 kcal/mol for Si-H in the gas phase [1.112]. The fragmented observations of adsorption from the molecular gas also speak against endothermic chemisorption. Finally, it would be surprising to have radically different behavior for the interactions of H_2 with silicon and with germanium. It appears safe to assume that chemisorption of hydrogen on silicon is in fact activated. Nevertheless, for so significant a system it would be important to have available direct experimental studies of the adsorption process from the molecular gas.

1.4.4 Activated Adsorption on Densely Packed Planes

a) Hydrogen and Nitrogen on W{110}, Re{0001}, and Ru{0001}

Polizzotti's field emission studies [1.25,26] of hydrogen on W{110} provide a good example of the kinetic inhibition of chemisorption on a smooth, low-index plane. At low temperatures, $T_s \simeq 40$ K, he isolated a weakly bound molecular state of hydrogen adsorbed on W{110}. On warming to 70 K the hydrogen evaporated, leaving behind a substantially clean surface, as already shown in Fig.1.6. From these observations it appears that a barrier amounting to >3.5 kcal/mol must be overcome for chemisorption of hydrogen on W{110}. Exposure to hydrogen with the surface at $T_s \geq 80$ K did bring about chemisorption on W{110}, but only after all the surrounding rough surfaces had been filled. Population with hydrogen presumably occurs by diffusion from the atomically rough regions, on which dissociative chemisorption of hydrogen is rapid.

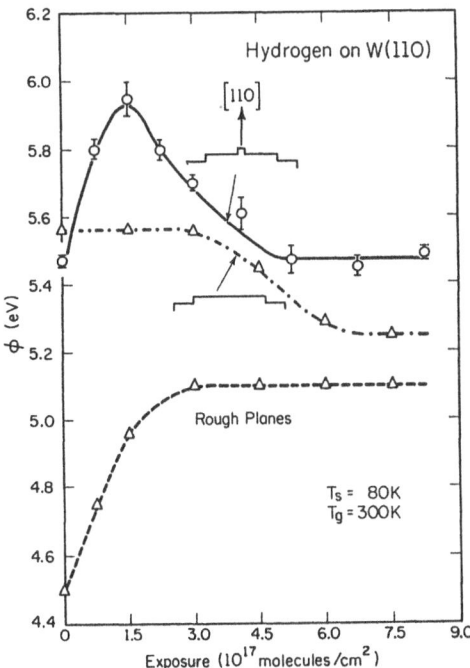

Fig.1.36. Chemisorption of hydrogen on smooth W{110}, and on W{110} with a small island of tungsten on it [1.28]. Change in work function ϕ provides a measure of amount adsorbed

This view was supported by experiments done by *Liu* [1.28], who compared the activity in hydrogen chemisorption of smooth {100} planes and {110} planes with a tungsten cluster on them. As indicated in Fig.1.36, the emission characteristics of the latter change immediately upon exposure to hydrogen, just as for rough planes, suggesting that the clusters catalyze the dissociation of hydrogen on the surface. Recent experiments by *Wang* and *Gomer* [1.113] on much larger planes formed by thermal annealing confirm the view that hydrogen chemisorption on W{110} is limited by processes on the surrounding rough regions. *Wang* and *Gomer* find that chemisorption occurs from the edges inward, at a rate much slower than would be expected if diffusion over the {110} plane itself were the limiting step. It is worth noting parenthetically that there is no indication of slow hydrogen chemisorption on Mo{110} [1.29].

The effect of gas temperature upon the sticking coefficient of hydrogen on W{110} has been examined by *Steinbrüchel* and *Schmidt* [1.114]. They find a slight *decrease* as the temperature of the beam is raised. However, they also report a significant diminution as the angle of incidence deviates from the normal, and conclude that there is some barrier to chemisorption that can be overcome by molecules with a higher momentum perpendicular to the surface. The temperature dependence speaks against this, however, and it is likely that the angular dependence is actually brought about by the presence of defects at which dissociation occurs.

Chemisorption is inhibited on the {0001} plane, the basal plane of rhenium. This is strikingly documented by the behavior of nitrogen on rhenium examined by *Liu* [1.27, 28]. When a rhenium field emitter cleaned and smoothed by field evaporation at low temperatures is exposed to nitrogen, the emission characteristics of the basal plane remain unchanged if gas and surface are at room temperature, as in Fig.1.7. Chemisorption does occur on the atomically rough regions around {0001}. If the surface is kept at 80 K, molecular nitrogen adsorbs on Re{0001}. It is completely removed on warming to $T_s \simeq$ 150 K, leaving behind a clean surface. If nitrogen is brought into contact with a rhenium surface at 550 K, then chemisorption takes place on the basal plane at a rate comparable to that on the rough surfaces of rhenium. Much the same end result is achieved by adsorbing nitrogen on the rhenium emitter at room temperature, pumping off the gas, and then warming to 550 K. These experiments suggest that chemisorption of nitrogen on Re{0001} is activated and that this plane fills by diffusion from the periphery, where dissociative chemisorption is rapid.

Observations on Re{0001} by *Grunze* et al. [1.115] are in agreement with these results. They find that at low temperatures, molecular nitrogen is bound on the clean surface with its molecular axis perpendicular to the plane. The dissociation probability for these weakly held molecules is quite low, and is believed to depend on the presence of defects. On a reasonably perfect {0001} plane, dissociative chemisorption was found to be slightly activated. *Asscher* et al. [1.116] in *Somorjai's* laboratory have examined the synthesis of ammonia on rhenium crystals of various orientations. In this reaction, dissociative chemisorption of nitrogen may be the limiting step, and it is of interest that reactivity increases in the sequence Re{0001}, {10$\bar{1}$0}, {11$\bar{2}$0}, {11$\bar{2}$1}; on {0001}, ammonia production is barely detectable, as would be expected from the chemisorption behavior for nitrogen established previously.

Chemisorption of hydrogen on Re{0001} also appears to be slow. In Fig.1.37, work function change obtained by *Liu* [1.28, 29] on Re{0001} are compared with changes on rough rhenium surfaces after hydrogen exposure at different surface temperatures. At 80 K there is no detectable change in emission from the basal plane until the rough surfaces are extensively covered. Only then does chemisorption occur on {0001}. As the surface temperature is raised, the initial rate of hydrogen chemisorption on the basal plane increases, and by 200 K the initial delay in filling the basal plane disappears. This behavior, so similar to what has been found on W{110}, makes it likely that chemisorption of hydrogen is activated on rhenium {0001}.

40

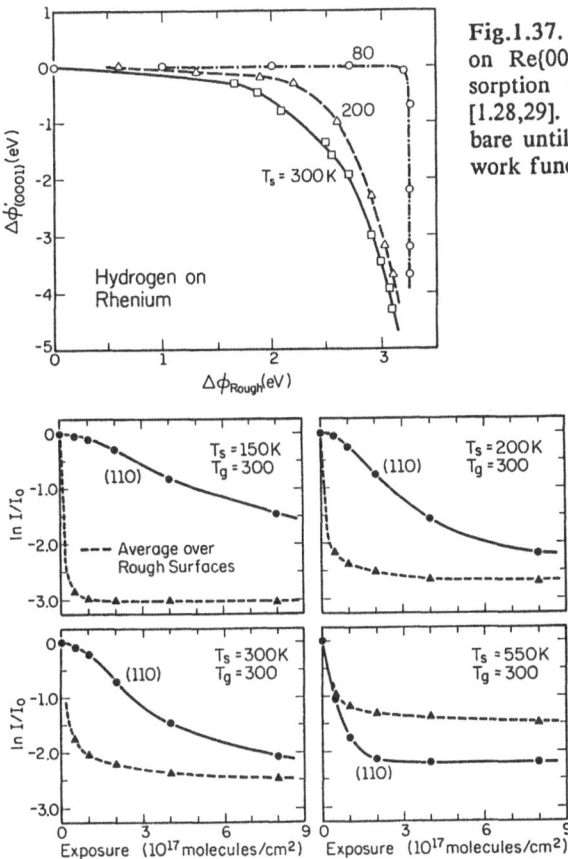

Fig.1.37. Chemisorption of hydrogen on Re{0001}, compared with chemisorption on rough rhenium surfaces [1.28,29]. At T_s = 80 K, {0001} stays bare until rough surfaces are filled. (ϕ: work function)

Fig.1.38. Chemisorption of nitrogen on W{110} and on rough tungsten surfaces, at different surface temperatures [1.28,29]. (I: field emission current)

That chemisorption of nitrogen also is activated on W{110} was demonstrated by *Liu* [1.28,29]. Shown in Fig.1.38 are comparisons of the electron emission from W{110}, and from the atomically rough tungsten planes, when the surface at different temperatures is exposed to nitrogen. As the surface temperature T_s increases, the initial rate of nitrogen chemisorption on W{110} increases. At $T_s \simeq 550$ K, chemisorption on W{110} and on the rough planes occurs at comparable rates. If the {110} plane is kept at 80 K, molecularly bound γ-nitrogen forms. Warming the surface slowly removes this molecular adsorption, but the W{110} is not restored to cleanliness - it is substantially covered with chemisorbed nitrogen. Conversion into the chemisorbed state occurs either during the initial exposure, or more likely during the subsequent warming, by conversion from the γ-states. *Liu* and *Ehrlich* [1.29] conclude that an activation energy of $\simeq 7$ kcal/mol limits the chemisorption of nitrogen on W{110} for surface temperatures in the range 300 to 550 K. Diffusion of nitrogen on W{110} has been studied by *Polak* [1.117], and is too slow to account for the observed

41

rates of filling. The {110} plane does not populate from the periphery, as is sometimes claimed by analogy with the behavior of hydrogen; rather it occurs by dissociation on {110} itself.

Measurements of the angular dependence of nitrogen evolution by *Cosser* et al. [1.118] are in reasonable accord with the conclusions from field emission studies. In Fig.1.39 the desorption intensities observed by *Cosser* et al. from tungsten {310} are compared with their measurements from tungsten {110}. The measurements from {310} correspond to a cosine distribution; the experiments on {110} are fit by $\cos^n \theta$, with $3 \leq n \leq 4$, and *Cosser* et al. conclude that chemisorption on W{110} is activated, occurring over a barrier of $\simeq 4$ kcal/mol. Rather different results have been obtained recently by *Pfnür* et al. [1.119] in detailed molecular beam studies of nitrogen chemisorption on a W{110} plane at 800 K. At incident beam energy less than 10 kcal/mol, the initial sticking coefficient $s(0)$ is fairly constant. However, as shown in Fig.1.40, at higher translational energies $s(0)$ rises sharply up to energies of $\simeq 25$ kcal/mol, where the sticking coefficient becomes fairly constant.

The initial sticking coefficient does not scale with momentum normal to the surface, as might have been expected from the angular distribution of the desorption flux measured by *Cosser* et al. [1.118]. In the supersonic beam studies, the sticking coefficients are not greatly affected by the angle of incidence. *Pfnür* et al. [1.119] do not find any significant effects of vibrational excitation on the rate. They interpret the failure of normal momentum scaling as due to scrambling of paral-

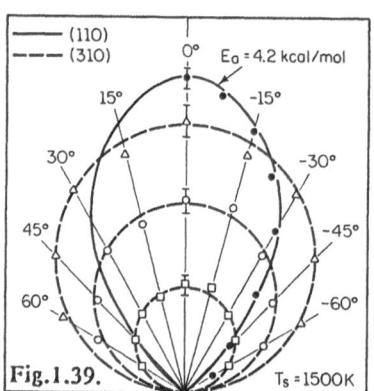 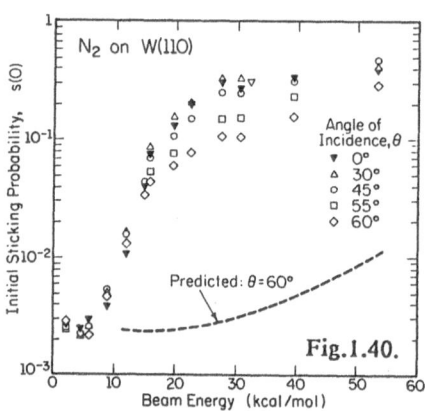

Fig.1.39. Angular distribution of nitrogen desorbed from W{310} and W{100} [1.118]. Measurements from W{310} obtained at N_2 exposures of $\sim 0.4 \cdot 10^{-6}$ Torr·s ▢; $0.85 \cdot 10^{-6}$ Torr·s ○ ; $8.5 \cdot 10^{-6}$ Torr·s △

Fig.1.40. Rate of nitrogen chemisorption on W{110} at increasing translational energies [1.119]. Dashed curve gives predictions for $\theta = 60°$ from data at $0°$ assuming $s(0)$ scales with normal component of translational energy

lel and perpendicular molecular motions, which they expect if the incident molecule experiences strong chemical interactions before it reaches the barrier. It should be noted that the value of 22 kcal/mol which *Pfnür* et al. derive for the mean barrier height from the measured energy dependence is far greater than the barrier found either by *Cosser* et al. [1.118] or by *Liu* and *Ehrlich* [1.29]. It appears as if the experiments with supersonic beams are exploring phenomena different from those examined in previous experiments. Nevertheless, it is comforting that all the work is in agreement in finding chemisorption of nitrogen on W{110} activated.

There are indications that chemisorption of nitrogen on ruthenium may be similar in its structural selectivity to what has been observed on rhenium. On Ru{0001}, *Danielson* et al. [1.120] found no chemisorption of nitrogen in studies using Auger spectroscopy to examine the adsorbed layers. They did observe the decomposition of ammonia on this plane, and report it is activated, occurring over a barrier of 12 kcal/mol.

The studies on W{110} and Re{0001} are important as they provided early, clearcut examples for activated chemisorption on smooth, low index planes of metals for which chemisorption elsewhere is rapid and appears to occur without activation, via precursur processes. It is the atomic arrangement of the surface which clearly dictates the course of the chemisorption process and forces dissociation to occur over a barrier on the densely packed planes.

b) Hydrogen on Platinum {111}

The first hint of unusual behavior in the chemisorption of hydrogen on platinum {111} was provided by the work of *Smith* and *Palmer* [1.121], at General Atomic, who studied the oxidation of deuterium on a {111}-oriented film formed inside their vacuum system. Deuterium from a molecular beam was allowed to interact with the Pt{111} surface immersed in an oxygen ambient; from the dependence of the rate of D_2O production upon the beam temperature, *Smith* and *Palmer* concluded that chemisorption of D_2 was activated and occurred over a barrier of 1.8 kcal/mol. However, these experiments were not done under ultrahigh vacuum conditions and there was no independent documentation of the cleanliness of the surfaces, so there remained some doubt about these interesting results.

A few years later, *Bernasek* and *Somorjai* [1.122] did modulated molecular beam studies of HD formation on smooth and stepped Pt{111}. Only on the stepped surface was HD detected, and they concluded that dissociation at steps was four orders of magnitude faster than at the terrace sites. This result was in apparent conflict with the measurements of *Lu* and *Rye* [1.123], who examined the adsorption of hydrogen, as well as the formation of HD in the mixed adsorption of

H_2 and D_2, on platinum {110}, {211}, {111}, and {110}. They report chemisorption on {110} only a factor of roughtly 20 faster than on {111}. At surface temperatures below 570 K they observed that the equilibration of H_2 with D_2 occurred with an activation energy between 1.9 and 1 kcal/mol, the barrier varying with orientation; on {111}, the activation energy for equilibration amounted to 1.2 kcal/mol. *Lu* and *Rye* interpreted these small barriers as indicating that surface diffusion of atomically bound hydrogen was the limiting step in the statistical mixing of isotopes.

Christmann et al. [1.124] carefully reexamined the chemisorption of hydrogen on platinum {111} and found an initial sticking coefficient of 0.1. In a second study, *Christmann* and *Ertl* [1.125] also looked at chemisorption on stepped {111} planes. On these, the initial sticking coefficient for hydrogen chemisorption was only a factor of four higher than on the flat platinum {111}. On both smooth and stepped {111} planes the formation of HD from H_2 and D_2 again was activated, occurring over a barrier of 2.5 kcal/mol on both; however, on stepped surfaces the rate of exchange was an order magnitude greater than on the smooth planes. *Christmann* and *Ertl* report no significant dependence of the rate of chemisorption upon temperature, and conclude that chemisorption is not activated. It is interesting to note, however, that their sticking coefficients drop rapidly with increasing coverage, as would be expected for slow, activated chemisorption.

At roughly the same time *Wachs* and *Madix* [1.126] reanalyzed the results of *Bernasek* and *Somorjai* [1.122] and discovered that the rate of hydrogen chemisorption on stepped {111} surfaces was only a factor of 20 higher than on flat {111} planes. At this stage, there appeared to be fairly general agreement that dissociative chemisorption did occur on platinum {111}, albeit at a smaller rate than on rougher surfaces, and that chemisorption on {111} was not activated.

The picture changed again a few years later, with a careful examination of the H_2-D_2 exchange on platinum {111} and {332} by *Salmerón* et al. [1.39] in *Somorjai's* lab. On {332}, a stepped plane, the sticking coefficient deduced from the modulated beam measurements of HD production did not show a temperature dependence, and *Salmerón* et al. concluded that chemisorption is not activated. The situation is different on Pt{111}. The variation of the sticking coefficient with surface temperature deduced from the rate of HD formation is in keeping with an activation barrier of 1.5 kcal/mol. However, the activation energy deduced from the dependence of the sticking coefficient upon beam temperature, or from its dependence upon the angle of incidence of the beam, amounted to 0.5 kcal/mol or less. *Salmerón* et al. conclude that on the flat {111} plane of platinum, hydrogen chemisorption is activated, and this result is in agreement with subsequent studies of the same surface. *Poelsema* et al. [1.127] in Jülich measured the scattering of thermal energy helium beams to follow the adsorption

of hydrogen on Pt{111} surfaces whose perfection had been characterized by the same technique. As the temperature of the {111} surface
was increased over the range from 90 to 300 K, the initial sticking
coefficient was observed to increase as well. For example at 90 K, s(0)
= 0.045, while at 240 K the value had risen to 0.06. To fit their own
measurements as well as the results of *Salmerón* et al. [1.39] taken over
temperatures from 500 to 900 K, *Poelsema* et al. [1.127] had to
abandon a simple exponential relation for the temperature dependence
of the sticking coefficient.

In these studies a sensitive dependence of the rate upon the presence of defects was noted and *Poelsema* et al. interpret their results as
due entirely to reaction at isolated defects. For such a model [1.128]
the sticking coefficient will depend upon the characteristic distance x_s
covered by physically adsorbed molecules diffusing over the surface
during their lifetime, and s(0) can be represented as proportional to

$$s(0) \sim s_m x_s \exp(-E_c/kT_s) , \qquad (1.21)$$

where E_c is the barrier to dissociation into the chemisorbed state at the
defects. Inasmuch as the lifetime of the molecules is proportional to
$\exp(E_d/kT)$, *Poelsema* et al. write

$$s(0) = CT_s^\alpha \exp[(E_d - E_c)/kT_s] , \qquad (1.22)$$

where T_s^α represents the joint temperature dependence of the diffusivity as well as of the condensation efficiency s_m, and α is some small
number between 1.25 and 1.5. They find good agreement with the
experimental data from 90 to 900 K, shown in Fig.1.41a, using $\alpha = 1.4$
and $E_d - E_c = 0.42$ kcal/mol. It is worth emphasizing that the activation
energy observed in these experiments is attributed entirely to transport
and reaction of the precursor molecules at defects where the chemisorption actually occurs. Once hydrogen is chemisorbed, it is presumed

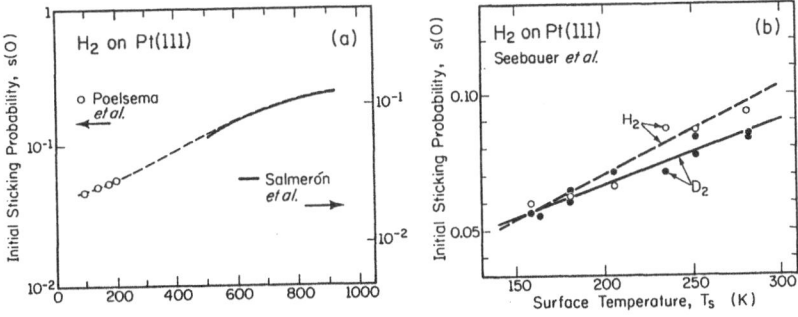

Fig.1.41. Chemisorption rate of hydrogen on Pt{111} at different surface temperatures. (a) Fit of (1.22) to experiments of *Poelsema* et al. [1.127] and of *Salmerón* et
al. [1.39]; (b) Measurements for H_2 and D_2 at low temperatures [1.130]

mobile enough to cover the entire {111} surface, even though defects make up only a small fraction of the total number of sites. More extensive measurements by *Comsa* [1.129] have quantified many of these results. Independent studies of chemisorption rates for H_2 and D_2 have also been done by *Seebauer* et al. [1.130], using laser desorption to detect the amount chemisorbed. As appears in Fig.1.41b, they find a 70% increase in the rate of chemisorption for hydrogen and a 50% increase for deuterium on raising the surface temperature from 160 to 280 K, in reasonable agreement with the previous work at Jülich.

Views about the chemisorption of hydrogen on platinum {111} seem to have come full circle: just as in the earlier studies of *Bernasek* and *Somorjai* [1.122], current work [1.127,129] points to defects as the sites where chemisorption occurs, the flat {111} itself being unreactive. It will be important to subject this very interesting system to further studies in order to reveal the details of the chemisorption process. At the least it is now reasonably clear that the close packed plane of platinum is relatively unreactive toward hydrogen compared with other, rougher planes. There does not yet seem to be any indication of a similar structural effect on other platinum metals. Field emission studies of hydrogen chemisorption have been made on rhodium [1.25,26] and iridium {111} [1.28,29]. Unlike hydrogen chemisorption on W{110}, the initial filling of Rh{111} or Ir{111} is not inhibited compared to that on the high index planes. This may just mean that diffusion of hydrogen is rapid on these metals even at low temperatures, and exploration of macroscopic surfaces would be interesting.

c) Hydrogen on Nickel {111} and {110}

Just as for platinum, the first indication that chemisorption of hydrogen on Ni{111} might be activated was provided by work at General Atomic. *Palmer* et al. [1.131] looked at the evolution of HD from a {111} oriented nickel film deposited on mica and maintained at 700 K. From the dependence of the rate of HD formation upon the temperature of the incident gas, which was varied from 300 to 1800 K, they deduced an activation energy to chemisorption of 2 kcal/mol. The angular dependence of the gas evolved was also in keeping with activated chemisorption, showing the peaked form expected from (1.17). Because of the uncertain state of cleanliness of the surface in this work these findings were soon relegated to the background. In fact, the {111} plane of nickel received considerable attention in chemisorption studies as a simple system with a well-defined surface structure [1.132].

Christmann et al. [1.133] examined hydrogen adsorption on Ni{111}, {110}, and {100} by a variety of techniques. Extensive observations of the kinetics of chemisorption were reported only on Ni{100}, for which the sticking coefficient was found to follow the relation $s(\Theta) = s(\Theta)(1-\Theta)$, with $s(0) \simeq 0.25$. This strong dependence upon the coverage Θ is surprising; it suggests either direct chemisorption without a precursor, or else activated chemisorption. On the other planes the initial sticking coefficient was of the same general magnitude as on {100}. *Lapujoulade* and *Neil* [1.134] as well as *Andersson* [1.135] also examined the initial rate of hydrogen chemisorption on nickel {100}, the former reporting $s(0) \sim 0.06$; the later measurements were an order of magnitude higher.

A really detailed examination of hydrogen chemisorption on nickel was reported by *Winkler* and *Rendulic* [1.136], the first in a series of significant studies. On Ni{111} they found a small initial sticking coefficient, shown in Fig.1.42, amounting to only 0.05. On a stepped {111} plane the rate was higher, with $s(0) = 0.24$, and on Ni{110}, $s(0)$ was equal to 0.96. The dependence upon coverage is also interesting. On {111} and on the stepped {111} plane, the rate diminishes rapidly as the surface fills with hydrogen. On {110} the rate initially does not depend upon coverage. *Rendulic* and *Winkler* [1.137] then examined the effect of temperature upon the rate of hydrogen chemisorption on Ni{111}. Changing the temperature T_s of the surface had little effect in the range $138 < T_s < 300$ K; they found $ds/dT_s < 10^{-5}$, with the gas at 300 K. Changing both gas and surface temperature simultaneously produced an increase in the rate, but this was still small, with $ds/dT = 3.5 \times 10^{-5}$. *Rendulic* and *Winkler* surmised that there might be a small activation barrier to chemisorption, possibly between 50 and 100 cal/mol. In subsequent measurements of the angular distribution of H_2 desorbing from {111}, *Steinrück* et al. [1.138] at Graz were able to fit their results with a highly peaked $\cos^{4.6}\theta$ law. Desorption of hydrogen from {110} gave a $\cos^{1.2}\theta$ distribution, close to that found for CO from the same surface. The results for {111}, interpreted as in

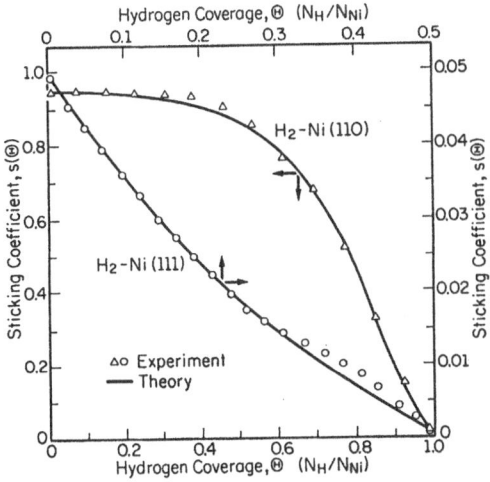

Fig.1.42. Rate of hydrogen chemisorption as a function of coverage on Ni(111) and Ni(110) [1.136]

Sect.1.3.3, clearly suggest activated chemisorption, but the authors decided that "the result of ds/dT = $3.5 \times 10^{-5} K^{-1}$ for the β-states practically excludes the existence of an activation barrier."

The facts about the activated chemisorption of hydrogen on Ni{111} first clearly emerged from the very extensive report by *Robota* et al. [1.40] in *Ertl's* laboratory. With both gas and surface at room temperature, s(0) is only 0.05 on Ni{111}, compared to almost unity on Ni{110}. Raising the translational energy perpendicular to the surface increased the sticking coefficient on Ni{111}, as in Fig.1.43, until s(0) = 0.4 at 2.8 kcal/mol. On Ni{110} the translational energy had no perceptible effect. Also, on this plane the rate of chemisorption (Fig.1.44) is initially independent of concentration, whereas on {111} the sticking coefficient drops rapidly as the amount adsorbed increases. The authors conclude that chemisorption of hydrogen on {111} is activated and occurs directly from the gas phase over a barrier of $\simeq$ 2 kcal/mol, in remarkably good agreement with the early report of *Palmer* et al. [1.131]; chemisorption on {110}, however, is not activated. Measurements of the angular distribution of H_2 desorbing from Ni{110} and {111} were in keeping with this view and with the previous work at Graz: desorption from {110} occurs with a $\cos\theta$ distribution; evolution from {111} is best described by $\cos^n\theta$, with n$\simeq$5. All these results are consistent with activated chemisorption on {111} that depends upon translational energy perpendicular to the surface to overcome the dissociation barrier. There is only one somewhat surprising result: neither

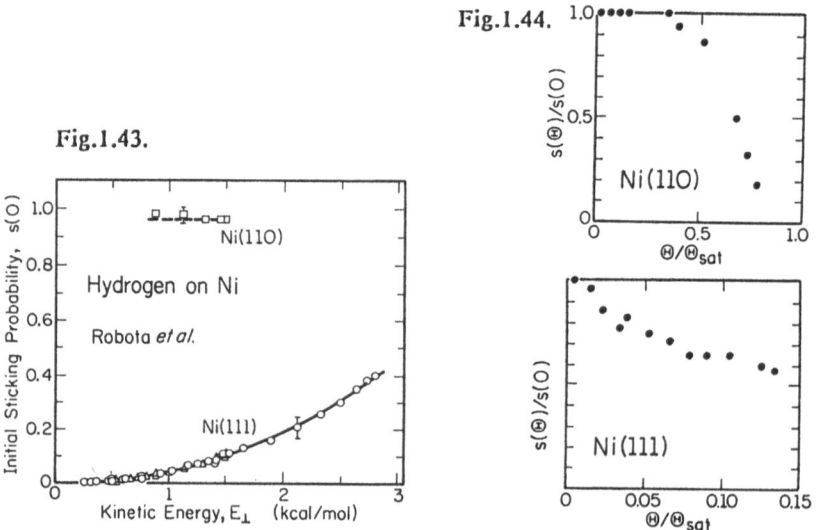

Fig.1.43. Effect of translational energy on the sticking coefficient of hydrogen on Ni{111} and Ni{110} [1.40]. ($E_\perp$: mean translational energy normal to the surface)

Fig.1.44. Concentration dependence of sticking coefficient for D_2 on Ni(110) and Ni(111) [1.40]

on Ni{111} nor on {110} is there an effect of surface temperature upon the value of the initial sticking coefficient. For {111} that observation is consistent with direct chemisorption from the gas. For {110}, *Robota* et al. [1.40] believe that precursors on the surface do not make an important contribution to the rate, but that still leaves the lack of coverage dependence noted earlier as something of a puzzle.

Further studies done at Graz are in good agreement with the results from *Ertl's* laboratory. *Steinrück* et al. [1.139,140] measured the angular dependence of adsorption and desorption from Ni{111} and found a $\cos^{4.5}\theta$ dependence for both. A $\cos\theta$ distribution accounted nicely for their measurements on {110}. Increasing the temperature of a Maxwellian molecular beam incident upon Ni{110} at $T_s = 220$ K diminished the initial sticking coefficient, starting from a value larger than 0.5 at $T_g = 90$ K; D_2 gave essentially the same values as H_2. On Ni{111}, however, raising the gas temperature *raised* the initial rate of chemisorption, as is shown in Fig.1.45, with no difference between H_2 and D_2. Surface temperature was found to have only a very small effect, raising the rate slightly. The authors conclude that on Ni{111} chemisorption occurs over a barrier of 1-2 kcal/mol, with no activation required on Ni{110}. It is of interst that *Lee* and *DePristo* [1.141] have recently done classical trajectory calculations for the dissociative chemisorption of H_2 on Ni{100}, {110}, and {111}. They find that to reproduce the significant difference in reactivity reported in experiments on {111} and {110} it is necessary to allow for the promotion of 4s to 3d electrons.

There are still some differences in the experimental results obtained by different groups on nickel {111}. While there is general agreement that the rate of chemisorption on {111} increases with beam energy, the detailed dependence found differs, as is apparent in Fig.1.46. *Hayward* and *Taylor* [1.142] conclude from their measurements that the rate of chemisorption does *not* scale with momentum normal to the surface, as suggested in other studies. Furthermore, they interpret the results in Fig.1.46 as indicating a threshold energy for reaction. Also contrary to previous work are the results of *Russell* et al. [1.143] at Pittsburgh, shown in Fig.1.47. They detect a significantly lower rate of chemisorption for D_2 than H_2 on Ni{111} at 87 K, in

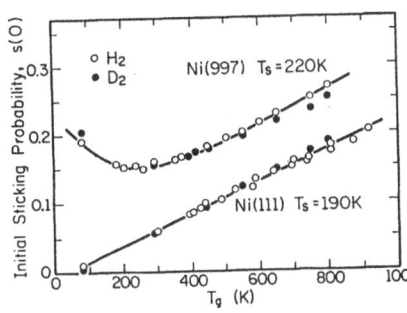

Fig.1.45. Dependence of chemisorption rate for hydrogen on nickel upon gas temperature T_g [1.140]

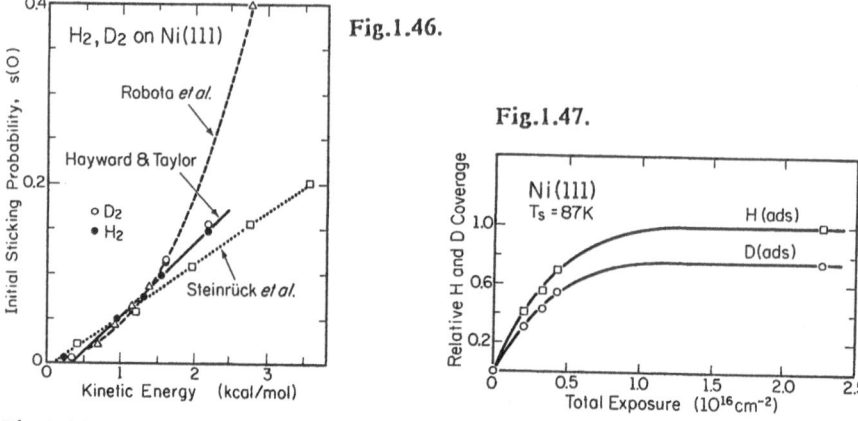

Fig.1.46.

Fig.1.47.

Fig.1.46. Comparison of different measurements of the translational energy dependence of the sticking coefficient for hydrogen on Ni{111} [1.142]

Fig.1.47. Isotope effect in chemisorption of hydrogen on Ni{111} at 87 K [1.143]

contradiction to the expectations for simple translational activation. There is also some difference in the results obtained at higher coverage, where an additional β_1-state desorbs. *Russell* et al. [1.144] find an angular distribution for the evolution of this state close to $\cos\theta$. They suggest as one possible explanation for this that at high surface coverages, vibrational excitation of the incoming gas might be effective in promoting chemisorption; if this is so, then in the desorption of β_1-H_2, the excess energy should go into internal excitations rather than translation. A rather different conclusion is reached by *Rendulic* et al. [1.145]. They find a strongly peaked angular distribution in the desorption of both β_2- and β_1-hydrogen, illustrated in Fig.1.48. However, they also report a very considerable effect of defects on Ni{111} upon chemisorption. By sputtering the surface they were able to change the angular distribution of the β_1-state from a $\cos^n\theta$ dependence with $n = 4.9$ to one with $n = 1$. They attribute this change to unactivated chemisorption on defects created during the sputtering.

Very interesting results have also been obtained on another reasonably smooth plane of nickel, the {100}, for which *Christmann* et al. [1.133] had already noted a surprising sensitivity of $s(\Theta)$ to coverage. In work with nozzle beams of H_2 and D_2, *Hamza* and *Madix* [1.146] find no effects of surface temperature upon the initial rate of chemisorption. However, the initial sticking coefficient increases as the translational energy of the incident molecules was increased. From measurements at different angles, *Hamza* and *Madix* conclude that it is again the normal component of the translational energy which is effective in overcoming the barrier of 1.2 kcal/mol. Inasmuch as the chemisorption of D_2 occurs somewhat more slowly than that of H_2, they believe that tunneling through the barrier may be significant. As the amount of gas on the surface was increased, the sticking coefficient on {100} de-

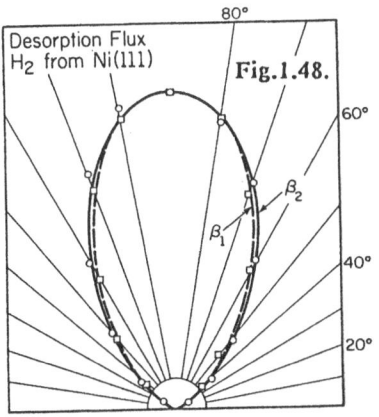

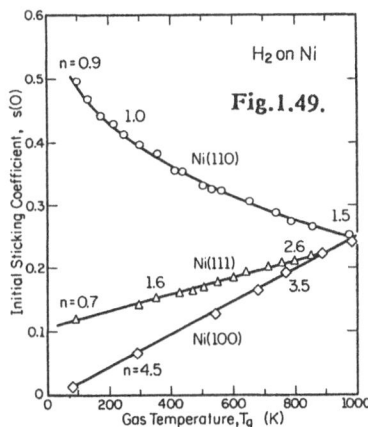

Fig.1.48. Angular dependence of hydrogen desorption from Ni(111) [1.145]. Both β_1- and β_2-state give a distribution that differs from a $\cos\theta$ dependence

Fig.1.49. Effect of gas temperature T_g upon the sticking coefficient $s(0)$, and upon the angular dependence of the sticking coefficient, for hydrogen on different planes of nickel [1.41]. The parameter n describes the angular dependence through the relation $\cos^n\theta$

creased steadily. All these results taken together certainly indicate that hydrogen chemisorption on Ni{100} is translationally activated.

The various studies of hydrogen chemisorption on nickel surfaces have been summarized and extended recently by *Rendulic* et al. [1.41]. As shown in Fig.1.49, they have carried out further measurements of the dependence of the sticking coefficient on Ni{111}, {110}, and {100} upon the angle of incidence. The small dependence of the sticking coefficient on {111} upon the temperature of the surface, which has been noted earlier, leads them to suggest possible contributions from a precursor process. Results on {110} are more interesting. On this plane the angular dependence of the sticking coefficient becomes more peaked, that is the exponent n in the function $\cos^n\theta$ used to represent the angular distribution becomes larger as the gas temperature increases. This is contrary to the trend expected for chemisorption by passage of fast molecules over a one-dimensional barrier. From the angular distribution at 950 K, *Rendulic* et al. deduce a chemisorption barrier of 0.8 kcal/mol, and they rationalize the variation with T_g as due to competition between two processes: unactivated chemisorption via a precursor at low gas temperatures, and directly from the gas at high temperatures. A similar sharpening of the angular distribution of the sticking coefficient at higher gas temperatures is also found on Ni{100}, but of course for this plane the rate of chemisorption rises with T_g. *Rendulic* et al. [1.41] again invoke the idea of two competing reaction paths: the dominant process is direct chemisorption from the gas over an activation barrier of $\simeq 1.4$ kcal/mol; at low temperatures, however, there may be some contributions from an unactivated precursor process.

51

Only a few years ago chemisorption on nickel was believed to occur in the way usual on the majority of transition metals. It is evident now that this is not the case on two smooth planes, Ni{111} as well as Ni{100}. On these, chemisorption is activated, and impinging molecules require translational energy to become dissociatively chemisorbed. There are hints that activated processes may also contribute on the much rougher Ni{110}. Just as in the chemisorption of alkanes discussed earlier, there are also suggestions that at lower temperatures the activated chemisorption of hydrogen on Ni{111} and {100} faces competition from nonactivated adsorption channels. There obviously is much that still must be done to explore the chemisorption kinetics of hydrogen on nickel. It is already clear, however, from the available material that on densely packed planes of fcc, bcc, and hcp metals dissociative chemisorption cannot proceed easily.

1.4.5 Slow Chemisorption of Nitrogen

a) Nitrogen on Platinum Group Metals

Chemisorption from molecular nitrogen is limited to a few of the transition elements. In the sixties and earlier, this high selectivity was ascribed to the extraordinarily strong bonding in the nitrogen molecule (dissociation energy D = 226 kcal/mol). It was presumed that on nickel and on the platinum metals chemisorption of nitrogen from the molecular gas was endothermic; that is, the energy of the bond formed in chemisorption between nitrogen and the surface was smaller than D/2, and chemisorption was limited by thermodynamic factors. If nitrogen atoms were formed on the surface, by decomposition of weakly bonded molecules containing nitrogen atoms, such as NH_3 or NO, their recombination with other nitrogen atoms and evolution as a molecular gas would be inhibited by the barrier to surface diffusion, as was illustrated in Fig.1.12. Diffusion of nitrogen atoms on atomically rough surfaces is known to be a slow process at low temperatures; on tungsten, for example, the activation energy for nitrogen atom diffusion can be as high as 35 kcal/mol [1.147]. Nitrogen layers, even if endothermic, could therefore be maintained in a metastable state on the surface up to moderate temperatures.

This interpretation of the very limited chemisorption of molecular nitrogen had to be completely revised in the seventies, as a consequence of observations by *Wilf* and *Dawson* [1.33] on platinum. In studies with carefully cleaned platinum wires, they were able to chemisorb nitrogen by long exposures of the sample to molecular nitrogen. As indicated in Fig.1.50, the rate of chemisorption was zero-order in the pressure. The layer so formed was quite stable at room temperature, and desorbed by second-order kinetics with an activation energy of 19 kcal/mol. *Wilf* and *Dawson* were able to enhance the rate of chemisorption by first activating the nitrogen on a hot tungsten

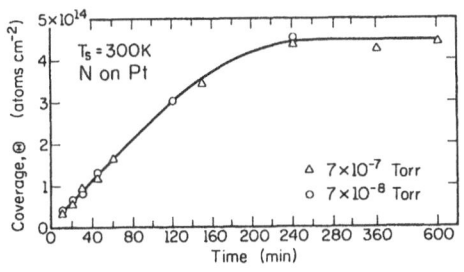

Fig.1.50. Chemisorption of nitrogen on platinum [1.33]

filament [1.148] in line of sight with the platinum surface. They concluded that chemisorption of nitrogen from the molecular gas occurred at active sites on the surface, with either dissociation or diffusion from the sites limiting the rate of chemisorption.

The important fact in these experiments is that dissociative chemisorption from the molecular gas occurs and is thermodynamically favorable at room temperatures. The frequently reported lack of nitrogen chemisorption on platinum, and possibly on other metals as well, must therefore be attributed to kinetic limitations, that is, to slow chemisorption. Rather similar results were obtained for nitrogen chemisorbing on nickel, another metal which was known not to chemisorb from the molecular gas. *Grunze* et al. [1.34] in 1979 found that long exposures of N_2 ($\simeq 10^6$ Langmuir) to Ni{110} at 591 K brought about chemisorption; the adsorbed material desorbed in two stages, with activation energies of 26 and 43 kcal/mol. Although chemisorption in these experiments may have been aided by the presence of excited species formed in the ion pump, there is little doubt that chemisorbed nitrogen layers do form in the chemisorption of NO or NH_3, and that these layers are stable at elevated temperatures [1.35]. This conclusion emerges from the desorption energies for N_2 reported for various metal surfaces and listed in Table 1.1. The desorption energies are too high to be attributed to a diffusion barrier.

Table 1.1. Desorption Energies E_{des} for Nitrogen

Surface	E_{des} [kcal/mol]	Investigator
Ni{111}	52	*Conrad* et al. [1.149]
	50	*Benziger* and *Preston* [1.150]
Ni{110}	43, 26	*Grunze* et al. [1.34]
Pd{331}	33	*Davies* and *Lambert* [1.151]
Pt	19	*Wilf* and *Dawson* [1.33]
Pt{111}	28	*Comrie* et al. [1.152]
	23	*Schwaha* and *Bechtold* [1.153]
Ir{110}	36–45	*Ibbotson* et al. [1.154]

It appears therefore that chemisorption on nickel and on the platinum metals is activated and occurs only if an appreciable barrier can be overcome. At the moment there is no information available concerning the magnitude of the activation energy involved, but from the lack of reactivity toward nitrogen by platinum metals at room temperature, the barrier should be on the order of magnitude of 10 kcal/mol or higher. It would certainly be interesting to subject the interactions between nitrogen and the platinum metals to the kind of scrutiny given to the interactions of hydrogen with nickel.

b) Chemisorption of Nitrogen on Iron

The interaction of nitrogen with iron is one of the classic examples of activated chemisorption. *Emmett* and *Brunauer* in 1934 [1.38] already wrote that "the rate of nitrogen adsorption is of the right magnitude to be the slow step in the catalytic synthesis of NH_3." Studies in *Beeck's* laboratories [1.155] indicated that chemisorption on iron films was slow, but it remained for *Bozso* et al. [1.36, 37] in *Ertl's* laboratory to carry out the first quantitative studies under ultrahigh vacuum conditions. Their work, summarized by the results in Fig. 1.10, revealed that on Fe{110} and {100} chemisorption is indeed activated and occurs over a barrier of 7 and 5 kcal/mol, respectively. From the dependence of the rate on surface temperature *Bozso* et al [1.36] deduced an activation energy essentially equal to zero on Fe{111}. Nevertheless, the rate of chemisorption on this surface at 500 K was only $\simeq 20$ times the rate on {100}. On all the surfaces studied, nitrogen forms an atomically bound state, with desorption energies of 58 kcal/mol on {100} and 51 kcal/mol on {111} and {110}. At low temperatures ($T_s = 140$ K) and rather high pressures, a molecularly bound α-state of nitrogen with an adsorption energy of 7.5 kcal/mol was also identified on Fe{111}.

Considerably more work has been done since these early studies to characterize chemisorption on Fe{111}. Even though this reaction is strictly speaking not activated, we will nevertheless consider nitrogen chemisorption on Fe{111}, as it turns out to be quite surprising. *Ertl* et al. [1.156] continued their investigation of nitrogen on iron with a detailed characterization of the kinetics of the molecular α-state at low temperatures, and with further measurements on the β-nitrogen layer atomically adsorbed on iron {111} at higher temperatures. They conclude that the condensation coefficient into the α-state, which acts as a precursor to atomically held nitrogen on the surface, is about a factor of 100 lower than usual. Furthermore, the prefactor for desorption out of the α-state is orders of magnitude higher than for conversion into the chemisorbed state. It is these two factors that conspire to give very low values for the observed sticking coefficient; at zero coverage, $s(0) = 2.2 \cdot 10^{-6} \exp(-E_a/kT)$, with $E_a = -0.8$ kcal/mol. *Böheim* et al. [1.157] have pointed out that energy transfer from the gas into the phonon modes of the solid is likely to be ineffective if the translational energy is comparable to the activation barrier leading to the chemisorbed state. They also report introductory experiments with a supersonic nitrogen beam impinging on Fe{110}. In view of the barrier of 7 kcal/mol previously found for this chemisorption process, it is not surprising that no chemisorbed nitrogen was detected in these limited experiments.

Grunze and his collaborators [1.158] have since subjected the molecular adsorption of nitrogen on Fe{111} to close scrutiny. By carrying out adsorption at ever lower temperatures, most recently at 64 K, they have been able to identify three molecular entities on the surface. Of these, α-nitrogen, the precursor to the atomically chemisorbed state, is made up of molecules π-bonded to the surface and lying flat on it [1.115]. At temperatures below 80 K, an additional γ-state has been observed; its desorption energy has been estimated at roughly 6 kcal/mol, and XPS [1.159] as well as HREELS [1.160] measurements suggest that the molecular axis of the γ-molecules is perpendicular to the surface. Molecules bound in this state have been observed to convert into the α-state; that is, they can act as precursors to the α-state. At still lower temperatures and at high γ-populations, an additional molecular state of binding, δ, has recently been isolated [1.158]; it has a desorption energy of < 5 kcal/mol and at temperatures below 100 K, nitrogen molecules in this δ-state are precursors to the formation of γ-nitrogen.

At higher temperatures, where the dissociative chemisorption of nitrogen becomes important, the γ-state populates directly from the gas phase and we therefore need not consider the δ-state further. A schematic potential diagram which summarizes the current state of understanding [1.161] in the system nitrogen-Fe{111} is given in Fig.1.51. It should be noted that despite the presence of the weakly bound molecular states γ and δ, it appears from recent studies summarized by *Grunze* et al. [1.161] that the dominant role in the chemisorption of

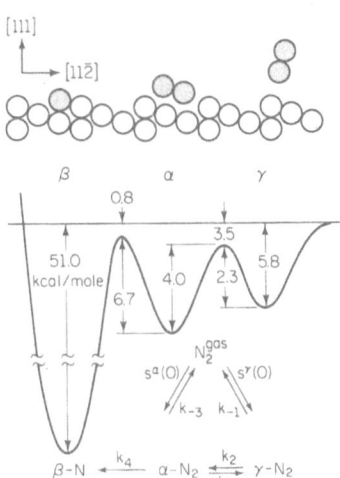

Fig.1.51. Schematic of configuration and potential energy for nitrogen on Fe{111} [1.161]

nitrogen on Fe{111}, that is in the formation of the β-state, is played by α-nitrogen. At temperatures $T_s \geq 214$ K, direct adsorption of nitrogen from the gas into α predominates in filling this precursor to dissociative chemisorption, as was already recognized by *Ertl* et al. [1.156].

After this fairly detailed outline of the energetics and kinetics of the molecular processes of nitrogen on the {111} plane of iron, two facts must be emphasized: (1) chemisorption from the gas phase is *not* activated. The activation energy for chemisorption determined by *Ertl* et al. [1.156] is -0.8 kcal/mol. (2) All the experiments upon which this detailed scheme has been built were done by controlling the surface temperature, so that only the precursor processes have been sampled.

Just recently, *Rettner* and *Stein* [1.162] have reported an experiment in which nitrogen from a supersonic beam was allowed to chemisorb on Fe{111}. Their measurements (Fig.1.52) suggest that translational energy increases the sticking coefficient of N_2. This is a bit unexpected, as there is not supposed to be a barrier to the dissociative chemisorption of nitrogen on Fe{111} from the gas phase. On running the nozzle source at 2000 K instead of 300 K, but at comparable translational energies, *Rettner* and *Stein* find uniformly higher rates at the higher temperatures. At 2000 K some atomization of molecular nitrogen has been reported at low pressures [1.148]. Atomization is probably negligible under the conditions of these experiments, however, and *Rettner* and *Stein* conclude that vibrational excitation of the nitrogen molecules enhances their sticking on Fe{111}.

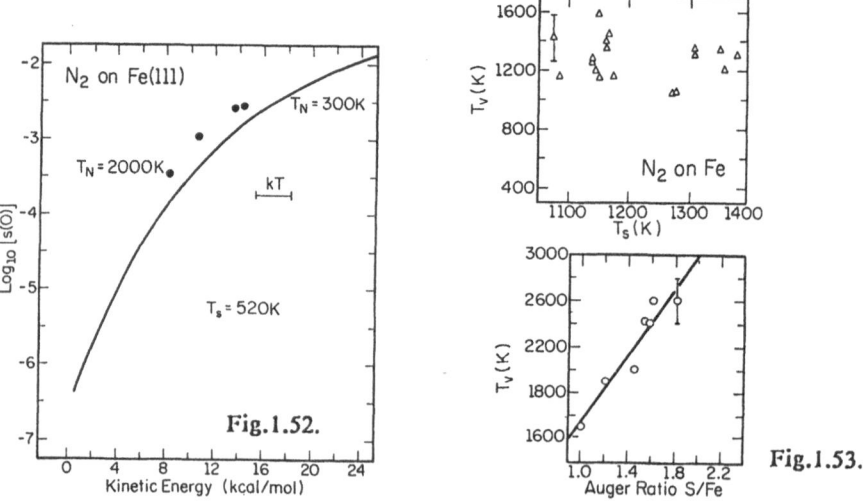

Fig.1.52. Effect of translational energy and nozzle temperature upon chemisorption of nitrogen from a supersonic beam on Fe{111} [1.162]

Fig.1.53. Vibrational temperature of nitrogen molecules evolved after permeation through iron diaphragm [1.163]. Top: Effect of surface temperature. Bottom: Effect of sulfur contamination

Again this would not have been predicted from previous studies. *Thorman* and *Bernasek* [1.163] determined the vibrational states of nitrogen molecules evolved from polycrystalline iron membranes after permeation by studying their electron beam induced fluorescence. For clean iron surfaces they found vibrational populations in keeping with the temperature of the surface, as shown in Fig.1.53. Although it is generally accepted that the results of permeation studies cannot automatically be equated to desorption measurements, and *Thorman* and *Bernasek* did not work on {111} surfaces, the effectiveness of the vibrational excitation observed in the molecular beam experiments is unexpected. It appears once more that the reaction channels for dissociative chemisorption probed in higher energy beam experiments may be rather different than the channels important under the usual thermal conditions. Chemisorption on iron studied with ordinary thermal beams of nitrogen would now be most interesting to round out the experiments in which chemisorption has been studied at different surface temperatures.

1.4.6 A Surprise: Oxygen on W{110}

The chemisorption of oxygen on tungsten has been of scientific interest since the days of *Langmuir* [1.164], and it has long been known that this is a rapid process, which occurs with a high initial sticking coefficient. Of the many studies devoted to the kinetics of oxygen chemisorption, the fairly recent work of *Wang* and *Gomer* [1.165] is of special interest: it affords a comparison of the behavior of the {110} and {100} planes of tungsten. As appears from the data in Fig.1.54, the {100} plane is more reactive; the initial sticking coefficient of oxygen is close to unity on this surface, whereas on a {110} plane at 300 K only half the incident molecules stick. The effects of surface temperature and of gas temperature upon the initial rate of chemisorption on W{110} are plotted in Fig.1.55. When the surface is maintained at temperatures $T_s \leq 200$ K, raising the temperature of the gas decreases the

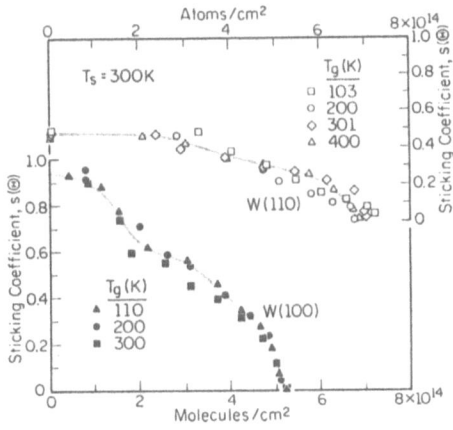

Fig.1.54. Oxygen chemisorption on tungsten at low temperatures [1.165]

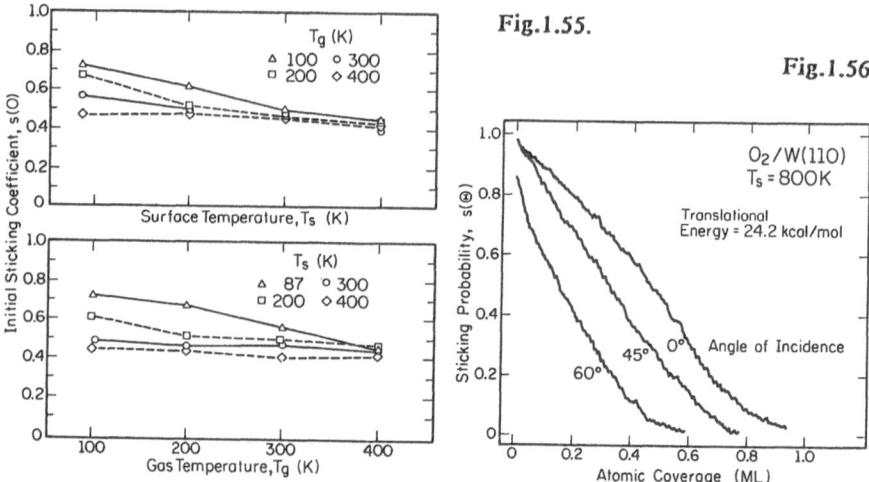

Fig.1.55.

Fig.1.56.

Fig.1.55. Effects of gas and of surface temperature upon oxygen chemisorption on W{110} at low temperatures [1.165]

Fig.1.56. Dependence of sticking coefficient upon coverage for oxygen molecules with 24.2 kcal/mol translational energy on W{110} [1.42]

initial rate of chemisorption; similarly, at gas temperatures $T_g \le$ 400 K, an increase in the temperature of the surface also lowers the rate. The qualitative dependence of the rate of oxygen chemisorption on W{110} upon temperature and upon coverage appears to conform to what is expected for a standard precursor process. *Wang* and *Gomer*, however, suggest that at low coverages, direct chemisorption rather than precursor processes appear to play the dominant role. Regardless of that, what is impressive about the chemisorption of oxygen on W{110} is its speed. With the surface at 87 K, for example, and the beam at 100 K, the sticking coefficient on the bare surface exceeds 0.7.

With this as a background, the results of the recent supersonic beam studies of oxygen chemisorption on W{110} reported by *Rettner* et al. [1.42] come as a considerable surprise. In Fig.1.11 is plotted the initial sticking coefficient found in this work as a function of the translational energy normal to the {110}, maintained at T_s = 800 K. At very low incident energies, the sticking coeffient seems to decrease as the energy is raised, but above 0.7 kcal/mol the sticking coefficient rises with increasing translational energy, reaching a value close to unity at an energy of 9 kcal/mol. At high incident energies the coverage dependence of the sticking coefficient, shown in Fig.1.56, is typical of what is expected for direct or activated chemisorption, falling monotonically as the amount of oxygen on the surface increases. From the increase in s(0) with translational energy, *Rettner* et al. deduce a mean barrier to chemisorption of 2.8 kcal/mol, but they postulate that precursor processes account for the behavior of the sticking coefficient at low incident energies.

Additional information about the behavior of oxygen on tungsten is clearly desirable. It will be interesting to test the effect of beam temperature at constant translational energy upon the rate of chemisorption, in order to sort out possible contributions from internal molecular motions. At the highest source temperature, $T \simeq 900$ K, atomization of oxygen is certainly negligible, but higher vibrational states of the molecule may be significant in accelerating the rate of chemisorption. Even without additional information, it appears already that chemisorption of oxygen at high translational energies occurs in a different fashion from chemisorption at lower energies. This is evident from the supersonic beam measurements [1.42] done at normal energies below 0.7 kcal/mol, and by comparison of the present studies with the earlier work using thermal beams [1.165], in which increasing the energy of the gas brings about a diminution of the sticking coefficient. At low surface and gas temperatures, the majority of the oxygen molecules incident upon W{110} are chemisorbed. We expect the orientation of the incoming molecules to play some role in direct chemisorption from the gas; chemisorption may, for example, be favored if the molecular axis is parallel to the interface. The effect of high translational energies may just be to force the oxygen molecules into a configuration on the surface from which chemisorption can occur more easily.

1.5 Summary

The material reviewed here is too diverse and fragmented to permit a few simple conclusions. One impression, however, should be immediate: activated chemisorption is not an isolated curiosity. Rather, it is a very widespread phenomenon, of interest in its own right, but also of interest for the insights it may yield about the mechanism of chemisorption in general. H.S. Taylor's original views seem to have had much merit.

The amount of detail available even about the most intensely examined system undergoing activated adsorption is still small. There are large gaps in our knowledge, and straightforward studies by presently available techniques could considerably enrich the current understanding of activated chemisorption. Molecular beam studies of hydrogen chemisorption on silicon, for example, are long overdue. Measurements to establish the activation energy for dissociative chemisorption of nitrogen on platinum metals would be highly desirable, even though they may prove more difficult. Concerted efforts to explore the mechanism of chemisorption on densely packed metal surfaces should also be rewarding.

It must be admitted, however, that we are at present seriously limited by the techniques available for probing the pathways important in overcoming chemisorption barriers. Much new information about the kinetics of chemisorption is being generated by using supersonic molecular beams to study surface encounters. Such experiments have demonstrated that translational energy is effective in promoting chemisorption. However, there are indications in several systems that dissociative chemisorption can also occur via alternative, lower energy paths which may be dominant in ordinary thermal environments. It would be immensely important to have the capability of doing state-selected chemistry in the systems examined above, in order to come to grips with the role of internal molecular degrees of freedom in affecting the kinetics of activated chemisorption. Given the continuing progress in tunable lasers, there is little doubt that such chemisorption studies from molecules in specified states are not far off.

One topic that has hardly been mentioned in this review is the theoretical effort to understand activated chemisorption. Theory, in the hands of *Lennard-Jones*, played a vital role in providing a conceptual foundation for activated chemisorption. With the amount of experimental information about the chemisorption process growing, the time should soon be appropriate for a more concerted theoretical effort to delineate the important material parameters that make it necessary to overcome an energy barrier in order to accomplish chemisorption in some systems, and not in others. At the moment, the more empirical approach of doing trajectory studies on trial potential surfaces at least offers the more immediate prospect of providing sounder foundations for the interpretations of experiments.

Acknowledgments. In the preparation of this review I have benefited greatly from discussions with present and former coworkers, and with friends at various laboratories. I am especially indebted to S. Brass, A. Cassuto, R.S. Chambers, G. Comsa, T. Engel, R. Liu, M. Grunze, and T.-C. Lo. This work was made possible by the support of the National Science Foundation under Grant NSF DMR 84-20751.

References

1.1   I. Langmuir: J. Am. Chem. Soc. **38**, 2221 (1916)
1.2   W. Frankenburg, A. Hodler: Trans. Faraday Soc. **28**, 229 (1932)
1.3   H.S. Taylor: J. Am. Chem. Soc. **53**, 578 (1931)
1.4   J.E. Lennard-Jones: Trans. Faraday Soc. **28**, 333 (1932)
1.5   A.J. Allmand, R. Chaplin: Trans. Faraday Soc. **28**, 223 (1932)
1.6   H.S. Taylor: Trans. Faraday Soc. **28**, 136 (1932)
1.7   J.K. Roberts: Proc. Roy. Soc. A **152**, 445 (1935)
      J.K. Roberts: *Some Problems in Adsorption* (University Press, Cambridge 1939)
1.8   O. Beeck, A.E. Smith, A. Wheeler: Proc. Roy. Soc. A **177**, 62 (1940)
1.9   O. Beeck: Rev. Mod. Phys. **17**, 61 (1945)
      O. Beeck: Disc. Faraday Soc. **8**, 118 (1950)
      O. Beeck, W.A. Cole, A. Wheeler: Disc. Faraday Soc. **8**, 314 (1950)
      O. Beeck: Adv. Catalysis **2**, 151 (1950)
      A. Wheeler, in: *Structures and Properties of Solid Surfaces*, ed. R. Gomer, C.S. Smith (Univ. of Chicago Press, Chicago 1953) Chap.XIII

1.10 R. Gomer, R. Wortman, R. Lundy: J. Chem. Phys. **26**, 1147 (1957)
R. Wortman, R. Gomer, R. Lundy: J. Chem. Phys. 27, 1099 (1957)
1.11 R. Gomer, J.K. Hulm: J. Chem. Phys. **27**, 1363 (1957)
1.12 R. Gomer: *Field Emission and Field Ionization* (Harvard Univ. Press, Cambridge 1961)
1.13 J.A. Becker, C.D. Hartman: J. Phys. Chem. **57**, 157 (1953)
1.14 G. Ehrlich: J. Phys. Chem. **60**, 1388 (1956)
1.15 G. Ehrlich: J. Chem. Phys. **34**, 29 (1961)
1.16 G. Ehrlich: J. Chem. Phys. **34**, 39 (1961)
1.17 L. Apker: Ind. Eng. Chem. **40**, 846 (1948)
1.18 D.A. King, in: *Chemistry and Physics of Solid Surfaces*, Vol.II, ed. R. Vanselow (CRC Press, Boca Raton 1979)
D.A. King, M.G. Wells: Proc. Roy. Soc. (London) A **339**, 245 (1974)
1.19 C. Kemball: Proc. Roy. Soc. A **207**, 539 (1951)
1.20 D.D. Eley, D.R. Rossington, in: *Chemisorption*, ed. W.E. Garner (Butterworth Scientific Publ., London 1957) p.137
1.21 K. Tamaru: J. Phys. Chem. **61**, 647 (1957)
1.22 T.A. Delchar, G. Ehrlich: J. Chem. Phys. **42**, 2686 (1965)
1.23 T.E. Madey, J.T. Yates Jr.: Nuovo Cimento Suppl. **5**, 483 (1967)
1.24 P.W. Tamm, L.D. Schmidt: Surf. Sci. **26**, 286 (1971)
1.25 R.S. Polizzotti: *Structure-Sensitive Chemisorption*, Rpt. R-646, Coord. Sci. Lab., Univ. of Illinois-Urbana, UILU-ENG 74-2210 (April 1974)
1.26 R.S. Polizzotti, G. Ehrlich: J. Chem. Phys. **71**, 259 (1979)
1.27 R. Liu, G. Ehrlich: J. Vac. Sci. Technol. **13**, 310 (1976)
1.28 R. Liu: *Chemisorption on Perfect Surfaces and Structural Defects*, Rpt. R-790, Coord. Sci. Lab., Univ. of Illinois-Urbana, UILU-ENG 77-2237 (Oct.1977)
1.29 R. Liu, G. Ehrlich: Surf. Sci. **119**, 207 (1982)
1.30 C.N. Stewart, G. Ehrlich: Chem. Phys. Lett. **16**, 203 (1972)
1.31 M. Balooch, M.J. Cardillo, D.R. Miller, R.E. Stickney: Surf. Sci. **46**, 358 (1974)
1.32 G.C. Bond: *Catalysis by Metals* (Academic, New York 1962) Chap.5
1.33 M. Wilf, P.T. Dawson: Surf. Sci. **60**, 561 (1976)
1.34 M. Grunze, R.K. Driscoll, G.N. Burland, J.C.L. Cornish, J. Pritchard: Surf. Sci. **89**, 381 (1979)
1.35 For a review see G. Ertl, in: *The Nature of the Surface Chemical Bond*, ed. T.N. Rhodin, G. Ertl (North Holland, Amsterdam 1979) Chap.5
1.36 F. Bozso, G. Ertl, M. Grunze, M. Weiss: J. Catalysis **49**, 18 (1977)
1.37 F. Bozso, G. Ertl, M. Weiss: J. Catalysis **50**, 519 (1977)
1.38 P.H. Emmett, S. Brunauer: J. Am. Chem. Soc. **56**, 35 (1934)
1.39 M. Salmerón, R.J. Gale, G.A. Somorjai: J. Chem. Phys. **70**, 2807 (1979)
1.40 H.J. Robota, W. Vielhaber, M.C. Lin, J. Segner, G. Ertl: Surf. Sci. **155**, 101 (1985)
1.41 For a review of the work at Graz see K.D. Rendulic, A. Winkler, H. Karner: J. Vac. Sci. Technol. A **5**, 488 (1987)
1.42 C.T. Rettner, L.A. Delouise, D.J. Auerbach: J. Chem. Phys. **85**, 1131 (1986)
1.43 J.A. Appelbaum, D.R. Hamann: Rev. Mod. Phys. **48**, 479 (1976)
1.44 G. Ehrlich: J. Chem. Phys. **31**, 1111 (1959)
1.45 This material is drawn largely from R.S. Chambers, G. Ehrlich: Surf. Sci. **186**, L535 (1987)
1.46 For a considerably more detailed discussion see A. Cassuto, D.A. King: Surf. Sci. **102**, 388 (1981)
1.47 A. Frennet, G. Liénard: J. chim. Phys. **67**, 598 (1970)
1.48 S.G. Brass, G. Ehrlich: Phys. Rev. Lett. **57**, 2532 (1986)

1.49 J.W. Gadzuk, J.K. Nørskov: J. Chem. Phys. **81**, 2828 (1984)
J.W. Gadzuk, S. Holloway: Chem. Phys. Lett. **114**, 314 (1985); Phys. Scr. **32**, 413 (1985); J. Chem. Phys. **84**, 3502 (1986)
J.W. Gadzuk: J. Chem. Phys. **86**, 5196 (1987); J. Vac. Sci. Technol. A **5**, 492 (1987)
1.50 S. Holloway, J.W. Gadzuk: J. Chem. Phys. **82**, 5203 (1985); Surf. Sci. **152/153**, 838 (1985)
S. Holloway: J. Vac. Sci. Technol. A **5**, 476 (1987)
1.51 An excellent review of the relevant kinetic theory is given by G. Comsa, R. David: Surf. Sci. Rpts. **5**, 145 (1985)
1.52 W. van Willigen: Phys. Lett. **28A**, 80 (1968)
1.53 G. Comsa, R. David, B.-J. Schumacher: Surf. Sci. **95**, L210 (1980)
1.54 J.C. Polanyi, W.H. Wong: J. Chem. Phys. **51**, 1439 (1969)
1.55 J.C. Polanyi: Science **236**, 680 (1987)
1.56 R. Suhrmann, H.J. Busse, G. Wedler: Z. Phys. Chem. NF **47**, 1 (1965)
1.57 G. Maire, J.R. Anderson, B.B. Johnson: Proc. Roy. Soc. (London) A **320**, 227 (1970)
1.58 A. Frennet: Catalysis Rev. - Sci. Eng. **10**, 37 (1974)
1.59 C.N. Stewart Jr.: *Dynamics of Dissociative Chemisorption - Methane on Rhodium*, Rpt. R-638, Coord. Sci. Lab., Univ. of Illinois-Urbana, UILU-ENG 73-2242 (Dec.1973)
1.60 S.G. Brass: *Dissociative Chemisorption of Methane on Transition Metals*, Rpt. R-939, Coord. Sci. Lab., Univ. of Illinois-Urbana, UILU-ENG 82-2205 (Feb.1982)
1.61 J.T. Yates Jr., T.E. Madey: Surf. Sci. **28**, 437 (1971)
1.62 S. Hellwig, J.H. Block: Surf. Sci. **29**, 523 (1972)
1.63 B.J. Hopkins, G.R. Shah: Vacuum **22**, 267 (1972)
1.64 R.A. Shigeishi: Surf. Sci. **51**, 377 (1975); **72**, 61 (1978)
1.65 C.N. Stewart, G. Ehrlich: J. Chem. Phys. **62**, 4672 (1975)
1.66 N.B. Slater: *Theory of Unimolecular Reactions* (Cornell Univ. Press, Ithaca 1959)
1.67 H.F. Winters: J. Chem. Phys. **62**, 2454 (1975)
1.68 H.F. Winters: J. Chem. Phys. **64**, 3495 (1976)
1.69 L.H. Jones, R.S. McDowell: J. Mol. Spectrosc. **3**, 632 (1959)
1.70 G. Hubbert, T.G. Kyle, G.J. Troup: J. Quant. Spectrosc. Radiat. Transfer **9**, 1469 (1969)
1.71 J.T. Yates Jr., J.J. Zinck, S. Sheard, W.H. Weinberg: J. Chem. Phys. **70**, 2266 (1979)
1.72 S.G. Brass, D.A. Reed, G. Ehrlich: J. Chem. Phys. **70**, 5244 (1979)
1.73 S.G. Brass, G. Ehrlich: Surf. Sci. **187**, 21 (1987)
1.74 S.G. Brass, G. Ehrlich: J. Chem. Phys. **87**, 4285 (1987)
1.75 T.-C. Lo: *Dynamics of Dissociative Chemisorption - Methane on Tungsten Single Crystal Surfaces*, Ph.D. Thesis, Dept. of Physics, Univ. of Illinois at Urbana-Champaign (Feb.1983)
1.76 C.T. Rettner, H.E. Pfnür, D.J. Auerbach: Phys. Rev. Lett. **54**, 2716 (1985)
1.77 C.T. Rettner, H.E. Pfnür, D.J. Auerbach: J. Chem. Phys. **84**, 4163 (1986)
1.78 E. Kemble: *The Fundamental Principles of Quantum Mechanics* (McGraw-Hill, New York 1937) Sect.21j
1.79 T.-C. Lo, G. Ehrlich: Surf. Sci. **179**, L19 (1987)
1.80 Numerical integration was used rather than the approximation in [1.79] to avoid problems pointed out to us by Dr. B.D. Kay, Sandia
1.81 A.V. Hamza, H.-P. Steinrück, R.J. Madix: J. Chem. Phys. **85**, 7494 (1986)
1.82 H.P. Steinrück, A.V. Hamza, R.J. Madix: Surf. Sci. **173**, L571 (1986)
1.83 A.V. Hamza, R.J. Madix: Surf. Sci. **179**, 25 (1987)

1.84 A.V. Hamza, H.-P. Steinrück, R.J. Madix: J. Chem. Phys. **86**, 6506 (1987)

1.85 P.D. Szuromi, W.H. Weinberg: Surf. Sci **149**, 226 (1985)

1.86 M.B. Lee, Q.Y. Yang, S.L. Tang, S.T. Ceyer: J. Chem. Phys. **85**, 1693 (1986)

1.87 S.T. Ceyer, J.D. Beckerle, M.B. Lee, S.L. Tang, Q.Y. Yang, M.A. Hines: J. Vac. Sci. Technol. A **5**, 501 (1987)

1.88 A.F.H. Ward: Proc. Roy. Soc. A **133**, 506 (1931)

1.89 T. Kwan: Adv. Catalysis **6**, 67 (1954)

1.90 J. Pritchard, F.C. Tompkins: Trans. Faraday Soc. **56**, 540 (1960)

1.91 J. Pritchard: Trans. Faraday Soc. **59**, 437 (1963)

1.92 C.S. Alexander, J. Pritchard: J. Chem. Soc., Faraday Trans. I **68**, 202 (1972)

1.93 J. Pritchard, T. Catterick, R.K. Gupta: Surf. Sci. **53**, 1 (1975)

1.94 M. Balooch, R.E. Stickney: Surf. Sci **44**, 310 (1974)

1.95 A. Gelb, M.J. Cardillo: Surf. Sci. **59**, 128 (1975); **64**, 197 (1977); **75**, 199 (1978)

1.96 G. Comsa, R. David: Surf. Sci. **117**, 77 (1982)

1.97 G.D. Kubiak, G.O. Sitz, R.N. Zare: J. Chem. Phys. **83**, 2538 (1985)

1.98 J.T. Law: J. Phys. Chem. **59**, 543 (1955)

1.99 R.M. Dell: J. Phys. Chem. **61**, 1584 (1957)

1.100 M.J. Bennett, F.C. Tompkins: Trans. Faraday Soc. **58**, 816 (1962)

1.101 Yu.I. Belyakov, T.N. Kompaniets: Sov. Phys. - Tech. Phys. **17**, 674 (1972)

1.102 Yu.I. Belyakov, N.I. Ionov, T.N. Kompaniets: Sov. Phys. - Solid State **14**, 2567 (1973)

1.103 J.T. Law, E.E. Francois: J. Phys. Chem. **60**, 353 (1956)

1.104 J.T. Law: J. Chem. Phys. **30**, 1568 (1959)

1.105 M.J. Datsiev, Y.I. Belyakov: Sov. Phys. - Tech. Phys. **15**, 166 (1971)

1.106 Ch. Kleint, B. Hartmann, H. Meyer: Z. Phys. Chem. **250**, 315 (1972)

1.107 K.-D. Brzóska, Ch. Kleint: Thin Solid Films **34**, 131 (1976)

1.108 G. Schulze, M. Henzler: Surf. Sci. **124**, 336 (1983)

1.109 H. Froitzheim, H. Lammering, H.L. Günter: Phys. Rev. B **27**, 2278 (1983)

1.110 B.M. Rice, I. NoorBatcha, D.L. Thompson, L.M. Raff: J. Chem. Phys. **86**, 1608 (1987)

1.111 L.M. Raff, I. NoorBatcha, D.L. Thompson: J. Chem. Phys. **85**, 3081 (1986)
      I. NoorBatcha, L.M. Raff, D.L. Thompson: J. Chem. Phys. **83**, 1382 (1985)

1.112 Note, however, that the bond dissociation energy for H_3Si-H is 90.3 kcal/mol.
      D.F. McMillen, D.M. Golden: Annu. Rev. Phys. Chem. **33**, 493 (1982)

1.113 S.C. Wang, R. Gomer: Surf. Sci. **141**, L304 (1984)

1.114 Ch. Steinbrüchel, L.D. Schmidt: Phys. Rev. B **10**, 4209, 4215 (1974)

1.115 M. Grunze, M. Golze, J. Fuhler, M. Neumann, E. Schwarz, in: *Proc. 8th Int'l Congr. on Catalysis*, Berlin 1984, Vol.IV (Dechema, Frankfurt/M. 1984) p.133

1.116 M. Asscher, J. Carrazza, M.M. Khan, K.B. Lewis, G.A. Somorjai: J. Catalysis **98**, 277 (1986)

1.117 A. Polak, G. Ehrlich: J. Vac. Sci. Technol. **14**, 407 (1977)
      A.J. Polak: *Desorption and Surface Diffusion - Nitrogen on Tungsten {110}*, Rpt. R-789, Coord. Sci. Lab., Univ. of Illinois-Urbana, UILU-ENG 77-2236 (Oct.1977)

1.118 R.C. Cosser, S.R. Bare, S.M. Francis, D.A. King: Vacuum **31**, 503 (1981)

1.119 H.E. Pfnür, C.T. Rettner, J. Lee, R.J. Madix, D.J. Auerbach: J. Chem. Phys. **85**, 7452 (1986)

1.120 L.R. Danielson, M.J. Dresser, E.E. Donaldson, J.T. Dickinson: Surf. Sci. **71**, 599 (1978)

1.121 J.N. Smith Jr., R.L. Palmer: J. Chem. Phys. **56**, 13 (1972)

1.122 S.L. Bernasek. G.A. Somorjai: J. Chem. Phys. **62**, 3149 (1975)

1.123 K.E. Lu, R.R. Rye: Surf. Sci. **45**, 677 (1976)

1.124 K. Christmann, G. Ertl, T. Pignet: Surf. Sci. **54**, 365 (1976)

1.125 K. Christmann, G. Ertl: Surf. Sci. **60**, 365 (1976)

1.126 I.E. Wachs, R.J. Madix: Surf. Sci. **58**, 590 (1976)

1.127 B. Poelsema, L.K. Verheij, G. Comsa: Surf. Sci. **152/153**, 496 (1985)

1.128 G. Ehrlich: J. Phys. Chem. Solids 1, 3 (1956)

1.129 G. Comsa: personal communication

1.130 E.G. Seebauer, A.C.F. Kong, L.D. Schmidt: Surf. Sci. **176**, 134 (1986)

1.131 R.L. Palmer, J.N. Smith, H. Saltsburg, D.R. O'Keefe: J. Chem. Phys. **53**, 1666 (1970)

1.132 J. Lapujoulade, K.S. Neil: J. Chem. Phys. **57**, 3535 (1972)

1.133 K. Christmann, O. Schober, G. Ertl, M. Neumann: J. Chem. Phys. **60**, 4528 (1974)

1.134 J. Lapujoulade, K.S. Neil: Surf. Sci. **35**, 288 (1973)

1.135 S. Andersson: Chem. Phys. Lett. **55**, 185 (1978)

1.136 A. Winkler, K.D. Rendulic: Surf. Sci. **118**, 19 (1982)

1.137 K.D. Rendulic, A. Winkler: J. Chem. Phys. **79**, 5151 (1983)

1.138 H.P. Steinrück, A. Winkler, K.D. Rendulic: J. Phys. C **17**, L311 (1984)

1.139 H.P. Steinrück, K.D. Rendulic, A. Winkler: Surf. Sci. **154**, 99 (1985)

1.140 H.P. Steinrück, M. Luger, A. Winkler, K.D. Rendulic: Phys. Rev. B **32**, 5032 (1985)

1.141 C.-Y. Lee, A.E. DePristo: J. Chem. Phys. **85**, 4161 (1986)

1.142 D.O. Hayward, A.O. Taylor: Chem. Phys. Lett. **124**, 264 (1986)

1.143 J.N. Russell Jr., S.M. Gates, J.T. Yates Jr.: J. Chem. Phys. **85**, 6792 (1986)

1.144 J.N. Russell Jr., I. Chorkendorff, A.-M. Lanzillotto, M.D. Alvey, J.T. Yates Jr.: J. Chem. Phys. **85**, 6186 (1986)

1.145 K.D. Rendulic, A. Winkler, H.P. Steinrück: Surf. Sci. **185**, 469 (1987)

1.146 A.V. Hamza, R.J. Madix: J. Phys. Chem. **89**, 5381 (1985)

1.147 G. Ehrlich, F.G. Hudda: J. Chem. Phys. **35**, 1421 (1961)

1.148 S.B. Nornes, E.E. Donaldson: J. Chem. Phys. **44**, 2968 (1966)

1.149 H. Conrad. G. Ertl, J. Küppers, E.E. Latta: Surf. Sci. **50**, 296 (1975)

1.150 J.B. Benziger, R.E. Preston: Surf. Sci. **141**, 567 (1984)

1.151 P.W. Davies, R.M. Lambert: Surf. Sci. **110**, 227 (1981)

1.152 C.M. Comrie, W.H. Weinberg, R.M. Lambert: Surf. Sci. **57**, 619 (1976)

1.153 K. Schwaha, E. Bechtold: Surf. Sci. **66**, 383 (1977)

1.154 D.E. Ibbotson, T.S. Wittrig, W.H. Weinberg: Surf. Sci. **110**, 294 (1981)

1.155 O. Beeck, A. Wheeler: J. Chem. Phys. 7, 631 (1939)

1.156 G. Ertl, S.B. Lee, M. Weiss: Surf. Sci. **114**, 515 (1982)

1.157 J. Böheim, W. Brenig, T. Engel, U. Leuthäusser: Surf. Sci. **131**, 258 (1983)

1.158 For a review, see M. Grunze, G. Strasser, M. Golze, W. Hirschwald: J. Vac. Sci. Technol. A **5**, 527 (1987)

1.159 H.-J. Freund, B. Bartos, R.P. Messmer, M. Grunze, H. Kuhlenbeck, M. Neumann: Surf. Sci. **185**, 187 (1987)

1.160 L.J. Whitman, C.E. Bartosch, W. Ho: J. Chem. Phys. **85**, 3688 (1986)

1.161 M. Grunze, G. Strasser, M. Golze: Appl. Phys. A **44**, 19 (1987)

1.162 C.T. Rettner, H. Stein: J. Chem. Phys. **87**, 770 (1987)

1.163 R.P. Thorman, S.L. Bernasek: J. Chem. Phys. **74**, 6498 (1981)

1.164 I. Langmuir: Chem. Rev. 6, 451 (1929)

1.165 C. Wang, R. Gomer: Surf. Sci. **84**, 329 (1979)

2. Physisorbed Rare Gas Adlayers

Klaus Kern and George Comsa

Institut für Grenzflächenforschung und Vakuumphysik
Kernforschungsanlage Jülich, D − 5170 Jülich, Fed. Rep. Germany

Within the Born-Oppenheimer approximation, the motion of a rare gas atom in front of a substrate surface can be described with a three dimensional potential energy relief, depending on two coordinates parallel $V(x, y)$ and one perpendicular to the surface $V(z)$. Since the electron affinity of the rare gases is negative and their ionization energy is large compared to the work function of almost all substrates, the electronic configuration of the rare gas atom is only slightly perturbed upon adsorption. Thus, rare gases interact with substrates mainly via induced dipole forces (van der Waals interaction), which can be modelled by a superposition of a long-range van der Waals attractive term (decaying as z^{-3}) and a short range repulsive term. The corresponding adsorption potential is shown in Fig.2.1; the minimum of $V(z)$ determines the adsorption energy in the limit of zero coverage. Particles which are bound to surfaces via these dispersion forces are called physisorbed. Typical physisorption energies range from several to hundreds of meV.

At low coverages, the state of the adsorbate is determined by the magnitude of the potential modulation $V(x,y)$ and the temperature. For temperatures which are large compared to the modulation energy, the adparticle can move freely on the surface. However, if the temperature is small compared to the modulation, the adparticle is fixed to its

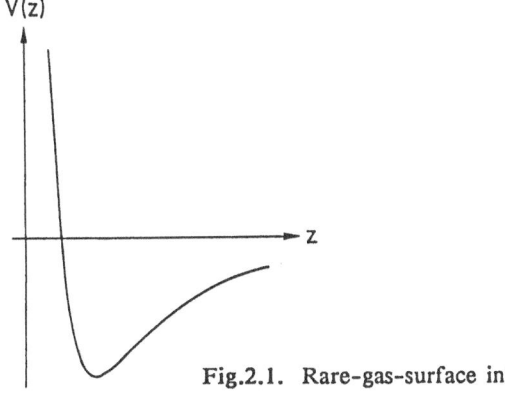

Fig.2.1. Rare-gas-surface interaction potential

adsorption site, i.e., the adsorption is localized. With increasing coverage, however, the mutual interactions between the adatoms gain importance. These interactions can be either direct or substrate mediated, repulsive or attractive in nature. It is the competition between these lateral adatom interactions and the potential modulation $V(x,y)$ which gives rise to a variety of phase transitions in the adsorbed layer. Phase transition is defined here as a change of the pair correlation function by changing the long range order, i.e., the structure, of the adlayer.

Physisorbed rare gas layers have attracted much interest during the last decade. This is not only because their investigation gives access to fundamental processes like condensation, melting or polymorphism, in low dimensional systems (2D) exhibiting enhanced fluctuation effects, but also because, from a theoretical point of view, the simple interaction forces allow first principle calculations.

Peierls [2.1] and *Landau* [2.2] discovered as early as the 30's that solids in two dimensions exhibit a number of remarkable properties. They demonstrated that for a 2D-solid the pair correlation function diverges algebraically at any finite temperature, i.e., strictly speaking, there should be no crystalline order in two dimensions. This rather spectacular observation remained almost unnoticed for decades. Today's interest in 2D-matter started with the remarkable studies of *Wegner* [2.3], *Jancovici* [2.4] and *Kosterlitz* and *Thouless* [2.5] in the late 60's and early 70's. In particular, *Kosterlitz* and *Thouless* [2.5] proposed a new criterion for crystalline order in two dimensions, "topological long-range order". Even if true long-range positional order does not exist, the crystalline axes could still be well defined, i.e., the angle that the bond between two adjacent atoms makes with some reference axis is an order parameter, and one can have bond orientational long-range order. Provided the shear modulus, μ, is finite, a two dimensional arrangement of atoms with long-range orientational order and algebraic positional order exhibits enough short- to medium-range order for a local crystalline structure to be defined. However, we should mention that the Peierls-Landau instability is almost negligible as long as we deal with crystallites less than say $\simeq 200$ nm in diameter. Indeed, as pointed out by *Abraham* [2.6], a 2D-solid has to be as large as $\simeq 10^{27}$ cm^2 in order to lose crystalline correlation equal to about one lattice spacing.

The resistance to shear is probably the best definition of a solid. *Kosterlitz* and *Thouless* [2.5] argued that at low temperatures there are no free dislocations present and hence the 2D-system is a "crystalline solid". In two dimensions, a dislocation is a defect in a perfect crystal in which half an extra row of atoms is added. At a certain temperature T_m, however, the free energy for spontaneous formation of free dislocations becomes negative, and the 2D-system will no longer resist shear, because an arbitrarily small shear stress will cause the free dislocations to move. Based on this concept, *Kosterlitz* and *Thouless* [2.5], *Halperin* and *Nelson* [2.7], and *Young* [2.8] developed a theory of two dimensional melting, initiating much of the current interest in 2D-solids. In contrast to 3D-systems, where melting is definitely a first order transition, they predicted that in two dimensional systems it has to be a second order one.

In addition, fascinating properties of 2D-solids arise from the fact that in experiment, any 2D-solid needs to be supported by a substrate. The physisorbed atoms form a modulated structure on the lattice of the substrate surface. The modulation arises from the competing interactions which favor different periodicities; the "natural" nearest-neighbor distance of the nonsupported adlayer differs in general from the lattice periodicity which the substrate potential tries to impose. The physisorbed adatoms form in general ordered structures. These may be commensurate with the substrate surface in certain ranges of coverage and temperature. However, when coverage and/or temperature is varied, the adlayer may contract or expand and become incommensurate with the supporting substrate. According to the present understanding of theory [2.9] these kind of phase transitions are driven by the spontaneous formation of misfit dislocations, so called domain walls.

The interest in adsorbed layers is not restricted to the understanding of fundamental two dimensional physics. As such films increase in thickness, they finally approach three dimensional behavior. The manner in which thin films grow is of great practical importance in various processes like adhesion, lubrication or production of submicron electronic devices. Investigation of the evolution from 2D to 3D behavior of rare gases is the most straightforward experiment to uncover the fundamental principles governing the growth properties.

After discussing the present status of the experimental techniques, we will examine structure, phase diagrams, phase transitions and lattice dynamics of physisorbed layers of the heavy rare gases Kr and Xe on a single crystal Pt{111} surface.

2.1 Experimental Techniques

2.1.1 General Remarks

The above discussion raises one main question: how can one probe the structure and dynamics of physisorbed adlayers with enough sensitivity and without disturbing these delicate systems? The most direct way to get such information is via a scattering experiment, the basic principle of which is illustrated in Fig.2.2. An incoming beam of probe particles

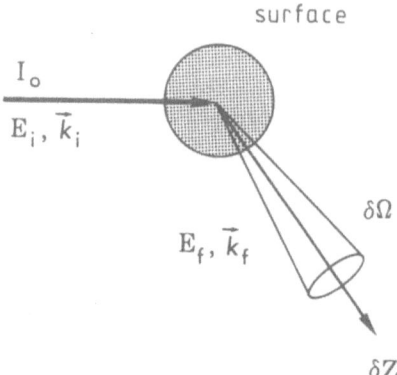

surface

Fig.2.2. Schematic sketch of a surface scattering experiment

I_o

$E_i, \vec{k}_i$

$\delta\Omega$

$E_f, \vec{k}_f$

δZ

(e.g., electrons, photons, neutrons or atoms) of wavevector k_i and energy E_i, impinges on a target, and the scattered intensity at wavevector k_f and energy E_f is measured in the solid angle element $\delta\Omega$. The complete information which can be deduced in such a scattering experiment, i.e., the behavior of the target particles in space and time, is contained in the double differential cross section:

$$\delta Z = I_0 \frac{d^2\sigma}{d\Omega dE_f} \delta\Omega \, \delta E_f \qquad (2.1)$$

where δZ is the number of probe particles in the energy interval δE_f scattered into the solid angle element $\delta\Omega$, and I_0 the intensity of the incoming probe particles. Except for trivial factors, the scattering cross section is determined by the energy exchange ($\hbar\omega$) and the momentum exchange (Q).

$$\frac{d^2\sigma}{d\Omega dE_f} \simeq S(Q,\hbar\omega)$$

$$\hbar\omega = E_f - E_i \, , \qquad Q = k_f - k_i \qquad (2.2)$$

The selection of the probe particles is determined by the characteristic energy, time and length scales of the phenomena to be investigated. In Table 2.1, we have collected a few characteristic parameters for physisorbed films of rare gases. In order to gain information on these different processes in a scattering experiment it is favorable when energy and momentum of the probe particles simultaneously match the characteristic parameters of the surface processes (energy, $\hbar\omega$, time, $1/\omega$, and length, $1/Q$). For particle waves (electrons, neutrons and atoms) the de Broglie relation gives $\lambda = \hbar/(2mE)^{1/2}$ whereas for electromagnetic waves (photons) the dispersion relation is $\lambda = \hbar c/E$. In Fig.2.3 we illustrate this wavelength-energy dispersion for the different probe particles. Obviously, the wavelength-energy range accessible to electrons, neutrons and He atoms, coincides with the energy and time-scales which are characteristic for collective excitations and atomic

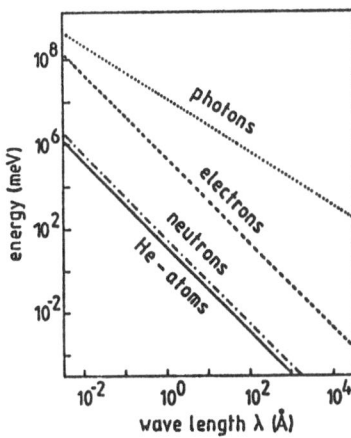

Fig.2.3. Wavelength-energy dispersion of non-relativistic particles (electrons, neutrons, He atoms) and electromagnetic waves (photons)

Table 2.1 Characteristic parameters of processes in physisorbed rare gas films.

energy-scale		
	adsorption energy	5-300 meV
	phonons	0.1-10 meV
time-scale		
	elemental step of diffusion	> 10^{-12} s
	phonon lifetime	> 10^{-11} s
length-scale		
	lattice parameter	3-5 Å
	correlation length	10-5000 Å
	phonon wavelength	≥ 1 Å
	(at the zone boundary)	

movements on the surface. On the other hand, for electromagnetic waves this coincidence is absent, generally $\hbar\omega$ and Q do not match.

The most convenient way to obtain structural information is via a diffraction experiment: a beam of well-defined wavelength impinges on a target, and the elastically scattered intensity is measured by an appropriate detector. For diffraction from a periodic two dimensional arrangement of atoms the structure factor $S(Q, \hbar\omega = 0)$ denotes:

$$S(Q,0) = \sum_m \delta(Q^{\parallel} - G_m^{\parallel}).$$ (2.3)

The terms $Q^{\parallel} = (Q_x, Q_y)$ and $G_m^{\parallel} = (G_{mx}, G_{my})$ are the momentum exchange vector and the reciprocal lattice vector in the surface plane of the two-dimensional structure, respectively. Thus, the Laue-condition for diffraction from a 2D-lattice reads

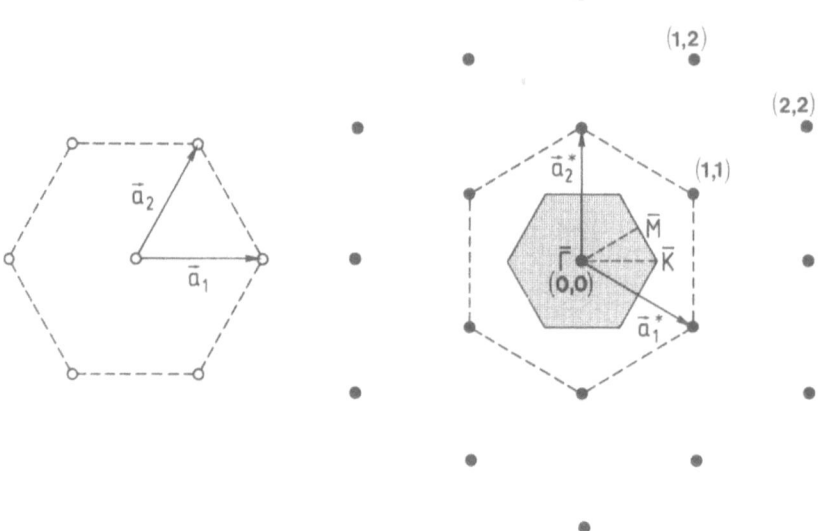

Fig.2.4. Real lattice (left) and reciprocal lattice (right) of a triangular 2D-net

$$Q^{\parallel} = G_m{}^{\parallel} \tag{2.4}.$$

In Fig.2.4, we show as an example the real and reciprocal space for the densest possible arrangement in a periodic 2D-lattice, the triangular lattice, which is the most widespread structure of ordered rare-gas adlayers and also the structure of the {111}-face of fcc crystals. The high symmetry points in the reciprocal lattice are denoted by $\bar{\Gamma}$, $\bar{M}$, and $\bar{K}$; the hatched area is the first Brillouin zone.

2.1.2 Probe Particles

The most stringent requirement for probe particles when used in studies of adsorbed overlayers is an adequate surface sensitivity. From this point of view, thermal He atoms are the most appropriate probe particles. Because of their large cross section, thermal atoms of small energy (<100 meV) interact with the outermost layer only; there is no penetration into deeper layers. The classical turning point of thermal energy He atoms is usually about 0.3 nm above the ion cores of the outermost layer.

In the energy range used in low energy electron diffraction (LEED), 10-1000 eV, the information depth is between 0.4 and 1 nm. This surface sensitivity of low energy electrons is mainly due to the large cross section for inelastic scattering. For high energy electrons (RHEED), X-rays, or neutrons, the information depth is normally much larger. Surface sensitivity in these cases can be obtained either by using grazing incidence- and exit-angles (high energy electrons and X-rays) or by using substrates with large surface to volume ratio like powdered samples to discriminate against background scattering (neutrons and X-rays). In the case of neutrons or X-rays, particular adsorbate/substrate combinations are chosen in which the scattering cross-sections of the adsorbate nuclei are much larger than those of the substrate.

a) Electrons

Electrons are the most widely used surface analytical probe particles [2.10]. Clean as well as adsorbate covered surfaces are routinely investigated by techniques like LEED, Auger electron spectroscopy (AES) or electron energy loss spectroscopy (EELS). The main concern when an electron beam is incident on a rare gas adlayer is the large cross section for electron stimulated desorption; its effect has to be carefully checked. Whereas LEED and AES are currently applied to the study of physisorbed phases at low temperatures, EELS has not been used so far. The reason is of a technical rather than fundamental nature. Although the momentum-energy range (Fig.2.3) covered by low energy electrons is suited for studies of collective lattice excitations, the energy resolution ($\simeq 4$ meV for these electrons) [2.11] is not sufficient to resolve these excitations in physisorbed films. This energy resolution is, however, good enough to study the internal vibrational modes of physisorbed molecules. Some experimental arrangements are currently set up to investigate this interesting subject.

In a typical experimental arrangement for the study of physisorbed films by electron diffraction [2.12,13], the LEED-optics (channelplate detector) is optimized for low electron beam currents in the nanoampere range in order to avoid electron stimulated desorption or local heating of the weakly bound films. The momentum resolution of a typical LEED-system is about 0.1 nm^{-1} (FWHM). However, very recently, instruments capable of very high resolution have been developed [2.14,15]. Another very promising improvement of LEED, invented recently by *Telieps* and *Bauer*, is the low energy electron microscopy (LEEM) [2.16]. In this technique, the surface is imaged with diffracted low-energy electrons. The lateral resolution obtainable is about 10 nm. Although this technique has been used so far only for the study of phase transitions on clean surfaces and metal layers, it should also be applicable to physisorbed films.

In recent years, the RHEED technique has been very popular in characterizing the type of multilayer growth [2.17]. However, the soundness of the interpretation of RHEED-patterns (streaks → perfectly flat surface; sharp spots → 3D-crystallites on the surface) has been severely challenged of late [2.18].

b) Neutrons

Among the probes currently used in condensed matter studies, neutrons have a unique status. Since the wavelength and energy of thermal neutrons are comparable with interatomic distances in condensed phases and with characteristic energies of most crystal atom motions, neutrons are the most versatile probe of bulk matter. Their application to surface studies, however, requires the use of high surface area materials, like Grafoil or powdered MgO, because of the weak scattering of neutrons by matter. These kind of substrates certainly have some disadvantages; they tend to have surface inhomogeneities and orientational ordering effects cannot be studied. On the other hand, there are also some advantages of the weak scattering: experiments can be done at high ambient pressures and on technologically important samples like raney-nickel or zeolites.

Typical momentum resolution obtained with neutron spectrometers are 0.05-0.1 nm^{-1} whereas the energy resolution can be as good as 1 μeV [2.19]. This extreme energy resolution allows, for instance, the measurement of rotational tunneling energies of adsorbed molecules, not accessible so far to any other method. Numerous physisorption systems have been investigated with neutron scattering. A large amount of information on the atomic arrangement in the adsorbed layer, distance of the layer from the substrate, intramolecular structure of the adsorbed species, collective excitations within the adsorbed layer, rotational and vibrational spectroscopy of adsorbed molecules, etc., has been obtained. An extended review on this subject has been given by *Thomas* [2.19]. A particularly interesting application of the technique, the incoherent quasielastic scattering, has been used recently by *Bienfait* [2.20] to study the diffusive motion of CH_4 films adsorbed on MgO to gain insight into the surface melting of these films.

c) X-Ray Photons

The recent rapid development of surface structural tools based on X-ray photons as probe particles is linked to the development of bright X-ray sources with the advent of synchrotron radiation. However, as already noted, X-rays only interact weakly with condensed matter, and thus it is necessary to develop techniques which allow a separation between surface and bulk scattering. In the case of clean surfaces, this can only be done by using grazing angles of incidence; a technique which has been developed recently by *Eisenberger* and *Marra* [2.21]. In the case of adsorbed layer studies, however, glancing angles may be avoided by choosing a strongly scattering adsorbate on a weakly scattering substrate. Because the scattering cross-section scales with the square of the atomic number, the heavier noble gases adsorbed on the basal plane of graphite, for instance, are an appropriate choice. With a typical X-ray scattering spectrometer for physisorbed film studies (beam energy $\simeq$ 10 keV), which has been installed at Stanford Synchrotron Radiation Laboratory [2.22], the adsorbed layer to substrate scattering ratio for a commensurate Kr monolayer on the basal plane of graphite was $\simeq 600/150$.

The important advantage of the use of synchrotron X-ray diffraction lies in the fact that the scattering can be interpreted in the kinematic approximation and that momentum resolution of the order 0.001 nm^{-1} can be obtained routinely. This very high resolution allows the study of spatial correlations from 1 nm to 1000 nm via line shape analysis of the diffraction peaks [2.23]. Also, like neutrons, X-rays are not restricted to low ambient pressures as electrons or He atoms are.

The main limitation in the use of X-rays in the study of thin physisorbed films stems from the discrepancy between the energy of the X-ray photons and the energy of collective phenomena in the adlayers. Although *Burkel* and coworkers [2.24] have recently measured optical phonons in pyrolytic graphite and in Be single crystals by means of inelastic scattering of X-rays, this technique will not be applicable to physisorbed films in the near future. The present limit for the absolute energy resolution is about 8 meV, which is twice as large as that of EELS.

d) Helium Atoms

The recent progress in generation of highly monochromatic He nozzle beams [2.25,26] and their combination with ultrahigh vacuum techniques has favored the development of several novel surface analytical tools based on the interaction of thermal He atoms with solid surfaces. He atom scattering is one of the oldest surface probes [2.27]. However, the lack of an appropriate He-beam source was the main hurdle in the development of this now very powerful analytical tool. For a long time the Knudsen (effusion) cell was the only means for producing molecular beams. The Maxwellian effusive beams have low intensity ($I_0 \simeq 10^{14}$ particles s^{-1}sr^{-1} and low monochromaticity ($\Delta v/v = \Delta\lambda/\lambda = 0.95$). Monochromaticity improvement by means of mechanical velocity selectors reduces the already low intensity to a level almost unacceptable for the requirements of surface analysis. The major break-through was the development of high pressure nozzle sources. The effect achieved by the invention of these sources is only comparable to that of the laser technology: simultaneous increase of intensity and monochromaticity by several orders of magnitude. Indeed, intensities of 10^{19} particles s^{-1}sr^{-1} and monochromaticities of $\Delta v/v = \Delta\lambda/\lambda \simeq 0.007$ are obtained routinely today (Fig.2.5). The beam monochromaticity is the result of the large number of collisions in the hydrodynamic expansion of a gas: the larger the number of collisions, the higher the monochromaticity;

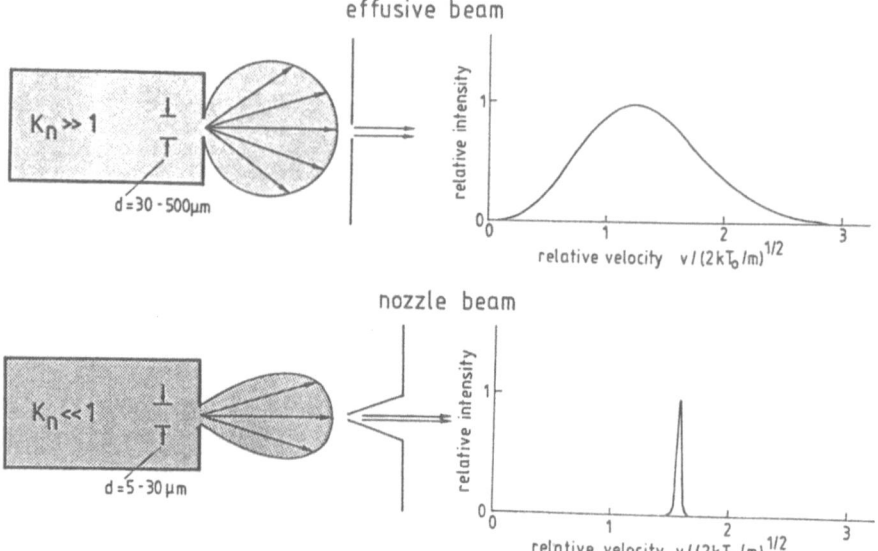

Fig.2.5. Comparison of effusive and nozzle-beam systems

i.e., the smaller the half width of the velocity distribution. Accordingly, high monochromaticity is obtained by using large stagnation gas pressures and/or low stagnation temperature. The latter increases the number of collisions because, at low relative velocities, the collision cross-section of atoms increases. Due to quantum effects, this increase is particularly pronounced for helium [2.28].

For a 20 meV He-beam, a velocity spread of $\Delta v/v \simeq 0.007$ corresponds to an energy spread smaller than 0.3 meV (FWHM), a value close to the energy resolution in optical spectroscopies. Note that even in the case of He the beam monochromaticity cannot be increased by cooling beyond a certain limit. For stagnation temperatures less than $\simeq 30$ K, the onset of clustering may destroy the beam properties [2.29].

The major characteristics of the He-beam as surface analytical tool are connected with the nature of the He-surface interaction potential. At distances not too far from the surface, the He atom is weakly attracted due to dispersion forces. At a closer approach, the electronic densities of the He atom and of the surface atoms overlap, giving rise to a strong repulsion. The classical turning point for thermal He is a few Angstroms in front of the outermost surface layer. It is this interaction mechanism which makes the He-beam an outstanding surface tool, sensitive exclusively to the outermost layer. The low energy of the He atoms and their inert nature ensures that He scattering is a completely nondestructive surface probe. This is particularly important when delicate phases, like rare-gas layers, are investigated.

The de Broglie wavelength of thermal He atoms is comparable with the interatomic distances in adsorbed overlayers. Thus, from measurements of the angular positions of the diffraction-peaks the size and orientation of the 2D unit cell, i.e., the structure of the outermost layer can be straightforwardly determined. Analysis of the peak intensities yield the potential corrugation, which usually reflects in a direct way the geometrical arrangement of the atoms within the 2D unit cell [2.30].

The energy of thermal He atoms is comparable to the energies of collective excitations in overlayers. Thus, in a scattering experiment, the He atom may exchange an appreciable part of its energy with the surface. This energy can be measured in time-of-flight experiments with a resolution $\simeq 0.3$ meV, more than 10 times better than with EELS [2.31]. Thus, surface phonon dispersion curves of rare-gas layers can be mapped out by measuring energy loss spectra at various momentum transfers in different crystallographic directions. This is a definite advantage of inelastic He-scattering over inelastic neutron scattering. (In view of the random orientation of powdered samples, which have to be used in neutron scattering, only average phonon density of states, but not dispersion curves can be obtained.) The range of energy transfer that can be covered by thermal He-atoms is limited at the low end by the present maximum resolution of $\simeq 0.3$ meV and at the high end by the nature of the scattering mechanism. The He-beam surface interaction time being finite, the upper limit for the detectable phonon energy is about 40 meV. So far, only modes with a component perpendicular to the surface have been clearly detected; this seems to be less a fundamental than a technical problem.

Besides the inelastic component, a certain fraction of the elastically scattered He atoms are always found between the coherent diffraction peaks. We will refer to this scattering as diffuse elastic scattering. This diffuse intensity is attributed to scattering from defects and impurities. Accordingly, this diffuse elastic scattering provides valuable information on the degree and nature of surface disorder. It can be used, for example, to study the growth of thin films [2.32] or to deduce information on the size, nature and orientation of surface defects [2.33]. Very recently, the peak shape analysis of the diffuse elastic component has also been used to study the diffusive motion of surface atoms [2.34].

Another remarkable way to use He-scattering for the study of adsorbed layers is based on the large total cross section for diffuse He-scattering of isolated adsorbates (e.g., $\Sigma_{Xe}^{He} \simeq 1.1$ nm^2 for $E_{He} = 18$ meV). This large cross section is attributed to the long-range attractive interaction which causes He atoms to be scattered out of the (0,0) beam. This extreme sensitivity of He-beams allows the extraction of important information concerning the lateral distribution of adsorbates, mutual interactions between adsorbates, dilute-condensed phase transitions in 2D, adatom mobilities, etc. [2.35], simply by monitoring the attenuation of the $(0,0)$ He-beam, i.e., the specular beam. This technique also allows the detection of impurities (including hydrogen!) in the per mille range, a level hardly attainable with almost all other methods.

The type of information that can be derived by He atom scattering from physisorbed adlayers is summarized in Table 2.2, together with the type of scattering involved.

In Fig.2.6 we show a schematic sketch of the high resolution He-scattering spectrometer used in the authors' laboratory for studies of thin physisorbed films [2.36]. The scattering geometry is fixed with $\theta_i + \theta_f = 90^\circ$, where θ_i and θ_f are the incident and outgoing angles, respectively. The vacuum system consists of four main units: the three-chamber nozzle-beam generator, the scattering chamber (with sample holder, LEED, CMA-Auger, and ion gun facilities), the pseudorandom chopper chamber, and the three-chamber detector unit. The scattering chamber has a base pressure in the low 10^{-11} mbar range which rises to about 10^{-9} mbar (He partial pressure) during He-beam operation. The extensive differential pumping serves to reduce the He-partial base pressure in the detector chamber to 10^{-15}-10^{-16} mbar. As a detector we use a commercial Extranuclear mass spectrometer with electron bombardment ionizer; the residual He-pressure in the detector chamber results in a count rate of about 40 counts/s at 5mA emission

Table 2.2 Types of He atom scattering from overlayers and information which can be inferred

Type of scattering	Energy and parallel momentum exchange	Information
diffraction	$\hbar\omega = 0$ $Q^{\parallel} > 0 \ (= G_m)$	layer structure, orientational ordering, correlation lengths, distance of layer from substrate, random step densities
specular	$\hbar\omega = 0$ $Q^{\parallel} = 0 \ (= G_0)$	thermodynamics, impurity and defect densities, lateral distribution of adsorbates, adatom mobility
inelastic	$\hbar\omega \neq 0$ $Q^{\parallel} > 0$	collective excitations within adsorbed layer, dynamical coupling between layer and substrate
diffuse elastic	$\hbar\omega = 0$ $Q^{\parallel} > 0 \ (\neq G_m)$	layer perfectness; growth mode; size, nature and orientation of surface defects; surface diffusion

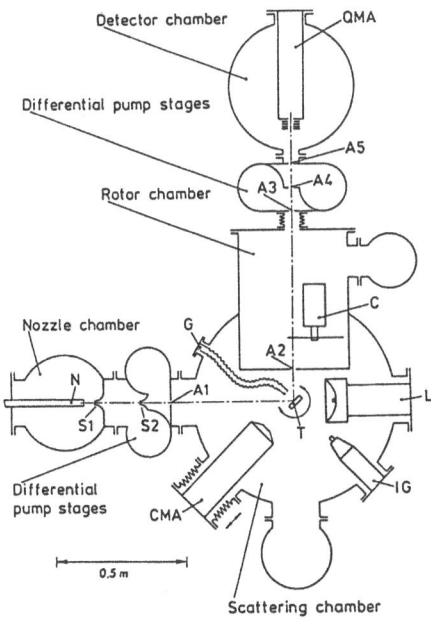

Fig.2.6. Schematic diagram of a high resolution He time-of-flight spectrometer. (N: nozzle beam source, S1,2: skimmers, A1-5: apertures, T: target, G: gas doser, CMA: Auger spectrometer, IG: ion gun, L: LEED, C: Magnetically suspended pseudorandom chopper, QMA: detector, quadrupole mass analyzer with channeltron)

current. This figure has to be compared with 10^5 counts/s, representing the signal of the first order He-diffraction beam (E_{He}=18 meV) from a Kr monolayer adsorbed on Pt {111} at 25 K.

In most experiments a 18 meV He-beam, produced by cooling the nozzle with liquid nitrogen, is used. The particle flux impinging on the target is about $2\cdot10^{19}$ He atoms $s^{-1}sr^{-1}$ with a velocity spread $\Delta v/v$ = 0.007 (He stagnation pressure: 150 bar). The beam divergence as well as the acceptance angle of the detector are 0.2°. Accordingly, the overall momentum resolution of the apparatus is about 0.1 nm^{-1}, corresponding to a transfer width for the higher order peaks around 20-30 nm. Information on correlation lengths up to 100 nm can be inferred by instrumental response function deconvolution of diffraction peaks monitored with appropriate statistics.

The energy distribution of the scattered He atoms is obtained by a cross-correlation analysis of the time-of-flight (TOF) spectra of the He atoms upon passing a pseudorandom chopper [2.36,37]. At a flight path of 790 mm the overall resolution of the spectrometer for a 18 meV beam with $\Delta v/v$ = 0.007 is 0.4 meV.

The experiments reported here illustrate the 2D phase transitions and the multilayer growth of rare-gas films and are performed on a Pt{111} crystal surface as substrate. The crystal is mounted on a manipulator which allows independent polar and azimuthal rotation, as well as tilting. The temperature can be regulated between 25 and 1800 K by means of liquid He cooling and/or electron bombardment heating.

The temperature of the crystal is controlled by chromel-alumel thermocouple calibrated in situ by Xe vapor pressure isotherms monitored with surface phonon spectroscopy [2.36,38]. The Pt crystal is cleaned by repeated cycles of argon and xenon ion bombardment and annealing to 1200 K. The defect density of the clean Pt-surface is less then 10^{-3} (average terrace width 200-300 nm) as probed with elastic He scattering.

2.2 Solid–Solid Transitions in Two Dimensions

2.2.1 Commensurability

Atoms adsorbed on a periodic substrate can form ordered structures. These structures can be either in or out of registry with the structure of the substrate. It is convenient to describe this ordering by relating the Bravais lattice of the adlayer to that of the substrate surface. *Park* and *Madden* [2.39] have proposed a simple vectorial criterion to classify the structures. Let a_1 and a_2 be the basis vectors of the adsorbate and b_1 and b_2 those of the substrate surface; these can be related by

$$\begin{bmatrix} a_1 \\ a_2 \end{bmatrix} = G \begin{bmatrix} b_1 \\ b_2 \end{bmatrix} \tag{2.5}$$

with the matrix

$$G = \begin{bmatrix} G_{11} & G_{12} \\ G_{21} & G_{22} \end{bmatrix} . \tag{2.6}$$

$|a_1 \times a_2|$ and $|b_1 \times b_2|$ are the unit cell area of the adlayer and substrate surface, respectively; det G is the ratio of the two areas. The relation between the two ordered structures is classified by means of this quantity as follows:

i) *det G = integer*
the structure of the adlayer belongs to the same symmetry class as that of the substrate and is in registry with the latter; the adlayer is termed *commensurate*.

ii) *det G = irrational number*
the adlayer is out of registry with the substrate; the adlayer is termed *incommensurate*.

iii) *det G = rational number*
the adlayer is again in registry with the substrate. However, whereas in i) all adlayer atoms are located in equivalent high symmetry adsorp-

tion sites, here only a fraction of adatoms is located in equivalent sites; the adlayer is termed *high-order commensurate*.

In Fig.2.7, we show a simple one-dimensional model illustrating this classification. The periodicity of the substrate surface is provided by a sinusoidal potential of period **b** and the adlayer by a chain of atoms with nearest neighbor distance **a**.

Assuming that the structural mismatch between adlayer and substrate is not too large ($\leq 15\%$), the "commensurability" is determined by the relative interaction strength h/u_c, which is the ratio of h, the lateral adatom interaction in the layer, to u_c, the modulation of the adsorbate-substrate potential parallel to the surface. It is not the whole adsorbate-substrate interaction energy, as often stated in the literature, which enters the determining ratio, but the diffusional barrier u_c. When this diffusional barrier is large compared to the lateral attraction, commensurate structures will be formed. On the other hand, when the lateral adatom interactions dominate, incommensurate structures will be favored. Only when the competing interactions are of comparable magnitude, may both registry and out of registry structures be stabilized by the complex interplay of these interactions thus leading to the occurrence of phase transitions between the various structures (Table 2.3).

We have recently measured the energetics of the adsorption of the heavy rare-gases Ar, Kr and Xe on the Pt{111} surface by means of He-scattering. In Table 2.4 we summarize the quantities pertinent to

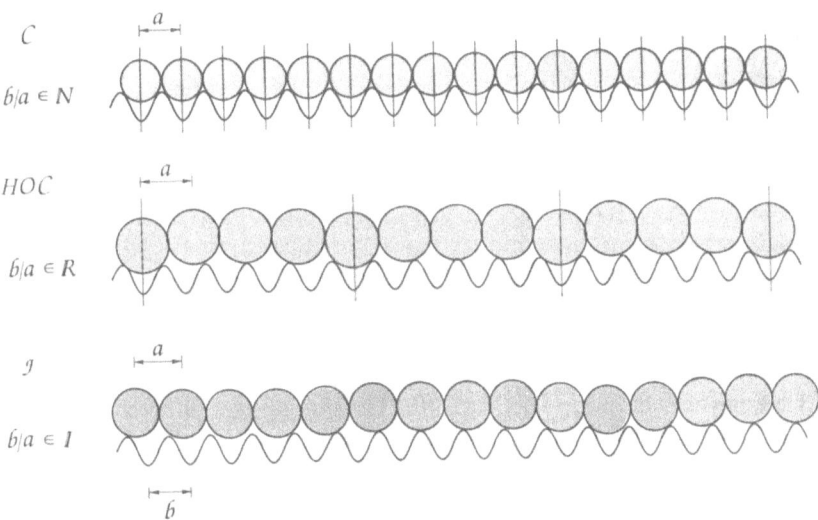

Fig.2.7. Simple one dimensional model of adsorbed adlayers. The substrate is represented by a sinosoidal potential of period b and the adlayer by a chain of atoms with nearest neighbor distance a. (C: commensurate, HOC: high order commensurate, I: incommensurate)

Table 2.3. Interaction strength h/u_c and commensurability

$h \ll u_c$	commensurate structure only
$h \gg u_c$	incommensurate structures only
$h \simeq u_c$	commensurate and incommensurate structures, corresponding phase transitions

Table 2.4. Characteristic energies of rare-gas adsorption on Pt{111} in the monolayer range

	Xe	Kr
isoteric heat q_{st} [meV] at $\theta \simeq 0$	277	128
lateral attraction h_ℓ [meV]	43	26
diffusional barrier u_c [meV]	~30	~10-20

Kr and Xe/Pt{111}. From inspection of the relevant quantities in Table 2.4 we can deduce that rare-gas monolayers on Pt{111} appear to be suited to study structural 2D-solid-solid transitions.

2.2.2 Fundamentals of the Theory Describing the Commensurate-Incommensurate Transition in 2D

The basic ideas of all modern theories of the Commensurate-Incommensurate transition are contained in the 1D model of *Frank* and *van der Merwe* (FvdM) developed in 1949 [2.40]. A linear chain of atoms with nearest neighbor distance **a** is placed in a sinusoidal potential of amplitude V and periodicity **b** representing the substrate. The mutual interaction of the atoms in the chain is assumed to be harmonic and characterized by a spring constant K. The calculations of FvdM show that for slightly different lattice parameters of chain and substrate, i.e., for a weakly incommensurate adlayer, the lowest energy state is obtained for a system which consists of large commensurate regions separated by regions of poor fit (Fig.2.8). The regions of poor lattice fit are called misfit dislocations, solitons or domain walls.

To be more quantitative we write the Hamiltonian for the linear chain problem:

$$H = \sum_n \frac{K}{2} \left(x_{n+1} - x_n - a \right)^2 + \sum_n V \cdot \left[1 - \cos\left(\frac{2\pi}{b} x_n \right) \right] \qquad (2.7)$$

where x_n is the position of the nth atom. By introducing the phase ϕ_n, which describes the positional deviation of an atom from its commensurate position

$$x_n = nb + \frac{b}{2\pi} \phi_n$$

and by approximating the discrete values ϕ_n by a continuous function $\phi(n)$, the Hamiltonian becomes

$$H = \int \left[\frac{Kb^2}{8\pi^2} \left(\frac{d\phi}{dn} - 2\pi\delta \right) + V[1 - \cos(p\phi)] \right] dn \qquad (2.9)$$

with $\delta = (a-b)/b$ being the natural misfit between adlayer and substrate and p the commensurability. The phase function ϕ for which the Hamiltonian is minimized satisfies the sine-Gordon equation:

$$\frac{d^2\phi}{dn^2} = p\,A\,\sin(p\phi) \qquad (2.10)$$

with $A = 2\pi\,(V/K)^{1/2}/b$ a solution of which is the solitary wavepacket distortion, or the so-called soliton:

$$\phi(n) = \frac{4}{p}\,\arctan[\exp(pn\cdot A^{1/2})] \,. \qquad (2.11)$$

The soliton solution is shown in Fig.2.8; it describes a domain wall located at n = 0 separating two adjacent commensurate regions. The soliton superlattice is a compromise between the elastic energy in the chain which favors the unperturbed incommensurate phase, and the

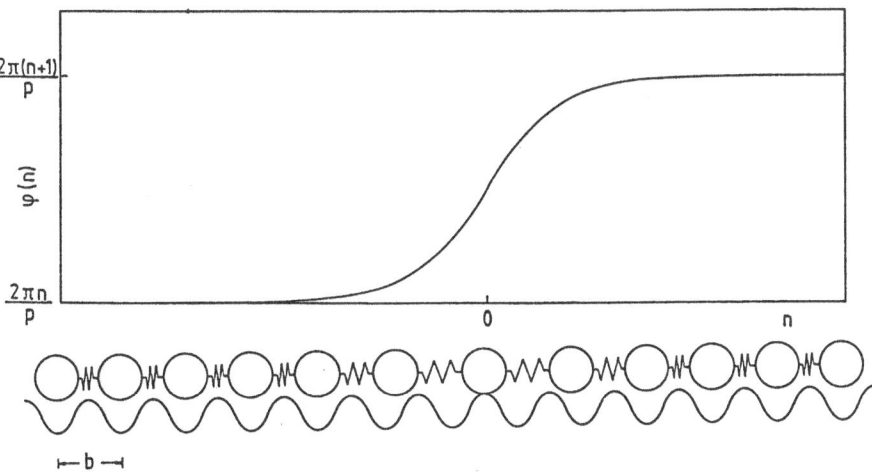

Fig.2.8. Soliton solution (domain wall) of the sine-Gordon equation. The domain wall separates two adjacent commensurate regions with $\phi = 2\pi(n+1)/p$ and $\phi = 2\pi n/p$

interaction energy with the substrate. The width of the domain wall in this model is $L_0 = 1/(pA^{1/2})$.

In two dimensional systems, walls are lines. In a triangular lattice with its C_{3v} symmetry, there are three equivalent directions. Therefore, domain walls can cross. Using Landau theory, *Bak, Mukamel, Villain,* and *Wentowska* (BMVW) [2.41] have shown that it is the wall crossing energy, Λ, which determines the symmetry of the weakly incommensurate phase and the nature of the phase transition.

i) $\Lambda < 0$, i.e., attractive walls. A hexagonal network of domain walls (HI) will be formed at the C–I transition because the number of wall crossings has to be as large as possible (Fig.2.9a). This C–HI transition should be first order.

ii) $\Lambda > 0$, i.e., repulsive walls. The number of wall crossings has to be as small as possible, i.e., a striped network of parallel walls (SI) will be formed in the incommensurate region (Fig.2.9b and c). The C–SI transition should be continuous. The striped phase is expected to be stable only close to the C–I transition. At larger misfits the hexagonal symmetry should be recovered in a first order SI⟷HI transition.

The FvdM model as well as the BMVW model neglect thermal fluctuation effects; both are T = 0 K theories. *Pokrovsky* and *Talapov* (PT) [2.42] have studied the C–SI transition including thermal effects. For T ≠ 0 K the domain walls can meander and collide, giving rise to an entropy-mediated repulsive force between meandering walls of the form $F \propto T^2/\ell^2$, where ℓ is the distance between nearest neighbor walls. Because of this inverse square behavior, the inverse wall separation, i.e., the misfit m, in the *weakly incommensurate phase* should follow a power law of the form:

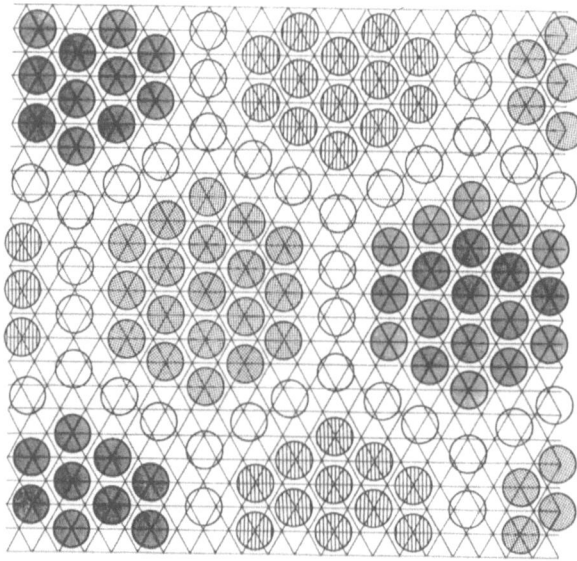

Fig.2.9a

Fig.2.9b

Fig.2.9c

↑   ↑
h   *sh*

Fig.2.9. Schematic diagram showing a) hexagonal and b) striped domain wall arrangements (in both pictures super light walls are shown). In incommensurate layers where the monolayer is compressed with respect to the commensurate lattice, domain walls are either heavy or super heavy c)

$$m = \frac{l}{\ell} \simeq \sqrt{1 - \frac{T}{T_c}} \; . \tag{2.12}$$

2.2.3 The C-I Transition of Monolayer Xe on Pt{111}

In a recent He diffraction study [2.43] we have shown that the adsorption system Xe/Pt{111} is dominated by the existence of a $(\sqrt{3} \times \sqrt{3})$ R30⁰ commensurate phase (Fig.2.10). The C-phase has been found to be stable in an extended temperature (62-99 K) and coverage range ($\leq 1/3$).

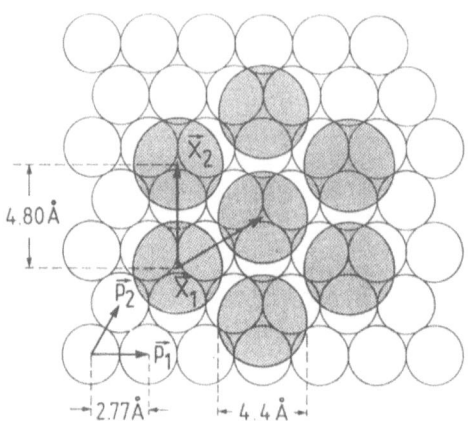

Fig.2.10. Commensurate $(\sqrt{3}\times\sqrt{3})R30^\circ$ Xe monolayer adsorbed on Pt{111}

The maximum coverage in this $(\sqrt{3} \times \sqrt{3})$ R30^0 commensurate structure is obviously θ_{Xe} = 1/3 (θ_{Xe}=1 corresponds to $1.5\cdot10^{15}$ atoms/cm^2, the density of Pt atoms in the {111} plane). Only one third of the adsorption sites are occupied, i.e., there exist three energetically degenerate commensurate sublattices. The commensurate Xe-lattice being expanded by about 9% with respect to the "natural" Xe-lattice, the coverage can be increased beyond θ_{Xe} = 1/3. Obviously, above this limit the adatoms cannot all occupy preferred adsorption sites, and the adlayer becomes incommensurate. Alternatively, due to anharmonic effects [2.43,44], the Xe-adlayer becomes incommensurate upon decreasing the temperature below $\simeq$ 62 K at constant coverage ($\theta_{Xe}\leq$1/3).

Before discussing the experimental results of the C-I transition of Xe on Pt{111} in detail, let us have a look at the diffraction pattern expected from a striped (SI) and from a hexagonal uniformly (HI) compressed phase. The structures with corresponding schematic diffraction patterns are shown in Fig.2.11. The diffraction patterns for the (n,n) and (n,2n) diffraction orders are shown for fully relaxed phases.

We discuss first the basic crystallography of the incommensurate phase as deduced from the measured patterns [2.45]. Fig.2.12 shows the $(2,2)_{Xe}$ and $(1,2)_{Xe}$ diffraction features obtained from a Xe layer of coverage $\theta_{Xe} \simeq$ 0.30 during the C-I transition at 54 K. The plots have been obtained by monitoring series of azimuthal scans (i.e., constant Q scans in the reciprocal space). The comparison with Fig.2.11 shows that the incommensurate Xe layer on Pt{111} is a striped phase (SI) with a uniaxial compression in the $\overline{\Gamma M}$ direction. Indeed, a three-peak structure for the $(2,2)_{Xe}$ diffraction feature, with the doublet located at $Q^{2,2}_{comm}$ + 0.048 $\mathring{A}^{-1}$ and the singlet peak located at $Q^{2,2}_{comm}$ + 0.190 $\mathring{A}^{-1}$

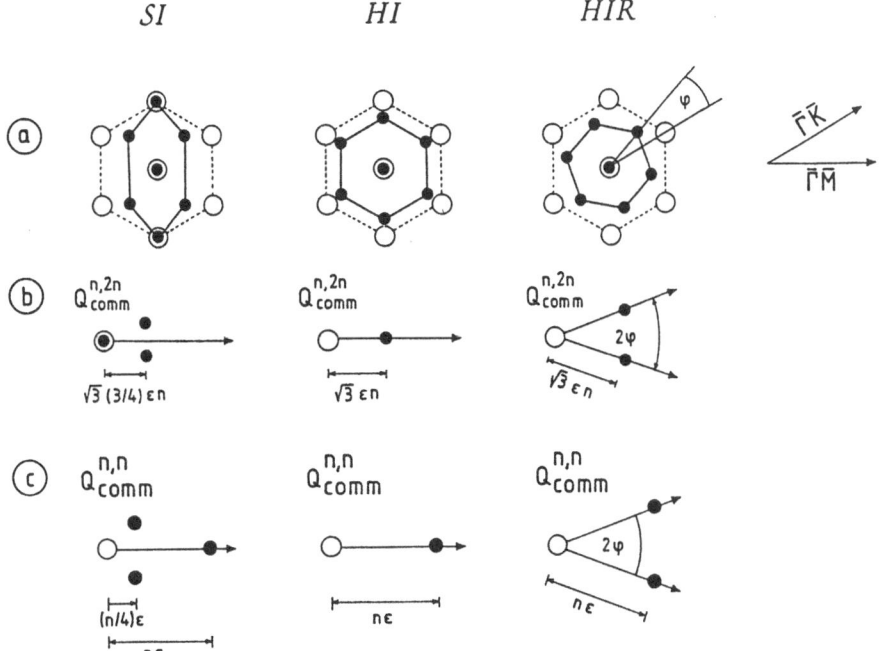

Fig.2.11. a) Real lattice and b,c) schematic representation of the (n,n) and (n,2n) diffraction features of various incommensurate structures. SI: striped incommensurate, HI: hexagonal incommensurate, and HIR: hexagonal incommensurate rotated. All phases are assumed to be fully relaxed. $\circ$ denotes the $(\sqrt{3} \times \sqrt{3})R30°$ commensurate and $\bullet$ the incommensurate structures

is observed (with $Q^{2,2}_{comm} = 3.02\ \text{Å}^{-1}$); whereas the $(1,2)_{Xe}$ pattern consists of a single peak at the commensurate position and a shallow doublet with the maximum intensity at about $Q^{1,2}_{comm} + 0.13\ \text{Å}^{-1}$ (with $Q^{1,2}_{comm} = 2.62\ \text{Å}^{-1}$). The observed incommensurability deduced from the well-defined polar location of the peaks in Fig.2.12a is $\epsilon = 0.95\ \text{Å}$ and corresponds to an interrow distance in the $\bar{\Gamma}\text{M}$ direction of $d_{SI} = 3.91\ \text{Å}$. This results in a misfit $m = 1-d_{SI}/d_C = 0.059$, where $d_c = 4.80 \cdot \cos 30°\ \text{Å}$ is the inter-row distance of the commensurate Xe structure in the same direction. From the measured polar and azimuthal peak widths in Fig.2.12 we can also estimate average domain sizes of the incommensurate layer. For the $\bar{\Gamma}\text{K}$ direction, i.e, parallel to the walls, we obtain $\simeq 350\ \text{Å}$ and for the perpendicular $\bar{\Gamma}\text{M}$ direction $\simeq 50\ \text{Å}$.

The analysis in the last paragraph has shown that the incommensurate Xe layer on Pt{111} at misfits of about 6% is a striped phase with the domain walls strongly relaxed, i.e., a uniaxially compressed layer. Indeed, for less relaxed domain walls, depending on the extent of the wall relaxation and on the nature of the walls (heavy or superheavy) additional satellites in the (n,n) diffraction patterns should

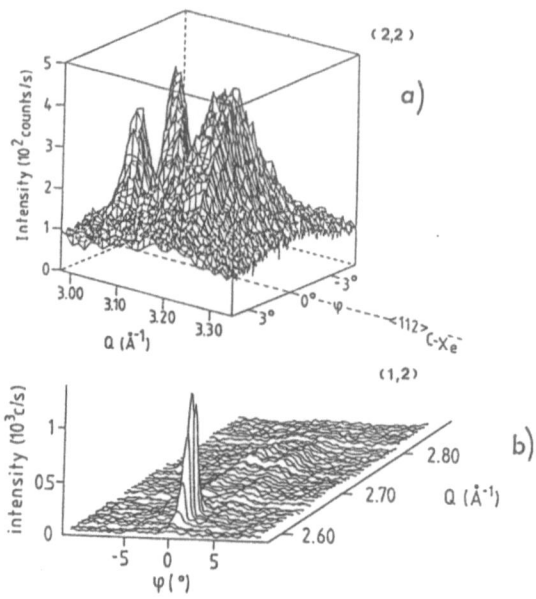

Fig.2.12. 3D-diffraction plot of the a) $(2,2)_{Xe}$ and b) $(1,2)_{Xe}$ diffraction features during the C-I transition at T = 54 K. Q denotes the wave vector in the $\bar{\Gamma}\bar{M}$- and $\bar{\Gamma}\bar{K}$-direction, respectively, while ϕ denotes the azimuthal angle

appear. In the case of a weakly incommensurate layer (misfits below $\simeq 3\%$) we observe an additional on-axis peak at $Q^{2,2}_{comm} + \epsilon/2$ in the (2,2) diffraction pattern. In order to determine the nature of the domain walls, we have calculated the structure factor for the different domain wall types as a function of the domain wall relaxation [2.46] following the analysis of *Stephens* et al. [2.47]. The observed additional on-axis satellite at $Q^{2,2}_{comm} + \epsilon/2$ in the weakly incommensurate phase is consistent with the occurrence of superheavy striped domain walls, and the observed peak intensities can be reproduced with a domain wall width of $\lambda \simeq 3$-5 Xe inter-row distances.

In Fig.2.13, we have analyzed the data of the C-SI transition in the weakly incommensurate phase with a least-squares fit of a power law form $m = m_o(1-T/T_c)^\beta$; the best fit parameters are $T_c = 61.7$ K, $m_o = 0.18$ and $\beta = 0.51 \pm 0.04$. The value of β is in good agreement with the Pokrovsky-Talapov prediction. Only data points up to misfits of about 4% have been included in the fit. The cutoff at $\simeq 4\%$ has been chosen in accordance with *Erbil* et al. [2.48], who have found the $\beta = 1/2$ power law to be only valid in this range for bromine intercalated graphite. For larger values, the misfit variation with reduced temperature is roughly linear; in this region the inter-wall distance is of order of the wall width and the PT-theory not applicable.

In contrast to the C-I transition of Kr on graphite [2.49], which is in fact a melting transition due to the instability of the hexagonal

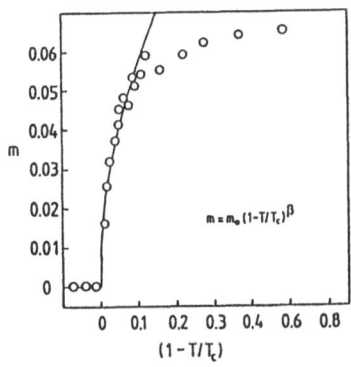

Fig.2.13. $\overline{\text{IM}}$-uniaxial misfit m versus reduced temperature during the C-SI transition. The solid line represents the power law fit

weakly incommensurate phase with respect to the formation of free dislocations (reentrant melting), the C-SI transition of Xe/Pt{111} is a solid-solid transition with the incommensurability simply related to the domain wall density. According to *Coppersmith* et al. [2.50], striped structures are stable if the number of sublattices, p (here, 3), is larger than $8^{1/2}$, whereas for hexagonal structures the criterion is p > 7.5±1.5. As mentioned, the critical exponent $\beta = 0.51\pm0.04$ deduced from the data in Fig.2.13 is in good agreement with the $\beta = 1/2$ prediction of Pokrovsky and Talapov. The Pokrovsky-Talapov model may essentially be applied to a substrate of uniaxial symmetry, although the original model calculations are performed for an isotropic substrate; thus, it should be applicable to the isotropic Pt{111} substrate. However, in a recent study, *Haldane* and *Villain* [2.51], pointed out that in the case of rare-gas monolayers on metal surfaces, substrate induced electric dipole interactions might be responsible for the square root law. Moreover, they inferred that even in the case of an insulating substrate (no induced dipole forces) the square root behavior should be valid, but only for very small misfits (m<0.001 !). At present, it is difficult to make a choice between the thermal fluctuation mechanism of Pokrovsky and Talapov and the substrate induced dipole mechanism of Haldane and Villain. However, it is worth noting that the experimental range of validity of the square root law in 2D-striped domain wall phases has been found to be much larger (a factor of $\simeq$30) than the limit given by Haldane and Villain, for "insulating" substrates (Br intercalated graphite [2.48]) as well as for metal substrates (Xe/Cu{110} [2.52], Xe/Pt{111} [2.45]).

The width of the superheavy domain walls in the striped phase, as obtained from an analysis of the satellite peak intensities, amounts to 3-5 inter-row distances (FWHM). With increasing incommensurability, the total length of the domain walls is expected to increase, while the wall thickness is expected to remain constant [2.53], giving rise to smaller and more numerous commensurate domains. For misfits larger than 3-4%, i.e., where the inter-wall distance becomes less than three times the wall width, the diffraction pattern of the striped incommensurate layer can no longer be distinguished from an unaxially compressed layer.

The most direct implication of the existence of a striped phase in Xe layers on Pt{111} is that the wall crossing energy is substantially positive. This is at variance with observations made for Kr layers on graphite [2.49], where the crossing energy was always found to be negative or at least only slightly positive (so that the entropy gain due to the free breathing of the honeycomb lattice is sufficient to stabilize the hexagonal symmetry). *Gooding* et al. [2.54] have studied the influence of the substrate potential modulation on the different wall energies. They found that for large potential modulations striped arrays of discommensuration might have lowest energy. This goes along with the large potential modulation observed for the Xe/Pt{111} system [2.55]. The extended misfit range (0<m<7.2%) in which the striped structure appears to be stable, is somewhat puzzling in view of recent theoretical results by *Halpin-Healy* and *Kardar* [2.56]. They have studied the occurrence of striped structures in the "striped helical Potts lattice gas model". Their results reveal a strong correlation between the extent of the striped phase regime and the wall thickness. Striped structures in an extended coverage range should appear only for "sharp" domain walls: with increasing wall thickness this range is expected to shrink substantially. The energy cost due to the wall repulsion seems to be too large for thick walls. They conclude that the wall width of 4-5 inter-rows in Kr monolayers on graphite [2.57] might be responsible for the absence of a striped phase in this system. The wall width in Xe layers on Pt{111} is similar; the coverage range in which the striped phase is found to be stable corresponds in Halpin-Healy and Kardars calculations to wall widths of 1-2 inter-rows.

When increasing the misfit above 6.5%, an additional on-axis peak at $Q^{1,2}_{comm} + \sqrt{3}\epsilon$ appears in the (1,2) diffraction spots (Fig.2.14). This marks the transition from the striped to the hexagonal incommensurate phase. Diffraction patterns composed of a peak at $Q^{1,2}_{comm}$ and a doublet at $Q^{1,2}_{comm} + (3/4)\sqrt{3}\epsilon$ originating from a SI phase, and an on-axis peak

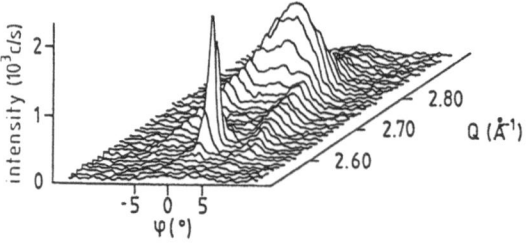

Fig.2.14. 3D-diffraction plot of the $(1,2)_{xe}$ diffractional feature of an incommensurate Xe layer on Pt(111) at misfit of 7.0% (θ_{Xe} = 0.35, T = 25 K.)

at $Q_{comm}^{1,2}$ + $\sqrt{3}\epsilon$ originating from a HI phase are observed in the misfit range 6.5% ≤ m ≤ 7.3%, with the HI peak intensity progressively increasing with coverage. We thus conclude that the SI phase transforms at misfits of ≃6.5% to a HI phase in a first order transition. Xe on Pt{111} appears to be the first 2D system fully consistent with the BMVW theory, i.e., the first system displaying the full sequence of C → SI → HI transitions with increasing incommensurability. Of course, the BMVW theory is a T = 0 K theory neglecting thermal fluctuation effects. However, *Halpin-Healy* and *Kardar* [2.58] have recently studied the various domain wall phases in the framework of a generalized helical Potts model, including finite temperature effects and two species of domain walls. Their results were in general agreement with the BMVW theory; in particular, they pointed out that the C → SI → HI sequence only occurs when assuming repulsive heavy and superheavy wall crossings. This is confirmed by the Xe/Pt{111} system.

2.2.4 Can High Order Commensurate Adlayers be Distinguished from Incommensurate Ones?

In the commensurate structure discussed in the last paragraph, all adatoms are located in equivalent high symmetry sites. Here, we will focus on overlayers where only a fraction of the adatoms occupies equivalent sites, layers which have been defined as high order commensurate (HOC). *Aubry* has studied the occurrence of such HOC phases at T = 0 K in a one-dimensional model [2.59]. His study revealed the unexpected result that upon variation of the chemical potential the overlayer will lock successively into all high order commensurate wavevectors and that there are no incommensurate phases in between. Thus, the wavevector varies continuously, but nonanalytically; a form which is known as a devil's staircase. The notion of a devil's staircase is somewhat academic, because finite temperature effects have been neglected in Aubry's theory. Indeed, a layer is only locked, i.e., can be regarded as high order commensurate, if the energy gain due to the occupation of the potential minima by a fraction of the adatoms is large compared to the temperature.

Until very recently there has been no convincing experimental evidence for the existence of high order commensurate physisorbed layers. This appeared to support the widespread belief that "experimentally it is impossible to distinguish between a HOC structure and an incommensurate structure" (*Per Bak* in [Ref.2.60, p.587]). This statement is certainly true if the only accessible experimental information is the ratio of the adlayer and substrate wavevectors from a diffraction experiment. Indeed, because one can always find one rational number within the confidence range of any experimental irrational number, i.e., the wavevector ratio supplied by the most refined experiment is always compatible with a high order commensurate phase. We will show here, however that there are at least four other criteria which allow an unequivocal distinction between high order commensurate and incommensurate. In a recent letter [2.61], we have proposed and demonstrated two criteria based on thermal expansion and adlayer buckling, respectively:

i) *Thermal expansion*

An incommensurate "floating" rare-gas layer is expected to thermally expand very much like the corresponding rare-gas bulk crystal, whereas a commensurate "locked" layer has to follow by definition the substrate at which it is locked. The thermal expansion of rare-gas solids being at least ten times larger than that of substrates normally used, the distinction between HOC and I becomes straightforward. Indeed, the thermal expansion criterion is a very sharp criterion, because it requires that the "locking" is strong enough to withstand temperature variatons over a sufficiently large range ($\geq 10K$) to allow for reliable thermal expansion measurements.

ii) *Commensurate buckling*

The locking of a high order commensurate adlayer is due to a fraction of the adlayer being located at high symmetry, high bonding substrate sites. These stronger bound atoms are located "deeper" in the substrate surface than the others; the adlayer is periodically buckled. This buckling superstructure should give rise to additional satellites in a diffraction experiment. Due to the extreme sensitivity of He-scattering to the surface topography these satellites can be detected by high resolution He-diffraction.

There are two additional criteria which, in principle, could be used to distinguish between I and HOC. The present state of the experimental techniques, however, is the main hurdle in applying them. Neither in-plane phonons, nor adsorbed layers of sufficient spatial coherence are yet accessible.

iii) *Lattice dynamical criterion*

A basic property of all commensurate layers (C as well as HOC) is that they have discrete rather than continuous character and that infinitesimal displacements of the whole layer cost a nonzero amount of energy; i.e., the commensurate layers are "locked" on the substrate. According to the general theorems of lattice dynamics, the broken translational invariance requires that the longitudinal and transverse phonon branches of HOC monolayers are optical modes (finite frequency at the $\bar{\Gamma}$ point). On the other hand, in incommensurate layers the monolayer can be shifted with respect to the substrate by an arbitrary vector without change in energy. Due to this reestablished translational invariance the in-plane vibrational modes become acoustic (zero frequency at the $\bar{\Gamma}$ point).

iv) Bragg peak singularities

In 3D crystals with perfect long-range order, the structure factor $S(Q)$ exhibits delta-function singularities, i.e., the Bragg peaks. As already stated, there is no real long-range positional order for incommensurate monolayers with continuous symmetry. Due to the algebraic decay of positional correlation, as proved by Jancovici, there is only a power law singularity of the form

$$S(Q) \simeq \sum_{G} \left| Q^{\|} - G_m \right|^{-2+\eta_G} . \qquad (2.13)$$

Since long range positional correlations in a commensurate monolayer are recovered by pinning to the substrate, the structure factor from a commensurate 2D layer exhibits delta function singularities at the diffraction conditions. Thus, a careful peak shape analysis could be used to distinguish between HOC and I.

2.2.5 The I-HOC Phase Transition of Monolayer Kr on Pt{111}

Figure 2.15a shows a series of polar He scans of the $(1,1)_{Kr}$ diffraction peak taken at 25 K along the $\bar{\Gamma}M_{Kr}$ direction of Kr monolayers adsorbed on a Pt{111} surface at coverages between 0.5 and 0.95 ML. The sequence is characteristic for a first order phase transition from a hexagonal solid phase with wavevector $Q = 1.769$ Å^{-1} ($d_{Kr} = 4.10$ Å) to one with $Q = 1.814$ Å^{-1} ($d_{Kr} = 4.00$ Å), below and above 0.8 ML, respectively. During the phase transition, the intensity diffracted from one phase increases at the expense of the other.

The question concerning the incommensurate "floating" versus high order commensurate "locked" nature of the two Kr-phases has been addressed by looking at their thermal expansion behavior and by searching for superstructure satellites. In Fig.2.15b the measured Kr-Kr interatomic spacing versus temperature is shown for submonolayer films of coverage 0.5 ML and 0.95 ML. The difference is striking. The low coverage phase shows a variation with temperature, very much like bulk Kr (dashed), and is thus an incommensurate "floating" phase. In contrast, the lattice parameter of the high coverage phase is - like that of the Pt substrate (solid) - constant within experimental error in the same temperature interval; accordingly, this Kr-phase is high-order commensurate "locked".

This assignment is supported by inspection of Fig.2.16a, where polar scans (He-beam energy 12 meV) in the $\bar{\Gamma}K_{Kr}$ direction of the

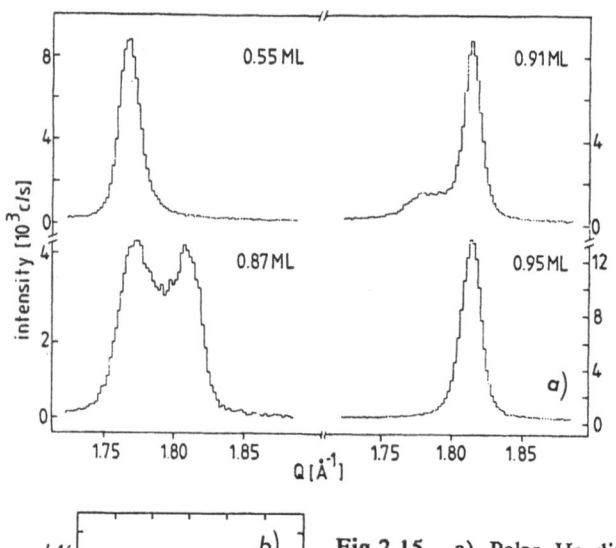

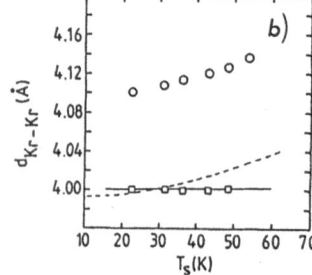

Fig.2.15. a) Polar He-diffraction scans of the $(1,1)_{Kr}$-diffraction order from Kr monolayers on Pt{111} at various Kr submonolayer coverages at T = 25 K. b) Kr-layer spacing vs temperature for the (□) high (0.95 ML) and (o) low (0.5 ML) coverage phase; temperature dependency of the lattice spacing of (—) Pt substrate, and of (---) bulk Kr

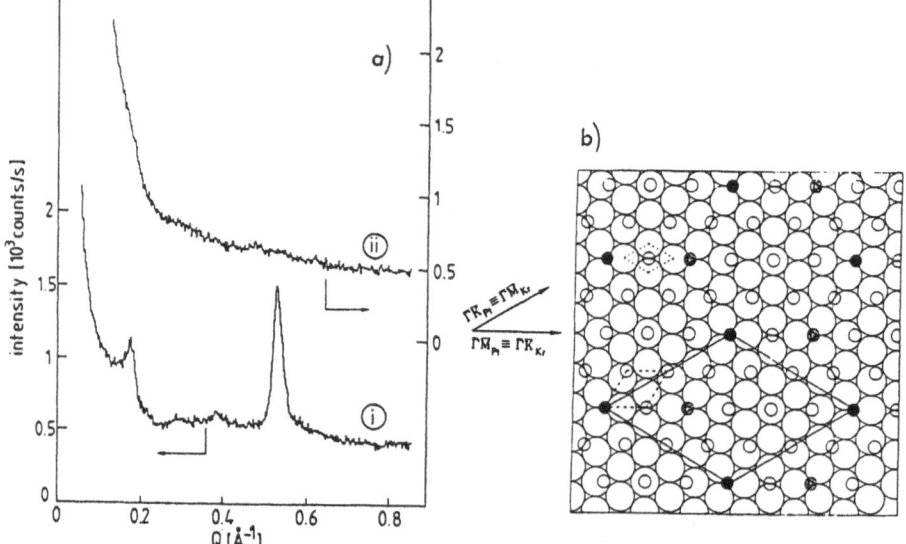

Fig.2.16. a) Polar He-diffraction scans of Kr monolayers in the vicinity of the specular peak (Q = 0 Å⁻¹); i) high (0.95 ML) and ii) low (0.5 ML) coverage phase, taken along the $\bar{\Gamma}\bar{K}_{Kr}$-azimuth. b) Schematic representation of the high coverage phase of Kr on Pt{111}; small circles represent Kr-atoms (d_{Kr-Kr}=4.00 Å) and large circles the Pt substrate atoms (d_{Pt-Pt}=2.77 Å)

"floating" and of the "locked" Kr layer are shown. The scans differ substantially: the locked scan clearly evidences the presence of a superstructure, while the floating one does not. The superstructure peak at $Q = 0.532 \pm 0.022$ $\text{\AA}^{-1}$ corresponds to 1/5 of the Pt substrate principal lattice vector. The origin of the superstructure peak is illustrated on the right hand of Fig.2.16b where the "locked" Kr phase on the Pt{111} surface is schematically shown. The Kr layer is rotated by 30° with respect to the substrate and its translational position is fixed by locating the central Kr atom in a preferred three-fold hollow site (say fcc). Obviously, the Kr atoms in fcc sites (filled small circles) form a hexagonal (5 x 5)R0° superstructure, which is responsible for the diffraction satellite at $Q_{Pt}^{1,1}/5$. Note that the superstructure is aligned with the substrate lattice while the Kr layer as a whole is rotated by 30°; this is the reason why the superstructure satellite is seen in the $\overline{\Gamma M}_{Pt} = \overline{\Gamma K}_{Kr}$ $= \overline{\Gamma M}_{superstructure}$-direction. The particular ratio between the lattice parameters of adlayer and substrate $\sqrt{3}\, d_{Pt}/d_{Kr} = 6/5$ produces an additional peculiarity: the same number of Kr atoms are located in hcp and fcc hollow sites. Thus the (5 x 5)R0° superstructure of the locked Kr atoms has a two-atomic basis. A simple counting in Fig.2.15b shows that one sixth of the Kr atoms are locked in a hollow site (fcc or hcp). This fraction appears to be sufficient to hinder the Kr layer from expanding freely over more than 25 K.

The correlation between negligible thermal expansion and the presence of a periodic adlayer buckling in the high coverage phase of Kr on Pt{111} unequivocally demonstrates the high order commensurate character of this phase. The superstructure peak in Fig.2.16a originates certainly from the layer buckling due to a fraction of Kr atoms located in high symmetry sites of the substrate. Indeed, the Kr layers being oriented along high symmetry axes of the substrate, there is no reason for the occurrence of mass density wave satellites (see below) which might complicate the interpretation [2.62].

2.2.6 Rotational Epitaxy of Monolayers

In the one-dimensional chain model of Frank and van der Merwe, the longitudinal misfit strain of the incommensurate layer is minimized by the formation of misfit dislocations. As demonstrated above, the formation of misfit dislocations - domain walls - also minimizes the total energy of an incommensurate 2D-layer close to the C-I transition (i.e., at small misfits). In two dimensional incommensurate overlayers, however, longitudinal as well as transverse strains are present. Since transverse strains have lower energy than longitudinal strains in a 2D-layer (transverse phonons are softer than longitudinal ones), the interconversion of these strains may minimize the total energy of the overlayer by rotating the adlayer out of the symmetry axes of the substrate.

Indeed, *Novaco* and *McTague* [2.63] have shown that for monolayers far from commensurability these rotations are energetically favorable and the rotation angle ϕ follows a simple relation:

$$\cos \phi = \frac{1 + (1+m)^2(1+2\eta)}{(1+m)[2 + \eta + \eta(1+m)^2]} \quad \eta > (1+m)^{-1}$$

$$\phi = 0 \quad \eta < (1+m)^{-1} \tag{2.14}$$

with $\eta = (c_\ell/c_t)^2 - 1$, c_ℓ and c_t being the longitudinal and transverse velocities of sound in the monolayer and m the misfit. For a Cauchy solid we have $c_\ell/c_t = 3^{1/2}$. Novaco and McTague also showed that this rotational epitaxy involves the creation of mass density waves, MDW, (also known as static distortion waves, SDW), i.e., there exists a periodic variation in the position of monolayer atoms from their regular lattice sites. These MDW bear relevant similarities to the charge density waves (CDW) of layered crystals [2.64]. Indeed, it is the combination of rotation and small displacive distortions of the adatom net which allows the adlayer to minimize its total energy in the potential relief of the substrate. In a diffraction experiment, these mass density waves should give rise to additional satellites.

Fuselier et al. [2.65] have introduced an alternative concept to explain the adlayer rotation: the "coincident site lattice". They pointed out that energetically more favorable orientations are obtained for rotated high-order commensurate structures. The larger the fraction of adatoms located in high symmetry sites, the larger the energy gain and the better the rotated layer is locked. It turned out that the predictions of the coincident site lattice concept for the rotation angle versus misfit agrees well with the Novaco-McTague predictions.

The Novaco-McTague rotational epitaxy has been observed in numerous adsorbate systems including Ar [2.66], Ne [2.67], and Kr [2.57] on graphite, Na/Ru{100} [2.68], K/Cu{100} [2.69] and K/Pt{111} [2.70]. It has also been observed for Xe monolayers on Pt {111} at large incommensurabilities [2.71]. As already shown in the last paragraph, the striped incommensurate phase of Xe on Pt{111} transforms above 6.5% misfit to a hexagonal incommensurate phase. When increasing the misfit above 7.2% the striped phase disappears and only the diffraction peaks characterizing the hexagonal phase are observed. However, these diffraction peaks at $Q^{1,2}_{comm} + \sqrt{3}\epsilon$ and $Q^{2,2}_{comm} + 2\epsilon$ (in the (1,2) and (2,2) diffraction patterns, respectively) start to split azimuthally with increasing misfit (Fig.2.17a). This obviously characterizes a Novaco-McTague rotated phase.

Figure 2.17b is a plot of the rotation angle ϕ as a function of the average misfit during the HI-HIR transition; the black dots are the measured data. The dashed line is the Novaco-McTague linear response

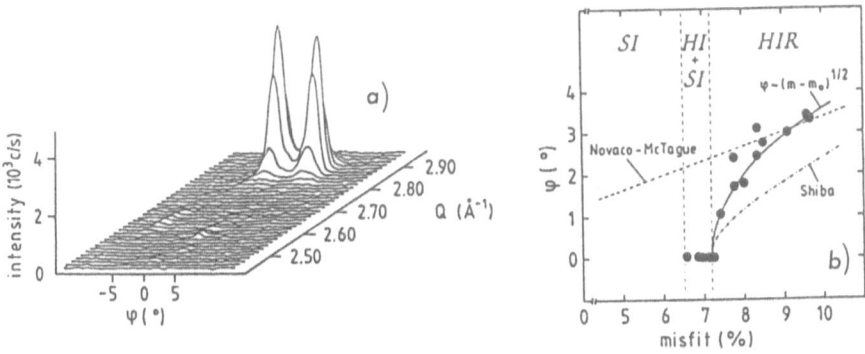

Fig.2.17. a) 3D-diffraction plot of the $(1,2)_{xe}$ diffraction feature of an incommensurate Xe layer on Pt{111} at a misfit of 9.6% ($\theta_{Xe}=0.41$, T=25 K). b) Rotation angle ϕ of Xe monolayers on Pt(111) versus misfit m

theory for a Cauchy solid (2.14). *Shiba* [2.72] noted that with increasing misfit a crossover from the domain wall regime (small incommensurabilities) to the modulation regime (large incommensurabilities) should occur, and that there should be a finite misfit for the onset of rotation. His curve for a Cauchy solid is also drawn in Fig.2.17b (dashed-dotted) for the value of his parameter ℓ which causes the HI-HIR transition to occur at a critical misfit of 7.2% (here $\ell{\approx}10$). The parameter ℓ in Shiba's theory is analogous to Frank and van der Merwe's ℓ, i.e., is the distance between the domain walls in units of Xe inter-row distances in the $\bar{\Gamma}M$ direction at the transition. Shiba's theory gives a qualitative account of the overall variation of rotation angle versus misfit but no quantitative account. Actually, the data are fitted well by a power law of the form $\phi \simeq (m-m_0)^{1/2}$ with $m_0 = 0.072$, which is shown as a solid line in Fig.2.17b. A similar power law behavior has been observed for the rotation angle of Kr layers on graphite [2.57] and in Cs intercalated graphite [2.73].

The observation of a finite critical misfit for the onset of rotation is also consistent with the approximate analysis of *Villain* [2.74] and the analytic treatment of *Gordon* and *Villain* [2.75]. Both have calculated the energy associated with a small rotation of a system of parallel walls near the C-I transition, and found that the domain wall rotation should take place in the incommensurate regime (i.e., a finite misfit) and not at the C-I transition. The Novaco-McTague model calculations [2.63] have been performed in the linear response approximation of the adsorbate-substrate interaction. Close to the C-I transition, however, this approximation fails and no rotation has to be expected for realistic values of the Lamé coefficients of the adlayer.

It is worthwhile to address here the question concerning the physics behind the rotational epitaxy; mass density waves [2.63] or high-order commensurability [2.65]? In a He-diffraction study from rotated Xe monolayers on Pt{111} we have, indeed, observed satellite peaks and assigned them to a high order commensurate superstructure [2.32]. However, *Mirta Gordon* [2.62] pointed out that these satellites could be due to the MDW. Here, we will show that, indeed, both MDW as well as commensurate buckling satellites are present in the rotated Xe monolayers on Pt{111}. The distinction between the two types of satellites is straightforward. As pointed out by Gordon, the MDW satellites should be subject to the following relation:

$$Q \simeq (8\pi/d_{Xe}^R)\,(m/\sqrt{3})\,(1 + m/8) \qquad (2.15)$$

where Q is the wavevector of the satellites, m the misfit and d_{Xe}^R the lattice constant of the rotated Xe layer. For not too large misfits, the MDW satellites should appear in the same direction as the principal reciprocal lattice vector of the Xe layer, i.e. in the $\overline{\Gamma M}_{Xe}$ direction. On the other hand, the commensurate buckling, according to its peculiar structure [Ref.32,Fig.2.3] should have its maximum amplitude in the $\overline{\Gamma K}_{Xe}$ direction. Moreover, these commensurate buckling satellites should only be present at the particular coverage where the high order commensurability becomes favorable, in the present case at monolayer completion (m=9.6%) [2.32], whereas the MDW satellites should be present in the entire misfit range where the Xe layer is rotated (7.2%-9.6%).

In Fig.2.18a we show polar He diffraction scans in the vicinity of the specular beam, with the scattering plane oriented along the $\overline{\Gamma M}_{Xe}$ direction, for rotated Xe layers of misfit 8.3% and 9.5%, respectively. In both scans satellite peaks are present. The dispersion of these peaks is shown in Fig.2.18b, and compared with Gordon's prediction for the MDW given above. The data follow qualitatively the predicted depen-

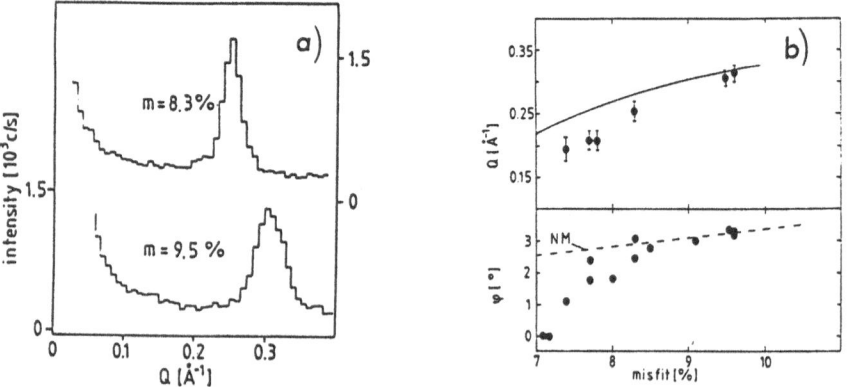

Fig.2.18. a) Polar He-diffraction scans of rotated Xe monolayers on Pt{111} in the vincinity of the specular peak (Q=0 Å^{-1}) taken along the $\overline{\Gamma M}_{Xe}$ azimuth, at m = 8.3% and 9.5%, respectively. b) Dispersion of the mass density wave (MDW) satellites with misfit m. The solid line is Gordon's relation (2.15)

dency; the agreement becomes quantitative at misfits larger than $\simeq 8\%$. The reason for the better agreement at large misfits is due to the fact that Gordon's analysis of the MDW (similar to Novaco-MacTague's model calculations) have been performed in the linear response approximation of the adsorbate-substrate interaction. As discussed in connection with Fig.2.17b above, this approximation is only justified at larger misfits. The measured intensities of the MDW satellites vary between 10^{-2} of the strongest Xe layer diffraction peak at small misfits to about 10^{-1} at large misfits. These intensities can be correlated with the amplitudes of the MDW. By means of [2.62,Eq.6-8] and by using realistic values for the Xe/Pt{111} potential (as listed in [2.76]) MDW amplitudes of the order 0.1 Å are obtained.

In Fig.2.19a we show scans like in Fig.2.18a but now measured in the $\bar{\Gamma}K_{Xe}$ direction for rotated Xe layers of misfits 8% and 9.6%. At variance with the scans in the $\bar{\Gamma}M_{Xe}$ direction (Fig.2.18a), a satellite

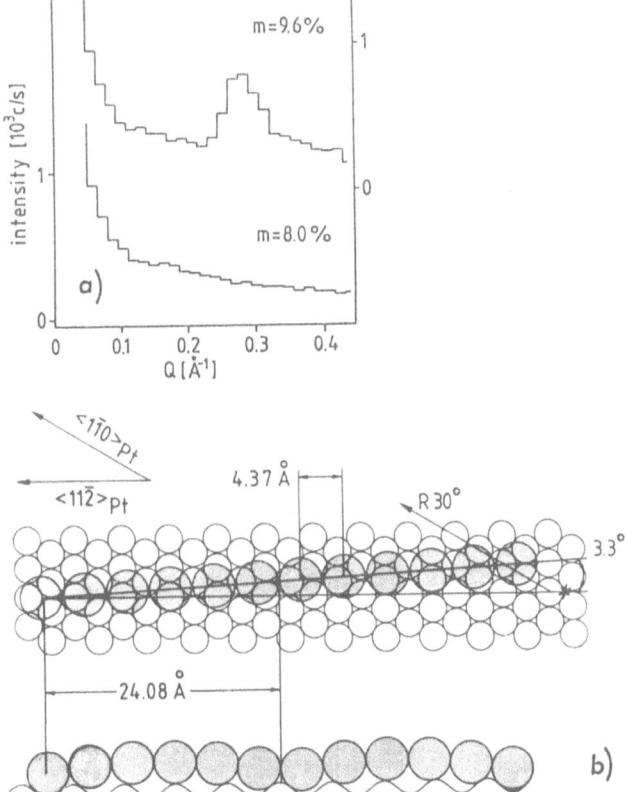

Fig.2.19. a) Polar He-diffraction scans of rotated Xe-monolayers on Pt{111} in the vicinity of the specular peak (Q=0 Å^{-1}) taken along the $\bar{\Gamma}K_{Xe}$-azimuth, at m = 8% and 9.6%, respectively. b) Upper and side view of a 3.3° rotated domain of the complete monolayer. The Xe domain is represented schematically by a chain of twelve atoms

peak is observed only for the complete Xe monolayer (m=9.6%). The location of this satellite peak at 0.28 Å^{-1} corresponds to a buckling period of 23 Å and can be ascribed to a high order commensurate structure shown in Fig.2.19b, as described in detail in [2.32]. Being present only at a particular misfit, this peak does not originate from a MDW. Thus, only at monolayer completion (m=9.6%) does the rotated Xe layer lock in the substrate, i.e., becomes a high order commensurate phase.

Thus the present answer to the question addressed is that both mass density waves and high order commensurability may be involved in rotational epitaxy.

2.3 Multilayer Growth of Rare Gases

2.3.1 Dynamical Coupling Between Adlayer and Substrate

The first systematic theoretical and experimental exploration of the dynamics of rare gas monolayers on metal surfaces was performed on Ag{111} [2.77]. The lattice dynamical calculations were based on a simple model - Barker pair potentials - to model the lateral adatom interactions and a rigid holding substrate. The calculations have supplied dispersion curves fully adequate to account for the available experimental data.

As expected from any model involving only central forces between the adatoms and a rigid substrate, the three modes of the monolayer decouple, and the perpendicular mode is dispersionless, i.e., the motion of the adatoms perpendicular to the surface acts like an Einstein oscillator. Because the perpendicular surface atom motions dominate the inelastic He cross sections, this dispersionless mode has, indeed, been observed in the experiments [2.77].

It is noteworthy that the most general conclusion emerging from this first systematic exploration, has been that "coupling between adatom and substrate atom motions is potentially more important than modest variations in the nature of the adatom-adatom potential" [2.77]. *Hall* et al. [2.78] extended their exploration by allowing the substrate atoms to move and by focussing on the coupling between the substrate and adlayer modes. As expected, the results of the calculations show that near the zone boundary $\bar{\text{M}}$ (the $\bar{\Gamma}\bar{\text{M}}$ direction has been explored), where the substrate phonon frequencies are well above those of the adlayer, the influence of the substrate adlayer coupling is small. Near the zone center $\bar{\Gamma}$ the anomalies introduced by the coupling are twofold:

1. A dramatic hybridization splitting around the crossing between the dispersionless adlayer mode and the substrate Rayleigh wave (and a less dramatic one around the crossing with the $\omega = c_\ell Q^{\parallel}$ line - due to

the Van Hove singularity in the projected bulk phonon density of states);

2. A substantial line width broadening of the adlayer modes in the whole region near $\bar{\Gamma}$ where they overlap the bulk phonon bands of the substrate: the excited adlayer modes may decay by emitting phonons into the substrate; they become leaky modes. These anomalies were expected to extend up to trilayers even if more pronounced for bi- and in particular for monolayers. More recent experimental data of *Gibson* and *Sibener* [2.79] qualitatively confirm these predictions, at least for monolayers. The phonon line widths appear to be broadened around $\bar{\Gamma}$ up to half of the Brillouin zone. The hybridization splitting could not be resolved, but an increase of the inelastic transition probability centered around the crossing with the Rayleigh wave and extending up to 3/4 of the zone has been observed and attributed to a resonance between the adatom and substrate modes.

Recent measurements performed on Ar, Kr and Xe layers on Pt{111} [2.80] with a substantially higher energy resolution ($\Delta E_I \leq 0.4$ meV) have now confirmed the theoretical predictions on the coupling effects within almost every detail (except for the hybridization around the Van Hove singularity, which has not been seen in spite of substantial effort). The sequence of He TOF spectra in Fig.2.20a taken along the $\bar{\Gamma}\bar{M}$ direction of the superstructure of the high-order commensurate (5x5) complete Kr-monolayer identical to the $\bar{\Gamma}\bar{M}_{Pt}$ = $\bar{\Gamma}\bar{K}_{Kr}$ directions (Sect.2.2.5) at 25 K gives a vivid picture of the coupling effects. The last spectrum θ_i = 37° taken near the zone boundary M exhibits a unique, sharp loss E $\simeq$ -3.7 meV resulting from the creation of an Einstein Kr-monolayer phonon (perpendicular Kr-Pt vibration); its width corresponds to the instrumental width of $\Delta E_I \simeq$ 0.38 meV; as expected there is no linewidth broadening near the zone boundary. On the other hand, the main peak in the first spectrum (θ_i=40°) taken near the $\bar{\Gamma}$-point located at E $\simeq$ -3.9 meV and which corresponds also to the creation of a Kr monolayer phonon is broadened by more than 0.5 meV. Of particular interest is the small peak at E $\simeq$ -3.1 meV, close to the position of the Pt substrate Rayleigh wave. The next two spectra θ_i = 39.5° and 39° taken always closer to the crossing between the Pt substrate Rayleigh wave and the Kr Einstein mode demonstrate strikingly the effect of the hybridization of the two modes: the originally tiny Pt-peak increases dramatically, while the Kr-peak is pushed slightly toward larger energies. After passing the crossover, the higher energy loss disappears abruptly. As predicted [2.78] the two features in the doublet only have comparable intensity quite near the crossover.

In Fig.2.20b we show the dispersion curve of the Kr-monolayer obtained from a large number of spectra like those in Fig.2.20a. The

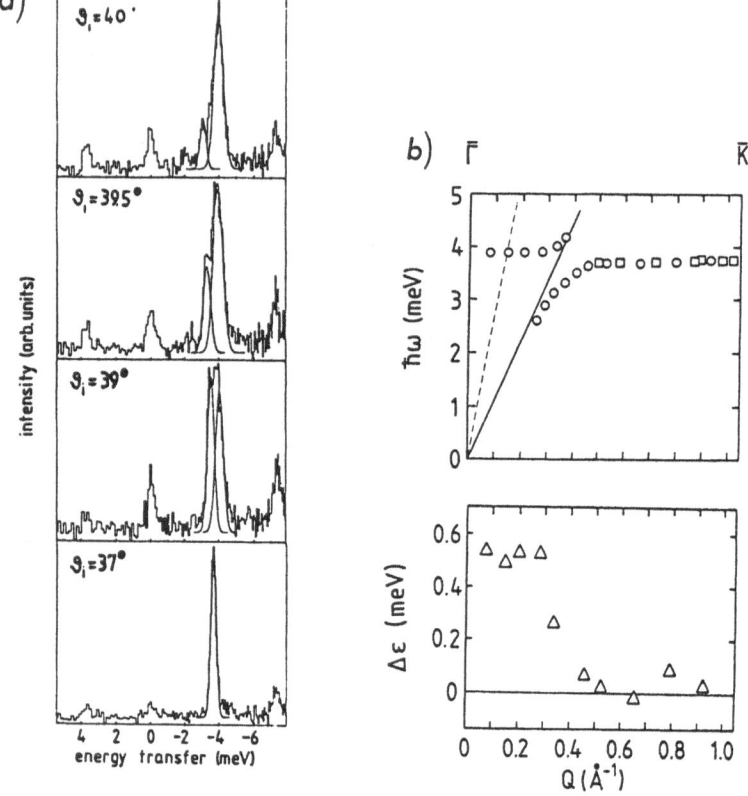

Fig.2.20. a) He energy loss spectra from a Kr monolayer taken along the $\overline{\Gamma K}_{Kr}$-azimuth. With decreasing angle phonons with larger wave vector are probed.
b) Experimental dispersion curve of the Kr monolayer and measured linewidth broadening $\Delta \epsilon$ of the Kr creation phonon peaks. The solid line in the dispersion plot is the clean Pt{111} Rayleigh phonon dispersion curve and the dashed line the longitudinal bulk band edge of the Pt{111} substrate both in the $\overline{\Gamma M}_{Pt}$ azimuth which is coincident with the $\overline{\Gamma K}_{Kr}$ azimuth

hybridization splitting around the crossover with the substrate Rayleigh wave (solid line) is clearly observed. The predicted tiny frequency upshift around the $\overline{\Gamma}$-point due to the coupling to the substrate vibrations is also seen.

The observed line width broadening is also shown in Fig.2.20b. As a measure of the broadening the quantity $\Delta \epsilon = [(\delta E)^2 - \Delta E_I^2]^{1/2}$, with δE the full width at half maximum of the major loss feature and ΔE_I the intrinsic instrumental broadening ($\Delta E_I = 0.38$ meV in the present experiment) is plotted as a function of the wave vector. For the ML a broadening larger than 0.5 meV is seen, and - as predicted - confined to the region near $\overline{\Gamma}$, where the adlayer mode overlaps the bulk bands of the substrate.

2.3.2 Layer-by-Layer Evolution of the Lattice Dynamics

In Fig.2.21a we show some characteristic He energy loss spectra which have been measured under identical scattering conditions ($\theta_i = 42°$, $E_i = 18.4$ meV) from Xe-films 1,2,3 and 25 monolayers thick adsorbed on Pt{111}. The kinematical conditions have been chosen to sample energy losses (phonon creation) with small momentum transfer, i.e., to probe the phonon dispersion near the zone center. Due to the high energy resolution, the different losses of the four films are clearly resolved. This makes it possible to straightforwardly determine the completion of each of the first two monolayers within a few percent. This kind of information, which can hardly be obtained with this accuracy by other methods, is very useful, for instance, in deducing thermodynamic properties of multilayer adsorption [2.38].

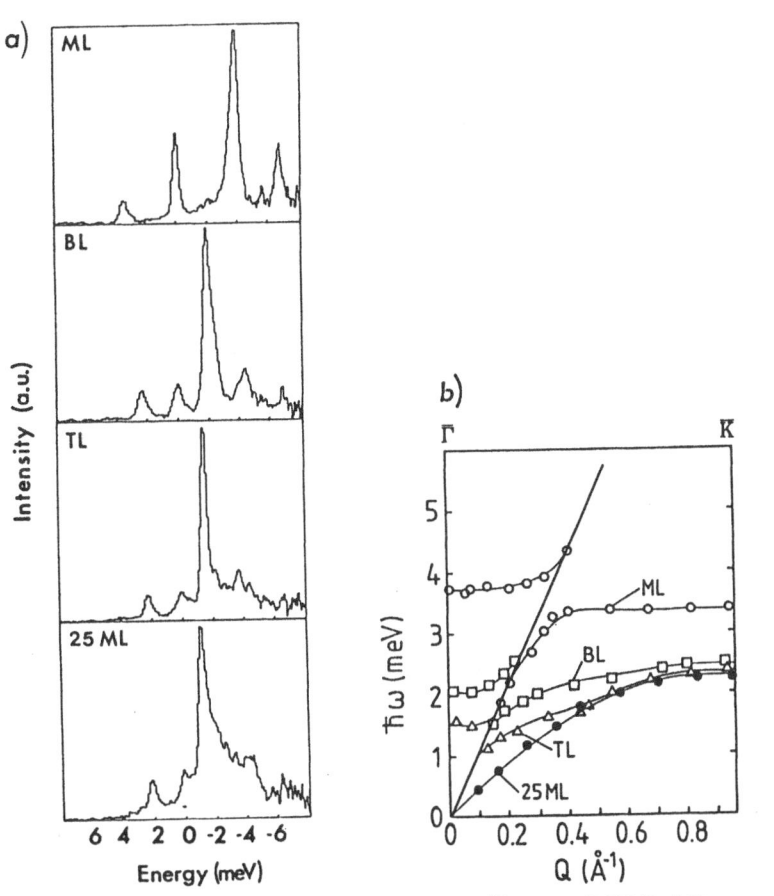

Fig.2.21. a) Energy loss spectra from Xe films on Pt{111}: ML: monolayer, BL: bilayer, TL: trilayer and 25 ML film
b) Experimental dispersion curves of the various Xe films along the $\overline{\Gamma K}_{Xe}$ azimuth

By varying the scattering angle, complete phonon dispersion curves for each film have been obtained. This is exemplified in Fig.2.21b, which is a reduced zone plot of the phonon dispersion along the $\overline{\Gamma K}$-azimuth of the Xe layers. The layer-by-layer evolution of the surface lattice dynamics with increasing film thickness is obvious. From the initial Einstein-mode for the monolayer a well developed Rayleigh mode (25 ML), characteristic for semi-infinite crystals is approached upon increase of the film thickness. It is noteworthy, that the phonon anomaly, due to the dynamical coupling between substrate Rayleigh wave and adlayer mode, is also present in the bi- and even the trilayer films. It is only the Q-range of the anomaly which becomes smaller, and its location shifts towards the zone center following the location of the intersection between the Xe and the substrate Rayleigh mode. Linewidth broadening, due to radiative damping into the substrate bulk bands has been found to be still substantial for bilayer films, while the trilayer shows no evidence for additional broadening. These results demonstrate that the influence of the substrate on adsorbed multilayer films is extended over several layers rather than being restricted to the first layer only.

2.3.3 Growth Mode and the Scale of Substrate Strength

The mode of nucleation and initial growth of thin films is a matter of longstanding interest. The basic question posed as long as 60 years ago is: given a known substrate/adsorbate, can the growth mode be predicted using atomistic principles only? From thermodynamics we know that when we coat a substrate s with a deposite d under an atmosphere v with which d coexists, three categories of initial film formation can occur [2.81]. These types of film growth can be classified using the surface tension σ. For a droplet of deposit d adsorbed on substrate s in equilibrium we have:

$$\sigma_{sv} = \sigma_{ds} + \sigma_{dv} \cos\theta \qquad (2.16)$$

where σ_{sv}, σ_{ds} and σ_{dv} are the surface tensions of the substrate-vapor, deposit-substrate, and deposit-vapor interfaces and θ is the contact angle (see Fig.2.22a for definition). Equation (2.16) is known as *Young's* equation [2.82].

The deposit wets the substrate completely when $\sigma_{sv} = \sigma_{ds} + \sigma_{dv}$, i.e., $\theta = 0$; the growth occurs layer-by-layer (Fig.2.22b). This growth mode is also known as Frank-van der Merwe or type 1 growth. When $\sigma_{sv} > \sigma_{ds} + \sigma_{dv}$ the deposit grows in a few layers on top of which 3D-islands are formed (Fig.2.22c). This growth mode is also known as partial wetting, Stranski-Krastanov or type 3 growth. When $\sigma_{sv} < \sigma_{ds} + \sigma_{dv}$, i.e., θ is finite, small 3D-cluster are nucleated directly on top of

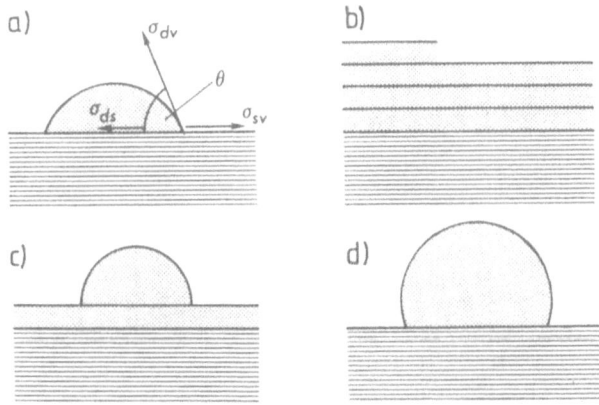

Fig.2.22. Schematic representation of the various growth modes.
a) definitions, b) complete wetting, c) partial wetting, and d) incomplete wetting

the substrate surface which eventually agglomerate into a continuous film (Fig.2.22d). This growth mode is also known as imcomplete wetting, Volmer-Weber or type 2 growth.

It has been suggested by *Sullivan* [2.83], that the determining parameter for the growth mode is the relative substrate strength u/h, which is the ratio of u, the adsorbed-substrate interaction, to h, the lateral adsorbate-adsorbate interaction. *Pandit, Schick* and *Wortis* [2.84], based on this idea, have developed a lattice-gas model of adsorption generating detailed phase diagrams of multilayer-adsorption (including wetting behavior, wetting transitions, roughening transitions, melting transitions, etc.) which qualitatively scale with the ratio u/h. Their results are too extensive as to be reviewed here. However, the basic result was that incomplete wetting should take place for low, partial wetting for intermediate, and complete wetting for large relative substrate strengths.

Shortly after, experiments designed to test the theory showed that, in contrast to the predictions, at low temperatures complete wetting is restricted to a very narrow intermediate range of substrate strengths [2.85]. Both, small as well as large u/h values resulted in incomplete wetting behavior

The reentrant wetting behavior (at large u/h) has been explained by the incompatibility between the crystal structure of the monolayer and that of the bulk adsorbate [2.86-88]. Strong substrates tend to compress the adsorbate monolayer beyond the density of close packed planes of the bulk solid to be grown. Since the next adsorbing layers tend to grow epitaxially on the monolayer they are also compressed and the stress in the layer will grow linearly with increasing thickness, preventing complete wetting. For weak substrates the argument is

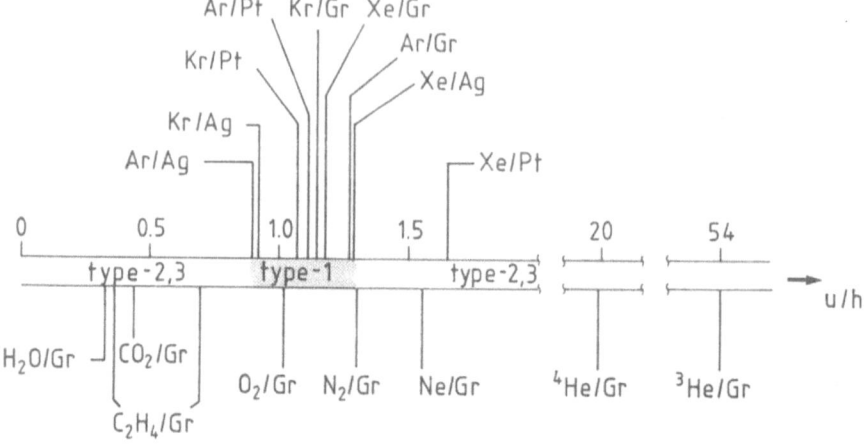

Fig. 2.23. Adsorbate-substrate systems ordered on the scale of substrate strengths. Systems exhibiting complete wetting are marked by bars above the center line, while those showing incomplete wetting are indicated below

similar, but now the density in the monolayer is lower than in the bulk solid. Thus, only for intermediate substrate strengths does a compatibility between the structure of the monolayer and that of the bulk adsorbate allow a uniform layer-by-layer growth.

However, this is not the whole truth as can be seen from Fig.2.23, where we show the adsorbate sytems investigated so far ordered on the scale of substrate strengths. The isosteric heat of adsorption is taken for u and the 0 K cohesive energy of the bulk phase of the adsorbing gases for h. Systems exhibiting complete wetting are marked by bars above the center line while those showing incomplete wetting are indicated below. Whereas most systems are in agreement with the reentrant wetting behavior, Xe/Pt{111} exhibits complete wetting although it is located well within the incomplete wetting range. As will be shown in the next section, this anomaly is probably due to registry effects in the Xe/Pt{111} system which result in a neglibible incompatibility between the bulk adsorbate and the full monolayer.

2.3.4 Epitaxial Layer Growth of Xe on Pt{111}

The He energy loss spectra shown in Fig.2.21a not only contain lattice dynamics information as discussed in Sect.2.3.2; they also contain direct information concerning the growth characteristics of the Xe multilayers. This information is in the diffuse elastic peak, i.e., in the peak at zero energy exchange. This peak originates from scattering at impurities and defects and its intensity is a sensitive measure of surface disorder. From the comparison with spectra taken from surfaces of known disorder, we can infer that the monolayer is well ordered and the multilayers even better. For the 25 ML thick film in

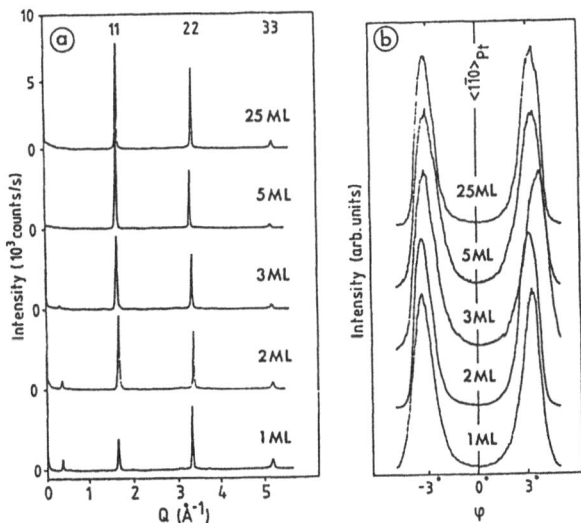

Fig.2.24. a) Polar and b) azimuthal diffraction patterns of Xe films of indicated thickness. The incident plane for the polar patterns was oriented through the left-hand peaks in the azimuthal patterns. All polar patterns are plotted on the same scale; the azimuthal patterns are normalized

Fig.2.21a, the diffuse elastic peak has nearly vanished. This shows that the 25 ML film is very flat, and thus that Xe on Pt{111} exhibits complete wetting. This goes along with the layer-by-layer evolution of the surface phonon dispersion discussed in Sect. 2.3.2.

The structure of the Xe multilayers has been characterized by measuring polar and azimuthal He-diffraction scans, shown in Fig.2.24. As already emphasized in Sect.2.2.6, at monolayer completion the Xe monolayer on Pt{111} is a Novaco-McTague rotated layer with rotation angle $\phi = \pm 3.3°$. The azimuthal plots in Fig.2.24 show that all consecutive layers growing on top of the first layer are likewise rotated by $\phi = \pm 3.3°$. Lattice constant and average domain size, as deduced from the polar diffraction plots, are also unchanged with increasing film thickness, i.e., $d_{Xe} = 4.33 \pm 0.03$ Å and average domain size $\simeq 300$ Å, respectively. Thus, the consecutive layers grow epitaxially on the preceding ones. Within experimental confidence there is no mismatch between the nearest neighbor distances in the monolayer ($d_{Xe}^{R-ML} = 4.33$ Å, T=25 K) and in the bulk Xe ($d_{Xe}^{b} = 4.34$ Å, T=25 K). This structural compatibility, which leads to an unstrained layer-by-layer growth appears to be a direct result of registry forces. Indeed, Xe/Pt{111} being a "strong substrate" system, the monolayer lattice parameter would be expected, in the absence of registry forces, to be compressed well beyond the bulk value. However, as shown in Sect. 2.2.6, at misfits of 9.6% the rotated monolayer locks into an energetically favorable high order commensurate structure. It is this high order commensurate locking of a fraction of adatoms which, by counterbalancing the tendency of the strong substrate to compress the monolayer lattice beyond the bulk value, allows for an unstrained layer-by-layer growth.

Note that the small peak at low Q-values in Fig.2.24a is a MDW satellite already shown in Fig.2.18a for smaller misfit. The very gradual disappearance of the mass density waves with increasing film thickness, again emphasizes that the influence of the substrate on adsorbed multilayer films extends over several layers.

2.4 Conclusion

Rare-gas monolayers on Pt{111} exhibit a marvellous diversity of 2D phases and phase transitions; most of these have been predicted theoretically, but never been seen completely on other substrates so far. Phases and transitions orginate as a function of temperature and coverage from the interplay between interadatom forces and the corrugation of the holding potential. The exploration of these phenomena is essential for the understanding of the fundamentals of 2D-layer structure and dynamics as well as for the epitaxial growth of thicker layers; the initial, decisive step of this growth appears to be determined by the adatom-substrate interaction.

The experimental exploration of the details of the structure and dynamics of all these phases is made possible by the use of high-resolution thermal He scattering. Energy and momentum resolution; sensitivity for disorder, impurities and defects; and non-destructivity are the distinctive features of thermal He scattering, making it an ideal tool for this kind of investigations.

Acknowledgements. The essential contribution of Rudolf David, Peter Zeppenfeld and Bob Palmer to the results presented here is gratefully acknowledged, as are enlightening discussions with John Black, Sam Fain, Harald Ibach, Doug Mills and Steven Sibener. The careful typing of Maria Kober is also gratefully acknowledged.

References

2.1   R.E. Peierls: Helv. Phys. Acta 7, 81 (1934) Suppl.II
2.2   L.D. Landau: Phys. Z. Sowjetunion II, 26 (1937)
2.3   F. Wagner: Z. Phys. 206, 465 (1967)
2.4   B. Jancovici: Phys. Rev. Lett. 19, 20 (1967)
2.5   J.M. Kosterlitz, D.J. Thouless: J. Phys. C 6, 1181 (1973)
2.6   F.F. Abraham: Physics Reports 80, 339 (1981)
2.7   B.I. Halperin, D.R. Nelson: Phys. Rev. B 19, 2457 (1979)
2.8   A.P. Young: Phys. Rev. B 19, 1855 (1979)
2.9   J. Villain, M.B. Gordon: Surf. Sci. 125, 1 (1983), and references therein

2.10 G. Ertl, J. Küppers: *Low Energy Electrons and Surface Chemistry* (Verlag Chemie, Weinheim 1985)

2.11 H. Ibach: J. Vac. Sci. Technol. A **5**, 419 (1987)

2.12 M.D. Chinn, S.C. Fain: J. Vac. Sci. Technol. **14**, 314 (1977)

2.13 J.A. Venables, J.L. Senguin, J. Suzanne, M. Bienfait: Surf. Sci. **145**, 345 (1984)

2.14 J.A. Martin, M.G. Lagally: In *Scanning Electron Microscopy*, ed. by O. Johari (Scanning Electron Microscopy, Inc., Chicago 1985) Vol.4

2.15 K.D. Gronwald, M. Henzler: Surf. Sci. **117**, 180 (1982)

2.16 W. Telieps, E. Bauer: Ultramicroscopy **17**, 57 (1985)

2.17 J.L. Seguin, J.Suzanne, M. Bienfait, J.G. Dash, J. Venables: Phys. Rev. Lett. **51**, 122 (1983)

2.18 D.E. Savage, M.G. Lagally: Phys. Rev. Lett. **55**, 959 (1985)

2.19 R.K. Thomas: Prog. Solid. St. Chem. **14**, 1 (1982)

2.20 M. Bienfait: Europhys. Lett. **4**, 79 (1987)

2.21 P. Eisenberger, W.C. Marra: Phys. Rev. Lett. **46**, 1081 (1981)

2.22 D.E. Moncton, G.S. Brown: Nucl. Instr. Meth. **208**, 579 (1983)

2.23 E.D. Specht, R.J. Birgeneau, K.L. D'Amico, D.E. Moncton, S.E. Nagler, P.M. Horn: J. de Phys. Lett. **46**, L561 (1985)

2.24 E. Burkel, J. Peisl, B. Dorner: Europhys. Lett. **3**, 957 (1987)

2.25 G. Brusdeylins, H.D. Meyer, J.P. Toennies, K. Winkelmann: *Rarefied Gas Dynamics*, ed. by J.L. Potter (AIAA, New York 1977) Vol.II

2.26 R. Campargue: J. Phys. Chem. **88**, 4466 (1984)

2.27 O. Stern: Naturwiss. **17**, 391 (1929)

2.28 J.P. Toennies, K. Winkelmann: J. Chem. Phys. **66**, 3965 (1977)

2.29 K. Kern, R. David, G. Comsa: Rev. Sci. Instr. **56**, 369 (1985)

2.30 T. Engel, K.H. Rieder: *Structural Studies of Surfaces with Atomic and Molecular Beam Diffraction*, Springer Tracts Mod. Phys., Vol.91 (Springer, Berlin, Heidelberg 1982)

2.31 J.P. Toennies: J. Vac. Sci. Technol. A **2**, 1055 (1984)

2.32 K. Kern, R. David, R.L. Palmer, G. Comsa: Phys. Rev. Lett. **56**, 2823 (1986)

2.33 A.M. Lahee, J.R. Manson, J.P. Toennies, Ch. Wöll: Phys. Rev. Lett. **57**, 471 (1986)

2.34 J.W. Frenken: private communication

2.35 G. Comsa, B. Poelsema: Appl. Phys. A **38**, 153 (1985)

2.36 R. David, K. Kern, P. Zeppenfeld, G. Comsa: Rev. Sci. Instr. **57**, 2771 (1986)

2.37 G. Comsa, R. David, B.J. Schumacher: Rev. Sci. Instr. **52**, 789 (1981)

2.38 K. Kern, R. David, R.L. Palmer, G. Comsa: Surf. Sci. **175**, L669 (1986)

2.39 R.L. Park, H.H. Madden: Surf. Sci. **11**, 188 (1968)

2.40 F.C. Frank, J.H. Van der Merwe: Proc. Roy. Soc. London Ser. A **198**, 216 (1949)

2.41 P. Bak, D. Mukamel, J. Villain, K. Wentowska: Phys. Rev. B **19**, 1610 (1979)

2.42 V.L. Pokrovsky, A.L. Talapov: Sov. Phys. JETP **51**, 134 (1980)

2.43 K. Kern, R. David, R.L. Palmer, G. Comsa: Phys. Rev. Lett. **56**, 620 (1986)

2.44 M.B. Gordon, J. Villain: J. Phys. C **18**, 3919 (1985)

2.45 K. Kern, R. David, P. Zeppenfeld, R.L. Palmer, G. Comsa: Solid State Commun. **62**, 361 (1987)

2.46 P. Zeppenfeld, K. Kern, R. David, G. Comsa: Phys. Rev. B (1988) to be published

2.47 P.W. Stephens, P.A. Heiney, R.J. Birgeneau, P.M. Horn, D.E. Moncton, G.S. Brown: Phys. Rev. B **29**, 3512 (1984)

2.48 A. Erbil, A.R. Kortan, R.J. Birgeneau, M.S. Dresselhaus: Phys. Rev. B **28**, 6329 (1983)

2.49 S.C. Fain, M.D. Chinn, R.D. Diehl: Phys. Rev. B **21**, 4170 (1980);
D.E. Moncton, P.W. Stephens, R.J. Birgeneau, P.M. Horn, G.S. Brown: Phys. Rev. Lett. **46**, 1533 (1981)

107

2.50 S.N. Coppersmith, D.S. Fisher, B.I. Halperin, P.A. Lee, W.F. Brinkman: Phys. Rev. B **25**, 349 (1982)

2.51 F.D.M. Haldane, J. Villain: J. de Phys. **42**, 1673 (1981)

2.52 M. Jaubert, M. Glachant, M. Bienfait, G. Boato: Phys. Rev. Lett. **46**, 1679 (1981)

2.53 F.F. Abraham, W.E. Rudge, D.J. Auerbach, S.W. Koch: Phys. Rev. Lett. **52**, 445 (1984)

2.54 R.J. Gooding, B. Joos, B. Bergersen: Phys. Rev. B **27**, 7669 (1983)

2.55 K. Kern, R. David, P. Zeppenfeld, G. Comsa: Surf. Sci. **195**, 353 (1988)

2.56 T. Halpin-Healy, M. Kardar: Phys. Rev. B **34**, 318 (1986)

2.57 K.L. D'Amico, D.E. Moncton, E.D. Specht, R.J. Birgeneau, S.E. Nagler, P.M. Horn: Phys. Rev. Lett. **53**, 2250 (1984)

2.58 T. Halpin-Healy, M. Kardar: Phys. Rev. B **31**, 1664 (1985)

2.59 S. Aubry: In *Springer Ser. Solid-State Sci.*, Vol.8, ed. by A.R. Bishop, T. Schneider (Springer, Berlin, Heidelberg 1978) p.264

2.60 P. Bak: Rep. Prog. Phys. **45**, 587 (1982)

2.61 K. Kern, P. Zeppenfeld, R. David, G. Comsa: Phys. Rev. Lett. **59**, 79 (1987)

2.62 M.B. Gordon: Phys. Rev. Lett. **57**, 2094 (1986)

2.63 A.D. Novaco, J.P. McTague: Phys. Rev. B **19**, 5299 (1979)

2.64 J.A. Wilson, A.D. Yoffe: Adv. Phys. **18**, 193 (1969)

2.65 C.R. Fuselier, J.C. Raich, N.S. Gillis: Surf. Sci. **92**, 667 (1980)

2.66 C.G. Shaw, S.C. Fain, M.D. Chinn: Phys. Rev. Lett. **41**, 955 (1978)

2.67 S. Callisti, J. Suzanne: Surf. Sci. **105**, L255 (1981)

2.68 D.L. Doering, S. Semancik: Phys. Rev. Lett. **53**, 66 (1984)

2.69 T. Aruga, H. Tochihara, Y. Murata: Phys. Rev. Lett. **52**, 1794 (1984)

2.70 G. Pirug, H.P. Bonzel: Surf. Sci. **194**, 159 (1988)

2.71 K. Kern: Phys. Rev. B **35**, 8265 (1987)

2.72 H. Shiba: J. Phys. Soc. Jpn. **48**, 211 (1980)

2.73 T. Clarke, N. Caswell, P.M. Horn: Phys. Rev. Lett. **43**, 2018 (1979)

2.74 J. Villain: Phys. Rev. Lett. **42**, 36 (1978)

2.75 M.B. Gordon, J. Villain: J. Phys. C **15**, 1817 (1982)

2.76 K. Kern, P. Zeppenfeld, R. David, G. Comsa: In *The Structure of Surfaces II*, ed. by J.F. Van den Veen, M.A. Van Hove, Springer Ser. Surf. Sci., Vo.11 (Springer, Berlin, Heidelberg 1988) p.488

2.77 K.D. Gibson, S.J. Sibener, B.M. Hall, D.L. Mills, J.E. Black: J. Chem. Phys. **83**, 4256 (1985)

2.78 B.M. Hall, D.L. Mills, J.E. Black: Phys. Rev. B **32**, 4932 (1985)

2.79 K.D. Gibson, S.J. Sibener: Faraday Discuss. Chem. Soc. **80**, 14 (1985)

2.80 K. Kern, P. Zeppenfeld, R. David, G. Comsa: Phys. Rev. B **35**, 886 (1987)

2.81 J.A. Venables, G.D.T. Spiller, M. Hanbücken: Rep. Prog. Phys. **47**, 399 (1984)

2.82 T. Young: Philos. Trans. **95**, 65 (1805)

2.83 D.E. Sullivan: Phys. Rev. B **20**, 3991 (1979)

2.84 R. Pandit, M. Schick, M. Wortis: Phys. Rev. B **26**, 5112 (1982)

2.85 M. Bienfait, J.L. Senguin, J. Suzanne, E. Lerner, J. Krim, J.G. Dash: Phys. Rev. B **29**, 983 (1984)

2.86 R.J. Murihead, J.G. Dash, J. Krim: Phys. Rev. B **29**, 5074 (1984)

2.87 D.A. Huse: Phys. Rev. B **29**, 6985 (1984)

2.88 F.T. Gittes, M. Schick: Phys. Rev. B **30**, 209 (1984)

3. Infrared Spectroscopy of Semiconductor Surfaces

Y.J. Chabal

AT&T Bell Laboratories, Murray Hill, NJ 07974, USA

Traditionally, photons have not been used as extensively as particles to study single crystal surfaces. While photons can effectively excite electrons at the surface, they cannot be confined to a few monolayers. Electromagnetic radiation is therefore not intrinsically surface sensitive. This fact is reflected in this review series, in the past six years only two articles have dealt with purely photon based techniques for single crystal surfaces [3.1,2]. Two factors have, however, contributed to a renewed interest in photon probes, particularly surface infrared spectroscopy (SIRS): 1) the development of faster and more powerful computers and faster and more sensitive IR detectors, and 2) the need to acquire information which cannot be readily derived from other experimental techniques. The dramatic improvement of high performance fast scanning interferometers results from the first and the recent predictive ability of first-principle calculations [3.3,4] is an example of the second. As a result, a number of laboratories around the world are now incorporating SIRS to investigate metal and semiconductor surfaces.

Although the present review specifically deals with semiconductor surfaces, it should be emphasized that the pioneering work of Francis and Ellison [3.5] and the theoretical description of Greenler [3.6] for *metal* surfaces remain the basis of SIRS. In fact, SIRS has been used to study adsorbates on metal surfaces much more extensively than on semiconductor surfaces. Several reviews have been published on this subject [3.7].

In surface infrared spectroscopy of semiconductor surfaces, one measures the vibrational response of adsorbates and the electronic absorption associated with surface states. SIRS is complementary to other vibrational techniques such as Electron Energy Loss spectroscopy (EELS), Inelastic Atom Beam Scattering (IABS) and Raman scattering and to electronic probes such as photoemission and inverse photoemission spectroscopies. The features that make SIRS attractive are its high resolution (<1 cm^{-1}), polarization properties and sensitivity to vibrational modes or electronic transitions located in the range of transparency of the substrate (i.e. below the gap and above the lattice absorption). At semiconductor surfaces, vibrational lines can be very sharp ($\simeq 1$ cm^{-1}) and cannot be resolved by EELS. For light adsorbates such as H, the modes are too high in frequency to be probed by IABS and too weak Raman scatterers to be detected by Raman spectroscopy. Furthermore, the surface state absorption can be relatively sharp ($\simeq 500$ cm^{-1}) and close to the Fermi level, making it difficult to resolve with direct and inverse photoemission.

The vibrational spectrum associated with an adsorbate is not only a direct identification of the chemical nature of the adsorbed molecule but also a measure of the orientation of the molecule and an indication of the structure of the bonding site. From line shape studies, dynamics such as determination of vibrational lifetime or dephasing mechanisms can often be extracted. Furthermore, it appears that the time resolution of SIRS may be good enough to study kinetic phenomena such as diffusion or reaction rates by this technique. Finally, the electronic absorption spectrum is a measure of the substrate surface structure to the extent that a reliable calculation can be performed. The correlation between the presence of surface states with specific coverages of an adsorbate and the simultaneous knowledge of the substrate/adsorbate vibrational spectrum can help shed light on the spatial origin of such states.

The purpose of this review is to give the basics of SIRS as it applies to semiconductor surfaces, rather than an exhaustive review of all the work done in the field [3.8]. We focus on the study of semiconductor/vacuum interfaces with a monolayer or submonolayer coverage of simple adsorbates. In Sect.3.1, explicit expressions are given in terms of the macroscopic dielectric response of the various media to calculate the electric fields and the optical absorption at the surface. The absorption is then related to the microscopic parameters of the adsorbate monolayer and various simplifying assumptions are discussed. The results of Sect.3.1 are necessary to give a quantitative basis for the choice of experimental geometries and parameters (Sect.3.2), and to provide a framework for quantitative analysis of the data (Sect.3.3). Section 3.2 spells out, for instance, various advantages of multiple internal reflection (MIR), a technique pioneered by Harrick in the sixties [3.9,10]. Among the less obvious attributes of MIR is its ability to probe vibrational modes polarized parallel to the surface with *as much sensitivity as* modes polarized perpendicular to the surface. Section 3.3 presents examples selected mostly for their tutorial aspect. The first deals with an experimentally very well-defined system, i.e. the most stable hydride phase on Si{100}, for which a rigorous and quantitative analysis of the data can be performed (Sect.3.3.1). Based on the knowledge extracted about this phase (monohydride), semiquantitative information can then be obtained on all of the hydride phases on

Si{100} and Ge{100} (Sect.3.3.2), and interesting dynamics inferred for the monohydride phase (Sect.3.3.3). Next, water decomposition on Si{100} is summarized as an instructive example of simple chemistry at surfaces (Sect.3.3.4). A few studies involving the electronic absorption of clean semiconductor surfaces are also included (Sect.3.3.5). Finally, recent data on organometallic decomposition are presented (Sect.3.3.6). From these illustrations, a quantitative assessment of the technique can be made. Section 3.4 deals more directly with the limitations of the technique and addresses the areas of expected growth such as *kinetic, in situ* measurements in conventional and laser-induced chemical vapor deposition (CVD) processes.

3.1 Theoretical Framework

Quantitative analysis of optical reflection spectra requires a microscopic theory of the surface optical response. Since first-principle treatments are generally not possible, a simple three-layer model with step function changes of dielectric constant has been used to approximate the "vacuum/adsorbate layer/substrate" system and to deduce the field strength at the adsorbate position. In this section, the results for a thin adsorbate layer with anisotropic response are first presented. A simple Lorentz oscillator model is then used to relate the measured reflectance to microscopic quantities such as the adsorbate oscillator effective charge. In the case of electronic absorption, the system "vacuum/active layer/substrate" is considered where the active layer is now the substrate surface region encompassing the surface states. Finally, the assumption of sharp boundary conditions (step function change of dielectric constants) is discussed, particularly as it relates to nonabsorbing semiconductor/vacuum interfaces.

3.1.1 Macroscopic Theory (Three-Layer Model)

The model used is shown in Fig.3.1. To make it more concrete and relevant to the experimental geometries used most commonly, a beam traveling in a semiconductor substrate is taken incident in the x-z plane onto the substrate/vacuum interface separated by a thin (d$\ll\lambda$) adsorbate layer. The vacuum is characterized by a real dielectric constant, ϵ^v. Both the substrate and the active adsorbate layer are characterized by complex dielectric constants, $\tilde{\epsilon}^s$ and $\tilde{\epsilon}$, respectively. The present derivation follows the work of McIntyre and Aspnes [3.11,12]. Since we are interested in submonolayer coverages of atoms or molecules oriented on the semiconductor surface, the active adsorbate layer is taken to be anisotropic but with the principal dielectric axes along x,y,z. The dielectric tensor is therefore diagonal with principal values

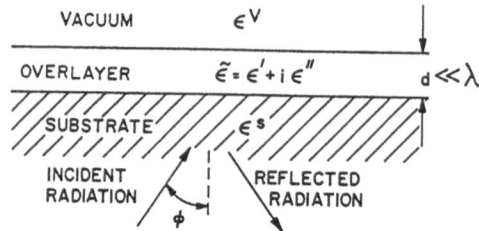

Fig.3.1. Model used to describe the response of a thin absorbate layer at a semiconductor/vacuum interface. The radiation is shown to be incident from the substrate side as is the case for total *internal* reflection spectroscopy. Note that to describe an *external* reflection geometry, the substrate is interchanged with vacuum ($\epsilon^s = 1$) and the vacuum interchanged with substrate [3.14]

$\tilde{\epsilon}_x, \tilde{\epsilon}_y, \tilde{\epsilon}_z$. For nonmagnetic media, the relationship between the electric and magnetic fields and induction is [3.13]:

$$D_i = \tilde{\epsilon}_i E_i \quad \text{with} \quad i=x,y,z \tag{3.1}$$

$$\mathbf{B} = \mathbf{H} . \tag{3.2}$$

The essence of the derivation, reported previously in detail [3.14], is to solve Maxwell's equations subject to the boundary conditions:

D_{normal} continuous (no surface free charge) , $\tag{3.3a}$

$E_{tangential}$ continuous , $\tag{3.3b}$

$B = H$ continuous (nonmagnetic media) $\tag{3.3c}$

at each interface. A linear approximation, when expanding in powers of d/λ (i.e., $\exp(\pm k_z d) \simeq 1 \pm i k_z d$) greatly simplifies the expressions. It is an excellent approximation since the effective thickness for adlayers or surface states is of order a few Ångstroms and the infrared wavelengths are of order a few 10^5 Å.

There are two cases depending on the polarization of the incident radiation, as shown in Fig. 3.2, namely 1) *s-polarization*: electric field perpendicular to the plane of incidence (strictly tangential to the surface), and 2) *p-polarization*: electric field parallel to the plane of incidence (with both normal and tangential components). For both polarizations, the propagation wave vector (x-component) is conserved in all three media (note that quantities without superscripts characterize the active layer):

$$k_x = k_x^s = k_s^v = \frac{\omega}{c} \sqrt{\epsilon^s} \sin\phi . \tag{3.4}$$

The quantities calculated (Fig.3.2) are

$$\tilde{r} \equiv E_r^s / E_t^s (d \neq 0) \tag{3.5a}$$

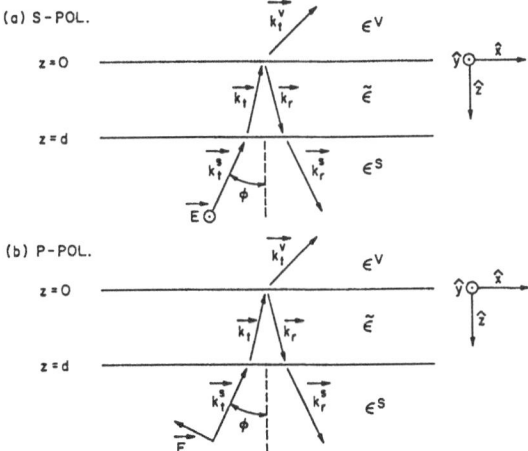

(a) S-POL.

(b) P-POL.

Fig.3.2. Schematic representation of the electric field polarization and of the propagation vectors in the various layers for (a) s-polarization, and (b) p-polarization. The cartesian axes are defined with z=0 at the vacuum/overlayer interface and (xz) the plane of incidence. The subscripts t and r refer to transmitted and reflected radiation, respectively. $\hat{x}$, $\hat{y}$ and $\hat{z}$ are unit vectors [3.14]

and a reference value

$$\tilde{r}^o \equiv E_r^s/E_t^s(d=0) \ , \tag{3.5b}$$

where the tilde is used to emphasize that these quantities are in general complex. In practice, the reference surface does not have to be the *clean* surface but only a surface with no vibrational or electronic absorption in the frequency range of interest. Since the layer dielectric constant is essentially 1 away from the absorption region, the effective thicknesses $(d_x, d_y$ and $d_z)$ in these transparent regions can be set to zero. Experimentally, the reflectivities $R \equiv |\tilde{r}|^2$ and $R^0 \equiv |\tilde{r}^o|^2$ are measured for both surfaces. The reflectance, R/R^0, is then obtained. The absorption associated with the active layer is best expressed as

$$\frac{\Delta R}{R} \equiv 1 - \left| \frac{\tilde{r}}{\tilde{r}^0} \right|^2 . \tag{3.6}$$

The results are:

1) For s polarization, where

$$k_z^j = \frac{\omega}{c} \sqrt{\epsilon_y^j - \epsilon^s \sin^2\phi} \quad \text{with} \quad j = v,s,layer , \tag{3.7}$$

$$\left. \frac{\Delta R}{R} \right|_{s\text{-pol}} = 4 \frac{\omega}{c} d_y \sqrt{\epsilon^s} \cos\phi \ \text{Im} \left\{ \frac{\epsilon_y^v - \tilde{\epsilon}_y}{\epsilon_y^v - \epsilon^s} \right\} . \tag{3.8}$$

Using $\tilde{\epsilon}_y \equiv \epsilon'_y + i\epsilon''_y$ and $\omega = 2\pi c \tilde{\nu}$ and assuming a nonabsorbing substrate (ϵ^s real), (3.8) becomes

$$\left.\frac{\Delta R}{R}\right|_{s\text{-pol}} = 8\pi\tilde{\nu}\sqrt{\epsilon^s}\, d_y \cos\phi \left[\frac{\epsilon''_y}{\epsilon^s - \epsilon^v_y}\right]. \tag{3.9}$$

2) For p polarization, where

$$k_z{}^j = \frac{\omega}{c}\sqrt{\epsilon_x{}^j}\left[1 - \frac{\epsilon^s}{\epsilon_z{}^j}\sin^2\phi\right] \quad \text{with} \quad j = v, s, \text{layer} \tag{3.10}$$

$$\left.\frac{\Delta R}{R}\right|_{p\text{-pol}} = 8\pi\tilde{\nu}\sqrt{\epsilon^s}\cos\phi$$

$$\cdot \text{Im}\left\{\frac{(\epsilon_x{}^v - \tilde{\epsilon}_x)d_x - \left[\dfrac{\epsilon_x{}^v}{\tilde{\epsilon}_z} - \dfrac{\tilde{\epsilon}_x}{\epsilon_z{}^v}\right]d_z\,\epsilon^s\sin^2\phi}{(\epsilon_x{}^v - \epsilon^s) - \left[\dfrac{\epsilon_x{}^v}{\epsilon^s} - \dfrac{\epsilon^s}{\epsilon_z{}^v}\right]\epsilon^s\sin^2\phi}\right\}, \tag{3.11}$$

where the tilde is again used to emphasize that $\tilde{\epsilon} \equiv \epsilon' + i\epsilon''$ will always be complex. Note that the subscripts of ϵ^v have been kept so that (3.11) can be used for *external* reflection ($\epsilon^s = 1$) on a nonisotropic substrate. In order to simplify the expression and cast it in a form similar to (3.9) for s-polarization, we now relax the assumption that the last medium be anisotropic. This is a justified assumption since for *internal* reflection this medium is vacuum.

Thus, setting $\epsilon_x{}^v = \epsilon_z{}^v = \epsilon^v$ and $\tilde{\epsilon}_x = \epsilon'_x + i\epsilon''_x$, $\tilde{\epsilon}_z = \epsilon'_z + i\epsilon''_z$, we can rewrite (3.11):

$$\left.\frac{\Delta R}{R}\right|_{p\text{-pol}} = 8\pi\tilde{\nu}\sqrt{\epsilon^s}\cos\phi$$

$$\cdot \frac{\left[\dfrac{\epsilon^s}{\epsilon^v}\sin^2\phi - 1\right]d_x\epsilon_x{}'' + \epsilon^s\epsilon^v\sin^2\phi\, d_y\,\dfrac{\epsilon_z{}''}{\epsilon_z{}'^2 + \epsilon_z{}''^2}}{(\epsilon^s - \epsilon^v)\left[\dfrac{\epsilon^s + \epsilon^v}{\epsilon^v}\sin^2\phi - 1\right]} \tag{3.12}$$

The point of writing out (3.9) and (3.12) explicitly is to gain insight into the relevant experimental parameters. These are the relative strengths of the electric fields responsible for the absorptions and the role of the various components of $\tilde{\epsilon}$ on the spectra.

If the spectrometer source is completely unpolarized and rotation of a linear polarizer is used to provide s- and p-polarized radiation

114

with unit strength, then the three relevant normalized field intensities are the coefficients of ϵ''_y, $\epsilon^{v2}/(\epsilon_z'^2+\epsilon_z''^2)\,\epsilon_z''$ and ϵ_x'' in (3.9,12);

for s-polarization:
$$I_y = \frac{4\epsilon^s \cos^2\phi}{\epsilon^s - \epsilon^v}, \tag{3.13}$$

for p-polarization:

$$I_z = 4\epsilon^s \cos^2\phi \;\; \frac{(\epsilon^s/\epsilon^v)\sin^2\phi}{(\epsilon^s-\epsilon^v)\left[\dfrac{\epsilon^s+\epsilon^v}{\epsilon^v}\sin^2\phi - 1\right]} \tag{3.14}$$

$$I_x = 4\epsilon^s \cos^2\phi \;\; \frac{(\epsilon^s/\epsilon^v)\sin^2\phi - 1}{(\epsilon^s-\epsilon^v)\left[\dfrac{\epsilon^s+\epsilon^v}{\epsilon^v}\sin^2\phi - 1\right]} \tag{3.15}$$

which are identical to equations (17-19) derived in [3.9] for a vacuum/substrate interface [3.15]. For small absorptions typical of monolayer films at a semiconductor/vacuum interface, $\epsilon_z'^2+\epsilon_z''^2 \simeq \epsilon^{v2} = 1.0$, and for $\epsilon^s \gg \epsilon^v$, (3.13-15) show that *all three intensities are roughly equal* to $4\cos^2\phi$. For a typical and practical incidence angle $\phi=45°$ the normalized field intensities are roughly 2. That is, all components of a surface dipole can be probed *with equal sensitivity*. This is not the case for external reflection for which the tangential field amplitudes are attenuated by approximately $(\epsilon^s)^{-1/2}$. The sensitivity to dipoles parallel to the surface is therefore ϵ^s times weaker than to perpendicular dipoles.

The second aspect deals with the active layer absorption itself and its spectral signature and will be dealt with in some detail in the next section. For now, we rewrite (3.9 and 12) as

$$\left.\frac{\Delta R}{R}\right|_{s\text{-pol}} = \frac{2\pi\nu}{\sqrt{\epsilon^s}\,\cos\phi}\,(I_y)(\epsilon''_y d_y)\,, \tag{3.16}$$

$$\left.\frac{\Delta R}{R}\right|_{p\text{-pol}} = \frac{2\pi\nu}{\sqrt{\epsilon^s}\,\cos\phi}\left[(I_x)(\epsilon_x'' d_x) + I_z\left(\frac{\epsilon^{v2}}{\epsilon_z'^2 + \epsilon_z''^2}\right)(\epsilon_z'' d_z)\right] \tag{3.17}$$

Note that the $1/\cos\phi$ factor arises from the sampling area and the $(\epsilon^s)^{-1/2}$ factor from the index matching with the active layer as discussed in detail by *Harrick* [3.16]. Equations (3.16,17) show that while the "parallel" spectra (x and y components) depend only on the absorption Im$\{\tilde\epsilon\}$ of the layer, the perpendicular spectrum (z component) depends on Im$\{1/\tilde\epsilon\}$ and will therefore be affected by both the real and

imaginary parts of $\tilde{\epsilon}$. As a result, a spectral *shift* will occur for resonant absorption, the magnitude of which depends on the optical
response of the layer (Sect. 3.1.2).

3.1.2 Microscopic Model of the Active Layer

The previous macroscopic considerations are useful only if $\tilde{\epsilon}$ can be
expressed in terms of *relevant* microscopic quantities such as the
dynamic charge of the adsorbate mode and the electronic screening
within the adsorbate layer. Since ab initio calculations are in general
difficult, we present here the parametrized Lorentz oscillator response
of the layer [3.17], which can be directly related to ab initio calculations only if every parameter can be explicitly calculated. In this
model, assuming a layer formed of identical oscillators, the dielectric
function is:

$$\tilde{\epsilon}_i = \epsilon_{\infty i} + \frac{v_{pi}^2}{v_{oi}^2 - v^2 - i\Gamma_i v} \qquad i=x,y,z \qquad (3.18)$$

where ϵ_∞ is the electronic screening of the layer, v_o the resonant frequency (in cm^{-1}) and Γ the natural linewidth (in cm^{-1}). The plasma
frequency (in cm^{-1}) of the oscillator, v_p, is related to the effective
charge e* and effective mass m* of the oscillator by

$$v_{pi}^2 = \frac{N_i}{\pi} \frac{e_i^{*2}}{m_i^* c^2} \qquad (3.19)$$

where N_i is the number of oscillators per unit volume which contribute to the ith component of the response. A useful definition of the
coverage N_{si}, the number of oscilators per unit area contributing to the
ith component of the absorption, is [3.18]:

$$N_{si} = N_i d_i \qquad (3.20)$$

We thus have

$$v_{pi}^2 d_i = \frac{N_{si}}{\pi} \frac{e_i^{*2}}{m_i^* c^2} . \qquad (3.21)$$

From (3.16-18, 21) it is clear that at least for the parallel components x and y, $\Delta R/R$ does not depend on d explicitly. However, for
the z component, the quantity $\epsilon_z'^2 + \epsilon_z''^2$ in the denominator requires
the knowledge of the effective thickness. Following *Mahan* and *Lucas*
[3.19], and as pointed out by others [3.20], we can write for a layer of
oscillators

$$d_i = \frac{4\pi N_{si}}{\displaystyle\sum_{i \neq j} (1/r_{ij})_i{}^3} \tag{3.22}$$

where the denominator is the dipole sum which depends on the specific arrangement of the adsorbate at the surface. In general, it can be written $\Sigma = CN_s{}^{3/2}$, where $C = 9.0336$ for a square lattice [3.21].

We can now work out estimates for the spectral shift occurring for the z component of p-polarized radiation mentioned at the end of Sect.3.1.1. That is, the resonance of $\text{Im}\{1/\tilde{\epsilon}_z\} = \epsilon_z{}''/(\epsilon_z{}'^2 + \epsilon_z{}''^2)$ occurs, using (3.18), at $(\tilde{\nu}_0{}^2 + \tilde{\nu}_p{}^2/\epsilon_{\infty z})^{1/2}$. For small shifts, the magnitude of the shift from $\tilde{\nu}_0$ is therefore $\Delta\tilde{\nu} \simeq \tilde{\nu}_p{}^2/(2\epsilon_{\infty z}\tilde{\nu}_0)$.

If the coverage, effective charge and mass of an adsorbate are such that $\Delta\tilde{\nu}$ is larger than the linewidths, then one must be careful to remove that shift before comparing the data to calculated resonant frequencies of ϵ_x, ϵ_y and ϵ_z. Such data analysis will be relevant to the analysis of H on Si{100} presented in Sect.3.3.1.

3.1.3 Model for Electronic Absorption

For electronic absorption, a quantitative interpretation of the $\Delta R/R$ spectra relies more directly on theory. A model similar to the Lorentz oscillator model must be formulated entirely by theory since the exact form of $\tilde{\epsilon}$ is not known a priori and cannot be directly measured experimentally. A more serious problem than for vibrational spectroscopy is the choice of a reference surface. The phase under study is usually the clean surface and the reference phase has some electronic absorption associated with the adsorbate in part of the broad frequency region necessarily covered. Detailed measurements for a variety of reference phases are performed to alleviate this difficulty. Another problem is the determination of the effective thickness of the active layer. In principle, this parameter is frequency dependent. Equations (3.9 and 12) are often recast in terms of effective dielectric functions which incorporate the quantity d. A recent review of the optical properties of semiconductor surfaces nicely summarizes these difficulties [3.22]. The formal solutions will be discussed in Sect.3.1.4 and examples given in Sect.3.3. Specific geometries for experimental studies will be discussed in Sect.3.2.3.

3.1.4 Discussion of the Assumption of Sharp Boundaries

As pointed out clearly by Feibelman [3.23], the three-layer model gives an unphysical description of the nature of induced surface charges on a *microscopic scale*. That is, the electric field relevant to the vibrational excitation of the adsorbate modes is not necessarily well described by a step function discontinuity in the dielectric function. For substrate/adsorbate modes, for instance, the relevant electric fields may not be the same as those on the vacuum side of the interface.

For electric fields polarized in the plane of the surface (x and y components), there is no surface charge induced and microscopic effects can be ignored as long as the substrate absorption is negligible. That is, since the parallel component of the electric field is continuous throughout the interface, the value of the field at the adsorbate can be accurately calculated from macroscopic quantities, ϵ^s and ϵ^v, even if the precise position of the absorbing dipole is not known.

The problem arises for fields polarized perpendicular to the surface plane (z component). For infrared frequencies (i.e., below the substrate plasma frequency), Feibelman defined the *surface* response function by an integro-differential equation

$$d_\perp(\omega) \equiv \frac{1}{(1-\epsilon^s/\epsilon^v)} \int_{z^s}^{z^v} dz' z' \frac{d}{dz'}(\epsilon_z) \tag{3.23}$$

where $\epsilon_z = E_z{}^s/E_z{}^v$. The values of z^s and z^v are chosen such that *all* the surface absorption occurs between these two values. For instance, if we make the same assumptions as in the three-layer model, i.e. all the absorption occurs between $z = 0$ and $z = d$, then (3.23) becomes, making use of the continuous nature of D_z

$$d_\perp(\omega) = d \frac{1/\tilde{\epsilon} - 1/\epsilon^v}{1/\epsilon^s - 1/\epsilon^v} \tag{3.24}$$

clearly showing that $d_\perp(\omega) = 0$ if $\tilde{\epsilon} = \epsilon^v$. The physical meaning of $d_\perp(\omega)$ is discussed in [3.23] where Feibelman shows that $\mathrm{Re}\{d_\perp(\omega)\}$ is the centroid of the surface position, i.e. the centroid of the induced charge which is responsible for the surface absorption. Moreover, in the case of semiconductor/vacuum interfaces for which there is *no bulk absorption*, $\mathrm{Im}\{d_\perp(\omega)\}$ is the surface power absorption.

When surface photoeffects are small and for frequency well below the substrate plasma frequency, Feibelman concludes that the assumption of step function dielectric constants is good as far as the computation of the fields is concerned. The problem arises, within the three-layer model, with the computation of the effective thickness d when one is dealing with submonolayer coverages.

In conclusion, for the vibrational study within the gap of weakly absorbing layers at semiconductor/vacuum interfaces, the three-layer model is well justified to calculate the electric fields. Calculations of the power absorption as they relate to geometry optimization are again well justified within the three-layer model. The aspect of the model which should be dealt with carefully is the determination of the effective thickness relevant for the layer response to perpendicular fields (z components). Fortunately, one can see from (3.17,18,21) that for weakly absorbing layers, $\epsilon_z" \ll \epsilon_z' \simeq 1$, an error in d_z only affects the spectral shift calculation and is negligible as far as *absorption intensities* are concerned.

The situation is not so simple for electronic absorption, particularly above the gap, that is when the substrate is absorbing. With regard to this problem, Bagchi et al. [3.24] have solved the integro/differential equations by using E_x, E_y and D_z (instead of E_z) as variables. Since D_z is continuous at the surface, the integral definition of the dielectric function

$$E_z(z) = \int dz' \ \epsilon_{zz}^{-1}(z,z') \ D_z(z') \tag{3.25}$$

can be written

$$\epsilon_{\alpha\beta}(z,z') = \delta_{\alpha\beta} \delta(z-z')\epsilon^0(z) + \Delta\epsilon_{\alpha\beta}(z,z')\delta_{\alpha\beta} \tag{3.26}$$

where the first term of (3.26) is Fresnel-like and the second represents the surface. That is, the surface is characterized by the diagonal tensor $\Delta\epsilon = (\Delta\epsilon_{xx}, \Delta\epsilon_{yy}, \Delta\epsilon_{zz})$. The equations are now local, since all the fields are continuous at the surface. Comparison with the previous theory gives, assuming an isotropic surface layer response $\tilde{\epsilon}$

$$\overline{\Delta\epsilon_{xx}} \equiv \int dz \ dz' \ \Delta\epsilon_{xx}(z,z') = d(\tilde{\epsilon}-\epsilon^s) \ , \tag{3.27}$$

$$\overline{\Delta\epsilon_{zz}} \equiv \int dz \ dz' \left[\epsilon_{zz}^{-1}(z,z') - \frac{\delta(z-z')}{\epsilon^0(z)} \right] = d(\tilde{\epsilon}^{-1} - 1/\epsilon^s) \tag{3.28}$$

with $\int \epsilon_{zz}(z,z')\epsilon_{zz}^{-1}(z',z'')dz' = \delta(z-z'')$ \hfill (3.29)

Computational efforts are now in progress to calculate the surface response as realistically as possible [3.22]. However, no theory can yet account for all the aspects of the surface response when both the substrate and surface layer have electronic absorption. For instance, recent calculations neglect local field effects [3.25] or simplify the problem by assuming that the atomic polarizability at the surface is the same as that in the bulk [3.26]. Nevertheless, these calculations are useful to interpret the measured surface electronic absorptions.

3.1.5 First-Principle Calculations

First-principle calculations of surface vibrational modes are difficult because they require that the computation of the surface total energy be accurate enough to yield reliable values for small displacements of the atoms about their equilibrium position. For metal substrates, density functional calculations of several atomic layer thick slabs have been used successfully within the Linear Augmented Plane Wave (LAPW) approximation to treat simple adsorbates such as H [3.27, 28]. This method is powerful for ordered systems with small unit cells. There is no restriction on the size of the electronic wave function so that it is useful to handle delocalized effects. A new technique combining density functional and scattering theory is being developed to tackle the case of *isolated* molecules at surfaces [3.29]. None of these have been applied to semiconductor surfaces yet, partly because the unit cells are often larger and cluster approximations are well justified for the more localized semiconductor surfaces.

In a cluster calculation, a finite number of atoms is chosen to represent the surface. The subsurface dangling bonds are made chemically equivalent to the bulk environment by attaching hydrogen or methyl groups [3.30]. Typically, up to four layers of the unit cell are considered. The symmetry of the ordered surface is used to fix some of the atoms in the deeper layers within certain planes or directions in order to reduce the number of possible degrees of freedom. The first step of the calculation is to calculate the surface geometry by minimizing the total energy with respect to the positions of all the atoms subject to symmetry constraints. This is usually determined with as large a cluster as necessary to simulate the constraining effects of the subsurface layers in a realistic manner.

The success of such calculations depends on the choice of a basis set (or basis functions) to represent the relevant electronic wavefunctions. The basis set typically consists of s-, p- and d-type atomic orbitals on each center and the number of such functions necessary to reproduce molecular geometries reliably has been well documented for a wide variety of systems [3.31]. The effects of larger basis sets on the calculated geometries can usually be checked by using smaller clusters. Electron correlation effects are usually considered by means of configuration interaction or by perturbation theory though their effects on molecular geometries of systems such as H on Si{100} are small. Once the geometry has been determined, the complete force field and vibrational frequencies can be evaluated using a smaller cluster but with a larger basis set. The complete quadratic force field can usually be evaluated easily and cubic or higher-order terms are also computed if necessary. Such calculations were done for H on Si{100} using efficient analytical gradient techniques [3.32]. In general, the absolute calculated frequencies are about 10% higher than experimentally measured [3.32]. However, in a systematic study of gas phase molecules, Raghavachari was able to establish that the *frequency splitting* among the modes involving a similar motion (e.g., symmetric and antisymmetric stretches) could be reproduced reliably [3.33]. In particular, in the case of the monohydride structure for H on Si{100} such a frequency splitting was found to be remarkably independent of the basis set or the level of theory used [3.34].

Finally, the dynamic dipole moments for the various normal modes can be computed using as large a basis set as possible and including electron correlation effects. For H on Si{100}, the correlation was introduced by means of Moller-Plesset perturbation theory carried out to fourth order [3.35]. The changes in the dipole moments are evaluated by taking finite steps long the direction of the normal mode under study and calculating the resulting dipole moments. A linear fit of the data about the equilibrium position yields the value $\Delta\mu/\Delta d$. The ratio of the dynamic dipole moments associated with modes that are close in energy can thus be obtained accurately [3.34]. The absolute dynamic dipole moment can also be extracted by multiplying the quantity $\Delta\mu/\Delta d$ by the extremum of the atomic displacements for the mode under study.

The capability of computing reliably such quantities as frequency splittings and relative dynamic dipole moments is essential to the understanding of vibrational spectra of *adsorbate-substrate* modes. In contrast to adsorbate *internal modes*, these modes have no gas phase equivalent because the surface cannot be well described by one or two atoms. The importance of ab initio calculations will be revealed in the study of H on Si{100} described in Sects.3.3.1, 2.

3.2. Experimental Geometries and Techniques

To design an experiment where the vibrational spectrum of submonolayer coverage of molecules or atoms at semiconductor surfaces can be probed sensitively, we separate the problem into two cases. The first involves the study of modes with a resonant frequency within the optical gap of the substrate, i.e. higher than any bulk phonon absorption and lower than bulk electronic absorption. Note that samples of high enough resistivity are used so that the free carrier absorption is also negligible in the region of interest. For the first case, the formalism developed in Sect.3.1 is appropriate where ϵ^v and ϵ^s are real and isotropic. The second case involves the study of modes with a resonant frequency in a region where the substrate absorbs sufficiently to prevent the use of internal reflection. In this case, the formalism developed in Sect.3.1 cannot be used as such but it can be modified to give semiquantitative information.

3.2.1 Vibrational Modes in Substrate Optical Gap

For conventional sources (e.g., globar), the surface infrared spectroscopist is usually in the regime where the signal-to-noise ratio, S/N, is proportional to the total radiation intensity incident on the detector, I. Although the limitation for *transmission* spectroscopy with broadband interferometers in the 2000 to 3000 cm^{-1} region can be the A/D limit, this is not true of *surface* spectroscopy for which some radiation is lost due to the requirements of small solid angle and spot size at the sample. Furthermore, considering that the capabilities of A/D's will improve in the near future beyond the 16 bit and 120 kHz specifications, it is appropriate to assume that the S/N is proportional to I. Based on this assumption, Greenler [3.36] pointed out that the quantity ΔR = R(0)-R(d) had to be optimized rather than $\Delta R/R$ = 1-R(d)/R(0).

To decide between external and internal reflection geometries to study a thin active layer at a semiconductor/vacuum interface, we first consider the case of a *single* reflection. The results of Sect.3.1 are applicable to *external* reflection simply by setting $\epsilon^s = 1$ (vacuum) and ϵ^v = substrate dielectric constant, and vice versa for *internal* reflection.

For s-polarization we have

$$\Delta R\bigg|_{\text{s-pol}} = \frac{2\pi\bar{\nu}}{\sqrt{\epsilon^s}\,\cos\phi} \left|r^0\right|^2 I_y(e_y{}''d_y) \tag{3.30}$$

where r^0 is defined by [3.14]

$$r^0 = \frac{\sqrt{\epsilon^s}\cos\phi - \sqrt{\epsilon_y{}^v - \epsilon^s\sin^2\phi}}{\sqrt{\epsilon^s}\cos\phi + \sqrt{\epsilon_y{}^v - \epsilon^s\sin^2\phi}} \tag{3.31}$$

and I_y is defined in (3.13) and ϵ''_y in (3.18 and 21).
For p-polarization we have:

$$\Delta R\bigg|_{\text{p-pol}} = \frac{2\pi\bar{\nu}}{\sqrt{\epsilon^s}\,\cos\phi} \left|r^0\right|^2 \left[I_x(\epsilon_x{}''d_x) + I_z\left(\frac{\epsilon^{v2}}{\epsilon_z{}'^2 + \epsilon_z{}''^2}\right)(\epsilon_z{}''d_z)\right] \tag{3.32}$$

where r^0 is defined by [3.14]

$$\hat{r}^0 = \frac{\epsilon_x^v \sqrt{\epsilon^s}\cos\phi - \epsilon^s \left[\epsilon_x^v \left(1 - \frac{\epsilon^s}{\epsilon_z^v}\sin^2\phi\right)\right]^{1/2}}{\epsilon_x^v \sqrt{\epsilon^s}\cos\phi + \epsilon^s \left[\epsilon_x^v \left(1 - \frac{\epsilon^s}{\epsilon_z^v}\sin^2\phi\right)\right]^{1/2}} \tag{3.33}$$

and I_x and I_z are defined in (3.14,15) and $\tilde{\epsilon}_z$ in (3.18,21,22). Note that $|\hat{r}^0|^2 = 1$ for both polarizations in the case of total *internal* reflection, $\phi \geq \sin^{-1}(\epsilon^v/\epsilon^s)^{1/2}$.

Figure 3.3 shows ΔR for both polarizations in the case of a single internal (a) and external (b) reflection. The parameters, $N_s = 5\cdot10^{14}$ cm^{-2}, $\epsilon_{\infty z} = 1.3$, $[(e^*/e)/(m^*/m)^{(1/2)}] = 0.05$, $\Gamma = 5$ cm^{-1} and $v_0 = 2100$ cm^{-1}, are typical of H on Si surfaces (very weakly absorbing monolayer). For the purpose of estimating the sensitivity, we assume the active layer to be isotropic. Figure 3.4 shows the more general case of a strongly absorbing monolayer. The only parameter changed is $[(e^*/e)/(m^*/m)^{(1/2)}] = 0.5$.

Figures 3.3 and 3.4 first show that for a *single* reflection, internal reflection gives a factor of 2 to 10 better S/N depending on polarization and strength of the active layer absorption. Second, they show that the external reflection spectrum is more complex due to interference between the negative x and positive z components in p-polarized

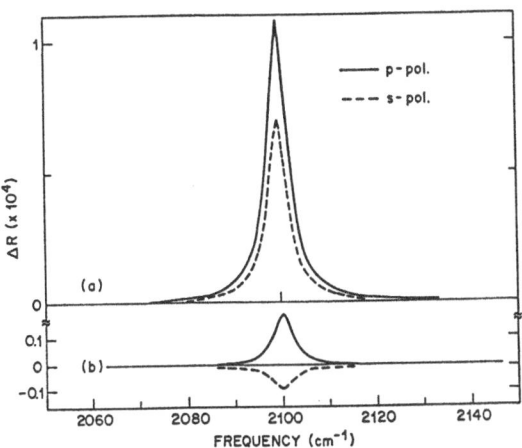

Fig.3.3. Model calculations of $\Delta R \equiv R(0)-R(d)$ in the case of (a) *internal* reflection, and (b) *external* reflection for both p- and s-polarizations. The parameters used are $\epsilon^v = 1$, $\epsilon^s = 11.7$, $v_{0x} = v_{0y} = v_{0z} = 2100$ cm^{-1}, $\Gamma_x = \Gamma_y = \Gamma_z = 5$ cm^{-1}, $N_{sx} = N_{sy} = N_{sz} = 5\times10^{14}$ cm^{-2}, $m_x^* = m_y^* = m_z^* = 1$ a.m.u., $(e^*/e)_x = (e^*/e)_y = (e^*/e)_z = 0.05$, $\epsilon_{\infty z} = 1.3$, $\phi = 45^0$. This represents the case of a very weakly absorbing, isotropic monolayer [3.14]

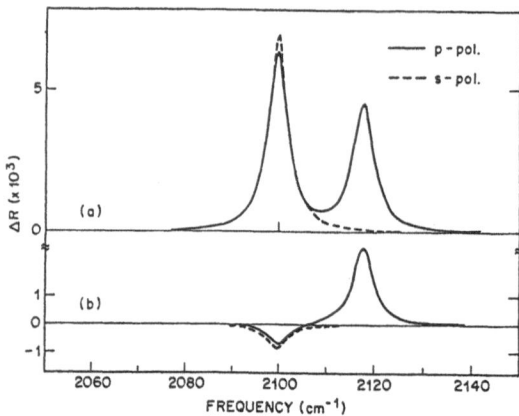

Fig.3.4. Model calculation of ΔR in the case of (a) *internal* reflection, and (b) *external* reflection for both polarizations. The parameters used are identical to those used for Fig.3.3 except that $(e^*/e)_x = (e^*/e)_y = (e^*/e)_z = 0.5$. This represents the case of a strongly absorbing, isotropic monolayer [3.14]

spectra. The physical interpretation of these changes of sign is that for external reflection the thin active layer can act as a reflection or anti-reflection coating, depending on the polarization, angle of incidence, and layer optical response. Third, the frequency shift of the z component, negligible for weak absorbers such as Si-H, is apparent in Fig. 3.4, i.e. for a strongly absorbing monolayer. We note that the shift is given by the expression derived at the end of Sect.3.1.2: $\Delta \nu \sim \nu_p^2/(2\epsilon_{\infty z}\nu_0) = 18$ cm^{-1} for the strongly absorbing monolayer.

The above results, along with similar calculations of ΔR for a range of angles of incidence show that: 1) single internal reflection always gives a better S/N than a single external reflection by a factor of 2 to over 10; 2) the spectrum obtained by internal reflection is simple (simple addition of the positive components of the field intensities I_x, I_y and I_z); and 3) I_x, I_y and I_z are very close in magnitude for a wide range of angles so that all the components of vibrational modes can be probed with equal sensitivity. Analysis of the data can also be done accurately even for highly diverging radiation and poor knowledge of the angle of incidence. This is not true of external reflection for which positive or negative contributions can occur and where the parallel components are *always* small.

For *multiple* internal reflection, a thin (~1 mm) plate with beveled edges is used. In practice, only a fraction of the incident radiation, f, couples because the spot size at the focus is usually larger than the coupling edge. In addition there are reflection losses at the input and output edges [3.9], so that only a fraction, T_0, of the radiation intensity reaches the detector. As a result, the measured ΔR for N reflections (on one side of the plate only) is:

$$\Delta R(N) = f\, T_0\, N\, \Delta R(1) \tag{3.34}$$

124

where $\Delta R(1)$ is given by (3.30) or (3.32). For a 5 mm spot size and a Si plate (t=0.05 cm, ℓ=5 cm), $\Delta R(50refl) = 6.25\Delta R(1)$. Since $\Delta R(1)$ is between a factor of 2 and over an order of magnitude larger than that for external reflection, multiple internal reflection practically gives one to two orders of magnitude better S/N than external reflection. Having established this point, we refer the reader to [3.9] and to Sect.3.2.3b for a review of experimental MIR geometries.

3.2.2 Vibrational Modes in Substrate Absorption Region

Since it is not possible in this case to use multiple *internal* reflection because of the substrate absorption, the alternatives are external reflection or transmission through a very thin sample. For external reflection, it is possible to calculate the expected $\Delta R/R$ from (3.8) and (3.11) by letting $\epsilon_y^v, \epsilon_x^v, \epsilon_z^v$ be complex. Unfortunately, there are no simple explicit expressions corresponding to (3.9 and 12). The optimum parameters (e.g. angle of incidence) depend strongly on the substrate medium and the nature of the overlayer. They can be determined numerically. In practice, normal reflection or transmission have been used for which the equations are greatly simplified and the sensitivity to tangential absorption is large [3.37, 38], as reviewed in [3.22].

Some physical intuition is gained by working out the two limits of very weakly and very strongly absorbing substrates. For the weakly absorbing substrates, the equations derived in Sect.3.1 are approximately valid and it is found that the largest $\Delta R|_{p-pol}$ is obtained for grazing incidence (e.g., $80° < \phi < 87°$ for a silicon substrate). Note, however, that ΔR is negative even for the z component and that the x and y components are small. The latter components are best probed near normal incidence, either by reflection or transmission through a thin sample.

The other limit implies $|\tilde{\epsilon}^v| \gg |\tilde{\epsilon}| \simeq \epsilon^s = 1$ (for vacuum). This is the same assumption made for metal substrates, for which the following expressions are obtained [3.18]

$$\Delta R\bigg|_{s-pol} = 8\pi\tilde{\nu}\,\cos\phi\,|r^0|^2\,\text{Im}\left\{\frac{\tilde{\epsilon}_y - \tilde{\epsilon}_y^{\,v}}{1 - \tilde{\epsilon}_y^{\,v}}\right\}d_y \qquad (3.35)$$

and

$$\Delta R\bigg|_{p-pol} \simeq 8\pi\tilde{\nu}\,|r^0|^2\,\sin^2\phi\,\left[\cos\phi\left(1 + \frac{1}{|\tilde{\epsilon}^v|^2}\frac{\sin^2\phi}{\cos^2\phi}\right)\right]^{-1}\,\text{Im}\left\{-\frac{1}{\tilde{\epsilon}_z}\right\}d_z$$

$$(3.36)$$

where $\epsilon^{v\prime\prime} \gg \epsilon^{v\prime}$ was assumed for simplicity. Equations (3.31,33) show that $|r^0| \simeq 1$ for both s and p polarizations. However, (3.13,15) emphasize that the y and x components of the fields are very small. Therefore, as apparent in (3.35, 36), only the z component of the active layer dielectric constant can be probed with grazing incidence reflection. The optimum angle (in radian) is:

$$\phi = \pi/2 - (3^{1/4}/|\epsilon^v|^{1/2}) .$$ (3.37)

The general conclusion of this section is that for absorbing substrates, external reflection must be used for which the sensitivity to modes parallel to the surface is substantially reduced and for which the maximum sensitivity to modes perpendicular to the surface is achieved by grazing incidence. Transmission is best for modes parallel to the surface on a weakly absorbing substrate.

As a result, one may want to investigate alternatives depending on the particular system under study. For instance, if the Si-O modes are of interest on a silicon substrate, standard multiple internal reflection on a Si plate is ill advised since the modes under study occur at a frequency ($\sim$900 cm^{-1}) where the Si substrate absorbs. However, epitaxial Si can be grown on Ge single crystals which are transparent above 700 the particular system under study. For instance, if the Si-O modes are of interest on a silicon substrate, standard multiple internal reflection on a Si plate is ill advised since the modes under study occur at a frequency ($\sim$900 cm^{-1}) where the Si substrate absorbs. However, epitaxial Si can be grown on Ge single crystals which are transparent above 700 cm^{-1}. Multiple internal reflection using a Ge plate with a thin ($\sim$20Å) Si layer is particularly attractive if modes parallel to the surface need to be studied [3.39]. Alternatively, a single internal reflection can be used through a very thin Si prism. The small path length through the bulk Si gives an easily detectable reflected intensity and the internal reflection gives a better sensitivity to both the parallel and perpendicular modes.

It is clear that depending on the particular system under study a variety of approaches are possible. A quantitative estimate of the sensitivity of the various approaches can be done by means of (3.30, 32, 33) for transparent or weakly absorbing substrates or of (3.8, 11, 33) in general.

3.2.3 Experimental Apparatus

a) Surface Infrared Spectrometer

The main drawbacks of SIRS have long been the lack of sensitivity and the narrow spectral range. The sensitivity can be increased substantially depending on the substrate under study by maximizing the interaction of the intrared beam with the surface. The spectral range can be increased by using interferometers instead of grating spectrometers. In the next section, we will show how a multiple internal reflection geometry may be used advantageously to increase the interaction with semiconductor surfaces. In this section, we describe an example of an interferometer and ultra-high vacuum (uhv) system used to perform IR studies of semiconductor surfaces.

In a Michelson interferometer, the source radiation is modulated by a moving mirror in one arm of the interferometer. In one mirror scan, which typically takes a fraction of a second, a region of 4000 cm^{-1} can be covered with one beam splitter and one detector. However, all of the source energy impinges on the detector at all times, tending to saturate it and posing dynamic range problems. This apparent drawback can be turned to an advantage since increasing the interaction with the surface often results in a decrease of overall throughput.

The interferometer shown in Fig. 3.5 is a nitrogen-purged Nicolet interferometer equipped with a Ge on KBr beam splitter and a standard globar source. While the ultimate resolution is 0.06 cm^{-1}, signal-to-noise ratio considerations in surface studies often restrict it to 0.1 cm^{-1}. The workable frequency region is 400 to 8000 cm^{-1}. The windows of most uhv systems are double O-ring sealed alkali halides (e.g. CsI, transparent from 200 cm^{-1} to the uv) and present therefore

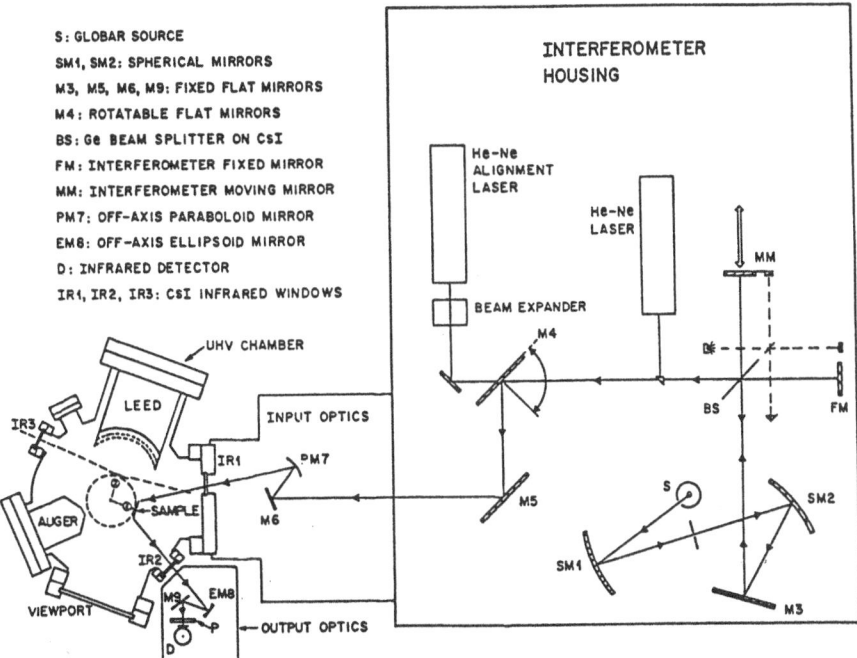

Fig.3.5. Schematic diagram of the experimental apparatus used by the author [from Y.J. Chabal, E.E. Chaban, S.B. Christman: J. Electron Spectr. Rel. Phenom. 29, 35 (1983)]

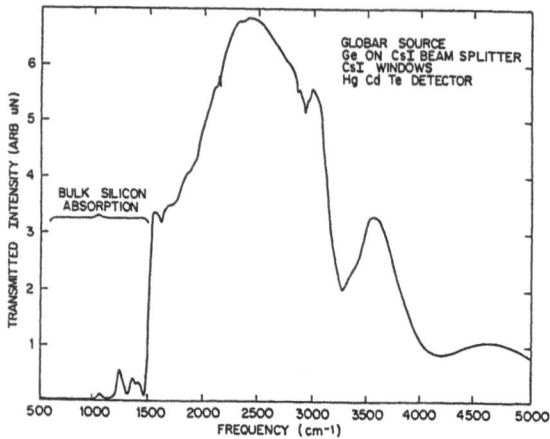

Fig.3.6. Typical transmission spectrum obtained with the apparatus shown in Fig.3.5 and a broad band HgCdTe detector. The spectral shape is the result of the globar source spectrum, the "Ge on CsI" beam splitter efficiency, the absorption of the reflective optics, the absorption of the silicon internal reflection plate and the response of the detector. Without the silicon sample, the spectrum would extend to 650 cm^{-1}. The region labelled "bulk silicon absorption" shows the effect of traversing 7.6 cm of silicon.

no restriction to the spectral range [3.40]. This range is reduced, however, by the particular detectors used, and by the intrinsic bulk absorption of the samples in a multiple internal reflection geometry. Thick silicon, for instance, is opaque below 1500 cm^{-1} (Fig.3.6) and germanium below 700 cm^{-1}, due to lattice absorption [3.41]. For the measurements presented here, optimum sensitivity was obtained by using a HgCdTe detector to study the Si-D, Ge-D and Ge-H vibrations and an InSb detector to study the Si-H and O-H vibrations. A typical sensitivity of $\Delta R/R = 1 \cdot 10^{-4}$ per reflection is achieved with 1 cm^{-1} resolution by averaging 2048 scans (6 minutes).

b) Sample Geometry

As shown in Sect.3.2.1, the multiple internal reflection (MIR) geometry is the key to the sensitivity of SIRS for semiconductor surfaces. In the configuration used by the author, the radiation is focused onto the beveled face of the sample shown in Fig.3.7, enters the material and bounces back and forth between the two sides of the sample before exiting at the other end. The technique was developed several decades ago by Harrick [3.42] and first applied to the study of H on semiconductor surfaces in uhv by Becker and Gobeli in 1963 [3.43]. The important feature of MIR is the fact that *no absorption* occurs upon total internal reflection except for that of the modes under study. The only loss is due to the poor match of the focused beam and the beveled input coupling edge, resulting typically in an order of magni-

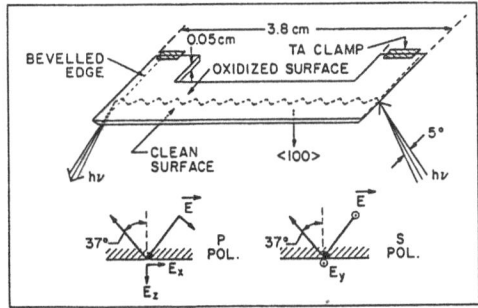

Fig.3.7. Sketch of the sample cut in a U–shape to allow easy clamping without interfering with the bevelled edges and to give a uniform resistive heating of the portion probed by the IR beam. Radiation is incident on the input bevelled edge and is totally internally reflected 50 times on the surface under study with an internal angle of incidence of 37°. The other surface is kept oxidized so that no hydrogen or deuterium chemisorption can take place. The adsorbate induced spectrum is obtained by subtracting from the spectrum obtained after a given exposure that of the clean surface. The electric field components at the surface for the two polarizations are shown schematically at the bottom [3.65]

tude lower throughput. Equation (3.34) showed that the net sensitivity can be increased linearly with the number of reflections. The only limits are the physical constraints on the size of the sample such as requirements for uniform heating, general handling and preparation ease, and plain geometric constraints. For example, the author used samples (3.8x1.5x0.05 cm³) cut in a U–shape. By clamping the sample at the top edges of the U, a uniform resistive heating could be achieved (Fig.3.7). The maximum temperature gradient recorded at the side edges was 50 K during annealing. However, a stability better than 10 K was achieved over 90% of the sample. The coupling edges were polished at 40° with respect to the surface normal and the radiation was externally incident at 30°, resulting in a 37° internal angle of incidence for Si. Equations (3.13-15) show that this internal angle gives roughly equal weight to the various components of the electric field. The number of reflections on the front surface alone is 50. In general, a 45° bevel for the coupling edges is a better choice because it avoids problems associated with light traveling down the length of the plate without interacting with the surface. We refer the reader to [3.9, Chap.4, p.138] for a thorough discussion of the geometries of internal reflection elements.

The polarization properties of MIR spectroscopy make it a powerful tool to determine quantitatively the orientation of molecules on single crystal surfaces. The bottom part of Fig.3.7 shows the field distribution for the two different polarizations; while p–polarized radiation can probe modes both parallel (x direction) and perpendicular (z direction) to the surface, s–polarized radiation is only sensitive to modes parallel (y direction) to the surface. As shown in section 3.1, the amplitude of the fields within the Si-H or Ge-H region can be accurately estimated using a three-layer model with the intermediate layer characterized by a nonisotropic dielectric constant. In the case of single domain surfaces for which all of the bonds are aligned along a particular direction within the surface plane, the dielectric constant is not azimuthally isotropic and complementary information can be obtained by performing similar measurements on a sample rotated by 90°. Thus all the components of the dynamic dipole moments can be extracted from the polarized measurements. The actual orientation of the bonds can then be determined from these measured moments if the electronic screening and the relative strengths of the modes are known. As described in Sect.3.3, the screening can often be experimentally measured while ab initio calculations are necessary to estimate the relative dipole moments.

c) Other Techniques

Two other approaches have been developed to study semiconductor surfaces besides the standard reflection spectroscopy: photoacoustic spectroscopy to detect hydrocarbon internal modes on Si and sapphire [3.44] and photothermal displacement spectroscopy [3.45] to measure the surface electronic absorption. Both measure the *absorption* as opposed to the *change* in reflection. Practically, this means that a reference signal is not needed unless specific comparisons have to be made. This is a decisive advantage in the case of the complicated chemical systems for which there is no good reference state of the surface. The in situ study of organometallic decomposition on silicon is such an example and will be discussed in Sect.3.3.6.

3.3 Selected Examples

In this section, selected examples are presented which do not represent an exhaustive list of all infrared spectroscopy performed at semiconductor interfaces [3.10,46]. Rather, they have been chosen for their tutorial aspect. First, the detailed geometry of the Si{100}-(2x1)H system is deduced from surface infrared spectroscopy and ab initio cluster calculations. In this case, the results derived in Sect.3.1 can be fully used to make connection with first principle calculations. Next, the nature of the H-saturated Si{100} and Ge{100} surfaces is determined from infrared spectroscopy on a semiquantitative basis. Some dynamics are inferred for the well-characterized Si{100}-(2x1)H system from both the IR data and molecular dynamics simulations. Semiquantitative information is also obtained for water decomposition on Si{100}. Electronic absorption on Si{100}, Ge{100} and Si{111} is then considered. Finally, recent data on in situ studies of organo-metallic decomposition are presented.

3.3.1 Structure of the Si{100}-(2x1)H System

The exposure of the Si{100}-(2x1) surface to atomic hydrogen first leads to the development of a sharp, well-defined (2x1) LEED pattern before becoming a (1x1) pattern at saturation coverage. The general understanding of this phenomenon is that (a) the clean {100} surface reconstructs by forming dimers and (b) the addition of one monolayer of H saturates the dangling bonds of each dimer unit as shown in Fig.3.8. A (2x1) unit cell is maintained with one H per Si, i.e. 2H per unit cell. There are therefore six vibrational modes associated with this this structure, two of which involve the stretch of the Si-H bonds. Since the other four, involving the SiH bend and frustrated translation, are too low in frequency to be detected with conventional surface IR spectroscopy, we focus on the two stretch modes occurring around 2100 cm^{-1}, from which accurate *structural* information can be learned [3.34]. A nominally flat Si{100} surface displays two domains rotated by 90° with (2x1) symmetry. Since the concentration of the two domains is equal, the surface is on average isotropic in 2-dimensions and polarization studies cannot be performed efficiently. However, Kaplan noted that clean surfaces cut a few degrees off the {100} plane around the $\langle 0\bar{1}1 \rangle$ axis gave only one domain of (2x1) symmetry [3.47]. Figure 3.9 shows the resulting structure anticipated when H is added to such stepped surfaces [3.34]. The importance of this geometry is that the dielectric constants of the Si-H layer can be associated with distinct modes: contributions to $\epsilon_{\langle 011 \rangle}$ come only from H adsorbed at steps, while contributions to $\epsilon_{\langle 100 \rangle}$ and $\epsilon_{\langle 0\bar{1}1 \rangle}$ primarily come from perpendicular and parallel components of H adsorbed on terraces (Fig.3.9). Figure 3.10a shows that these modes are the symmetric (perpendicular to surface) and antisymmetric (parallel to surface) stretches of the "monohydride". Note that unless the small perpendicu-

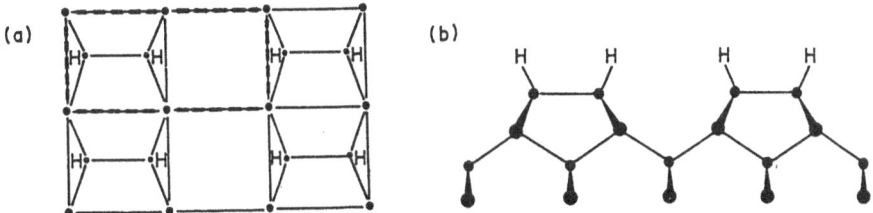

Fig.3.8. (a) Top view and (b) side view of a portion of a uniform monohydride phase on Si{100} or Ge{100}. The resulting 2x1 unit cell is outlined in dashed lines. Dimerization of the top Si or Ge layer is depicted by reduction of the Si-Si or Ge-Ge distance [3.14]

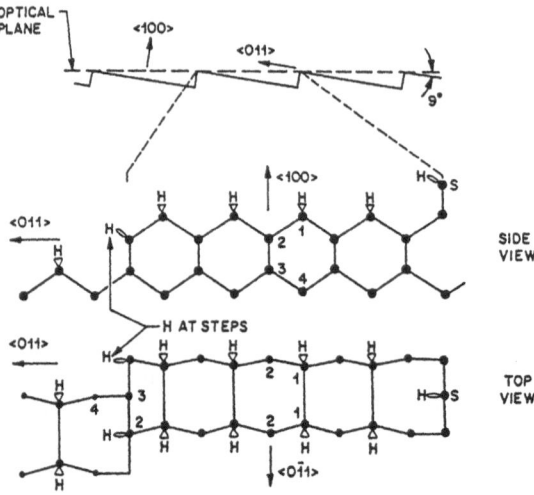

Fig.3.9. Schematic representation of a vicinal surface cut at 9⁰ from the (100) plane about the ⟨011⟩ axis, i.e. the optical plane with respect to which the probing electric field vectors (E_x, E_x or E_z) are described makes a 9⁰ angle with the (100) terraces. The atomic representation of the vicinal surface including one terrace and two steps is given below. The projection of H absorbed at all available dangling bonds of the clean surface (assuming that the dimers are not broken) is also shown schematically [3.14,34]

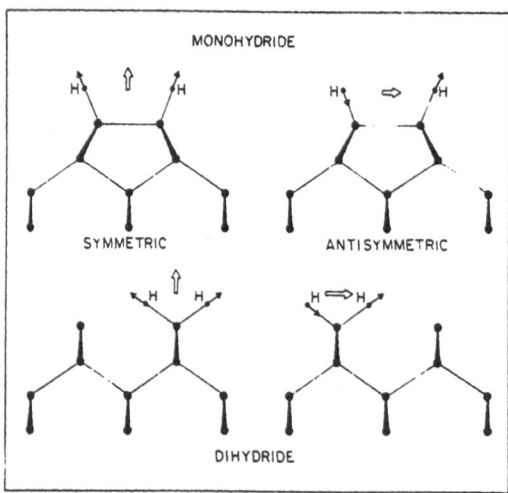

Fig.3.10. Scale drawing of (a) the monohydride, and (b) the dihydride structures. The arrows (not to scale) represent the direction of H displacements (solid arrows) and the *polarization* (double arrows) of the two normal modes involving the stretching of SiH bonds. The different magnitudes of the double arrows schematically indicate that the net dipoles associated with each normal modes are different for the two structures, with $\mu_\parallel < \mu_\perp$ for the monohydride and $\mu_\parallel > \mu_\perp$ for the dihydride.

132

lar component of the mode associated with H at steps has a resonance at the same frquency as the symmetric stretch of the monohydride, its contribution to $\epsilon_{\langle 100 \rangle}$ can be sorted out. As shown in Fig.3.11 the use of two samples oriented in orthogonal directions with respect to the probing radiation (cases I and II) makes it possible for s- and p-polarizations to probe (I) $S_I : \epsilon_{\langle 0\bar{1}1 \rangle}$ and P_I: $\epsilon_{\langle 0\bar{1}1 \rangle}$ and $\epsilon_{\langle 100 \rangle}$, and (II) S_{II}: $\epsilon_{\langle 0\bar{1}1 \rangle}$ and P_{II}: $\epsilon_{\langle 011 \rangle}$ and $\epsilon_{\langle 100 \rangle}$. As a result, the three components of ϵ can be measured unambiguously.

Results of such measurements showed that the monohydride modes exhibit a resonance at 2087 cm^{-1} in $\epsilon_{\langle 0\bar{1}1 \rangle}$) and at 2099 cm^{-1} in $\epsilon_{\langle 100 \rangle}$ at 300 K, while $\epsilon_{\langle 011 \rangle}$ is dominated by a strong resonance at 2087 cm^{-1} and much weaker contributions elsewhere in the spectrum [3.34]. Based on this knowledge, the data of Fig.3.12 show that upon annealing to 625K, *only* the resonances associated with the monohydride on the terraces remain; all the resonances associated with H at

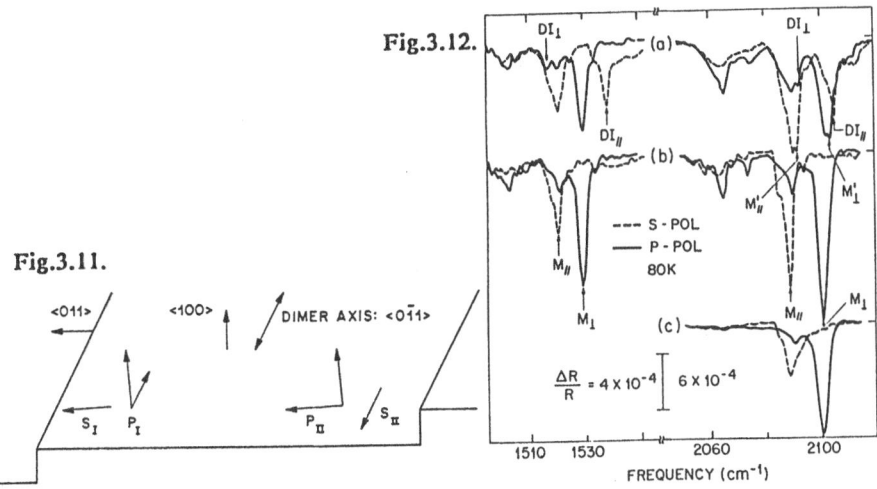

Fig.3.12.

Fig.3.11.

Fig.3.11. Schematic representation of the vicinal surface with a {100} terrace. Note that the relevant electric field directions, labelled S and P are not all exactly collinear with the ⟨011⟩, ⟨100⟩ and ⟨0$\bar{1}$1⟩ axes due to the angle made between the optical plane and the {100} terraces. Straighforward projection of the field vectors along the crystal axes is performed for quantitative data analysis [3.50]

Fig.3.12. Surface infrared spectra associated with (a) the (3x1) phase prepared by H-saturation at 375 K, (b) the (2x1) phase obtained upon a 475 K anneal, and (c) the (2x1) phase obtained upon a 625 K anneal. On the left-hand side of the figure, spectra resulting from pure D exposures are shown. On the right-hand side of the figure, spectra resulting from pure H exposures are shown. The sample is cut 5° off the {100} plane. The s-polarization (dashed lines) has an electric field exactly along the ⟨0$\bar{1}$1⟩ axis. The p-polarization (solid lines) has components 5° off the ⟨100⟩ and ⟨011⟩ directions. The data are taken at 80 K resulting in narrower lines and a small ($\approx$3 cm^{-1}) blue shift from the room temperature frequencies (Sect.3.3.3). The resolution is 0.5 cm^{-1} for th SiH region and 1 cm^{-1} for the SiD region [3.49]

steps disappear. To the extent that the monohydride, characterized only by two sharp modes with distinct frequencies and polarizations, is the most stable hydride on Si{100}, it represents a *model* system which can be reproducibly prepared and for which a quantitative analysis is possible.

In this analysis, it is important to determine the origin of the frequency shift of the resonance in $\epsilon_z = \epsilon_{\langle 100 \rangle}$. As mentioned in Sects.3.1,2, it can arise from extended dynamical effects, i.e. inter-unit cell dynamical coupling. The resulting shift $\Delta \tilde{\nu} \simeq \tilde{\nu}_p^2 / 2 \tilde{\nu}_0 \epsilon_{\infty z}$ will be detected if $\Delta \tilde{\nu} \geq 0.5$ cm^{-1}, the spectral resolution. A shift between the antisymmetric (parallel modes) and symmetric (perpendicular mode) can also arise because of *local* dynamical effects. The latter is sometimes referred to as "chemical" shift because it involves the coupling of normal modes *within* the unit cell and is therefore a signature of the particular chemical bonding within the cell. The two effects must be distinguished if, as is the case for this system, the data are compared to cluster calculations performed on a *single* unit cell.

Experimentally, the inter-unit cell dynamical coupling (extended) can be measured by means of isotopic mixture experiments. As more D is substituted for H at constant coverage, the Si-H normal modes will approach their "isolated frequency" value, i.e. the value devoid of dynamical coupling. If all the splitting is due to extended coupling, then it should decrease to zero as the D concentration increases. If on the other hand it is due to local effects (intra-unit cell), the splitting will remain constant as long as enough unit cells are occupied purely by H. Then, as one of the H is replaced by D, the remaining H will be characterized by a new vibrational frequency, usually falling in between that of the two normal modes. Results of such experiments showed [3.34] that the extended dynamical coupling only accounts for ~3 cm^{-1} for H and ~2 cm^{-1} for D. The rest of the splitting, 9 cm^{-1} for H and 7.5 cm^{-1} for D, is due to local effects as confirmed by the appearance of a new isolated frequency for very low H concentration at ~2092 cm^{-1} (i.e., in between the parallel and perpendicular components).

For a surface cut 9° from the {100} plane, the numbers of the terrace and step sites are $5.4 \cdot 10^{14}$ cm^{-2} and $1.36 \cdot 10^{14}$ cm^{-2}, respectively. Since the Si-H effective mass ($m^* \simeq 1$ amu) and the resonant frequencies ($\tilde{\nu}_{0\parallel} = 2087$ cm^{-1}, $\tilde{\nu}_{0z} = 2096$ cm^{-1}) are known, we are left with three unknowns: $e_\parallel^*$, $e_\perp^*$ and $\epsilon_{\infty z}$. The data provide us with three independently determined quantities

$$\int \epsilon_{\langle 011 \rangle} \, d\tilde{\nu} \qquad\qquad (3.38)$$

from s-polarization in configuration II,

$$\int \frac{\epsilon^{v2}}{\epsilon'_{\langle 100 \rangle}{}^2 + \epsilon''_{\langle 100 \rangle}{}^2} \, \epsilon''_{\langle 100 \rangle} \, d\tilde{\nu} \simeq \frac{1}{\epsilon_{\infty z}{}^2} \int \epsilon''_{\langle 100 \rangle} \, d\tilde{\nu} \tag{3.39}$$

from p-polarization in configuration I, and

$$\Delta\tilde{\nu} \simeq \tilde{\nu}_p{}^2 (2\tilde{\nu}_0 \epsilon_{(\infty z)})^{-1} \tag{3.40}$$

from isotopic mixture experiments.

Fitting the data with the measured $\Delta\tilde{\nu} = 3$ cm^{-1} yields $(e_\parallel{}^*/e) \simeq 0.11$, $(e_z{}^*/e) \simeq 0.14$ and $\epsilon_{\infty z} = 1.4$. Note that the values of e^* reported in [3.7, 14] are too small by a factor of $(4\pi)^{1/2}$. The unscreened ratio

$$r \equiv \frac{\int \epsilon''_{\langle 0\bar{1}1 \rangle} \, d\tilde{\nu}}{\int \epsilon''_{\langle 100 \rangle} \, d\tilde{\nu}} \tag{3.41}$$

can then be obtained, $r \simeq 0.35 \pm 0.1$. We note that the large error in r is mostly due to the uncertainty in $\epsilon_{\infty z}$. Here $\Delta\tilde{\nu}$ is used to determine $\epsilon_{\infty z}$. Because it is so small and known only within 0.5 cm^{-1}, it leads to a $\pm 30\%$ error in r. In summary, the local (or "chemical") splitting is $+9$ cm^{-1} for H, $+7.5$ cm^{-1} for D and the unscreened intensity ratio is r $= 0.35 \pm 0.1$.

In the cluster calculations, a large cluster (Si_9H_{14}) was used to calculate the geometry by minimizing the cluster total energy with respect to bond lengths and bond angles. From this geometry, the vibrational frequencies and intensities could be calculated using a smaller cluster (Si_6H_{14}). Although the absolute value of vibrational frequencies is too high by about 10%, it was found that the splitting between similar normal modes (e.g. symmetric and antisymmetric stretch modes) could be calculated accurately (within a few cm^{-1}). The results for the monohydride and monodeuteride gave a symmetric stretch frequency of, respectively, 11 cm^{-1} and 9 cm^{-1} higher than the antisymmetric stretch frequency, in good agreement with the data. The calculated intensities associated with each normal mode gave the ratio

$$\mu_\parallel/\mu_\perp = 0.59 \, , \quad \text{i.e.} \quad r = (0.59)^2 = 0.35 \tag{3.42}$$

again in good agreement with the data.

In closing, it is important to stress the limitations of SIRS if no theoretical input is available. For instance, if $\Delta\tilde{\nu}$ is too small to be observable, then $\epsilon_{\infty z}$ cannot be determined unless either μ_{par} or μ_{perp} is known independently. Unfortunately, the bare dipole moments are rarely accessible experimentally. As a result, the geometry (bond length and angle) cannot be obtained by SIRS experiments alone. In the case of the monohydride, the geometry was determined by cluster calculations and then used to calculate quantities accessible to the experiment. In particular, the angle made by the Si-H bond with the surface normal was found to be $\alpha = 20°$. Such a value could *not* be obtained from the data alone, i.e. using $\tan\alpha = \mu_\parallel/\mu_\perp$, because the calculations showed that the back projections of $\mu_\parallel$ and $\mu_\perp$ along the Si-H bond do not give the *same* value for the two dynamic dipole moments.

3.3.2 H-saturated Si{100} and Ge{100} Surfaces

Despite its limitations in determining precise adsorbate geometries, SIRS is a powerful tool to unravel complex structures at surfaces. The H-saturated phases on Si{100} and Ge{100} are good examples because both surfaces display a number of coexisting phases which cannot be detected by other probes such as EELS due to poor frequency resolution [3.48]. We refer the reader to [3.49, 50] for details of the summary presented here.

Let us consider the Si{100} and Ge{100} surfaces saturated with H atoms at room temperature. H/Si{100} displays a reasonably sharp (1x1) LEED pattern while H/Ge{100} displays weak 1/2-order beams, remnant of a (2x1) LEED pattern. Prior to the SIRS work, the accepted view was that Si{100}-(1x1)H was made up of a uniform *dihydride* phase with all the dimers broken and that the Ge{100}-(2x1)H had only partial dihydride coverage. The initial key findings of SIRS were that at saturation 1) the Si{100}-(1x1)H surface was characterized by strong modes characteristic of *monohydride* with additional modes present, and 2) the Ge{100}-(2x1)H was characterized by a spectrum identical to that obtained for lower coverages for which a very sharp (2x1) pattern is present. Thus it was apparent that the monohydride structure was an important part of the H-saturated Si{100} surface and the *dominant* part of the Ge{100} surface.

The next important observation was that a Si{100} surface saturated with H atoms while maintained at 380 K displayed a sharp (3x1) LEED pattern; yet it was characterized by an IR spectrum *identical* to that of the Si{100}-(1x1)H surface, prepared by H-saturation at room temperature. These two observations pointed to the fact that the Si{100}-(1x1)H surface was in fact a phase with local (3x1) arrangement and no long range order. The problem was then reduced to interpreting the SIR spectra associated with the (3x1) phase only. As for the Si{100}-(2x1)H case, single domain structures could be prepared on crystals cut vicinal to the {100} plane; thus, polarized SIRS could be profitably used. The data in Fig.3.12 were taken on a sample at 80 K because the bands are sharper (Sect.3.3.3) and different spectral components could be more easily resolved.

Annealing the sample to 475K (Fig.3.12b) produced two very well defined changes: 1) the LEED pattern changed from a (3x1) to a (2x1), and 2) the IR spectrum displayed the loss of two lines only, labeled $DI_\perp$ and $DI_\parallel$ in Fig.3.12a. Intermediate annealing confirmed that the intensity ratio of these two lines remained constant as they became weaker. Analysis of the data shows that 1) the mode perpendicular to the surface is now at lower frequency than that parallel to the surface, i.e. $\Delta_H \equiv \tilde{\nu}_{\langle 100 \rangle}^\circ - \tilde{\nu}_{\langle 0\bar{1}0 \rangle}^\circ = -12.5$ cm^{-1} for H; 2) the magnitude of the splitting between these two modes is larger for deuterium, i.e. $\Delta_D = -21$ cm^{-1}; and 3) the intensity of $DI_\parallel$ is substantially larger than that of $D_\perp$, r$\simeq$3 assuming $\epsilon_{\infty z} \simeq 1.4$. The unambiguous assigment of these lines to the dihydride structure (Fig.3.10b) relied again on cluster calculations [3.49] and isotopic mixture experiments. The theoretical predictions for the dihydride structure were $\Delta_H = -9$ cm^{-1}, $\Delta_D = -22$ cm^{-1} and r$\simeq$3. In addition, the calculations showed that the "isolated" frequency of the dihydride structure should be +7 cm^{-1} *higher* than the isolated frequency of the monohydride structure, and +5 cm^{-1} for the corresponding isolated deuterium modes. The results of isotopic mixture experiments (Fig.3.13) clearly show that DI is 6.5 cm^{-1} higher in frequency than M. For this diluted H concentration, no lines were observed at $DI_\parallel$ and $DI_\perp$ or $M_\parallel$ and $M_\perp$; only the isolated frequencies remained, beautifully confirming that both the dihydride and monohydride structures are simultaneously present in the (3x1) phase.

The excellent resolution (1 cm^{-1}) and S/N of the data make it possible to resolve a small shift ($\sim$2 cm^{-1}) of the wave number of the monohydride modes whenever the dihydride is observed (see M′ in Fig.3.13 and M′$_\perp$ in Fig3.12a). As discussed at the end of [3.49], this small shift arises from small back-bond relaxation of the monohydride structure caused by the presence of a neighboring dihydride. This observation was taken as supporting evidence that a dihydride structure was always formed right next to a monohydride structure and the model shown in Fig.3.14 was proposed to account for the observations on the {100} terraces of H-saturated Si{100}.

The other unlabeled features in the spectra of Fig.3.12a,b were shown to arise from H (or D) at steps and defects. The presence of these features in Fig.3.12b indicates that H is more strongly bound at steps and defects than in a dihydride configuration. However, the removal of these features upon annealing to 625K shows that they are less stable than H in a monohydride configuration [3.51].

Based on the above semiquantitative knowledge of the various hydride phases on Si{100} and their associated IR spectra, we now summarize the qualitative information obtained from IR studies of the H/Ge{100} system [3.50] for which no calculations were performed.

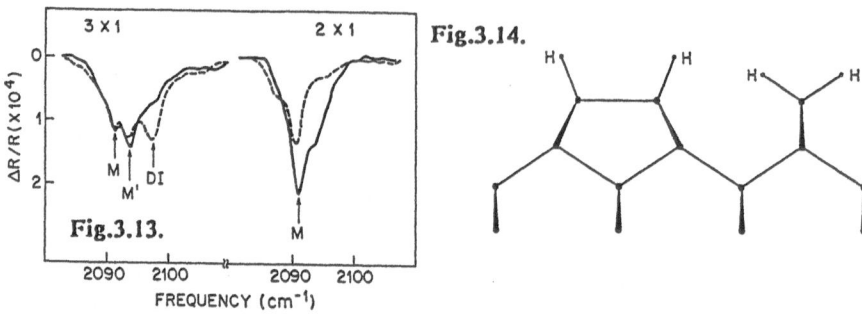

Fig.3.14.

Fig.3.13. Surface infrared spectra of the SiH stretch region obtained upon saturation exposure of an isotopic mixture at 375K ((3x1) phase) and after a subsequent 475K anneal ((2x1) phase). Conditions are therefore identical to those yielding Fig.3.12a,b except that hydrogen represents only 7.5% of the total saturation coverage (measured from IR intensities). The data are taken at room temperature on a Si(100) 5° sample with the s-polarization (dashed lines) and p-polarization (solid lines) defined as in Fig.3.9. The resolution is 1 cm^{-1}. M, M' and DI correspond to the isolated frequencies of pure monohydride, monohydride with a neighboring dihydride, and dihydride, respectively [3.49]

Fig.3.14. Model (drawn to scale) of the (3x1) unit cell. The top layer Si and H atoms are all along the $\langle 0\bar{1}1 \rangle$ axis, i.e. parallel to the step edges for the vicinal samples [3.49]

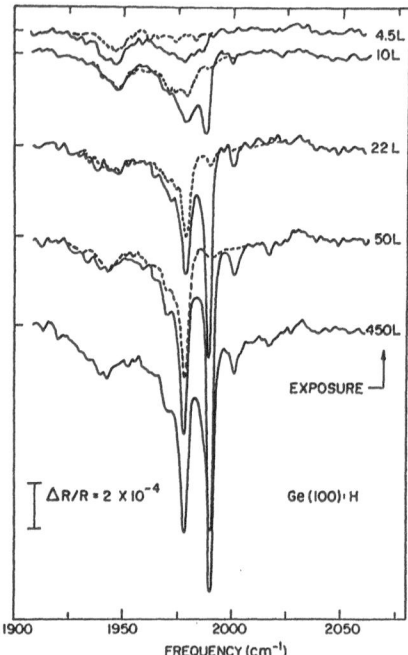

Fig.3.15. Surface infrared spectra as a function of H coverage on a nominally flat Ge{100} surface at room temperature. The exposures are quoted in terms of *molecular* hydrogen exposures, using a 2000K tungsten ribbon (1x2 cm^2) placed 4 cm in front of the sample to dissociate the molecular hydrogen. Atomic hydrogen exposures are approximately a factor of 100 lower. The dashed lines represent the results for s-polarization and the solid lines for p-polarization. The resolution is 2 cm^{-1}. There is no observed absorption for any exposure up to 450 L in the 750 to 1000 cm^{-1} region where the noise level is approximately 5x10^{-5} per reflection [3.50]

Figure 3.15 shows that the IR spectrum is dominated at all coverages (except for $\theta \leq 0.1$ ml) by two modes at 1979 cm^{-1} and 1991 cm^{-1}, polarized parallel and perpendicular to the surface respectively. This

138

observation, along with the (2x1) LEED pattern and the absence of scissor mode in the 750 to 1000 cm^{-1} region (characteristic of the dihydride structure) at all coverages, indicates that the main hydride phase is monohydride, even at saturation coverage. Dihydride is *not* formed on Ge{100} at any coverage [3.52]. As expected, the faint (2x1) LEED pattern at saturation shows that the (2x1) unit cells have a poor *long range* arrangement due to etching of the surface by H atoms.

For the purpose of this review, we note without showing the data that although the splitting $\Delta_H = 12$ cm^{-1} is identical to that of the monohydride on Si{100}, its origin is quite different. Isotopic mixture experiments show that the two modes *continuously* shift with increasing D concentration at saturation coverage to the value of 1980.5 cm^{-1}. The extended dynamical interaction (inter-unit cell) is therefore dominant for this system [3.50].

3.3.3 Dynamics of H on Si{100}

In this section, we outline the procedures by which dynamical information can be obtained from SIRS. In the favorable case of the Si{100}-(2x1)H phase for which a *single* structure (the monohydride) can be prepared and is stable over a large range of temperatures, analysis of the line shape is warranted. Such an analysis for the monohydride lines is particularly interesting as variations in width and line positions with temperatures are observed. From the spectra, changes in concentration of monohydride to defect sites with temperature can be ruled out. Temperature dependent inhomogeneous broadening is negligible and line broadening and shift must be due to interaction with phonons. Such anharmonic coupling to phonons can be calculated by molecular dynamics simulations [3.53].

In general, the main drawback of such simulations is that they rely on fitted parameters, i.e. they are not first principle, and they are classical calculations. In the case of the Si{100}-(2x1)H system, ab initio calculations of the force fields were available. Both diagonal and off-diagonal stretch and bend force constants, including non-negligible cubic and quartic terms, were calculated from first principles. Quantum effects were included by means of a novel stochastic method described in [3.54]. The molecular dynamics calculations could therefore be performed without adjustable parameters.

The main results are: 1) the lifetime and dipole broadening are negligible ($\sim 10^{-3}$ cm^{-1} for H and $\sim 10^{-2}$ cm^{-1} for D), and (2) the dominant broadening mechanism is due to dephasing, i.e. to anharmonic coupling between the Si-H bending mode and the Si-H stretch. We refer the reader to [3.53] for details. Because the silicon substrate lattice was taken to be harmonic, the simulations made no predictions on line shifts (4.1 cm^{-1} red shift from 40 K to 500 K for H). The anharmonic extension of the calculations is important as shifts can come from several mechanisms, including dephasing, thermal expansion and energy transfer mechanisms.

In conclusion, it appears that dynamical information can be obtained from line shape analysis in favorable cases. As data are compiled for a number of different systems for which the *structure* is *well known* and for which inhomogeneous broadening can be ruled out, interesting dynamical information will be obtained. Using the present calculations as a benchmark, it appears that the vibrational lifetime of adsorbates on semiconductors is long enough (10^{-7} to 10^{-9} s) that *direct* measurements of the lifetimes using tunable pulsed lasers may be possible.

3.3.4 H_2O on Si{100}

In this section, we summarize the findings of SIRS for H_2O adsorption on Si{100} to give an example of SIRS contribution to understanding simple chemistry at surfaces. Conflicting interpretations motivated a SIRS study: UPS data indicated that water adsorbed molecularly [3.55] while EELS studies showed that water was dissociated into H and OH [3.56]. The UPS group proposed that the probe electrons in EELS could dissociate molecular water and that the samples used in EELS experiments contained a high density of defects (steps) that would dissociate water.

As shown in Fig.3.16, the Si-H and O-H modes of dissociated water are clearly observed upon exposure at room temperature

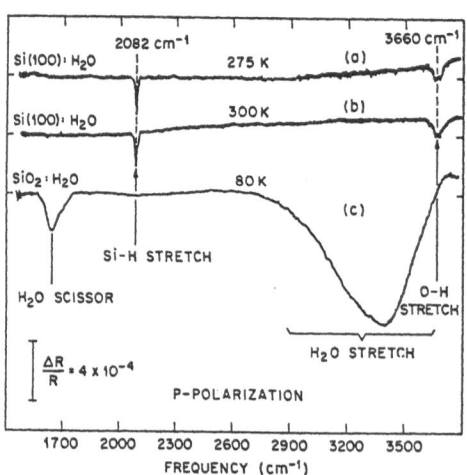

Fig.3.16. Surface infrared spectra obtained upon exposure of clean Si{100}-(2x1) to (a) 0.5 L water at $T_s = 275K$, and (b) 10 L water at $T_s = 300K$. Curve (c) is obtained upon exposure of an oxidized Si{100} surface (native oxide) to 10 L water at $T_s = 80K$. Data were taken at the exposure temperatures as indicated on each spectrum [3.58]

[357, 58]. Molecular water could only be observed by condensing water on an oxide surface or on a layer with predissociated water (water was found to dissociate on a *clean* Si{100} surface at 80 K).

By using vicinal samples with steps along the $\langle 0\bar{1}1 \rangle$ direction as was done for the H/Si{100} studies, it was possible to test whether dissociation can take place at steps: monitoring possible Si-H modes polarized along the $\langle 011 \rangle$ direction would indicate the presence of H at steps. Although the data could not rule out the presence of some H at steps, they clearly showed that steps are not *saturated* with hydrogen [3.51]. Since the substrate temperature was too low for H to diffuse on the surface, it was concluded that water dissociation takes place on the terraces themselves.

3.3.5 Electronic Absorption of Si{100} and Si{111}

a) Si{100}

As evident from direct and inverse photoemission studies, clean semiconductor surfaces have states in the gap, both below and above the Fermi level, E_F. These states are associated with dangling bonds and usually disappear when impurities such as H, O or H_2O are adsorbed on the surface. When the symmetry and energetics of these states are such that vertical transitions are allowed from the filled to empty states, optical absorption takes place. An important advantage of optical spectroscopy is the polarization information which helps determine the symmetry and spatial orientation of surface states. In general, however, the interpretation is less straightforward than that of direct or inverse angle resolved photoemission since the convolution of the filled and empty states is probed (joint density of states) rather than only one or the other. SIRS is therefore complementary to the above electron spectroscopies.

Results for the clean Si{100} surface are shown in Fig.3.17. The change in reflectance $\Delta R/R = (R^o - R^{clean})/R^o$ is plotted as a function of energy for three different temperatures. Here R^o is the reflectivity after the surface has been exposed to oxygen, hydrogen or water. The spectra are characterized by a lack of sharp features and by a temperature dependent threshold and intensity around 0.45 eV. The lack of sharp features and the weak absorption indicates that there are no direct transitions in this frequency range.

It is in general hard to prove the uniqueness of an interpretation of such a broad spectrum. In [3.59] the functional and temperature dependences of $\Delta R/R(v)$ below 0.75 eV were used to determine the nature of the transition. As suggested previously by Mönch et al. [3.60], a model where *indirect* transitions (phonon-assisted) are involved was put forth. It was found that the absorption and creation of

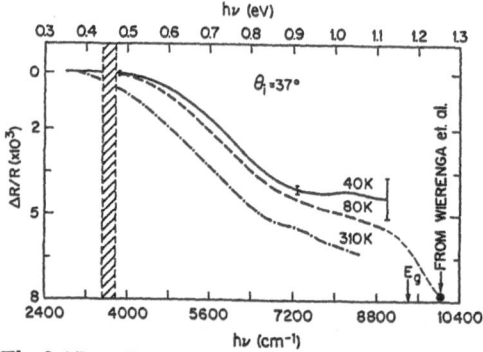

Fig.3.17. Change of reflectivity after surface cleaning vs. optical frequency for three different temperatures. The background surfaces were obtained by exposing the clean surface to hydrogen, oxygen, or water. The internal incident angle is 37⁰. The point labeled Wierenga is taken from the reflectometric studies in the 1.0 to 3.0 eV region of P.E. Wierenga, M.J. Sparnaay, and A. van Silfhout: Surf. Sci. **99**, 59 (1980), and corrected for the field strength at the surface. The resolution is 8 cm⁻¹. The hatched area corresponds to a frequency range where no signal is modulated by the interferometer [3.59]

0.06 eV phonons could account for the temperature dependence of the threshold at 0.44 eV. Combining the results of UPS [3.61,62] it was then suggested that the indirect transitions occurred from the top of the bulk valence band projected at Γ to empty surface at J'. The origin of the difficulties to interpret optical absorption is apparent: transitions may be direct or indirect, and may involve surface states as well as bulk states projected onto the surface plane. The quantitative interpretation of electronic absorption data is therefore rarely available without independent information from other experiment.

b) Si{111}

Purely qualitative interpretation can be useful. For instance, Fig.3.18 shows the electronic absorption associated with the clean Si{111}-(7x7) and with the laser-annealed Si{111}-(1x1) surfaces. The background surface in this case is the H-saturated surface [3.63]. Without any knowledge of the states involved in the transition at 0.3 eV for the (7x7) surface, the spatial origin of such states may be inferred by noting that 1) such electronic absorption is completely absent in the laser-annealed (1x1) surface, and 2) the sharp peak at 0.3 eV on the (7x7) is quenched by low coverages of hydrogen. STM studies have shown that the main difference between the (7x7) and laser-annealed (1x1) surfaces is the absence of deep holes on the latter [3.64]. Furthermore, the extensive IR work [3.63,65] on the H/Si{111} system demonstrated that H is preferentially chemisorbed in the deep corner holes of the (7x7) surface at low coverages. The origin of the 0.3 eV IR absorption can therefore be assigned to the corner holes of the (7x7) surface.

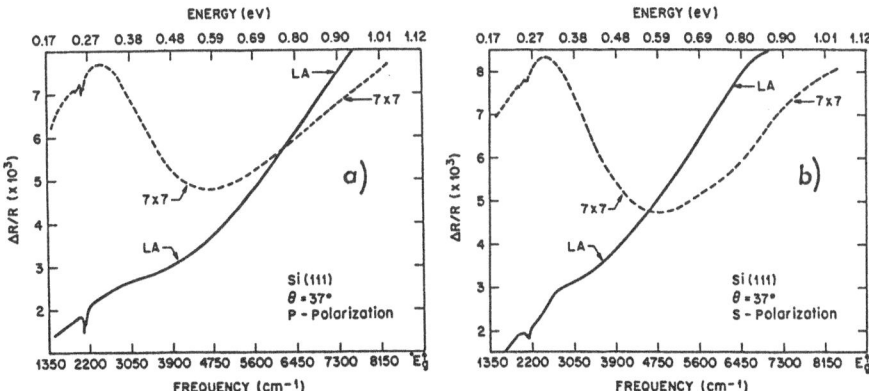

Fig.3.18. Change in reflectivity (note the direction of the abscissa is reversed from that of Fig.3.17) of clean Si{100}-(7x7) (dashed line) and laser-annealed Si{111}-(1x1) (solid line) for (a) p-polarization and (b) s-polarization. The background surfaces are the H-saturated surfaces. Note the weak and sharp features around 2100 cm⁻¹ corresponding to Si-H vibrational absorption of the *reference* surface.

Such information is useful for inelastic STM, UPS and inverse UPS studies on this system. At worst, it is another fingerprint of the quality of the Si{111}-(7x7) surface during IR studies. It emphasizes the need for broadband spectroscopy even when a narrow vibrational mode is under study, since the latter "rides" on top of this large absorption.

A particularly nice example of the power of polarized SIRS can be found in the study of the cleaved Si{111}-(2x1) surface. Using a MIR geometry, Chiarotti et al. [3.66] established early on the presence of a strong optical absorption at 0.45 eV on this surface. Later, by means of external reflection, they extended the range of this study to above the indirect band gap, uncovering a broad and weaker surface absorption band between 1.0 and 3.5 eV [3.37]. These data were an important consideration when Pandey theoretically reexamined the reconstruction of the Si{111}-(2x1) surface. He concluded that the surface atoms are arranged in a chainlike structure, with strong π bonding [3.67], instead of being buckled as was previously thought. This model also implied that the strong absorption at 0.45 eV should be highly anisotropic, with a maximum for the electric field polarized parallel to the surface and along the chain axis, i.e. along the $\langle \bar{1}10 \rangle$ axis [3.68]. In contrast, the buckling model would give a maximum absorption for a field orthogonal to the $\langle \bar{1}10 \rangle$ axis.

Two studies were performed to test the model, an external reflection absorption experiment by *Chiaradia* et al. [3.69] and a photothermal displacement spectroscopy (PTDS) study by Olmstead at al. [3.70]. The results are shown in Figs.3.19,20. In the reflection experiment, the radiation is normally incident on the surface (i.e. E-field parallel to the

143

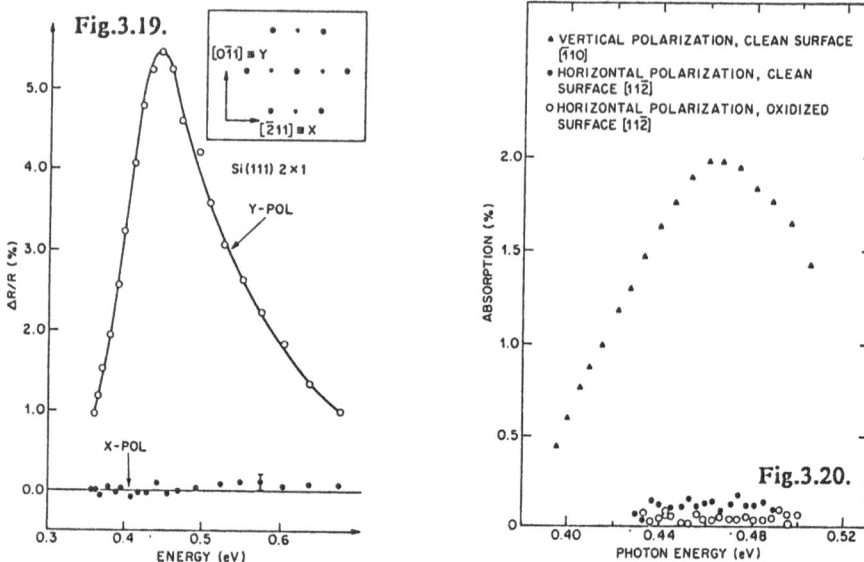

Fig.3.19. Differential reflectivity spectra of a Si{111}-(2x1) single domain surface, for light polarizations along the ⟨211⟩ and ⟨01̄1⟩ directions (curves labeled x and y, respectively). The error bar of ± 1x10⁻³ is also shown. The inset represents a sketch of the LEED pattern (with integer and half-order spots) and the main crystallographic directions in the {111} plane [3.69]

Fig.3.20. Si{111}-(2x1) surface-state absorption spectrum obtained by photothermal displacement spectroscopy [3.70]. Saturation oxidation was obtained after ~1 h exposure at 10⁻⁷ Torr oxygen

surface) with the polarization continuously rotated from E//⟨01̄1⟩ to E//⟨2̄11⟩. In Fig. 3.19, the differential reflectivity (reflectance) is plotted for the two extreme polarizations. The absorption at 0.45 eV is maximum for E//⟨01̄1⟩ and falls to zero as E approaches the ⟨2̄11⟩ direction. In the PTDS experiment, radiation from a Kr⁺-pumped F-center laser is normally incident on the surface. The deflection of the He-Ne laser radiation with a 45° angle of incidence is then recorded as a function of pump frequency. The absorption at 0.46 eV, plotted in Fig.3.20, is again clearly maximum for E//⟨01̄1⟩. Polarized differential reflectivity studies were later extended to higher energy to explore the dependence of the 1-3.5 eV absorption [3.38]. The results could once again be reconciled only with the Pandey model. Both the reflectivity and PTDS data constitute one of the strongest supports for a surface reconstruction model that was in drastic departure from the conventional wisdom. For completeness we list in [3.71, 37] studies performed on Ge{111} and on III-V semiconductors. We refer the reader to some reviews on the subject [3.22, 72].

3.3.6 Chemistry at Semiconductor Surfaces

In Sects.3.3.1-5 the examples have focused on simple systems for which the surface cleanness and morphology could be well controlled. It is clear, however, that SIRS is a versatile tool to study more complicated but technologically very important systems. For instance, little is known about chemically prepared semiconductor surfaces and initial stages of metal deposition by chemical vapor deposition (CVD) or photoinduced CVD. In this section, we want to make the reader aware of some beginning efforts in this complex field.

Very recently, silicon surfaces prepared by wet chemistry were found to have a very low surface recombination velocity [3.73]. SIRS showed the presence of a featureless band at 2100 cm^{-1} which was interpreted as a monolayer of Si-H at the surface. Since this interpretation was questioned [3.74], the surface was reexamined in our laboratory [3.75]. The spectrum observed after pirhana oxidation followed by a HF etch is shown in Fig.3.21. It clearly contains very sharp and distinctive features, which are not seen on the more featureless spectra obtained after atomic H exposure of clean Si{111} [3.63]. This spectrum is very reproducible but has recently been interpreted [3.75]. It is characterized by a strong mode normal to surface at 2087 cm^{-1} and three bands with strong tangential components at 2077, 2110 and 2141 cm^{-1}. When placed in a vacuum system for subsequent analysis by electron techniques, hydrocarbons are seen to adsorb on the surface and the SiH vibration band is modified. Continuous monitoring of the surface from atmosphere to uhv is therefore very important and can be done with SIRS.

With regard to the decomposition of organo-metallic gases, recent work has focused on the laser-assisted decomposition of Fe(CO)$_5$ on Si{111} [3.76]. Monitoring the carbonyl absorption by SIRS as a function of temperature and laser irradiation was instrumental in uncover-

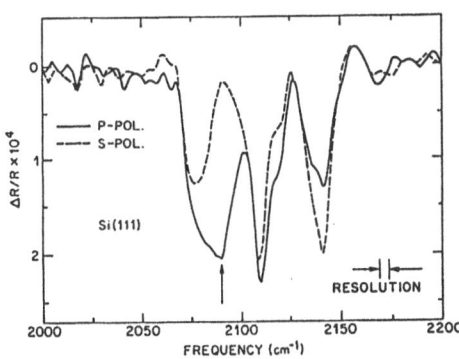

Fig.3.21. Surface infrared spectra of the piranha-etched (H$_2$SO$_4$:H$_2$O$_2$ cleaned) and subsequently HF-etched Si(100) surface in the Si-H stretch frequency range. The reference surface is obtained after flashing to 500°C in vacuum (P $\simeq$ 10^{-7} Torr)

ing the mechanism and energetics of $Fe(CO)_5$ decomposition. Work is actively being done to study the initial stages of Al deposition upon exposure of Si{111} to tri-butylaluminum [3.75].

3.4 Problems and Future Directions

While the previous section shows that chemical, structural and dynamical information of adsorbates at semiconductor surfaces can be obtained from SIRS and proper theoretical treatments, the technique faces severe limitations which must be addressed. In particular, the poor sensitivity to vibrational modes occurring in a frequency range where the substrate absorbs (usually the low frequency region) is very restrictive. Since most of the adsorbate-substrate modes occur in this region, the substrate absorption rules out the use of multiple internal reflection geometry and photoacoustic detection in this range. Emission spectroscopy [3.77] for which the sample acts as the IR source may be helpful if the largest solid angle compatible with the results of Sect.3.2.2 can be used. An interesting possibility regarding this problem is the use of synchrotron radiation in the range 100 to 1000 cm^{-1}. For the geometrical requirements of surface studies (small spot size and low divergence), the synchrotron radiation is estimated to be one to two orders of magnitude brighter than a conventional source, depending on the frequency [3.78]. Such an increase in intensity would make it possible to lose energy to the substrate in a total internal reflection configuration and yet have enough throughput for sensitive detection.

The second problem associated with the study of semiconductor surfaces is that of free carrier absorption when the substrate temperature is high. The ability to measure IR spectra on hot substrates is important for in situ studies of CVD processes. Emission spectroscopy is by far the most sensitive means of obtaining surface vibrational spectra for hot samples and might be performed without cooling the spectrometer if the samples are hot enough (i.e. $T_{sample} \gg 300$ K).

The ability of SIRS to probe semiconductor surfaces under a variety of conditions is important because it makes it possible to use the technique for in situ measurements of technologically important processes. For instance, we have shown in Sect.3.3.6 that work is beginning on laser-induced photochemistry, plasma etching and thermal chemistry at semiconductor surfaces, which are processes that require the understanding of the microscopic mechanisms of surface reactions. Thanks to the improvement in fast scanning interferometers, SIRS appears to be a viable real-time probe for kinetic measurements requiring subsecond time resolution [3.79, 80]. Resolutions of 25 ms are possible [3.81] using standard techniques. Better time resolution can be achieved for repetitive processes either by nonstandard interferometric techniques or with grating spectrometers [3.82]. We feel that relatively large efforts will be made to tackle kinetic measurements in the next few years. Such measurements are particularly pertinent to the relevant technological processes at semiconductor interfaces.

3.5 Conclusions

In this chapter, the basic formalism necessary to design and analyze SIRS studies of semiconductor surfaces has been presented. The general conclusion is that the use of multiple internal reflection is more advantageous except in some cases for which external reflection or emission spectroscopy is better suited.

The various examples presented have emphasized the features of SIRS that make it a unique probe. These are:

1) High resolution (~0.25 cm^{-1}) necessary to resolve sharp modes (~1 cm^{-1}) at semiconductor surfaces and to differentiate between various kinds of modes (e.g. H at steps or in monohydride structure, isolated monohydride and monohydride next to a dihydride).

2) Polarization properties, making it possible to probe and distinguish modes parallel to the surface. In contrast to MIR SIRS, specular EELS is *not* sensitive to the parallel modes. The combination of high resolution and polarization capabilities have made SIRS a much superior tool compared to EELS in the study of H on semiconductor surfaces.

3) The compatibility of SIRS with ambient pressures and electric fields sets it apart as a powerful in situ probe. Other photon probes such as Raman scattering have not shown enough sensitivity so far to be used in a versatile way.

4) The possibility of time-resolved measurements of important kinetic phenomena on the ms time scale.

Acknowledgments The author is grateful to K. Raghavachari, D.R. Hamann, J.C. Tully and J.E. Reutt for useful conversations and to S.B. Christman, E.E. Chaban and M.E. Sims for technical assistance for most of the experiments reviewed here.

References

3.1 G.A. Bootsma, L.J. Hanekamp, O.L.J. Gijzeman: In *Chemistry and Physics of Solid Surfaces IV*, ed. by R. Vanselow, R.F. Howe, Springer Ser. Chem. Phys., Vol.20 (Springer, Berlin, Heidelberg 1982) p.77

3.2 A. Campion: In *Chemistry and Physics of Solid Surfaces VI*, ed. by R. Vanselow, R.F. Howe, Springer Ser. Surf. Sci., Vol.5 (Springer, Berlin, Heidelberg 1986) p.261

3.3 R. Biswas, D.R. Hamann: Phys. Rev. Lett. 56, 2295 (1986)

3.4 P.J. Feibelman, D.R. Hamann: Surf. Sci. 173, L582 (1986)

3.5 S.A. Francis, A.H. Ellison: J. Opt. Soc. Am. 49, 131 (1959)

3.6   R.G. Greenler: J. Chem. Phys. **44**, 310 (1966)
3.7   F.M. Hoffmann: Surf. Sci. Reports 3, 107 (1983);
      E. Schweitzer, A.M. Bradshaw: In *Advances in Spectroscopy: Spectroscopy of Surfaces*, ed. by R.E. Hester (Wiley, New York 1988) Chap.3
      Y.J. Chabal: Surf. Sci. Rpt. **8**, 211 (1988)
3.8   The proceedings of 5th Int'l Conf. on Vibrations at Surfaces, published in J. Electron. Spect. Rel. Phenom. **44,45** (1987) contain good references for other work done in the field.
3.9   N.J. Harrick: *Internal Reflection Spectroscopy* (Wiley, New York 1967; 2nd printing Harrick Scientific Corp., Ossining, NY 1979)
3.10 F.M. Mirabella, N.J. Harrick: Internal Reflection Spectroscopy: Review, Supplement (Harrick Scientific Corp., Ossining, NY, 1985);
      F.M. Mirabella, Appl. Spectrosc. Rev. **21**, 45 (1985)
3.11 J.D.E. McIntyre, D.E. Aspnes: Surf. Sci. **24**, 417 (1971)
3.12 D.E. Aspnes: private communication
3.13 L.D. Landau, E.M. Lifshitz: *Electrodynamics of Continuous Media* (Pergamon, New York 1960) Chap.XI
3.14 Y.J. Chabal: Vibrational properties of semiconductor surfaces and interfaces, in *Semiconductor Interfaces: Formation and Properties*, ed. by G. Le Lay, J. Derrien, N. Boccara, *Springer Proc. Phys.* **22**, 301 (Springer, Berlin, Heidelberg 1987)
3.15 See [Ref.3.9, p.27]
3.16 See [Ref.3.9, pp.42-44]
3.17 F. Wooten: *Optical Properties of Solids* (Academic, New York 1972)
3.18 H. Ibach, D.L. Mills: *Electron Energy Loss Spectroscopy, Surface Vibrations* (Academic, New York 1982) pp.93-98
3.19 G.D. Mahan, A.A. Lucas: J. Chem. Phys. **68**, 1344 (1978)
3.20 Z. Schlesinger, L.H. Greene, A.J. Sievers: Phys. Rev. B **32**, 2721 (1985) footnote 17
3.21 The value of C depends on the particular arrangement of the oscillators. For a planar triangular lattice, C=8.8904, as discussed in [3.18].
3.22 P. Chiaradia: Optical properties of surfaces and interfaces, in *Semiconductor Interfaces: Formation and Properties*, ed. by G. Le Lay, J. Derrien, N. Boccara, *Springer Proc. Phys.* **22**, 290 (Springer, Berlin, Heidelberg 1987)
      M.A. Olmstead: Surf. Sci. Rpts. **6**, 159 (1986)
3.23 P.J. Feibelman: Progress Surf. Sci. **12**, 287 (1982)
3.24 A. Bagchi, A.K. Rajagopal: Solid State Commun. **31**, 127 (1979)
      A. Bagchi, R.G. Barrera, A.K. Rajagopal: Phys. Rev. **B20**, 4824 (1979)
3.25 A. Selloni, P. Marsella, R. Del Sole: Phys. Rev. **B33**, 8885 (1986)
      A. Selloni, R. Del Sole: Surf. Sci **168**, 35 (1986)
3.26 W.L. Mochan, R.G. Barrera: Phys. Rev. Lett. **56**, 2221 (1986)
3.27 P.J. Feibelman, D.R. Hamann: Surf. Sci. **179** 153 (1986); ibid **182**, 411 (1987)
3.28 S.R. Chubb, W.E. Pickett: Phys. Rev. Lett. **58**, 1248 (1987)
3.29 P.J. Feibelman: Phys. Rev. Lett. **58**, 2766 (1987)
3.30 E.G.A. Redondo, W.A. Goddard: J. Vac. Sci. Technol. **21**, 344 (1982)
3.31 W.J. Hehre, L. Radon, P.V.R. Schleyer, J.A. Pople: *Ab Initio Molecular Orbital Theory* (Wiley, NY, 1986)
3.32 J.A. Pople, R. Krishnan, H.B. Schlegal, J.S. Binkley, Inst. J. Quant. Chem. Symp. **13**, 225 (1979); ibid **15**, 269 (1981)
3.33 K. Raghavachari, J. Chem. Phys. **81**, 2717 (1984)
3.34 Y.J. Chabal, K. Raghavachari, Phys. Rev. Lett. **53**, 282 (1984)
3.35 R. Khrishnan, M.J. Frisch, J.A. Pople, J. Chem. Phys. **72**, 4244 (1980) and references therein
3.36 R.G. Greenler, J. Vac. Sci. Technol. **12**, 1410 (1975)

3.37 S. Nannarone, P. Chiaradia, F. Ciccacci, R. Memeo, P. Sassaroli, S. Selci, G. Chiarotti, Solid State Commun. 33, 593 (1980); P. Chiaradia, G. Chiarotti, F. Ciccacci, R. Memeo, S. Nannarone, P. Sassaroli, S. Selci, Surf. Sci. 99, 70 (1980)

3.38 S. Selci, P. Chiaradia, F. Ciccacci, A. Cricenti, N. Sparvieri, G. Chiarotti, Phys. Rev. B31, 4096 (1985)

3.39 W.C.M. Claassen, J. Dieleman: "An in-situ infrared study of the interaction of oxygen, fluorine plasmas with Si, SiO₂ surfaces", Topical Meeting on Microphysics of Surfaces, Beams, Adsorbates, Technical Digest Series 1987, Vol.9 (Optical Society of America, Washington, DC 1987)

3.40 P. Hollins, J. Pritchard: J. Vac. Sci. Technol. 17, 665 (1980)

3.41 R.J. Collins, H.Y. Fan: Phys. Rev. 93, 674 (1954)

3.42 N.J. Harrick: Phys. Rev. Lett. 4, 224 (1960)

3.43 G.E. Becker, G.W. Gobeli: J. Chem. Phys. 38, 2942 (1963)

3.44 G.S. Higashi, L.J. Rothberg: Appl. Phys. Lett. 47, 1288 (1985)

3.45 M.A. Olmstead, N.M. Amer, S. Kohn, D. Fournier, A.C. Buccara: Appl. Phys. A 32, 141 (1983)
M.A. Olmstead, N.M. Amer: J. Vac. Sci. Technol. B1, 751 (1983)

3.46 For example, some other pioneering work not mentioned so far:
D.B. Novotny: J. Vac. Sci. Technol. 9, 1447 (1972)
E.D. Palik, R.J. Holm, A. Stella, H.L. Hughes: J. Appl. Phys. 53, 8454 (1982)
H.J. Stein, P.S. Peercy: Phys. Rev. B22, 6233 (1980)

3.47 R. Kaplan: Surf. Sci. 93, 145 (1980)

3.48 F. Stucki, J.A. Schaefer, J.R. Anderson, G.J. Lapeyre, W. Göpel: Solid State Commun. 47, 795 (1983)

3.49 Y.J. Chabal, K. Raghavachari: Phys. Rev. Lett. 54, 1055 (1985)

3.50 Y.J. Chabal: Surf. Sci. 168, 594 (1986)

3.51 Y.J. Chabal: J. Vac. Sci. Technol. A3, 1448 (1985)

3.52 Curiously, the SIRS data are ignored by very recent EELS work on H/Ge(100) that claims that complete dihydride formation is achieved: L. Papagno, X.Y. Shen, J. Anderson, G. Schirripa Spagnolo, G.J. Lapeyre, Phys. Rev. B34, 7188 (1986)

3.53 J.C. Tully, Y.J. Chabal, K. Raghavachari, J.M. Bowman, R.R. Lucchese: Phys. Rev. 31, 1184 (1985)

3.54 A. Nitzan, J.C. Tully: J. Chem. Phys. 78, 3959 (1983)

3.55 D. Schmeisser, F.J. Himpsel, G. Hollinger: Phys. Rev. B27, 7818 (1983); D. Schmeisser, Surf. Sci. 137, 197 (1984)

3.56 H. Ibach, H. Wagner, D. Bruchmann: Solid State Commun. 42, 457 (1982); Appl. Phys. A29, 113 (1982)

3.57 Y.J. Chabal: Phys. Rev. B 29, 3677 (1984)

3.58 Y.J. Chabal, S.B. Christman: Phys. Rev. B29, 6974 (1984)

3.59 Y.J. Chabal, S.B. Christman, E.E. Chaban, M.T. Yin: J. Vac. Sci. Technol. A1, 1241 (1983)

3.60 W. Mönch, P. Koke, S. Krueger: J. Vac. Sci. Technol. 19, 313 (1981)

3.61 F.J. Himpsel, D.E. Eastman: J. Vac. Sci. Technol. 16, 1297 (1979); for a review see D.E. Eastman: J. Vac. Sci. Technol. 17, 492 (1980)

3.62 R.I.G. Uhrberg. G.V. Hansson, J.M. Nicholls, S.A. Flödstrom: Phys. Rev. B24, 4684 (1981)

3.63 Y.J. Chabal, G.S. Higashi, S.B. Christman: Phys. Rev. B28, 4472 (1983)

3.64 R.S. Becker, J.A. Golovchenko, G.S. Higashi, B.S. Schwartzentruber: Phys. Rev. Lett. 57, 1020 (1986)

3.65 Y.J. Chabal: Phys. Rev. Lett. 50, 1850 (1983)

3.66 G. Chiarotti, S. Nannarone, R. Pastore, P. Chiaradia: Phys. Rev. B4, 3398 (1971)

3.67 K.C. Pandey: Phys. Rev. Lett. 47, 1913 (1981)

3.68 R. Del Sole, A. Selloni: Solid State Commun. **50**, 825 (1984); Phys. Rev. **30**, 883 (1984)

3.69 P. Chiaradia, A. Cricenti, S. Selci, G. Chiarotti: Phys. Rev. Lett. **52**, 1145 (1984)

3.70 M.A. Olmstead, N.M. Amer: Phys. Rev. Lett. **52**, 1148 (1984)

3.71 M.A. Olmstead, N.M. Amer: Phys. Rev. B **29**, 7048 (1984)

3.72 P. Chiaradia, A. Cricenti, G. Chiarotti, F. Ciccacci, S. Selci: In *The Structure of Surfaces*, ed. by M.A. Van Hove, S.Y. Tong, Springer Ser. Surf. Sci. Vol.2 (Springer, Berlin, Heidelberg 1985) p.66
M.A. Olmstead: Surf. Sci. Rpts. **6**, 159 (1986)
S. Selci, F. Ciccacci, G. Chiarotti, P. Chiaradia, A. Cricenti: J. Vac. Sci. Technol. **A5**, 327 (1987)

3.73 E. Yablonovitch, D.L. Allara, C.C. Chang, T. Gmitter, T.B. Bright: Phys. Rev. Lett. **57**, 249 (1986)

3.74 B.R. Weinberger, G.G. Peterson, T.C. Eschrich, H.A. Krasinski: J. Appl. Phys. **60**, 3232 (1986)

3.75 V.A. Burrows, Y.J. Chabal, G.S. Higashi, S.B. Christman: Appl. Phys. Lett. (August 1988) to be published

3.76 J.R. Swanson, C.M. Friend, Y.J. Chabal: J. Chem. Phys. **87**, 5028 (1987)

3.77 S. Chiang, R.G. Tobin, P.L. Richards, P.A. Thiel: Phys. Rev. Lett. **52**, 648 (1984)

3.78 W.D. Duncan, G.P. Williams: Appl. Opt. **22**, 2914 (1983)
G.P. Williams: Int'l J. Infrared Submillim. Waves **5**, 829 (1984)

3.79 V.A. Burrows, S. Sundaresan, Y.J. Chabal, S.B. Christman: Surf. Sci. **160**, 122 (1985); ibid **180**, 110 (1987)

3.80 M.A. Chesters: J. Electron. Spect. Rel. Phenom. **38**, 123 (1986)

3.81 J.E. Reutt, D.J. Doren, Y.J. Chabal, S.B. Christman: Phys. Rev. Lett. (October–December 1988) and J. Vac. Sci. Technol. (May/June 1989)

3.82 B.A. Hayden: unpublished

4. Surface Phonons — Theory

A.A. Maradudin

Department of Physics
University of California, Irvine, CA 92717, USA

In this review I should like to summarize what in my view have been some of the significant advances in the theoretical study of surface phonons in the 1980s, and to point to some directions in which the field can go in the next few years.

In speaking of surface phonons, I have in mind vibrational surface excitations studied on the basis of discrete, atomistic models of crystals. Consequently, surface acoustic waves, which are generally discussed on the basis of continuum mechanics and which have an intrinsic interest of their own, will not be considered here.

Although my emphasis will be on the theoretical side of the subject, experimental resuts will be described when they introduce or otherwise bear on some particular theoretical point.

During the past six years several reviews have appeared that deal with various aspects of the physics of surface phonons, including the theory of these excitations [4.1-7]. The reader is referred to them for material that limitations of space and time prevented from being included in this review.

4.1 Discoveries and Advances

Significant discoveries and advances in the theoretical and experimental study of surface phonons were made during the past seven years. In this section several of them are described briefly.

4.1.1 Advances in Experimental Studies of Surface Phonons

Although theoretical studies of surface phonons date from the pioneering work of *Lifshitz* and *Rosenzweig* [4.8] and of *Rosenzweig* [4.9] in the late 1940s, and although many theoretical investigations in this field were carried out in the succeeding three decades [4.10], there has been a flowering of activity in it commencing in 1980. I think it is indisputable that a major reason for this was the introduction of two experimental methods for the determination of the dispersion curves of surface phonons on clean and adsorbate covered surfaces.

At the beginning of the time period to which this survey is devoted the use of helium atom scattering for the measurement of surface phonon dispersion curves had just been introduced and used for the measurement of dispersion curves of surface phonons on the surfaces of alkali-halide crystals [4.11]. Since then, the technique has been applied to the study of surface phonons on metal surfaces [4.12,13] and on layered compounds [4.3], to the study of surface phonons in layers of rare gas atoms adsorbed in an incommensurate fashion on metal substrates [4.14-17], and to the observation of surface optical phonons in NaF [4.18] and LiF [4.19]. The energy of the surface phonons that can be studied by this method is limited by the contribution from multiphonon scattering processes, which varies from crystal to crystal. However, surface phonon energies as high as 60 meV have been measured by helium atom scattering. The energy resolution of the method has been improved to the point that it is now in the range of 0.2 to 0.5 meV. One can begin to think about measurements of surface phonon line-shapes in the not too distant future.

During the same period a second experimental technique has been introduced for the determination of the dispersion curves of surface phonons on the clean surfaces of metals and on adsorbate covered surfaces, namely electron energy loss spectroscopy [4.20]. In this method a beam of electrons of energy E_i and wave vector $\mathbf{k}$ impinges on a crystal surface, excites or de-excites a phonon of frequency ω and wave vector $\mathbf{q}_{\parallel}$, and emerges with energy E_s and wave vector $\mathbf{k}_s$. If we take into account that $\hbar\omega$ is always negligible compared to E_i and E_s, the surface phonon dispersion curve $\omega=\omega(\mathbf{q}_{\parallel})$ is obtained from the pair of equations that express the conservation of energy and the components of the wave vectors parallel to the surface [4.20]:

$$\left| E_i - E_s \right| = \hbar\omega \tag{4.1}$$

$$k(\sin\theta_i - \sin\theta_s) = q_{\parallel}, \tag{4.2}$$

where $k = |\mathbf{k}_i| = |\mathbf{k}_s|$, and θ_i and θ_s are the polar angles of $\mathbf{k}_i$ and $\mathbf{k}_s$, respectively, measured from the normal to the surface. At each value of θ_i for a fixed θ_s (i.e. for each value of $q_{\parallel}$) the intensity of the outgoing beam is measured as a function of $E_i - E_s$, which yields the loss spectrum. A peak occurs in this spectrum when the conditions for exciting or de-exciting a surface phonon are satisfied. The energy resolution of this method is now about 7 meV for the large angle scattering required for mapping out surface phonon dispersion curves. Surface phonon dispersion curves have been obtained by this method for Ni{001} [4.21,22], TaC{001} [4.23], NbC{001} [4.24], Ni{001}-p(2x2)O [4.25], and Ni{001}-c(2x2)O [4.26]. This method is not limited by the energy of the surface phonon and is thus well suited to the study of high frequency phonons associated with adsorbed layers of atoms.

The existence of two experimental methods for measuring surface phonon dispersion curves provides us with a way of testing models used in calculating dynamical properties of surfaces, and of estimating the magnitude of surface relaxation if the variation of atomic force constants with interatomic separation is known from an independent calculation. In the case of adsorbate covered surfaces, these methods can provide information about the symmetry of the adsorption site, and together with the results of quantum chemistry calculations can give information about the adsorbate-substrate distance.

4.1.2 New Computational Methods

Several new approaches to the calculation of the dynamical properties of clean and adsorbate covered surfaces have been developed, and other methods that had been developed earlier were significantly improved since the start of the present decade. All of them are based on phenomenological models of interatomic forces, and most of them have been directed at the calculation of quantities that are important for the interpretation of experimental data bearing on the determination of surface phonon dispersion curves. In describing them, it is convenient to divide them into methods that are applicable to clean surfaces and those that are applicable to adsorbate covered surfaces.

A) Clean Surfaces

a) *Spectral Densities*

The quantities important for the interpretation of experimental data on surface phonon dispersion curves are the spectral densities defined by

$$\rho_{\alpha\beta}(\kappa\kappa'|\mathbf{k}_{\parallel}\omega) = \sum_{j} e_{\alpha}(\kappa|\mathbf{k}_{\parallel}j)e_{\beta}^{*}(\kappa'|\mathbf{k}_{\parallel}j)\delta(\omega-\omega_{j}(\mathbf{k}_{\parallel})) \tag{4.3}$$

that appear in the expressions for the contribution from one-phonon processes to the cross section for atom-surface scattering and to the electron energy loss spectrum. Knowledge of these functions helps to identify such features in the experimental spectra as surface phonons or resonances.

In this expression $\mathbf{k}_{\parallel}$ is a two-dimensional wave vector in the plane of the crystal surface, and is confined to the surface first Brillouin zone; the equilibrium positions of the atoms in the semi-infinite crystal are given by the vectors $\mathbf{x}(\ell\kappa) = \mathbf{x}_{\parallel}(\ell) + \mathbf{x}(\kappa)$, where $\mathbf{x}_{\parallel}(\ell) = \ell_{1}\mathbf{a}_{1} + \ell_{2}\mathbf{a}_{2}$ is a translational vector of the two-dimensional Bravais lattice that defines the periodicity of the crystal in directions parallel to its surface, and $\mathbf{x}(\kappa)$ ($\kappa=1,2...$) is a vector in the basis associated with each lattice site of the Bravais lattice that generates the semi-infinite crystal; $\mathbf{a}_1$ and $\mathbf{a}_2$ are the two noncollinear primitive translation vectors of the Bravais lattice, and ℓ_1 and ℓ_2 are any two integers that we denote collectively by ℓ; $\omega_j(\mathbf{k}_\|)$ is the frequency of the jth normal mode for the wave vector $\mathbf{k}_\|$, and $e_\alpha(\kappa|\mathbf{k}_\| j)$ is the corresponding unit eigenvector. The latter are obtained from the equation

$$\omega_j{}^2(\mathbf{k}_\|)e_\alpha(\kappa|\mathbf{k}_\| j) = \sum_{\kappa'\beta} D_{\alpha\beta}(\kappa\kappa'|\mathbf{k}_\|)e_\beta(\kappa'|\mathbf{k}_\| j), \tag{4.4}$$

where $D_{\alpha\beta}(\kappa\kappa'|\mathbf{k}_\|)$ is the partially Fourier transformed dynamical matrix of the crystal:

$$D_{\alpha\beta}(\kappa\kappa'|\mathbf{k}_\|) = \frac{1}{(M_\kappa M_{\kappa'})^{1/2}} \sum_{\ell'} \Phi_{\alpha\beta}(\ell\kappa;\ell'\kappa')\exp\{-i\mathbf{k}\cdot[\mathbf{x}_\|(\ell) - \mathbf{x}_\|(\ell')]\} .$$

$$\tag{4.5}$$

In (4.5) the $\{\Phi_{\alpha\beta}(\ell\kappa;\ell'\kappa')\}$ are the elements of the atomic force constant matrix, and M_κ is the mass of the κth kind of atom. With a proper choice of the $\{\mathbf{x}_\|(\ell)\}$ and the $\{\mathbf{x}(\kappa)\}$, the $\{\Phi_{\alpha\beta}(\ell\kappa;\ell'\kappa')\}$ can describe the effects of the relaxation of the interplanar spacings near the surface, of surface reconstruction, and of the presence of layers of adsorbed atoms. The spectral density $\rho_{\alpha\beta}(\kappa\kappa'|\mathbf{k}_\|\omega)$ has a peak at each frequency for which $e_\alpha(\kappa|\mathbf{k}_\| j)$ and $e_\beta(\kappa'|\mathbf{k}_\| j)$ are nonzero.

The function $\rho_{\alpha\beta}(\kappa\kappa'|\mathbf{k}_\|\omega)$ can be calculated conveniently from the expression

$$\rho_{\alpha\beta}(\kappa\kappa'|\mathbf{k}_\|\omega) = \frac{\omega}{\pi i} [U_{\alpha\beta}(\kappa\kappa'|\mathbf{k}_\|\omega-i0) - U_{\alpha\beta}(\kappa\kappa'|\mathbf{k}_\|\omega+i0)] , \tag{4.6}$$

where $U_{\alpha\beta}(\kappa\kappa'|\mathbf{k}_\| z)$ is the Green's function for the semi-infinite crystal,

$$U_{\alpha\beta}(\kappa\kappa'|\mathbf{k}_\| z) = \sum_j \frac{e_\alpha(\kappa|\mathbf{k}_\| j)e_\beta{}^*(\kappa'|\mathbf{k}_\| j)}{z^2 - \omega_j{}^2(\mathbf{k}_\|)}, \tag{4.7}$$

and satisfies the time-independent equation of motion

$$z^2 U_{\alpha\beta}(\kappa\kappa'|\mathbf{k}_\| z) - \sum_{\kappa''\gamma} D_{\alpha\gamma}(\kappa\kappa''|\mathbf{k}_\|)U_{\gamma\beta}(\kappa''\kappa'|\mathbf{k}_\| z) = \delta_{\alpha\beta}\delta_{\kappa\kappa'}. \tag{4.8}$$

A simple and commonly used method for calculating $U_{\alpha\beta}(\kappa\kappa'|\mathbf{k}_\| z)$, and hence $\rho_{\alpha\beta}(\kappa\kappa'|\mathbf{k}_\|\omega)$, is to replace the semi-infinite crystal by a crystal slab consisting of a finite number of atomic layers, typically from 10 to 70, calculate the $\{e_\alpha(\kappa|k_{\parallel}j)\}$ and $\{\omega_j(k_{\parallel})\}$ for values of $k_{\parallel}$ inside the surface first Brillouin zone, and then carry out the sum on j in (4.7).

Although the slab method is perhaps the simplest available at the present time for the calculation of the frequencies of surface phonons, and a method that is still often used for the calculation of local densities, such as $\rho_{\alpha\beta}(\kappa\kappa'|k_{\parallel}\omega)$, it has come in for some criticisms in recent years. These are of two major types. The first is that in the limit as $k_{\parallel} \to 0$ the decay length of the surface phonons into the crystal from each surface, which is of the order $2\pi/|k_{\parallel}|$, becomes comparable with the thickness of the slab. The surface phonons localized to one surface therefore "feel" the surface phonons localized to the other surface, and their frequencies are shifted from the values they would have on the surface of a semi-infinite crystal. Accuracy in obtaining surface phonon frequencies is therefore compromised in the long wavelength limit by the use of the slab method. More serious is the following feature of the slab method. If one works with a slab of, say, twenty atomic layers, one obtains sixty normal mode frequencies for each value of $k_{\parallel}$ in the simplest structures. If a histogram is then used to construct a local density of states corresponding to a given value of $k_{\parallel}$, there can be too few frequencies to reproduce fine structure in the density of states such as peaks associated with the presence of surface resonance modes. Thicker slabs, and consequently larger dynamical matrices, may be required to obtain enough modes for each value of $k_{\parallel}$ for such spectral densities to be computed accurately enough to reveal such features.

However, it would be a mistake, I think, to write off the slab method for the calculation of $\rho_{\alpha\beta}(\kappa\kappa'|k_{\parallel}\omega)$. Its simplicity makes it attractive, and recent advances in the design of computers makes working with very large dynamical matrices quite feasible. In fact, calculations employing crystal slabs of up to 105 layers have been reported [4.27].

In parallel with, and to some extent stimulated by, the awareness of the difficulties that can arise in the use of the slab method for the calculation of local spectral densities, there has been an increase of interest in Green's function approaches to such calculations. Three approaches of this type have been developed or have been widely used during the past four years. In the first, the surface is treated as an extended defect in an infinite crystal [4.28,29]. The interactions among atoms on opposite sides of a fictitious plane that passes through the crystal but contains no atoms itself are set equal to zero by subtracting the corresponding terms from the equations of motion of the crystal. The dynamical Green's function for the cut crystal, $U_{\alpha\beta}(\kappa\kappa'|k_{\parallel}z)$, is then solved for in terms of the Green's function for the infinite crystal $G_{\alpha\beta}(\kappa\kappa'|k_{\parallel}z)$,

$$U_{\alpha\beta}(\kappa\kappa'|\mathbf{k}_{\parallel}z) = G_{\alpha\beta}(\kappa\kappa'|\mathbf{k}_{\parallel}z) +$$

$$\sum_{\kappa''\mu}\sum_{\kappa\nu} G_{\alpha\beta}(\kappa\kappa'|\mathbf{k}_{\parallel}z)\Delta D_{\mu\nu}(\kappa''\kappa|\mathbf{k}_{\parallel})U_{\nu\beta}(\kappa\kappa'|\mathbf{k}_{\parallel}z) ,$$

$$(4.9)$$

where $G_{\alpha\beta}(\kappa\kappa'|\mathbf{k}_{\parallel}z)$ is the solution of

$$z^2 G_{\alpha\beta}(\kappa\kappa'|\mathbf{k}_{\parallel}z) - \sum_{\kappa''\gamma} D_{\alpha\gamma}^{\;0}(\kappa\kappa''|\mathbf{k}_{\parallel})G_{\gamma\beta}(\kappa''\kappa'|\mathbf{k}_{\parallel}z) = \delta_{\alpha\beta}\delta_{\kappa\kappa'} ,$$

$$(4.10)$$

and $\overset{\leftrightarrow}{D}_{\alpha\beta}^{\;0}(\mathbf{k}_{\parallel})$ and $\overset{\leftrightarrow}{\Delta D}_{\alpha\beta}(\mathbf{k}_{\parallel})$ are defined in terms of the force constants of the infinite crystal $\{\Phi_{\alpha\beta}^{\;0}(\ell\kappa;\ell'\kappa')\}$, and the addition to them $\{\Delta\Phi_{\alpha\beta}(\ell\kappa;\ell'\kappa')\}$ associated with the cutting of the bonds crossing the fictitious plane,

$$D_{\alpha\beta}^{\;0}(\kappa\kappa'|\mathbf{k}_{\parallel}) = \frac{1}{(M_{\kappa}M_{\kappa'})^{1/2}}\sum_{\ell'} \Phi_{\alpha\beta}^{\;0}(\ell\kappa;\ell'\kappa')\exp\{-i\mathbf{k}_{\parallel}\cdot[\mathbf{x}_{\parallel}(\ell) - \mathbf{x}_{\parallel}(\ell')]\}$$

$$(4.11a)$$

$$\Delta D_{\alpha\beta}(\kappa\kappa'|\mathbf{k}_{\parallel}) = \frac{1}{(M_{\kappa}M_{\kappa'})^{1/2}}\sum_{\ell'} \Delta\Phi_{\alpha\beta}(\ell\kappa;\ell'\kappa')\exp\{-i\mathbf{k}_{\parallel}\cdot[\mathbf{x}_{\parallel}(\ell)-\mathbf{x}_{\parallel}(\ell')]\}.$$

$$(4.11b)$$

The term $\Delta D_{\alpha\beta}(\kappa\kappa'|\mathbf{k}_{\parallel})$ is nonzero only for κ and κ' within one or two atomic layers of the fictitious plane, on both sides of it, so that the use of matrix partitioning methods considerably simplifies the solution of (4.9). Since the creation of a pair of free surfaces in this way in a crystal in which the atomic displacements satisfy cyclic boundary conditions is equivalent to creating a crystal slab, this approach is in fact a sophisticated slab method. Its advantage is that the Green's function of the infinite crystal, $G_{\alpha\beta}(\kappa\kappa'|\mathbf{k}_{\parallel}z)$, can be calculated with great accuracy to yield Green's functions of the cut crystal with an accuracy that could be achieved in a slab calculation only with slabs of several hundred or even thousand layers.

A second method for obtaining $U_{\alpha\beta}(\kappa\kappa'|\mathbf{k}_{\parallel}z)$ proceeds by solving (4.8) directly, by regarding it as a one-dimensional, finite difference equation in the κ's [4.30]. The solution is carried out by a combination of analytic and numerical methods, and yields spectral densities with a great deal of fine structure revealed (Fig.4.1).

Finally, the surface Green's function matching method of *Garcia-Moliner* [4.31], which was originally constructed for the study of the surface properties of media in the continuum approximation, has been applied to the determination of the dynamical Green's function of a semi-infinite crystal [4.32-35]. Results in agreement with experiment were obtained for ionic crystal surfaces [4.34] and for metal surfaces

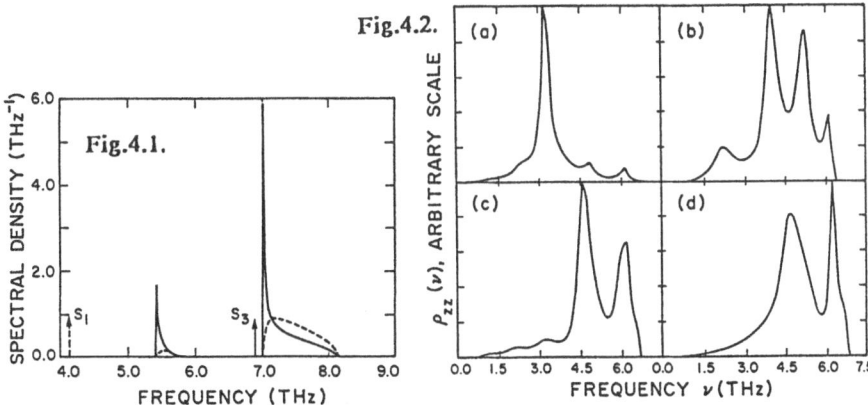

Fig.4.1. Spectral densities $\rho_{xx}(\kappa\kappa|\mathbf{k}_\parallel|\omega)$ (solid curves) and $\rho_{zz}(\kappa\kappa|\mathbf{k}_\parallel|\omega)$ (dotted curves) for atoms in the surface layer of a nearest neighbor central force model of an fcc crystal bounded by a {111} surface. The wave vector $\mathbf{k}_\parallel$ is at the $\overline{K}$ point of the surface Brillouin zone. The arrows indicate the frequencies of surface phonons that contribute delta function peaks to the spectral densities [4.30]

Fig.4.2. The spectral density $\rho_{zz}(\nu)$ versus ν for atoms in (a) the surface layer; (b) one layer into the crystal; (c) two layers into the crystal; and (d) three layers into the crystal, for a third neighbor interaction model of tungsten bounded by a {001} surface [4.37]

[4.35]. Its equivalence to the first of the Green's function methods described above has been shown [4.33].

At the beginning of the present decade, the first calculations of local densities of phonon states, $\rho_{\alpha\beta}(\kappa\kappa'|\mathbf{k}_\parallel|\omega)$, by another Green's function method, namely the continued fraction-recursion method, were barely a year old [4.36]. Although it has been used in calculations of surface phonons on clean and adsorbate covered surfaces from time to time (Fig.4.2) [4.37-39], it does not seem to have entered the armamentarium of surface phonon theory as a conventional weapon. Some reasons for this may be the slow convergence of the method when it is applied to adsorbate covered surfaces [4.39], ambiguities in the termination of the continued fraction [4.39] (Fig.4.3), and the existence of alternative methods that yield accurate results even for adsorbate covered layers. This method works well when there are vibrational modes that lie above the bulk phonon bands - high frequency surface optical phonons induced by an adsorbate layer. However, when there are sharp resonances at frequencies below the maximum bulk phonon frequency, the convergence of the continued fraction representation for Green's functions from which the spectral densities are obtained decreases significantly, and accurate results are hard to obtain by its use.

An interesting demonstration of the predictive value of phenomenological force constant models is provided by the experimental search

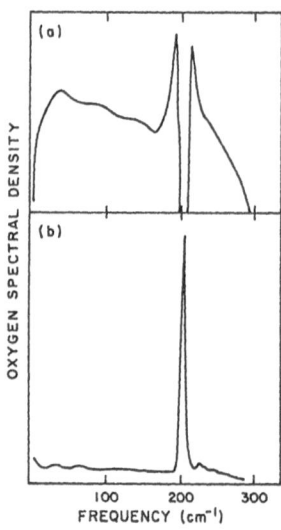

Fig.4.3. Spectral densities for parallel motion of the oxygen at the $\overline{\Gamma}$ point of the surface Brillouin zone for a c(2x2) adlayer of oxygen on Ni{001}, calculated on the basis of two different terminations of the continued fraction representation of these spectral densities. In (a) it is assumed that the coefficients in the continued fraction oscillate to infinite order; in (b) it is assumed that they converge to unique limiting values [4.39]

for the surface mode S_6 along the $\overline{\Gamma X}$ direction on the Ni{001} surface. This is a mode that was expected to scatter electrons. The intensity of this scattering had been estimated to be only 1/25 that of the scattering from the mode S_4, which had been observed by *Lehwald* et al. [4.40], but the mode S_6 was not seen in the same experiment. Recently, a rigid ion, multiple scattering calculation based on slab eigenvectors for the modes S_4 and S_6 was used to predict the actual intensities of electrons scattered from these modes [4.41]. It was found that the ratio of the S_4 to S_6 scattering intensities is very sensitive to the energy of the incident electrons at the experimental scattering angle. There are three ranges of electron energies in the experimentally accessible range at which the intensity of scattering from the mode S_6 is comparable to that from S_4; there are other electron energies at which this intensity is substantially lower. The original experiments were done at one of the latter energies. When the experiment was repeated at an appropriate energy, the mode S_6 was found [4.41]. When a contraction of 2.6% in the spacing between the outermost two atomic layers was included in the calculations, the theoretical and experimental scattering intensities were brought into very good agreement without changing the mode frequencies significantly.

b) *Surface Phonons and Resonances*

If one is interested in the dispersion curves of surface phonons alone, and not in the continuum of bulk phonons perturbed by the presence of the surface, the slab and several Green's function methods generally give more information than one needs. Many years ago *Gazis* et al. [4.42] developed a method for obtaining the dispersion curves of surface phonons on crystal surfaces by using the discrete analogue of

158

the method used by Lord Rayleigh in the continuum limit to obtain the dispersion curve of the surface acoustic waves that today bear his name. They wrote the equations of motion of the atoms in a semi-infinite crystal, including the atoms in the surface layers, in the form they would have if they were part of an infinite crystal. They assumed solutions of these equations that decayed exponentially with increasing distance into the interior of the crystal from the surface and obtained the decay constants in these solutions as functions of $k_{\parallel}$ and ω. These solutions were then substituted into the equations that arise because the equations of motion of the atoms in the surface layers are different from those of the atoms in the bulk, and obtained the relation between ω and k thereby. Recently, the analogous problem for surface spin waves [4.43] and edge magnons [4.44] was solved for crystals by the use of the expansion of the spin operators in terms of Gottlieb functions, $\{\phi_s(m)\}$ (s=0,1,2...). These are a complete set of polynomials multiplied by the square root of an exponentially decreasing weight function that are defined at the integer points m=0,1,2... and are orthonormal with respect to summation over this range. The attractiveness of expansions in these functions in surface problems is that the exponentially decaying weight function mimics the exponential decay of the displacement field in a surface mode, and the boundary conditions are incorporated into the equations of motion and do not have to be treated separately. This method has now been applied to the study of surface phonons by *Latkowski* and *Black* [4.45]. Its ease of use and the fact that it focuses on surface modes alone should gain this method wider acceptance in the future. Yet another approach to such calculations, namely the "matching method" introduced by *Feuchtwang* [4.46] in 1967 for the calculation of surface phonon dispersion curves, has been revived in recent work by *Szeftel* and his colleagues [4.47,48] and used in the study of surface resonance modes as well.

c) *Anharmonic Effects*

The methods just described for the calculation of dynamic correlation functions of atoms in crystal suffaces are all based on the harmonic approximation. For the analysis of either helium atom scattering experiments or electron energy loss experiments carried out at elevated temperatures, where anharmonic effects should be large, it is important to be able to calculate such correlation functions with the effects of anharmonicity included. Recently, such calculations have been carried out by the methods of molecular dynamics applied to a slab of a finite number of layers of atoms in an fcc crystal bounded by a pair of {111} surfaces in which the atoms interact through a Lennard-Jones pair potential [4.49]. Peaks in the corresponding spectral densities associated with surface phonons are visible (Fig.4.4), together with their shift in position and broadening with increasing temperature.

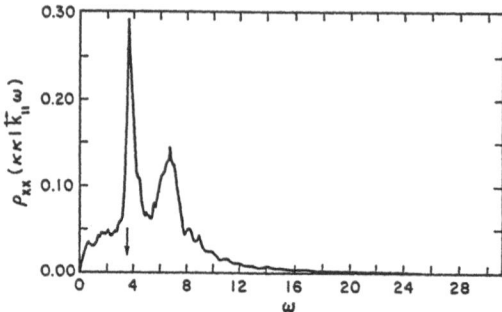

Fig.4.4. A plot of the spectral density $\rho_{xx}(\kappa\kappa|k_\parallel\omega)$ versus ω for atoms in the top layer of a 21-layer slab of an fcc Lennard Jones crystal bounded by {001} surfaces, calculated by molecular dynamics. The wave vector $k_\parallel$ is one-fifth the distance to the boundary of the surface Brillouin zone in the $\langle 100 \rangle$ direction; the temperature is about one-half the melting temperature. The frequency of the Rayleigh wave peak is denoted by an arrow [4.49]

d) *First Principles Calculations of Surface Phonon Dispersion Curves*

It should be clear from the preceding discussion that the vast majority of theoretical calculations of surface phonon properties carried out to date have been based on force constant models of the substrate. In these models the atomic force constants in the interior of the crystal are generally obtained from bulk phonon dispersion curves measured by neutron spectroscopy. The atomic force constants coupling atoms in the surface layers of the crystal generally have to be modified from their values in the bulk in order to fit experimental dispersion curves for surface phonons. The use of such models has been very helpful in interpreting experimental data concerning the structure and dynamics of clean and adsorbate covered metal surfaces. However, there are at least two drawbacks to the use of such models. The first is that by their nature they are not predictive: the parameters of a model must be obtained from experimental data - sometimes the same data that are being analyzed by the use of the model. The second is that such models may not be unique: more than one model can sometimes give an equally good fit to experimental data [4.50-52].

Consequently, one of the most significant developments in the theory of surface phonons during the past seven years has been the appearance of first principles calculations of the dispersion curves of surface phonons on low index metal surfaces. The first such calculations, for Na{001} and subsequently for K{001}, were carried out by *Beatrice* and *Calandra* [4.53-55], who expanded the electronic ground state energy of 15-layer slabs of Na{001} to second order in the pseudopotential describing the electron-ion interaction [4.56,57], and then re-expanded the result in powers of the displacements of the ion cores from their rest positions to obtain the electronic contribution to the atomic force constants. Combined with the ionic contribution, the latter permit the dynamical matrix of the crystal slab, and hence the surface mode frequencies, to be obtained. The unperturbed electronic subsystem in this work consisted of noninteracting electrons confined to a finite region in the direction normal to the surfaces of the slab by infinite potential barriers. The effects of a multilayer relaxation of the spacing between the atomic layers [4.58] on the normal mode frequencies of the slab were also studied in this work. Subsequently, *Eguiluz* et al. have used a similar approach to calculate the frequencies of surface phonons on the {001} surface of Na [4.59] and on the {110} surface of Al [4.60]. The chief differences between their work and that of *Beatrice* and *Calandra* are that the wave functions of the unperturbed electronic subsystem were calculated selfconsistently on the basis of the *Kohn-Sham* equations [4.61] of the local density approximation of density functional theory, and the effects of exchange and correlation were built into the calculation of the density response function that appears in the electronic contribution to the dynamical matrix. The results of the selfconsistent calculation of the surface phonon frequencies differ in some significant ways from those obtained on the basis of the infinite barrier model, particularly in the case of Al. In the calculations resported in [4.59] surface relaxation was not taken into account. However, in their subsequent calculations [4.60], *Eguiluz* et al. took surface relaxation into account in a first principles fashion, and found that it can have a significant effect on the frequencies of surface phonons, particularly on those associated with the Al{110} surface, for which the surface relaxation is large. A typical result of such calculations, for a 23-layer slab of Al{110}, is presented in Fig.4.5. Surface phonons are clearly seen below the continuum of bulk modes, and in the gaps in the continuum of bulk modes.

A somewhat different approach to the first principles calculation of the dispersion curves of surface phonons on metal surfaces was taken by *Ho* and *Bohnen* [4.62], who also studied Al{110}. In their calculations, a few of the interplanar force constants at the surface of Al{110} were obtained from the distortion energy produced when the atoms in the top layer of a slab were displaced in patterns correspond-

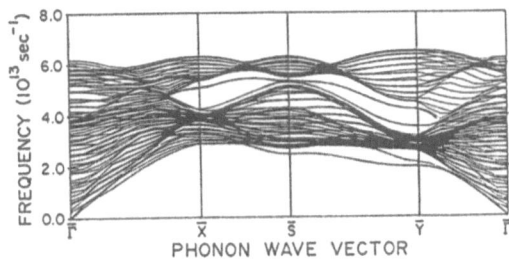

Fig.4.5. Dispersion curves of phonons in a 23-layer slab of Al{001} [4.60]

ing to wave vectors at the $\overline{X}$ and $\overline{Y}$ points on the boundary of the surface Brillouin zone. The distortion energy was calculated by a pseudopotential-based total energy calculation, and was used to obtain the values of the phonon frequencies at the $\overline{X}$ and $\overline{Y}$ points. A force constant model was then fitted to these frequencies in order to obtain dispersion curves along symmetry lines in the surface Brillouin zone.

Such calculations are still in their infancy. The calculations cited indicate their feasibility at the same time that they reveal their difficulty. We stand at the threshold of an era in which parameter-free determinations of surface dynamical properties from first principles will become as commonplace as the determination of bulk electronic structures of solids are today, but we still have a long way to go before that goal is achieved.

B) Adsorbate Covered Surfaces

The properties of systems of one or more layers of an atomic or molecular species deposited on a crystalline substrate of a different species are of interest to physicists and chemists alike. Such systems provide examples of two-dimensional systems whose properties, particularly their phase transitions, are of great current interest to theorist and experimentalist alike. At the same time, many chemical processes of fundamental and technological interest take place at surfaces coated with layers of adsorbed or physisorbed atoms, for instance the first stages of oxidation. The study of the vibrational properties of adatoms on different, ideal faces of crystals can assist in understanding other properties of such systems.

Within the past four years, several theoretical methods have been developed for the study of the vibrations of adsorbate covered surfaces. Three types of systems have been studied. The first is a single adparticle on an otherwise ideal surface. The second is a layer of adparticles that forms a lattice that is in perfect registry with the substrate. The third is an adlayer or several adlayers forming a lattice that is incommensurate with the substrate. In the first and third cases, the combined system of adparticle(s) and substrate no longer has any periodicity in directions parallel to the interface.

One can subdivide the vibrational modes of such systems into two types. Modes of the first type have frequencies above the maximum frequency of the bulk modes of the substrate. From such modes something can be learned about the nature of the adparticles, their binding to the surfaces, and their positions relative to the substrate. Modes of the second type have frequencies at and below the maximum frequency of the bulk modes of the substrate. From such modes one can not only learn something about the adparticles, but also about the geometry and interatomic forces in the first few layers of the substrate.

a) A Single Adparticle

It is an interesting point that the sociological differences between chemists and physicists, reflected in the interest the former display in molecules and the interest the latter display in crystals, also manifest themselves in their approaches to the theoretical study of the dynamical properties of crystal surfaces, particularly adsorbate covered surfaces.

One of the methods used for the calculation of the vibrational frequencies of a single atom adsorbed on a metal substrate replaces the infinite substrate by a finite cluster of atoms, and the system of adatom and substrate cluster is treated as a large molecule whose normal mode frequencies are calculated by standard methods [4.63]. Calculations of this type for H, O and S adatoms on Ni{111} and Ni{001} surfaces [4.64] gave well-converged results for the modes whose frequencies lie above the maximum frequency of the substrate, with clusters of 57 nickel atoms. The adatom-substrate and the substrate-substrate interactions were obtained from quantum-chemical calculations [4.65].

An alternative to such finite cluster calculations for an isolated adatom is the use of the real-space continued fraction/recursion method [4.38,66]. This method yields the spectral density $\rho_{\alpha\alpha}(\ell\kappa;\ell\kappa|\omega)$ for the motion of atom $(\ell\kappa)$ in the direction α. Localized modes show up as delta functions above the maximum frequency of the substrate in this spectral density; resonance modes show up as broader peaks below the maximum frequency of the substrate. Mean square displacements $\langle u_{\alpha}^2(\ell\kappa)\rangle$ of the adatom and of the atoms of the substrate can also be calculated from $\rho_{\alpha\alpha}(\ell\kappa;\ell\kappa|\omega)$.

A method that could be used to advantage for such calculations, but which has not up to the present time [4.67], is the one proposed many years ago by *Wagner* [4.68] for the study of the vibrations of crystals containing defects with internal degrees of freedom. It could be used in the case that molecules, not just atoms, are adsorbed on a crystal surface. Its application in the present context requires knowledge of the dynamical Green's function for a semi-infinite crystal or for a slab, but these are being calculated almost routinely at the present time. Both resonance and localized modes would be accessible by this method.

b) A Periodic Array of Adatoms

Several methods have been developed for calculating the vibrational modes of a periodic array of atoms on a substrate. The most accurate of these at the present time is the Green's function method developed by *Mills* and his colleagues [4.69,70] and used in their studies of the lattice dynamics of ordered oxygen overlayers on Ni{100} and Ni{111}. In this method, the dynamical Green's function $U_{\alpha\beta}(\kappa\kappa'|k_{\parallel}z)$ is obtained for a semi-infinite crystal with an ordered overlayer of adatoms on its surface from the solution of the equations of motion (4.8) it satisfies. The discontinuity of the Green's function across the branch cut it possesses along the real frequency axis yields a spectral density that has peaks at the frequencies of the vibrational modes for a given value of the two-dimensional wave vector characterizing the vibrations of the semi-infinite system. High frequency, adsorbate-induced surface optical modes and resonances can be detected clearly by this method.

The dispersion curves of the vibration modes of the {0001} surface of graphite containing a monolayer of krypton and xenon in registry with the substrate have been calculated by *de Wette* and his colleagues [4.71,72]. Finely divided graphites - graphitized carbon blacks and exfoliated graphites - have been widely used for a long time as substrate materials for adsorption studies because the exposed surfaces in these materials are predominantly the well-characterized {0001} basal plane surfaces. There is increasing experimental interest in the study of a variety of physical properties of gases adsorbed on graphite surfaces. The work of *de Wette* et al. is the first in which the dynamics of the graphite substrate have been taken into account. The results have been used in calculations of mean square displacements of the adsorbed atoms and of the thermodynamic functions of the adsorbate/substrate system.

More recently, *Marchese* et al. [4.73] have used molecular dynamics simulations to study the effect of temperature on the dispersion curves of surface phonons in a registered xenon monolayer physisorbed on graphite {0001}. At low temperatures, the results are in good agreement with the results of lattice dynamical calculations [4.71,72,74]. The phonon frequencies are shifted to lower values with increasing temperature, and their lifetimes decrease.

Simpler approximate methods of obtaining the vibrational frequencies of periodic arrays of adatoms on metal surfaces have been developed by *Black* [4.75] and by *Lloyd* and *Hemminger* [4.76,77]. In the former work the vibration frequencies were obtained by finding the normal modes of a layer of adparticles and one or two layers of the substrate. This is a kind of slab calculation for the adsorbate/substrate system. The method was applied to the case of oxygen p(2x2) and c(2x2) overlayers on Ni{001}, and to a c(2x2) overlayer of CO in the "on top" location on Ni{001}. Good agreement between the results of these calculations and of exact Green's function calculations [4.69] is found. *Lloyd* and *Hemminger* [4.76,77] have studied the modes of periodic overlayers using a finite cluster of atoms. The cluster is bounded by square boundaries parallel to the surface and, as in the work of

Black [4.75], a finite number of substrate atoms is used. Periodic boundary conditions are introduced by coupling the atoms on one side of the cluster to atoms on the opposite side of the cluster. These periodic boundary conditions allow the dispersion of the various vibrational modes of interest to be studied. In application to a p(2x2) oxygen overlayer on Ni{001} this approach led to improvements in the earlier models of *Rahman* et al. for this system [4.69], and showed that the experimental data can be explained (Fig.4.6) without invoking lateral interactions between neighboring adsorbate atoms, as is required in the case of the c(2x2) oxygen overlayer on Ni{001}.

This cluster approach has subsequently been applied by *Banse* et al. [4.78] to the study of the effects of vacancy defects in c(2x2) oxygen overlayers on Ni{001} on the vibrational spectra measured by the inelastic scattering of low energy electrons.

It should be emphasized that the use of such cluster methods can yield accurate values of only the frequencies of the high frequency surface modes associated with an adlayer, i.e. the modes whose frequencies lie above the continuum of bulk vibrations of the substrate. This is because these modes are so highly spatially localized that the substrate plays only a small role in the determination of their frequencies. If it is desired to obtain accurate results for the vibrational spectral densities in the frequency range allowed the bulk modes of the substrate, methods such as the slab method or Green's function method must be used.

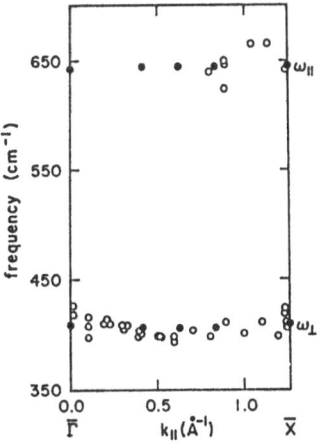

Fig.4.6. Experimental dispersion curves for p(2x2) oxygen on Ni(001) (open circles) are compared to theoretical results obtained for a finite cluster of atoms. Dispersion is along the $\langle 110 \rangle$ direction, and $\bar{\Gamma}$ and $\bar{X}$ are defined with respect to the substrate surface Brillouin zone [4.76]

4.2 Surface Phonons on Complex Crystals and Structures

Not all theoretical and experimental studies of surface phonons have been devoted to these excitations on low index faces of simple metals, clean or adsorbate covered, or the surfaces of ionic crystals. In this section we consider surface phonons on much more complex crystals and crystal structures.

4.2.1 Surface Phonon Anomalies Caused by the Electron–Phonon Interaction

Measurements of surface phonon dispersion curves, for instance by helium atom scattering, have begun to reveal aspects of surface lattice dynamics that are unexpected on the basis of the underlying bulk lattice dynamics [4.79]. Microscopic mechanisms that are forbidden, or hindered, in the bulk for symmetry reasons may be allowed or enhanced at the surface by the lowering of the symmetry that is associated with a surface. An example of this is provided by the surface phonon anomalies induced by the electron–phonon interaction. These have been divided into Type I anomalies, which appear due to a surface enhanced electron–phonon interaction and have no counterpart in the bulk dispersion curves; and Type II anomalies, which are deeply modified at the surface in systems that already show anomalies in the bulk phonon dispersion curves [4.79].

An example of a Type I anomaly may be provided by the strongly localized surface optical phonon observed by helium atom scattering from the {0001} surface of the layered compound GaSe [4.3]. The frequency of this mode lies well below the corresponding bulk optical phonon band. The large softening of a surface branch that this represents is suggestive of a change in the electronic structure of the compound occurring at the surface.

The surface phonon anomalies observed on the {111} surface of Cu, Ag, and Au [4.80] consist of a strong softening of the quasi-longitudinal acoustic resonance mode. A theoretical study [4.81] of the Ag{111} anomaly shows that this is due to a softening of in-plane surface force constants. The microscopic origin of this softening is thought to lie in a phonon induced s–d hybridization of the electronic states that is connected with the large quadrupolar deformability of silver ions [4.82-84]. However, an alternative explanation, that seems to explain the data better for all three metals, attributes the softening of the quasi-longitudinal resonance to a weakening of s–d hybridization at the surface in these metals [4.85].

In the case of the Au{111} surface an additional anomaly is observed in the lowest frequency surface phonon propagating in the ⟨110⟩ direction. This consists of a discontinuity in its dispersion curve at about half the distance to the zone boundary. It has been attributed to a strong hybridization between a pseudo-surface mode that can exist in the ⟨110⟩ direction on the {111} surface of an fcc crystal, and the quasi-longitudinal resonance mode and Rayleigh wave [4.86].

Type II anomalies have been observed in theoretical surface phonon dispersion curves calculated for TiN [4.87] on the basis of a cluster deformation model of the lattice dynamics of transition metal compounds [4.88] and a Green's function approach [4.29]. The Rayleigh wave dispersion curve shows a weak anomaly at about one-half the distance to the surface Brillouin zone boundary in the ⟨100⟩ direction, while the quasi-longitudinal resonance branch shows a large anomaly at two-thirds the distance to the zone boundary along the same direction (Fig.4.7). The Rayleigh wave anomaly would not be expected in a mode that has peeled off from the bottom of the bulk transverse acoustic band, since no anomaly occurs in the bulk transverse acoustic branches along symmetry directions. It occurs because of the elliptical polarization of the Rayleigh wave that couples the transverse acoustic to longitudinal acoustic components of the displacement field, which implies a relatively strong hybridization of the Rayleigh wave with the quasi-longitudinal resonance. The appearance of an anomaly at the zone boundary or at one-third or two-thirds of the distance to the zone boundary depends on whether the nearest, the second-nearest, or the third-nearest neighbor nitrogen-nitrogen quadrupolar interaction is dominant, respectively. Near the surface the second-nearest neighbor interaction can dominate the third-nearest neighbor interaction because of the smaller number of third-nearest neighbors relative to the number of second-nearest neighbors, so that the anomaly in the Rayleigh wave dispersion curve is shifted to smaller wave vectors. The quasi-longitudinal resonance is primarily of bulk

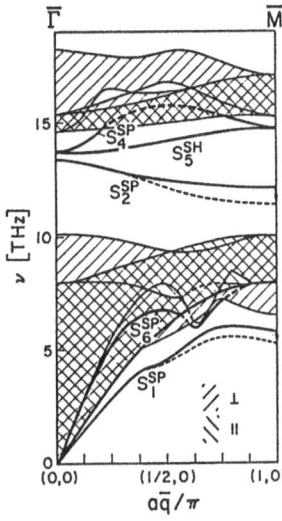

Fig.4.7. The surface phonon dispersion curves in the ⟨001⟩ direction on the TiN{001} surface [4.87]

character and keeps its bulk position. A competing model for the lattice dynamics of transition metal compounds, namely *Weber's* double-shell model [4.89], does not yield an anomaly in the Rayleigh wave dispersion curve on TaC [4.90], a compound very similar to TiN. Thus, when experimental surface phonon dispersion curves for TiN become available, they may be able to distinguish between competing lattice dynamical models that seem to reproduce bulk phonon dispersion curves equally well.

In an experimental study of surface phonons on the {001} surface of TaC by high resolution electron energy loss spectroscopy, *Oshima* et al. [4.23] have observed a surface optical mode in the gap between the acoustic and optical branches of the phonon density of states. This is called a microscopic surface optical mode because the amplitudes of vibration in it are localized within a few atomic layers of the surface in contrast with the more deeply penetrating surface optical modes obtained from macroscopic theory, that is from the solution of Laplace's equation. The existence of such modes was predicted in 1968 [4.91,92] but sixteen years were to pass before they were first observed (Fig.4.8).

Similar results have recently been obtained by electron energy loss spectroscopy for the {001} surface of NbC [4.24].

Theoretical calculations of surface phonons on the {001} surface of NaCl structure crystals in which the atoms interact with nearest neigh-

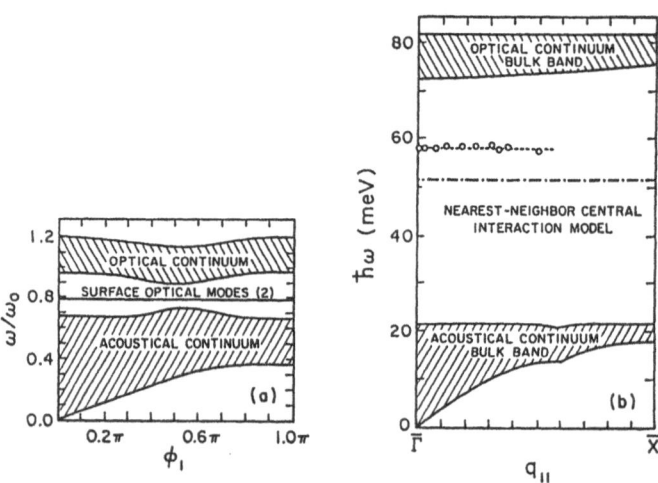

Fig.4.8. (a) Theoretical dispersion curves in the ⟨100⟩ direction for surface phonons on the {001} surface of an 18-layer slab of an NaCl-type crystal in which nearest and next-nearest neighbor atoms interact, through central forces [4.91]

(b) Experimental dispersion curve along the line from $\bar{\Gamma}$ to $\bar{X}$ on the TaC{001} surface. The open circles represent the data points. The dash-dotted line represents the frequency of a surface phonon calculated from the model of a linear diatomic chain with a free end [4.23]

bor forces only have been carried out on the basis of a slab of 15 atomic layers by *Oshima* [4.93]. Allowing for changes (stiffening) in the in-plane atomic force constants in the outermost plane of atoms at each face of the slab leads to the appearance of two new surface optical modes with frequencies above the top of the bulk optical band, in which the atomic motion is predominantly parallel to the surface, and a pair of new surface optical modes with frequencies just above the bulk acoustic band. Such modes have been observed experimentally in both TaC and NbC [4.23,24].

4.2.2 Surface Phonon *Kohn* Anomalies

Rayleigh wave dispersion curves along the ⟨110⟩ direction on the {111} surface of platinum, measured by helium atom scattering, reveal several anomalous kinks [4.13] (Fig.4.9). In addition, a longitudinal branch is found at higher frequencies that is similar to the quasi-longitudinal resonance modes observed in Cu, Ag, and Au. The anomalies in the Rayleigh wave dispersion curve occur at different wave vectors in the surface Brillouin zone from those at which the bulk phonons have anomalies. These surface phonon anomalies are interpreted as being due to a surface *Kohn* anomaly [4.94]. Such anomalies are expected to be strongly enhanced in going from three- to two-dimensional systems. They are weakened when the surface temperature is increased from 160 K to 400 K. This is expected, since the sharpness of a *Kohn* anomaly is related to the sharpness of the Fermi surface, which is smeared out at high temperatures. Similar experiments on Pt{111}

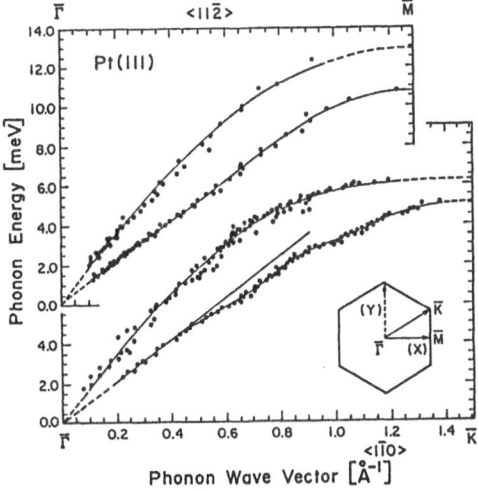

Fig.4.9. Measured surface phonon dispersion curves for Pt{111}. The upper two curves are for the $\overline{\Gamma}$-$\overline{M}$ (⟨112⟩) direction, and the bottom two curves are for the $\overline{\Gamma}$-$\overline{K}$ (⟨110⟩) direction. The curves show the best fit Fourier expansions. The solid line in the lowest set of data corresponds to a group velocity of 11.1 meV·Å [4.13]

surfaces covered with hydrogen fail to show distinct anomalies in the Rayleigh wave dispersion curve. This is consistent with the possibility that the anomaly on the clean surface is due to a surface electronic state which is strongly affected by the absorbate.

An explanation for the anomaly that does not invoke surface *Kohn* anomalies is suggested by the observation of a sharp discontinuity in the Rayleigh wave dispersion curve observed on Au{111} [4.85]. This anomaly has been attributed to the hybridization of a pseudo-Rayleigh wave with the Rayleigh wave and the longitudinal mode [4.86]. However, it is felt that this is not the explanation for the anomalies observed on Pt{111} since the longitudinal and Rayleigh modes show relative frequencies more characteristic of Cu and Ag, where such discontinuities are not observed.

4.2.3 Molecular Crystals

Recently, theoretical studies of surface phonons on molecular crystals have begun to be carried out [4.95, 96]. Such calculations are more difficult than the same calculations for ionic crystals or metals, because of the librational degrees of freedom possessed by the molecules constituting these crystals, in addition to the translational degrees of freedom, even if the intramolecular vibrations are neglected. The static structures and the phonon frequencies of the {001}, {110}, and {111} surfaces of solid α-N_2 were recently studied by the slab method at zero temperature [4.96]. It was found that the molecular center of mass positions, orientations, and bond lengths for molecules in the surface layers are significantly different from their bulk values. Several types of surface vibrations were found in this work. In the region of the external vibrations, surface modes are more or less hybrid modes between phonons and librations. They are grouped into the following four categories: (1) modes primarily consisting of surface phonons; (2) modes primarily consisting of surface librations; (3) hybrid modes of surface phonons and surface librations; and (4) a class of mixed modes in which the surface phonons and surface librations are mixed with bulk phonon and librational modes. In the intramolecular vibration range the surface modes fall into two categories, the pure surface modes and mixed modes. The richness of this mode structure is due to the additional orientational and internal degrees of freedom associated with the internal molecular structure, that are absent from atomic crystals. The fact that accurate intermolecular potentials are known for many molecular crystals makes it possible to study surface vibrations on them with fewer adjustable parameters than, say, for metallic surfaces. More calculations of this kind can be expected in the future.

4.2.4 Crystal Surfaces with High Miller Indices

Two groups have reported calculations of surface phonons and resonance modes on crystal surfaces with high Miller indices. *Armand* and
Masri have studied the {117} face of an fcc crystal on the basis of a
nearest neighbor, central force model of the crystal, using the method
of Green's function generating functions to obtain surface spectral
densities [4.97] (Fig.4.10). This surface contains steps, and stepped surfaces are of interest in catalysis problems. Steps act as preferential sites
for the adsorption of reactive gases, and play an important role in the
nucleation of the two-dimensional adsorbed structures, while the
adsorption on the terraces behaves as in the case of a low index
surface. *Black* and *Bopp* [4.27] have studied the {100}, {110},...{332}
surfaces on a nearest neighbor model of an fcc crystal, as well as the
{51$\bar{1}$}, {71$\bar{1}$}, {91$\bar{1}$}, and {553} surfaces. A slab calculation of the frequencies of surface and resonance modes was carried out. As the
Miller indices increase, larger numbers of layers in the slab are
required to obtain convergent frequencies. Slabs containing as many as
105 layers were used. The number of surface modes increases as the
step and terrace sizes increase. In the wave vector direction up the
stairs, the surface mode frequencies decrease as the step and terrace
size increase.

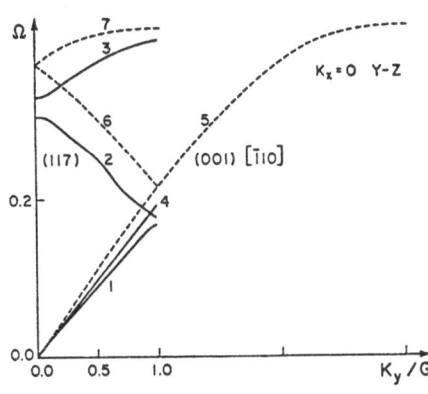

Fig.4.10. Surface phonon dispersion
curve for the fcc (117) vicinal surface
along the y-direction with a sagittal
polarization: localized mode (curve 1);
resonances (curves 2 and 3); bottom of
the bulk bands (curve 4); dashed lines:
localized mode for the {001} surface in
the (110) direction (curve 5) and
vibrational structures obtained by
three-folding curve 5 (curves 5-7)
[4.97]

4.2.5 Reconstructed Surfaces

Many crystalline surfaces on ionic crystals, semiconductors, and metals
undergo surface reconstruction, in which the surface unit cell is
enlarged relative to the one that is obtained by simply terminating the
bulk crystal at the surface in question. The problem of determining
why a given surface undergoes a particular reconstruction is one of
considerable current interest in the study of surface structures. We will
not be concerned with this question here but will focus on the surface
phonons that occur on a reconstructed surface.

It is straightforward to predict the qualitative features of the dispersion curves of such surface phonons. The enlarged surface unit cell of a reconstructed surface translates into a correspondingly decreased surface first Brillouin zone. Then, if the portions of the dispersion curves of the surface phonons on the unreconstructed surface that lie outside this smaller Brillouin zone are "folded" back into it by displacements through suitable translation vectors of the two-dimensional reciprocal lattice defined by the reconstructed surface, the "switching on" of the surface reconstruction will open up gaps in the resulting dispersion curves at the center of the new Brillouin zone and at points on its boundary.

Allan and *Mele* [4.98] have carried out calculations of the dispersion curves of surface phonons on the Si{001} (2x1) surface. The calculations are based on coupling a tight-binding Hamiltonian for the surface electronic structure to a nearest neighbor, central force elastic Hamiltonian. The electronic terms in this Hamiltonian are obtained by fitting bulk Si electronic bands, and the parameters in the elastic part of the Hamiltonian are chosen to minimize the ground state energy of the crystal at the correct lattice constant and to reproduce the frequency of the zone center optical phonon. The bulk phonon spectra obtained from this model are in excellent agreement with the experimental curves. To obtain the (2x1) reconstruction of the Si{001} surface, pairs of Si atoms are coupled in rows along the ⟨110⟩ direction on both {001} surfaces of a ten-layer slab, and the slab is then equilibrated by the use of the Hellman-Feynman theorem. The equilibrated surface structure thus obtained consists of rows of asymmetric (tilted) dimers on the surface. It is essentially the one obtained by *Chadi* [4.99,100], which is very similar to the one obtained in the selfconsistent pseudopotential calculations of *Yin* and *Cohen* [4.101], which is in qualitative agreement with the results of LEED experiments [4.102]. Phonon dispersion curves were then computed for the ten-layer slab. The results reveal a number of surface phonon branches below the bulk acoustic mode bands, above the bulk optical mode bands, and in gaps in the bulk phonon spectrum projected onto the surface Brillouin zone. Several of the branches in the spectrum can be interpreted simply, and yield structural information about the bonding properties of the surface. It is suggested that electron energy loss spectroscopy could be used to observe several of the calculated branches of surface modes, particularly those lying in gaps in the projected bulk phonon spectrum.

The calculations of *Allan* and *Mele*, however, were interpreted on the basis of an incomplete projected bulk phonon spectrum, which was partially responsible for the large number of surface modes observed. These calculations were corrected by *Mazur* and *Pollmann* [4.103,104], and although the number of surface modes found decreased, an "amazingly large number of localized surface modes" was still obtained.

The effects of hydrogenation on the surface phonons on Si{001} (2x1) were studied subsequently by *Allan* and *Mele* [4.105]. These calculations were also interpreted on the basis of an incomplete projected bulk phonon spectrum. They were corrected in [4.104].

More recently, the low frequency phonon spectrum of the Si{111} (2x1) surface has been calculated by *Alerhand* and *Mele* [4.106]. Their results show a low frequency, largely dispersionless surface mode in addition to the Rayleigh wave. It arises from strong electronic polarization effects associated with the quasi-one-dimensional nature of the structure of this reconstructed surface, which is assumed to consist of strongly π-bonded zigzag chains of surface atoms that run parallel to each other [4.107]. These results appear to explain the recent experimental observation of a nearly dispersionless surface mode on this surface by *Harten* et al. [4.108].

Recently, *Masri* and *Tasker* [4.109] have studied the structure and dynamics of a calcium doped magnesium oxide {001} surface that has undergone reconstruction. This work represents the first theoretical prediction of a segregation-induced surface reconstruction in an oxide.

4.2.6 Amorphous Media

The only existing theory of surface phonons on the surface of an amorphous solid has been constructed by *Laughlin* and *Joannopoulos* [4.110,111]. Silicon dioxide was chosen as a prototype because of its technological importance and because of the availability of experimental data on porous glass. However, the methods and ideas used are applicable to any amorphous material.

The model of a surface on an amorphous material used by *Laughlin* and *Joannopoulos* is a nearest neighbor Bethe lattice in which a bond has been broken. A Bethe lattice is a nonperiodic branching structure that contains no rings of bonds. The local geometry of a Bethe lattice is chosen to be the same as that of the crystal it is modeling, i.e. the building block of the Bethe lattice considered by *Laughlin* and *Joannopoulos* is an SiO_4 tetrahedron. The breaking of a bond produces two fundamental types of surface, one terminating on a silicon atom, and a similar surface with an oxygen atom attached. One advantage of using a Bethe lattice as a model for topological disorder is that vibrational Green's functions for this structure can be calculated exactly, in analytic form. It does have the drawback, however, of lacking long wavelength excitations, in particular low frequency acoustic phonons. Local densities of states for both a silicon-terminated and an oxygen-terminated surface were calculated from these Green's functions. They show features that can be interpreted as surface states associated with particular types of motion of surface atoms, e.g. bond stretching or wagging motions of these atoms. Infrared absorption and Raman scattering spectra of the Bethe lattice with a surface have also been calculated by *Laughlin* and *Joannopoulos*, and compared with the results of such calculations for bulk Bethe lattices to reveal surface features in these spectra. The results have been used to interpret experimental infrared absorption and Raman spectra of porous Vycor glass and bulk fused silica [4.112].

It is likely that future work on surface phonons on amorphous media will be based on the prescription for generating amorphous structures suggested by *Wooten* and *Weaire* [4.113], and that the local densities of states for surface atoms will be computed by the continued fraction/recursion method [4.114]. The computations will be heavier than in the case of calculations based on a Bethe lattice, but the results should be more realistic, becaues the structures generated by the *Wooten-Weaire* scheme are much closer to those of real amorphous media than are those provided by the Bethe lattice.

4.2.7 An Incommensurate Array of Adatoms

There is a great deal of interest at the present time in the vibrational properties of monolayers or multilayers of adatoms that form periodic structures in two dimensions whose periods differ in an incommensurate fashion from the periods of the substrate on which they are deposited. Such incommensurate phases are believed to occur in the phase diagrams at full monolayer coverage of krypton and xenon adsorbed on graphite {0001} [4.115]. Monolayers of argon, krypton, and xenon deposited on Ag{111} are incommensurate with the substrates [4.116].

Helium atom scattering experiments on these systems show a single, dispersionless mode. When additional layers of the rare gas atoms are deposited on Ag{111} an evolution of this dispersionless vibrational mode into a dispersive Rayleigh wave is observed (Fig.4.11) [4.15].

Many of the experimental results can be understood on the basis of a simple model in which the rare gas atoms interact with nearest neighbor forces, and the first layer of rare gas atoms is tied to a smooth, rigid substrate by perpendicular forces determined from the frequency of the dispersionless monolayer mode [4.16]. In a better treatment of the coupling of the phonons of the overlayer(s) to the phonons of the substrate, the latter was modeled by an elastic continuum [4.117]. In this case it is found that when the adsorbate modes lie inside the bulk bands of the substrate they are leaky modes, that is they radiate energy into the substrate and acquire a finite lifetime as a result. This lifetime has been measured for the case of a monolayer of

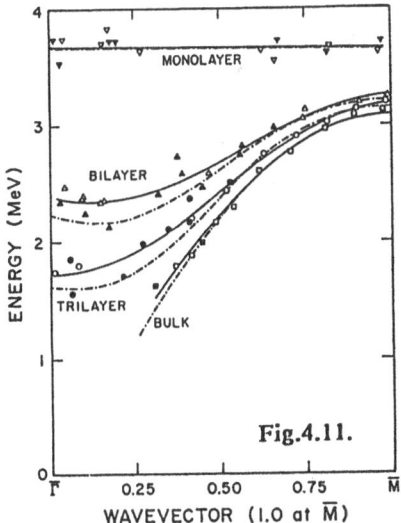

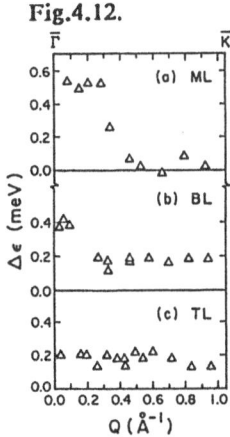

Fig.4.11.

Fig.4.12.

Fig.4.11. Experimental surface phonon dispersion curves for argon monolayers, bilayers, trilayers, and bulk (25 layers) on Ag{111}, obtained by inelastic He scattering. The open squares correspond to phonon loss events, while the closed squares correspond to phonon gain events. The full lines are a fit to the data, while the dash-dot lines are results of calculations based on nearest neighbor interactions between adatoms [4.16]

Fig.4.12. Measured linewidth broadening $\Delta\epsilon$ of the Kr creation phonon peaks. (a) monolayer; (b) bilayer; (c) trilayer [4.17]

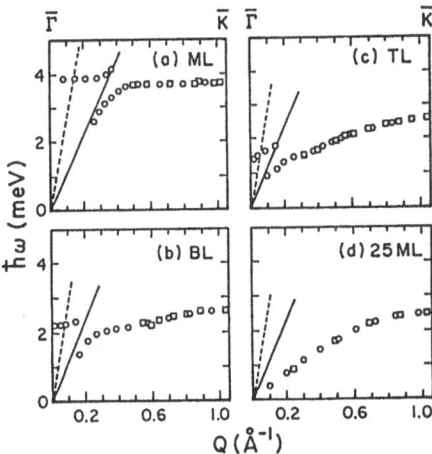

Fig.4.13. Experimental surface phonon dispersion curves of Kr films along the $(\overline{\Gamma K})_{Kr}$ azimuth. (a) monolayer; (b) bilayer; (c) trilayer; (d) 25 monolayers. The circles and squares represent phonon creation and annihilation events, respectively. The solid line is the clean Pt{111} Rayleigh phonon dispersion curve, the dashed line is the longitudinal bulk band edge of the Pt{111} substrate, both in the $(\overline{\Gamma M})_{Pt}$ azimuth which is coincident with the $(\overline{\Gamma K})_{Kr}$ azimuth [4.17]

175

Kr on Pt{111} by *Kern* et al. [4.17]. It disappears when the Kr mode stops overlapping the bulk phonons (Fig.4.12). When the modes of the overlayer(s) cross the Rayleigh wave of the substrate, hybridization of these modes is expected. This hybridization has been observed recently in experiments carried out by *Kern* et al. [4.17] (Fig.4.13).

In none of the theoretical work mentioned has the incommensurateness of the substrate and the adlayer(s) been taken into account. It is feasible to deal with this lack of registry in the high temperature limit by an extension of the method of molecular dynamics. Preliminary results for nonregistered phases of krypton and xenon on graphite {0001} obtained by such calculations have been reported [4.118].

4.3 Some Directions for Future Research

To conclude this review, I should like to point to some theoretical and experimental problems concerning surface phonons that in my view merit study in the future.

The first is the calculation of surface phonon dispersion curves and the associated eigenvectors from first principles by a variety of approaches, so that the advantages and shortcomings of each for this purpose can be judged. My personal preference is for methods that can yield individual atomic force constants, which are more informative than interplanar force constants despite the greater computational effort required in such calculations. The extension of these calculations to include anharmonic force constants would be desirable.

At the phenomenological level, the construction of interatomic potentials for atoms in the surface layers of a crystal that are noncentral, many-body potentials, and are capable of predicting static surface structures and harmonic and anharmonic dynamical properties of surface atoms would be welcome indeed. They would constitute the starting point for perturbation-theoretic studies of surface dynamics as well as for nonperturbative molecular dynamics simulations of such properties. They would also provide the starting point for theoretical calculations of surface free energies as functions of the surface orientation for predictions of equilibrium crystal shapes (Chaps.13,15).

Among the types of anharmonic surface properties that are susceptible to perturbative calculation, I mention the line shapes of the surface phonon contribution to the cross sections for the inelastic scattering of He atoms from surfaces. The resolution of such experiments is now reaching such levels that these line shapes may soon be measured. A comparison of experimental and theoretical results would yield useful information about anharmonic interactions in the surface layers of a crystal, where they are expected to be larger than in the bulk due to the larger excursions of the atoms about their equilibrium positions.

Path-integral quantum Monte-Carlo calculations would yield non-perturbative results for anharmonic surface phonon properties at low temperatures where molecular dynamics simulations are no longer valid.

While on the subject of anharmonic effects, I should like to point out that the effects of anharmonicity on surface phonons need not be limited to the small frequency shifts and lifetimes they acquire from that source and which can be treated by perturbation theory. They can give rise to fundamentally different kinds of surface waves, namely surface acoustic solitons. Some first steps in the direction of studying such waves have been taken by *Sakuma* and his coworkers [4.119-122], but the methods they use seem to me to be unnecessarily cumbersome, and the last word on the subject has yet to be spoken.

An interesting problem in the theory of the dynamical properties of adsorbate covered surfaces is that of determining the surface waves on them when the adlayer is incommensurate with the substrate. This is a much more difficult problem than that of the dynamics of recon-structed surfaces because unlike the situation in the latter case the combined system of adlayer and substrate here no longer has any translational periodicity parallel to their interface.

The study of surface phonon mediated phase transitions occurring at the surface of a crystal before they occur in the bulk should also be worthwhile. An example of such a transition might be surface ferro-electricity. Another might be surface superconductivity. A theoretical analysis of such transitions requires the use of anharmonic interactions, perhaps in the form of the on-site, quartic, core-shell displacement contribution to the crystal potential energy introduced by *Bilz* et al. [4.123] in their studies of the ferroelectric transition in the perovskites. A self-consistent phonon calculation of the frequencies of surface phonons as functions of temperature could reveal a surface instability at a temperature lower than the one at which such an instability occurs in the bulk.

To conclude this listing of some directions for future research into surface phonons I mention two experimental problems.

Although symmetry and other invariance conditions prevent the atoms in crystals of the diamond structure from having nonzero trans-verse effective charges, so that these crystals display no reststrahl absorption at the Raman frequency, the same conditions allow atoms in the surface layers of these crystals to acquire nonzero transverse effec-tive charges. Estimates [4.124] suggest that the magnitudes of these surface induced effective charges can be of the order of 0.5 e for the Si{111} (2x1) surface, where e is the magnitude of the electronic charge. Some of the surface modes associated with the surfaces of these crystals should therefore be infrared active. The experimental determination of the associated infrared absorption spectrum should provide very useful information about the dynamical properties of the surfaces of the elemental semiconductors, since it would be a continuous spectrum rather than a line spectrum due to the loss of translational symmetry in the direction normal to the surface. Since it is a surface effect, this absorption should be very weak. However, there have been significant advances in recent years in the use of infrared absorption techniques for the study of surface excitations (Chap.3), and it may be hoped that the effect described will eventually be seen.

Finally, the availability of experimental surface phonon dispersion curves along other than symmetry directions in the surface first Brillouin zone could provide information about atomic force constants that do not appear in the expressions for the frequencies along symmetry directions.

Acknowledgment. This work was supported in part by the National Science Foundation, Grant No. DMR 85-17634.

References

4.1   G. Benedek: Surf. Sci. **126**, 624 (1983)
4.2   F.W. de Wette: Comments on Solid State Phys. **11**, 89 (1984)
4.3   J.P. Toennies: J. Vac. Sci. Technol. A **2**, 1055 (1984)
4.4   G. Benedek: Physica **127** B, 59 (1984)
4.5   R.F. Wallis, J. Vac. Sci. Technol. A **3**, 1422 (1985)
4.6   J. Szeftel: Surf. Sci. **152/153**, 797 (1985)
4.7   J.E. Black, in: *Topics in Current Physics, Vol.41: Structure and Dynamics of Surface 1*, ed. W. Schommers, P. von Blanckenhagen (Springer, Berlin 1986), p.153
4.8   I.M. Lifshitz, L.M. Rosenzweig: Zh. Eksp. Teor. Fiz. **18**, 1012 (1948)
4.9   L.M. Rosenzweig, Tr. Fiz. Otdel. Fiz.-Mat. Fakul'teta Khark. Gos. Univ. **2**, 19 (1950) (English translation of this paper available as Tech. Rpt. 75-44 from Dept. of Physics, Univ. of California, Irvine)
4.10   Reviews of this work can be found in A.A. Maradudin, E.W. Montroll, G.H. Weiss, I.P. Ipatova: *Theory of Lattice Dynamics in the Harmonic Approximation* (Academic, New York 1971); and in A.A. Maradudin, R.F. Wallis, L. Dobrzysnski: *Surface Phonons and Polaritons* (Garland STPM Press, New York 1980)
4.11   G. Brusdeylins, R.B. Doak, J.P. Toennies: Phys. Rev. Lett. **44**, 1417 (1980)
4.12   U. Harten, J.P. Toennies, Ch. Wöll: Farad. Discuss. Chem. Soc. **80**, 1 (1985)
4.13   U. Harten, J.P. Toennies, Ch. Wöll, G. Zhang: Phys. Rev. Lett. **55**, 2308 (1985)
4.14   B.F. Mason, B.R. Williams: Surf. Sci. **139**, 173 (1984)
4.15   K.D. Gibson, S.J. Sibener, Phys. Rev. Lett. **55**, 1514 (1985)
4.16   K. Gibson, S.J. Sibener, B. Hall, D.L. Mills, J.E. Black: J. Chem. Phys. **83**, 4256 (1985)
4.17   K. Kern, P. Zeppenfeld, R. David, G. Comsa: Phys. Rev. B **35**, 886 (1987)
4.18   G. Brusdeylins, R. Rechsteiner, J.G. Skofronick, J.P. Toennies, G. Benedek, L. Miglio: Phys. Rev. Lett. **54**, 466 (1985
4.19   G. Bracco, R. Tatarek, S. Terreni, F. Tommasini: Phys. Rev. B **34**, 9045 (1986)

4.20 H. Ibach, D.L. Mills: *Electron Energy Loss Spectroscopy and Surface Vibrations* (Academic, New York 1982)

4.21 S. Lehwald, J.M. Szeftel, H. Ibach. T.S. Rahman, D.L. Mills: Phys. Rev. Lett. **50**, 518 (1983)

4.22 M. Rocca, S. Lehwald, H. Ibach, T.S. Rahman: Surf. Sci. **138**, L123 (1984)

4.23 C. Oshima, R. Souda, M. Aono, S. Otani, Y. Ishizawa: Phys. Rev. B **30**, 5361 (1984); Solid State Commun. **57**, 283 (1986)

4.24 C. Oshima, R. Souda, M. Aono, S. Otani, Y. Ishizawa: Phys. Rev. Lett. **56**, 240 (1986)

4.25 H. Ibach, D. Bruchmann: Phys. Rev. Lett. **44**, 36 (1980)

4.26 J.M. Szeftel, S. Lehwald, H. Ibach, T.S. Rahman, J.E. Black, D.L. Mills: Phys. Rev. Lett. **51**, 268 (1983)

4.27 J.E. Black, P. Bopp: Surf. Sci. **140**, 275 (1984)

4.28 A.A. Maradudin, J. Melngailis: Phys. Rev. **133**, A1188 (1964)

4.29 G. Benedek: Surf. Sci. **61**, 603 (1976)

4.30 J.E. Black, T.S. Rahman, D.L. Mills: Phys. Rev. B **27**, 4072 (1983)

4.31 F. Garcia-Moliner: Ann. Physique **2**, 177 (1977)

4.32 F. Garcia-Moliner, G. Platero, V. Velasco: Surf. Sci. **136**, 601 (1984)

4.33 G. Platero, V.R. Velasco, F. Garcia-Moliner, G. Benedek, L. Miglio: Surf. Sci. **143**, 243 (1984)

4.34 A.C. Levi, G. Benedek, L. Miglio, G. Platero, V.R. Velasco, F. Garcia-Moliner: Surf. Sci. **143**, 253 (1984)

4.35 G. Platero, V.R. Velasco, F. Garcia-Moliner: Surf. Sci. **152/153**, 819 (1985)

4.36 M. Mostoller, U. Landman: Phys. Rev. B **20**, 1755 (1979)

4.37 J.E. Black, B. Laks, D.L. Mills: Phys. Rev. B **22**, 1818 (1980)

4.38 J.E. Black: Surf. Sci. **105**, 59 (1981)

4.39 T.S. Rahman, J.E. Black, D.L. Mills: Phys. Rev. B **25**, 883 (1982)

4.40 S. Lehwald, J.M. Szeftel, H. Ibach, T.S. Rahman, D.L. Mills: Phys. Rev. Lett. **50**, 518 (1983)

4.41 M.-L. Xu, B.M. Hall, S.Y. Tong, M. Rocca, H. Ibach, S. Lehwald, J.E. Black: Phys. Rev. Lett. **54**, 1171 (1985)

4.42 D.C. Gazis, R. Herman, R.F. Wallis: Phys. Rev. **119**, 533 (1960)

4.43 S.E. Trullinger: J. Math. Phys. **17**, 1884 (1976)

4.44 A.A. Maradudin, S.L. Moss, S.L. Cunningham: Phys. Rev. B **15**, 4490 (1977)

4.45 J. Latkowski, J.E. Black: (unpublished)

4.46 T.E. Feuchtwang: Phys. Rev. **155**, 731 (1967)

4.47 J. Szeftel, A. Khater, F. Mila: Saclay Preprint No. DPhG/PAS/87.01

4.48 F. Mila, J. Szeftel: Saclay Preprint No. DPhG/SPAS/87-13

4.49 A.R. McGurn, A.A. Maradudin, R.F. Wallis, A.J.C. Ladd: Phys. Rev. B **37**, 3964 (1988)

4.50 J. Szeftel, S. Lehwald, H. Ibach, T.S. Rahman, J.E. Black, D.L. Mills: Phys. Rev. Lett. **51**, 268 (1983)

4.51 D.L. Mills: Phys. Today **40**, 568 (1987)

4.52 J.E. Black, D.A. Campbell, R.F. Wallis: Surf. Sci. **115**, 161 (1982)

4.53 C. Beatrice, C. Calandra: Phys. Rev. B **28**, 6130 (1983)

4.54 C. Calandra, A. Catellani, C. Beatrice: Surf. Sci. **148**, 90 (1984)

4.55 C. Calandra, A. Catellani, C. Beatrice: Surf. Sci. **152/153**, 814 (1985)

4.56 W.A. Harrison: *Pseudopotentials in the Theory of Metals* (Benjamin, New York 1966)

4.57 E.G. Brovman, Yu.M. Kagan: Zh. Ekspr. i Teor. Fiz. **52**, 557 (1967) [Soviet Physics JETP **25**, 365 (1967)]

4.58 R.N. Barnett, U. Landman, C.L. Cleveland: Phys. Rev. B **28**, 1667 (1983)

4.59 A.G. Eguiluz, A.A. Maradudin, R.F. Wallis, in: *Phonon Physics*, ed. J. Kollar, N. Kroo, N. Menyhard, T. Siklos (World Scientific, Singapore 1985), p.604

4.60 A.G. Eguiluz, R.F. Wallis, A.A. Maradudin: Bull. Am. Phys. Soc. **32**, 683 (1987) (UCI Dept. of Physics Techn. Rpt. 87-48); Phys. Rev. Lett. **60**, 309 (1988)

4.61 W. Kohn, L.J. Sham: Phys. Rev. **140**, A1133 (1965)

4.62 K.M. Ho, K.P. Bohnen: Phys. Rev. Lett. **56**, 934 (1986)

4.63 E.B. Wilson, J.C. Decuis, P.C. Cross: *Molecular Vibrations* (McGraw-Hill, New York 1955)

4.64 J.E. Black, P. Bopp, K. Lutzenkirchen, M. Wolfsberg: J. Chem. Phys. **76**, 6431 (1982)

4.65 T.A. Upton, W. Goddard III: *Chemistry and Physics of Solid Surfaces*, Vol.III, ed. R. Vanselow, W. England (CRC Press, Boca Raton, Florida 1980); Phys. Rev. Lett. **46**, 1635 (1981)

4.66 J.E. Black: Surf. Sci. **100**, 555 (1980)

4.67 The approach used by F.O. Goodman, Surf. Sci. **116**, 573 (1982), however, is close to that of [4.68]

4.68 M. Wagner: Phys. Rev. **131**, 2520 (1963)

4.69 T.S. Rahman, J.E. Black, D.L. Mills: Phys. Rev. B **25**, 883 (1982)

4.70 T.S. Rahman, D.L. Mills, J.E. Black: Phys. Rev. B **27**, 4059 (1983)

4.71 E. de Rouffignac, G.P. Alldredge, F.W. de Wette: Chem. Phys. Lett. **69**, 29 (1980)

4.72 E. de Rouffignac, G.P. Alldredge, F.W. de Wette: Phys. Rev. B **24**, 6050 (1981)

4.73 M. Marchese, G. Jacucci, M.L. Klein: Surf. Sci. **145**, 364 (1984)

4.74 G.G. Gardini, S.F. O'Shea: unpublished work referenced in [4.73]

4.75 J.E. Black: Surf. Sci. **116**, 240 (1982)

4.76 K.G. Lloyd, J.C. Hemminger: Surf. Sci. **143**, 509 (1984)

4.77 K.G. Lloyd, J.C. Hemminger: J. Chem. Phys. **82**, 3858 (1985)

4.78 B.A. Banse, K.G. Lloyd, J.C. Hemminger: J. Chem. Phys. **86**, 2986 (1987)

4.79 G. Benedek, M. Miura, W. Kress, H. Bilz: Surf. Sci. **148**, 107 (1984)

4.80 R.B. Doak, U. Harten, J.P. Toennies: Phys. Rev. Lett. **51**, 578 (1983)

4.81 V. Bortolani, A. Franchini, F. Nizzoli, G. Santoro: Phys. Rev. Lett. **52**, 429 (1984)

4.82 J.A. Moriarty: Phys. Rev. B **26**, 1754 (1982)

4.83 L.D. Marks, V. Heine, D.J. Smith: Phys. Rev. Lett. **52**, 656 (1984)

4.84 H. Bilz, W. Weber, in: *Proc. Int'l Conf. on Latent Image in Photography, Trieste 1983*

4.85 U. Harten, J.P. Toennies, Ch. Wöll: Faraday Discuss. Chem. Soc. **80**, 137 (1985)

4.86 V. Bortolani, G. Santoro, U. Harten, J.P. Toennies: Surf. Sci. **148**, 82 (1984)

4.87 G. Benedek, M. Miura, W. Kress, H. Bilz: Phys. Rev. Lett. **52**, 1907 (1984)

4.88 M. Miura, W. Kress, H. Bilz: Z. Phys. B **54**, 103 (1984)

4.89 H. Frölich: Phys. Lett. A **35**, 325 (1971)
W. Weber, H. Bilz, U. Schröder: Phys. Rev. Lett. **28**, 600 (1972)
W. Weber: Phys. Rev. B **8**, 5082 (1973)

4.90 G. Lakshmi, W. Weber: Verhandl. DPG (VI) **10**, 617 (1975), summary only

4.91 R.F. Wallis, D.L. Mills, A.A. Maradudin: In *Localized Excitations in Solids*, ed. by R.F. Wallis (Plenum, New York 1968) p.403

4.92 A.A. Lucas: J. Chem. Phys. **48**, 3156 (1968)

4.93 C. Oshima: Phys. Rev. B **34**, 2949 (1986)

4.94 W. Kohn: Phys. Rev. Lett. **2**, 393 (1959)
E.J. Woll Jr., W. Kohn: Phys. Rev. **126**, 1693 (1962)

4.95 G.S. Pawley, S.L. Chaplot: phys. stat. solidi (b) **99**, 517 (1980)

4.96 K. Kobashi, R.D. Etters: J. Chem. Phys. **82**, 4341 (1985)

4.97 G. Armand, P. Masri: Surf. Sci. **130**, 89 (1983)

4.98 D.C. Allan, E.J. Mele: Phys. Rev. Lett. 53, 826 (1984)

4.99 D.J. Chadi: Phys. Rev. Lett. 43, 43 (1979); J. Vac. Sci. Tech. 16, 1290 (1979)

4.100 R.M. Tromp, R.G. Smeenk, F.W. Saris, D.J. Chadi: Surf. Sci. 133, 137 (1983)

4.101 M.T. Yin, M.L. Cohen: Phys. Rev. B 24, 2303 (1981)

4.102 B.W. Holland, C.B. Duke, A. Paton: unpublished work referenced in [4.98]

4.103 A. Mazur, J. Pollmann: Phys. Rev. Lett. 57, 1811 (1986)

4.104 D.C. Allan, E.J. Mele: Phys. Rev. Lett. 57, 1812 (1986)

4.105 D.C. Allan, E.J. Mele: Phys. Rev. Lett. B 31, 5565 (1985)

4.106 O.L. Alerhand, E.J. Mele: Phys. Rev. Lett. 59, 657 (1987)

4.107 K.C. Pandey: Phys. Rev. Lett. 47, 1913 (1981); 49, 233 (1982)

4.108 U. Harten, J.P. Toennies, Ch. Wöll: Phys. Rev. Lett. 57, 2947 (1986)

4.109 P. Masri, P.W. Tasker: Surf. Sci. 149, 209 (1985)

4.110 R.B. Laughlin, J.D. Joannopoulos, in: *Lattice Dynamics*, Ed. M. Balkanski (Flammarion Sciences, Paris 1978), p.294

4.111 R.B. Laughlin, J.D. Joannopoulos: Phys. Rev. B 17, 4922 (1978)

4.112 R.B. Laughlin, J.D. Joannopoulos, C.A. Murray, K.A. Hartness, T.J. Greytak: Phys. Rev. Lett. 40, 461 (1978)

4.113 F. Wooten, D. Weaire: J. Non-Cryst. Solids 64, 325 (1984)
F. Wooten, K. Winer, D. Weaire: Phys. Rev. Lett. 54, 1392 (1985)

4.114 R. Haydock. V. Heine, M.J. Kelly: J. Phys. C 5, 2845 (1972); 8, 2591 (1975)

4.115 J.A. Venables, P.S. Schabes-Retchkiman: J. Phys. (Paris) Suppl.C4 38, 105 (1977)

4.116 J. Unguris, L.W. Bruch, E.R. Moog, M.B. Webb: Surf. Sci. 87, 415 (1979); 109, 522 (1981)

4.117 B. Hall, D.L. Mills, J.E. Black: Phys. Rev. B 32, 4932 (1985)

4.118 B. Firey, F.W. de Wette, in: *Phonon Physics*, ed. J. Kollar, N. Kroo, N. Menyhard, T. Siklos (World Scientific, Singapore 1985), p.624

4.119 T. Sakuma, Y. Kawanami: Phys. Rev. B 29, 869 (1984)

4.120 T. Sakuma, Y. Kawanami: Phys. Rev. B 29, 880 (1984)

4.121 T. Sakuma, T. Miyazaki: Phys. Rev. B 33, 1036 (1986)

4.122 T. Sakuma, O. Saito: Phys. Rev. B 35, 1294 (1987)

4.123 R. Migoni, H. Bilz, D. Bauerle: Phys. Rev. Lett. 37, 1155 (1976)

4.124 E. Evans, D.L. Mills: Phys. Rev. B 5, 4126 (1972)

5. Interpretation of the NEXAFS Spectra of Adsorbates Using Multiple Scattering Calculations

John A. Horsley[1]

Department of Chemistry and Zettlemoyer Center for Surface Studies
Lehigh University, Bethlehem, PA 18015, USA

Near edge x-ray absorption fine structure (NEXAFS) spectroscopy is proving to be a powerful tool for investigating the geometric and electronic structure of chemisorbed molecules [5.1]. The polarization dependence of NEXAFS resonances gives information on the orientation of the adsorbed molecules with respect to the surface, and the position of certain resonances can be used to measure intramolecular bond lengths [5.2]. However, except for diatomic or pseudodiatomic adsorbates, the near edge region presents a complex structure that is difficult to interpret in many cases. For a full interpretation it is often necessary to compare the experimental NEXAFS spectrum with a calculated spectrum for some model of the adsorbed species.

Calculations based on the ab initio Hartree-Fock potential can be used to obtain photoionization cross sections for molecules, but the representation of the continuum states presents severe problems and so far the application of such methods has been restricted mainly to diatomic molecules [5.3]. In contrast, using multiple scattering theory the continuum states are no more difficult to calculate than the bound states. In general, multiple scattering theory provides an efficient way of calculating both bound and continuum states and leads to results that are accurate enough to be used as a guide to the interpretation of experimental NEXAFS spectra of complex molecules and adsorbed species.

In this review I first give a brief history of the application of multiple scattering theory to the calculation of NEXAFS spectra for isolated molecules, and then discuss multiple scattering calculations of the NEXAFS spectra of simple adsorbates (atoms and diatomic molecules). I then illustrate the application of these techniques to more complex adsorbates through a number of examples.

---

[1] Present address: Catalytica, 430 Ferguson Dr., Bldg.3, Mountain View, CA 94043, USA

5.1 Multiple Scattering Calculations of Near Edge Spectra of Molecules

Multiple scattering theory has been successfully used for a number of years to calculate bound states for molecules and clusters using the multiple scattering $X\alpha$ (MS-$X\alpha$) method of Johnson and coworkers [5.4]. In the MS-$X\alpha$ method the one-electron Schrödinger equation is solved for a model potential known as the "muffin-tin" potential. Each atom is enclosed in a sphere of a given radius and the whole molecule or cluster is enclosed by an outer sphere which is tangential to the atomic spheres. The potential is spherically averaged within the atomic spheres and outside the outer sphere, and is taken to be constant in the intersphere region. For the exchange potential Slater's local statistical ($X\alpha$) potential [5.5] is used

$$V_{exch} = -6\alpha \, [3/8 \, \pi \, \rho(x)]^{1/3} \tag{5.1}$$

where $\rho(x)$ is the charge density and α is a parameter usually chosen on the basis of atomic calculations. The initial potential is normally obtained from a superposition of atomic charge densities and the final potential is then calculated by an iterative self-consistent field procedure.

In his original derivation of the secular equation for the multiple scattering model for molecules and clusters, *Johnson* [5.6] pointed out that it is also possible to set up the molecular secular equations for positive energies (corresponding to continuum states) as well as negative energies (corresponding to bound states). The multiple scattering $X\alpha$ method was extended to the calculation of continuum states independently by *Dill* and *Dehmer* [5.7] and *Davenport* [5.8]. The first calculations of NEXAFS spectra for molecules by this method were carried out by *Dehmer* and *Dill* [5.9] for N_2 and CO. At the same time *Davenport* [5.8] reported calculations of the uv photoionization cross sections for the same molecules.

In the K-edge spectrum of N_2, a shape resonance was calculated by Dehmer and Dill at ~11 eV above the ionization threshold, within 1 eV of the experimental resonance peak. The agreement was surprisingly good because the potential used was for the ground state of the neutral molecule, with no allowance for relaxation effects due to the presence of the core hole. However, the use of the neutral molecule potential was compensated by using a value of $\alpha = 1$ in the $X\alpha$ exchange potential, which is higher than the more usual value of $\alpha \approx 0.7$. This gives rise to a deeper potential and to some extent simulates the effect of the core hole. *Davenport* [5.8] obtained the same shape resonance in the uv photoionization spectrum of N_2 at somewhat higher energy than experiment, using the ground state potential and the more usual value of α. He suggested that instead of using a higher value of α it would be better to include the effect of the core hole directly by using the so-called "transition state" potential in which half an electron is removed from the appropriate core orbital. As we shall see, this potential does indeed lead to very good agreement with experiment.

Dehmer and *Dill* also calculated the cross sections for transitions to the unoccupied bound states below the ionization threshold in N_2, including the principal $1s \rightarrow \pi^*$ transition and transitions to a number of Rydberg levels. By combining the bound state and continuum calculations they were able to reproduce accurately the whole K-edge region in N_2.

Following these pioneering calculations the MS-Xα method was used to calculate the K-edge spectra of various transition metal complexes by *Kutzler* et al. [5.10]. The approach was similar to that used by Dehmer and Dill. Separate calculations were performed on bound and continuum states and the calculated cross sections were convoluted with a Lorentzian to account for the broadening due to the short core-hole lifetime and the instrumental resolution. The calculations used the full core-hole potential. The agreement with experiment was good and all the major features in the spectra were reproduced. In the continuum the principal resonance for these transition metal complexes was interpreted as a shape resonance produced by the confinement of the upper state wavefunction by the surrounding ligands, in contrast to previous interpretations in terms of a transition to an atomic-like *p*-orbital.

A multiple scattering method for the calculation of NEXAFS spectra for molecules and clusters, similar in many respects to the MS-Xα method, has been developed by *Durham* et al. [5.11]. The method is based on a muffin-tin potential which is constructed by superposition of the separate atom potentials. Slater's Xα exchange potential is used, with $\alpha = 1$. The cluster is divided into shells surrounding the central excited atom, the atoms in each shell having the same (within ~1 a.u.) radial distance from the central atom. The multiple scattering equations are first solved within each shell and then the multiple scattering between shells is added, leading to a very efficient method of computation. The advantage of this method is that very large clusters can be handled without enormous computational effort, because of the non-self-consistent potential and the overall efficiency of the shell-structure approach. However, many properties, such as the charge distribution, are poorly described by a non-self-consistent potential and so the method may be less accurate in many cases than the MS-Xα approach used by *Dehmer* and *Dill* [5.9] and *Kutzler* et al. [5.10]. The method has been used to interpret the K-edge structure in metal complexes and metalloproteins and has been used by *Norman* et al. [5.12] to calculate NEXAFS spectra of adsorbed atomic oxygen on Ni{100}. This calculation will be discussed in the next section.

5.2 Atomic and Diatomic Adsorbates

The MS-Xα method and the related multiple scattering method of Durham et al. can in principle be used to calculate the NEXAFS spectra of adsorbates by performing calculations on a cluster model consisting of the adsorbate and the neighboring surface atoms. The use of cluster models in the calculation of XPS and UPS spectra of adsorbates by the MS-Xα and other methods is now well established [5.13]. The size of the clusters required to represent the surface appears to depend on the particular system that is being studied, but in many cases a simple cluster consisting only of the nearest neighbor metal atoms appears to be adequate to reproduce the main features of the spectra.

The K-edge spectra of atomic oxygen on Ni{100} in various adsorption sites have been calculated by *Norman* et al. [5.12] using cluster models and the multiple scattering method of *Durham* et al. [5.11] outlined above. The calculated spectra were fully tested for convergence in the size of the cluster. It was found that about 30 Ni atoms were required to obtain convergence, i.e. the spectra were not dominated by scattering from just the nearest neighbor surface atoms. A test was also made to determine whether the spectra could be reproduced by including just the single-scattering terms and ignoring multiple scattering. It was found that for single-scattering the results changed significantly as each successive shell of neighbors was added and were not converged even with ten shells of neighbors. The agreement with the fully converged multiple scattering calculation was also poor, even with a large number of shells, indicating that the single scattering approximation is not adequate for this system.

Calculated NEXAFS spectra for oxygen in a number of different adsorption sites were compared with experiment. The calculation for the fourfold hollow site with the O atom 0.9 Å above the substrate gave good agreement with the experimental NEXAFS spectra, the agreement for the other possible sites being significantly worse. It therefore appears possible to discriminate between different adsorption sites for atomic adsorbates by means of calculated NEXAFS spectra. The situation is less clear for molecular adsorbates, however, because their NEXAFS spectra are dominated by intramolecular scattering resonances and so differences in adsorbate-substrate scattering for different sites will be less easy to detect.

NEXAFS spectra for NO adsorbed on Ru{0001} were calculated by *Chou* et al. [5.14] using the MS-Xα method with clusters representing various possible adsorption sites. In this case there were no experimental spectra with which the calculations could be compared, although NEXAFS spectra for NO on Ni{100} have been published [5.15]. The potential used was obtained by imposing a muffin-tin form on a self-consistent potential for ground state NO calculated by the discrete variational method (DVM) [5.16]. The calculated NEXAFS spectra were, as expected, dominated by the intramolecular π^* and σ^* resonances. The spectra of the N K-edge, however, showed additional weak features at higher energy that were attributed to scattering from substrate atoms. These features were missing from the calculated spectra for the O K-edge. The σ^* shape resonance was calculated to be at 14.1 eV above threshold for the gas phase molecule, which is in poor agreement with the experimental value (4.0 eV), probably because the ground state NO potential was used and the effect of the core hole was not taken into account. The σ^* shape resonance was calculated to shift ~3 eV to higher energy on chemisorption, which is not in agreement with the experimental results for NO on Ni{100}, where almost no shift in the σ^* shape resonance was observed on chemisorption. It is well known that a change in the radius of the outer sphere in the muffin-tin potential can cause shifts in the position of the bound states and continuum resonances. However, an increase in the outer sphere radius leads to a shift to lower energies, not higher, so the calculated shift cannot be attributed to changes in the muffin-tin geometry. In addition, as the NO was moved closer to the Ru cluster the σ^* shape resonance shifted to even higher energy and broadened considerably. It would appear that the change in position and shape is connected with the interaction of the NO with the Ru cluster. The discrepancy between the results of the cluster calculations and the experimental results for NO on Ni{100} is puzzling and further calculations of the effect of chemisorption on the shape resonance position are needed in order to shed light on this question.

Experimental NEXAFS spectra indicate that in general the adsorbate-substrate interaction has very little *direct* influence on the position of a continuum resonance [5.15]. However, there will be an *indirect* influence because changes in the intramolecular bond lengths due to chemisorption give rise to shifts in the continuum resonance positions. This suggests that it may be reasonable to omit the substrate atoms from the continuum state calculation and perform the calculation on the isolated molecule, using the geometry of the chemisorbed species. Bound states are more directly affected by absorbate-substrate bonding and so it is necessary to use a cluster containing the substrate atoms to calculate these states.

Using this approach, *Horsley* et al. [5.17] have calculated the K-edge NEXAFS spectrum for CO on Ni{100} with the simplest cluster model for an "on top" site, namely CO bonded to a single nickel atom. The NiCO cluster was used in the calculations of the π^* bound state resonance and the σ^* shape resonance was calculated using the isolated CO molecule with a bond length slightly larger than the gas phase

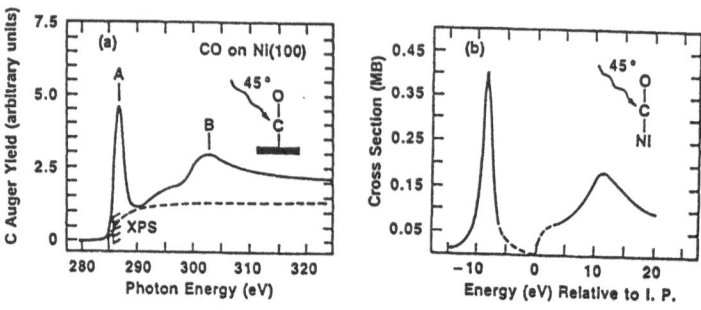

Fig.5.1 a. Experimental NEXAFS spectrum at the C K-Edge for CO on Ni{100} with an x-ray incidence angle of 45°. The background absorption is indicated by the dashed line. b. Calculated NEXAFS spectrum of the C K-edge for CO bonded to a single Ni atom. The incident radiation makes an angle of 45° with respect to the Ni-C-O axis. The Rydberg region (not calculated) is indicated by the dashed line

value. The calculated carbon K-edge spectrum for 45° incident radiation is shown in Fig.5.1, together with the experimental carbon K-edge spectrum of CO on Ni{100}. Apart from an arctangent-like background, which is present in the experimental spectrum but not in the calculated spectrum, the agreement between experiment and calculation is remarkably good for such a simple cluster model. The calculated separation between the bound and continuum resonances is somewhat greater than experiment. As discussed below, the shape resonance position is sensitive to the amount of overlap of the atomic spheres in the muffin-tin potential. In the calculation shown in Fig.5.1 a 30% sphere overlap was used. With the more usual value of 20% the separation between the resonances is in significantly better agreement with experiment.

The origin of the arctangent-like background is not clear. The background is not present in the K-edge spectrum of gas phase CO. It has tentatively been attributed to mixing of the conduction band states in the metal with the adsorbate continuum states [5.17]. Clearly, a more accurate theoretical model is required in order to understand the origin of this feature.

5.3 Computational Method for Complex Molecules and Clusters

Before going on to consider calculations on more complex adsorbates, I shall first describe in some detail the computational method that has proved successful for these species. A similar computational method was used in the NiCO cluster calculations outlined above.

The potential for the continuum state calculations is the MS-Xα "transition state" potential, which corresponds to a configuration in which half an electron is removed from the appropriate core orbital..

For calculations on the bound state resonances, half an electron may be added to the appropriate unoccupied orbital. However, for the sake of consistency it is probably better to use the same potential for both bound and continuum states. This has been found to lead to calculated NEXAFS spectra in good agreement with experiment for benzene and thiophene (see following section). It has the additional advantage that only one SCF calculation needs to be carried out for both bound and continuum states. The eigenvalues of the unoccupied levels in the transition state calculation are taken as the positions of the bound state resonances relative to the ionization threshold. Again, this is consistent with the continuum state calculations, where the (positive) energies of the continuum states are also referenced to the muffin-tin zero of energy. This was the procedure used by *Dehmer* and *Dill* [5.9] in their calculation of the NEXAFS spectrum of N_2 (although with the neutral ground state potential).

In the transition state calculations the core hole is localized on a particular atom. This means that if there are a number of equivalent (by symmetry) atoms in the molecule, one of them is chosen to provide the site for the core hole and is separated from the others. This involves breaking the symmetry of the molecule so that a point group of lower symmetry than the full point group of the molecule, or cluster, is used in the calculations. It has been established for some time that broken-symmetry calculations lead to better results than full-symmetry calculations for processes involving the removal of a core-electron [5.18,19].

To ensure that the potential has the correct asymptotic form, it must be modified by substituting a "Latter tail" [5.20] of the form -2/r (in Rydbergs) for the calculated potential at large distances.

In MS-Xα calculations it has been found that allowing the atomic spheres to overlap to some extent often leads to better agreement between calculated and experimental quantities. An overlap of about 20% has been found to produce the best agreement with experiment in many cases (see for example the MS-Xα calculation on ethylene by *Rösch* et al. [5.21] and on benzene by *Case* et al. [5.22]. The experience of the author has been that a sphere overlap of 20% also produces the best agreement between calculated and experimental NEXAFS spectra, provided that the transition state potential is used. Shape resonance positions are fairly sensitive to the extent of overlap but with an overlap of 20% the calculated resonance position is usually within 1 eV of experiment.

Bound states and continuum states require separate calculations. Oscillator strengths for each bound state transition must be converted into cross sections using the relationship given by *Dehmer* and *Dill* [5.9]:

$$\sigma(E) = (\pi e^2 h/mc) \, df/dE = (\pi e^2 h/mc)f \, dn/dE \qquad (5.2)$$

where f is the oscillator strength and n is the principal quantum number of the level. The height of the resonance is given by $\sigma(E)$ and the width by $(dn/dE)^{-1}$. The latter quantity is not well defined for valence levels so some value for the resonance width must be chosen and the height adjusted accordingly. The obvious choice is to use the observed peak width in the experimental NEXAFS spectrum. If no experimental width is available, then some "typical" resonance width in the range 1-2 eV can be used. For the purpose of producing a simu-lated NEXAFS spectrum for comparison with experiment the bound state resonances are represented by Lorentzian (or Gaussian) functions with the chosen linewidth and height $\sigma(E)$ from the oscillator strength conversion formula.

Oscillator strengths are most efficiently calculated for a muffin-tin type of potential using the acceleration form of the transition matrix element

$$M = \langle f | \nabla V | i \rangle \qquad (5.3)$$

because in the intersphere region the potential is constant and $\nabla V=0$, so the matrix element becomes simply the sum of terms over the radi-ally symmetric atomic and outer sphere regions. Expressions for these terms were derived by *Noodleman* [5.23], who calculated the oscillator strengths for valence level transitions in a number of metal complexes, with mixed success. However, for the core-orbital transitions observed in NEXAFS spectra, which are essentially localized on a particular atom, *Noodleman's* method does an excellent job of reproducing the experimental results and has been used for all the calculations reviewed in the following section.

For the continuum states the acceleration form of the matrix element is again the most efficient for calculation of the absorption cross section. This method was used by *Davenport* [5.8] to calculate the uv photoionization cross sections for the valence levels of N_2 and CO. Davenport's program has been modified to allow the use of a core orbital as the initial state.

5.4 More Complex Adsorbates

5.4.1 Ethylene and Ethylidyne on Pt{111}

Multiple scattering calculations have been used to clear up an apparent anomaly in the NEXAFS spectrum of ethylene chemisorbed on Pt{111} [5.14,17]. Surface vibrational spectroscopy indicates that the most probable configuration of the chemisorbed ethylene on Pt{111} is the so-called "di-σ" configuration in which the ethylene is bonded to two

190

adjacent surface atoms and has rehybridized to sp^3 [5.25]. The π-bonded configuration where the midpoint of the C-C bond is directly over a surface Pt atom was considered to be unlikely from a comparison of the vibrational frequencies of the surface species with those of π-bonded ethylene in Zeises salt.

The NEXAFS spectrum of ethylene on Pt{111} at 70K is shown in Fig.5.2, for perpendicular and grazing incident radiation. It can be seen that the bound state resonance completely disappears for perpendicular incident radiation. This means that the ethylene $π^*$-orbital retains its identity in the adsorbed molecule, which suggests a π-bonded configuration rather than a sp^3 hybridized configuration [5.24]. However, in the calculated NEXAFS spectrum for a cluster representing ethylene di-σ bonded to two Pt atoms, shown in Fig.5.3, the bound state resonance shows the same behavior as in the experimental spectrum, with negligible intensity for radiation perpendicular to the C-C bond. An orbital contour plot of the upper state of the main peak in the bound state resonance reveals that the orbital is essentially an antibonding combination of a $π^*$ ethylene orbital and the d-orbitals on the Pt atoms. The experimental NEXAFS spectrum is therefore consistent with a di-σ bonded configuration. In fact, the calculated NEXAFS spectrum for a cluster representing π-bonded ethylene agrees rather less well with experiment than the calculation for the di-σ bonded cluster. A C-C bond length of 1.49 Å has been deduced for the chemisorbed ethylene, using the relationship between $σ^*$ resonance position and bond length in gas phase hydrocarbons. This is close to the typical

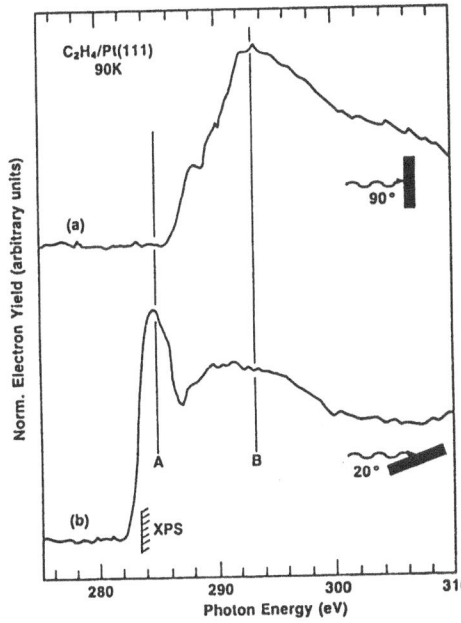

Fig.5.2. Experimental NEXAFS spectra at the C K-edge of ethylene adsorbed on Pt(111) at 90 K for incidence angles of 20° and 90°

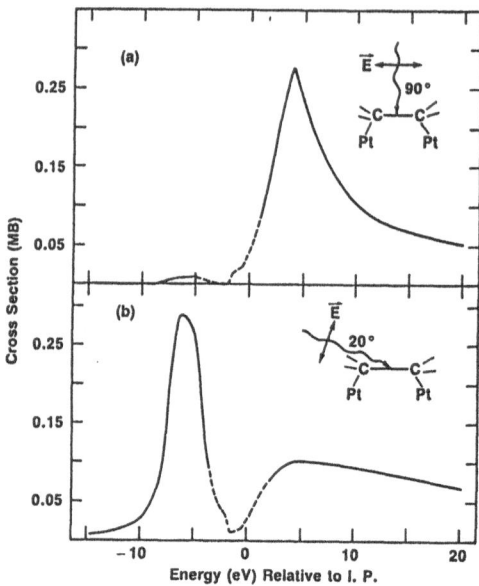

Fig.5.3. Calculated K-edge NEXAFS spectra for $Pt_2C_2H_4$ di-σ bonded cluster at incidence angles 90° and 20° with respect to the C-C axis. The cross section shown is the calculated integrated cross section divided by 4π

C-C single bond distance, which would also indicate sp^3 hybridized ethylene rather than π-bonded ethylene, for which the bond length increase should be much smaller.

At room temperature on Pt{111} the adsorbed ethylene loses hydrogen, leaving the ethylidyne species (CCH_3). In the NEXAFS spectrum of this species [5.17] the σ^* shape resonance shows the expected behavior for a molecule with the C-C axis perpendicular to the surface, but there is a prominent bound state resonance for both perpendicular and grazing incidence. A MS-Xα calculation for a cluster consisting of ethylidyne bonded to three Pt atoms showed that the bound state resonances correspond to transitions to antibonding combinations of the Pt d-orbitals and p-orbitals on the adjacent carbon atom. For grazing and perpendicular incidence the appropriate carbon p-orbitals are respectively perpendicular and parallel to the surface plane. Calculated NEXAFS spectra for this cluster reproduce the prominent bound state resonances for both directions of the incident radiation.

5.4.2 Benzene on Pt{111}

For the diatomic or pseudodiatomic adsorbates considered so far the NEXAFS spectra have a very simple structure consisting of at most one bound state resonance and one continuum resonance. The K-edge spectra of more complex molecules usually show several resonances and are more difficult to interpret. Without a full interpretation it is often difficult to relate the K-edge spectrum of the gas phase species to the NEXAFS spectrum of the adsorbed species in an unambiguous way, and changes in the molecular geometry on chemisorption cannot be measured with confidence. Multiple scattering calculations can lead to the assignment of the various resonances so that an unambiguous interpretation of the spectrum of the adsorbate can be made.

A good example of the use of a multiple scattering calculation to shed light on a complex spectrum is provided by the benzene molecule. The experimental gas phase carbon K-edge spectrum of benzene, obtained by high-energy electron energy loss spectroscopy, and a calculated K-edge spectrum obtained from a MS-Xα calculation, are shown in Fig.5.4 [5.26]. Peaks A and B correspond to transitions to empty π^* states, the e_{2u} and b_{2g} π^* levels. Peaks C and D correspond to three σ^* resonances, with peak D consisting of two overlapping resonances. The calculated NEXAFS spectrum for polarized radiation incident at 90° and 20° to the plane of the ring is shown in Fig.5.5. Although no surface metal atoms were included in this calculation, the spectra in Fig.5.5 can be taken to represent the spectrum of chemisorbed benzene for perpendicular and grazing incidence, if the interaction with the surface does not produce significant shifts in the resonances. With the calculated NEXAFS spectra as a guide, the gas phase and surface spectra can be related unambiguously. Figure 5.6 shows the spectra of benzene in the gas, solid, and monolayer (on Pt{111}) states. The pol-

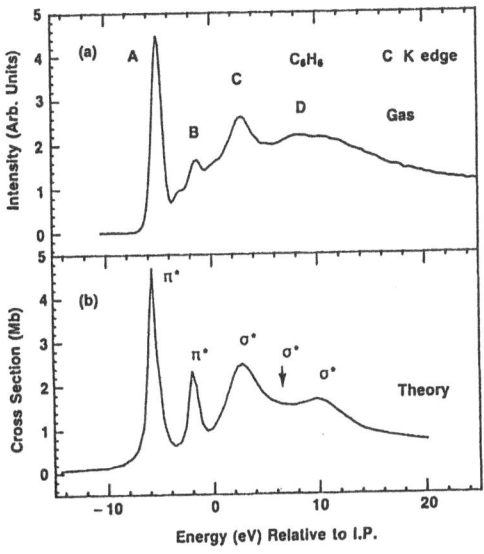

Fig.5.4. Comparison of the experimental and calculated gas phase carbon K-edge spectra of benzene. The calculated spectrum has been shifted so that the ionization threshold matches that determined experimentally by XPS

193

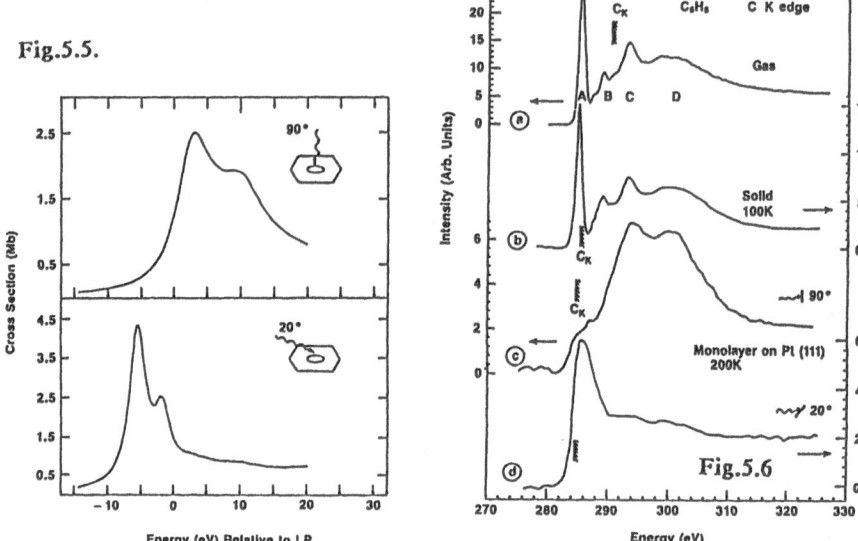

Fig.5.5.

Fig.5.6

Fig.5.5. Calculated carbon K-edge spectrum of benzene for photoabsorption of linearly polarized light incident at a) 90° and b) 20° to the plane of the ring

Fig.5.6 a. Carbon K-shell spectrum of gaseous benzene recorded by electron energy loss of 2.8 keV electrons. b. Carbon K shell NEXAFS spectrum of solid benzene recorded by partial electron yield from a multilayer condensate on Pt{111} at 100K. c. Partial electron yield NEXAFS spectrum at normal incidence of monolayer benzene on Pt{111} at 100K after heating the sample in Pt.b to 200 K. d. Monolayer benzene NEXAFS spectrum recorded at glancing (20°) x-ray incidence

arization dependence of the monolayer resonances together with the results of the multiple scattering calculation indicate that the two peaks observed at perpendicular incidence are indeed σ^* resonances that correspond to the resonances C and D in the gas phase spectrum. Resonances A and B appear to broaden and merge into a single resonance in the spectrum of the adsorbed species.

Sette et al. [5.27] have shown that for noncyclic aliphatic hydrocarbons (and similar noncyclic molecules) a given shape resonance can usually be associated with a given bond, or type of bond, in the molecule. In the case of benzene, however, the σ^* resonances correspond to states that are delocalized over the whole ring. The delocalized states can be regarded as arising from the interaction of a set of localized σ^* antibonding states, each of these states being localized on one of the C-C bonds in the ring, rather as the delocalized bonding states are built up from localized C-C bonding states by *Jorgensen* and *Salem* [5.28]. This suggests that an *intensity-weighted average* of the positions of the observed resonances should be used in the correlation of the shape resonance positions with bond length for benzene and other aromatic molecules. The intensity-weighted average energy of peaks C and D (8.8 ± 0.5 eV relative to the ionization threshold) is in good agreement with the predicted position, 8.5 ± 0.5 eV, for a single σ^* shape resonance corresponding to a bond length of 1.4 Å.

5.4.3 Saturated Cyclic Hydrocarbons

Similar delocalized σ^* resonances are seen in the carbon K-edge spectra of gas phase and chemisorbed saturated cyclic hydrocarbons, such as cyclopropane, cyclobutane and cyclohexane [5.30]. In the case of cyclopropane and cyclobutane the σ^* resonances can be identified with states that are "outside" and "inside" combinations of carbon p-orbitals, where the p-orbitals are all pointing toward the center of the ring or are all perpendicular to this direction. The assignments are shown for cyclopropane and cyclobutane in Fig.5.7. The assignments are based on the calculated polarization dependence of the resonances in a MS-Xα calculation. There are as yet no experimental NEXAFS spectra for monolayer cyclopropane and cyclobutane, although a spectrum for multilayer cyclobutane, which shows no polarization dependence, has been reported. Analogous delocalized σ^* resonances have

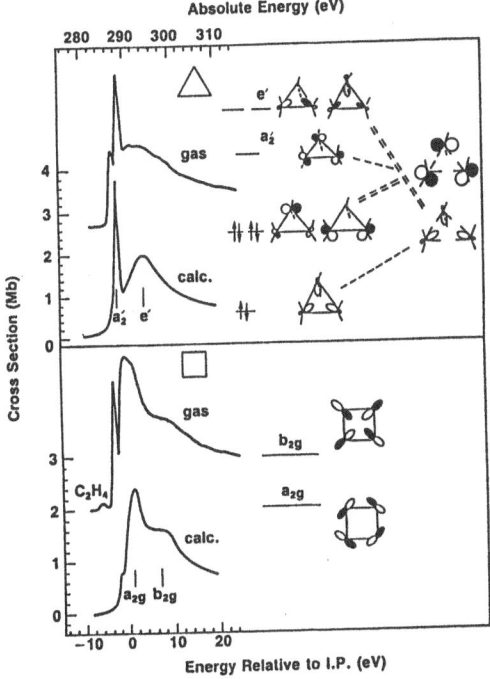

Fig.5.7. Gas phase carbon K-edge spectra for cyclopropane and cyclobutane recorded by electron energy loss of 2.8 keV electrons. The antibonding states that correspond to the upper states of the principal observed resonances are also shown

195

been assigned in the carbon K-edge spectrum of cyclohexane, based on MS-Xα calculations. Both multilayer and monolayer NEXAFS spectra for cyclohexane on Pt{111} have been reported. The observed polarization dependence of the σ^* resonances indicates that the molecules lie parallel to the surface in the monolayer state, and there is partial orientation of the molecules even in the multilayer state.

In the K-edge spectra for the cyclic hydrocarbons shown in Fig.5.7 a fairly intense bound state resonance can be seen below the σ^* resonances. MS-Xα calculations indicate that these resonances correspond to transitions to states of mixed Rydberg-valence character, the valence component being C-H antibonding. The orientation of the carbon p-orbitals in this antibonding combination is perpendicular to the plane of the ring, so the resonances have (C-H) π^* character, and their calculated polarization dependence is the opposite of that of the σ^* resonances. The corresponding resonance for cyclohexane in the monolayer state does indeed show a polarization dependence that is the opposite of the σ^* resonance polarization dependence. The MS-Xα calculations appear to overestimate the contribution of the Rydberg component of this state and underestimate the valence contribution, because the calculated intensities are significantly lower than the observed intensities for the mixed states.

Similar mixed-character C-H resonances have been observed in the K-edge spectra of other hydrocarbons and analogous O-H and N-H resonances have been observed in the K-edge spectra of H_2O and NH_3 [5.31]. In the case of ethylene C-H resonances have been identified both in the gas phase or for ethylene chemisorbed on Ag{100} and Ag{110} [5.32]. The K-edge spectrum of polymeric hydrocarbons such as polyethylene also exhibit strong C-H resonances [5.32], which have been assigned with the help of MS-Xα calculations. The identification of C-H resonances is of some importance as it allows NEXAFS studies of the orientation of C-H bonds and of hydrogenation and dehydrogenation reactions.

5.4.4 Thiophene and Platinum Metallacycle

Perhaps the most important contribution that multiple scattering calculations can make to the interpretation of NEXAFS spectra of chemisorbed species is to help identify an unstable reaction intermediate in a surface reaction. Often, appropriate model compound references are not available to provide spectra that can be compared with the spectrum of the intermediate. In these cases multiple scattering calculations on cluster models of the intermediate may provide the best supporting evidence for the attribution of the observed spectrum to a particular species.

A good example of the use of multiple scattering calculations to help identify a surface intermediate is provided by the proposed metallacycle intermediate in the desulfurization of thiophene on Pt{111} [5.34]. Before discussing the evidence for this intermediate, however, we must first examine the carbon K-edge spectrum of thiophene.

The K-edge spectrum of thiophene has been discussed in some detail by *Hitchcock* et al. [5.34] and assignments made on the basis of multiple scattering calculations. The spectrum is somewhat more complex than that of benzene because of the presence of the C-S bond in thiophene. The calculated spectrum for thiophene is shown in Fig.5.8, together with the gas phase K-edge spectrum. The assignments of the calculated resonances are also shown. There are two σ^* shape resonances in the continuum that are delocalized over the ring, similar to the σ^* resonances in benzene. There are two bound state π^* resonances, the resonance at higher energy overlapping a σ^* resonance assigned to the C-S bond. The first peak in the experimental spectrum corresponds to the first π^* resonance, and the second to overlapping π^* and C-S σ^* resonances with approximately equal contributions from each.

Stöhr et al. have used NEXAFS spectroscopy to monitor the desulfurization of thiophene on Pt{111} [5.33]. For thiophene on Pt{111} at 180 K (corresponding to monolayer coverage) the first π^* peak in the spectrum is completely absent for radiation incident perpendicular to the surface plane, showing that the thiophene lies flat on the surface. As the temperature is increased the σ^* resonances undergo an evolution, which is shown in Fig.5.9. In the 180 K spectrum the first peak can be assigned to the C-S σ^* resonance and the two resonances at higher energy are the delocalized σ^* resonances associated with the thiophene ring. Upon heating the C-S σ^* resonance diminishes and eventually disappears, while a new peak appears at lower energy. The changes in the resonances associated with the thiophene ring are relatively minor, however. These results, together with changes observed in the sulfur $L_{2,3}$ edge resonances, led Stöhr et al. to propose the formation of a species with a carbon skeleton similar to thiophene in which the sulfur atom is replaced by a surface Pt atom. The proposed species would be an intermediate in the desulfurization process, undergoing decomposition to butadiene at higher temperatures.

There are no suitable organometallic reference compounds that could provide supporting evidence for the proposed intermediate, but multiple scattering calculations on the metallacycle can be carried out without difficulty. Figure 5.10 shows the calculated carbon K-edge spectrum for the platinum metallacycle for polarized radiation incident perpendicular to the plane of the ring [5.35], together with the corresponding thiophene spectrum for comparison. The differences between

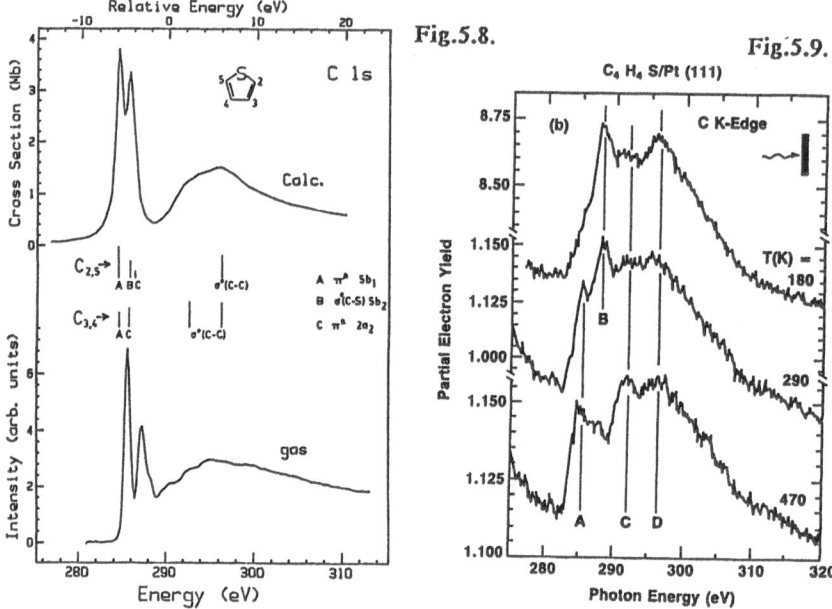

Fig.5.8. Comparison of calculated and experimental gas phase carbon 1s spectra of thiophene. The scale for the calculated cross section refers to the value per carbon atom, averaged over the $C_{2,5}$ and $C_{3,4}$ sites

Fig.5.9. NEXAFS spectra at normal x-ray incidence for the C K-edge of thiophene on Pt{111} after annealing to various temperatures

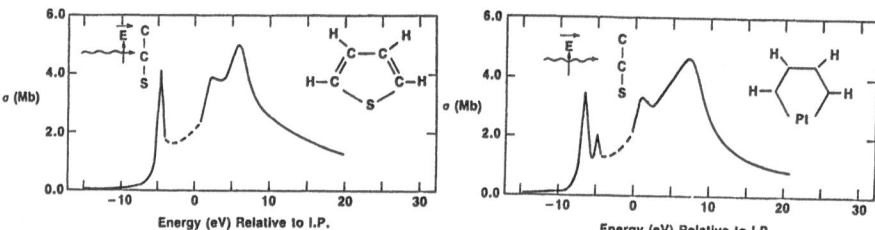

Fig.5.10. Calculated carbon K-edge spectra for thiophene and platinum metallacycle for radiation incident perpendicular to the plane of the ring

the calculated spectra for thiophene and the metallacycle largely parallel the changes in the experimental spectra for the surface species shown in Fig.5.9. The C-S σ^* resonance in thiophene is replaced by two Pt-C σ^* resonances, the more intense resonance lying at significantly lower energy than the C-S σ^* resonance. The change in the relative intensities of the delocalized σ^* resonances, although minor, also reproduces the observed change fairly accurately. The calculation shows that the delocalized σ^* resonances do serve as a "fingerprint" for the particular ring skeleton found in thiophene and the metallacycle. The fact that these resonances do persist in the NEXAFS spectrum of the intermediate while the changes in the region of the C-S σ^* resonance follow the predicted changes for the formation of Pt-C bonds, provides strong support for the existence of the metallacycle intermediate. Other candidates would be expected to show much greater changes in the region of the delocalized σ^* resonances (above 290 eV), a prediction that can be tested by further calculations.

5.5 Conclusions

At present, multiple scattering calculations are the only method for calculating both bound and continuum states of complex molecules and clusters with reasonable accuracy and efficiency. Using cluster models for the adsorbate-substrate system, multiple scattering calculations can provide interpretation of the NEXAFS spectra of adsorbates that enable an assignment to be made among various possible adsorption sites or bonding configurations. They can be used to relate unambiguously the spectrum of the gas phase molecule to the spectrum of the chemisorbed molecule, leading to information on changes in geometry on chemisorption. Finally, and perhaps most importantly, multiple scattering calculations can provide calculated reference spectra that can help to identify unstable intermediates in surface reactions where no experimental reference spectra are available.

Future development of the method will probably focus on obtaining a more accurate reproduction of the adsorbate NEXAFS spectrum. The effect of including the substrate atoms in the calculation of adsorbate continuum states needs to be investigated more thoroughly. The arctangent-like background in the NEXAFS spectra of molecular adsorbates also requires explanation. In many cases multiple scattering calculations, although predicting resonance positions with good accuracy, do not do a particularly good job of reproducing resonance widths. The calculated resonances are often considerably narrower than experiment even for gas phase molecules. For surface adsorbates there is an additional broadening due to adsorbate-substrate interaction which has not yet been reproduced by a calculation.

Even with the simple models used so far, however, significant progress has been made. Multiple scattering calculations are certain to become increasingly useful as NEXAFS spectra for more complex adsorbates (including polymers) are obtained.

Acknowledgments. The author is grateful to the Department of Chemistry, Lehigh University and the Zettlemoyer Center for Surface Studies for hospitality. Support was provided by the U.S. Department of Energy under Contract No. DE-AC22-85PC80014 and Contract No. DE-AC02-83ERI3090.

References

5.1   J. Stöhr: Z. Physik **B61**, 439 (1985)

5.2   J. Stöhr, F. Sette, A.L. Johnson: Phys. Rev. Lett. **53**, 1684 (1984)

5.3   R.R. Lucchese, G. Raseev, V. McKoy: Phys. Rev. **A25**, 2572 (1982)
R.R. Lucchese, V. McKoy: Phys. Rev. **A28**, 1382 (1983)
M.G. Smith, R.R. Lucchese, V. McKoy: J. Chem. Phys. **79**, 1360 (1983)

5.4   K.H. Johnson: Adv. Quantum Chem. **7**, 143 (1973)

5.5   J.C. Slater, T.M. Wilson, J.H. Wood: Phys. Rev. **179**, 28 (1969)

5.6   K.H. Johnson: J. Chem. Phys. **45**, 3085 (1966)

5.7   D. Dill, J.L. Dehmer: J. Chem. Phys. **61**, 692 (1974)

5.8   J.W. Davenport: Phys. Rev. Lett. **36**, 945 (1976)

5.9   J.L. Dehmer, D. Dill: J. Chem. Phys. **65**, 5327 (1965)

5.10 F.W. Kutzler, C.R. Natoli, D.K. Misemer, S. Doniach, K.O. Hodgson: J. Chem. Phys. **73**, 3274 (1980)

5.11 P.J. Durham, J.B. Pendry, C.H. Hodges: Comput. Phys. Commun. **25**, 193 (1982)

5.12 D. Norman, J. Stöhr, R. Jaeger, P.J. Durham, J.B. Pendry: Phys. Rev. Lett. **51**, 2052 (1983)

5.13 R.P. Messmer: In *The Nature of the Surface Chemical Bond*, ed. by T.N. Rhodin, G. Ertl (North Holland, Amsterdam 1979)

5.14 S.H. Chou, F.W. Kutzler, D.E. Ellis, P.L. Cao: Surf. Sci. **164**, 85 (1985)

5.15 J. Stöhr, R. Jaeger: Phys. Rev. **B26**, 4111 (1982)

5.16 E.J. Baerends, D.E. Ellis, P. Ros: Chem. Phys. **2**, 41 (1973)

5.17 J.A. Horsley, J. Stöhr, R.J. Koestner: J. Chem. Phys. **83**, 3146 (1985)

5.18 P.S. Bagus, H.F. Schaeffer III: J. Chem. Phys. **56**, 224 (1972)

5.19 G. Loubriel: Phys. Rev. **B20**, 5339 (1979)

5.20 R. Latter: Phys. Rev. **81**, 385 (1955)

5.21 N. Rösch, W.G. Klemperer, K.H. Johnson: Chem. Phys. Lett. **23**, 149 (1973)

5.22 D.A. Case, M. Cook, M. Karplus: J. Chem. Phys. **73**, 3294 (1980)

5.23 L. Noodleman: J. Chem. Phys. **64**, 2343 (1976)

5.24 R.J. Koestner, J. Stöhr, J.L. Gland, J.A. Horsley: Chem. Phys. Lett. **105**, 332 (1984)

5.25 H. Steininger, H. Ibach, S. Lehwald: Surf. Sci. **117**, 685 (1982)

5.26 J.A. Horsley, J. Stöhr, A.P. Hitchcock, D.C. Newbury, A.L. Johnson, F. Sette: J. Chem. Phys. **83**, 6099 (1985)

5.27 F. Sette, J. Stöhr, A.P. Hitchcock: J. Chem. Phys. **81**, 4906 (1984)

5.28 W.L. Jorgensen, L. Salem: *The Organic Chemists Book of Orbitals* (Academic, New York 1973)

5.29 E. Keulen, F. Jellinek: J. Organomet. Chem. **5**, 490 (1966)
B. Rees, P. Coppens: Acta Crystallogr. Sect. **B29**, 2516 (1973)

5.30 A.P. Hitchcock, D.C. Newbury, I. Ishii, J. Stöhr, J.A. Horsley, R.D. Redwing, A.L. Johnson, F. Sette: J. Chem. Phys. **85**, 4849 (1986)

5.31 R. Jaeger, J. Stöhr, T. Kendelewicz: Surf. Sci. **134**, 547 (1983)

5.32 J. Stöhr, D.A. Outka, K. Baberschke, D. Arvanitis, J.A. Horsley: Phys. Rev. B **36**, 2976 (1987)

5.33 J. Stöhr, J.L. Gland, E.B. Kollin, R.J. Koestner, A.L. Johnson, E.L. Muetterties, F. Sette: Phys. Rev. Lett. **53**, 2161 (1984)

5.34 A.P. Hitchcock, J.A. Horsley, J. Stöhr: J. Chem. Phys. **85**, 4835 (1986)

5.35 J.A. Horsley: unpublished calculations

6. Near-Edge X-Ray Absorption Fine Structure Spectroscopy: A Probe of Local Bonding for Organic Gases Solids, Adsorbates, and Polymers

D.A. Outka and J. Stöhr

IBM Almaden Research Center, 650 Harry Rd.
San Jose, CA 95120- 6099, USA

The near edge X-ray absorption fine structure (NEXAFS) of the carbon K-shell is shown to be a sensitive probe of certain types of bonds and functional groups in organic molecules, regardless of whether the molecules are: isolated in the gas phase, condensed, chemisorbed on a surface, or a long chain polymer. To demonstrate this principle, the K-shell excitation spectra of molecules containing similar bonds and/or functional groups are presented and compared in the gas phase, the solid state and adsorbed on surfaces. From a systematic study of molecules with increasing size, guidelines are established which allow the understanding of the NEXAFS spectra of macromolecules and polymers based on a simple building block picture. As examples we discuss the spectra of single layer Langmuir-Blodgett and thin polymer films on surfaces and show that they can be interpreted in like manner to those of simple isolated molecules. We also discuss limitations of the building block approach arising from conjugation effects in molecules or band structure formation in three dimensional periodic solids.

6.1 Development of NEXAFS: An Overview

Inner shell excitation spectroscopy of molecules has been an active research field since the early 1970's [6.1,2]. Most of these studies involved low-Z molecules and were carried out at the K-shell of carbon, nitrogen and oxygen using inelastic electron energy loss spectroscopy (IEELS). For the study of free molecules the IEELS technique offers an attractive alternative to conventional X-ray absorption spectroscopy and yields almost identical results [6.1]. The increasing interest in understanding the properties and structure of molecules bonded to surfaces led to the development of a technique in the early 1980's [6.3,4] which allowed the recording of K-shell spectra of chemisorbed molecules. This technique, now known as the near edge X-ray absorption fine structure (NEXAFS) utilizes high intensity monochromatic and polarized synchrotron radiation for the K-shell excitation process and electron [6.5] or photon [6.6] detection techniques to monitor the X-ray absorption probability. In terms of instrumentation NEXAFS utilizes the same equipment as the earlier developed surface X-ray absorption fine structure (SEXAFS) technique [6.7] which was first demonstrated in 1978 [6.8].

In NEXAFS spectroscopy an inner shell electron is excited to unoccupied electronic states near the ionization limit. This spectral region is immediately below that of SEXAFS which extends from approximately 30 eV to several hundred eV above the ionization limit. The interpretation of the near edge structures differs from SEXAFS in that the final states are determined by the molecular potential of the valence electrons and cannot be described as simple backscattering events off the cores of neighbor atoms. For example, in a simple one-electron picture of the near-edge core excitation event, the final states can be considered to be unoccupied molecular orbitals. By examining the presence and character of these unoccupied molecular orbitals, near-edge spectroscopy provides a probe of the bonding in molecules. Among the capabilities of near-edge core level excitation spectroscopy are: the ability to detect the presence of certain bonds in molecules (e.g., C-C, C=C, C≡C, C-O, C=O, C≡O, and C-H) [6.9]; to determine the orientation of molecules in ordered systems [6.4, 10]; and to estimate the lengths of bonds between low-Z atoms [6.9, 11].

This chapter will primarily concern itself with the first mentioned ability of near-edge spectroscopy: to detect and distinguish between various types of bonds. To accomplish this goal, we will present a variety of spectra of increasingly larger molecules and outline systematics which enable the understanding of the spectra of complex molecules such as polymers. Our discussion will mostly concentrate on *organic* molecules and intermediates since except for a few simple cases (e.g., CO, N_2 etc.) most low-Z molecules fall into this category. It should be mentioned, however, that a considerable amount of work has been carried out on intermediate and high-Z *inorganic* molecules and metal complexes using near-edge X-ray absorption spectroscopy [6.12-19].

The near-edge K-shell excitation spectra of dozens of molecules have now been measured either in the gas phase using IEELS [6.2] or absorbed on surfaces using NEXAFS [6.20]. Over the past few years as the understanding of near-edge spectroscopy has increased, these studies have been the subject of several reviews. In early NEXAFS studies the primary interest was simply to demonstrate that the spectral features were, indeed, related to molecular bonding. Thus one of the first reviews concerned itself principally with the spectra of diatomic or pseudo-diatomic molecules such as carbon monoxide and methoxy groups (CH_3O-) [6.21]. By the time of this early review, the spectra of a few larger molecules such as benzene had been measured but were incompletely understood. The next major development was the realization that certain near-edge features could be used to estimate bond lengths between low-Z atoms in both gas phase and chemisorbed molecules [6.22-24]. Since that time there has been a steady progression in the study and understanding of the near-edge spectra of larger and more complex molecules. This includes an understanding of the spectra of both more complex bonding units such as cyclic [6.25] and heterocyclic [6.26-28] hydrocarbons, aromatic rings [6.29] and of the spectra of polyfunctional molecules [6.20,30,31]. Most recently, NEXAFS spectroscopy has been applied to long chain molecules such as in LB (Langmuir-Blodgett) films [6.32,33] and polymers [6.34,35]. The present study continues this progression to larger molecules by presenting new spectra for a variety of single-layer LB films and polymer samples, showing that they can be interpreted by comparison to the spectra of simpler monofunctional molecules. In addition, several comparisons are presented showing the similarities between the spectra of chemically similar molecules in the gas phase, in the condensed state, chemisorbed on surfaces, and in a polymer. This demonstrates that near-edge spectroscopy is a probe of the *local* chemical bonding in a molecule, largely unaffected by changes in molecular size, the presence of other functional groups in a molecule or the phase of the molecule, i.e. gas, solid or chemisorbed. Finally, we outline limitations of the simple building block approach used here for the understanding of the spectra of large molecules. At present, the NEXAFS spectra of three dimensional periodic solids such as diamond and graphite can only be partly understood in terms of the spectra of the individual sp^3 and sp^2 C-C bonds or ring-like building blocks such as benzene and cyclohexane. Rather, band structure effects modify the spectra to a degree that a one-to-one correlation between local bonds and the observed near-edge structures is lost.

6.2 Experimental Details

As mentioned above core level excitation spectroscopy is experimentally investigated by two principal techniques. In the gas phase, IEELS is chiefly employed [6.36] while for surface and solid state studies the technique used is usually NEXAFS which sometimes is referred to as XANES (X-ray absorption near edge structure) [6.37]. The IEELS technique is usually operated under conditions of high (several keV) primary beam energy and small momentum transfer such that its results approximate those of a NEXAFS experiment [6.36].

Some of the NEXAFS spectra presented in this paper are new and were measured on beam line I-1 at the Stanford Synchrotron Radiation Laboratory with use of a grasshopper monochromator (1200 lines/mm holographic grating). Details of the beam line and data aquisition technique have been previously published [6.4, 21]. All NEXAFS spectra of chemisorbed and condensed molecules were carried out under ultra-high-vacuum conditions (about $1 \cdot 10^{-10}$ Torr) by means of partial electron yield detection. The spectra were normalized by the spectra of the clean surface [6.38]. The LB and polymer films on Si substrates were measured by means of total electron yield detection in a chamber which was evacuated without baking to about $1 \cdot 10^{-8}$ Torr. The spectra were divided by the signal from a gold grid reference monitor in order to correct for the transmission function of the monochromator [6.38]. The energy scale was calibrated using the carbon contaminants on the optical surfaces of the monochromator which have two major absorption peaks at 284.7 and 291.0 eV. No radiation damage effects were observed in the samples at the photon flux ($\simeq 2 \cdot 10^9$ photons s^{-1}) used in the present experiments.

The spectra of several single layer LB films on silicon substrates are reported in this work. To prepare the films, the silicon wafers were first dipped in Buffered Oxide Etch (J.T. Baker, Phillipsburg, NJ) in order to remove the native oxide layer along with contaminants. Since the LB films were prepared in air, however, the oxide is expected to have partially regrown. Immediately preceding the transfer of the LB monolayer, the wafer was further cleaned with a solution of No Chro Mix Water (Barnstead, Nanopure grade) and dried in nitrogen flow. Several different LB films were measured including: cadmium arachidate ($Cd(CH_3(CH_2)_{18}CO_2)_2$), cadmium 10,12-di-yne-hexacosanate ($Cd(CH_3(CH_2)_{12}-C \equiv C-C \equiv C-(CH_2)_8CO_2)_2$), (referred to as DIAC) and a semifluorinated cadmium nonadecanate, ($Cd(CF_3(CF_2)_7(CH_2)_{10}CO_2)_2$), (referred to as SFN), all on Si{111} surfaces. The SFN was synthesized by Dr. R.J. Twieg at IBM Research Laboratory in San Jose and the DIAC was a gift of Dr. B. Tieke of Ciba Geigy; and the LB monolayers were prepared on a commercially available film balance (Joyce Loebl) with a solid teflon trough instead of the originally supplied glass tank. The solutions used to prepare the LB films were $2.5 \cdot 10^{-4}$ mol·l^{-1} CdCl$_2$ except for the DIAC for which $1.0 \cdot 10^{-3}$ mol·l^{-1} CdCl$_2$ was used. The pH was maintained between 7 and 8 with a Na$_2$CO$_3$ buffer except for SFN for which the pH was held at 8.4 using NH$_4$OH, instead. The transfer pressure used was 30 mN·m^{-1} except for SFN for which a transfer pressure of 40 mN·m^{-1} was used.

The polymer samples examined in this study were thin films spun onto Si{111} wafers at $2 \cdot 10^3$ rpm, room temperature, and from a toluene solution. Among the polymer spectra reported in this chapter are polystyrene, $[-CH_2-CH(C_6H_5)-]_n$, (referred to as PSTY); polymethylmethacrylate, $[-CH_2-C(CH_3)(CO_2CH_3)-]_n$, (referred to as

PMMA); and polybutadiene $[-CH_2-CH = CH-CH_2-]_n$, (referred to as PBD). The concentrations of the solutions were: $2.706 \cdot 10^{-2}$ g·ml^{-1} for the polystyrene; $1.256 \cdot 10^{-2}$ g·ml^{-1} for PMMA; $3.85 \cdot 10^{-2}$ g·ml^{-1} for the low molecular weight polybutadiene; and $3.59 \cdot 10^{-2}$ g·ml^{-1} for the high molecular weight polybutadiene. The films were then heated to 340 K for 4 h under N_2 flow.

6.3 Results and Analysis

In this section, the core level excitation spectra of a variety of molecules ranging from small isolated gas phase molecules to polymers will be presented and compared. The spectra are organized to demonstrate a few general guidelines to use in the qualitative interpretation of core level excitation spectra. The emphasis is on showing how the spectra of small molecular building blocks in the form of diatomics and/or larger functional groups can be used to understand the spectra of molecules with increasing complexity and size.

6.3.1 The Signatures of Individual Bonds

Characteristic resonances or signatures for selected small molecules with C-H, C-O and C-C bonds are shown in Fig.6.1. The figure also compares IEELS spectra of free molecules with angle integrated NEXAFS spectra of the same chemisorbed molecules. Two important points can be made from inspection of the spectra. First, each molecule has its characteristic NEXAFS spectrum, second, the spectra of free and chemisorbed molecules are very similar.

Gas and Chemisorbed Molecules

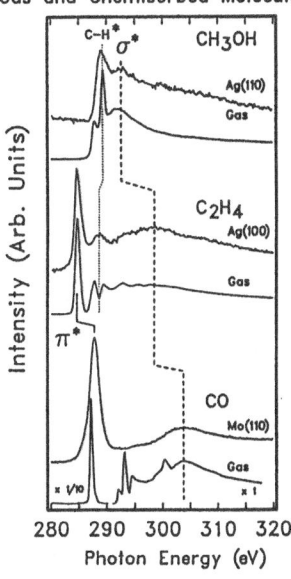

Fig.6.1. Comparison of IEELS spectra of various simple gas-phase molecules with angle-averaged NEXAFS spectra of the same molecules chemisorbed on metal surfaces. The spectra demonstrate that different molecules have characteristic K-shell excitation spectra and that the spectra of free and chemisorbed molecules are remarkably similar. The main resonances are labelled according to the symmetry of the corresponding antibonding molecular orbitals which are the final states of the observed K-shell excitations.

The dominant resonances observed in the spectra in Fig.6.1 can be directly associated with transitions to antibonding molecular orbitals. They can be labelled according to the molecular symmetry of the final state orbital and are called π^* and σ^* resonances, for short. In particular, the absence or presence of the π^* resonance signifies the hybridization of the intramolecular bond, i.e. whether the molecule is π bonded or not. For example, triple bonded CO exhibits a prominent C-O π^* resonance while single bonded methanol does not. Also, the σ^* resonance position is a direct measure of the intramolecular bond length [6.9, 11]. This is evident from the large shift of the C-O σ^* resonance position between methanol (C-O bond length 1.425Å) and CO (1.128Å). If present, the strongest resonance is associated with π^* C-O or C-C bonds. As clearly revealed by the methanol and ethylene spectra, C-H resonances can also be observed [6.39].

The difference between the spectra of free and chemisorbed molecules lies mainly in the disappearance of certain sharp resonances, i.e. the peaks between 290 eV and 302 eV for CO. These resonances are known to originate from transitions to Rydberg states below the 1s ionization potential (296.2 eV) and a two-electron excitation (at 301 eV) involving π orbitals [6.40]. Upon chemisorption, these resonances vanish because the Rydberg orbitals are significantly affected by the bond to the surface owing to their large spatial extent. The overlap with metal states reduces the life-time of the Rydberg state and the resonances are broadened beyond recognition. Also, multielectron excitation energies and intensities are often affected by the surface bond. In CO, for example, the double excitation at 301 eV in the gas phase spectrum disappears upon chemisorption because of the interaction of the π orbitals with the surface. This is also indicated by the significant broadening of the π^* resonance at 288 eV. In contrast, the resonances associated with molecular antibonding orbitals survive the chemisorption process because of their localized nature. Surprisingly, they not only survive, but in case of an unchanged intramolecular geometry, they even remain at the same excitation energies. These effects are the foundation of NEXAFS spectroscopy since they make possible the understanding of the spectra of chemisorbed molecules from those of free molecules.

6.3.2 Assembly of Diatomics to Functional Groups

As discussed above the NEXAFS resonances of diatomic and pseudo-diatomic molecules can be directly linked to the π and σ bonds in the molecule. In a simple picture larger molecules can be viewed as an assembly of diatomic building blocks. One may therefore envision that the NEXAFS spectra of complex molecules are simply a superposition of those of the diatomic building blocks. The localized nature of the NEXAFS excitation suggests that this simple approach should work.

A beautiful example of the building block picture is provided by the K-shell excitation spectrum of acetonitrile (CH_3CN) [6.41] and its building blocks hydrogen cyanide (HCN) [6.42] and ethane (C_2H_6) [6.43]. As shown in Fig.6.2, the acetonitrile spectrum contains the C-C resonance of ethane and the $C \equiv N$ π^* and σ^* resonances of HCN. There is also an indication of C-H and/or Rydberg resonances in the CH_3CN spectrum in the region 288-292 eV, similar to those in HCN and C_2H_6. Figure 6.2 also shows that the spectrum of the free acetonitrile molecule is almost identical to that of a condensed acetonitrile multilayer and that of acetonitrile chemisorbed on Ag{110}. As discussed for the diatomic or pseudo-diatomic molecules shown in Fig.6.1 the K-shell excitation spectrum of a molecule is almost unaffected by extra-molecular interactions, provided the internal structure of the molecule is preserved. In our example, both the weak chemisorption bond to the Ag{110} surface and the inter-molecular van der Waals interactions in the solid do not change the intramolecular structure.

To test the building block picture further a series of complex alcohols containing C-H, C-O, C-C single, double or triple bond, were examined [6.30]. For these studies methanol (CH_3OH), n-propanol, ($CH_3CH_2CH_2OH$), allyl alcohol ($CH_2 = CHCH_2OH$), and propargyl alcohol ($CH \equiv CCH_2OH$) were condensed on a cold substrate at 120 K so that ice layers were formed. The spectra are shown in Fig.6.3.

Two major features are observed for methanol and are characteristic of the alcohol group. These are a C-H π^* resonance at 289 eV and a C-O σ^* resonance at 293 eV [6.39]. As shown in Fig.6.1, a similar spectrum for methanol is also observed in the gas phase [6.44]. The NEXAFS features associated with the alcohol group are expected to be relatively invariant since the alcohol group varies little structurally from molecule to molecule. In accordance, using methanol as a prototype for the alcohol group, all of the alcohols exhibit the NEXAFS features of this group.

The NEXAFS spectra of the complex alcohols also exhibit the features of the other local bonds in the molecule. For example, the NEXAFS spectrum of propargyl alcohol has a sharp $C \equiv C$ π^* resonance at 285.7 eV and a broad $C \equiv C$ σ^* resonance at 310 eV just like the spectrum of gaseous acetylene [6.9]. Likewise, allyl alcohol has a sharp $C = C$ π^* resonance at 284.4 eV and a $C = C$ σ^* resonance at 300 eV like the 284.8 eV and 302 eV peaks in the NEXAFS of ethylene [6.9]. All of the complex alcohols have a C-C bond, and therefore a peak in the region of 293 eV is expected for the C-C σ^* resonance by comparison to the NEXAFS spectrum of gaseous ethane. Such a peak is present in the case of the complex alcohols, although it over-

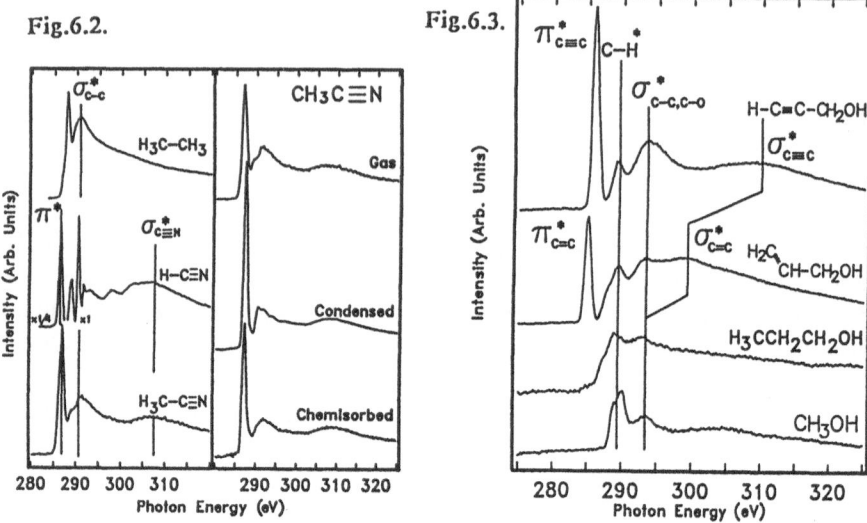

Fig.6.2.

Fig.6.3.

Fig.6.2. Demonstration of the building block principle of K-shell excitation spectra. (a) Gas-phase IEELS spectra of ethane (C_2H_6) [6.43] hydrogen cyanide (HCN) [6.42] and acetonitrile (CH_3CN) [6.41]. The spectrum of acetonitrile can be understood as a supposition of those of the other two molecules. (b) Comparison of K-shell excitation spectra of acetonitrile as a free molecule, condensed as a thick multilayer and chemisorbed as a monolayer on Ag{110}. The insensitivity of the spectra to the phase of the molecule indicates that the main resonances arise from transitions between states that are strongly localized on the molecule.

Fig.6.3. Carbon K edge NEXAFS spectra of condensed propargyl alcohol, allyl alcohol, n-propanol, and methanol at an x-ray incidence angle of 20° [6.30]. These spectra show that the NEXAFS spectra of many molecules are simply a combination of features due to the individual bonds comprising the molecule.

laps with that of the C-O σ^* resonance and is not distinctly resolved. From the above studies it appears that the NEXAFS spectra of these polyfunctional molecules can be described as a superposition of the features expected of the individual diatomic building blocks comprising the molecule.

If the above results were extended to aromatic rings we would expect to find only one σ^* resonance in the NEXAFS spectrum of benzene (C_6H_6), corresponding to the unique C-C bond length of 1.395 Å. The NEXAFS spectrum of benzene, however, exhibits two pronounced σ^* resonances. This is clearly revealed by Fig.6.4 where the NEXAFS spectra for ethylene (C_2H_4) and benzene chemisorbed on Pt{111} are compared. Both spectra exhibit a similar polarization dependence of the first peak A which is readily identified as a π^* resonance. Since in both molecules the π^* orbitals are oriented perpendicular to the C-C bonds, the fact that the π^* resonance is observed at

208

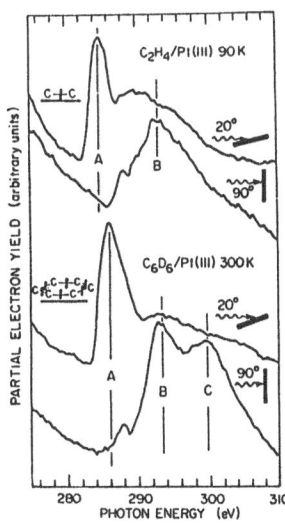

Fig.6.4. Carbon K-edge NEXAFS spectra of ethylene [6.52] and benzene [6.53] monolayers on Pt{111}. Peak A is a C-C π^* resonance and peaks B and C σ^* resonances. The splitting of the σ^* resonances B and C in benzene is caused by resonance interaction of the σ^* antibonding orbitals in the molecule. The indicated orientation on the surface is consistent with the polarization dependence of both the π^* (peak A) and σ^* (peaks B and C) resonance intensities.

grazing incidence (E along surface normal) unambiguously shows that both molecules lie down on the surface. Surprisingly, the benzene spectrum exhibits two σ^* resonances at normal X-ray incidence, rather than the expected one, as observed for ethylene. The presence of two σ^* resonances is not specific to chemisorbed benzene on Pt{111}, but the same two resonances are also observed for free and condensed benzene molecules. Hence they signify a breakdown of the simple picture which associates each σ^* resonance with a specific bond.

The explanation of the NEXAFS spectrum of benzene has been provided by a multiple scattering Xα calculation which shows that multiple resonances may result from "resonance" interactions between localized molecular states [6.29]. In the case of aliphatic hydrocarbons or for the complex alcohols discussed earlier, there is only a weak interaction between localized antibonding σ^* states corresponding to a given type of bond. In contrast, for benzene and other aromatic molecules there is a strong interaction between the localized σ^* states, producing a set of delocalized σ^* states which are significantly separated in energy. In fact, for benzene there are several π^* and σ^* antibonding orbitals, giving rise to two strong transitions to π^* and σ^* states, respectively. In the spectrum in Fig.6.3 the second π^* resonance of benzene is not resolved but it is in fact responsible for the increased width of the π^* resonance relative to ethylene.

Another example where the interaction between adjacent bonds leads to a splitting of resonances is shown in Fig.6.5 which shows the π^* region in the C K edge spectra of various unsaturated carboxylic acids and alcohols [6.30]. Here the splitting is a result of the conjugation of C-C π^* bonds. The top two spectra are of propargyl alcohol and propanoic acid which have C≡C and C=O π^* bonds, respectively.

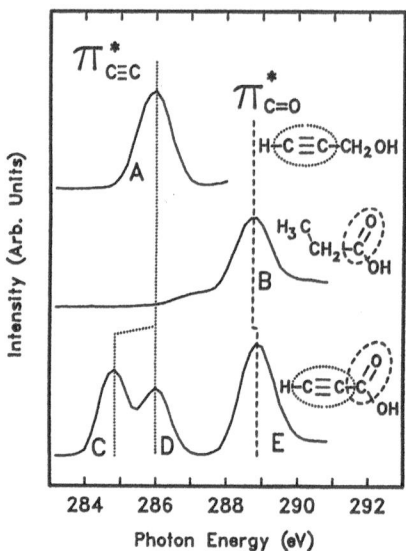

Fig.6.5. This figure shows the effects of conjugation on core level excitation spectra. (a) is the NEXAFS spectrum of condensed propargyl alcohol; (b) is the NEXAFS spectrum of condensed propanoic acid; and (c) is the NEXAFS spectrum of condensed propiolic acid [6.30]. The top two spectra show the π^* resonances of molecules with isolated π^* bonds which are not conjugated. The π^* bonds of the molecule whose spectrum is shown at the bottom are conjugated, however, which is manifest as a splitting of the C≡C π^* resonance into two peaks.

Since the π bonds are isolated they are unaffected by conjugation. Note that in this case the C≡C π^* resonance in the top spectrum is a single peak. In the bottom spectrum of propiolic acid, however, there is interaction between the C≡C π bond and the C=O π bond. This conjugation splits the C≡C π resonance into two components at 284.8 and 286 eV. The angular dependence of these resonances [6.30] shows that the peak labelled "C" is that component of the C≡C bond which is parallel to the C=O π bond, while the peak labelled "D" is that component of the C≡C bond which is perpendicular to the C=O π bond [6.30]. The former is shifted down in energy by the effects of conjugation.

The above examples show how the spectra of diatomics develop into those of larger functional groups. In the absence of conjugation a simple building block picture can be used to assemble the NEXAFS spectra from those of the diatomics. Conjugation between adjacent bonds spoils this simple picture in that it leads to a splitting of resonances as shown for the σ^* resonance in benzene and the π^* resonance in propiolic acid. However, the knowledge of this effect allows one to understand and predict the spectra of larger functional groups.

6.3.3 The Fingerprints of Functional Groups

In our effort to understand and predict the NEXAFS spectra of macromolecules and polymers it appears reasonable to take the building block approach one step further. Above we have discussed the step from diatomics to larger functional groups, e.g. a benzene ring. We could now use the spectra of functional groups as our new, larger, building stones and assemble polyfunctional molecules. The reason behind taking progressive steps rather than trying to directly link the spectra of macromolecules to diatomics is that we can ignore effects like conjugation. We simply accept the fingerprints of certain functional groups as given and use them as new building blocks. Within this picture the signatures of functional groups are taken to depend only upon the local intra-group bonding and not upon external bonds. Before we discuss the assembly of functional groups we shall give examples of some of the important functional constituents of polymers.

A common feature of most polymers is the presence of a hydrocarbon chain backbone consisting of sp^3 bonded carbons. Examination of the core level excitation spectra of a series of saturated hydrocarbon molecules of varying length and geometric form and under various conditions shows that they all exhibit certain common features. Figure 6.6 compares the core excitation spectra of molecules which are composed of sp^3 hybridized C-C bonds including gaseous hexane [6.45]; condensed cyclohexane; a cadmium arachidate LB film which largely consists of a hydrocharbon chain [6.32]; and a polyethylene polymer [6.34]. The spectra of these samples are rather similar, each exhibiting three principal resonances. The first resonance at 288 eV has previously been assigned to a transition to C-H antibonding molecular orbitals [6.34, 45]. The second peak at approximately 293 eV is assigned to the C-C σ^* resonance, and the third peak at approximately 300 eV has also been associated with the C-C σ bonding because of its angular dependence [6.32]. Thus, Fig.6.6 shows that these hydrocarbons have similar spectra. There are a few differences between the spectra which can be partly attributed to angular effects for some of the ordered samples or the presence of other functional groups. For example, the carboxylate group ($-CO_2$) in the arachidate and the CH_3 groups in the gaseous hexane introduce minor differences in the respective spectra from that expected of hydrocarbon chains containing only CH_2 groups [6.45]. Despite these minor differences these three peaks can be considered a fingerprint for a hydrocarbon chain and will be repeatedly observed in the spectra of the various polymer films discussed later.

Spectra for another important functional group, benzene, are shown in Fig.6.7. Among the spectra shown are gaseous benzene, condensed benzene, benzene chemisorbed on a Mo{110} surface, and PSTY. The spectrum for chemisorbed benzene has been angle averaged. The π^* resonance is broadened and both the π^* and σ^* resonances are shifted due to the interaction with the surface which causes a slight change in the molecular structure [6.46]. In each spectrum the characteristic resonances of a benzene ring are apparent. The most prominent feature is the $\pi_{\phi 1}^*$ transition (peak 1) at approximately 285 eV which is readily apparent in the spectra of each of the molecules just listed. Two other resonances (peaks 4 and 5) are also apparent in each of the benzene containing spectra: $\sigma_{\phi 1}^*$ and $\sigma_{\phi 2}^*$ at 293

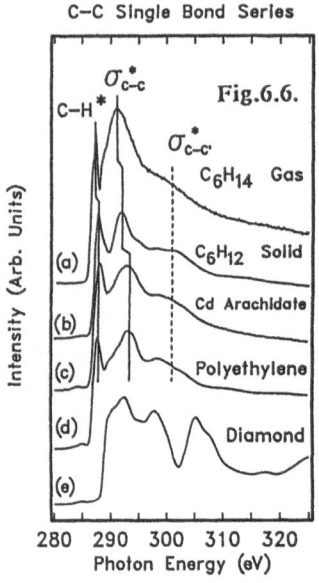

C–C Single Bond Series

Fig.6.6.

σ^*_{c-c}

C–H*

$\sigma^*_{c-c'}$

C_6H_{14} Gas

(a)

C_6H_{12} Solid

(b)

Cd Arachidate

(c)

Polyethylene

(d)

Diamond

(e)

Intensity (Arb. Units)

280 290 300 310 320
Photon Energy (eV)

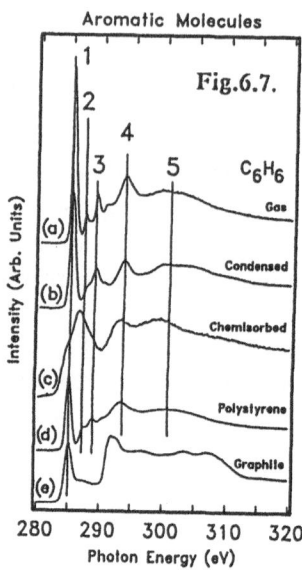

Aromatic Molecules

Fig.6.7.

1
2
4
3   5 C_6H_6

Gas

(a)

Condensed

(b)

Chemisorbed

(c)

Polystyrene

(d)

Graphite

(e)

Intensity (Arb. Units)

280 290 300 310 320
Photon Energy (eV)

Fig.6.6. This figure shows the similarity between the core level excitation spectra of various hydrocarbon molecules. (a) is the IEELS spectrum of gaseous hexane [6.45]; (b) is the NEXAFS spectrum of condensed cyclohexane; (c) is the NEXAFS spectrum of Cd arachidate [6.32]; (d) is the NEXAFS spectrum of polyethylene [6.34]; and (e) is the NEXAFS spectrum of diamond. The spectra of the first four molecules are quite similar and exhibit three characteristic resonances which are labelled.

Fig.6.7. This figure shows the similarity between the core level excitation spectra of various benzene-containing molecules. (a) is the IEELS of gaseous benzene [6.43]; (b) is the NEXAFS spectrum of condensed benzene; (c) is the angle-averaged NEXAFS spectrum of benzene chemisorbed on Mo{110} at 90 K [6.46]; (d) is the NEXAFS spectrum of polystyrene; and (e) is the angle-averaged NEXAFS spectrum of graphite. The first four spectra are quite similar exhibiting several characteristic resonances. The assignments for the resonances are: (1) is $\pi^*_{\phi 1}$; (2) is C-H*; (3) is $\pi^*_{\phi 2}$; (4) is $\sigma^*_{\phi 1}$; and (5) is $\sigma^*_{\phi 2}$.

and 300 eV, respectively. The two remaining resonances, peak 2 (C-H*) and peak 3 ($\pi^*_{\phi 2}$), are weaker in intensity so not always apparent in the spectra of the more complex systems. Nevertheless, the prinicpal resonances for benzene are unmistakable in each of the benzene containing molecules.

The fingerprint guideline is not limited to C-C bonds but applies to other types of bonds as well. For example, functional groups containing C-O bonds such as carboxylic acids and esters have characteristic features in their core level excitation spectra. Figure 6.8 compares the spectra of a variety of molecules containing the carboxylate group ($-CO_2-$) ranging from gaseous formic acid, [6.47] gaseous methyl formate [6.31], and the polymer, polymethylmethacrylate. Each of these spectra exhibits three peaks which are characteristic of the

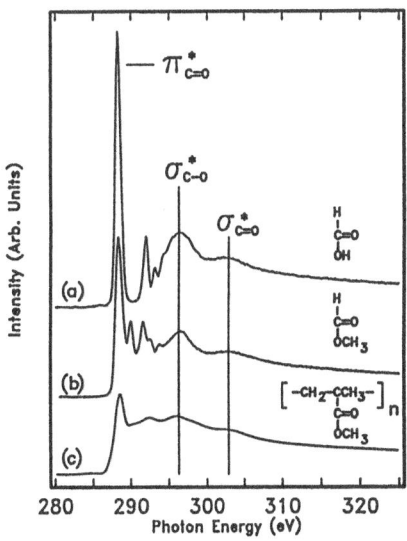

Fig.6.8. This figure shows the similarity between the core level excitation spectra of various molecules containing the carboxylate group (-CO₂-). (a) is the IEELS of gaseous formic acid [6.47]; (b) is the IEELS of geseous methyl formate [6.31]; and (c) is the NEXAFS spectrum of polymethylmethacrylate, PMMA. All of these spectra contain the three labelled resonances which are characteristic of the carboxylate group.

carboxylate group. The first peak is the sharp C=O π^* resonance at approximately 289 eV. The second peak is the C-O σ^* resonance at 296 eV and the third peak is the C=O σ^* resonance at 302 eV. There are also a few notable differences between these spectra which can be explained. For example, the IEELS of gases usually exhibit small sharp peaks immediately below the ionization limits which are largely Rydberg in character. This accounts for the small peaks near 291 eV in the spectra of formic acid and methyl formate (Fig.6.8). In addition, PMMA contains a hydrocarbon backbone which contributes to the spectrum above 287 eV making the features of the carboxylate group less prominent. Despite these minor differences the spectra are quite consistent with the superposition principle.

6.3.4 Assembly of Functional Groups to Macromolecules

A second general principle to use in interpreting core level excitation spectra is that, for polyfunctional molecules, the spectra are a superposition of the features from each of the functional groups in the molecule. This is true, provided the functional groups do not interact chemically to change the geometry and character of the functional groups.

An example which demonstrates both, the assembly of diatomic C-C building blocks to a functional group and the further assembly of this group to a polymer is shown in Fig.6.9. This figure shows the spectra of gaseous 2-butene, and two polybutadiene polymers. The 2-butene spectrum exhibits all resonances expected form a superposition of those of ethane (C-C σ^* and C-H resonances; see Fig.6.2) and ethylene (C=C σ^* and π^* and C-H resonances; see Fig.6.1), in accordance with the building block principle. As expected from the building

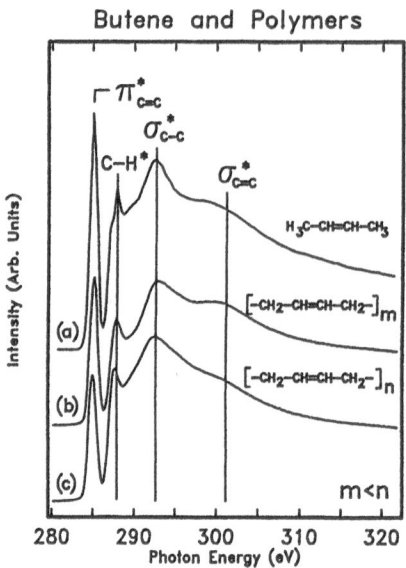

Butene and Polymers

Fig.6.9. This figure shows how changes in the proportion of C=C π^* bonds in a polymer is manifest in the core level excitation spectrum. (a) is the IEELS of gaseous butene [6.54]; (b) is the NEXAFS spectrum of a low molecular weight polybutadiene polymer; and (c) is the NEXAFS spectrum of a high molecular weight polybutadiene polymer. Overall, these spectra are all quite similar as expected since they all contain the same types of bonds. The concentration of C=C π bonds differs among the polymer samples, however, with (c) having a lower concentration which is manifest in the NEXAFS as smaller C=C π^* and σ^* resonances.

block picture, the spectra of 2-butene and polybutadiene which consists of 2-butene building blocks are very similar. The difference between the two polymer sepctra is caused by a difference in molecular weight, which affects the concentration of C=C π bonds. The lower spectrum is of a sample with higher molecular weight and clearly fewer C=C π bonds. This is reflected in the spectra by smaller intensities for the C=C π^* and σ^* resonances compared to the middle spectrum which is of a polymer of lower molecular weight.

To show that the core level excitation spectrum of a polyfunctional molecule is a superposition of the spectral features of its constituent functional groups, let us consider several hydrocarbon molecules containing various additional functional groups. Figure 6.10 shows the NEXAFS spectra of DIAC which is a LB film assembled from molecules containing a hydrocarbon chain and two C≡C bonds. Since the hydrocarbon chain comprises a majority of the molecule, the NEXAFS spectrum resembles to a large extent that of the saturated hydrocarbons previously discussed such as cadmium arachidate shown at the bottom of Fig.6.10. The NEXAFS spectrum of DIAC, in particular, shows the three resonances characteristic of hydrocarbon chains which were identified in the spectra of Fig.6.6. The C≡C bonds of DIAC are apparent

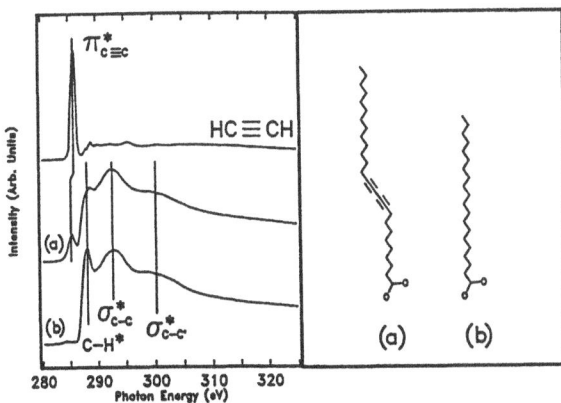

Fig.6.10. This figure shows that the core level excitation spectrum for the LB film containing DIAC is a superposition of the features from its constituent functional groups. At the top is shown the IEELS of gaseous acetylene [6.55]; (a) is the NEXAFS spectra of DIAC; and (b) is the NEXAFS spectrum of cadmium arachidate at an x-ray incident angle of 50° from the surface normal [6.32]. At the right are shown the structural formulas for (a) and (b). The spectra of DIAC is a superposition of the π^* resonances from the C≡C bonds at approximately 285.5 eV and the three resonances above 287 eV characteristic of the hydrocarbon chain.

by the C≡C π^* resonance which occurs at approximately 285.5 eV as in gaseous acetylene shown at the top of Fig.6.10. This is the only feature in the spectrum of DIAC which is clearly attributable to the C≡C group, because the other features of the acetylene spectrum are weak compared to the π^* resonance and they overlap with the strong peaks attributed to the hydrocarbon chain. Thus the NEXAFS spectrum of DIAC is consistent with a superposition of the core level excitation features expected of the hydrocarbon chain and of the C≡C bonds.

As with the fingerprint guideline, the superposition principle is not restricted to functional groups consisting of only C-C bonds. For example, the spectrum of a LB film assembled from a semifluorinated acid (SFN) which consists partly of a normal hydrocarbon chain and partly of a fluorinated hydrocarbon chain, looks like a superposition of the spectrum of a normal hydrocarbon and a fluorocarbon (Fig.6.11). The spectrum of SFN differs from the previous example of LB films, however, in that the spectral features of the hydrocarbon chain are *less* prominent than that of the other parts of the molecule, because the features due to the fluorine-carbon bonds are so intense. For example, the hydrocarbon part of the molecule is only apparent in the NEXAFS spectrum by the C-H* resonance at approximately 289 eV. This is the only peak which is not overlaid by the features from the fluorine-carbon bonds. The other major resonances in the spectrum of SFN are quite similar to those of simple fluorocarbons such as *per*-fluorohexane (Fig.6.11). In particular, the C-F σ^* resonance at 292 eV, the CF_2^* resonance at 294 eV, the C-C σ^* resonance at 297 eV, and the C-C' σ^*

215

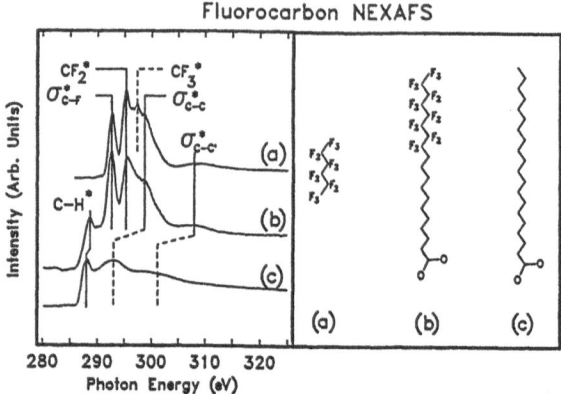

Fig.6.11. This figure shows that the core level excitation spectrum for the LB film containing SFN is a supersposition of the features from its constituent functional groups. (a) is the IEELS of *per*-fluorohexane [6.56]; (b) is the NEXAFS spectrum for SFN; and (c) is the NEXAFS spectrum of cadmium arachidate at an x-ray incidence angle of 50° from the surface normal [6.32]. At the right are shown the structural formulas for (a), (b), and (c). The spectrum of SFN is a superposition of the C-H* resonance from the hydrocarbon chain at 288 eV and the four resonances characteristic of the -CF$_2$- group above 291 eV.

resonance at 308 eV are directly comparable between the spectra of gaseous C$_6$F$_{14}$ and SFN. The assignments for the various C-F bonds were based upon a previous detailed study of fluorocarbons by *Hitchcock* et al. [6.56]. The spectrum of SFN does differ from that of *per*-fluorohexane in regard to the absence of a small peak at approximately 296 eV. This is associated with the CF$_3$ group, however, which comprises a larger component of *per*-fluorohexane than of SFN, so is expected to be weaker in the latter. Thus the NEXAFS spectrum of SFN in Fig.6.11 is quite consistent with the superposition principle.

6.3.5 Limitations of the Building Block Approach

With some limitations, the fingerprint guideline can also be extended to molecular solids such as diamond and graphite. For example, the spectrum of diamond is presented in Fig.6.6 along with that of various hydrocarbons. Diamond is not really a hydrocarbon, but it does have tetrahedral C-C bonds like one. The NEXAFS spectrum of diamond reflects some of these similarities and differences. For example, diamond does not have C-H bonds, so the C-H* resonance is absent from the NEXAFS spectrum for diamond in Fig.6.6. Diamond also has C-C single bonds so the prominent C-C σ* resonance should be retained in the NEXAFS spectrum of diamond. However, the region above 290 eV in diamond is no longer dominated by one dominant peak but instead several peaks are observed. This indicates a breakdown of the building block picture. The reasons for the loss of correspondence

216

between local bonds and resonances lies in the formation of electronic energy bands by interaction of local states in a periodic potential. As shown by *Müller* et al. [6.48] the near-edge fine structure in such cases is determined by the appropriate angular momentum projection of the electronic density of states. In a scattering picture this is equivalent to summing over all scattering paths of the excited photoelectron in the lattice. Thus the long range periodicity of the lattice leads to strong contributions from scattering events involving distant neighbors which significantly distort the structure arising from scattering processes involving only the nearest neighbor atoms.

Graphite is another example of a solid whose core level excitation spectrum exhibits a resemblance to that of a simple molecule, benzene, without being completely explainable in terms of the molecular spectrum. The spectrum of graphite is shown in Fig.6.7 along with the spectra of various benzene-containing molecules. All of the spectra in Fig.6.7 contain a π^* resonance at approximately 285 eV which reflects the local character of the π^* bonds, or in the case of graphite the flatness of the π^* conduction bands. On the other hand, in the σ^* region, the spectrum of graphite has considerable structure in the range between 291 and 305 eV which is attributed to formation of σ^* energy bands. Even if all features in the graphite and diamonds spectra are not readily understandable in terms of local bonds, the π^* resonance may be used to clearly distinguish the two forms of carbon from each other. This characteristic feature can be utilized, for example, for determining whether carbon films are graphitic or diamond-like.

6.4 Conclusions and Future Prospects

Considerable progress has been achieved in just the past few years in the interpretation of near-edge core level excitation spectra. For example, spectral features which were not understood, but quite apparent, in spectra taken less than five years ago such as the C-H* resonances [6.34,45] and the multiple C-C σ^* resonances of aromatic systems [6.28,29] can now be identified. Because of this advancement much more information can be obtained regarding thin organic films or intermediates in surface reactions from near-edge core level excitation spectroscopy. For example, the orientations of the hydrocarbon chains of ordered, *single-layer*, LB films can now be determined from an analysis of the angular dependence of resonances in NEXAFS spectra [6.32,33]. Such information is difficult to obtain by other techniques in the monolayer limit for LB films, particularly on non-metallic substrates such as silicon. A further advancement in the understanding of near-edge spectroscopy is the size of molecules which are now being studied. Four years ago, the most complex molecules being studied with near-edge spectrosopy were monofunctional and the size of benzene. Now polyfunctional molecules and even polymers can be readily characterized using near-edge spectroscopy.

There are still aspects of near-edge spectra which are not understood, however. For example, some weaker peaks are not understood in presently available spectra, such as the high energy C-C′ σ^* resonance observed for long hydrocarbon chain molecules (Fig.6.6). The σ^* spectral region becomes even less understood in the upper limit of molecular size, for example, in three dimensional solids such as graphite and diamond. In this case the effects of electronic-band formation, i.e. multiple scattering processes involving third and higher-shell neighbors, and even single scattering EXAFS-like features complicate the spectra. Thus continued experimental and theoretical work is needed to completely understand some aspects of the spectra of molecules and to extend near-edge spectroscopy to three dimensional solids.

Limitations still exist with regard to the experimental acquisition of data, especially for NEXAFS which requires synchrotron storage rings as a source of X-rays. The major limitations are the signal-to-noise ratio and instrumental resolution. For gas phase IEELS and bulk NEXAFS studies, the signal to noise ratio is adequate, but the analysis of small concentrations such as a submonolayer on a surface is often difficult. Increased photon fluxes from improved synchrotron radiation sources and use of fluorescence detection [6.6, 49, 50] will eventually alleviate these problems. High resolution spectra from IEELS are already available for selected systems [6.51]. Higher resolution X-ray absorption spectra, however, will require a new generation of monochromators which are now under construction.

Acknowledgements. We would like to thank A. Hitchcock and I. Ishii of McMaster University for providing copies of their gas phase IEELS spectra; J. Swalen and J. Rabe of IBM Almaden Research Center for providing the LB films; T. Russell of IBM Almaden Research Center for providing the polymer samples; and B. Hermsmeier (University of Hawaii), H.H. Rotermund (Fritz-Haber Institut), P. Stevens, J. Solomon and R.J. Madix (Stanford University) and J.P. Roberts, A.C. Liu and C.M. Friend (Harvard University) for help with some of the measurements and helpful discussions. This work was done in part at the Stanford Synchrotron Radiation Laboratory which is supported by the office of Basic Energy Sciences of DOE and the Division of Materials Research of NSF.

References

6.1   C.E. Brion: Comments At. Mol. Phys. 16, 249 (1985)
6.2   A.P. Hitchcock: J. Electron Spectrosc. Related Phenomena 25, 245 (1982)
6.3   J. Stöhr, K. Baberschke, R. Jaeger, T. Treichler, S. Brennan: Phys. Rev. Lett. 47, 381 (1981)
6.4   J. Stöhr, R. Jaeger: Phys. Rev. B 26, 4111 (1982)
6.5   J. Stöhr, N. Noguera, T. Kendelewicz: Phys. Rev. B 30, 5571 (1984)

6.6 D.A. Fischer, U. Döbler, D. Arvanitis, L. Wenzel, K. Baberschke, J. Stöhr: Surf. Sci. 177, 114 (1986)

6.7 For a review see: J. Stöhr: In *X-ray Absorption: Principles, Applications, Techniques of EXAFS, SEXAFS and XANES*, ed. by D.C. Koningsberger, R. Prins (Wiley, New York 1988) p.443

6.8 P.H. Citrin, P. Eisenberger, R.C. Hewitt: Phys. Rev. Lett. 41, 309 (1978); J. Stöhr, D. Denley, P. Perfetti: Phys. Rev. B 18, 4132 (1978); J. Stöhr: Jpn. J. Appl. Phys. 17, Suppl. 17-2, 217 (1978)

6.9 F. Sette, J. Stöhr, A.P. Hitchcock: J. Chem. Phys. 81, 4906 (1984)

6.10 J. Stöhr, D.A. Outka: Phys. Rev. B 36, 7891 (1987)

6.11 J. Stöhr, F. Sette, A.L. Johnson: Phys. Rev. Lett. 53, 1684 (1984)

6.12 U.C. Srivastava, H.L. Nigam: Coord. Chem. Rev. 9, 275 (1972)

6.13 R.A. Bair, W.A. Goddard III: Phys. Rev. B 22, 2767 (1980)

6.14 A. Bianconi, M. Dell'Ariccia, P.J. Durham, J.B. Pendry: Phys. Rev. B 26, 6502 (1982)

6.15 T.A. Smith, J.E. Penner-Hahn, M.A. Berding, S. Doniach, K.O. Hodgson: J. Am. Chem. Soc. 107, 5945 (1985)

6.16 J.E. Penner-Hahn, T.A. Smith, B. Hedman, K.O. Hodgson, S. Doniach: J. Phys. (Paris) C8, 1197 (1986)

6.17 S.S. Hasnain, L. Alagna, N.J. Blackburn, R.W. Strange: J. Phys. (Paris) C8, 1129 (1986)

6.18 J. Wong, F.W. Lytle, R.P. Messmer, D.H. Maylotte: Phys. Rev. B 30, 5596 (1984)

6.19 M. Belli, A. Scafati, A. Bianconi, S. Mobilio, L. Palladino, A. Reale, E. Burattini: Solid State Commun. 35, 355 (1980)

6.20 J. Stöhr, D.A. Outka: J. Vac. Sci. Technol. A 5, 919 (1987)

6.21 J. Stöhr: In *Chemistry and Physics of Solid Surfaces V*, ed. by V.R. Vanselow and R. Howe, Springer Ser. Chem. Phys, Vol.35 (Springer, Berlin, Heidelberg 1984)

6.22 F. Sette, J. Stöhr: In *EXAFS and Near Edge Structure III*, ed. by K.O. Hodgson, B. Hedman, J.E. Penner-Hahn, Springer Proc. Phys., Vol.2 (Springer, Berlin, Heidelberg 1985) p.250

6.23 J. Stöhr: Z. Physik B 61, 439 (1985)

6.24 J. Stöhr: In *The Structure of Surfaces*, ed. by M.A. Van Hove, S.Y. Tong, Springer Ser. Surf. Sci., Vol.2 (Springer, Berlin, Heidelberg 1985) p.140

6.25 A.P. Hitchcock, D.C. Newbury, I. Ishii, J. Stöhr, J.A. Horsley, R.D. Redwing, A.L. Johnson, F. Sette: J. Chem. Phys. 85, 4849 (1986)

6.26 J. Stöhr, J.L. Gland, E.B. Kollin, R.J. Koestner, A.L. Johnson, E.L. Muetterties, F. Sette: Phys. Rev. Lett. 53, 2161 (1984)

6.27 A.P. Hitchcock, J.A. Horsley, J. Stöhr: J. Chem. Phys. 85, 4835 (1986)

6.28 D.C. Newbury, I. Ishii, A.P. Hitchcock: Can. J. Chem. 64, 1145 (1986)

6.29 J.A. Horsley, J. Stöhr, A.P. Hitchcock, D.C. Newbury, A.L. Johnson, F. Sette: J. Chem. Phys. 83, 6099 (1985)

6.30 D.A. Outka, J. Stöhr, R.J. Madix, H.H. Rotermund, B. Hermsmeier, J. Solomon: Surf. Sci. 185, 53 (1987)

6.31 I. Ishii, A.P. Hitchcock: J. Electron Spectrosc. Related Phenomena 46, 55 (1988)

6.32 D.A. Outka, J. Stöhr, J.P. Rabe, J. Swalen: J. Chem. Phys. 88, 4076 (1988)

6.33 D.A. Outka, J. Stöhr, J.P. Rabe, J. Swalen, H.H. Rotermund: Phys. Rev. Lett. 59, 1321 (1987)

6.34 J. Stöhr, D.A. Outka, K. Baberschke, D. Arvanitis, J.A. Horsley: Phys. Rev. B 36, 2967 (1987)

6.35 C.A. Kovac, J.L. Jordan, R.A. Pollak: Mat. Res. Soc. Symp. Proc. 72, 247 (1986)

6.36 C.E. Brion, S. Daviel, R.N.S. Sodhi, A.P. Hitchcock: AIP Conf. Proc. 94, 429 (1982)

6.37 A. Bianconi: In *X-ray Absorption: Principles, Applications, Techniques of EXAFS, SEXAFS and XANES*, ed. by D.C. Koningsberger, R. Prins (Wiley, New York 1988) p.573

6.38 D.A. Outka, J. Stöhr: J. Chem. Phys. **88**, 3539 (1988)

6.39 The assignment of the resonances at 289 eV and 293 eV differs from that originally given in [6.44] for methanol. The 289 eV resonance is assigned as a transition to C-H derived antibonding orbitals, as in the case of hydorcarbons [6.25,45]. The 293 eV resonance is assigned as a C-O σ^* resonance as discussed in [6.9]

6.40 H.Agren, R. Arneberg: Phys. Scripta **30**, 55 (1984)

6.41 I. Ishii, A.P. Hitchcock: private communication

6.42 A.P. Hitchcock, C.E. Brion: J. Electron Spectrosc. Related Phenomena **15**, 201 (1979)

6.43 A.P. Hitchcock, C.E. Brion: J. Electron Spectrosc. Related Phenomena **10**, 317 (1977)

6.44 G.R. Wight, C.E. Brion: J. Electron Spectrosc. Related Phenomena **4**, 25 (1974)

6.45 A.P. Hitchcock, I. Ishii: J. Electron Spectrosc. Related Phenomena **42**, 11 (1987)

6.46 J.P. Roberts, A.C. Liu, C.M. Friend, J. Stöhr: unpublished data

6.47 I. Ishii, A.P. Hitchcock: J. Chem. Phys. **87**, 830 (1987)

6.48 J.E. Müller, J.W. Wilkins: Phys. Rev. B. **29**, 4331 (1984);
J.E. Müller: In *EXAFS and Near Edge Structure III*, ed. by K.O. Hodgson, B. Hedman, J.E. Penner-Hahn, Springer Proc. Phys. 2, p.7 (Springer, Berlin Heidelberg 1986)

6.49 J. Stöhr, E.B. Kollin, D.A. Fischer, J.B. Hastings, F. Zaera, F. Sette: Phys. Rev. Lett. **55**, 1468 (1985)

6.50 D. Arvanitis, U. Döbler, L. Wenzel, K. Baberschke, J. Stöhr: J. Physique (Paris) **C8**, 173 (1986)

6.51 A.P. Hitchcock, C.E. Brion: J. Electron Spectrosc. Related Phenomena **18**, 1 (1980)

6.52 R.J. Koestner, J. Stöhr, J.L. Gland, J.A. Horsley: Chem. Phys. Lett. **105**, 332 (1984)

6.53 A.L. Johnson, E.L. Muetterties, J. Stöhr: J. Am. Chem. Soc. **105**, 7183 (1983)

6.54 A.P. Hitchcock, S. Beaulieu, T. Steel, J. Stöhr, F. Sette: J. Chem. Phys. **80**, 3972 (1984)

6.55 A.P. Hitchcock, C.E. Brion: J. Electron Spectrosc. Related Phenomena **22**, 283 (1981)

6.56 A.P. Hitchcock, P. Fisher, R. McLaren: *Proc. of a NATO Advanced Study Institute on Giant Resonances in Atoms, Molecules and Solids* (Les Houches, France, June 1986)

7. Surface Kinetics with Near Edge X-Ray Absorption Fine Structure

John L. Gland

Corporate Research Laboratory, Exxon Research and Development Co.
Annandale, NJ 08801, USA

This review is focused on the use of soft X-ray absorption methods for characterizing the rates of surface reactions. In order to measure rates of surface reactions it is first necessary to characterize the adsorbed species which are involved in the reaction. A group of methods which together are generally capable of resolving key structural aspects of adsorbed species will be briefly introduced. A brief introduction will then be made for a number of methods used for determining rates of surface reactions. Transient near edge X-ray absorption fine structure (NEXAFS) measurements based on electron detection in vacuum will be introduced, and several examples of rate measurements will be discussed in order to highlight the strength of this technique. A brief introduction to fluorescence yield near edge-spectroscopy (FYNES) in the soft X-ray region will then be given. FYNES is an interesting new surface method which can be used to characterize adsorbed species even in the presence of reactive gas. Transient FYNES characterization of the kinetics of CO displacement by hydrogen will then be discussed to highlight the utility of transient surface methods capable of characterizing surface reactions in the presence of reactive environments.

Surface reactions play a key role in a large number of important technological areas ranging from catalysis to corrosion, from chemical vapor deposition to lubrication. However, detailed understanding of surface reaction rates and mechanisms has been limited in many cases by our inability to measure the rates of key surface reactions directly. In addition, detailed characterization of the structure and bonding of adsorbed species (reactants and products) has been difficult, particularly for adsorbed molecules containing more than three or four atoms. In recent years, the development and extension of a number of surface spectroscopies has significantly enhanced our ability to characterize the structure and bonding of adsorbed species on model single crystal surfaces even for fairly complex adsorbed species [7.1,2].

In order to develop a detailed understanding of a surface process, several important aspects of the system must be characterized. Initially the structure and composition of the surface should be established. The structure and bonding of adsorbed reactants, products, and stable surface intermediates should be established so that chemical analogies from inorganic chemistry can be used if available. The important reaction steps should then be isolated from the overall reaction scheme so that their kinetic behavior can be characterized in detail.

A combination of surface methods is generally required to establish the detailed structure and bonding of adsorbed species on single crystal surfaces. The combination of techniques used varies depending on the system being studied as well as the spectroscopic preference of the investigator. We have found that the critical aspects of structure and bonding for many surface reaction systems can be characterized using a standard combination of surface methods. Initially, low energy election diffraction (LEED) and Auger electron spectroscopy (AES) are used to characterize the symmetry and cleanliness of the clean single crystal surface.

Adsorbed species are characterized using temperature programmed desorption (TPD) to establish overall decomposition pathways along with stoichiometries and thermal stabilities of adsorbed intermediates [7.1,3]. Hydrogen TPD intensities are especially useful in determining overall hydrogen surface stoichiometries since many surface methods cannot detect hydrogen.

X-ray photoemission spectroscopy (XPS) intensities depend linearly on the number of atoms present so that core level intensities can be used to determine the concentration of surface atoms [7.4]. XPS peak positions depend on the chemical environments of the photoemitting atom. Thus core level shifts can be used to characterize oxidation state changes and in particular can often be used to detect dissociation of molecular species on surfaces.

High resolution electron energy loss spectroscopy (HREELS) is an attractive method for obtaining vibrational spectra of adsorbed species [7.5]. Vibrational spectra are a sensitive probe of the structure and bonding of adsorbed molecules and intermediate species. Sufficient resolution is available so that deuterium substitution often allows direct mode assignments to be made. Unfortunately the 60 to 80 cm^{-1} resolution available generally precludes resolution of transitions involving other isotopic species like ^{18}O. Otherwise, mode assignments and structural information are obtained by comparison with inorganic complexes and cluster complexes where unambiguous structures can be determined by X-ray scattering. In general, the vibrational information available from HREELS furnishes critical information concerning the structure and bonding of adsorbed species as evidenced by the widespread use of this fairly recent technique.

NEXAFS can furnish critical additional information concerning the structure and bonding of adsorbed species as indicated by other chapters in this collection and several recent reviews [7.6,7]. Since the resonant transition observed between core levels and the lowest unoccupied molecular orbitals obey dipole selection rules, the polarized soft X-rays from synchrotron light sources can generally be used to establish the orientation of adsorbed molecules and intermediate species. In addition, the separation between and number of low lying unoccupied molecular orbitals is a sensitive function of bond distances and molecular structure.

Taken together, this combination of four techniques can be used to establish the structure and bonding of adsorbed molecules ranging from diatomics like CO [7.8,9] to more complex molecules like thiophene [7.10] and pyridine [7.11]. Establishing the structure and bonding of reactant molecules is a necessary first step in developing a detailed understanding of surface reactions. With structural information established for the important species involved in a surface process, a number of techniques is availble for measuring surface reaction rates. Several of the methods available for measuring rates of various types of surface reaction steps will be briefly introduced in the following section.

Surface reactions can be divided into two general classes which require different strategies. Reaction rates can be characterized by measuring the rate of product appearance in the gas phase if desorption is not rate limiting. A wide variety of reactor types and detection methods is available for studying reactions under these conditions [7.12]. For instance, catalytic reaction rates are often characterized using gas chromatographic or mass spectrometric detection of products from batch or differential reactors. For single crystal surfaces, temperature programmed desorption is a widely used method for characterizing both the kinetics of appropriate surface reactions and the kinetics of the desorption process when desorption is rate limiting [7.1,3]. A wide range of molecular beam based kinetic methods is also available for characterizing the transient behavior of surface reactions on well-characterized surfaces [7.13]. All these methods generally rely on detection of products which desorb rapidly from surfaces.

A second important class of reaction steps involves chemical reactions where rapid desorption does not occur. In this situation, no gas phase products are avilable so surface analysis methods must be used to measure the reaction rates. In order to characterize surface reactions of this type, the methods employed must be capable of distinguishing between the reactants and products and also capable of measuring the rate of concentration change for the important species.

7.1 An Overview of Methods for Characterizing Surface Reactions

A number of methods has been employed to measure the rates of surface reactions involving exclusively adsorbed species. Methods which rely on electron or particle detection are generally limited to vacuum environments. A brief introruction to laser induced desorption (LID), transient secondary ion mass spectroscopy (SIMS), transient ultraviolet photoemission spectroscopy (UPS) and high resolution electron energy loss spectroscopy (HREELS) will be made in the following section. Vacuum based NEXAFS methods which rely on electron detection will be discussed in a later section in more detail. Available methods which can be or could be used to measure reactions involving adsorbed species in the presence of reactive environments will then be reviewed briefly. These techniques include infrared reflection-absorption spectroscopy (IRAS), dynamic work function change (DWFC) measurements, and second harmonic generation (SHG). Fluorescence yield near-edge spectroscopy (FYNES), a soft X-ray absorption technique which can be used in the presence of reactive gases will be discussed in a later section.

Laser-induced desorption (LID) has recently been used successfully to characterize surface diffusion and a number of surface reactions [7.14-18]. This method relies on mass spectrometric detection of molecular species rapidly desorbed from a small spot on the surface heated by a pulse of laser light. Mass spectrometric detection results in good capability for distinguishing between similar chemical species. Reliable surface concentration measurements can be made by integrating the amount of desorbed material. Unfortunately it is generally not possible to desorb dissociatively adsorbed species so that surface reactions which link adsorbed intermediates may not be accessible.

Secondary ion mass spectroscopy (SIMS) is an interesting surface technique since it is often very sensitive to adsorbed species as well as being capable of discriminating between similar chemical species [7.19-23]. Transient SIMS has been used both in a temperature programmed and isothermal mode to characterize several surface reactions. Unfortunately the intensities of the signals observed may be difficult to relate directly to the concentration of adsorbed species without extensive calibration procedures.

Transient ultraviolet photoemission measurements have also been used to characterize surface reaction rates [7.24-26]. However, difficulties can arise because it is sometimes difficult to energy or angularly separate valence transitions for adsorbed species or mixtures of adsorbed species (reactants and products).

High-resolution electron energy loss spectroscopy (HREELS) can usually distinguish between similar adsorbed species, since the vibrational spectrum depends strongly on molecular structure and bonding. Several surface reactions have been characterized using spectrometers designed specifically for transient measurements [7.27-29]. Potential problems include difficulties with tuning as work functions change during reaction, and difficulties with concentration determination because vibrational intensities often depend on the order in the adsorbed layer, orientation of the adsorbed molecules, dipole-dipole coupling, and other factors.

Infrared reflection absorption spectroscopy (IRAS) has been developed as a method for obtaining vibrational spectra from adsorbed monolayers on well-characterized surfaces in the presence of reactive environments. Recently improved instrumentation has made it possible to acquire vibrational spectra fairly rapidly in favorable situations [7.30]. The combination of excellent ability to discriminate between similar species, reasonable time resolution, as well as ability to obtain spectra in the presence of reactive environments suggests that IRAS holds substantial promise as a method for studying surface reactions. As mentioned previously, vibrational intensities are not easily related to surface concentration but also depend on nonlocal effects such as order in the adsorbed layer and dipole coupling in the adsorbed overlayer. Therefore care must be exercised when measuring surface concentrations with vibrational methods.

Dynamic work function changes (DWFC) measurements can also be made in the presence of reactive environments using the Kelvin probe method [7.31,32]. DWFC measurements have been particularly useful for the simpler chemical systems since the ability to discriminate between adsorbates is somewhat limited. Since the work function change is often a complex function of coverage even for simple adsorbates, fairly extensive standardization procedures are often required before kinetic measurements can be made.

Second-harmonic generation (SHG) is a recently developed laser-based method which can be used to characterize the nonlinear optical response of adsorbed species. Since SHG is a photon-in, photon-out method, it can also be used to characterize adsorbates in reactive environments. In fact SHG has been used to characterize surface reactions on liquid water surfaces [7.33] as well as reactions on single crystal surfaces [7.34,35]. Because SHG is a laser-based technique, the technique holds exciting prospects for characterization of surface reactions on short time scales. Studies of fast phase transitions are discussed in a recent review [7.36].

7.2 Transient Near-Edge X-ray Absorption Fine Structure (NEXAFS) as a Probe of Surface Reactions

Resonant transitions between adsorbate core levels and the lowest unfilled molecular orbitals have been used successfully to characterize the orientation and bonding in carbon, nitrogen, oxygen and sulfur-containing adsorbates on metal surfaces [7.6, 7, 37-39]. Since the transitions probe the lowest unfilled valence orbitals, NEXAFS can easily distinguish between similar adsorbed species. Each type of atom in a molecule can also be probed independently since the core levels for different atomic species occur at very different energies. For instance, the carbon edge occurs neat 285 eV, the nitrogen edge occurs near 410 eV, while the oxygen edge occurs at 543 eV. Thus the behavior of each type of atom in a molecule can be probed independently. This was illustrated recently by independently characterizing the behavior of carbon and sulfur during thiophene dissociation on platinum [7.10].

The ability to distinguish between similar adsorbed species is also enhanced by our ability to control the orientation of the polarized light relative to the surface. Since NEXAFS transitions obey dipole selection rules, specific incidence angles can be selected for optimized detection of a specific species. X-ray absorption is also a nondestructive probe of adsorbed surface species, thus surface reactions can be quenched if required so that the surface species can be characterized in detail using other surface spectroscopies. Reaction quenching studies are important in some cases to develop a more detailed picture of all the surface species present during reaction. In summary, NEXAFS is an attractive method for characterizing surface reactions because it combines the ability to discriminate between similar chemical species based on the sensitivity of the valence levels with the element specificity of a core level method.

In the following section we will discuss the experimental setup required for NEXAFS measurements in vacuum. Several NEXAFS studies of surface reactions in vacuum environments will then be briefly reviewed. The reactions include desorption of CO from the Pt{111} surface [7.40], reorientation of pyridine on the Ag{111} surface [7.41], and dehydrogenation of ethylene to form ethylidyne on the Pt{111} surface [7.42].

7.2.1 Soft X-ray Absorption Using Electron Detection

Two primary components are required for soft X-ray absorption studies. A soft X-ray beamline is required to focus and monochromatize the radiation produced by the synchrotron light source [7.43]. An experimental end station is required which contains: 1) the sample, 2)

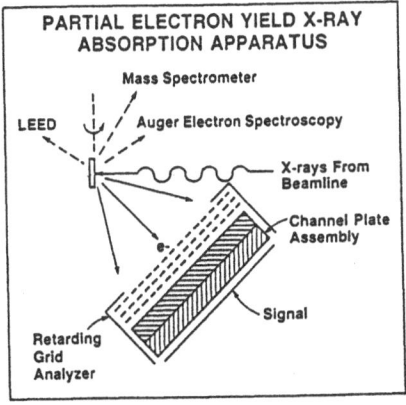

PARTIAL ELECTRON YIELD X-RAY ABSORPTION APPARATUS

Mass Spectrometer

LEED

Auger Electron Spectroscopy

X-rays From Beamline

Channel Plate Assembly

Signal

Retarding Grid Analyzer

Fig.7.1. A schematic diagram of key components for a typical "end station" for soft X-ray absorption measurements.

facilities for preparing the surface and characterizing adsorbed species, and 3) a detector system to measure the electrons emitted following X-ray adsorption. The schematic diagram shown in Fig.7.1 outlines the key components of an end station for vacuum NEXAFS The electron detector system illustrated was designed for use during kinetic studies and has a large acceptance angle to enhance the total signal collected. In the system illustrated in Fig.7.1, the sample surface can also be rotated to face an Auger electron spectrometer to verify sample cleanliness, a LEED system to check the surface order, and a mass spectrometer so the TPD spectra can be used to characterize adsorbed species. Using this type of end station, NEXAFS spectra can be obtained from species which are adsorbed on an in situ characterized surface. The thermal behavior of the adsorbate can also be verified during spectroscopic experiments by acquiring TPD spectra as required.

The electron detector illustrated is equipped with a retarding field analyzer since the signal-to-background in NEXAFS spectra can be substantially improved by using a cutoff filter set about 20% below the appropriate Auger energy [7.44,45]. Basically this filter rejects the large secondary electron peak at low energy. During kinetic experiments it is also quite important to obtain an accurate measurement of the incident soft X-ray flux since the intensity of incident X-ray beam decays as the beam current in the storage ring decays. Accurate I_0 data can be collected using a number of methods, however, it is often convenient to measure the current to ground from a highly transmitting grid placed near the sample. If sufficient current is not available for direct current measurement, photoelectrons from the grid can be amplified by an electron multiplier. I_0 is then used to normalize the NEXAFS data so that the results are independent of ring current or fluctuations in incident intensity.

7.2.2 CO Desorption from the Pt{111} Surface

Isothermal CO desorption from the Pt{111} surface was characterized recently using transient NEXAFS as part of a program to establish the viability of this technique [7.40]. The following discussion describes the methods used and the results obtained during this initial set of experiments on Beamline I-1 at SSRL.

Figure 7.2 compares a spectrum from the clean Pt{111} surface with a spectrum from the CO saturated surface. Subtraction of these two normal incidence spectra highlight the large π resonance of the adsorbed CO. The isothermal desorption experiments discussed in this section were performed at normal incidence while monitoring the intensity of the π resonance. The surface was first saturated with CO at 170 K so that the intensity of the π transition could be determined for a CO saturated surface as an internal coverage standard. The sample was then rapidly heated to an isothermal desorption temperature in the 360 to 440 K range. The intensity of the π resonance was monitored continuously as the CO desorbed from the surface. After about 1000 seconds the sample temperature was rapidly raised to 600 K to desorb all remaining CO so that a second internal standard could be established for zero CO coverage. An Auger spectrum was then taken to verify that the surface was clean after the desorption experiment.

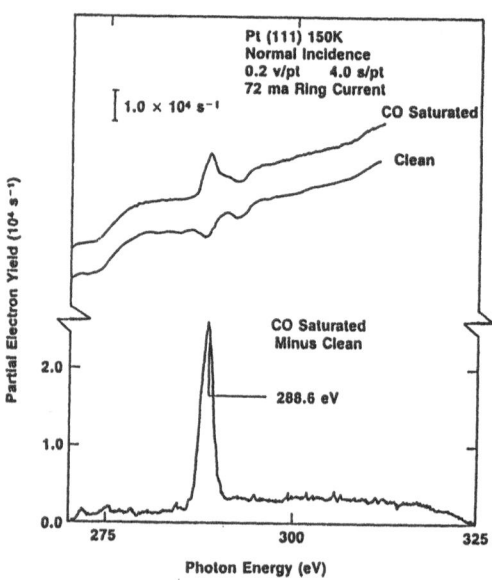

Fig.7.2. The clean and CO saturated spectra in the top panel illustrate that adsorbed CO has a large π resonance at normal incidence. The dips in the clean spectrum are caused by carbon contamination on the beam line optics. The difference spectrum in the lower panel indicates that a saturated CO monolayer results in a 2.5×10^{-4} cts/s signal for a ring current of about 70 mA.

228

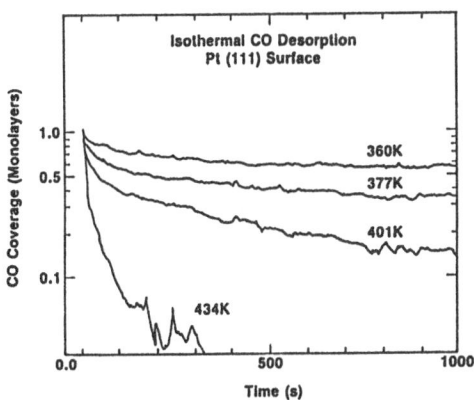

Fig.7.3. A logarithmic plot of the CO coverage as a function of time for a series of desorption temperatures. The slopes of these curves are the logarithm of the desorption rate. The nonlinear behavior of this plot indicates that the desorption rate is not a simple first order process.

A series of isothermal desorption experiments are shown in Fig.7.3 to illustrate typical experimental results. The intensity of the π resonance is plotted as a function of time for several temperatures. The rate of CO desorption is simply the slope of the θ versus time plot. Thermal activation of the desorption process is clearly indicated since the rate of desorption increases with increasing temperature. These data were taken with an incident photon flux of about 5×10^9 photons per second with 50 mA in the ring. Desorption rates in the range from about 10^{-2} to 10^{-5} monolayers per second could be determined based on these data. More rapid rates could be easily measured with increased incident photon flux or by repeating experiments until an appropriate signal-to-noise level is achieved. Such repetitive adsorption or desorption experiments reaction studies might be performed using a modulated gas sources or molecular beam system. Background pressures limited our ability to characterize slower reaction rates since readsorption of background CO causes difficulty for desorption rates lower than about 10^{-5} monolayers per second.

Desorption from a single adsorbed state would be expected to result in a simple first order desorption. The nonlinear behavior of the log of CO coverage versus time plots shown in Fig.7.3 suggests that these desorption kinetics are more complex. Complex behavior for CO desorption from the Pt{111} surface has been observed previously and attributed to the existence of at least two desorption peaks (processes) [7.46,47]. The energetics of CO desorption from the Pt{111} have been determined using a number of methods [7.46-49]. For CO coverages below 0.5 monolayers the heats of desorption based on these studies agree with previous values of about 28 kcal/mole. At higher coverages our results suggest that the heat of desorption is substantially smaller than 28 kcal/mole. Decreased heats of desorption have been observed previously for coverages near saturation and have been attributed to repulsive interactions in the adsorbed overlayer. More extensive desorption rate measurements are required to establish a detailed model for this desorption process since the kinetics are fairly complex.

7.2.3 Pyridine Reorientation of the Ag{111} Surface

A recent letter by *Bader* et al. reported a NEXAFS study of pyridine on the Ag{111} surface performed with the SX-700 monochromator at BESSY in Berlin [7.41]. As part of this work they measured NEXAFS in real time during continuous adsorption of pyridine at 100K as shown in Fig.7.4. They found that the normalized intensity of pyridine's π resonance for normal incidence increases dramatically between 4 and 5.5 L and decreases for larger exposures as indicated in Fig.7.4. This change in π intensity indicates a reorientation of the pyridine molecule with increasing surface coverage. These adsorption measurements were done on a time scale of about 3×10^{-4} monolayers per second. A detailed analysis of spectra taken after fixed exposures indicates that pyridine is oriented $45^0 \pm 5^0$ relative to the surface at low coverages. For coverage just below one monolayer a phase transition occurs which results in a reorientation of the pyridine so that the ring is $70^0 \pm 5^0$ relative to the surface plane. Exposures above 5.5 L result in the formation of a disordered multilayer. This example highlights the unique ability of NEXAFS to monitor subtle changes in adsorbed monolayers in real time. Fairly detailed HREELS studies were performed for this

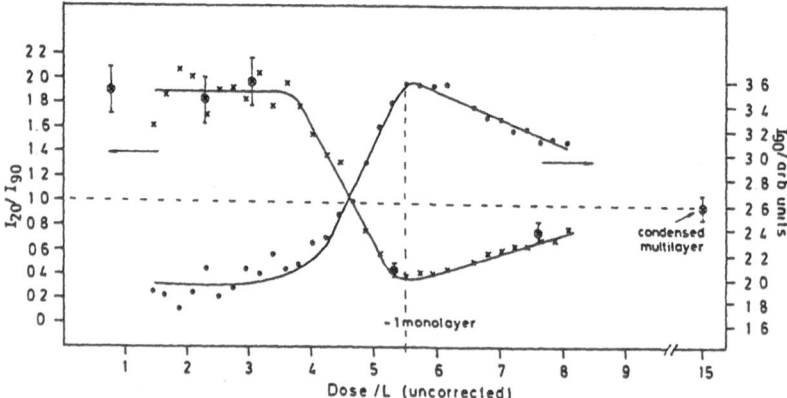

Fig.7.4. From [7.41], increasing pyridine coverage on the Ag{100} surface results in a reorientation of the pyridine molecules at high coverage. Normalized nitrogne K-edge π resonance intensity for normal indicence ($\Theta = 90^0$), I_{90} (full circle) and the intensity ratio I_{20}/I_{90} (crosses) as a function of dose during permanent adsorption of pyridine (~0.1 L/min) on Ag{111} at 100K. The I_{20}/I_{90} data points with error bar were independently measured at fixed pyridine doses. There is a phase transition near 4.5 L.

system in an attempt to establish molecular orientation [7.50]; however, disagreement with these NEXAFS results indicates difficulties with interpretation of the vibration intensities.

7.2.4 Ethylene Conversion to Ethylidyne on the Pt{111} Surface

Adsorbed ethylidyne on the Pt{111} surface has been characterized using a number of surface spectroscopies [7.51]. Recently the kinetics of ethylidyne formation have also been characterized [7.52,53]. *White's* group using TPD and transient SIMS [7.53] found that ethylidyne formation has an activation energy of 17.3 ± 1 kcal/mole and observed a fairly complex coverage dependence. During ethylidyne formation, ethylene loses a hydrogen, a second hydrogen migrates to form the CH_3 group and a metal carbon bond is formed. NEXAFS spectra taken of adsorbed ethylene and ethylidyne suggest that ethylidyne concentrations might be measured in the presence of coadsorbed ethylene [7.54,55]. The normalized spectra shown in the bottom panel of Fig.7.5 confirm that the π resonance of ethylidyne observed for normal incidence furnishes a spectroscopic window for the detection of

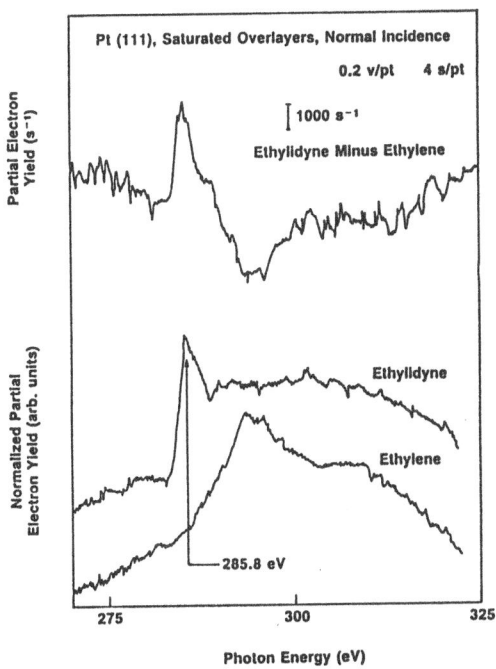

Fig.7.5. Spectra for ethylene and ethylidyne at normal incidence are illustrated in the lower panel. The difference spectrum in the upper panel indicates that a saturated overlayer of ethylidyne results in a signal of 3×10^{-3} cts/s for a ring current of about 60 mA.

adsorbed ethylidyne even in the presence of coadsorbed ethylene. These spectra were obtained as an intitial step in a recent detailed study of ethylidyne formation kinetics on the Pt{111} surface [7.42]. This kinetic study was performed on Beamline I-1 at SSRL.

During these transient NEXAFS experiments kinetic data were obtained by monitoring the intensity of ethylidyne's π resonance at normal incidence. Spectra were acquired periodically during the experimental sequence to check the position of the π resonance and to verify that the surface was not contaminated. Initially, the intensity of the π resonance was monitored for an ethylene-saturated surface to establish an internal coverage calibration. The temperature was then rapidly increased to the reaction temperature desired and the rate of ethylidyne formation was monitored directly. After about 1000 seconds the surface was rapidly heated to 360K to drive the reaction to completion. The intensity of the π resonance was determined again to establish a second internal coverage standard. A series of isothermal kinetic experiments is illustrated in Fig.7.6. The log of the ethylene coverage is plotted against time for a series of reaction temperatures. The linear behavior of the logarithmic plot suggests that the ethylene dehydrogenation reaction is first order in ethylene coverage. This simple kinetic rate law disagrees with earlier SIMS kinetic data [7.53], suggesting that nonlinear neutralization effects may have caused difficulty with the SIMS data. The rate of ethylidyne formation increases with increasing temperature, indicating that the surface reaction is thermally activated. An Ahrrenius plot of the isothermal rate data is

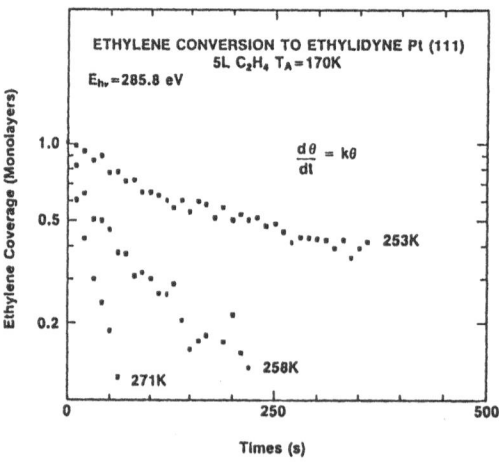

Fig.7.6. The logarithm of the ethylene coverage is plotted as a function of time for several isothermal reaction temperatures. No gas phase products are produced during ethylidyne formation in this temperature range so spectroscopic methods are required for characterization of the surface reaction rate. The linear behavior of the semilog plot indicates that the surface reaction is first order in ethylene surface coverage.

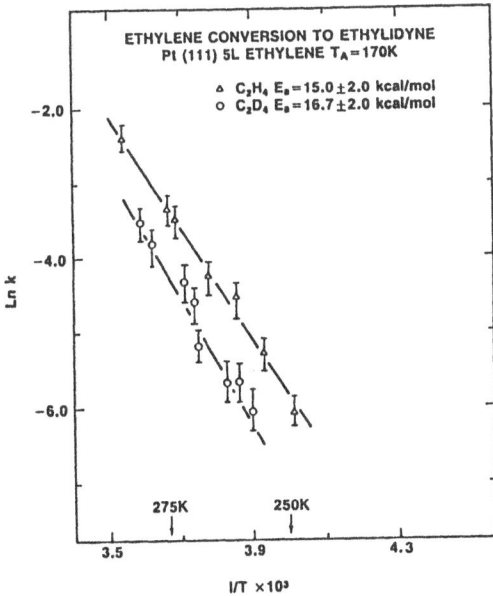

Fig.7.7. An Ahrrenius plot for ethylene conversion to ethylidyne on the Pt(111) surface. A kinetic isotope effect is observed for perdeutero ethylene conversion indicating that hydrogen is involved in the rate limiting step.

shown in Fig.7.7. The 15.0 ± 2 kcal/mole activation energy determined agrees with previous work [7.53]. As indicated in Fig.7.7 we also observe that perdeutero ethylene dehydrogenates more slowly than hydrogenated ethylene, suggesting that C-H bond activation is involved in the rate limiting step. The first order behavior observed indicates that a single ethylene molecule is involved in the rate limiting step. Taken together these kinetic results are beginning to provide a detailed mechanistic picture of an interesting surface reaction. This particular reaction is difficult to study using more conventional methods since the reaction occurs in a temperature range where no products desorb from the surface.

7.2.5 Summary: Transient NEXAFS Using Electron Detection

Transient NEXAFS is a promising technique for characterizing surface reactions. The chemical specificity provided by this technique allows similar chemical species to be easily differentiated. Adequate time resolution is available using synchrotron light sources so that surface reaction rates in the 10^{-2} to 10^{-5} monolayer/s ranges can be easily characterized. The examples reviewed reinforce the concept that for adsorbed species which have been characterized in detail using a combination of surface techniques, detailed kinetic data promise to provide important new insight into the chemical behavior of surface species.

7.3 Fluorescence Yield Near-Edge Structure (FYNES) in the Soft X-ray Region

Fluorescence detection of soft X-ray absorption has recently been developed as a method for detecting X-ray absorption above the carbon K-edge (285 eV) [7.44,56,57]. Fluorescence detection makes it possible to obtain X-ray absorption spectra in the presence of reactive gaseous environments since it is a photon-in, photon-out method [7.57]. Fluorescence yield (FY) measurements have been used extensively in the hard X-ray region to characterize both near edge structure and extended X-ray absorption fine structure (EXAFS) [7.58]. However, since the FY decreases abruptly with decreasing atomic number [7.59], FY has been considered to be poorly suited for detection of X-ray absorption for low z elements in the soft X-ray region. Because of a strong interest in characterizing surface species under reaction conditions, a long-term program to establish FY as a viable detection method in the soft X-ray region has been undertaken. Initial vacuum experiments focused on thiophene adsorption on the Ni{100} surface were performed at the sulfur K-edge (2472 eV) [7.60,61]. A more advanced detector for use at the carbon K-edge was then developed and used to characterize ethylene chemisorption on the Cu{100} surface in vacuum [7.62]. A more sophisticated apparatus capable of performing FY measurements at the carbon K-edge in the presence of up to 10 torr of gas is described in the following section [7.56].

7.3.1 Apparatus

The experimental apparatus consists of a multiple level vacuum chamber with a small high pressure reaction chamber on top where the FYNES experiments are performed [7.56,63]. The primary vacuum chamber is equipped with standard surface science instrumentation including facilities for Auger electron spectroscopy, thermal desorption, low energy electron diffraction and sputtering. A long travel manipulator is used to transfer the sample to the reaction chamber on the upper level. The reaction chamber can be isolated from the main chamber by a gate valve and pumped independently using a turbomolecular pump as indicated in Fig.7.8. The reaction chamber could also be isolated from the UHV soft X-ray beamline and synchrotron light source by a combination of two thin windows and a ballast region. The windows were 1000 Å boron and tin films, supported on a high transmission nickel grid mounted on the center of two 2 $^3/_4$" gate valves. Thus each window could be inserted or removed independently. The windows transmit over 30% of the radiation in the 300 eV energy range. The tin window has an absorption edge around 490 eV which absorbs second order radiation coming from the monochromator for primary energies

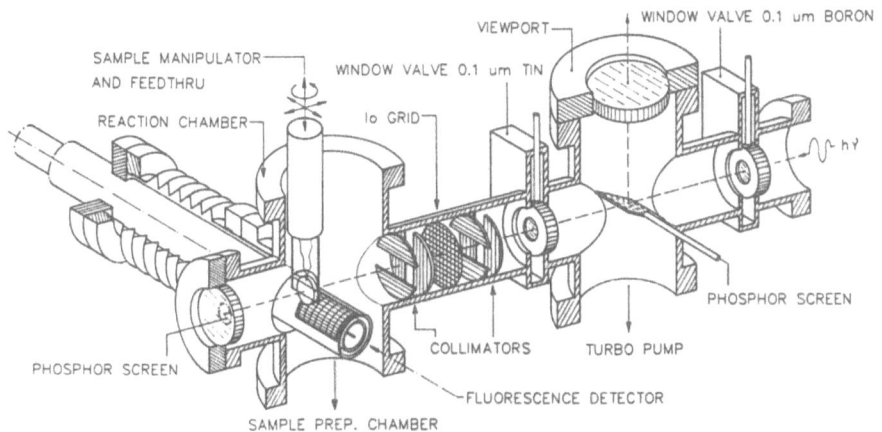

Fig. 7.8. A schematic of the apparatus for fluorescence yield near edge spectroscopy (FYNES) under reactive gases at elevated pressures. The drawing shows a cross section of the fluorescence detector and the combination of thin windows used to isolate the reaction chamber from the photon source.

above 245 eV in the carbon edge region. Each window could withstand a hydrogen pressure differential over 10 torr without any significant leakage to the vacuum side.

X-ray absorption by the sample was measured using a high resolution fluorescence detector optimized for carbon radiation [7.56]. The detector is a differentially pumped ultra high vacuum compatible proportional counter whose position could be varied in order to maximize the fluorescence signal, as shown in Fig. 7.8. The radiation from the sample passes through two 1 μm thick polypropylene windows (each 85% transmitting for C-K$_\alpha$ radiation) with a differnetially pumped region in between. The windows were mounted on 60% transmitting grids of cylindrical symmetry for optimum energy resolution and improved strength. The energy discrimination characteristics of the detector have proven to be crucial for our application, since excitation of the Ni-L$_\alpha$ edge by third order light occurs at about the same monochromator position needed for C-K$_\alpha$ electron excitation. In order to perform carbon-edge NEXAFS experiments we had to reject the signal coming from nickel by using a pulse height discriminator and setting a window around the carbon peak.

The experiments were done on the U1 beam line at the National Synchrotron Light Source, Brookhaven National Laboratory. We estimated the total photon flux to be about $1 \cdot 10^{11}$ photons per second at 300 eV photon energy for a ring current of 100 mA.

Spectra at glancing (30°) and normal incident photon geometries for a CO-saturated Ni{100} crystal in vacuum at 100 K are shown as insets in Fig. 7.9. These spectra were taken with about 300 mA in the

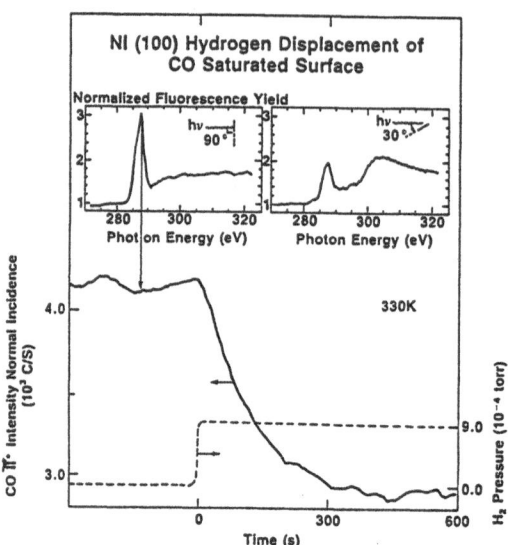

Fig.7.9. Displacement of a monolayer of chemisorbed CO by 9×10^{-4} torr of hydrogen occurs in less than 600 seconds at 330K. The inset spectra illustrate normalized CO spectra taken with fluorescence detection for both normal and glancing incidence light with a CO saturated surface at 100K in vacuum.

ring (100 points, 4 s/pt). They have been normalized by spectra from the clean surface, but a count rate of about 4000 cts/s was observed at the π resonance peak for normal incidence. These spectra illustrate the high sensitivity of our FY detection techniques. The spectra are qualitatively identical to those obtained using electron detection techniques [7.7]. However, the signal-to-background (STB), which is the ultimate parameter in determining the sensitivity of the technique, is close to 2 for the edge jump of a CO-saturated nickel surface, and goes as high as 10 for the π resonance peak at normal incidence. These STB are better than those obtained by any electron yield technique by more than an order of magnitude [7.62]. If we establish a criterion where structure determinations are possible for STB larger than 10%, this means that our fluorescence detection scheme should be able to detect coverages as low as 1% of a chemisorbed CO monolayer.

The lower panel in Fig.7.9 illustrates that chemisorbed CO is rapidly displaced by hydrogen pressures above 10^{-4} torr. This displacement reaction is quite surprising since CO is being displaced by more weakly adsorbed hydrogen. The discovery of this unexpected surface reaction emphasizes the important role that in situ surface methods can play in developing our understanding of surface reactions.

During these transient FYNES experiments the Ni{100} surface was first saturated with CO at the displacement temperature (330 K in Fig.7.9) and a NEXAFS spectrum was acquired. The monochromator was then moved to the peak of the CO π resonance and the entrance

236

windows were inserted. After monitoring the intensity of the π resonance for several hundred seconds, hydrogen was introduced into the reaction chamber to the desired pressure. As indicated in Figs. 7.9,10 the CO was rapidly displaced from the surface for hydrogen pressures above 10^{-4} torr. NEXAFS spectra for both polarizations were taken after the displacement reaction to insure that the CO was actually displaced from the surface. Auger spectra taken following the desorption cycle were used to check the cleanliness of the surface after the displacement.

A series of transient FYNES experiments is shown in Fig. 7.10 for a hydrogen pressure of 0.012 torr. The logarithm of the CO coverage is plotted as a function of time. The rate of CO displacement is simply the slope of a linear coverage plot. These transient FYNES results clearly allow us to make detailed kinetic measurements in the presence of hydrogen. Reaction rates in the 10^{-5} to 10^{-2} monolayer/s region are easily accessible using current technology. The increase in rate with increasing temperature indicates that the displacement reaction is thermally activated. A displacement reaction which is first order in CO coverage would result in a linear decrease in the logarithm of the CO coverage with time. More complex kinetics are suggested by these results. A more detailed discussion of the displacement kinetics and mechanism will be presented after a brief description of thermal desorption experiments done for this same displacement reaction for smaller hydrogen pressures.

In order to verify the unexpected FYNES result we have performed a series of temperature programmed desorption (TPD) experiments [7.64] which are summarized in Fig. 7.11. Rates of displacement were measured by integrating CO TPD spectra for a series of displace-

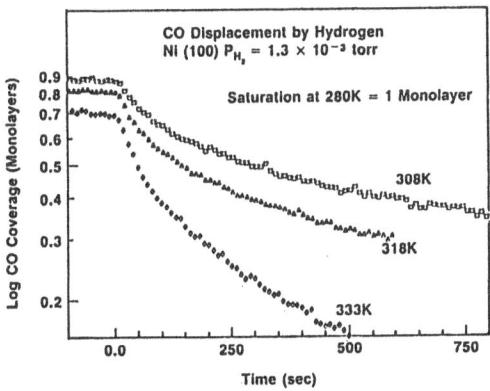

Fig. 7.10. The logarithm of CO coverage taken from transient FYNES experiments, as a function of displacement time for several temperatures. Reaction rates in the region 10^{-2} monolayer/s to 10^{-5} monolayer/s can be characterized using this in situ technique.

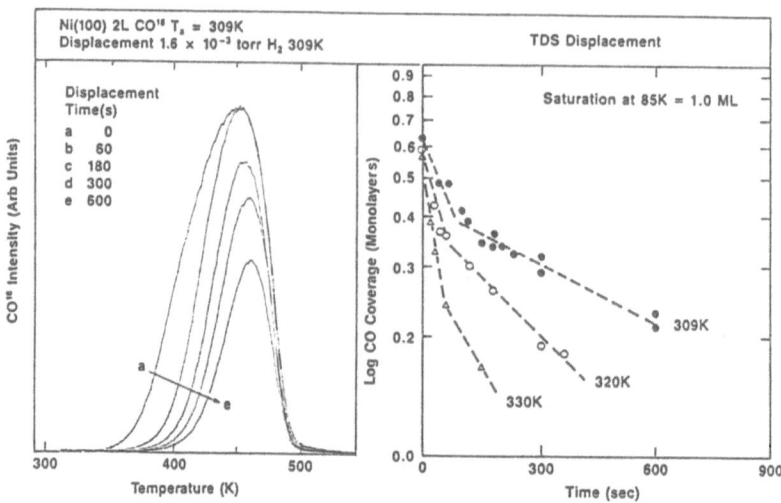

Fig.7.11. A representative series of CO TPD spectra following displacement by 1.6 x 10⁻⁴ torr of H_2 for the indicated time periods is shown in the left panel. The right panel presents the log of CO coverage versus displacement time for a series of displacement temperatures. Rates of displacement were measured by integrating CO TPD spectra for a series of displacement times.

ment times. Hydrogen pressures above 10^{-4} torr cause rapid displacement verifying that the incident soft X-ray radiation was not causing the observed displacement. CO desorption spectra from a series of displacement experiments at 309 K are shown in the left panel of Fig.7.11. The right panel presents a more extensive data set, including displacement experiments at 309, 320 and 330 K. The rates of displacement are positive order in CO coverage since larger rates are observed for larger CO coverages. Note that the displacement reaction initially removes the low temperature shoulder on the CO desorption peak. The low temperature CO shoulder remains when the coverage of undisplaced CO is above 0.3 to 0.4 monolayers (compare curves c and d in Fig.7.3). A transition in the CO displacement rate occurs in the same coverage range as indicated in the right panel of Fig.7.3. This evidence suggests that the displacement rate is *first* order in CO coverage and that the displacement reaction may depend on the energies or structure of the adsorbed CO overlayer. Similar behavior is observed for higher H_2 pressures in Fig.7.10. The rapid increase in rate with increasing temperature shown in the right panel of Fig.7.11 clearly indicates that the displacement process is thermally activated. Estimates indicate that the displacement reaction at high coverage has an activation energy of about 8 ± 2 kcal/mole for a hydrogen pressure of $1.6 \cdot 10^{-3}$ torr. A series of displacement experiments with deuterium indicates that the rate of CO displacement by deuterium is within 5% of the rate of CO displacement by H_2.

The experimental data clearly indicate that CO can be displaced by hydrogen from the Ni{100} surface. The displacement reaction is unexpected since in the 300 K temperature range, coadsorbed hydrogen does not cause a detectable modification of coadsorbed CO [7.63,65-67]. Yet the 7 to 12 kcal/mole activation energies observed for displacement are substantially smaller than the 25 to 30 kcal/mole heat of adsorption [7.68,69]. Several types of mechanism are consistent with the available data. Two of these mechanisms will be briefly discussed. Repulsive interactions in the hydrogen crowded adsorbed overlayer may decrease the strength of the Ni-CO bond for a small fraction of the adsorbed CO molecules. Thermal desorption of this small concentration of weakened CO is consistent with the observed data if CO adsorption strengths can be decreased to the 7-12 kcal/mole range for a small fraction of the adsorbed CO molecules. A direct energy transfer process between thermally activated CO and dissociating hydrogen is also consistent with the observed kinetic behavior. In this scenario, a substantial fraction of the hydrogen's 23 kcal/mole heat of adsorption would be transferred to an adjacent thermally activated Ni-CO bond. The present data do not allow us to determine the molecular details of the displacement reaction. Further experiments are planned to refine our mechanistic understanding of this process.

7.3.2 Summary: Soft X-ray Absorption Using Fluorescence Detection

Fluorescence detection is an exciting development for soft X-ray absorption studies on surfaces. With current light sources and detector systems adsorbed monolayers can easily be detected and characterized using the fluorescence detection method. This detection method allows adsorbed monolayers to be characterized in the presence of reactive gases. The application of FYNES as a transient method has been demonstrated in a recent study of CO displacement from the Ni{100} surface [7.56,63]. The observation that chemisorbed CO is displaced by more weakly adsorbed hydrogen was quite unexpected; this discovery highlights the importance of in situ surface spectroscopies. Surface reaction rates in the 10^{-2} to 10^{-5} monolayer/s range can be measured using current detector systems and light sources. Pressures up to 10 torr and sample temperatures up to 500 K can be accommodated with the current system design. A wide range of important surface reactions is currently accessible and can be characterized using this method in the near future.

7.4 Future Opportunities

Soft X-ray absorption has played an important role in characterizing adsorbed species and improving our understanding of bonding on surfaces. The additional flexibility provided by fluorescence detection promises to add substantially to our ability to characterize the structure and bonding of adsorbed species even in the presence of reactive environments. With current light sources and detector systems a wide range of important surface reactions can be characterized in detail using transient soft X-ray absorption methods. More intense light sources and more sophisticated monochromators promise to improve intensity by several orders of magnitude. The increased intensity will result in substantially improved time resolution and improved sensitivity. Improved intensities will be especially important for the development of the FYNES since pressure isolation windows cause a substantial decrease in intensity available at the sample.

More sophisticated soft X-ray detector systems also promise to improve the sensitivity of the FYNES technique. Rapid advances in solid state detectors and soft X-ray optics may result in detectors with higher sensitivity and energy resolution. FYNES instrumentation capable of characterizing surface reactions at higher temperatures and under higher pressures are now being developed. Pressures in the 1 atmosphere range and tempratues in the 1000 K range can be achieved with extensions of current technology. A wide range of important surface reactions will become accessible with these improvements. The opportunities to characterize adsorbed surface species in detail under reaction conditions will be enhanced substantially by FYNES. This capability for characterizing adsorbed species and reaction rates may play an important role in increasing our understanding of the factors which control surface reactions.

Acknowledgment. During development of the high pressure FYNES technique over the past three years it has been a real pleasure to collaborate with Dr. Daniel Fischer and Dr. Francisco Zaera. Their perseverance, insight and enthusiasm have made this method become a reality. The electron yield NEXAFS measurements were done in part at the Stanford Synchrotron Radiation Laboratory which is supported by the Office of Basic Energy Science at DOE and the Division of Materials Research at NSF. The FYNES measurements were done in part at the National Synchrotron Light Source at Brookhaven National Laboratory which is supported by DOE.

References

7.1   G.A. Somorjai: *Chemistry in Two Dimensions: Surfaces* (Cornell Univ. Press, Ithaca 1981)

7.2   M.W. Roberts, C.S. McKee: *Chemistry of the Metal-Gas Interface* (Clarendon, Oxford 1978)

7.3   R.J. Madix: CRC Crit. Rev. Solid State Mater. Sci. (USA) 7, 143 (1978)

7.4   P.K. Ghosh: *Introduction to Photoelection Spectroscopy* (Wiley, New York 1983)

7.5   H. Ibach, D.L. Mills: *Electron Energy Loss Spectroscopy* (Academic, New York, 1982)

7.6   J. Stöhr, Z. Phys. **B61**, 439 (1985)

7.7   J. Stöhr and R. Jaeger, Phys. Rev. **B26**, 4111 (1982)

7.8   F. Sette, J. Stöhr, E.B. Kollin, D.J. Dwyer, J.L. Gland, J.L. Robbins, A.J. Johnson, Phys. Rev. Lett. **54**, 935 (1985)

7.9   D.W. Moon, S. Cameron, F. Zaera, W. Eberhardt, R. Carr, S.L. Bernasek, J.L. Gland, D.J. Dwyer, Surf. Sci. **180**, L123 (1987)

7.10   J. Stöhr, J.L. Gland,. E.B. Kollin, R.J. Koestner, A.J. Johnson, E.L. Muetterties, F. Sette, Phys. Rev. Lett. **53**, 2161 (1984)

7.11   A.L. Johnson, E.L. Muetterties, J. Stöhr, F. Sette, J. Phys. Chem. **89**, 4071 (1985)

7.12   R.B. Anderson (ed.): *Experimental Methods in Catalytic Research* (Academic, New York 1968)

7.13   M.P. D'Evelyn, R.J. Madix: Surf. Sci. Rep. 3, 413 (1983)

7.14   J.P. Cowin, D.J. Auerbach, C. Becker, L. Wharton: Surf. Sci. **78**, 545 (1987)

7.15   D. Burgess Jr., I. Hussla, P.C. Stair, R. Viswanathan, E. Weitze: J. Chem. Phys. **79**, 5200 (1983)

7.16   M.G. Sherman, J.R. Kingsley, D.A. Dahlgren, J.C. Hemminger, R.T. McIver Jr.: Surf. Sci. **148**, L25 (1985)

7.17   R.B. Hall, A.M. DeSantolo, S.J. Bares: Surf. Sci. **161**, L533 (1985)

7.18   R.B. Hall: J. Phys. Chem. **91**, 1007 (1987)

7.19   M. Mohri, H. Kikabayashi, K. Watanabe, T. Kamashina: Appl. Surf. Sci. **1**, 170 (1987)

7.20   A. Benninghoven, R. Beckmann, D. Griefendorf, M. Shemmer: Appl. Surf. Sci. **6**, 288 (1980)

7.21   J.R. Creighton, J.M. White: Surf. Sci. **122**, L648 (1982)

7.22   L.A. Deluoise, N. Winograd: Surf. Sci. **159**, 199 (1985); **154**, 79 (1985)

7.23   P.L. Radloff, J.M. White: Acc. Chem. Res. **19**, 287 (1986)

7.24   G.W. Rubloff, Surf. Sci. **89**, 566 (1979)

7.25   F. Steinbach, J. Schutte: Rev. Sci. Instrum. **54**, 1169 (1983)

7.26   R. Haight, J. Baker, J. Stark, R.H. Storz, R.R. Freeman, P.H. Bucksbaurm: Phys. Rev. Lett. **54**, 1302 (1985)

7.27   L.J. Richter, B.A. Gurney, J.S. Villarrubia, W. Ho: Chem. Phys. Lett. **111**, 185 (1984)

7.28   L.J. Richter, W. Ho: J. Chem. Phys. **83**, 2569 (1985)

7.29   L.H. DuBois, T.H. Ellis, S.D. Kevan: J. Vac. Sci. Technol. **A3**, 1643 (1985)

7.30   V.A. Burrows, S. Sundaresan, Y.J. Chatal, S.B. Christman: Surf. Sci. **160**, 122 (1985)

7.31   H.A. Englehardt, P. Feulner, H. Pfnür, D. Menzel: J. Phys. **E10**, 1133 (1977)

7.32   P.R. Norton, P.E. Bindner, K. Griffiths, T.E. Jackson, J.A. Davies, J. Rustig: J. Chem. Phys. **80**, 3859 (1984)

7.33   G. Berkovic, Th. Rasing, Y.R. Shen: J. Chem. Phys. **85**, 7374 (1985)

7.34   H.W.K. Tom, C.M. Mate, X.D. Zhu, J.E. Crowell, T.F. Heinz, G.A. Somorjai, Y.R. Shen: Phys. Rev. Lett. **52**, 348 (1984)

7.35   H.W.K. Tom, X.D. Zhu, Y.R. Shen, G.A. Somorjai: Surf. Sci. **167**, 167 (1986)

7.36   H.W.K. Tom: Proc. Int'l Laser Soc. Meeting, Seattle, Washington, 1986 (American Inst. of Physics, in press)

7.37   J. Haase: Appl. Phys. A **38**, 181 (1985)

7.38   F. Sette, J. Stöhr: *EXAFS and Near Edge Structure III*, Springer Proc. Phys. 2, 250 (Springer, Berlin, Heidelberg 1985)

7.39   J. Stöhr, K. Baberschke, R. Jaeger, R. Treichler, S. Brennan: Phys. Rev. Lett. **47**, 381 (1981)

7.40 J.L. Gland, E.B. Kollin, D.A. Fischer, R. Carr, F. Zaera: in preparation
7.41 M. Bader, J. Haase, K.H. Frank, A. Puschmann, A. Otto: Phys. Rev. Lett. 56, 1921 (1986)
7.42 F. Zaera, D.A. Fischer, E.B. Kollin, R. Carr, J.L. Gland: J. Am. Chem. Soc., submitted
7.43 R.L. Johnson: In *Handbook on Synchrotron Radiation*, Vol.1b (North Holland, New York 1983) p.173
7.44 D.A. Fischer, U. Döbler, D. Arvanitis, L. Wenzel, K. Baberschke, J. Stöhr: Surf. Sci. 177, 114 (1986)
7.45 J. Stöhr, R.S. Bauer, J.C. McMenamin, L.I. Johannson, S. Brennan: J. Vac. Sci. Technol. 16, 1195 (1979)
7.46 G. Ertl, M. Neumann, K.M. Streit: Surf. Sci. 64, 393 (1977)
7.47 H. Steininger, S. Lehwald, H. Ibach: Surf. Sci. 132, 264 (1982)
7.48 T.H. Lin and G.A. Somorjai: Surf. Sci. 107, 573 (1981)
7.49 K. Horn and J. Pritchard: J. Physique 38, C4 (1977)
7.50 J.E. Demuth, K. Christmann, P.N. Sandra: Chem. Phys. Lett. 76, 201 (1980)
7.51 For a review, see H. Ibach, D.L. Mills: *Electron Energy Loss Spectroscopy and Surface Vibrations* (Academic, New York 1982) p.326
7.52 F. Zaera, G.A. Somorjai: In *Hydrogen in Catalysis: Theoretical and Practical Aspects*, ed. by Z. Paal, P.G. Menon (Dekker, New York 1987)
7.53 K.M. Ogle, J.R. Creighton, S. Akhter, J.M. White:, Surf. Sci. 169, 246 (1986)
7.54 R.J. Koestner, J. Stöhr, J.L. Gland, J.A. Horsley: Chem. Phys. Lett. 105, 332 (1984)
7.55 J.A. Horsley, J. Stöhr, R.J. Koestner: J. Chem. Phys. 83, 3146 (1985)
7.56 D.A. Fischer and J.L. Gland: Proc. Int'l Conf. of Soft X-Ray Optics and Technology (1986), in press
7.57 F. Zaera, D.A. Fischer, S. Shen, J.L. Gland: Surf. Sci., submitted
7.58 J. Jaklevic, J.A. Kirby, M.P. Klein, A.S. Robertson, G.S. Brown, P. Eisenberger: Solid State Commun. 23, 679 (1977)
7.59 M.O. Krause: J. Phys. Chem. Ref. Data 8, 307 (1979)
7.60 J. Stöhr, E.B. Kollin, D.A. Fischer, J.B. Hastings, F. Zaera, F. Sette: Phys. Rev. Lett. 55, 1468 (1985)
7.61 D.A. Fischer, J.B. Hastings, F. Zaera, J. Stöhr, F. Sette: Nucl. Instrum. Methods Phys. Res. A246, 561 (1986)
7.62 D.A. Fischer, U. Dobler, D. Arvanitis, L. Wenzel, K. Baberschke, J. Stöhr: Surf. Sci. 177, 144 (1986)
7.63 F. Zaera, D.A. Fischer, S. Shen, J.L. Gland: Surf. Sci., submitted
7.64 S. Shen, F. Zaera, D.A. Fischer, J.L. Gland: J. Chem. Phys., submitted
7.65 G.E. Mitchell, J.L. Gland, J.M. White: Surf. Sci. 131, 167 (1983)
7.66 D.W. Goodman, J.T. Yates Jr., T.E. Madey: Surf. Sci. 93, L135 (1980)
7.67 B.E. Koel, D.E. Peebles, J.M. White: Surf. Sci. 107, L367 (1981)
7.68 J.C. Tracy: J. Chem. Phys. 56, 2736 (1971)
7.69 J.C. Tracy, J.M. Burkstrand: CRS Crit. Rev. Solid State Sci. (USA) 4, no.3, 381 (1974)

8. Overview of Electron Microscopy Studies of the So-Called "Strong Metal-Support Interaction" (SMSI)

R.T.K. Baker

Chemical Engineering Department, Auburn University
Auburn, AL 36849-3501, USA

Tauster and coworkers [8.1,2] discovered that when the Group VIII noble metals supported on oxides from Groups IIA–IIB were heated in hydrogen at temperatures in excess of 775 K the systems exhibited unusual chemisorption properties. They attributed the suppression of H_2 and CO chemisorption capacity to the existence of a strong metal-support interaction (SMSI).

In the present chapter, an attempt is made to present an overview of the role that electron microscopy techniques have played in the identification of some of the structural and chemical features associated with this interaction. The most notable contributions are a) the finding that in the SMSI state the function of the metal is to provide a source of H atoms by dissociation of H_2, which are responsible for reducing the support to a lower oxide [8.3], and b) evidence indicating that migration of both metal and reduced oxide support species are involved in the process [8.4]. The details of these and other aspects will be discussed below.

8.1 Experimental

The majority of the experiments reported here were performed on model systems prepared according to the following procedure. Controlled amounts of metals were introduced onto the support surfaces by vacuum evaporation of spectrographically pure metal wire from a clean tungsten filament at a residual pressure of about 10^{-6} Torr. The amount of metal and distance separating the filament from the specimen were selected so as to produce an approximate monolayer coverage on the support surface. Films of titanium oxide, aluminum oxide, and silicon dioxide of about 35 nm thickness were produced by sputtering from the respective target materials onto rock salt. The rock salt was subsequently dissolved away in water, leaving the oxide films floating on the surface. Sections of the films were washed to remove all traces of salt, then mounted on stainless steel electron microscope grids. Carbon films of similar thickness were prepared by vacuum deposition from a carbon arc onto a Pyrex slide. The resulting carbon film was released from the slide by gradual immersion in water and finally mounted on a microscope grid.

Specimens were subsequently treated in hydrogen at temperatures in the range 425 to 1075 K and after cooling and passivation, they were examined in a Philips EM 300 transmission electron microscope. In other experiments described in this chapter, specimen behavior was observed directly by in situ scanning transmission electron microscopy during treatments in hydrogen at temperatures in the range 300 to 1000 K.

8.2 Results and Discussion

Up until recent years, the information obtained from transmission electron micrographs of supported metal particles has been used mainly to verify the average particle size measurements from X-ray and chemisorption techniques [8.5,6]. Occasionally, the microscopy data have also been the key factor in resolving apparent discrepancies in the values obtained from X-ray and chemisorption measurements [8.7]. It was the realization that the more detailed analysis of a particle size distribution available from a microscopy study might shed some light on the prevailing mechanism of particle sintering in a given system that has led to a renewed interest in the use of particle size distribution data.

We have used particle size distribution measurements in an attempt to correlate the relative strengths of the interactions between certain metals and supports [8.8,9]. Figure 8.1 shows a comparison of the size distributions of platinum particles on four support materials following treatment in hydrogen at 975K. It is evident that the distribution from platinum on titanium dioxide is narrower than for the other supports and suggests that the strength of the metal-support interaction follows the sequence: $Pt/TiO_2 > Pt/Al_2O_3 > Pt/SiO_2 \simeq Pt/C$. The notion that inhibition of particle sintering could be used as a general diagnostic for the attainment of the SMSI condition was quickly dispelled during subsequent studies of other supported metal systems. Examination of specimens of palladium supported on titanium dioxide and aluminum oxide shows that following reduction in hydrogen at temperatures from 425 to 1075 K, the palladium particles exhibit a greater degree of sintering on titanium dioxide than on aluminum oxide (Fig.8.2). This relationship existed despite the fact that the former system was in an "SMSI state" at temperatures in excess of 775 K [8.9].

Before leaving the area of particle size distributions, a cautionary remark should be made regarding the accuracy of results derived from

244

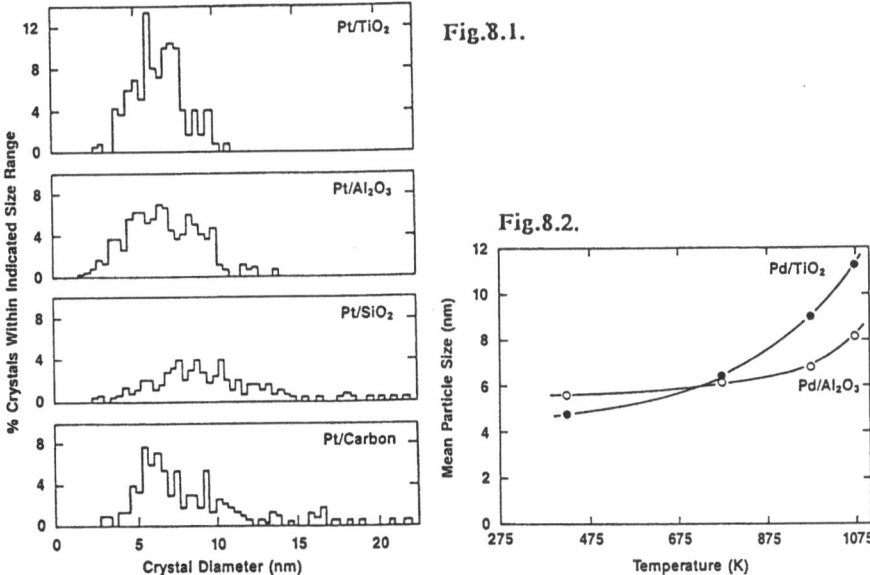

Fig.8.1.

Fig.8.2.

Fig.8.1. Particle size distribution curves for similar loadings of platinum on titania, alumina, silica and carbon after reaction in hydrogen at 975 K

Fig.8.2. Variation of mean palladium particle size with reduction temperature on titania and alumina

electron microscopy studies. Particle detectability and apparent size are found to be sensitive functions of defocus, and as a consequence of the elevation of particles in the specimen [8.10]. This aspect is particularly noticeable in particles below about 2 nm in size. For all sizes of particles, the effects of overlap, and orientation sensitivity of the contrast of the support material, can seriously affect size analysis. Moreover, certain combinations of particle/support orientation can make the particle undetectable [8.11]. These pitfalls are most severe for microscopy specimens that have been prepared from conventional powdered catalyst materials where the terrain of the surface is uneven and the metal particles are randomly orientated with respect to the direction of the electron beam. In contrast to thin film specimens, where the majority of metal particles lie in the same plane as the substrate and interpretation of the information contained in the transmission image is a less formidable task.

Many of the problems mentioned above can be surmounted by using some of the more sophisticated electron imaging techniques such as phase and diffraction contrast, which require the specimen to be tilted, and dark field contrast where both the specimen and illumination are tilted [8.10,11].

As a result of the pioneering efforts of several electron microscopy groups [8.12-14], workers in the field of heterogeneous catalysis

245

have come to recognize that the technique can yield fundamental information on such aspects as the morphology of particles, their chemical identity, and the nature of their interaction with a support medium.

The equilibrium shape of supported metal particles is dependent upon the nature of the forces present at the surfaces of the adjoining phases. If one assumes that a certain degree of atomic mobility exists in the surface layers of the metal particles at the reaction conditions, then the situation can be described according to Young's equation which relates the characteristic interfacial energies existing in the system. For a metal-gas support system at equilibrium:

$$\gamma_{gs} = \gamma_{ms} + \gamma_{mg} \cos\Theta \, , \tag{8.1}$$

where Θ is the contact angle between the metal particle and the support, γ is the surface energy, and the subscripts s, m and g refer to the support, metal and gas, respectively.

Two cases can be considered, illustrated schematically in Fig. 8.3.

a) $\gamma_{gs} > \gamma_{ms}$; $\cos\Theta > 0$, $\Theta < 90°$. \qquad (8.2a)

The metal particles will wet the support and tend to spread and cover the support, in order to minimize the surface energy of the support. In this case the particles would be expected to be cap-shaped, possibly faceted and in the extreme, spread to form thin flat structures, features associated with strong metal-support interactions.

b) $\gamma_{gs} \le \gamma_{ms}$; $\cos\Theta < 0$, $\Theta \ge 90°$. \qquad (8.2b)

Under these circumstances, the particles will be in a nonwetting condition and will tend to assume the most energetically favorable form of a sphere or a polyhedral configuration, characteristics indicative of a relatively weak interaction between the metal and support.

Examples of these two situations can be seen in the micrographs of Fig. 8.4. Figure 8.4a shows the appearance of platinum particles which have been deposited on a thin film of titanium dioxide and then

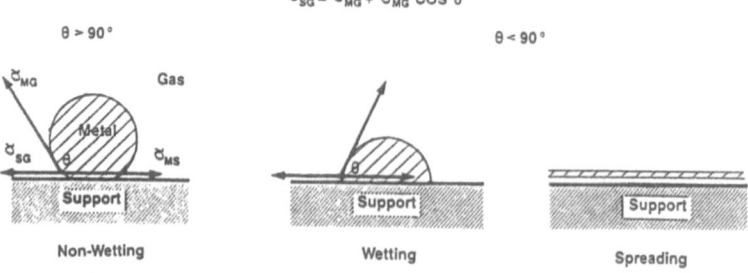

THE METAL-SUPPORT-GAS INTERFACE

$$\gamma_{SG} = \gamma_{MG} + \gamma_{MG} \cos\theta$$

Fig.8.3. The metal-support-gas interface

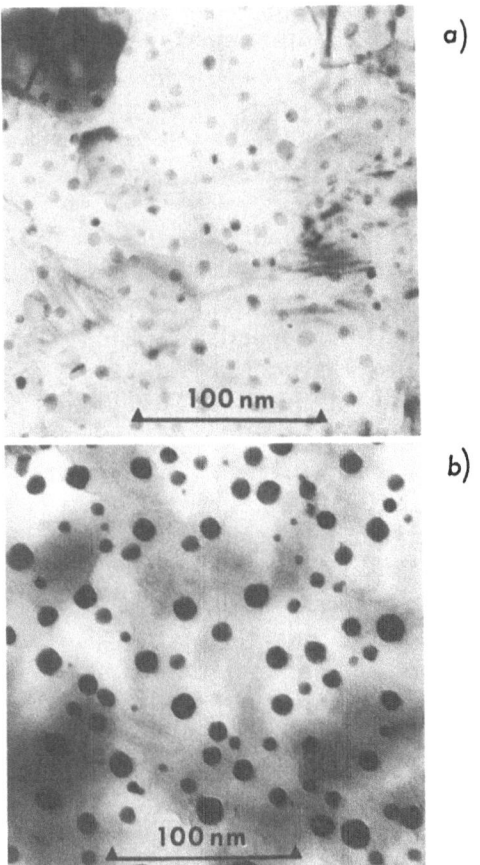

Fig.8.4a. Transmission electron micrograph of a platinum/titania specimen after heating in H_2 at 1025 K for one hour. b Transmission electron micrograph of a platinum/alumina specimen after heating in O_2 at 1025 K for one hour

treated in hydrogen at 1025 K for one hour. Inspection of the metal particles shows them to be predominantly hexagonal in outline and not only thin, but of uniform thickness across the particle, indicative of a "pillbox" morphology. At the same time the support is also found to undergo transformation from TiO_2 to a lower oxide, Ti_4O_7. Treatment of the sample in this condition with oxygen at 1025 K caused the support to revert back to TiO_2 and the platinum particles to increase in size and become more rounded in outline and dense, characteristics of a globular morphology (Fig.8.4b).

It is interesting to compare this behavioral pattern with that reported by *Ruckenstein* and *Chu* [8.15] for platinum/aluminum oxide samples. These workers found that in an oxygen environment the metal particles tended to spread out on the support, which they believed was due to partial oxidation of the platinum, and created a two-dimensional fluid layer of oxide in coexistence with a metallic core. They further suggested that reaction in hydrogen caused sintering of the particles as the oxide was converted to metal and that in this state the particles no longer wetted the aluminum oxide. In this case, it was tacitly assumed that the support did not undergo transformation in either gaseous environment.

The dramatic differences in morphological characteristics of platinum particles on these oxide supports can be rationalized by reference to the interfacial forces operative in each system [8.16] and provide an excellent example of how the nature of the support can influence the shape of metal/metal oxide particles dispersed on it.

When specimens of iron [8.17] and ruthenium [8.18] supported on titanium dioxide were heated in hydrogen at about 775 K, the metal particles underwent a transformation from discrete solid crystallites to toroidal-shaped structures. These results were discussed in terms of a model in which metal atoms migrate from the metal particle surface to freshly reduced regions of the oxide support adjacent to the particles. It was argued that if hydrogen was better able to reduce titania adjacent to the metal particles compared to titania beneath the particles, then a driving force would be established for transporting metal, i.e. the existence of a strong metal-support interaction is responsible for inducing the change in particle morphology [8.17]. It is not clear why this reorganization in particle shape is not observed with all titania supported metal-hydrogen systems, but it may be related to the ease of oxidation of the metal.

Metal particle location and orientation with respect to the support structure are further characteristics which can be established from examination of electron micrographs. For example, it is a simple operation to determine whether particles in a given system tend to be located at particular regions of the support such as edges, steps, pore mouths or other surface imperfections. The ability of small particles to preferentially collect at surface defects was exploited by *Bassett* [8.19] who used this feature to highlight the existence of monatomic steps on the surface of sodium chloride crystals by decorating the specimen with gold particles. *Hennig* [8.20] used the same approach to reveal the presence of very shallow pits in the surface of graphite crystals which were produced by expansion of vacancies during oxidation. Figure 8.5 is a micrograph of platinum supported on a network of alumina and titanium dioxide where it is possible to see that after treatment in hydrogen, platinum particles supported on islands of alumina have tended to move towards to edges of these regions, i.e. they have become located at the alumina/titanium dioxide interface.

The strength of the metal-support interaction may affect not only the metal particle size but also its surface structure, a feature which

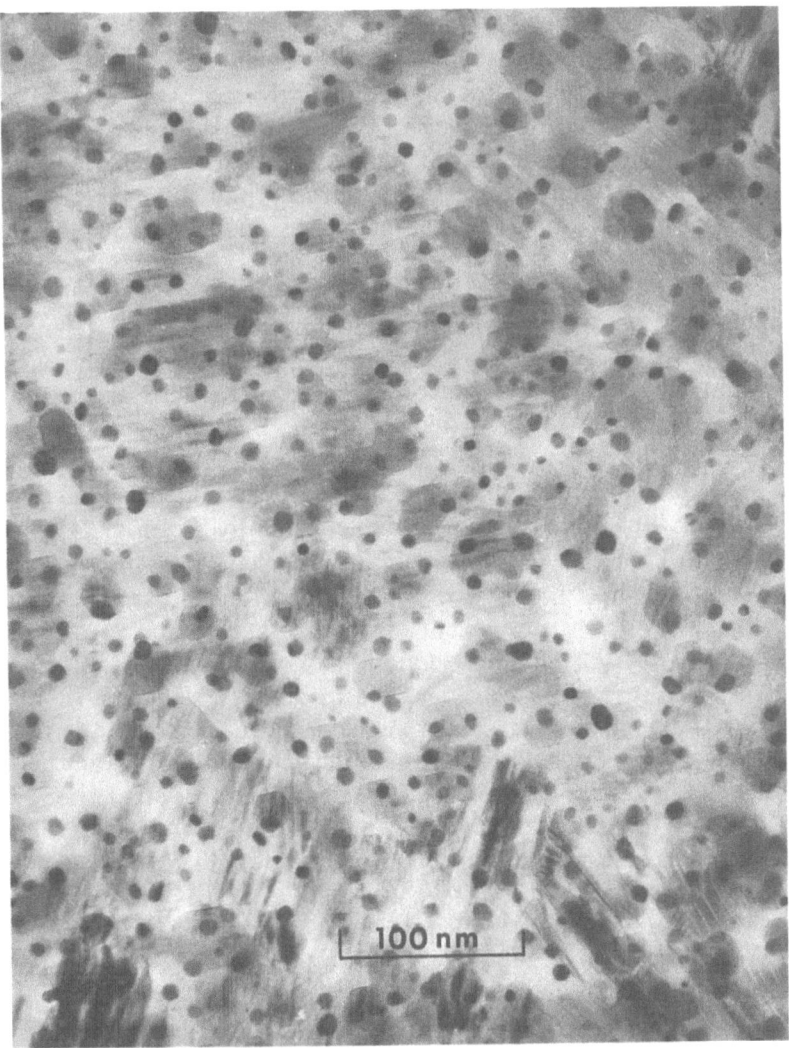

Fig.8.5. Electron micrograph showing the migration of platinum particles on an alumina island to the interface between alumina and an underlying titania film after reaction in hydrogen at 975 K

can have a profound influence on the performance of a supported metal catalyst. According to the nature of the support and the method of preparation of the metal-supported catalyst, thermodynamic equilibrium of the metal particles can be modified in such a way that the nature of the exposed faces and the ratio of particular atoms with different coordination numbers (corner, edge or face atoms) can be changed [8.21-24]. *Dalmai-Imelik* and coworkers [8.25] succeeded in preparing well-faceted nickel crystallites on silica by reduction of a nickel antigorite precursor. Using electron microscopy and electron

249

diffraction, they were able to establish the conditions for producing the metal crystallites in either the {100} or {110} orientations parallel to the support. They further demonstrated that the catalytic activity of the epitaxially grown nickel-supported catalysts for ethylene hydrogenation was much higher than that of randomly oriented catalysts.

The area of scanning transmission electron microscopy (STEM) represents a new generation of electron microscopes which have been developed by adding the functions of the scanning electron microscope to high resolution transmission electron microscopes. This technique offers important new ways of determining both the chemical nature and structural features of supported particles [8.26,27]. For example, it is possible to obtain diffraction information from particles as small as 2 nm in size, and to perform microanalysis of such particles using electron energy loss spectroscopy. Recently, *Jiang* and coworkers [8.28] compared the characteristics of nickel-iron particles supported on titanium and aluminum oxides following reduction in hydrogen at 713 and 773 K with a variety of techniques. Energy dispersive X-ray analysis carried out in the STEM showed that although alloy particles were formed on both supports, the extent of alloy formation was greater on the titania than on the alumina support, and was related to the fact that iron was more easily reduced to the metallic state on the titania. Another key finding was that under reducing conditions alloy particle growth occurred more readily on titania than on alumina, thereby dispelling the notion that suppression of sintering was an SMSI criterion. This conclusion was also reached from other studies based on the interaction of palladium with alumina and titanium oxide supports [8.9].

Up to this point, the discussion has centered around what transmission electron microscopy can reveal about the characteristics of the metal particles in a supported catalyst system. Electron diffraction can also supply valuable information about the support structure. *Baker* and coworkers [8.8] used this technique in conjunction with lattice spacing measurements to demonstrate that during reduction of the platinum/titanium oxide system in hydrogen the support underwent a structural and chemical transformation. It was found that in the presence of platinum at temperatures in excess of 825 K, the support was converted from TiO_2 to Ti_4O_7. In the absence of platinum under similar reducing conditions only the rutile form of TiO_2 was observed. Figure 8.6 is a high resolution electron micrograph of a platinum/titanium oxide specimen, which has been heated in hydrogen at 1100 K for one hour. The lattice spacings which correspond to various crystal planes of Ti_4O_7 are shown on the micrograph. It was claimed that platinum catalyzed the reduction of the support by providing a source of hydrogen atoms via dissociation of molecular hydrogen.

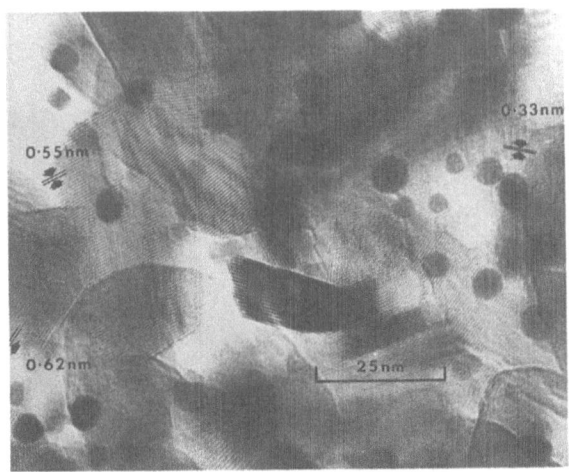

Fig.8.6. High resolution micrograph of platinum on titanium oxide reduced at 1100K showing lattice fringes of Ti_4O_7 and platinum particles on the substrate

In an attempt to establish that the ability of the metal to dissociate molecular hydrogen is a prerequisite for SMSI, this same group of workers studied the behavior of silver on titanium oxide [8.3]. This system was selected because silver is one of the metals which does not dissociate hydrogen and, as such, provides a rigorous test of the model. Figure 8.7 is an electron micrograph of a silver/titanium-oxide specimen which has been treated in hydrogen at 825 K for one hour. It can be seen that the particles are dense and globular in outline, and there are a number of examples of adjacent particles which have been caught in the act of coalescing (indicated by the arrows). Electron diffraction examination of this specimen showed that the support was in the rutile form of TiO_2, there being no evidence of a lower oxide state. Collectively, these features indicated that silver, probably because of its inability to dissociate molecular hydrogen, did not reduce the titanium oxide and as a consequence could not induce the SMSI state.

In a second series of experiments the concept of a geometrically designed catalyst system was employed to demonstrate that silver could undergo a SMSI interaction with the titanium oxide provided a second metal was present to perform the function of dissociating hydrogen. Figure 8.8 is a schematic representation of the specimen design and the subsequent change in appearance following reaction in hydrogen. Some of the previously treated silver/titanium-oxide specimens were coated with a monolayer of platinum in such a manner that only half of the specimen was exposed to the platinum flux. This combination was then reheated in hydrogen for one hour at 825 K, cooled and examined by transmission electron microscopy. Inspection of the bimetallic regions revealed some massive changes in the appearance of the specimen

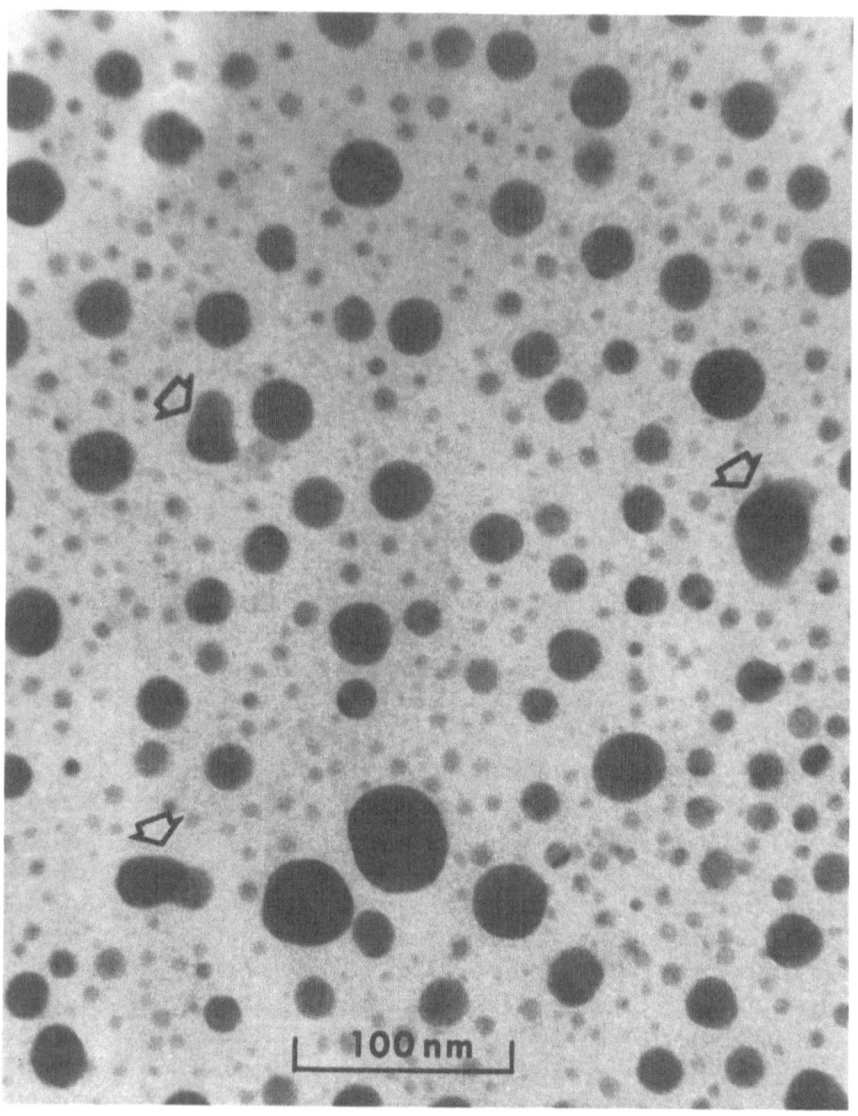

Fig.8.7. Electron micrograph of silver on titania after heating in H_2 at 825 K for one hour. (Arrows indicate particles in the process of coalescing)

compared to its starting conditions. It was impossible to distinguish from the morphological characteristics which particles were silver and which were platinum. All particles were faceted and relatively thin and the majority were less than 2.5 nm in size. Electron diffraction analysis showed that the support had been converted from the rutile structure of TiO_2 to a reduced state, Ti_4O_7. This support transformation was found to occur in regions devoid of platinum, although the extent of the change decreased dramatically as one moved away from the bime-

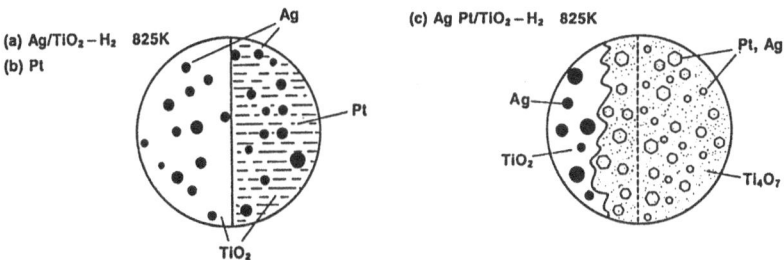

Fig.8.8. Geometrically designed model catalyst of silver and platinum on titania

tallic boundary. It was also interesting to find a corresponding grada-
tion in silver particle characteristics from thin faceted crystallites to
large dense globules. This pattern of behavior was rationalized in terms
of hydrogen spillover which was claimed to be responsible for the
reduction of TiO_2 in the vicinity of the platinum boundary, an effect
which would be expected to decrease with distance from the boundary.
In addition to providing support for the notion that in a SMSI interaction the metal catalyzed reduction of the support, this study also demonstrated that if a given metal did not dissociate hydrogen it could still
undergo such an interaction through the aid of a second metal capable
of performing this function.

Other workers [8.29] have carried out electron microscopy studies
of the rhodium/titanium-oxide system and attempted to characterize
the nature of the support as a function of reduction temperature. Analysis of the electron diffraction patterns of specimens treated in
hydrogen at 773 K showed that the support consisted of a mixture of
anatase and rutile phases of TiO_2. The support composition was determined by X-ray diffraction to be 70% anatase and 30% rutile under
these conditions. Reduction of similar specimens at 1073 K resulted in
a complete transformation of the support to the rutile phase, a result
which was found in the absence of the metal, which suggests that the
anatase to rutile phase change is not catalyzed by the rhodium particles. No evidence was found in these experiments for the formation of
crystalline or other suboxide phases by either electron diffraction or
X-ray diffraction.

It is possible that the difference in support structure reported by
these workers to that found by *Baker* and coworkers [8.3,8] arises from
variations in specimen handling procedures following reduction in
hydrogen. In the studies performed by the Exxon group, specimens
were always passivated prior to removal from the reactor tube. This
procedure, which is frequently used when transferring bulk catalysts to
and from a reactor unit, consists of replacing hydrogen by an inert gas
flow and cooling specimens to room temperature followed by exposure

to a flow of 2% oxygen-inert gas for one hour before finally transferring in air to the microscope. In a recent investigation *Dumesic* et al. [8.4] have highlighted the dangers which can be encountered in determining the correct chemical state of a specimen from a post-reaction examination if a careful passivation step is not performed. Electron diffraction analysis of nickel/titanium-oxide specimens undergoing reaction in the controlled atmosphere electron microscope [8.30] in the presence of 1.0 Torr hydrogen showed that at 1075 K the oxide transformed from TiO_2 to Ti_4O_7. Post-reaction examination of similarly prepared specimens which were treated in a flow reactor in 1 atm 10% H_2/Ar at about 1075 K exhibited the same structural conversion provided that the specimens were passivated before removal from the reactor. In experiments where this step was omitted, the diffraction pattern showed no evidence of Ti_4O_7 formation. Based on these findings, it is imperative to be acquainted with specimen treatment history before attempting to draw conclusions regarding reaction mechanisms from electron diffraction data.

Many of the difficulties associated with a post-reaction electron microscopy examination are readily surmounted by controlled atmosphere electron microscopy (CAEM) [8.30]. This technique enables one to study reactions between gases and solids at very high magnifications, while they are taking place under realistic conditions of temperature, pressure, and reaction time. Furthermore, since it is now possible to carry out this type of experiment in the scanning transmission electron microscope (STEM) one has the ability to perform in situ chemical analysis of the reacting specimen using electron diffraction, energy-dispersive X-ray (EDX) and electron energy loss spectroscopy (EELS). The key design feature in the technique is the ability to operate at high gas pressure in the specimen region while maintaining very low pressure in the rest of the microscope. This aspect is accomplished by differential pumping around the specimen region.

In an attempt to learn more about the SMSI phenomenon, *Baker* and coworkers [8.31] used controlled atmosphere electron microscopy to study the effects of depositing powdered titanium dioxide on nickel surfaces with respect to the formation of filamentous carbon from acetylene. The growth of this form of carbon is readily observable with controlled atmosphere electron microscopy [8.32] and is therefore an ideal reaction to probe the properties of a metal catalyst surface.

During this investigation it was found that if the titanium dioxide/nickel system was heated directly in acetylene to temperatures near 1000 K then growth of carbon filaments did not occur on regions of the nickel surface which were covered with the titanium dioxide. Above this temperature the oxide tended to spall and as bare metal was exposed to the gas phase, prolific filament growth occurred. In contrast, effective passivation of the metal surface was achieved at all temperatures by pretreating the samples in hydrogen at about 770 K. During this reduction treatment, the titanium oxide was observed to undergo a restructuring and eventually wet and spread over the nickel surface. This process could be reversed by heating these samples in oxygen at 845 K, which resulted in the reformation of titanium dioxide crystallites and reexposure of the nickel surface to the gas phase, which catalyzed filament formation during subsequent reaction in acetylene.

It was claimed that these results provided direct evidence that under reducing conditions, titanium oxide species were capable of migrating over metal surfaces and could be stabilized in the form of discrete titanium dioxide crystallites under oxidizing conditions. As such, it was believed that the results supported the notion that the collection of reduced titanium oxide species on metal surfaces could be the origin of strong metal-support interactions for titanium oxide supported metal particles.

In more recent studies, *Dumesic* and coworkers [8.4] have extended the studies of the nickel/titania system in an attempt to characterize the role of the metal in the reduction process, and in particular to determine whether the metal facilitates the migration of reduced titania species onto its surface during reaction in hydrogen. In this investigation they used in situ scanning transmission electron microscopy to continuously follow the physical and chemical changes of a model system consisting of overlapping nickel and titania films during reaction in hydrogen.

When such samples were heated in 1.0 Torr hydrogen, the nickel film was observed to transform into large globular particles at around 800 K. When the temperature was raised to 1000 K, light areas formed around some of the nickel located on the titania, suggesting that the oxide was being depleted in these regions. On continued reaction up to 1080 K, this attack became more intense, and deep holes, which were apparently devoid of titania, were produced behind many of the nickel particles. Figure 8.9 is a sequence showing the movement and coalescence of two nickel particles and the creation of trails in the titania film as a result of this action. From inspection of these and many other micrographs it is apparent that the size of the holes in the oxide is considerably larger than the nickel particles responsible for producing them. This aspect led to the conclusion that at least a fraction of the titania removed from the support was present within the metal particles. It was significant that the appearance of regions of the titania not in contact with nickel remained visually unchanged, even at temperatures up to 1200 K.

Corresponding electron diffraction analysis showed that at 1000 K the titania was present in the rutile form, whereas at 1080 K it was

255

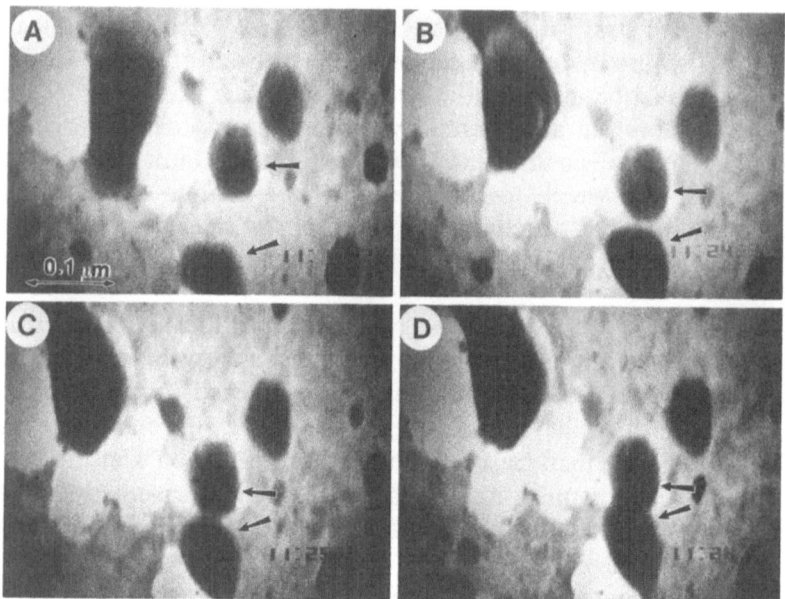

Fig.8.9. Sequence showing the interaction of nickel particles with a titania film in 1 Torr H_2 at 1070 K

completely converted to Ti_4O_7. It is possible that this reduced state was formed at 1000 K in the regions surrounding the nickel particles but was of such a minor concentration that its pattern was masked by the strong rutile signal.

The results of this investigation clearly demonstrate that migration of both titania and metal species may be involved in the initiation of the SMSI state during treatment in hydrogen at temperatures in excess of 775 K.

8.3 Summary

From examination of the profiles of supported metal particles, one can determine their morphological features such as size, shape and topography. These characteristics are directly dependent on the strength of the interaction between the particles and the support, e.g. the formation of large globular particles is indicative of a weak metal-support interaction, whereas the formation of faceted particles that are of uniform thickness and very thin is associated with a strong metal-support interaction.

The particle size distribution obtained from analysis of electron micrographs provides a direct measure of the growth characteristics of the supported particles. The shape of such a distribution is a function of the predominant sintering mechanism operative in a given system.

One would expect that the distribution derived from a system where particle migration was the overriding mechanism would follow a broad Gaussian-shaped curve skewed towards the large particle diameter side. Such a relationship would be indicative of a weak metal-support interaction. In contrast, for a system where particle growth occurred exclusively by the atomic migration mode a Lorentzian distribution may be found. This type of dependence would be expected for a system where a significant interaction existed between the metal and the support. The degree to which this dependence is maintained with increasing reaction temperature provides a measure of the relative strength between the metal and support.

A further aspect which can be obtained from an electron micrograph is the location of particles on the support surface. It is a simple task to determine whether there are preferred collection sites for particles on surfaces such as edges, steps, surface defects or grain boundaries. In a designed system consisting of overlapping support films, it is then possible to determine whether the metal particles tend to accumulate on one support or the other, or whether the support/support interface is the preferred collection site.

Using micro-diffraction techniques, it is now possible to obtain both structural and chemical information from individual supported metal particles. One can determine the structural relationship between the metal and support and establish whether the particles grow in a preferred orientation such as an epitaxial arrangement. Based on the chemical information, it is possible to establish if a new compound is formed between the metal and the support.

Finally, using conventional electron diffraction, one can follow structural changes in the support material which can lead to major changes in the nature of the metal-support interaction. This aspect has been cited as one of the factors responsible for the origin of the so-called strong metal-support interaction.

References

8.1   S.J. Tauster, S.C. Fung, R.L. Garten: J. Am. Chem. Soc. **100**, 170 (1978)

8.2   S.J. Tauster, S.C. Fung: J. Catal. **55**, 29 (1978)

8.3   R.T.K. Baker, E.B. Prestridge, L.L. Murrell: J. Catal. **79**, 348 (1983)

8.4   J.A. Dumesic, S.A. Stevenson, R.D. Sherwood, R.T.K. Baker: J. Catal. **99**, 79 (1986)

8.5   D. Cormak, R.L. Moss: J. Catal. **13**, 1 (1969)

8.6   C.R. Adams, H.A. Benesi, R.M. Curtis, R.G. Meisenheimer: J. Catal. **1**, 336 (1962)

8.7   G.B. McVicker, R.L. Garten, R.T.K. Baker: J. Catal. **54**, 129 (1978)

8.8   R.T.K. Baker, E.B. Prestridge, R.L. Garten: J. Catal. **56**, 390 (1979)

8.9   R.T.K. Baker, E.B. Prestridge, G.B. McVicker: J. Catal. **89**, 422 (1984)

8.10 P.C. Flynn, S.C. Wanke, P.S. Turner: J. Catal. **33**, 233 (1974)

8.11 M.M.J. Treacy, A. Howie: J. Catal. **63**, 265 (1980)

8.12 M.J. Stowell, T.J. Law, J. Smart: Proc. Roy. Soc. London A318, 231 (1970)

8.13 J.G. Allpress and J.V. Sanders: Surf. Sci. 7, 1 (1967)

8.14 S. Ino, S. Igawa: J. Phys. Soc. Jap. 22, 1365 (1965)

8.15 E. Ruckenstein, Y.F. Chu, J. Catal. 59, 109 (1979)

8.16 R.T.K. Baker: J. Catal. 63, 523 (1980)

8.17 B.J. Tatarchuk, J.J. Chludzinski, R.D. Sherwood, J.A. Dumesic, R.T.K. Baker: J. Catal. 70, 433 (1981)

8.18 R.T.K. Baker, J.J. Chludzinski: unpublished results

8.19 G.A. Bassett, Phil. Mag. 33, 1042 (1958)

8.20 G. Hennig: J. Inorg. Nucl. Chem. 24, 1129 (1962)

8.21 R. van Hardeveld, A. van Moonfort, Surf. Sci. 4, 396 (1966)

8.22 R. van Hardeveld, F. Hartog: Surf. Sci. 15, 189 (1969)

8.23 W. Romanowski: Surf. Sci. 18, 373 (1969)

8.24 J.J. Burton: Catal. Rev. 9, 209 (1974)

8.25 G. Dalmai-Imelik, C. Leclerq, A. Maubert-Muguet: J. Solid State Chem. 16, 129 (1976)

8.26 F. Delannay: Catal. Rev. Eng. Sci. 22, 141 (1980)

8.27 M.J. Yacaman: Appl. Catal. 13, 1 (1984)

8.28 X-Z. Jiang, S.A. Stevenson, J.A. Dumesic, T.F. Kelly, R.J. Casper: J. Phys. Chem. 88, 6191 (1984)

8.29 A.K. Singh, N.K. Pande, A.T. Bell: J. Catal. 94, 422 (1985)

8.30 R.T.K. Baker and P.S. Harris: J. Phys. E 5, 793 (1972)

8.31 R.T.K. Baker, J.J. Chludzinski, J.A. Dumesic: J. Catal. 93, 312 (1985)

8.32 R.T.K. Baker, M.A. Barber, P.S. Harris, F.S. Feates, R.J. Waite: J. Catal. 26, 51 (1972)

9. Theory of Desorption Kinetics

H.J. Kreuzer

Department of Physics, Dalhousie University
Halifax, NS B3H 3J5, Canada

Equilibrium properties of an isolated physical system are obtained theoretically by minimizing the total free energy of the system. A phenomenological approach employs the framework of equilibrium thermodynamics, whereas a microscopic model would be evaluated using the methods of equilibrium statistical mechanics. Let us next open the system to mass and energy exchange with its environment by pumping on it, heating it, or bombarding it with a laser or particle beam. The resulting nonequilibrium state will evolve in time leading to transport or kinetic processes whose time scales will depend on how fast and from where mass and energy are exchanged. To describe such processes at a macroscopic level one employs the methods of nonequilibrium thermodynamics. If a microscopic understanding is sought we must resort to the tools developed in nonequilibrium statistical mechanics. In this review we will look at the kinetics of adsorption and desorption of a gas at a solid surface using both approaches. For the most part we will concentrate on desorption.

The, theoretically, simplest way to induce desorption of an adsorbate is to disturb the initial equilibrium by rapidly pumping away the gas phase above the solid surface. To attain the final equilibrium state, with no adsorbate present, we must supply the adsorbed molecules with enough energy to break the surface bond. In this isothermal experiment, this energy would be drawn out of the thermal reservoir of the solid. Because rapid pumping is usually not feasible experimentally, one resorts to temperature programmed desorption in which one raises the temperature of the solid in a controlled manner, usually linearly, to increase the energy supply to the adsorbate. A variant of this technique involves raising the temperature rapidly to some higher value which is then held fixed during the ensuing desorption process. Thermal desorption induced by heating with a laser pulse is a further technique. Thermal desorption experiments and their interpretation have been reviewed by *Menzel* [9.1]. Desorption can also be induced by supplying energy to the adsorbate with an electron or ion beam [9.2] or resonantly coupling an infrared laser to some internal vibrational mode of an adsorbed molecule [9.3,4].

In our presentation of theoretical methods to deal with adsorption-desorption kinetics we will start in the next section by looking at the kinetics of a two-phase adsorbate using the methods of nonequilibrium thermodynamics. This not only provides us with an economic way of describing experimental data but also allows us to identify the phenomenological transport coefficients that must be calculated in a microscopic theory. Such a programme is then outlined in Sect.9.2 based on the master equation.

9.1 Nonequilibrium Thermodynamics of a Two-Phase Adsorbate

9.1.1 Preliminary Comments

At low coverage adsorbed particles are, on average, so far apart from each other that their mutual interaction is negligible. As a consequence, desorption is a first order process with the rate of desorption proportional to the number of adsorbed particles left to desorb. As coverage builds up, the interaction between adsorbed particles becomes important which leads, for attractive interactions, to clustering, and, eventually, for both attractive and repulsive interactions, to long range order. Here we are primarily concerned with gas-solid systems in which, for certain ranges of coverage and temperature, the adsorbate exhibits coexistence of a dilute, gas-like phase together with islands of a condensed phase, both, of course, being two-dimensional (2D). The 2D gas phase consists of two parts: one being adsorbed on the bare surface, the other on top of the condensed islands. It would lead to difficulties with the Gibbs phase rule if we were to treat them as separate phases. Rather we should deal with the 2D gas as one phase in an external potential that is different on the bare surface and on top of the islands.

Nonequilibrium thermodynamics deals with macroscopic quantities such as mass and energy densities and can only be applied to systems in local equilibrium [9.5]. This is to say that the time evolution of these variables must be slow on a microscopic time scale and their spatial variation must be small over a microscopic length scale. For a gas these are readily identified as the collision time t_c, and the mean free path $\ell = \bar{v}/t_c$ between collisions (where $\bar{v}$ is the average particle velocity). These concepts readily transcribe to highly mobile adsorbates. For localized adsorption, t_c and ℓ can be identified as the time needed and distance traveled before an isolated, adsorbed particle occupies a site next to another adsorbed particle. In a two-dimensional gas on a surface, ℓ will be of the order of a few ten lattice sites only. If we were to average over such small areas, the total number of particles found in them would be only a few hundred at the most, implying that the statistical fluctuations in the numbers obtained would be of order 10% and more. Such situations can only be satisfactorily described by microscopic models but not by nonequilibrium thermodynamics. We will, therefore treat the densities in the various phases and components as constant. Thus, the relevant extensive thermodynamic variables are the number of particles in the 3D gas phase, N_3, in the condensed phase of the adsorbate, N_1, in the 2D gas phase on the bare surface, N_2, and on top of the islands, $N_{2'}$. In addition, to deal with questions of energy transfer, we must introduce the internal energies, U_i, with i = 1, 2, 2', and 3, of the various phases and components.

Desorption of a two-phase adsorbate can proceed via several channels, namely (i) out of the 2D gas phase on the bare surface with a time constant t_d, (ii) out of the 2D gas phase on top of the condensed islands with a time constant $t_{d'}$, and (iii) directly out of the condensed phase with a time constant t_c. It is sometimes argued that the last process is negligible due to the fact that particles in this phase are more strongly bound by roughly the 2D latent heat. However, we should keep in mind that in many systems the entropy gain out of the condensed phase will outweigh the energetic disadvantage. We will see in particular that desorption from the condensate is crucial around monolayer coverage. On the other hand, as long as desorption proceeds predominantly via the 2D gas phase, the depletion of the condensate islands takes place by evaporating into the 2D gas phase, with a time constant t_{ev}, from where particles then desorb into the 3D gas phase. Two rather different situations may obtain: (i) if $t_{ev} \ll t_d$, evaporation is so fast that during the desorption process a quasi-equilibrium is maintained between the adsorbed phases. In such a situation, there will be a coverage regime where the desorption kinetics is roughly zero order. (ii) If $t_{ev} \gg t_d$, then evaporation is the slowest process in a chain and thus rate determining. With evaporation proceeding via the rim of the condensed islands, one expects roughly half order kinetics. We will see in the numerical examples that the desorption kinetics of two-phase adsorbates are much more interesting than that in the intermediate regime. A good review of the experimental situation has been given by *Menzel* [9.6].

There are a number of phenomenological models developed to explain zero and fractional order desorption from two-phase adsorbates [9.7-14]. Some authors have approached the problem more microscopically by starting from the rate equation as given by transition state theory and calculating the activation energies and entropies within the Bragg-Williams and quasichemical approximation of the lattice gas model [9.15-22].

Our aim here is not so much to fit particular data, but to examine the richness of desorption kinetics in relatively simple model systems, and to indicate how more sophisticated models can be set up to fit actual data such as the desorption kinetics of rare gases on metals and of metals on metals in order to extract rate constants.

9.1.2 General Formulation

To set up phenomenological rate equations for the adsorption-desorption kinetics of a two-phase adsorbate, it is expeditious to follow the Onsager approach to nonequilibrium thermodynamics [9.5,23-25]. We neglect diffusive processes in the two-dimensional gas phase and treat the densities as spatially constant (or averaged). To set up balance equations properly, we consider the total system, consisting of gas phase, adsorbate, and solid, as closed in a volume V_t and isolated with total energy U_t. As extensive variables we consider the particle numbers and the respective energy variables where the subscript s refers to the solid. Following Onsager we can write the macroscopic balance equations as

$$\frac{dX_i}{dt} = \sum_{j=1}^{10} L_{ij} \frac{\partial S}{\partial X_j}\bigg|_{U_t,V_t,N,N_s} .$$ (9.1)

Such a description is valid (i) in the linear regime, and (ii) as long as local equilibrium pertains. The latter condition in particular implies that

$$S(U_t, V_t N, N_s) = S_1(U_1, A_1, N_1) + S_2(U_2, A_2, N_2)$$
$$+ S_{2'}(U_{2'}, A_{2'}.N_{2'}) + S_3(U_3, V, N_3) + S_s(U_s, V_s, N_s) .$$

(9.2)

Thus we get in (1)

$$\frac{\partial S}{\partial N_j}\bigg|_{U_j,A_j(V)} = - \frac{\mu_j}{T_j} ,$$ (9.3)

$$\frac{\partial S}{\partial U_j}\bigg|_{N_j,A_j(V)} = \frac{1}{T_j} ,$$ (9.4)

introducing the chemical potentials μ_j and the temperatures T_j in the various phases and components.

Next we note that mass conservation in the system $dN_s/dt = 0$, $(d/dt)(N_1+N_2+N_{2'}+N_3) = 0$ and energy conservation $dU_t/dt = 0$ give us 30 conditions on the 100 phenomenological coefficients L_{ij}, because the coefficients in front of each thermodynamic force have to vanish independently. Onsager's reciprocity relations $L_{ij} = L_{ji}$ eliminate another 43 coefficients.

Under isothermal situations, the equations simplify considerably and can be written as

$$\frac{dN_1}{dt} = - \frac{L_{12}}{T} (\mu_2-\mu_1) - \frac{L_{12'}}{T} (\mu_{2'}-\mu_1) - \frac{L_{13}}{T} (\mu_3-\mu_1) ,$$ (9.5)

$$\frac{dN_2}{dt} = - \frac{L_{12}}{T} (\mu_1 - \mu_2) - \frac{L_{23}}{T} (\mu_3 - \mu_2) - \frac{L_{22'}}{T} (\mu_{2'} - \mu_2) , \qquad (9.6)$$

$$\frac{dN_{2'}}{dt} = - \frac{L_{12'}}{T} (\mu_1 - \mu_{2'}) - \frac{L_{22'}}{T} (\mu_2 - \mu_{2'}) - \frac{L_{2'3}}{T} (\mu_3 - \mu_{2'}) , \qquad (9.7)$$

$$\frac{dN_3}{dt} = - \frac{d}{dt} (N_1 + N_2 + N_{2'}) . \qquad (9.8)$$

Note that equilibrium conditions, i.e., $\mu_i = \mu_j$, imply vanishing fluxes.

It is our next task to relate the 6 Onsager coefficients in (9.5-8) to experimentally accessible quantities like sticking coefficients and heats of activation. We do this in turn for the individual processes.

a) Adsorption

To determine the coefficients L_{13}, L_{23}, and $L_{2'3}$ controling adsorption and desorption in the 3D gas phase, we assume that all other phases and components are in equilibrium except that the 3D gas phase has a small excess of particles ΔN_3. We can then write

$$\mu_3 = \bar{p}_3 + \left.\frac{\partial \mu_3}{\partial N_3}\right|_{T,V} \Delta N_3 , \qquad (9.9)$$

where $\bar{p}_3 = \bar{p}_2 = \bar{p}_1 = \bar{p}_2{}' = \bar{p}$ is the chemical potential in equilibrium, and the derivative is evaluated at the equilibrium point. We treat the 3D gas phase as ideal so that

$$\bar{p}_3 = k_B T \ln\left(\frac{N_3}{V} \lambda^3\right) , \qquad (9.10)$$

where

$$\lambda = \frac{h}{(2\pi m k_B T)^{1/2}} , \qquad (9.11)$$

is the thermal de Broglie wavelength. Also note that $\Delta N_3 = (V/k_B T) \cdot (P - \bar{P})$. We thus get for (9.5-7) under adsorption conditions

$$\left.\frac{dN_i}{dt}\right|_{ads} = - \frac{L_{i3}}{T} \frac{V}{N_3} (P - \bar{P}) , \qquad (9.12)$$

for $i = 1, 2, 2'$.

On the other hand, one usually writes the rate equation for adsorption as

$$\left.\frac{dN_i}{dt}\right|_{ads} = S_i A_i \frac{P - \bar{P}}{(2\pi m k_B T)^{1/2}} \qquad (9.13)$$

in terms of an excess flux of particles hitting a surface area A_i and sticking with a probability S_i. Thus we get, by comparing (9.12) and (9.13),

263

$$\frac{L_{i3}}{T} = - S_i \, A_i \, \frac{1}{(2\pi m k_B T)^{1/2}} \, \frac{N_3}{V} = - S_i \, A_i \, \frac{1}{\lambda^2 h} \, e^{\mu_3/k_B T} \, . \qquad (9.14)$$

Information about the dynamics of energy exchange between a particle hitting the surface and the solid is contained in the sticking coefficients S_1, S_2, and $S_{2'}$. As for their relative magnitude, one can argue that light gas particles, e.g. helium and neon, experience poor energy exchange on a bare surface of heavy atoms such as a transition metal, for reasons of mass mismatch, so that in such cases $S_{2'} \gg S_2$. On the other hand, sticking in a condensate island is not too probable, i.e. $S_1 \ll 1$, because a particle hitting an island of condensed phase will not instantaneously penetrate at the spot of collision but more likely skid along the top of the island, i.e. be part of the gas phase 2′, and attach itself at the rim of the island.

b) Desorption

Desorption will take place if there are in one or more of the adsorbate components too many particles, more than required by equilibrium. Let us first look at desorption from the bare surface. We write

$$\mu_2 = \bar{\mu}_2 + \frac{\partial \mu_2}{\partial N_2}\bigg|_{T,A_2} \Delta N_2 \, , \qquad (9.15)$$

but demand again $\bar{\mu}_3 = \bar{\mu}_2 = \bar{\mu}_1 = \bar{\mu}_{2'} = \bar{\mu}$. Thus from (9.6) we obtain

$$\frac{dN_2}{dt}\bigg|_{des} = \frac{L_{23}}{T} \, \frac{\partial \mu_2}{\partial N_2}\bigg|_{T,A_2} \Delta N_2 \, , \qquad (9.16)$$

with L_{23} given in (9.16) as a negative quantity, and $\Delta N_2 = N_2 - \overline{N}_2$. To evaluate (9.16) further, we must adopt a thermodynamic model for the 2D gas phase on the bare surface, such as a van der Waals gas, or a lattice gas, in the Bragg-Williams or quasi-chemical approximation. To keep matters transparent, we will treat the 2D gas phase as an ideal mobile gas, so that we can write

$$\mu_2 = k_B T \, \ln(N_2/q_2) \qquad (9.17)$$

where

$$q_2 = \frac{A_2}{\lambda^2} \, q_z \, e^{E_d/k_B T} \qquad (9.18)$$

is the single particle partition function in a surface potential of depth $-E_d$. Treating the latter as a harmonic oscillator (around the bottom of the well) we also have

$$q_z = (e^{h\nu_z/k_B T} - 1)^{-1} \, . \qquad (9.19)$$

From (9.17) we get

$$\frac{\partial \mu_2}{\partial N_2}\bigg|_{T,A_2} = \frac{k_B T}{N_2} = k_B \, \frac{T}{q_2} \, e^{-\bar{\mu}_2/k_B T} \qquad (9.20)$$

264

which with $\bar{\mu}_2 = \bar{\mu}_3$ replaced by (9.10,14), gives

$$\frac{dN_2}{dt}\bigg|_{des} = - S_2 \frac{k_B T}{h q_z} e^{-E_d/k_B T} \Delta N_2 \ . \tag{9.21}$$

Phenomenologically, this rate equation is usually written

$$\frac{dN_2}{dt}\bigg|_{des} = - r_{d2} \Delta N_2 \ , \tag{9.22}$$

so that we can identify the desorption rate constant as

$$r_{d2} = S_2 \frac{k_B T}{h q_z} e^{-E_d/k_B T} \ . \tag{9.23}$$

In the high temperature limit, $k_B T \gg h\nu_z$, we have $q_z \simeq k_B T/h\nu_z$, so that (9.23) reduces to the familiar form

$$r_{d2} = S_2 \, \nu_z \, e^{-E_d/k_B T} \ . \tag{9.24}$$

If both adsorption and desorption are present we must combine (9.13,21) to get for the relevant terms in (9.6,7)

$$- \frac{L_{23}}{T} (\mu_3 - \mu_2) = S_2 A_2 \frac{P}{(2\pi m k_B T)^{1/2}} - S_2 \nu_z e^{-E_d/k_B T} N_2 \ , \tag{9.25}$$

$$- \frac{L_{2'3}}{T} (\mu_3 - \mu_{2'}) = S_{2'} A_1 \frac{P}{(2\pi m k_B T)^{1/2}} - S_{21} \nu_z' e^{-E_d'/k_B T} N_{2'} \ . \tag{9.26}$$

The term involving L_{13}, i.e. adsorption on and desorption from the condensed phase, will be listed as r_{13} (9.58,59). To have a closed set of equations, one expresses P in terms of the particle density in the 3D gas phase via the ideal gas law. In practice, one would most likely keep P as an experimental control parameter.

Before closing this subsection, we note that (9.25,26) can be obtained more directly be relating $\bar{P}$ in (9.13) to $\bar{\mu}_3$ via the ideal gas law (9.10) as

$$\bar{P} = \frac{k_B T}{\lambda^3} e^{\bar{\mu}_3/k_B T} \ , \tag{9.27}$$

and equating the latter to μ_2 in (9.17). We can then write (9.25), and likewise (9.26), completely in terms of chemical potentials

$$- \frac{L_{23}}{T} (\mu_3 - \mu_2) = S_2 A_2 \frac{1}{h\lambda^2} (\mu_e{}^{\mu_3/k_B T} - \mu_2 e^{\mu_2/k_B T}) \ , \tag{9.28}$$

with μ_3 and μ_2 given by (9.9,17). respectively. Note that the chemical potentials change as a function of time.

c) Two-Dimensional Condensation and Evaporation

We next isolate the processes of condensation and evaporation between the condensed and dilute phases of the adsorbate. Let us first assume that there are slightly too many particles in the 2D gas phase on the bare surface so that

$$\mu_2 = \overline{\mu}_2 + \frac{\partial \mu_2}{\partial N_2}\bigg|_{T,A_2} \Delta N_2 , \qquad (9.29)$$

with all other chemical potentials being equal. Thus the first term in (9.5) gives, after using (9.17),

$$\frac{dN_1}{dt}\bigg|_{cond} = -\frac{L_{12}}{T} \frac{k_B T}{N_2} \Delta N_2 \qquad (9.30)$$

In analogy to (9.13), one can also write phenomenologically

$$\frac{dN_1}{dt}\bigg|_{cond} = S_c A_1^{\varsigma} f \Delta j_2 \qquad (9.31)$$

implying that condensation takes place when an excess flux of particles, Δj_2, from phase 2 sticks on the circumference of the condensed phase 1. The rate is proportional to the circumference rather than the area of phase 1 because particles in phases 1 and 2 are restricted to move in two dimensions. If phase 1 consists of n disjoint, disc-like islands of total area A_1, then $\varsigma = 1/2$ and $f = (n\pi)^{1/2}$. However, if phase 1 has any other, jagged circumference then $0.5 < \varsigma < 1$ and $f = f_o n^{1-\varsigma}$ with $f_o \geq 2\sqrt{\pi}$. We should note that neither ς nor f can be assumed constant throughout the coexistence region. In particular, close to the lower coexistence point, the number of islands will be growing. Midway through the coexistence region, islands will start to coalesce. Towards the upper coexistence point, part of the condensed phase will touch the outer boundary of the surface and thus will no longer be part of the interphase boundary to the 2D gas phase. These rather complex features are very difficult to mimic phenomenologically and should be kept in mind when discussing numerical results obtained by keeping ς and f constant throughout the coexistence region.

Assuming the 2D gas to be ideal, the excess flux according to (9.31) can be written as

$$\Delta j_2 = \frac{(N_2 - \overline{N}_2)}{A_2} \frac{\overline{v}}{\pi} , \qquad (9.32)$$

where

$$\overline{v} = \sqrt{\frac{\pi k_B T}{2m}} . \qquad (9.33)$$

We then can relate the particle density to the spreading pressure, Π, via the ideal gas law

$$N_2 - \bar{N}_2 = \frac{A_2}{k_B T}(\Pi_2 - \bar{\Pi}_2) \tag{9.34}$$

in analogy with (9.12). For a mobile 2D gas we can express the average speed, $\bar{v}$, in terms of the distance between adsorption sites, d, and the average jump frequency, τ_0, as

$$\bar{v} = \frac{d}{\tau_0} e^{-q/k_B T} \tag{9.35}$$

where q is the height of the barrier to diffusion. Equations (9.30-35) obviously determine L_{12}.

To include evaporation in conjunction with condensation we replace $\bar{N}_2$ in (9.32) via the ideal gas law and (9.17) so that (9.31) reads

$$\frac{dN_1}{dt}\bigg|_{\text{cond,ev}} = S_c A_1 \varsigma f \frac{\bar{v}}{\pi}\left[\frac{N_2}{A_2} - \frac{1}{\lambda^2} q_z e^{E_d/k_B T} e^{\bar{\mu}_2/k_B T}\right] \tag{9.36}$$

Next, we equate chemical potentials, $\bar{\mu}_2 = \bar{\mu}_1$, and are faced with the problem of writing down $\bar{\mu}_1$ in terms of N_1, A_1, and T for the condensed phase. How to proceed for a lattice gas model will be shown in the forthcoming paper [9.25]. Here we adopt a simple Einstein model, suitable for a liquid- or solid-like condensate. Thus we write

$$\bar{\mu}_1 = -w + k_B T \ln(1-e^{-h\nu/k_B T})^2 - k_B T \ln(q_z e^{E_d/k_B T}) . \tag{9.37}$$

The last term compensates a similar one in $\bar{\mu}_2$. They would be absent in a completely 2D theory of the adsorbate. However, because we eventually want to study adsorption and desorption, thus having to include coupling terms to the 3D gas phase, these correction terms must be included. Further note that in (9.37) w is the heat of evaporation per particle and ν is the Einstein frequency of the condensate. We will see later on that ν essentially determines the coverage at the lower coexistence point. Inserting (9.37) into (9.36), we get the terms with L_{12} in (9.5-7). It is important to note that the evaporation rate is zero order and, moreover, that evaporation proceeds through the same circumference as condensation.

Expression (9.36) is valid within the coexistence region. We note that keeping the factor $f = 2(n\pi)^{1/2}$, with the number of islands, n, fixed, assumes that there is no coalescence or breakup of islands throughout the adsorption-desorption process. This is a rather serious assumption, as one realizes by following adsorption from low coverage or desorption from high coverage. Let us assume in the former situation that we start adsorbing onto a clean surface. Below the coexistence region, expression (9.36) is zero because $A_1 = 0$. As we cross with the accumulated coverage into the co-existence region, we need, in addition to (9.36), a term to describe spontaneous condensation. Such questions of nucleation will not be addressed in this paper.

Turning to desorption, let us start with an initial coverage above the coexistence region where we have only one, namely the condensed, phase which, by definition, is homogeneous. Upon desorbing, this phase will be thinned out, but remain homogeneous on the large scale. Entering the coexistence region, two scenarios are possible: (i) if diffusion across the surface is much slower than desorption, the adsorbate will evolve far from equilibrium, remaining homogeneous on a large scale; and (ii) if diffusion is faster than desorption, a 2D gas phase will separate from condensed islands as the coexistence region is entered. In the following we will accept the second scenario, assuming that the time scale of phase separation is much faster than all other scales in the system.

Let us deal with evaporation and condensation between the condensed phase 1 and the gas phase 2' on top of it. Let us follow a gas particle on top of an island of the condensate. Being mobile, it will eventually hop over the rim of the island at which stage it has a finite chance of becoming part of the 2D gas phase on the bare surface; this process will be studied in the next subsection. Or else, it can attach itself to a terrace site to become part of the condensed phase. For the latter process, the first term in (9.5) obviously contains an expression like (9.36). However, there must be additional channels. Let us assume that the surface of the solid is completely covered by a monolayer of the condensed phase. The latter, thus, has no boundary and particle exchange with the second adlayer cannot proceed via the process discussed above. Rather, there must be a process by which particles in the first monolayer get squeezed out and into the second layer with a rate proportional to the area A_1 covered by the condensed phase. Rather than introducing such additional channels, we adopt an expression like (9.36) for the first term in (9.6) as well.

We can now discuss the terms in (9.5-7) describing adsorption on and desorption from the condensed phase directly to complete the discussion of the last subsection. Within the coexistence region, we can first assume that desorption out of the condensate is proportional to its area A_1. On the other hand, desorption might only occur from the rims of the condensate islands. Above the coexistence region, we assume that desorption is again a first order process. These rates are listed in (9.58 and 59).

d) Equilibration of the 2D Gas Phases

It remains to examine the last term in (9.5), which describes the exchange of particles between the 2D gas on the bare surface and that on top of the condensate islands. Phenomenologically, one would argue that an excess particle on top of the condensed phase will eventually come to its rim where it either becomes part of the condensate with a probability $S_{c'}$, introduced in (9.38), or it enters the gas phase on the bare surface. Because it is unlikely that the particle gets reflected, we can set the probability for the latter process to be $1-S_{c'}$. Thus, we would write

$$\left.\frac{dN_2}{dt}\right|_{eq} = (1-S_{c'})\, A_1 S'\, f'\, \Delta j_2 , \qquad (9.38)$$

from which, by comparison with (9.30,31), we get $L_{22'}$.

e) Equilibrium Properties

For vanishing fluxes, $dN_i/dt = 0$ in (9.5-7), the equality of five pairs of chemical potentials determines the equilibrium properties of the system. To determine the partial coverages $\theta_i = N_i/N_s$, we start from (9.25) and note that in the coexistence region, the densities in the 2D gas phase and in the condensed phase remain constant. The density in the latter, n_c, is less than the density $n_s = N_s/A$ of lattice sites. We write $A_2 = A-A_1$, $A_1 = N_1/n_c$, $\theta_{1c} = n_c/n_s$ and note that θ_{1c} is the coverage in the condensed phase at the upper coexistence point; it differs from the total coverage by $\theta_{2'}$. We get from (9.25)

$$\theta_2 = \frac{\overline{P}}{(2\pi m k_B T)^{1/2}}\, \nu_z^{-1}\, e^{E_d/k_B T}\, \frac{1}{n_s}\left[1 - \frac{\theta_1}{\theta_{1c}}\right]. \qquad (9.39)$$

Obviously, $\theta_2 = 0$ once $\theta_1 > \theta_{1c}$. Likewise (9.26) yields, for all $\theta_1 > 0$,

$$\theta_{2'} = \frac{\overline{P}}{(2\pi m k_B T)^{1/2}}\, \nu_{z'}^{-1} e^{E_{d'}/k_B T}\, \min(\theta_1/n_c,\, 1/n_s) \qquad (9.40)$$

Comparison of (9.39 and 40) yields for $\theta_1 < \theta_{1c}$

$$\theta_{2'} = B_o\, \frac{\theta_1 \theta_2}{\theta_{1c} - \theta_1} , \qquad (9.41)$$

where

$$B_o = \frac{\nu_z}{\nu_{z'}}\, e^{(E_{d'} - E_d)/k_B T} . \qquad (9.42)$$

Turning to (9.38) we get for $\theta_1 < \theta_{1c}$

$$\theta_2 = \frac{1}{\lambda^2 n_s}\, (1 - e^{-h\nu/k_B T})^2\, e^{-w/k_B T}\left[1 - \frac{\theta_1}{\theta_{1c}}\right] = B_1\left[1 - \frac{\theta_1}{\theta_{1c}}\right] \qquad (9.43)$$

Note that $(1-\theta_1/\theta_{1c})N_s$ is the number of sites available to the 2D gas phase on the bare surface. We must obviously insist that the density in the latter is much lower than that in the condensed phase, i.e., $\theta_2 \ll 1-\theta_1/\theta_{1c}$, implying from (9.43) that

$$B_1 = \frac{1}{\lambda^2 n_s}(1 - e^{(-h\nu/k_B T)})^2 \; e^{-w/k_B T} \ll 1 . \tag{9.44}$$

If systems parameters are such that (9.44) is not satisfied, we are not justified to treat phase 2 as ideal.

A comparison of (9.42 and 39) shows that in the temperature regime where phases 1 and 2 coexist, the pressure remains constant at

$$\overline{P} = \frac{h}{\lambda^3} \nu_z \; e^{-E_d/k_B T} \; (1 - e^{-h\nu/k_B T})^2 \; e^{-w/k_B T} , \tag{9.45}$$

while the coverage increases.

Lastly we get from (9.38) for $\theta_1 > 0$

$$\theta_{2'} = B_1 \, B_o \, \theta_1 \tag{9.46}$$

consistent with what we get by inserting (9.45) into (9.39). The diluteness of the gas phase on top of the condensed islands requires that $B_o B_1 \ll 1$. We can also express θ_1 in terms of the total coverage θ for $0 < \theta_1 < \theta_{1c}$ as

$$\theta_1 = \frac{(\theta - B_1)}{1 - B_1/\theta_{1c} + B_1 B_o/\theta_{1c}} . \tag{9.47}$$

We note that phase coexistence is given in the coverage regime

$$B_1 \leq \theta \leq \theta_{1c} + B_o B_1 . \tag{9.48}$$

9.1.3 Results

We start by listing our basic rate equations (9.5-7) in terms of the partial coverages $\theta_1 = N_i/N_s$ introduced above, where N_s is the number of adsorption sites on the surface. We get

$$\frac{d\theta_1}{dt} = r_{12} + r_{12'} + r_{13} , \tag{9.49}$$

$$\frac{d\theta_2}{dt} = -r_{12} + r_{23} + r_{22'} , \tag{9.50}$$

$$\frac{d\theta_{2'}}{dt} = -r_{12'} - r_{22'} + r_{2'3} . \tag{9.51}$$

The term r_{12} is given by (9.36), divided by N_s. We note that $r_{12} = 0$ outside the coexistence region. The rate (9.36) contains the number n of condensed islands via the factor $f = 2(n\pi)^{1/2}$ specified above (9.30). We assume that n does not change during desorption. If we, therefore, start the desorption process within the coexistence region, we can set

$$n = \frac{N_1(t = 0)}{N_i} = \frac{N_s}{N_i} \theta_1(t = 0) , \tag{9.52}$$

where N_i is the average number of particles per island. If we start desorption above the coexistence region, then we set $\theta_1(0)$ in (9.52) equal to the coverage at its upper limit. We thus get from (9.36), also using (9.37),

$$r_{12} = S_c f_o \left[\frac{\theta_1(0)n_s}{N_i}\right]^{1-\zeta} (\theta_1/\theta_{1c})^{\zeta} \frac{k_B T}{h} \lambda \left[\frac{\theta_2}{1-\theta_1/\theta_{1c}} - B_1 \Theta(\theta_1)\right],$$
(9.53)

where B_1 is given in (9.44). We recall that for disc-like islands $\zeta = 1/2$, and $f_o = 2\pi^{1/2}$ so that r_{12} is proportional to $(\theta_1)^{1/2}$. Also note the explicit dependence of r_{12} on the island size, N_i.

With similar arguments we get from (9.38)

$$r_{12'} = S_c' f_o' \left[\frac{\theta_1(0)n_s}{N_i}\right]^{1-\zeta} (\theta_1/\theta_{1c})^{\zeta} \frac{k_B T}{h} \lambda [\theta_{2'}\theta_{1c}/\theta_1 - B_o B_1 \Theta(\theta_1)],$$
(9.54)

where B_o was introduced in (9.42). Also, we get

$$r_{22'} = (1 - S_c') f_o' \left[\frac{\theta_1(0)n_s}{N_i}\right]^{1-\zeta} (\theta_1/\theta_{1c})^{\zeta} \frac{k_B T}{h} \lambda$$

$$\cdot \left[\frac{\theta_{2'}\theta_{1c}}{\theta_1} - \frac{B_o}{\lambda^2} \frac{\theta_2}{(1 - \theta_1/\theta_{1c}) \Theta(\theta_1)}\right].$$
(9.55)

From (9.25 and 26) we get

$$r_{23} = S_2 \left[\frac{1}{n_s} (1 - \theta_1/\theta_{1c}) \frac{P}{(2\pi m k_B T)^{1/2}} - \nu_z e^{-E_d/k_B T} \theta_2\right], \quad (9.56)$$

$$r_{2'3} = S_2' \left[\frac{1}{n_s} \frac{\theta_1}{\theta_{1c}} \frac{P}{(2\pi m k_B T)^{1/2}} \nu_z' e^{-E_{d'}/k_B T} \theta_{2'}\right]. \quad (9.57)$$

Finally, in the coexistence region, we get

$$r_{13} = S_1 \frac{1}{n_s} \frac{\theta_1}{\theta_{1c}} \left[\frac{P}{(2\pi m k_B T)^{1/2}} - \nu_z n_s B_1 e^{-E_d/k_B T}\right], \quad (9.58)$$

and above the coexistence region

$$r_{13} = S_1 \left[\frac{1}{n_s} \frac{P}{(2\pi m k_B T)^{1/2}} - \nu_1(\theta_1 - \theta_{2'})e^{-(E_d+w)/k_B T}\right]. \quad (9.59)$$

To begin the presentation of results let us first look at the rate equations (9.49-51) analytically, in special cases under desorption conditions, i.e., with $P = 0$. Suppose we neglect contributions from the gas phase "2'" on top of the islands by setting $\theta_{2'} = 0$ and also direct desorption out of the condensed phase by putting $r_{13} = 0$. The total coverage $\theta = \theta_1 + \theta_2$ then changes according to $d\theta/dt = r_{23}$ with r_{23} given in (9.56) as proportional to θ_2. Let us first assume that θ_2 does not change much during desorption implying from (9.50) that $r_{23} \simeq (r_{12})$. If furthermore $\theta_2 \ll \theta_1$ than (9.53) implies that $d\theta/dt \propto - \theta^{1/2}$. However, as θ decreases the first term in (9.53) comes into play leading to a transition from half order to fractional order desorption which eventually, as the condensed phase gets depleted, changes over to first order.

Let us next look at a system in which particle exchange between the 2D phases is faster than desorption so that, with $r_{12} \simeq 0$, a quasi-equilibrium is maintained on the surface. We can then replace θ_2 in (9.56) by expressions (9.43,47) to obtain

$$\frac{d\theta}{dt} \simeq - S_2 \nu_2 e^{-E_d/k_B T} \lambda^2 B_1 \left[1 - \theta_{1c}\theta - \frac{B_1}{1 + B_o B_1 - B_1/\theta_{1c}} \right].$$

(9.60)

In the high coverage regime we thus expect the desorption rate to increase linearly with decreasing coverage. It is followed by a regime of zero order kinetics down to the point where the condensed phase has disappeared, below which first order kinetics takes over again. Note, however, that in our numerical examples we will find that typically $r_{12} > r_{23}$, so that zero order desorption comes about in a more complicated fashion.

We now turn to numerical examples to substantiate the above discussion. We present our results both for isothermal desorption and for temperature programmed desorption. For the latter, an adsorbate is prepared at an initial temperature T_o with equilibrium partial coverages given by (9.43,46,47). The rate equations (9.49-51) are then solved with the temperature changing linearly in time, $T = T_o + \alpha t$ where α is typically between 1 and 100 K/s. Ideally, in an isothermal desorption experiment one keeps $T = T_o$ and generates nonequilibrium conditions by rapidly, i.e., on a time scale fast compared to desorption, pumping the gas phase away. As this is rarely possible to do, one rapidly increases the temperature from T_o to T, keeping it constant thereafter. We mimic this situation by a fast temperature programmed desorption between T_0 and T to be followed by isothermal desorption. One point of interest in our study is whether any nonequilibrium effects produced in this manner could be observed. For all our numerical results we assume that readsorption is negligible by putting $P = 0$ in the rates r_{13}, r_{23} and $r_{2'3}$.

For our numerical examples we have chosen the parameters of the order of those for Xe on metals but note that parameters typical for Ar on metals or for metals on metals do not introduce new qualitative features. Preexpotential frequency factors are varied from $1 \cdot 10^{11}$ to $1 \cdot 10^{14}$ s^{-1} and sticking coefficients are taken of the order of one. The

Einstein frequency, introduced in (9.36), is chosen to give a reasonable value for the coverage at the lower coexistence point given by (9.44,48). Little is known about the island sizes. Theoretical estimates, obtained by Monte Carlo simulations, seem of limited value as islands on real surfaces seem to grow at most to the size of the patches of perfect terraces surrounded by steps. We will vary the number of particles in an island, N_i, over many orders of magnitude.

In Fig.9.1 we study the island size dependence of the isothermal desorption rate starting from an initial coverage within the coexistence region. We have set direct desorption from the condensed islands equal to zero and chosen the sticking coefficients S_2 and $S_{2'}$ as unity. Let us recall that condensation-evaporation rates, r_{12} and $r_{12'}$, are proportional to $S_c/N_i^{1/2}$ so that a variation of N_i is equivalent to a variation of S_c. The upper curve is calculated for small islands with particle numbers, N_i, less than $1 \cdot 10^{18}$; it does not change, not even quantitatively, down to N_i as small as $1 \cdot 10^3$. In particular, it shows a pronounced zero order regime which, at the lower coexistence point, changes over to first order desorption from the 2D gas phase. We note that upon starting desorption at the highest coverage the rate jumps up dramatically. This is due to the fact that a small adjustment in the initial coverages, as given at the preparation temperature T_0, will change the overall rate dramatically. Indeed, such a jump is present, even if we follow the temperature rise from T_0 to T via a linear temperature ramp with $\alpha = 10$ K/s. If we increase N_i to more than $1 \cdot 10^{22}$ particles, the isothermal desorption characteristics change dramatically in that the zero order

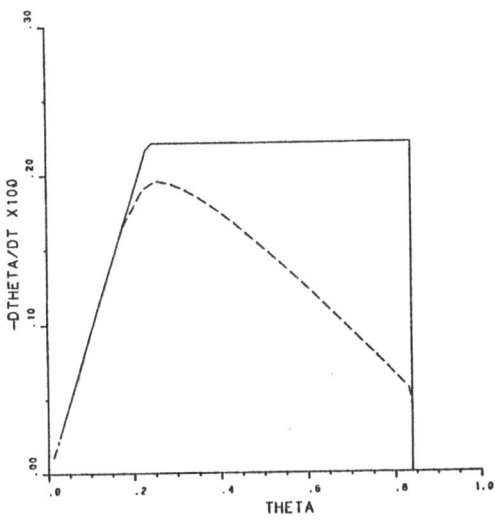

Fig.9.1. Isothermal desorption at $T = 80$ K reached from the preparation temperature $T_0 = 60$ K via a linear temperature ramp with a heating rate $\alpha = 10$ K/s. System's parameters are: $n_s = 0.2$ Å^{-2}; $S_2 = S_{2'} = S_c = 1$; $S_{c'} = 0.5$; $S_1 = 0$; $\theta_{1c} = 0.85$; $\nu_z = \nu_{z'} = 1 \times 10^{11}$ s^{-1}; $\nu = 1 \times 10^{12}$ s^{-1}; $E_d = 2400$ K; $E_{d'} = 2000$ K; $w = 400$ K. Solid curve $N_i = 1 \times 10^{10}$; dashed curve $N_i = 1 \times 10^{24}$

273

regime disappears. Starting again at a high coverage within the coexistence region, the rate initially increases because in this example desorption still proceeds through the 2D gas phase, with evaporation of the islands the rate limiting step. The increase in the rate can be attributed to an increase of the amount of 2D gas present due to a shrinking area occupied by the condensed islands. We point out once more that the unreasonably large island size, $N_i = 1 \cdot 10^{22}$, can be reduced by choosing the sticking coefficient S_c much less than one. In particular we had specified the 2D gas to be totally mobile and ideal. Thus, if the hopping speed in (9.34) is smaller by, say, m orders of magnitude, then the effective condensation coefficient is down by the same amount, so that to get the same prefactor, $S_c N_i^{1/2}$, we could choose N_i smaller by 2m orders of magnitude. In this sense the lower curve in Fig.9.1 has indeed physical significance. We note that following the time dependence of the partial coverages, $\theta_i(t)$, one finds that for $N_i > 1 \cdot 10^{22}$ pronounced nonequilibrium effects show up in that $\theta_2(t) \langle\langle \theta_2^{eq}$ and $\theta_1(t) \rangle\rangle \theta_1^{eq}$. In Fig.9.2 we show temperature programmed desorption traces for the two systems studied in Fig.9.1. For small islands we see the typical leading edge of zero order desorption, whereas, for large islands, the desorption traces obtained by starting at higher initial coverages lie initially below those starting at low coverage.

Let us at this stage point out the most likely pathway of desorption in the coexistence region. Although $\theta_{2'} << \theta_2$ we still find that $r_{23} << r_{2'3} < r_{12}$. Desorption thus proceeds by islands evaporating into the 2-phase via rates r_{12}. Subsequently these particles jump on top of the islands into the 2'-phase with rates $r_{22'}$, from which they finally desorb with rates $r_{2'3}$.

In Fig.9.3 we include direct desorption from the condensed islands via the rates given by (9.58,59) matching the two expressions at the upper critical point to prevent a jump in the overall desorption rate. The obvious effect for small island sizes is to change the zero order regime of Fig.9.1 into a first order regime. If, in the coexistence region, we have desorption from the rims of the islands, we should observe half order desorption. However, for the parameters used, the rate r_{13} is too small to affect the zero order result. We have argued above that the number of condensed islands, and thus their overall circumference, shrinks as the upper coexistence point is approached. If we mimic this by making the rate r_{13} proportional to $(A_1(A-A_1))^{1/4}$, rather than to $A_1^{1/2}$, we find little change to the first order result.

Next we vary the Einstein frequency characterizing the condensate via the chemical potential shown in (9.37). Via the parameter B_1 in (9.44), it shifts the lower coexistence point. The influence on the isothermal desorption rate, as shown in Fig.9.4, is to shift the changeover point between first and zero order kinetics. Changing the 2D heat of condensation, w, has a similar effect.

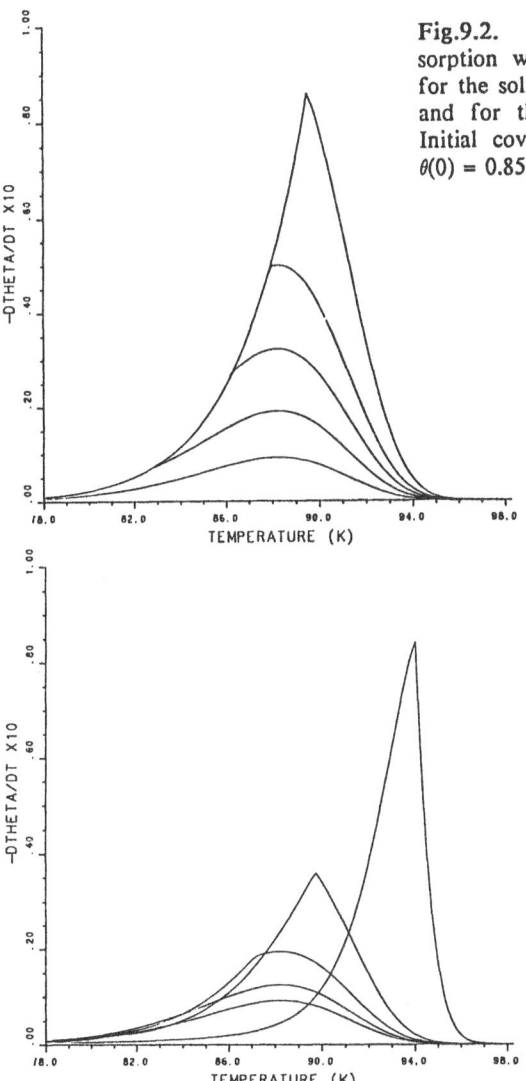

Fig.9.2. Temperature programmed desorption with a heating rate $\alpha = 0.5$ K/s for the solid curve of Fig.9.1 (upper panel) and for the dashed curve (lower panel). Initial coverages are from top to bottom $\theta(0) = 0.85, 0.60, 0.45, 0.30, 0.15$

We note that an increase in the vibrational frequency, ν_z, perpendicular to the surface, simply produces an increase in the desorption rate for small islands. However, for large islands, i.e., for $N_i \gg 1 \cdot 10^{22}$, which according to our arguments above is appropriate for a localized gas phase with a rather small hopping rate, reducing ν_z to about $1 \cdot 10^{11}$ causes (i) the appearance of half order desorption and a decline of the desorption rate for higher coverages as shown in Fig 9.5. These effects have been discussed analytically at the beginning of this section.

We note that the overshoot in the solid curve in Fig.9.5 is a real effect and is a result of heating the sample rapidly from the preparation temperature to the desorption temperature. In the depicted case

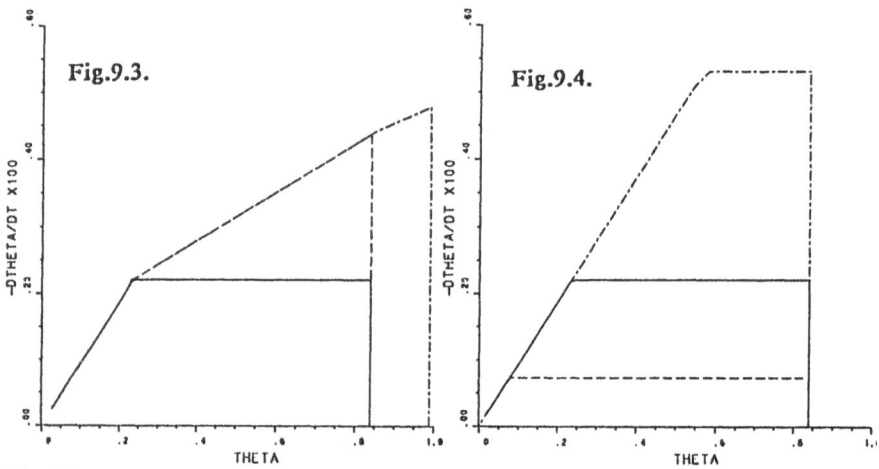

Fig.9.3. Isothermal desorption. All parameters appear as solid curve in Fig.9.1 (repeated for comparison) except $S_1 = 1$

Fig.9.4. Isothermal desorption as shown in Fig.9.1 for $N_i = 1 \times 10^{10}$ with $\nu = 2 \cdot 10^{12}$ s^{-1} (dash-dot), 1×10^{12} s^{-1} (solid) and $5 \cdot 10^{11}$ s^{-1} (dash)

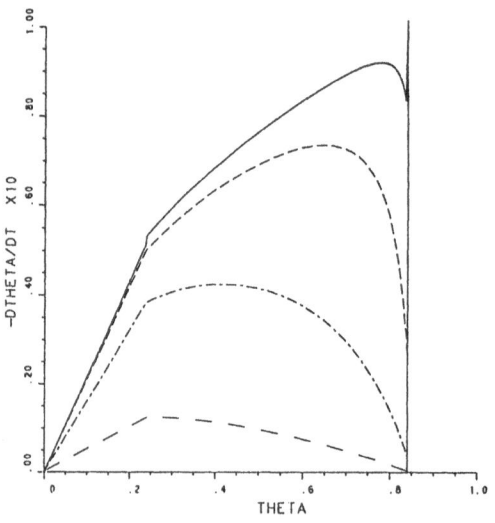

Fig.9.5. Isothermal desorption, as shown in Fig.9.1 for large islands, $N_i = 1 \times 10^{24}$ but with $\nu_{z'} = 10^{14}$ s^{-1}, and $\nu_z = 1 \times 10^{15}$ s^{-1} (solid), 1×10^{14} s^{-1} (dash), 1×10^{13} s^{-1} (dash-dot), and 1×10^{12} s^{-1} (long dash). Initial heating rate $\alpha = 100$ Ks^{-1}

the heating rate was 100 K/s. For faster rates the overshoot would be larger and sharper; only for heating rates less than 10 K/s does it disappear. Such features have been observed experimentally, but are usually suppressed as being due to problems with the temperature control.

276

9.1.4 A Simplified Model

In this section we want to simplify the above theory by concentrating on systems in which the surface phases remain in (local) equilibrium with each other. We also neglect the component 2' of the 2D gas phase on top of the condensed islands. Summing (9.5,6), we get, quite generally still, for the number of particles, $N_a = N_1 + N_2$, in the adsorbate

$$\frac{dN_a}{dt} = \sum_{i=1,2} \frac{L_{i3}}{T} (\mu_i - \mu_3) .$$ (9.61)

On the other hand, by looking at adsorption for a moment, we can introduce sticking coefficients via (9.13)

$$\frac{dN_a}{dt} = \sum_{i=1,2} S_i A_i \frac{(P - \overline{P})}{(2\pi m k_B T)^{1/2}} .$$ (9.62)

We can next use (9.27) to replace $\overline{P}$ by μ_3 and equate $\mu_3 = \mu_a$, because we assumed that $\mu_1 = \mu_2 = \mu_a$. To get an explicit expression for μ_a we treat the adsorbate as an immobile lattice gas within the Bragg-Williams approximation. This is not the best of models but suffices for our purposes. More sophisticated models can be incorporated similarly. In terms of the total coverage $\theta = \theta_1 + \theta_2$ we then get

$$\frac{dN_a}{dt} = \sum_{i=1,2} S_i \frac{A_i}{N_s} \frac{1}{(2\pi m k_B T)^{1/2}} \left[P - \frac{k_B T}{\lambda^3} H(\theta) \right]$$ (9.63)

Where, within the Bragg-Williams approximation, we have

$$H(\theta) = \frac{\theta}{1 - \theta} \frac{1}{q_2} e^{-cw\theta/k_B T}$$ (9.64)

outside the coexistence region, and

$$H(\theta) = \frac{1}{q_2} e^{-cw/2k_B T}$$ (9.65)

within it. Here c is the coordination number of the lattice. Using the expressions for the partial areas A_i, we get under desorption conditions with P = 0 below the coexistence region, i.e., for $\theta < \theta_{2c}$,

$$\frac{d\theta}{dt} = - S_2 \frac{k_B T}{n_s h \lambda^2 q_2} \frac{\theta}{1 - \theta} e^{-cw\theta/k_B T}$$ (9.66)

and above the coexistence region, i.e., for $\theta > \theta_{1c}$,

$$\frac{d\theta}{dt} = - S_1 \frac{k_B T}{n_s h \lambda^2 q_2} \frac{\theta}{1 - \theta} e^{-cw\theta/k_B T}$$ (9.67)

whereas within the coexistence region one finds

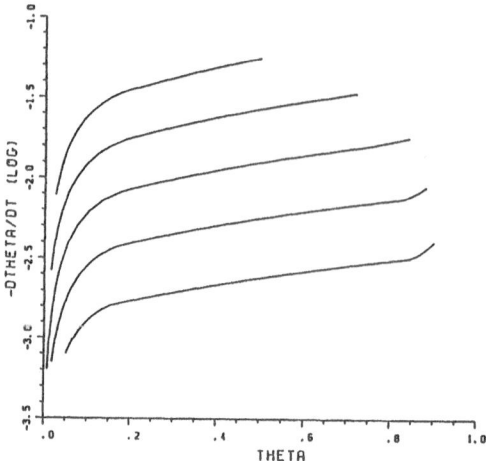

Fig.9.6. Isothermal desorption as calculated from (9.66-68). Parameters: $S_1 = 1.0$, $S_2 = 0.5$, cw = 375 K ($\theta_{1c} = 0.8$), $\nu_g = 6\cdot10^{10}$ s^{-1}, desorption tempertures from top to bottom 84 K, 82 K, 80 K, 78 K, 76 K

$$\frac{d\theta}{dt} = - \frac{k_B T}{n_s h \lambda^2 q_2} \left[(S_1 - S_2) \frac{\theta_1}{\theta_{1c}} + S_2 \right]. \qquad (9.68)$$

Thus, we see that this simple model produces zero order kinetics within the coexistence region if $S_1 = S_2$, in which case our model reduces to the standard one, for instance explored by *Nagai* [9.20-22]. On the other hand, if $S_1 > S_2$, one finds first order desorption within the coexistence region, whereas $S_1 < S_2$ would actually produce decreasing rates. In Fig.9.6 we show some isothermal desorption traces at various temperatures for parameters appropriate for Xe/Ru. Our model reproduces experimental data shown in [9.6] rather well. Deviations can be traced to the use of the Bragg-Williams approximation and can be easily removed by the use of more sophisticated ˊthermodynamic models.

9.2 Microscopic Approaches

A macroscopic description of adsorption-desorption kinetics based on non-equilibrium thermodynamics, as outlined above, introduces a number of parameters associated with the Onsager transport coefficients, such as sticking coefficients, activation energies and preexponential factors. For their calculation from first principles, one starts from the microscopic form of the total energy of the coupled gas-solid system as given by its hamiltonian $H = H_s + H_g + V$. Here H_s is the hamiltonian of the isolated solid accounting for its vibrational (phonon) and electronic degrees of freedom, H_g is the hamiltonian of the gas, and V contains the coupling between the solid and the gas. The latter contains a static part, namely the surface potential, V_s; gas particles

278

trapped in its bound states form the adsorbate. Neglecting electronic effects for the sake of the discussion here, we note that the thermal vibrations of the solid, with amplitudes u(t), induce a dynamic part in $V = V_s[r-u(t)]$. Here r is the distance of a gas particle above the (vibrating) surface of the solid. For small amplitudes we can expand V to get

$$V = V_s(r) - u(t) \cdot \frac{dV_s}{dr} + \dots \qquad (9.69)$$

The time dependence in the phonon-coupling term allows energy transfer between the solid and the gas necessary in adsorption and desorption. Trying to understand these processes we, of course, do not want to follow the atomic motion as given by the Schrödinger equation over time scales of 10^{-13} s and shorter, but attempt to extract the statistical evolution of an ensemble of gas particles. A successful approach to such a kinetic description is based on the master equation

$$\frac{dn_i}{dt} = \sum_j (W_{ij} n_j - W_{ji} n_i) . \qquad (9.70)$$

Here n_i is the probability that state "i" with energy E_i in the surface potential V_s is occupied; "i" labels both bound states (adsorbate) and continuum states (free gas). The first term on the right hand side of (9.70) gives the rate of increase in the occupation of state "i" due to transitions from all other states "j" which occur at a rate W_{ij}. The second term on the right hand side gives the rate at which particles leave state "i". The transition rates W_{ij} have been calculated from (9.69) using Fermi's golden rule; we note that the energy difference is made up by absorbing or emitting phonons depending whether the initial or final states have higher energy. The master equation has been used to calculate sticking coefficients, energy accommodation coefficients, desorption rates and energy distributions. Most theories are restricted to the low coverage regime where the interaction between adsorbed particles can be neglected. One notable exception is a mean field kinetic theory of multilayer helium adsorption and desorption for which such interesting phenomena like a compensation effect between the desorption energy and the preexponential factor have been predicted. We omit giving any details here as this subject has been recently surveyed extensively in a monograph on physisorption kinetics [9.28].

9.3. Outlook

In this paper we have presented a general approach to the adsorption-desorption kinetics based on the methods of nonequilibrium thermodynamics. For the most part we employed a simple model of a two-phase adsorbate in which the 2D gas phase is treated as ideal whereas the 2D condensed phase is described by an Einstein model. This allowed us to specify all formulae explicitly and to investigate the potential of the approach in a, hopefully, transparent way. Our numerical examples should demonstrate the richness of this simple model and in particular its ability to reproduce many details of experimental desorption data. This study should be convincing that (i) inspection of experimental data alone can easily lead to wrong conclusions, and in particular (ii) that the customary habit to attribute unexpected features to a possible coverage dependence of pre-exponential factors and desorption energies might, in some cases, be overly hasty and misleading.

What must be done in the future is to improve the thermodynamic model, e.g., by using the lattice gas model in the Bragg-Williams and quasi-chemical approximations or a 2D van der Waals model. This will ultimately allow fits of experimental desorption data, e.g., for rare gases on metals and metals on metals. It would be very interesting to have a "complete" set of experimental data on one particular gas-solid system including the temperature and coverage dependence of the chemical potential, sticking coefficients, heats of adsorption, and iso-thermal and temperature programmed desorption data. In this situation almost all parameters in the above equations are known and fits to the desorption data can be used to estimate such quantities as island size and to identify desorption pathways.

The same approach based on nonequilibrium thermodynamics can be extended to include dissociative adsorption-desorption phenomena and also surface reactions.

As to microscopic models of adsorption-desorption kinetics, we want to stress again that almost all of them are restricted to the low coverage regime. One would hope that theorists will finally attack the much more difficult, but also more interesting problem of finite cov-erage effects. perhaps along the lines of the mean field kinetic model of multilayer helium desorption.

Acknowledgement. This work was supported in part by a grant from the Natural Sciences and Engineering Council of Canada.

References

9.1   D. Menzel: In *Interactions on Metal Surfaces*, ed. by R. Gomer, Topics Appl. Phys., Vol.4 (Springer, Berlin, Heidelberg 1975)
        In *Chemistry and Physics of Solid Surfaces IV*, ed. by R. Vanselov, R Howe, Springer Ser. Chem. Phys., Vol.20 (Springer, Berlin, Heidelberg 1982)

9.2   *Desorption Induced By Electronic Transition.* ed. by N.H. Tolk, M.M. Traum, J.C. Tully, T.E. Madey, Springer Chem. Phys., Vol.29 (Springer, Berlin, Heidelberg 1983)

9.3   T.J. Chuang: Surf. Sci. Rep. 3, 1, 1983

9.4   P. Piercy, Z.W. Gortel, H.J. Kreuzer: In *Advances in Multi-photon Processes and Spectroscopy*, Vol.3, ed. by S.H. Lin (World Scientific, Singapore 1987)

9.5   H.J. Kreuzer: *Nonequilibrium Thermodynamics and its Statistical Foundations* (Oxford Univ. Press, Oxford 1981 and 1983)

9.6   D. Menzel: In *Kinetics of Interface Reactions*, ed. by M. Grunze, H.J. Kreuzer, Springer Ser. Surf. Sci., Vol.8 (Springer, Berlin, Heidelberg 1987)

9.7   J.R. Arthur, A.Y. Cho: Surf. Sci. 36, 641, 1974

9.8   J.R. Arthur: Surf. Sci. 38, 394, 1974

9.9   J.A. Venables, M. Bienfait: Surf. Sci 61, 667, 1976

9.10 M. Bienfait, J.A. Venables: Surf. Sci. 64, 425, 1977

9.11 G. Le Lay, M. Manneville, R. Kern: Surf. Sci. 65, 261, 1977

9.12 J. Suzanne, M. Bienfait: J. de Phys. 38, C4-93, 1977

9.13 M. Bertucci, G. Le Lay, M. Manneville, R. Kern: Surf. Sci. 85, 471, 1979

9.14 R. Opila, R. Gomer: Surf. Sci. 112, 1, 1981

9.15 D.L. Adams: Surf. Sci. 42, 12, 1974

9.16 P.K. Johansson: Chem. Phys. Lett. 65, 366, 1979

9.17 U. Leuthäuser: Z. Physik B 37, 65, 1980

9.18 V.P. Zhdanov: Surf. Sci. 111 (1981) L662; 148 (1984) L691; 171, L461, 1986

9.19 K. Nagai, T. Shibanuma, M. Hashimoto: Surf. Sci. 145, L459, 1984

9.20 K. Nagai: Phys. Rev. Lett. 54, 2159, 1985

9.21 K. Nagai: Surf. Sci. 176, 193, 1986

9.22 K. Nagai, A. Hirashima: Surf. Sci. 171, L464, 1986

9.23 W. Brenig, K. Schönhammer: Z. Physik B 24, 91, 1976

9.24 H. Müller, W. Brenig: Z. Physik B 34, 165, 1979

9.25 H.J. Kreuzer, S.H. Payne: Surf. Sci. (to be published)

9.26 T.L. Hill: *Statistical Mechanics* (McGraw-Hill, New York 1956)

9.27 A. Clark: *The Theory of Adsorption and Catalysis* (Academic, New York, 1970)

9.28 H.J. Kreuzer, Z.W. Gortel: *Physisorption Kinetics*, Springer Ser. Surf. Sci., Vol.1 (Springer, Berlin, Heidelberg 1986)

10. Fractals in Surface Science: Scattering and Thermodynamics of Adsorbed Films

Peter Pfeifer

Department of Physics, University of Missouri – Columbia
Columbia, MO 65211, USA

One of the reasons why fractals have received so much interest in the past few years is that they enable one to tackle a notoriously difficult problem in a spectacularly successful way. The problem is the characterization and understanding of strongly disordered systems. Strongly disordered means that disorder exists over many length scales. Indeed, numerous systems of outstanding interest are neither crystalline nor homogeneously (weakly) [10.1] disordered, so that classic concepts like unit cells, Voronoi cells, dislocations, surface steps, etc. are not applicable. The systems we have in mind are porous solids, rough catalyst surfaces, colloidal aggregates, and polymers, to name just a few. Often they are referred to as systems with complex geometry. To fix the ideas for the time being, one may think of them in terms of critical phenomena (percolation clusters, random walks...).

The second reason for the interest in fractals is that fractals provide a laboratory for physics and chemistry in nonintegral dimensions (fractal dimension). It turns out that physical phenomena in nonintegral dimensions are much richer than what one might expect from the physics in dimensions 1, 2, and 3. Various quantities distinct in nonintegral dimensions coincide or vanish for d-dimensional Euclidean spaces (d=1,2,3). The origin of this degeneracy of Euclidean spaces is that in addition to being scale invariant they are also translation invariant. So fractals teach us what happens when translation invariance is lost, but scale invariance is preserved [10.2].

Typically, a property f of a scale-invariant system takes the form

$$f = f(\text{geometry};...) = f(D, \tilde{D}, D_{top}, d;...) \tag{10.1}$$

where the quantities appearing as arguments are the system's fractal, spectral, topological, and embedding dimensions, respectively (to be explained in detail in Sect.10.1). This is, in a nutshell, the success story of fractal theory of disordered materials: system properties depend on the disorder (geometry) only through D and $\tilde{D}$. The fractal dimension D measures disorder in terms of the space-filling ability of the system ((10.2) in Sect.10.1). The spectral dimension $\tilde{D}$ measures disorder in terms of the system's branching or connectivity properties

(10.7). The other two dimensions D_{top} and d in (10.1), both integers, are not disorder indexes. D_{top}=1,2,3 specifies whether the system is a curve, surface, or volume, and d is the dimension of the Euclidean space in which the system is embedded (usually d=3; for d=2 see Sect.10.4). For completeness' sake it should be mentiond that, depending on the property f, yet other dimensions may occasionally occur. Conversely, there are properties that depend on D only or on $\tilde{D}$ only.

This chapter discusses five recent examples for (10.1) in the field of surface science. The examples are elastic scattering from systems with a fractal boundary (Sect.10.2,3, including He scattering from adsorbate islands on an ordered surface); inelastic scattering from diffusing molecules on a fractal surface (Sect.10.4); Henry's law of adsorption (Sect.10.5); BET condensation of a gas onto a fractal surface (Sect.10.6); and Bose-Einstein condensation in ^{4}He sub-monolayers on a fractal substrate (Sect.10.7). The last example shows that even the ideal gas is affected by nonintegral dimensionality.

The choice of topics is motivated by the fact that surface scattering and thermodynamics of adsorbed films lie at the heart of all studies of ordered and weakly disordered surfaces. So it is of interest to see how methods and results for flat surfaces carry over to strongly disordered surfaces. For scattering, one should not expect simple parallels because in scattering from ordered surfaces one is interested in the local arrangement of atoms or molecules, whereas in the fractal case one is interested in the large-scale structure of the surface. Some features of this complementarity are indicated in Table 10.1.

The reader who is interested in a more general picture of fractals (for instance ubiquity of fractal materials, models for their origin, properties of them) is referred to [10.3-11] for reviews, to [10.12-17] for conference proceedings, and to the seminal books by *Mandelbrot* [10.18,19]. References [10.3] and [10.10] are specifically oriented to surface problems and describe experimental methods, case studies, and applications such as optimization of chemical reaction rates on fractal surfaces. Given those earlier accounts, the present paper shows where we go from there and how much territory still remains to be explored.

Table 10.1 Local vs global structural information in various scattering modes

	Ordered or weakly disordered surfaces	Strongly disordered (fractal) surfaces
Momentum transfer q	$q > 1 Å^{-1}$	$q \ll 1 Å^{-1}$
Local structure	Bragg peaks	prefactors of power laws
Global structure	line shape of Bragg peaks	exponents of power laws

10.1 Fundamentals of Fractal Geometry

This section consists of two definitions, each followed by a lengthy series of remarks. The remarks not only illustrate the underlying concepts but often also describe important consequences and results.

10.1.1 Definition 1. Put down a monolayer of molecules (e.g. nitrogen) of radius r on the surface. If M(R), the number of molecules within distance R from a fixed point on the surface, grows as

$$M(R) \propto R^D \ (R \gg r) \tag{10.2}$$

for increasing R (Fig.10.1), then the exponent D is by definition the fractal dimension of the surface. Equation (10.2) is called the mass-radius relation. The lower and upper end of the R range in which (10.2) holds are called the inner and outer cutoffs of the fractal behavior.

Remarks:

1a. If the object on which the molecules are adsorbed is a straight line, then a sphere of radius $R \gg r$ cuts out R/r molecules, whence $M(R) \propto R^1$. If the object is a flat surface, then similarly R^2/r^2 molecules are cut out, whence $M(R) \propto R^2$. So D in (10.2) indeed has the natural meaning of dimension.

1b. The model curve in Fig.10.1 (cross section of some hypothetical surface) has $D = \ln 4 / \ln 3 = 1.26$. Indeed, the curve by construction consists of four identical pieces, each of which is a copy of the whole curve downscaled by a factor 3. So if we increase R by a factor 3, M(R) increases by a factor 4: M(3R)=4M(R), which is solved by $M(R) \propto R^{\ln 4 / \ln 3}$.

1c. The just mentioned property of the curve in Fig.10.1, that the "whole" can be decomposed into parts similar to the "whole", is called self similarity or scale invariance of the system. It is one of the basic properties of fractals in general. In fractals generated by random processes ("random fractals") it is satisfied in a statistical sense. The system looks the same at all magnifications. A weaker form of scale invariance is self-affinity [10.20], in which case the "whole" obtains from its parts by rescaling the parts differently in the horizontal and vertical direction. The roughening transition (Chap.14) falls into this

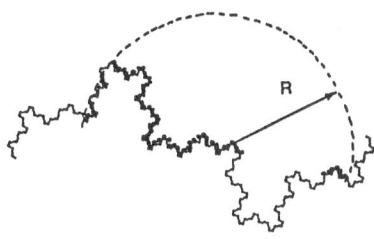

Fig.10.1. Illustration of the definition of D, (10.2). In this example, $D = \ln 4 / \ln 3$

285

class. Self-affine fractals have considerably more complicated properties. Unlike self-similar fractals, they have an intrinsic length scale, called crossover length, associated with them. Here we will consider only self-similar fractals. Many results, however, will also hold in the self-affine case if the system diameter does not exceed the crossover length.

1d. The curve in Fig.10.1 attains the value $D>1$ in (10.2) by winding back and forth. The larger D, the more winding (irregular) is the curve. So, for any system one has $D_{top} \leq D \leq d$ and $D-D_{top}$ is a measure of the disorder of the system. If $D=D_{top}$, the system is ordered or weakly disordered. The type of disorder that may exist in the case $D=D_{top}$ is described in Table 10.2 and Fig.10.2.

1e. "A good definition should be the hypothesis of a theorem" [10.21]. Equation (10.2) can easily be converted into a relation between experimentally measurable quantities. Suppose that the D-dimensional surface is carried by particles of variable diameter L and constant shape, then the number of molecules on a particle of diameter L scales like the number of molecules cut out by a sphere of radius L/2 on the largest particle, i.e. like L^D (10.2). A macroscopic volume holds $\propto L^{-3}$ particles of size L. So, the number of molecules (monolayer) per unit volume of adsorbent, as a function of adsorbent particle size L, is proportional to L^{D-3}[10.22]. Experimental examples for this power law are shown in Fig.10.3 [10.23,24]. There are many other ways in which

Table 10.2 Surface structure in terms of irregularities. In a continuum representation of the surface, irregularities are defined as points where the surface has zero radius of curvature. D' denotes the fractal dimension of the set of all irregularities. Notice the relation $D = \max\{D', D_{top}\}$.

	Irregularities		
	fill entire surface	do not fill surface	are absent
Value of D	$D > D_{top}$	$D = D_{top}$	$D = D_{top}$
Value of D'	$D' = D$	$D' < D$	$D' = 0$
Disorder	strong (fractal)	weak (subfractal) [10.3,10]	none
Example	Fig.10.1	Fig.10.2	sphere

Fig.10.2. Illustration of the fact that systems with $D=D_{top}$ may still be quite irregular. The curve is Cantor's staircase [10.18,19] and has $D=1$, $D'=\ell n2/\ell n3$

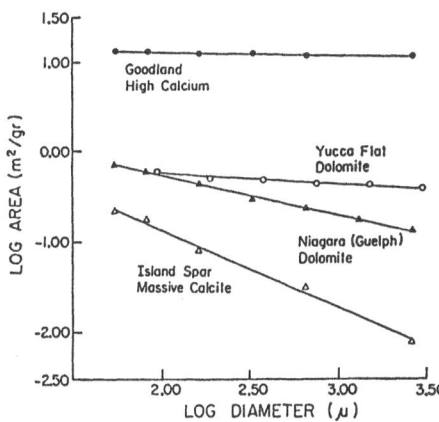

Fig.10.3. Ca/MgCO$_3$ systems exhibiting fractal surface structure through the L^{D-3} law discussed in the text (adsorbate: Kr). The D values from top to bottom are 2.97, 2.81, 2.58, and 2.16, with standard deviations less than 0.04 [10.23]

(10.2) affects physical phenomena on a fractal surface. Indeed, (10.2) implies that any physical property that depends on the number of distant neighbors from a given molecule necessarily depends on D. One example is the electronic energy transfer from an excited donor molecule to acceptor molecules by dipole-dipole interaction (donor and acceptors on the surface). On a substrate with fractal dimension D, the survival probability of the donor at time t, p(t), takes the form [10.25,26]

$$p(t) = \exp(-\text{const} \cdot t^{D/6}) \ . \tag{10.3}$$

A second example, with applications to corrosion, is the adsorbate-adsorbate interaction if the adsorbed species are ions (an electrically neutral layer of ions). The interaction is the screened Coulomb potential $\phi(R)$. *Debye-Hückel* theory on a fractal yields [10.27]

$$\phi(R) \propto \frac{1}{R^{(D+1)/4}} \exp\left(-\text{const} \cdot R^{(D-1)/2}\right) \ , \tag{10.4}$$

showing that screening is most effective when D is large.

1f. An alternative definition of D reads as follows. Divide space into cells (boxes) of side length r and count the number N(r) of boxes that intersect the surface. Then the relation which determines D is

$$N(r) \propto r^{-D} \ . \tag{10.5}$$

The two definitions, (10.2) and (10.5), are equivalent [10.22]. The one analyzes the space-filling ability of the surface by counting distant neighbors, the other by coverings of the surface. The interest in (10.5) comes from the fact that one may replace boxes intersecting the surface by molecules forming a monolayer, in which case (10.5) describes how the number of molecules per monolayer scales with the radius r of the molecules. It shows that a surface with D>2 has no well-defined area because N(r)r^2 ($\propto$ area measured with molecules of radius r) decreases with increasing r. From this viewpoint, (10.5) measures disorder by measuring the decreasing accessibility of the surface to molecules of increasing size. For experimental examples (adsorption data) for (10.5) see [10.28–30].

1g. Inaccessibility of the surface to molecules of a given size is related to the notion of pores. Consequently, Remark 1f suggests that a fractal surface should also have a characteristic pore size distribution. Indeed, if one defines $V_{pore}(r)$ as the volume around the solid that is inaccessible to spheres of radius r (Fig.10.4, cumulative volume of pores of radius $\leq r$), one finds [10.31]

$$V_{pore}(r) = C \left[\frac{D-D_{top}}{d-D_{top}} \right]^{d-D_{top}} r^{d-D} \tag{10.6}$$

for a general fractal in d dimensions (V_{pore} = volume in d dimensions). The constant C equals the prefactor in (10.5). The result (10.6) is an example of a property that depends on all three dimensions D, D_{top}, d and vanishes for Euclidean structures ($D=D_{top}$). It shows in what way pores disappear in the limit of zero disorder [10.32]. In Sect.10.6 we will use (10.6) to treat multilayer adsorption on a fractal surface. In general, one expects D and d to occur jointly, as in (10.6), whenever interactions of the fractal with its surroundings are at issue.

10.1.2 Definition 2.
Let a molecule perform a random walk (diffuse) on the surface. If the probability G(0,t) for the molecule to return to the origin at time t decreases as

$$G(0,t) \propto t^{-\tilde{D}/2} \tag{10.7}$$

with increasing t (Fig.10.5), then $\tilde{D}$ by definition is the spectral dimension (also called fracton dimension) of the surface.

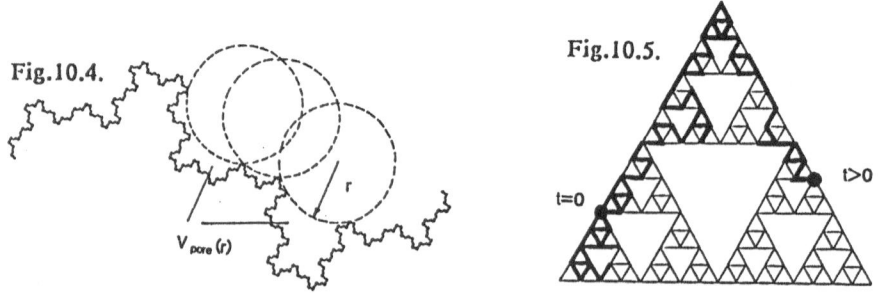

Fig.10.4.

Fig.10.5.

$V_{pore}(r)$

$t=0$

$t>0$

Fig.10.4. The pore volume $V_{pore}(r)$

Fig.10.5. A random walk on a curve with $D=\ell n3/\ell n2$ and [10.33,34] $\tilde{D}=2\ell n3/\ell n5$. The probability for return to the origin on this branched curve is smaller than on the nonbranched curve in Fig.10.1 (compare text)

288

Remarks:

2a. In the early studies of spectral dimension [10.33,34], emphasis was on the density of vibrational states (central forces) onto which the diffusion problem (10.7) can be mapped. This is where the name comes from. However, realistic descriptions of vibrational properties of fractals are considerably more complicated (noncentral forces, etc.) [10.36,37], so it has become customary to define $\tilde{D}$ exclusively in terms of diffusion.

2b. To see that $\tilde{D}$ has the meaning of dimension, recall the fundamental solution $G(r,t)$ of the diffusion equation in d-dimensional Euclidean space,

$$\frac{\partial}{\partial t} G(r,t) = K \nabla^2 G(r,t) \tag{10.8a}$$

$$G(r,0) = \delta(r) \tag{10.8b}$$

(K = diffusion coefficient). It has the familiar form

$$G(r,t) = (4\pi K t)^{-d/2} \exp[-|r|^2/(4Kt)] \tag{10.9}$$

and gives the probability density for a molecule at the origin at time 0, to be at position r at time t. Thus, for diffusion in Euclidean space, the probability of return to the origin is $G(0,t) \propto t^{-d/2}$ and depends on the dimension of the diffusion space indeed as written down in (10.7). The fractal case of the full $G(r,t)$ will be discussed in Sect.10.4.

2c. In general, the spectral dimension equals neither D_{top} nor D but lies in between:

$$D_{top} \leq \tilde{D} \leq D . \tag{10.10}$$

The model curve in Fig.10.5 (cross section of a multiply connected surface) is such a case. In Fig.10.1 on the other hand we have $\tilde{D}=D_{top}$ because any nonbranched system can be deformed, without changing $G(0,t)$, into a Euclidean space of dimension D_{top}, whence $\tilde{D}=D_{top}$ from (10.9). Thus $\tilde{D}$ depends on the ramification of the substrate (to analyze ramification it is often convenient to represent the fractal as a lattice of sites and bonds on which diffusion takes place). With increasing ramification, the escape probability of the molecule goes up; the probability of return to the origin goes down; and thus $\tilde{D}$ goes up. $\tilde{D}=D$, the upper limit in (10.10), is presumably realized for structures like *Menger*'s sponge [10.19]. Perhaps somewhat surprisingly, the list of models for which $\tilde{D}$ is nontrivial ($\neq D_{top}$) and known exactly is short. It consists of the Sierpinski gasket in Fig.10.5, relatives thereof [10.38], and $\tilde{D}=4/3$ for percolation clusters in any dimension [10.33].

2d. Equation (10.10) is the second major example, next to (10.6), for the removal of degeneracies of Euclidean spaces. For Euclidean spaces, one has $D_{top}=\tilde{D}=D$. So one expects, and does find, that many properties of diffusion on fractals exhibit an unusual behavior as compared to Euclidean diffusion. The most prominent examples are the mean-square displacement $\xi^2(t)$ as a function of time; the mean number of distinct visited sites, $S(t)$, as a function of time; and the mean diffusion time $\tau(L)$ to reach a target on a substrate of diameter L. They satisfy

$$\xi^2(t) \propto t^{\tilde{D}/D} \qquad (10.11)$$

$$S(t) \propto t^{\min\{1,\tilde{D}/2\}} \qquad (10.12)$$

$$\tau(L) \propto L^{\max\{1,2/\tilde{D}\}D} . \qquad (10.13)$$

The rule is that quantities which involve distances depend on both $\tilde{D}$ and D, whereas quantities which involve no metric depend on $\tilde{D}$ only. Equations (10.11-13) and their consequences for the kinetics of diffusion-controlled chemical reactions have been reviewed in considerable detail in [10.9,10,35,39,40], so we restrict the discussion to a few points.

Equation (10.11) shows that the classical diffusion law $\xi^2(t) \propto t$ in Euclidean spaces (independent of dimension!) no longer holds for fractals. For fractals ($\tilde{D}<D$), ξ^2 grows sublinearly with t, that is much more slowly. This is called anomalous diffusion. It follows that the diffusion coefficient K enters no longer as proportionality factor between ξ^2 and t. Instead, within a factor of unity one has

$$\xi^2(t) = (Kt/\ell_0^2)^{\tilde{D}/D} \ell_0^2 \qquad (10.11^*)$$

where ℓ_0 is the elementary step length of the random walk (inner cutoff of the fractal behavior) and K is the diffusion coefficient on the identical but ordered substrate [10.41]. In Sect.10.4 we will use (10.11^*) to estimate time scales for scattering from a diffusing adsorbate.

If we write (10.11) as $t \propto \xi^{2D/\tilde{D}}$ and interpret t as "mass" of the random walk trajectory within distance ξ from the origin, it follows from Definition 1 that the walk has the fractal dimension

$$D_{walk} = \min\{2D/\tilde{D},D\} . \qquad (10.14)$$

The reason for min{...} is that "mass" in (10.2) should only count the number of distinct visited sites within distance ξ. So if that number is proportional to t (nonrecurrent walk), then $D_{walk}=2D/\tilde{D}$. If that number is not proportional to t, then most sites within distance ξ are visited infinitely often (recurrent walk) and the walk fills the substrate, whence $D_{walk}=D$. Thus, (10.14) tells us that diffusion is recurrent if $\tilde{D} < 2$ and nonrecurrent for $\tilde{D} > 2$. It also proves (10.12): use (10.11,14) to evaluate $S(t) \propto [\xi(t)]^{D_{walk}}$. In fact, (10.12) is just another way of expressing recurrence vs. nonrecurrence. Similarly for (10.13).

For substrates in three-dimensional space, diffusion-controlled reactions are fastest for $\tilde{D}=3$ (whence D=3) if substrates with the same number of surface sites are compared; and for D=1 (whence $\tilde{D}=1$) if substrates with the same diameter L are compared [10.10]. While the detailed analysis of this question requires more theory than presented here, a qualitative picture can be obtained as follows. In the first case, the reaction rate is determined by S(t) which goes up when $\tilde{D}$ goes up (10.12). In the second case, the rate is determined by $\tau(L)$ which goes up when D goes down (10.13).

10.2 Small-Angle Scattering from Fractal Surfaces

Small-angle x-ray and neutron scattering as a method of fractal surface analysis was introduced by *Bale* and *Schmidt* in 1984 [10.42]. Practically at the same time, the first small-angle scattering studies of fractal aggregates appeared [10.43]. The method has become one of the major tools in fractal analysis since [10.44]. Indeed, it applies to a wide variety of systems and there is no competition from other scattering techniques because only x-rays and neutrons can probe strongly disordered surfaces.

For scattering from a three-dimensional solid with a surface of fractal dimension D ("surface fractal", Fig.10.6), the scattered intensity I(q) as a function of the length q of the scattered wave vector (momentum transfer $\hbar q$) obeys [10.42]

$$I(q) \propto q^{D-6} . \tag{10.15}$$

In contrast, for scattering from fractal aggregates where both the solid and the surface scale with fractal dimension D ("mass fractals"), the scattering law is [10.43]

$$I(q) \propto q^{-D} . \tag{10.16}$$

A unified description of (10.15) and (10.16) can be found in [10.45]. Here we are concerned only with the case (10.15). We describe the main ingredients of a rigorous derivation [10.46] of (10.15). The motivation for this is that both in the limit D→2 and in the limit D→3 there are a number of unresolved questions. For D=2, one would like to know the term next to q^{-4} in (10.15) in order to extend the method to subfractal and self-affine surfaces [10.47]. For D=3, where the surface becomes space-filling, the scattering law depends on how the

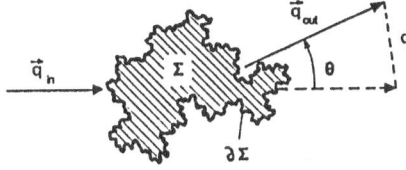

Fig.10.6. Elastic scattering ($|q_{in}| = |q_{out}|$) of x-rays or neutrons from a solid Σ with fractal surface $\partial\Sigma$. The scattered wave vector has length $q = (2\pi/\lambda) \sin(\theta/2)$ where λ is the wave length and θ is the scattering angle

limit D→3 is taken [10.48]. For convenience, we express everything in terms of x-rays.

We denote the scatterer by Σ (continuum representation, set of all occupied points) and its surface by $\partial\Sigma$. In Σ, a uniform electron density ρ_0 is assumed. The scattered intensity, in the usual single-scattering framework, is then given by

$$I(q) = I_0 \rho_0^2 \left| \int \rho(x) e^{-iq \cdot x} d^3 x \right|^2 \tag{10.17}$$

where q is the scattered wave vector, I_0 is the scattered intensity per electron, $\rho(x)=1$ if x is in Σ and $\rho(x)=0$ otherwise, and integration is over all space. So the starting point of the theory is the same as in x-ray crystallography except that here we are interested in the large-scale structure of Σ (small q values) where individual atoms are no longer resolvable. This is why here the scatterer can be represented by a uniform electron density. It reduces the determination of the correlation function $g(r)$, (10.19a), to a purely geometric problem, (10.19b):

$$I(q) = I_0 \rho_0^2 V \int g(r) e^{iq \cdot r} d^3 r \tag{10.18}$$

$$g(r) = \frac{1}{V} \int \rho(x) \rho(x+r) d^3 x \tag{10.19a}$$

$$= \frac{1}{V} \cdot \text{volume of the intersection of } \Sigma \tag{10.19b}$$

$$\text{with } \Sigma \text{ translated by } r$$

where V is the volume of Σ. Note that $g(r)$ is normalized so that $g(0)=1$. The quantity of experimental interest is the spherically averaged intensity $I(q)$, for which from (10.18) we obtain

$$I(q) = 4\pi I_0 \rho_0^2 V \int_0^\infty g(r) \frac{\sin qr}{qr} r^2 dr \tag{10.20}$$

where $g(r)$ is the spherical average of $g(r)$.

So far, it is not obvious yet that $g(r)$ and hence $I(q)$ should be particularly sensitive to the surface structure of Σ. This sensitivity is brought out by the following identity (proofs of Theorems 1-3 in this section will be presented elsewhere [10.46]).

10.2.1 Theorem 1. Let $\partial\Sigma_s$ denote the surface in Σ at a distance s from $\partial\Sigma$ (Fig.10.7). Note that $\partial\Sigma_s$ is smooth even if $\partial\Sigma$ is fractal. For each point x on $\partial\Sigma_s$, let $W_r(x)$ be the volume of all points outside Σ whose distance from x is $\leq r$ (Fig.10.7). Then, for an arbitrary scatterer Σ ($\partial\Sigma$ fractal or not),

$$g(r) = 1 - (4\pi V r^2)^{-1} \int_0^r ds \int_{\partial\Sigma_s} \frac{d}{dr} W_r(x) d^2 x \tag{10.21}$$

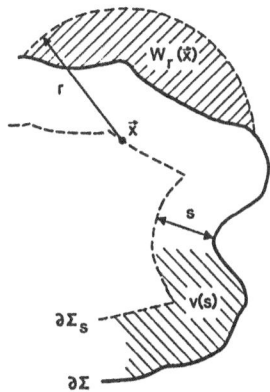

Fig.10.7. Illustration of the volumes $W_r(\mathbf{x})$ and v(s) occurring in Theorems 1 and 2 (Sect.10.2)

where $d^2\mathbf{x}$ is the surface element on $\partial\Sigma_s$.

The theorem says that we can get g(r) by integrating the volume $W_r(\mathbf{x})$ along the surface $\partial\Sigma_s$, and then integrating over all s. Compared with (10.19b), the achievement of (10.21) is that the spherical average is taken care of, that only local volumes are needed, and that everything is reduced to points within distance r from $\partial\Sigma$. We now use (10.21) to determine the small-r behavior of g(r). This is the key step in order to find the dependence on D when the surface $\partial\Sigma$ is fractal. Indeed, while in a lattice representation of Σ (infinite system) self-similarity shows up at ever larger distances, self-similarity in a continuum representation (finite system) shows up at even smaller distances. The following theorem gives the full account of the small-r behavior.

10.2.2 Theorem 2. Let v(s) be the volume of all points in Σ whose distance from $\partial\Sigma$ is $\leq r$ (Fig.10.7). If the surface has fractal dimension D, let $\mu_D(\partial\Sigma)$ be the D-dimensional Hausdorff measure of $\partial\Sigma$ (to be explained below), and if the surface is smooth (D=2) let S be its area. Then for $r\to0$

$$g(r) = 1 - \frac{1}{2Vr} \int_0^r v(s)ds + O(r)v(r) \quad (\partial\Sigma \text{ arbitrary}) \qquad (10.22a)$$

$$= 1 - \frac{1}{4V} \gamma_D \mu_D(\partial\Sigma)r^{3-D} + o(r)v(r) \quad (\partial\Sigma \text{ fractal}) \qquad (10.22b)$$

$$= 1 - \frac{S}{4V} r + o(r) \quad (\partial\Sigma \text{ smooth}) \qquad (10.22c)$$

where $\gamma_D = \pi^{(3-D)/2}[(4-D)\Gamma((5-D)/2)]^{-1}$ and Γ is the gamma function. Note $\gamma_2 = \gamma_3 = 1$.

We discuss (10.22) in several steps. Equation (10.22a) shows that for small r the integral over $\partial\Sigma_s$ in (10.21) reduces to $2\pi r v(s)$. This is not only important as a source for the results (10.22b,c) but also because from (10.22a) one can get terms beyond that by estimating v(s)

293

beyond its leading-order term. To see how (10.22b) arises, we go back to (10.5) and identify v(s) with the total volume of boxes of size s that intersect $\partial\Sigma$. Each box has the volume s^3 so that $v(s) \propto s^{-D} s^3$ by (10.5). This explains r^{3-D} in (10.22b) and r in (10.22c). To appreciate the prefactors, we note that v(s) is proportional to the content of the surface $\partial\Sigma$. If $D=2$, the content is the surface area S. But for $D>2$, we have seen in Remark 1f that the surface area is not well-defined (it is infinite if self similarity extends to arbitrarily small length scales). This is why the Hausdorff measure μ_D, the generalization to nonintegral dimensions of length/area/volume, enters for $D>2$. A precise definition can be found in [10.18,19]. Here it suffices to know that within a factor of order unity, $\mu_D(\partial\Sigma)$ equals the prefactor in (10.5). As it should, it has units of $(length)^D$ and satisfies

$$\mu_D(\partial\Sigma) \leq L^D \tag{10.23}$$

where L is the diameter of Σ.

From (10.22b,c) it is now straightforward to find the scattered intensity I(q) for large q. (Large q means $q \gg L^{-1}$, corresponding to that small r in (10.22) means r<<L. In practice, one has the upper bound $q<<1 \text{Å}^{-1}$ in Table 10.1, of course.) By integrating (10.20) twice by parts, substituting (10.22b,c) into the result, and doing Fourier asymptotics [10.42], one obtains:

10.2.3 Theorem 3. For q→∞, the scattered intensity is given by

$$I(q) = \pi I_0 \rho_0^2 \Gamma(5-D) \sin\left[\frac{D-1}{2}\pi\right] \gamma_D \mu_D(\partial\Sigma) q^{D-6} + o(q^{D-6})$$

$$(\partial\Sigma \text{ fractal}) \tag{10.24a}$$

and

$$I(q) = 2\pi I_0 \rho_0^2 S q^{-4} + o(q^{-4}) \quad (\partial\Sigma \text{ smooth}). \tag{10.24b}$$

Equation (10.24b) is the classical q^{-4} Porod law for small-angle scattering. Equation (10.24a) is the detailed version of the fractal law (10.15). It gives the absolute expression for the prefactor in (10.15). It is one of the few results in fractal theory where a complete expression for the prefactor is available. As a corollary, (10.24a) identifies the scenario for which I(q) vanishes [10.42] for D→3 (note $\sin[(D-1)/2\pi] \to 0$ for D → 3). The scenario is that the diameter L of the scatterer remains finite (10.23), [10.48].

10.3 Application: Small–Angle He Scattering from Adsorbate Islands on an Ordered Surface

The development in the previous section was done for a three-dimensional scatterer and correspondingly three-dimensional wave vectors. It

294

is not difficult to generalize this to scattering in arbitrary space dimension d. The fractal dimension D of the boundary $\partial\Sigma$ then satisfies $D \leq d$ and the correlation function $g(r)$, (10.22b), behaves like $1 - C \cdot r^{d-D}$, where C is a constant. The resulting generalization of the scattering law (10.15) reads [10.49]

$$I(q) \propto q^{D-2d} . \tag{10.25}$$

Let us now see how (10.25) can be applied to cases other than d=3. The idea is that if an ordered, single-crystal surface is partially covered with an adsorbate that arranges itself into islands (as opposed to forming a homogeneous layer), and if we could do an x-ray or neutron experiment in the plane that contains the islands (one layer thick), (10.25) would provide a powerful means to investigate the boundary structure of the islands. Indeed, scaling relations in the theory and simulations of island formation suggest that fractal island boundaries should be quite common. The desired in-plane scattering experiment can be realized by He scattering from the adsorbate (Fig.10.8) as follows. The scattering is assumed to be elastic. We decompose the wave vectors into components parallel and perpendicular to the surface:

$$\mathbf{q}_{in} = \mathbf{q}_{in,\|} + \mathbf{q}_{in,\perp}$$

$$\mathbf{q}_{out} = \mathbf{q}_{out,\|} - \mathbf{q}_{in,\perp}$$

$$q_\| = |\mathbf{q}_{in,\|} - \mathbf{q}_{out,\|}| . \tag{10.26}$$

Since the perpendicular components are independent of the momentum transfer parallel to the surface, the parallel components correspond to the desired in-plane scattering. Thus, if the adsorbate islands have a boundary with fractal dimension D, then

$$I(q_\|) \propto q_\|^{D-4} \tag{10.27}$$

from (10.25), d=2, and (10.26). With currently available q ranges for He scattering, it should be no problem to look at islands of 100 Å or more.

In the case where that island diameters vary over an appreciable range, (10.27) may have to be averaged over the diameter distribution,

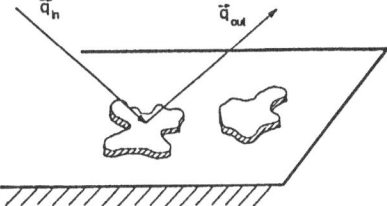

Fig.10.8. He scattering from adsorbate islands on an ordered surface

available from the line shape of Bragg peaks. Finally, we mention that small-angle He scattering is also able to detect adsorbate clusters that correspond to mass fractals, (10.16). For adsorbate mass fractals, the intensity will scale as $q_{\parallel}^{-D}$. Since $D \le 2$, the exponent here is ≥ -2, whereas the exponent in (10.27) is always ≤ -2. Thus, the two scattering laws do not overlap, just as (10.15) and (10.16) do not overlap.

10.4 Scattering from a Diffusing Adsorbate on a Fractal Surface

Here we discuss a recently proposed method of measuring the spectral dimension $\tilde{D}$ of a fractal surface [10.50]. It is clear from the relevance of $\tilde{D}$ for reaction dynamics described in Sect.10.1 that experimental methods of obtaining $\tilde{D}$ are of great applicational importance. Apart from the measurements in [10.40] (kinetics of exciton fusion), no experimental $\tilde{D}$ values have been measured as yet. For a proposal to determine $\tilde{D}$ from NMR experiments, see [10.51].

We pursue the question of whether quasi-elastic neutron scattering, which has been used to measure the diffusion coefficient K of adsorbates on ordered surfaces [10.52], is sensitive to anomalous diffusion on a fractal surface. The expression exploited by these experiments is [10.53]

$$S(q,\omega) \propto \frac{Kq^2}{\omega^2 + (Kq^2)^2} \tag{10.28}$$

(diffusion in d-dimensional Euclidean space, d=1,2,3) where $S(q,\omega)$ is the incoherent dynamic structure factor, i.e. the incoherently scattered intensity as a function of momentum transfer $\hbar q$ (in d dimensions) and energy transfer $\hbar\omega$ (Fig.10.9). So, the answer to the question seems to be negative since (10.28) does not depend on the dimension of the diffusion space at all. However, we know from (10.11) that a dynamical quantity may well exhibit its dependence on dimensionality only if we go to nonintegral dimensions. The dynamic structure factor has the same property.

Similarly to (10.18) for elastic scattering, the inelastic intensity $S(q,\omega)$ is given by [10.53]

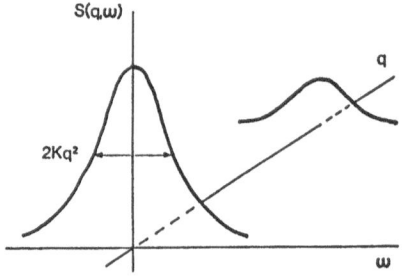

Fig.10.9. Measurement of the diffusion coefficient K of an adsorbate on a smooth surface by quasi-elastic scattering (diffusional line broadening). The line maximum equals $K^{-1}q^{-2}$ and the width equals $2Kq^2$ (10.28). Small q corresponds to large diffusion distances; small ω corresponds to large diffusion times.

$$S(q,\omega) \propto \int_{-\infty}^{\infty} dt \int d^d r \; G(r,t) e^{i(\mathbf{q}\cdot\mathbf{r}-\omega t)} \qquad (10.29)$$

where $G(r,t)$ is the adsorbate autocorrelation function, i.e. the probability density for a molecule at the origin at time 0, to be at position r at time t (recall Sect.10.1). In (10.29), the coordinates r are with respect to the d-dimensional Euclidean space in which the system is embedded and in which the scattering takes place. Of course, the embedding space will typically be three-dimensional.

Thus, the problem is to find $G(r,t)$ for the general case of anomalous diffusion. (For normal, Euclidean diffusion, $G(r,t)$ is given by (10.9) and leads to (10.28) upon substitution into (10.29)). The second moment of $G(r,t)$ obeys

$$\int r^2 G(r,t) d^d r = \xi^2(t) \propto t^{\tilde{D}/D} \qquad (10.30)$$

by (10.11). We also know that a random walk on a self-similar system is scale-invariant. This implies that, after spherical averaging, $G(r,t)$ is of the form [10.50]

$$G(r,t) = r^{-d} h(r/\xi(t)) \qquad (10.31)$$

where h is a system-specific function. The point is that (10.31) and (10.11) are already enough to infer the scaling behavior of $S(q,\omega)$: substitution of the two relations into (10.29) and simple changes of variables yield

$$S(q,\omega) \propto q^{-2\delta} f(\omega q^{-2\delta}) \qquad (10.32)$$

$$\delta \equiv \tilde{D}/D \qquad (10.33)$$

where the line-shape function f depends on the scaling function h in (10.31). The result (10.32) states that the line maximum decreases like $q^{-2\delta}$ with increasing q, and that the line width increases like $q^{2\delta}$ with increasing q. From either one, the ratio δ can be determined. Thus if we know D from some other experiment, quasi-elastic scattering indeed yields $\tilde{D}$. The relevant q range is as in small-angle scattering, namely $q \ll 1$ Å^{-1}. The energy range may be obtained from (10.11*). For example, for methane on a carbon surface, estimated energy transfers are >40 μeV [10.50].

Under special circumstances, it is possible to get the explicit form of the scaling function h and hence of the line shape f. For example, if the fractal is nonbranched, then [10.50]

$$h(u) = u^D \exp(-u^{2\delta}) \qquad (10.34)$$

(we omit all constants here). The Euclidean result (10.9) corresponds to the special case of $\tilde{D}=D=d$ in (10.34,31,11). *Banavar* and *Willemsen* [10.54] have shown that (10.34) holds whenever diffusion on the fractal satisfies the Chapman-Kolmogorov equation, i.e. is a Markov process (for yet another derivation of (10.34), see [10.55]). So one is led to the conclusion that (10.34) is universally valid. This conclusion, however, has recently been questioned [10.9]. If the doubts turn out to be valid, the possibility of non-Markovian diffusion would be highly unusual. It would mean that the diffusion process is no longer memoryless, i.e. that the future of the process depends no longer only on the present state but also on its past.

10.5 Henry's Law of Adsorption on a Fractal Surface

Much of the early work in fractal surface analysis was based on adsorption data [10.22,23,28,29]. Carefully determined monolayer values were used to see whether power laws such as in Fig.10.3 or (10.5) would be obeyed. These monolayer values were obtained from standard low-coverage analyses of adsorption isotherms (BET etc.). All these standard analyses rest of course on the implicit or explicit assumption that the surface is smooth. Consequently, an important question is whether one can consistently infer fractal behavior (D>2) from "smooth-surface" monolayer data. The intuitive answer is yes because at low coverage one expects the adsorbed molecules to "see" only the local surface structure, as opposed to the fractal structure at large scales. But what about the substrate potential in the vicinity of a fractal surface, and what about energetically inequivalent surface sites?

These and related questions concerning the adsorption at low coverage have been studied in detail in a paper [10.56] whose title we have borrowed for this section. It is a three-page self-contained paper. So we restrict ourselves here to giving the bottom lines and referring the reader to [10.56] for more details.

(a) Henry's law refers to the proportionality at low coverage between the amount adsorbed and the pressure of the gas (linear portion of the isotherm). For a smooth surface, the proportionality factor, called Henry's-law coefficient, equals $Sr\beta e^{-\beta E}$ where S is the surface area, r is the adsorbate radius, $\beta=(kT)^{-1}$ is the inverse temperature, and E is the effective substrate potential (a constant). For a fractal surface, Sr is replaced by the Hausdorff measure of the surface - $\mu_D(\partial\Sigma)$ in the notation of Sect.10.2 - times r^{3-D}. Notice the parallel to (10.22). Thus from the temperature dependence of Henry's-law coefficient one can determine $\mu_D(\partial\Sigma)r^{3-D}$, and by subsequent variation of the adsorbate size r one can get D.

(b) If the geometric disorder creates energetically inequivalent sites, preferential adsorption at strong bonding sites is negligible at temperatures T such that $kT\gg\Delta E$ where ΔE is the maximal difference in binding energy between different sites.

(c) Apart from Henry's-law coefficient, adsorption isotherms at low coverage are independent of D. In particular, they are always linear at sufficiently low pressure. Since in the determination of monolayer values Henry's-law coefficient is always a fitting parameter, it follows that fractal behavior may indeed be deduced from "smooth-surface" monolayer data.

10.6 BET Condensation of a Gas on a Fractal Surface

In the Henry's-law regime, adsorbate-adsorbate interactions are negligible. The opposite extreme is adsorption at very high coverage where attractive forces between the adsorbed molecules lead eventually (as the number of layers grows to infinity) to the formation of bulk liquid on the surface. At very high coverage, one expects adsorption isotherms to be different on a fractal surface than on an ordered surface because the presence of pores of all sizes (Fig.10.1) imposes systematic restrictions on the growth of multilayers as the pressure is increased. It will be convenient to consider the thermodynamic limit, i.e. to treat the surface as an infinite system.

The Brunauer-Emmett-Teller (BET) theory [10.57,58] of multilayer adsorption provides an interesting test case to explore such D dependences at high coverage. We focus on the BET model because of its time-honored experimental success in the analysis of low-coverage data, and because it constitutes a well-defined statistical mechanical model [10.57,58] which is exactly solvable for an ordered surface. Recalling [10.57] that BET theory predicts a film surface tension of $-\infty$ at infinite coverage on an ordered surface, a question of some interest is whether the same holds on a fractal surface.

On an ordered surface, the BET isotherm reads, as is well known,

$$\theta(x) = \frac{cx}{1-x} \frac{1}{1+(c-1)x} \tag{10.35}$$

where θ is the coverage (number of adsorbed molecules per surface site), c is the adsorption strength related to the energy difference between adsorption in the first layer and adsorption in higher layers, and x is the relative pressure (ratio of pressure to saturation pressure, $0 \leq x \leq 1$). Thus, $x \to 1$ corresponds to the condensation of bulk liquid on the surface. To obtain the fractal generalization of (10.35), there have been two different approaches. The approach by *Fripiat* et al. [10.60] uses the idea that a completed m-layer becomes increasingly smooth, and thus decreases in surface area available for further adsorption, as m increases. In the thermodynamic limit (no maximum layer thickness), it yields

$$\theta_F(x) = \frac{c}{1+(c-1)x} \sum_{m=1}^{\infty} m^{2-D} x^m .$$ (10.36)

The second approach [10.56] treats the problem as one of adsorption in pores of various sizes: in a single pore of radius r, there holds an m-layer BET isotherm [10.59] (adsorption restricted to maximum m layers) with $m \propto r$ and with an effective number of surface sites proportional to r^2. According to (10.6) the number of pores of radius r scales like r^{-D-1}. Adding all contributions from individual pores, one finds

$$\theta_C(x) = \frac{cx}{1-x} (D-2) \int_1^{\infty} m^{1-D} \frac{1-(m+1)x^m+mx^{m+1}}{1+(c-1)x-cx^{m+1}} dm .$$ (10.37)

Here the thermodynamic limit amounts to the absence of a largest pore radius. Numerical results for (10.37) are shown in Fig.10.10. In agreement with Sect.10.4, they are D-independent at very low pressures. At high pressures, the amount adsorbed at a given pressure decreases with increasing D, as expected from the spatial constraints for D>2.

Both (10.36) and (10.37) reduce to the classical formula (10.35) in the smooth-surface limit D→2. For (10.36) this is obvious, and for (10.37) one first has to integrate by parts (for D=2, the vanishing of the prefactor is counterbalanced by the nonintegrability of m^{1-D}). As will be seen below, both equations also diverge for x→1, in agreement with the fact that on an infinite fractal surface there is still room for infinitely thick layers (compare (10.6) for r→∞). But the two results are not identical. An instructive way to see this is to note that (10.36) can be factored into a part that depends only on c and a part that depends only on D. Not so (10.37). The decoupling of c and D in (10.36) can be interpreted as (10.36) treating the adsorption strength in

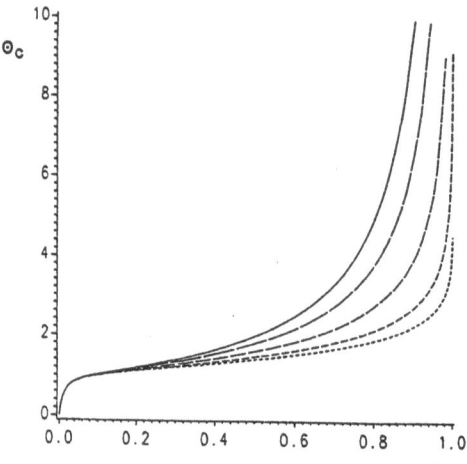

Fig.10.10. Fractal BET isotherm (10.37) on surfaces of dimension D=2.0, 2.2, 2.5, 2.8, and 3.0 (from top to bottom). Here c=100 [10.61]

an average way (average over all pores) rather than in each pore separately as (10.37) does. There are other reasons to believe that (10.37) is more accurate than (10.36). See [10.61] which will give the complete analytical and numerical analysis of the two theories. In view of this difference between (10.36) and (10.37) it is rather remarkable that they agree numerically quite well. Up to about x=0.9 the difference is only a few percent, even for D=3. Only for x→1 does (10.36) exceed (10.37) as much as by a factor of two (10.40).

Unlike many other properties of fractals we have seen in the previous sections, the isotherms (10.36) and (10.37) do not have the form of simple scaling laws. The reason is that these two equations describe adsorption over the entire pressure range, but are sensitive to D only at high x. Thus it is only in the condensation regime that one may expect, and does indeed find, a simple scaling behavior. For x→1 and 2≤D<3, the isotherms diverge as

$$\theta_F(x) \sim \Gamma(3\text{-}D) \frac{1}{(1\text{-}x)^{3\text{-}D}} \tag{10.38a}$$

$$\theta_C(x) \sim A(D) \frac{1}{(1\text{-}x)^{3\text{-}D}} \tag{10.38b}$$

where Γ is the gamma function and the prefactor $A(D)$ is given by

$$A(D) = (D\text{-}2) \int_0^\infty t^{1\text{-}D} \left[1 - \frac{te^{-t}}{1\text{-}e^{-t}} \right] dt \ . \tag{10.39}$$

For $x \to 1$ and $D = 3$ the isotherms diverge only logarithmically:

$$\theta_F(x) \sim \ell n \left(\frac{1}{1\text{-}x} \right) \tag{10.40a}$$

$$\theta_C(x) \sim \frac{1}{2} \ell n \left(\frac{1}{1\text{-}x} \right) . \tag{10.40b}$$

Thus the condensation process at the surface has all the features of a critical phenomenon. The critical exponents and amplitudes are uniquely determined by the fractal dimension. As a special case, the $1/(1\text{-}x)$ divergence of the classical BET isotherm is recognized as the fingerprint of D=2. For the interpretation of (10.40b) and the analytic continuation of (10.37) to formal D values larger than 3, see [10.62].

10.7 Bose-Einstein Condensation in Nonintegral Dimensions

The last two sections might leave one with the impression that thermodynamic properties of films (thin or thick) on a fractal surface are exclusively controlled by the fractal dimension D of the surface. This final section, which is a preview of [10.63], shows that the thermodynamics may also depend on the spectral dimension $\tilde{D}$. Thus, $\tilde{D}$ turns out to be a key quantity both for nonequilibrium (diffusion-controlled kinetics) and equilibrium (here: critical temperature, specific heat) properties.

The problem at issue is this: Given that a system of noninterating Bose particles at sufficiently low temperatures undergoes Bose-Einstein condensation ("superfluidity") in three but not in two dimensions [10.64], what is the situation in dimensions in between? Since the particles by definition have only kinetic energy, we have to solve the free-particle Schrödinger equation on a fractal and count the number of eigenstates as a function of energy ϵ. Knowing this density of states, we can then compute the partition function and everything becomes a standard exercise in statistical mechanics. The point is that the density of states $\rho(\epsilon)$ scales like

$$\rho(\epsilon) = A\epsilon^{(\tilde{D}-2)/2} , \tag{10.41}$$

where A is a constant ($\rho(\epsilon)d\epsilon$ = number of states with energy between ϵ and $\epsilon+d\epsilon$). This can be derived by exploiting the formal equivalence between the Schrödinger equation and the diffusion equation, and is already implicit in [10.33]. Since diffusion is governed by $\tilde{D}$, it is thus no surprise that quantum translational motion is also governed by $\tilde{D}$.

Now for an ideal Bose gas with density of states $\rho(\epsilon)$, the condition for a condensed phase (macroscopic population of the ground state) is that $N^*(T) < N$ where

$$N^*(T) = \int_0^\infty \frac{\rho(\epsilon)d\epsilon}{e^{\epsilon/(kT)}-1} \tag{10.42}$$

is the number of particles not in the ground state (at zero chemical potential), and N is the total number of particles. The equality $N^*(T_c)=N$ defines the critical temperature T_c (Fig.10.11). Equations (10.41,42) give

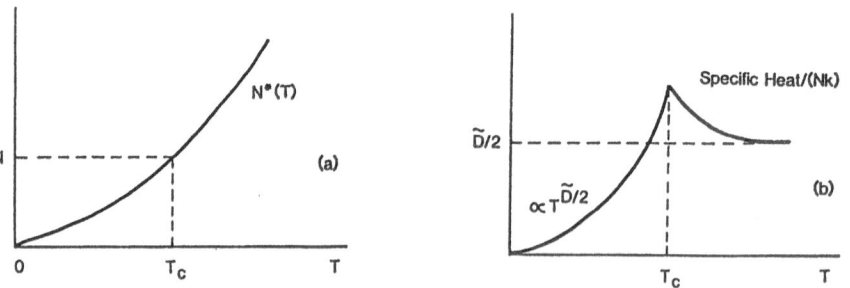

Fig.10.11. $\tilde{D}$ dependence of Bose-Einstein condensation. (a) Determination of the critical temperature T_c. (b) Specific heat at constant Hausdorff measure (volume, if $\tilde{D}=3$) as a function of temperature

$$A \int_0^\infty \frac{\epsilon^{(\tilde{D}-2)/2}}{\exp[\epsilon/(kT_c)]-1} = N \qquad (10.43)$$

for T_c. For $\tilde{D}=2$, the integral diverges, so that (10.43) cannot be realized at any finite density of the gas. For any $\tilde{D}>2$, however, condensation is possible. The no-condensation result for $\tilde{D}=2$ is seen to have its origin in $T_c \sim (\tilde{D}-2)N/(2A)$ for $\tilde{D} \to 2$, i.e. in that T_c goes to zero for $\tilde{D} \to 2$.

Since Bose-Einstein condensation is the zeroth order theory for superfluidity, and since this zeroth order theory depends on $\tilde{D}$, it appears very likely that also a theory for the real Bose gas, on a fractal surface, will have to depend on $\tilde{D}$. It may also be necessary to revise current theories [10.65,66] of superfluid films in porous Vycor glass if the fractal surface of Vycor glass [10.67] turns out to have a spectral dimension larger than two.

10.8 Conclusion

An attempt has been made to answer the question "Suppose we know that a given system is a fractal. So what?" This attempt has naturally led to a focus on theoretical developments. The developments have been aimed at realistic, experimentally testable results. So, not least, it is hoped that this chapter will be read as an invitation for numerous outstanding experiments.

Acknowledgments. The author is greatly indebted to M.W. Cole, M. Obert, and P.W. Schmidt for their collaboration on work quoted here. Acknowledgment is made to the donors of the Petroleum Research Fund, administered by the American Chemical Society, for partial support of this work.

References

10.1 J.M. Ziman: *Models of Disorder* (Cambridge Univ. Press, Cambridge 1979)
10.2 This also explains why critical phenomena in statistical mechanics are a rich source of fractals: it is precisely at a phase transition (critical point) that translation invariance is destroyed macroscopically
10.3 P. Pfeifer: Chimia 39, 120 (1985)
10.4 S.H. Liu: Solid State Phys. 39, 207 (1986)
10.5 H.J. Herrmann: Phys. Rep. 136, 153 (1986)
10.6 B.M. Smirnov: Sov. Phys. Usp. 29, 481 (1986)
10.7 I.M. Sokolov: Sov. Phys. Usp. 29, 924 (1986)
10.8 P. Meakin: CRC Crit. Rev. Solid State Mater. Sci. 13, 143 (1987)
10.9 S. Havlin, D. Ben-Avraham: Adv. Phys. 36, 695 (1987)
10.10 P. Pfeifer, in: *Preparative Chemistry Using Supported Reagents*, ed. P. Laszlo (Academic, New York 1987) p.13
10.11 D. Avnir, ed.: *The Fractal Approach to the Chemistry of Disordered Systems: Polymers, Colloids, Surfaces* (Wiley, New York 1988)

10.12 M.F. Shlesinger, B.B. Mandelbrot, R.J. Rubin, eds.: Proceedings of a Symposium on Fractals in the Physical Sciences, J. Stat. Phys. **36**, 519-921 (1984)

10.13 F. Family, D.P. Landau, eds.: *Kinetics of Aggregation and Gelation* (North-Holland, Amsterdam 1984)

10.14 R. Pynn, A. Skjeltorp, eds.: *Scaling Phenomena in Disordered Systems*, NATO Adv. Study Inst. B133 (Plenum, New York 1985)

10.15 N. Boccara, M. Daoud (eds.): *Physics of Finely Divided Matter*, Springer Proc. Phys., Vol.5 (Springer, Berlin, Heidelberg 1985)

10.16 H.E. Stanley, N. Ostrowsky (eds.): *On Growth and Form*, NATO Adv. Study Inst. E100 (Martinus Nijhoff, Boston 1986)

10.17 L. Pietronero, E. Tosatti, eds.: *Fractals in Physics* (North-Holland, Amsterdam 1986)

10.18 B.B. Mandelbrot: *Fractals: Form, Chance, and Dimension* (Freeman, San Francisco 1977)

10.19 B.B. Mandelbrot: *The Fractal Geometry of Nature* (Freeman, San Francisco 1982)

10.20 B.B. Mandelbrot: Physica Scripta **32**, 257 (1985)
B.B. Mandelbrot in: [10.17], p.3

10.21 J. Glimm as quoted in M. Reed, B. Simon: *Methods of Modern Mathematical Physics*, Vol.1 (Academic, New York 1972) p.259

10.22 P. Pfeifer, D. Avnir: J. Chem. Phys. **79**, 3558 (1983); **80**, 4573 (1984)

10.23 D. Avnir, D. Farin, P. Pfeifer: J. Colloid Interface Sci. **103**, 112 (1985)

10.24 For a comparison with reaction-rate power laws, see D. Farin, D. Avnir: J. Phys. Chem. **91**, 5517 (1987)

10.25 J. Klafter, A. Blumen: J. Chem. Phys. **80**, 875 (1984)

10.26 For experimental measurements (fluorescence decay of donor) compare P. Levitz, J.M. Drake: Phys. Rev. Lett. **58**, 686 (1987); D. Pines-Rojanski, D. Huppert, D. Avnir: Chem. Phys. Lett. **139**, 109 (1987); and references therein

10.27 R. Blender, W. Dietrich: J. Phys. A **19**, L785 (1986)

10.28 D. Avnir, D. Farin, P. Pfeifer: Nature (London) **308**, 261 (1984)

10.29 D. Farin, A. Volpert, D. Avnir: J. Am. Chem. Soc. **107**, 3368, 5319 (1985)

10.30 F.M. Gasparini, S. Mhlanga: Phys. Rev. **B33**, 5066 (1986)

10.31 P. Pfeifer, A. Salli: to be published; see also [10.22]

10.32 Equation (10.6) gives the leading order in r for $D>D_{top}$. For $D=D_{top}$, terms of order $r^{d-D'}$ become leading where D' is as in Table 10.2. So $V_{pore}(r)$ may, but need not, vanish for $D=D_{top}$.

10.33 S. Alexander, R. Orbach: J. Physique Lett. **43**, L625 (1982)

10.34 R. Rammal, G. Toulouse: J. Physique Lett. **44**, L13 (1983)

10.35 R. Rammal: J. Stat. Phys. **36**, 547 (1984)

10.36 I. Webman, in: [10.17], p.343

10.37 J.A. Krumhansl: Phys. Rev. Lett. **56**, 2696 (1986)

10.38 R. Hilfer, A. Blumen: J. Phys. A **17**, L537, L783 (1984)

10.39 G. Zumofen, A. Blumen, J. Klafter: J. Chem. Phys. **82**, 3198 (1985); A. Blumen, J. Klafter, G. Zumofen, in: [10.17], p.399

10.40 R. Kopelman: J. Stat. Phys. **42**,185 (1986); R. Kopelman, S. Parus, J. Prasad: Phys. Rev. Lett. **56**, 1742 (1986)

10.41 Similarly, the prefactors for (10.12,13) obtain by making the substitutions $t \to Kt/\ell_0^2$, $r \to K/\ell_0^2$, and $L \to L/\ell_0$

10.42 H.D. Bale, P.W. Schmidt: Phys. Rev. Lett. **53**, 596 (1984)

10.43 D.W. Schaefer, K.D. Keefer: Phys. Rev. Lett. **53**, 1383 (1984); S.K. Sinha, T. Freltoft, J. Kjems, in: [10.13], p.87

10.44 For recent reviews see J.E. Martin, A.J. Hurd: J. Appl. Cryst. **20**, 61 (1987); P.W. Schmidt: Makromol. Chemie (Macromol. Symp. Vol., 1988), in press

10.45 P. Pfeifer, in: *Multiple Scattering of Waves in Random Media and Random Rough Surfaces*, eds. V.K. Varadan, V.V. Varadan (Pennsylvania State University 1987), p.45

10.46 P. Pfeifer: to be published

10.47 For a tentative scattering law for self-affine surfaces see P.Z. Wong, J. Howard, J.S. Lin: Phys. Rev. Lett. 57, 637 (1986)

10.48 P. Pfeifer, P.W. Schmidt: Phys. Rev. Lett. 60, 1345 (1988)

10.49 J.K. Kjems, P. Schofield, in: [10.14], p.141

10.50 P. Pfeifer, A.L. Stella, F. Toigo, M.W. Cole: Europhys. Lett. 3, 717 (1987)

10.51 J.R. Banavar, M. Lipsicas, J.F. Willensen: Phys. Rev. B 32, 6066 (1985)
G. Jug: Chem. Phys. Lett. 131, 94 (1986)

10.52 J.P.Coulomb, M. Bienfait, P. Thorel: J. Physique 42, 293 (1981); Faraday Discuss. Chem. Soc. 81, 1 (1985)

10.53 W. Marshall, S.W. Lovesey: *Theory of Thermal Neutron Scattering* (Clarendon Press, Oxford 1971), Chapt.11

10.54 J.R. Banavar, J.F. Willemsen: Phys. Rev. B 30, 6778 (1984)

10.55 B. O'Shaughnessy, I. Procaccia: Phys. Rev. Lett. 54, 455 (1985)

10.56 M.W. Cole, N.S. Holter, P. Pfeifer: Phys. Rev. B 33, 8806 (1986)

10.57 W.A. Steele: *The Interaction of Gases with Solid Surfaces* (Pergamon, Oxford 1974)

10.58 For noteworthy retrospect see S. Brunauer: Langmuir 3, 3 (1987)

10.59 T.L. Hill: J. Chem. Phys. 14, 263 (1946)

10.60 J.J. Fripiat, L. Gatineau, H. Van Damme: Langmuir 2, 562 (1986)

10.61 P. Pfeifer, M. Obert: to be published

10.62 P. Pfeifer: to be published

10.63 P. Pfeifer, M.W. Cole: to be published; P. Pfeifer, M.W. Cole: Bull. Am. Phys. Soc. 32, 862 (1987)

10.64 R. Kubo: *Statistical Mechanics* (North-Holland, Amsterdam 1965)

10.65 M. Ma, B.I. Halperin, P.A. Lee: Phys. Rev. B 34, 3136 (1986)

10.66 P.B. Weichmann, M.E. Fisher: Phys. Rev. B 34, 7652 (1986)

10.67 A. Höhr, H.B. Neumann, P.W. Schmidt, P. Pfeifer, D. Avnir: Phys. Rev. B 38, 1462 (1988)

11. Critical Phenomena of Chemisorbed Atoms and Reconstruction — Revisited

Theodore L. Einstein

Department of Physics and Astronomy, University of Maryland
College Park, MD 20742, USA

In the theory of critical phenomena of lattice models, two dimensions holds special fascination for several reasons. In this low dimensionality, fluctuations are stonger, leading to more pronounced divergences near the transition and to grosser violations of mean field predictions. In contrast to three dimensions, there are several exactly-solved models, which provide invaluable benchmarks for understanding critical properties. One of the few places one can hope to find realizations of such models is at surfaces of materials, either by adsorbed atoms or by reconstruction [11.1] of the top layer(s). Regarding reconstruction, 2D does not mean planar: In many reconstructions, several layers near the surface are involved. All that is required for the transitions to be 2D is that displacements eventually decay exponentially as one goes deep into the bulk and that there be no new periodicity exhibited by the ordered state perpendicular to the surface, just parallel to it. On the other hand, we will not consider "ordinary" surface transitions, i.e., 2D critical behavior at the surface generated by a bulk, 3D transition [11.2].

In volume IV of this series [11.3] I reviewed such critical behavior for chemisorbed atoms. At that time, we at Maryland had just reported [11.4] exponents for p(2x2) O/Ni{111} which differed from those expected: Ising-like rather than 4-state Potts-like. When several explanations proved unsatisfying [11.3,5], we undertook a series of painstaking simulations of generic lattices with sizes comparable to those of defect-free regions on typical metal surfaces (~4000 sites). This finite size limits how close one can approach the transition; a key question, requiring explicit calculations, is whether one can in fact get close enough to gauge critical behavior. A goal was to anticipate with ideal data what good experiments would show. In the course of this work we learned many things useful for doing future experiments. These ideas have been published in four rather lengthy papers [11.6-9]. I am grateful to the editors for the second opportunity to present these findings in a unified and hopefully accessible format.

This chapter will begin with a brief review of important concepts. Ideally, the interested reader will look at the earlier review [11.3] or a general reference [11.10] since time forbids a complete recap here.

(Moreover, it may be interesting to find what we did not understand then!) Realistically, few will; thus, the following is intended to be self-contained. In Sect.11.2, I go through many of the findings from our analyses of simulations, stressing ideas that might guide experiments. Topics are highlighted at the start of most paragraphs to permit quick access to topics of special interest and to facilitate skimming. In Sect.11.3, I describe experimental progress since 1981. This section is lamentably shorter than I would have predicted in 1981, and also shorter than the detailed account of what might be expected, presented in the preceding section . Since the discussion is couched in terms of these expectations, readers who plowed through Sect.11.2 will be rewarded and those who did not, can look back. Section 11.4 presents my views on where the field is headed.

This review focuses on chemisorption and reconstruction, since reviews here in 1983 [11.11] and especially 1985 [11.12] emphasized the physisorption cases and since our lattice gas calculations seem more generally applicable. For a somewhat broader review, albeit for a very different audience, see [11.13]. Another somewhat more general and comprehensive review was recently written by *Unertl* [11.14].

11.1 Brief Recap

We specialize here on phase transitions of lattice gas models, typically studied by varying temperature T at fixed coverage. It is usually more convenient to use a reduced temperature $t \equiv (T-T_c)/T_c$. The ordered state at low T (t<0) leads to new sets of spots in a diffraction experiment at positons we shall denote $\{k_0\}$. If the lattice gas has N sites, the intensity of these new spots, as $t \rightarrow 0^-$, has the form (for N large)

$$I(k_0,T) \propto N^2(-t)^{2\beta} . \tag{11.1}$$

Here β is the exponent associated with long-range order, typically called the magnetization exponent based on the analogy with magnetic systems. Near T_c, there will be strong fluctuations, leading to additional, critical scattering with intensity.

$$I(k_0,T) = N\chi_\pm \left|t\right|^{-\gamma} \text{ for } t \gtrless 0 , \tag{11.2}$$

in analogy with the spin susceptibility, χ, in magnetic systems. The *inverse* width of the critical scattering is associated with the correlation length of these fluctuations:

$$\xi(T) = \xi_\pm \left|t\right|^{-\nu} \text{ for } t \gtrless 0 . \tag{11.3}$$

In addition, energy fluctuations lead to a divergence of the specific heat, having the form

$$C(T) = N \, C_{\pm} \left| t \right|^{-\alpha} + \text{const.} \tag{11.4}$$

For physisorption systems, particularly with grafoil substrates (so that there are many layers of adsorbate), this property can be measured directly sensitively; for chemisorption systems, such measurements are not feasible since the substrate contribution masks that of the surface. Contrary to a statement in [11.3], however, the specific heat can be measured for chemisorption or reconstruction systems, using a scattering experiment, by low-resolution monitoring of the energy-like anomaly in correlation functions [11.7], to be described below (Sect.11.3.3).

In principle, one should allow for different exponents α', γ', ν' below T_c. As a consequence of the scaling hypothesis (which says that ξ is the only important length in the critical regime), these primed exponents are expected to be the same as their unprimed counterparts. Hyperscaling further interrelates the exponents, so that only two are predicted to be independent. In 2D, these have the simple form

$$\alpha = 2 - 2\nu \quad \text{and} \quad \gamma = 2(\nu - \beta) . \tag{11.5}$$

The fascination of the exponents is that they are "universal", determined only by the spatial dimension (here 2) and the symmetry of the ordered state relative to the higher-T disordered state. They do *not* depend on any of the microscopic interactions of the system, which in contrast do determine T_c. In addition, the "critical amplitude ratios", C_+/C_-, χ_+/χ_-, and ξ_+/ξ_-, are universal quantities, although the individual amplitudes themselves are non-universal. Moreover, the correlation length ξ is not uniquely specified - nor, for that matter, is the order parameter and its fluctuations χ - and the critical amplitude ratio depends on which definition is used.

To predict the universality class of the disordering of a particular ordered state, one compares the expansion of the Landau-Ginsburg-Wilson Hamiltonian to that of well-studied magnetic systems [11.3,15]. In this framework, continuous transitions of adsorbed atoms are predicted to occur in just four distinct classes (see Table 11.1). Three can be described by q-state Potts models [11.16], with q = 2, 3, 4. These models can be described by the Hamiltonian

$$H = -J \sum_{\langle i,j \rangle} \delta_{s_i,s_j} \tag{11.6}$$

where the subscript denotes lattice sites, and the sum is over nearest neighbors. In essence, in this model each site can have one of q colors, called S_i from analogy to a spin (pointing to the q vertices of a regular polyhedron in a q-1 dimensional spin space). If neighboring sites are the same color, there is an (attractive) interaction. Otherwise, there is

Table 11.1. 2D Universality Classes, Exponents, and Examples

Universality Class	α	β	γ	ν	Sample pattern (full list in [11.3,15] + Example)
Ising (2-state Potts)	0(log)	1/8	7/4	1	c(2x2)/square, rectangular Cl/Ag{100}
X-Y with cubic anisotropy		non-universal!			(2x1)/centered rectangular: O/W{110} (2x1)/square: W{100}
3-state Potts	1/3	1/9	13/9	5/6	($\sqrt{3}$x$\sqrt{3}$)/triangular noble gas/graphite
4-state Potts	2/3	1/12	7/6	2/3	p(2x2)/triangular O/Ru{0001} [graphitic] (2x2)/honeycomb H/Ni{111}?
1st order	1	0	1	1/2	p(2x2)/honeycomb
chiral 3-st. Potts	1/3	1/9	13/9	5/6	(3x1)/centered rectangular

either no interaction or (with slight modification of H [11.16]) a repulsion (such that with random "coloration" the interaction is zero). The origin of the correspondence between the Potts and the lattice gas models does not arise from any microscopic mapping of one Hamiltonian into another. Instead, it has to do with the interaction between fluctuating ordered domains. Thus, for example, the ($\sqrt{3}$x$\sqrt{3}$) R 30° on a triangular lattice, which has three degenerate ground states (say A, B, C), is in the 3-state Potts universality class. Casual inspection [Ref.11.3, Fig.4] shows that the boundary between like domains will have one energy while the boundary between any pair of unlike domains will have a different energy; of course, it is the free energy rather than just the energy that matters. (Furthermore, closer inspection shows that there is a triaxial chirality involved, in that A-B, B-C, and C-A boundaries are different from B-A, C-B, and A-C. But this difference turns out not to change critical behavior [11.17,18]; it is "irrelevant" in a renormalization group sense.) The 2-state Potts model is just the familiar Ising model. This model applies to systems in which there are just two degenerate ordered states, i.e., each atom faces an "either-or" choice in some sense.

According to conventional Landau theory, the overlayer-induced spots must occur at high-symmetry positions of the surface Brillouin

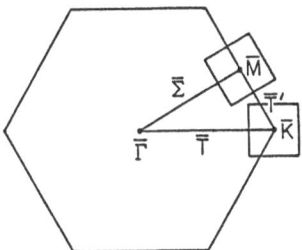

Fig.11.1 Surface Brillouin zone for a triangular substrate. Ordering with $(\sqrt{3} \times \sqrt{3})$ R30° and p(2x2) symmetry correspond to peaks at $\bar{K}$ and $\bar{M}$, respectively. These two positions for k_0 are the only ones at which, according to conventional Landau theory, continuous disordering can occur. The structure factor was computed along the high-sysmmetry lines and within the appropriate squares (adapted from [11.6])

zone (SBZ) in order for the transition to be continuous: k_0 in this case is at a corner and in other cases may be at a midpoint of an edge (Fig.11.1). Thus, e.g., the (7x7) reconstruction of Si{111} [11.19], c(4x2) CO/Ni{111} [11.20], and the c(2x8) reconstruction of Ge{111} [11.21] are expected to disorder discontinuously (first-order) if they do not go to an incommensurate ordered state (which apparently occurs for Ge [11.21]). As usual in 2D, one must be wary of mean-field like arguments. (E.g., it is well-known that conventional Landau analysis predicts incorrectly that 3- and 4-state Potts models do not disorder continuously because their expansions have cubic terms [11.15]). Below, we will consider the (3x1) on a centered rectangular lattice, which obviously violates this high-symmetry criterion but disorders continuously to an incommensurate disordered state (as required since k_0 is not a point of high symmetry). Finally, there is the possibility of observing critical-like behavior at a first-order transition. As an example, we have studied the p(2x2) on a honeycomb lattice. (For q > 4, Potts models have first-order transitions; this p(2x2) honeycomb has 8 degenerate ordered states.) For further details on Landau classification, see [11.3, 15].

Corrections to the pure power-law behavior of (11.1-4) become évident as one gets farther from T_c. Specifically, one speaks of corrections to scaling of the form

$$|t|^{-\lambda} \left[1 + D_{1,\pm} |t|^{\Delta_1} + D_{2,\pm} |t|^{\Delta_2} + ... \right]. \tag{11.7}$$

At least, there will be analytic corrections ($\Delta_1 = 1$, $\Delta_2 = 2,...$), due e.g., to ambiguities in defining t, as in the special case of the Ising model [11.22]. (If the definition of t happens to correspond to a "pure field", the leading correction may be of even higher order, e.g., $\Delta_1 = 2$ for the Ising model [11.23]). More often, there are weaker ("confluent") singularities, i.e., $\Delta_1 < 1$. These exponents are universal, but the amplitudes, the D's, are not, and some may vanish. In special cases, the leading correction can be logarithmic, eventually dwarfing "pure" behavior close enough to T_c. We have argued [11.24] that the effect of multiple scattering - sometimes claimed to obscure critical behavior in

LEED [11.25] - is simply to alter (not necessarily increase) the size of the D's. (For light adatoms, the atomic scattering function has a pronounced minimum at ~90° [11.26], so with normal incidence one can minimize scattering within the overlayer [11.4]. Hence, by varying the incident angle, one can check for such effects.)

The critical exponents described above are defined in terms of fixed field (here chemical potential) rather than fixed coverage, typically used in chemisorption experiments. If this "isochore" passes near a maximum in the phase diagram, there is no problem. At coverages away from such "saturation" values, one encounters Fisher renormalization [11.27], by which $\beta \rightarrow \beta/(1-\alpha)$ and similarly for γ and ν, while $\alpha \rightarrow -\alpha/(1-\alpha)$, see (11.15 and 16). Thus, the transition is smoothed.

In light of the importance attached to ξ as the only important length in the critical region, it is fruitful to catalog all significant lengths in discussing phase transitions. The characteristic size of defect-free regions of the substrate, L_s, limits the growth of ξ. Once ξ becomes comparable to L_s, we expect the power-law divergences to break down. Correspondingly, this criterion sets a lower limit $|t_{min}| \sim L_s^{-1/\nu}$ of closest approach to T_c; for smaller t, the data is contaminated by finite-size effects. With the notable exception of graphite, most substrates studied to date have L_s on the order of 100's of Å. Another length due to the instrument used for measurements is the characteristic distance L_i over which correlations are sampled coherently. In LEED, L_i is typically the transfer width of the instrument response function [11.28]. For conventional LEED systems $L_i \gtrsim 100$ Å, but high-resolution systems can achieve L_i's nearly an order of magnitude larger [11.29]. Atom scattering (Chap.14 and [11.30]) can also attain L_i's up to ~500 Å, but electrons tend to be easier to use. In the critical region, when $\xi \ll L_i$, one has the diffraction limit, while in the opposite limit one probes short-range order and finds energy-like anomalies. The corrections to scaling, which enter away from T_c when ξ is small, may be associated with small length scales such as the lattice spacing or the electron inelastic mean-free path (if one is considering multiple scattering).

More strikingly, there may be weak but relevant fields which lower the symmetry of the Hamiltonian. Close to T_c one then observes scaling associated with the lower symmetry system, but when ξ decreases to a size comparable to the characteristic length associated with the field, there is crossover to scaling appropriate to the higher-symmetry system. Such behavior was indeed invoked [11.31] to try to explain our O/Ni{111} exponents. The form of the crossover function is generally unknown, but in some cases the general form has been computed using renormalization group methods [11.32].

11.2 Results from Computed Structure Factors

11.2.1 General Features

As described in the introduction, we used Monte Carlo methods to produce ideal "data" for generic lattice gases [11.6-9]. Since much exact work existed already for Ising models, we considered models which had $(\sqrt{3} \times \sqrt{3})$ R 30° and p(2x2) ordered structures on triangular lattices [11.6,7]. We used a lattice with 3,888 sites (comparable to the size of defect-free regions on good metal surfaces) a hexagonal perimenter, and periodic boundary conditions. We fixed the chemical potential such that the coverage near T_c was close to saturation. For the $\sqrt{3} \times \sqrt{3}$ overlayer, the Hamiltonian was the simplest possible:

$$H = E_1 \sum_{\langle ij \rangle_1} n(r_i)\, n(r_j) \tag{11.8}$$

where $n(r_i) = 0, 1$ is the occupancy of the lattice site at r_i and the nearest neighbor repulsion E_1 was fixed at 1 to set the energy scale. For the p(2x2), a second neighbor repulsion E_2 is also needed; it was set to $E_1/2$. Further details are given in [11.6].

We computed the structure factor

$$S(k,T) = \left\langle \left| \sum_j n(r_j)\, \exp[i(k+k_0)\cdot r_j] \right|^2 \right\rangle \tag{11.9}$$

which is just proportional to the intensity one would measure in the kinematic limit. A series of contour plots in [11.6] nicely illustrates the narrowing and increase in intensity of S(k) as one approaches a continuous phase transition. The novice is encouraged to look at the two pages of plots (Figs.2 and 3 of [11.6]). The best way to analyze the wealth of data - not yet applied to any adsorption experiment - is to take cognizance of the phenomenologically predicted scaling form [11.10]

$$S(k,T) = a_1 |t|^{-\gamma} X_\pm \left(a_2 |t|^{-\nu} |k| \right) \tag{11.10}$$

for small $|t|$ and $|k|$; $X_\pm(y)$ are universal functions of $y = k\xi$, while a_1 and a_2 are system-dependent constants. Figure 11.2 shows such a plot for the $\sqrt{3} \times \sqrt{3}$ system with k running from the corner of the SBZ ($\bar{K}$ of Fig.11.1) toward the center ($\bar{\Gamma}$) using 3-state Potts values of γ and ν and with T_c picked to give the best scaling. To within statistical error, k's up to half way to the center can be made to scale. Multiple scatter-

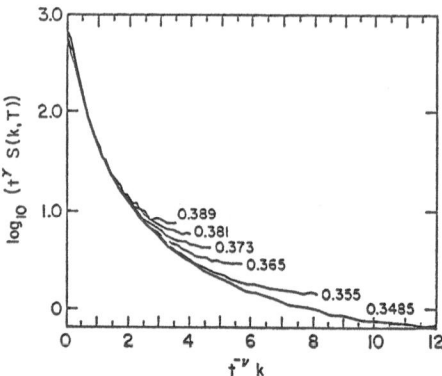

Fig. 11.2 Structure factor above T_c scaled according to (11.10) for the $\sqrt{3} \times \sqrt{3}$ case, assuming γ and ν of the 3-state Potts model and $T_c/E_1 = 0.338$, with k along $\bar{T}$, starting from $\bar{K}$ (from [11.6])

ing could decrease (or increase) this cut-off. Once the correlation length becomes comparable to the size of the system L_s, the universal behavior of (11.10) as well as (11.1-4) break down due to finite-size effects. Here, data within about 2% of T_c cannot be made to scale. Recall that since the amplitude of ξ is not universal, t_{min} will be the value of t, for a given size lattice, below which finite-size breakdown sets in.

11.2.2 Diffraction-Limit Results

Finding T_c: The more conventional analysis involves separating $S(k, T)$ into various measurables and using log-log plots (vs t) to estimate exponents. Adopting as the criterion for T_c that value which maximizes the linearity of the log-log plot over some thermal range, careful analysis in [11.6] shows that the convergence of T_c is faster than that of the effective exponents, so that the lack of knowledge of T_c does not unduly affect the ability to estimate exponents. In Sect. 11.2.3, we will discuss an alternate way to estimate T_c using a point of inflection. In this experiment one finds scaling over less than two decades of reduced temperature, usually only about one, so that it is hard to get a sense of convergence.

Finding χ, ξ: We first consider the diffraction limit [11.10] $k\xi \ll 1$ or $\xi \ll L_i$. In this limit, S separates into a contribution from long-range order below T_c and critical scattering peaking at T_c. In the following, $\chi(T)$ and $\xi(T)$ are determined as those values which, at each T, minimize a least-squares fit of $\{S(k, T)\}$ to a Lorentzian form, $\chi/(1 + \xi^2|k|^2)$. Near T_c, the fit is insensitive to the large-k cutoff. (Since this definition ignores possibly sizable non-Lorentzian behavior near T_c, the χ and ξ differ somewhat from conventional definitions but nonetheless have the same critical properties.) Below T_c one must also

314

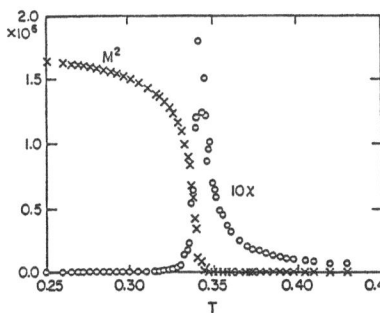

Fig.11.3 Temperature dependence of χ and of M^2 for the $\sqrt{3}\times\sqrt{3}$ case. Note the large critical amplitude ratio of χ above and beneath T_c; experimentally deducing exponents for critical scattering below T_c is difficult if not impossible. Note also the different dependence on the number of scatterers N, indicated in eqs.(11.1,2) (from [11.6])

worry about the δ-function contribution from long-range order, called M^2 in Fig.11.3 based again on the analogy with magnetic systems. With our periodic boundary conditions, this contribution only occurs at k_0, i.e., $k = 0$, but in an experiment (or simulation with other boundary conditions) this term would be spread over a width proportional to the inverse size, i.e., $N^{-1/2}$; this neighborhood of k_0 would have to be excluded from the least-squares fitting. Once T_c is estimated, (above T_c) one can include data at $k = 0$ (or $k \simeq 0$ in an experiment) to decrease noise. While one can imagine other schemes to extract χ and ξ, one must be careful not to weight large-k data excessively. For example, if one takes ξ^2 from the least-squares fit of $S^{-1}(k)$ vs k^2, $k\xi$ is not small, and the "wrong" critical behavior is observed [11.6].

Thermal cutoffs: The divergence of ξ will be limited by the size of defect-free regions, and when ξ becomes comparable to this size, scaling breaks down. Since in these experiments (or simulations) one has little data to spare, one should not remove a larger range around T_c than necessary. The procedure suggested in [11.3] - analyzing all the data that appear to scale - is dangerous in that one might choose T_c so that data near it scale when they really should not. One is particularly prone to this poor choice when the thermal range of data is small. An objective criterion is to exclude data which are more than some specific fraction of the maximum correlation length found. In our case, we used the fraction 1/2, but with non-periodic boundary conditions, the maximum ξ will be smaller, and this criterion might cause one to discard more data than necessary. Since this procedure involves non-universal features, it is not possible to give strict prescriptions. At the other limit, far from T_c, ξ decreases to the size of other lengths in the problem (e.g. the lattice spacing) and corrections to scaling become large. Again, the experimenter must exercise judgement. In our case, we could sometimes use data as far as 25% away from T_c in log-log plots.

Definition of t: Above T_c, we redefined [11.33] the reduced temperature t as $1 - (T_c/T)$. In some cases [11.33] this definition has led to smaller corrections to scaling - the two definitions differ by a term of order t^2. The difference in the exponents from the two definitions is on the order of the statistical errors. Some experts prefer this alternative defintion since it corresponds more closely to the form appearing in renormalization group calculations [11.34].

Results and statistical errors, $T < T_c$: In Table 11.2 are listed under "effective" the results of the log-log fits for $0.003 < t < 0.25$; the pure exponents are taken from Table 11.1. While T_c is non-universal, it was corroborated using several other more sophisticated numerical techniques. The level of accuracy is seen to be of order 10%, good enough to distinguish between universality classes but not much more. The error bars, of order 10% again, were based on standard error propagation techniques starting from 2-5% errors in the $S(k,T)$. Similar errors are to be expected from good experiments. Given the linearity of the fits (Fig.11.4), the estimate of the exponents did not change beyond the error bars when either upper or lower thermal cutoff was varied. The estimates are quite sensitive to the estimate of T_c: a 12% variation in the estimated T_c produced a 17% variation in the deduced γ.

Problems for $T < T_C$: Below T_c the contribution due to long-range order, called M^2, can be extracted from $S(k,T)$ by subtracting

Table 11.2. Summary of Computed Exponents

	$\sqrt{3}\times\sqrt{3}$			p(2x2)	
	Pure	Effective		Pure	Effective
kT_c/E_1	-	0.338±0.002		-	0.344±0.002
γ	1.44	1.25±0.07		1.17	1.13±0.06
ν	0.83	0.77±0.05		0.67	0.70±0.09
β (Δ=1)	0.111	0.104±0.010	(Δ=1)	0.083	0.083±0.009
(Δ=1)(min χ^2: $T_c = 0.335$)		0.087±0.010	(Δ=1) $T_c = 0.343$		0.078
no Δ, max linearity $T_c = 0.334$		0.078±0.010	no Δ $T_c = 0.343$		0.065
Δ=2/3 $T_c = 0.336$		0.111±0.019	$(\log\lvert t\rvert)^{1/8}$ $T_c = 0.343$		0.077±0.020

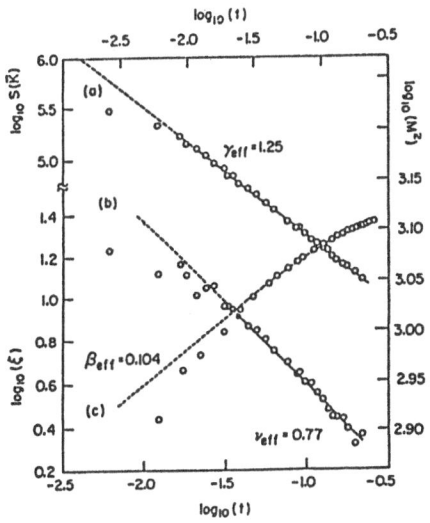

Fig.11.4 Log-log plots for the $\sqrt{3}\times\sqrt{3}$ case. a) χ above T_c, b) ξ above T_c of M^2 below T_c. The thermal scale at the top applies to c), that at the bottom to a) and b). The fit in c) includes a non-linear term in the fitting function (from Ref.11.13, adapted from [11.6])

$\lim\limits_{k\to 0}$ S(k, T), based on some fit (Fig.11.3). For large system size - larger than here - (11.1 and 2) suggest this subtraction is unnecessary, since the critical scattering is of order $1/N$ compared to M^2. Far from T_c, χ is small and need not be removed. In a plot of log M^2 vs log $|t|$, depicted in Fig.11.4, nonlinearities are clearly present; no choice of T_c will remove them. Evidently, corrections to scaling are larger below T_c than above. Since ratios of corresponding amplitudes of these corrections above and below T_c [e.g., D_{1+}/D_{1-} in (11.7)] are universal, we expect this behavior to be general. If larger defect-free regions were achievable, one could simply use data closer to T_c and discard more data at large $|t|$. Since this is not the case, one must cope with the corrections. Without additional information, the simplest approach is to assume analytic corrections ($\Delta_1=1$) and so fit M^2 to $(|t|+Dt^2)^{2\beta}$, but with T_c at the value determined above the transition. The results are listed in Table 11.2. For the $\sqrt{3}\times\sqrt{3}$ the β_{eff} is 6% too low, but within error bars, while for the p(2x2) there is fortuitous agreement. If, instead, T_c is freed and fit along with the slope, then in both cases the estimate of T_c declines (by 1% and 1/3%, respectively) and the estimate of β drops considerably (by 16% and 6%, respectively). Thus, in fitting exponents below T_c, without knowledge of T_c, one does better by using the estimate of T_c found from critical scattering *above* T_c. Lyle Roelofs had realized this already while analyzing data for O/Ni{111} [11.4], but we thought it was due to limited data below T_c. For the two models under study, Δ_1 is known in advance from theoretical work. For the 3-state Potts model, $\Delta_1 = 2/3$ [11.35]; with this ansatz, we estimate $\beta = 0.111 \pm 0.019$ (fortuitous exact agreement) and

$T_c = 0.336 \pm 0.002$. For the 4-state Potts model, the leading correction is logarithmic [11.16, 36], a general feature of classes of models at the last value of some parameter at which the transition is continuous. (For $q > 4$, Potts models have first-order transitions.) As seen in Table 11.2, fitting [11.36] to $|t|^{2\beta} \cdot (1+D[\ell n|t|]^{1/8})$ does not improve agreement of β with the pure value. For the Ising model, remarkably, there are no low-order non-analytic corrections to scaling [11.22]. Note that the amplitudes of the corrections to scaling are non-universal, and may even vanish. This happens, for instance, in the Baxter-Wu model, a member of the 4-state Potts class [11.16]. Finally, we tabulate the effective β when no corrections are included for our two systems; in both cases, the exponent is much too low.

Critical scattering below T_c unobservable: In our model systems, we can similarly compute the exponents γ' and ν' associated with critical scattering in the ordered regime. The former in particular shows strong nonlinearities in log-log plots. However, the amplitude of this critical scattering is quite small. We computed the critical amplitude ratio χ_+/χ_- as about 40 for both these models and for their Potts analogues. (For the Ising model, it is known to be nearly 38 [11.37]). For the correlation length, the corresponding ratio is about 4 [11.37] (somewhat lower for the $\sqrt{3}x\sqrt{3}$). Intuitively, one is tempted to use these ratios to interpret the more rapid onset of corrections to scaling below T_c: the (diverging) correlation length and susceptibility which manifest the critical region decrease much more rapidly below T_c than above as $|t|$ increases.

Asymmetries: Lattice symmetry (Fig. 11.1) dictates that the contours of $S(k,t)$ have only 3-fold and 2-fold symmetry, respectively, in our two systems, much lower than the analogous Potts systems. In a small $|t|$ expansion of $S(k,t)$ (around k_0), one can write

$$S(k,T) = \chi(T)\left[\left[1 - \xi(T)^2 |k|^2\right] + b(T)f(k_r, k_a) + ...\right], \qquad (11.11)$$

where k_r, k_a are the radial and azimuthal components of k (with respect to the center of the SBZ); $f = k_r^3 - 3k_a^2 k_r$ for the $\sqrt{3}x\sqrt{3}$ and $f = k_a^2 - k_r^2$ for the p(2x2). These definitions of χ and ξ differ slightly from those used earlier, but have the same critical properties. Analysis of $b(T)$, as explored in [11.6], gives interesting insights into the asymmetries in domain wall energies which complicate these lattice gases compared to their Potts analogues. With present system sizes and resolution, we do not believe these features can be well measured, but this prospect does offer a keen challenge to the experimentalist.

11.2.3 Energy-Like Limit

In the other limit, $k\xi \gg 1$, $X_{\pm}(y)$ of (11.10) also has a simple expansion [11.10] known to experts in critical phenomena for nearly two decades [11.38] but not applied to surface problems until recently:

$$X_{\pm}(y) = C_1 y^{-\gamma/\nu} [1 + C_2^{\pm} y^{-(1-\alpha)/\nu} \pm C_3 y^{-1/\nu} ...] . \qquad (11.12)$$

The leading term cancels the susceptibility prefactor of (11.10); the first and third terms are analytic, while the second has an energy-like anomaly. (Its first derivative is like a specific heat, while its integral is the [singular part of the] free energy.) Note that for an infinte system with $k \neq 0$, $k\xi \gg 1$ can always be achieved if one gets close enough to T_c; for $k \neq 0$, $S(k, T_c)$ must have an energy-like anomaly. This behavior is intimately related to the observation that two-site or multisite correlation functions, i.e., $\langle n(r)n(r+R) \rangle$ or $\langle n\ n\ n \rangle$, etc., as functions of T, have energy-like anomalies near T_c: $\mathcal{C}_1 \mp \mathcal{C}_2 |t|^{1-\alpha}$; more precisely, it is combinations of such correlation functions that are insensitive to the phase of the order-parameter, such as $\Sigma_r \langle n(r)n(r+R) \rangle$. Recalling the lattice gas Hamiltonian (11.8), we see that these correlation functions are, up to multiplicative constants, the constituents of the energy; this is the basis of the expansion in (11.12). (See [11.7,39] for a more formal derivation). While this discussion suggests that α might be measured in photoemission, core shifts, etc. [11.40], in general, the shifts seem to be too small and the precision inadequate.

LEED integrated intensity: If one measures the integrated intensity in LEED, these effects should be evident:

$$I(k_I, T) = \int_0^{k_I} dk\ k \int d\theta_k\ S(k, T)$$

$$\sim A_0 \mp B_{\pm} |t|^{1-\alpha} - A_1 t... \quad [T \gtrless T_c] \qquad (11.13)$$

The three coefficients depend on k_I; the A's come from the analytic part. The right side is valid only when k_I is "large enough", i.e., close to T_c and for moderately large integration radius. Again, the thresholds are non-universal, so our numerical studies give suggestive rather then definitive values. Once ξ grows comparable to the size of the defect-free regions, there will again be finite-size rounding; in this case that means that $I(k_I, T)$ will be analytic over a small neighborhood near T_c, which should be removed from fits. This behavior is illustrated in Fig.11.5. (The range of t over which this smearing occurs need not be the same as in the diffraction limit, since a different definition of ξ may be involved, with a different $\xi_{\pm}$.) At the other limit of the scaling range, there is no evidence that corrections to scaling

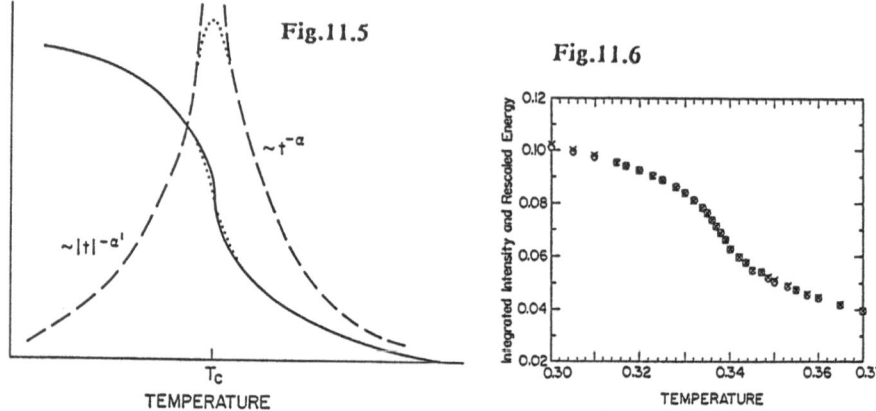

Fig.11.5 Schematic of the energy-like anomaly associated with positive α, as given in (11.13), seen by monitoring the integrated intensity around an overlayer-induced, extra spot in a diffraction experiment (solid line). The derivative with respect to temperature (dashed line) diverges like the specific heat, illustrating that the point of inflection of the solid curve offers a good estimate of T_c. The dotted lines indicate the finite-size rounding which occurs in real systems when the correlation length ξ exceeds the size of the defect-free regions L_s. Far away from T_c, when ξ << L_i, the solid curve is given by (11.1) [(11.2)] below [above] T_c, plus presumably large corrections to scaling (from [11.41])

Fig.11.6 Illustration of eq. (11.14): Structure factor integrated over 2.3% of the surface Brillouin zone [circles] vs. T, plotted with the rescaled energy [x's] for the √3x√3 case. The energy like behavior of the integrated structure factor near T_c is evident (from [11.7])

enter more quickly below T_c, although we have not considered this problem carefully.

Illustration of energy-like behavior. The key point to verify is that for adequate k_I, the intensity behaves like the energy:

$$I(k_I,T) = -u(k_I) E(T) + w(k_I) \qquad (11.14)$$

As k_I is increased, the largest $|t|$ at which (11.13) holds increases, but the size of $B_{\pm}/A_0$ decreases. Figure 11.6 coplots the left and right sides of (11.13) for k_I ~ 5/36 of the distance from the corner of the SBZ (k_0), to the SBZ center, i.e., 2.3% of the SBZ area. (The parameters u and w were determined by a least-squares fit.) For this temperature range and k_I, $I(k_I,T)$ gives the same information about the critical behavior as the [internal] energy, or for that matter, its directly measurable derivative, the specific heat. (This correspondence continues to hold over the finite-size-rounded region.) An important corollary is that *the point of inflection of $I(k_I,T)$ corresponds to the peak of the specific heat* [11.7,41]; this fact justifies the common LEED practice [11.42] of using the point of inflection as a "best estimate" of T_c. For the simple Ising model, we have, in fact, determined the A_i, $B_{\pm}$ exactly [11.7].

320

As a further test, we determined the effective α using the right side of (11.13) to fit $I(k_I, T)$ as well as the energy. For Potts models with continuous transitions, duality arguments show $B_+ = B_-$ [11.45] (and scaling dictates $\alpha' = \alpha$), reducing the number of fitting parameters. We also tried fits with $A_1 = 0$ (no linear term). Within statistical errors, similar values were found for fits of E and of $I(k_I, T)$ once the area was 2.3% of the SBZ and certainly once it was 4.6%. To gauge the finite-size-rounded region, we decreased the lower thermal cut-off until the effective α changed significantly; this change occurred below $|t| \sim 0.02$. The effective α decreased considerably when A_1 was included, giving better agreement with the pure value for the $\sqrt{3} \times \sqrt{3}$ but poorer agreement for the p(2x2)! To restate the key point, however: if k_I is large enough, the problems encountered in fitting $I(k_I, T)$ are no different from those in fitting the specific heat from, say, a direct calorimetric measurement [11.44, 45] (as performed on physisorption but not chemisorption systems) or from direct Monte Carlo calculations [11.46].

To summarize, there are several advantages to this approach. 1) High resolution and cumbersome deconvolution of an instrument response function are not required, so the measurement is comparatively easy. 2) Multiple scattering is intrinsically not a complication. 3) Values of α differ considerably between universality classes (especially in contrast to β). Some caveats are: 1) for Ising systems ($\alpha=0$ [logarithmic]) or first-order ($\alpha=1$), the energy anomaly is hard to distinguish from the analytic background. Indeed, for this reason, this method was not used much in 3D, where α is usually small [11.47]. When it is not certain *a priori* whether the disordering is Ising or first-order, this method does not really clarify matters. One way to proceed is to use a coverage well away from saturation. In the Ising case, one expects the logarithmic singularity to become a cusp. If the transition is first-order, one expects a broad smear characteristic of a coexistence region [1.48]. 2) If the thermal fitting range is small and too close to T_c, one might interpret the analytic behavior as evidence of an Ising transition. 3) Given the sensitivity of fits, it is risky to rely on a single exponent to attribute critical behavior to a particular universality class.

11.2.4 Melting to an Incommensurate Disordered Phase

Incommensurate overlayers are characterized by $S(k, T)$ peaking at positions in reciprocal space not simply related to the primitive lattice vectors (i.e., the overlayers' primitive vectors are not simply related to those of the substrate). This behavior says nothing about whether the adatoms are in registry with the substrate as in strong chemisorption, or whether a continuum of adsorption sites are occupied, as in physisorption (Chap.2 and [11.49]) or crystal growth (see, for example

Fig.11.7 a) The (3x1) phase on a centered rectangular lattice, with the interactions producing the ordered overlayer. b) The surface Brillouin zone of this substrate, along with the lines along which the structure factor was computed. Here k_0 is Δ_c, a point of low symmetry. From conventional Landau theory, one would (mistakenly) not expect this overlayer to disorder continuously (from [11.6])

[11.50]). On the other hand, it is possible to distinguish between a "floating" phase, in which there are algebraic decay of correlations and so novel behavior of $S(k,T)$ [11.51] and a disordered phase, for which decay is still exponential and the analysis of Sect.11.2.2 should apply.

We have studied an example of a transition to the latter by a (3x1) on a centered rectangular lattice. This pattern is illustrated in Fig.11.7a, with the pairwise interactions depicted. In particular, we focus on the case $E_1 = \infty$, $E_2 = 1 = -E_3$, which was explored by *Huse* [11.52] using existing exact results of *Baxter* [11.53]. In the corresponding SBZ, shown in Fig.11.7b, we see that the k_0 associated with order is not a high-symmetry point. (In fact, we had to concoct a Greek letter for it!) Simple considerations of interface energies for the 3 possible ordered domains (A, B, C) shows A|B, B|C, and C|A walls differ from B|A, C|B, and A|C walls. This "uniaxial chirality" is predicted [11.17] to alter the leading critical behavior, in contrast to the "triaxial chirality" [discussed after (11.6)] distinguishing the $\sqrt{3}$x$\sqrt{3}$ from the 3-state Potts model, which merely introduces a correction to the scaling term with a predicted exponent Δ (not to be confused with the SBZ labels!). A basic question is whether the (3x1) can disorder continuously in a single transition. As catalogued in [11.8,54], Landau-like arguments and several continuum studies suggest that a floating phase must intervene, but most numerical studies (including ours) support *Huse* and *Fisher's* [11.17] contention that such a single disordering transition can occur, but (perhaps) in a new universality class.

Scans of $S(k,T)$ near the transition are depicted in Fig.11.8. While the long-range-order spike always lies at $\bar{\Delta}_c$, the critical scattering peaks above $\bar{\Delta}_c$ (for our choice of chemical potential), except at T_c. Again, however, the critical scattering satisfies the scaling form of (11.10), but a novel feature is that the peak of $t^\gamma S(k,t)$ is not at $k = 0$. (Indeed, the shift of the peak, $q \sim |t|^{\bar{\beta}}$, with the prediction [11.17] $\bar{\beta} = \nu$.) The low symmetry of contours of $S(k,T)$ complicates the analysis; the reader is referred to [11.8] for details. One analysis yields $\gamma_{eff} =$

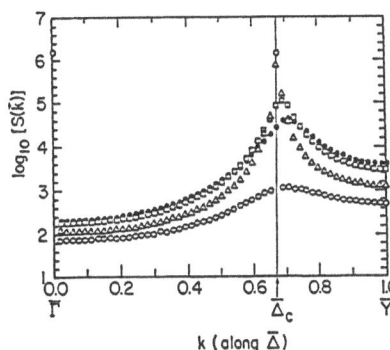

Fig.11.8 Plots of S(k,T) for a 36 x 108 lattice at temperatures about 5% below T_c (circles), at T_c (triangles), 5% above T_c (squares), and 10% above T_c (solid hexagons). Only at T_c does the critical scattering peak at Δ_c. For the chosen chemical potential, it peaks above it, for T ≷ T_c. The curves for T ⟩ T_c can be scaled as in (11.10), using 3-state Potts exponents (from [11.8])

1.35 ± 0.10 and ν_{eff} = 0.85 ± 0.10, consistent with 3-state Potts values (γ=13/9, ν=5/6). Another analysis gives γ_{eff} = 1.55 ± 0.12, and ν_{eff} = 0.87 ± 0.08 or 0.95 ± 0.09, depending on which direction in k-space is analyzed. Assuming 3-state Potts exponents, the critical amplitude ratios of χ and ξ are 160 ± 80 and 5.0 ± 1.7, much greater than for the $\sqrt{3}$x$\sqrt{3}$ case or for a 3-state Potts model. This behavior is consistent with predictions that the new universality class might have 3-state Potts exponents but a new scaling function. Similarly, in a low-resolution-type procedure following Sect.11.2.3, α_{eff} = 0.45 ± 0.05 (consistent with that analysis without a linear $A_1 t$ term) but B_+/B_- = 3.2 ± 1.3 (rather than unity). If there were an intervening floating phase, then the lower-T transition from the commensurate state should have the Pokrovsky-Talapov form [11.55,56] while the higher-T transition should show Kosterlitz-Thouless [11.55] behavior. In the former, the specific heat diverges as T_c is approached from below but remains finite during approach from above [11.56]. In the latter, the specific heat peaks above the transition in a non-divergent, non-singular way [11.57]. In such a floating-phase scenario, one would not expect that same sort of anomalous behavior above and below the transition. However, in this case one must be unusually wary of finite-size limitations (see [11.8]).

While [11.8] reports several other studies, one is of particular interest: an investigation of qξ as the transition is approached. In a commensurate phase, this limit should be zero since q "locks in", while in a floating phase it should be ∞ since correlations decay with a power law rather than exponentially. *Huse* and *Fisher* [11.17] predicted this limit to be a universal number, while our studies give a value dependent on chemical potential. Nonetheless, their prediction might be correct, and we - and a potential experimentalist - are just seeing a slow approach to the universal number due to the limited size of our system.

It would certainly be interesting to see an experimental study of such a transition. The system H/Fe{110} was thought to be an example

[11.58] until it was determined - after much misdirected theoretical work [11.59] - that H adsorbs in the quasi-threefold rather than bridge site [11.60]. More generally, incommensurate disordered phases are probably quite common in chemisorption systems, without being so identified. For example, we believe the region labelled "antiphase domains" in our study of O/Ni{111} [11.61,62] is such a case.

11.2.5 Critical Behavior at Temperature-Driven First-Order Transitions

In Table 11.1 there are, perhaps surprising to many, critical exponents for a first-order transition. For example, a leading experimentalist wrote that "critical scattering is an unambiguous hallmark of a second-order phase transition". To the contrary, there are cases in which a first-order phase transition will have all the qualitative features of a second-order one [11.63]. These cases can sometimes arise in the generic situation in which the terminal point of a line of first-order transitions (in, say, a temperature-chemical potential phase diagram) is for some reason not a usual [second-order] critical point. Such first-order transitions are called "temperature-driven" to distinguish them from "field-driven" transitions such as found in crossing the line of first-order transitions, where the more usual first-order discontinuous behavior is found. As a possible lattice gas illustration of this behavior, we have studied the p(2x2) on a honeycomb substrate. With its eight degenerate ground states, this model calls to mind the 8-state Potts model. (However, they are not in the same universality class; e.g., their order parameters have different dimensions.) Recall that for q > 4, q-state Potts models have first-order transitions [11.16], but these transitions are examples of the behavior just discussed. Indeed, Monte Carlo studies of 5- and 6-state Potts models showed disordering behavior barely distinguishable from that of the 4-state Potts model [11.64].

Critical behavior: For the p(2x2)/honeycomb, we proceed as before (see [11.9] for details) to compute S(k,T). Since this substrate is not Bravais, there are complications in k-space. The 6 integer-order spots $\bar{\Gamma}$ in the first ring are not equivalent to the specular spot $\bar{\Gamma}$. Also, the outer and inner half-integer, overlayer-induced spots are not connected by a reciprocal lattice vector and so are not equivalent [11.61]. Choosing as k_0 an inner such spot, we perform a log-log analysis and find $\gamma_{eff} = 0.86 \pm 0.20$, $\nu_{eff} = 0.55 \pm 0.15$, and (with a $\Delta = 1$ correction to scaling) $\beta_{eff} = 0.08 \pm 0.03$, compared to the discontinuity fixed-point values of 1, 1/2, and 0, respectively (see, for example, [11.65]). Using γ_{eff} and ν_{eff}, we find that S(k,T) satisfies the same scaling form of (11.10) as for the continuous transitions (Fig.11.9). (In a similar plot for the 8-state Potts model, we find best scaling for $\gamma = 0.85$ and $\nu =$

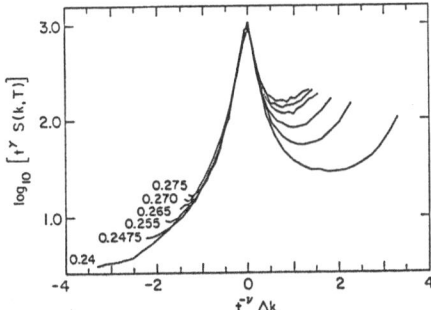

Fig.11.9 Structure γ factor for the p(2x2) on a honeycomb lattice, above T_c, scaled according to (11.10) with $\gamma = 0.86$, $\nu = 0.55$, and $T_c/E_1 = 0.233$. This temperature-driven first-order transition could not be readily distinguished from a continuous transition. The incipient divergences on the right are indicative of a divergent susceptibility at the outer integer-order spots, associated with the (vanishing) binding energy difference between the two Bravais triangular sublattices (from [11.9])

0.6.) The critical amplitude ratios of χ and ξ are much smaller than for the continuous transitions: 13^{-4}_{+11} and 1.1 ± 0.3, respectively.

Much of this work was motivated by attempts to understand our nemesis, p(2x2) O/Ni{111}. As discussed near the end of [11.3], *Schick* suggested an interpretation in terms of a Heisenberg model with corner cubic anisotropy [11.67]. Several aspects of the beam shape, both experimental and in these simulations, seem inconsistent with that explanation [11.9]. Indeed, in other searches of coupling-parameter space we were never able to find the Ising-like behavior of Schick's scenario [11.5].

Inequivalent beams: As noted above, the first ring of integer beams in this model are not equivalent to the specular beam. $S(\mathbf{k}, t)$ at these outer beams depends not only on the total coverage but also on the difference between the two triangular sublattices of the honeycomb net. This occupation difference is conjugate to a binding energy difference such as one would find between fcc-like and hcp-like sites on a close-packed (fcc or hcp) surface. As explored and depicted in [11.9], there is a susceptibility associated with this field leading to divergent $S(0, T_c)$ at the outer integer spots, even when, as here, the binding energy difference is zero. Unfortunately, it would be difficult to observe such critical fluctuations near an integer beam, but this type of behavior is more general.

In typical diffraction patterns there can be several inequivalent sets of adsorbate-induced spots. One can imagine (though not necessary be able to create in a physical system) a field (i.e., a binding energy difference) which favors one set over others. The susceptibility (or fluctuations) with respect to these fields can be different for each set of spots. If the field turns out to be relevant, then the different spots can have different exponents γ and β [11.10b]. (However, according to the scaling hypothesis [11.10], there is only one diverging length scale near T_c, so that the ξ's at each spot will have the same ν.) For the p(2x2)/honeyomb, the field associated with the difference between the inner and outer half-order spots is irrelevant, so that the two sets should differ only in amplitudes of corrections to scaling. However, for a p(2x2)/*square* [11.68], the field associated with the difference between (1/2, 1/2) and (1/2, 0) spots is relevant; *Enting* [11.69] predicts the universal relation $\beta_{1/2,1/2} = 4\beta_{1/2,0} - 1/4$. An analogous difference may have been seen for S/W{110} [11.70]. Even more generally, some diffraction features can in principle change discontinuously at a transition while others change continuously (in analogy to transitions where magnetization changes discontinuously while energy changes continuously, or vice versa).

11.2.6 Effects of Defects

In statistical mechanics it is common to make a somewhat arbitrary division of defects into two sorts: annealed and quenched. [11.71,10c] Annealed defects are in thermal equilibrium with the adatoms and, therefore, move around. The simplest case is vacancies, i.e., fixed coverage at a value below saturation. We have already noted that the result is smoothing of the transition, with Fisher renormalization of the exponents. The possibility of vacancies also leads to corrections to scaling, the form of which has been studied for the Ising model [11.23].

In the other limit, the defects are fixed in position [11.72]. In Monte Carlo simulations, this behavior is readily modeled by fixing a number of vacancies (or extra atoms) in random positions and not propagating them. By fixing the occupancy at certain positions, these quenched defects limit the growth of the correlation length ξ, thereby introducing a new length scale of order $N_D^{-1/2}$, where N_D is the defect density. The effect is essentially the same as having a finite-size system in a simulation, or a mean terrace width in an experiment. We have seen that below some $|t|_{min}$, singular behavior is no longer observed. Thus, we expect $|t|_{min} \simeq N_D^{1/2\nu}$ once the defects dominate, a relation we plan to verify numerically and experimentally.

11.3 Experimental Progress Since ISISS-1981

In spite of the many examples of ordered chemisorbed overlayers appearing in catalogues [11.73], the search has been disconcertingly difficult for relatively simple ones to which to apply theoretical concepts of critical phenomena [11.74]. A chemisorption system with a good $\sqrt{3}\times\sqrt{3}$ transition has proved particularly elusive; e.g., a recent study of Al/Si{111} found first order behavior [11.75]. (Such transitions are frequently observed in physisorption.) However, a recent careful study [11.45] of the specific heat of H/graphite with a.c. calorimetry shows striking sensitivity to the chemical potential (i.e., to where on the phase boundary the transition occurs).

11.3.1 4-State Potts Systems: O/Ru{0001}

Very recent reports of *Piercy* and *Pfnür's* LEED study of the disordering of p(2x2) O/Ru{0001} [11.76] are, thus, particularly heartening. The integrated intensity of an overlayer-induced spot was measured with a Faraday cup subtending 2.3% of the SBZ. First they determine T_c from the point of inflection as 754 ± 0.5 K. Then with this T_c, removing a Debye-Waller factor, using a linear term, and fixing $B_+ = B_-$ (duality) and $\alpha' = \alpha$ (scaling), they find $\alpha = 0.60 \pm 0.04$. The error bars are determined from variances resulting from changing T_c by ± 1 K and excluding a window around T_c of 0 to 0.01 T_c. If either of the two equalities are not forced, the results are not changed significantly. The upper cutoff was set at $|t| = 0.03$, when the spot width became comparable to the cup aperture. (Note that this criterion is not sure to apply in all cases; other length scales associated with corrections to scaling may introduce nonlinearities to the log-log plot for smaller $|t|$.) As in our simulations, α_{eff} is below the pure value, but here less so, perhaps due to the smaller t's in the fitting range. Proceeding to the diffraction limit, Piercy and Pfnür use a small Faraday cup to measure the peak intensity below T_c (determined above). The log-log plot vs t is linear from 0.002 to 0.2; surprisingly, no nonlinear term is needed and the slope gives $2\beta = 0.17^{+0.03}_{-0.001}$, in excellent agreement with the pure value 1/6. Above T_c, they fit to a delta function plus Lorentzian, convoluting with a low T_c profile to represent the instrument response function. From log-log plots of χ and ξ vs t, they determine $\gamma = 1.08 \pm 0.07$ and $\nu = 0.68 \pm 0.03$ (compared with pure values of 7/6 and 2/3). The data are linear out to near 0.1, the largest t plotted. Finite-size rounding sets in around t = -0.004, corroborated by the non-vanishing of the delta-function component for t < 0.004.

At $\theta = 1/2$, preliminary measurements are made of a (2x2) phase composed of three rotationally related domains of p(2x1). This model is expected to be the universality class of the Heisenberg model with a face-centered anisotropy, known to have a first-order transition [11.77]. Nonetheless, they find Ising-like exponents: $\alpha = 0.01 \pm 0.02$ and $\beta = 0.13 \pm 0.02$. While one might be able to reinterpret the integrated intensity to mean $\alpha \simeq 1$, consistent with first-order critical scattering, the large value of β is mysterious.

11.3.2 Ising Systems

Even the search for a good Ising system has taken longer than was expected at ISISS-1981. An obvious category to study are overlayers with c(2x2) order on square lattices. The first attempt at Maryland, O/Ni{100} [11.78], proved unsatisfactory since upon disordering, the adatoms dissolve into the substrate, introducing uncontrollable bulk degrees of freedom. The next study was Cl/Ag{100} [11.79]. As the c(2x2) is heated to disordering, chlorine desorbs irreversibly, presumably due to very strong nearest neighbor repulsion. In this case, however, one can observe a reversible transition to the ordered state by increasing θ at fixed T. (For O/Ni{100}, there is a coexistence regime with p(2x2) on the low-coverage side of the pure c(2x2) phase.) To extract β, $\ell n I/\theta^2$ was plotted vs $\ell n(\theta-\theta_c)/(1/2 -\theta_c)$, where θ_c is the coverage at the transition. The θ^{-2} was included to isolate the critical variation from the non-singular dependence on density, while the normalization of the coverage was chosen so that the fully ordered state at $\theta = 1/2$ corresponds to T = 0. In this case one expects that (in the diffraction limit) $I \simeq (\mu-\mu_c)^{2\beta}$ as $\mu \rightarrow \mu_c^+$, where μ_c is the (unknown) chemical potential at the transition. Near the transition

$$\theta - \theta_c \simeq (\mu-\mu_c)^{1-\alpha} + const \cdot (\mu-\mu_c) \tag{11.15}$$

so that (for $\alpha > 0$)

$$I \simeq (\theta-\theta_c)^{2\beta/(1-\alpha)}. \tag{11.16}$$

Equation (11.16) is just an example of Fisher renormalization. By maximizing the linearity of the log-log plot, Taylor et al. estimated $\beta/(1-\alpha)$ = 0.12 ± 0.03; the value of θ_c, with error bars < 2%, was consistent with the point of inflection of I vs θ. The exponent is consistent with the Ising values of β = 1/8 and α = 0. However, this experiment was hardly ideal in that data in the disordered region were not adequate to permit corroborating extractions of γ and ν, and since less than a decade of abscissa scale was available. It is noteworthy that no non-linearity appeared.

Another possible realization of the Ising model is a p(2x1) on a rectangular lattice. *Wang* and *Lu* studied O/W{112} as an example [11.80]. They analyze data over the range $0.002 \lesssim |t| \lesssim 0.11$ and quote exponents β = 0.13 ± 0.01, γ = 1.79 ± 0.14, and ν = 1.09 ± 0.11. Following a procedure popularized for physisorption [11.81], they do not use a log-log plot nor exclude the nonlinear, finite-size contaminated data near T_c. Instead, they take note of the fact that in addition to "smearing" the transition, finite size shifts the effective T_c. Since a real system contains a distribution of sizes of defect free regions, one convolutes the fitting function, e.g., $|t|^{2\beta}$, with a Gaussian distribution of T_c's. This procedure, for which no explicit justification has been published, implicitly parameterizes an unknown correction to scaling: It assumes that the shift in T_c is more important than the smearing of the transition. It assumes that corrections to scaling will be smaller if the reduced temperature is defined with respect to T_c of the finite-size system rather than the infinite system. The instrument response function is completely neglected, a reasonable approximation for synchrotron radiation ($L_i \sim 10^4$ Å), but not for LEED. In our opinion, it is not the best method; if used it should be checked with the procedures presented in Sect. 11.3.

Two recent measurements of Ising behavior have captured theorists' attention [11.82]. *Kim* and *Chan* [11.83] studied the physisorption system CH_4/graphite foam with a.c. calorimetry. Evidence for Ising-like behavior include 1) logarithmic divergence of the specific heat, 2) a critical amplitude ratio (of the specific heat and so of energy-like anomalies) of unity, and 3) behavior of the phase boundary near its maximum (the width of the coexistence region going as $(-t)^\beta$).

In the other case, *Campuzano* et al. [11.84] studied the (1x2) reconstruction of the rectangular Au{110} surface. The experimenters used a novel mirror electron microscopy (MEMLEED) system [11.85], which promises ultrahigh resolution with instrument widths approaching 10^4 Å, although in this research more conventional resolution two orders of magnitude smaller was used. The reconstruction-induced spot was fit to a Gaussian plus a Lorentzian. Above T_c, from log-log plots over $0.04 \leq t \leq 0.2$ they quote $\gamma = 1.75 \pm 0.03$ and $\nu = 1.02 \pm 0.02$. Below T_c, a log-log plot of the intensity of the Gaussian vs t, for $0.006 \leq |t| \leq 0.1$, is remarkably straight; the estimate of β is 0.13 ± 0.022. The quoted exponents are in excellent agreement with pure Ising values, but we believe the error bars of the critical-scattering exponents are rather optimistic, with $\pm$ 10% or so more in keeping with our model calculations and with other cited experiments. For example, we did a log-log fit of the exact Ising correlation length over the same thermal range and found T_c 0.5% too high and $\nu_{eff} = 0.89$. Extracting data points from the published figures and trying some alterations of the fit, we produced similar values of ν_{eff}. A subsequent integrated-intensity measurement [11.86] of a reportedly better gold surface had a T_c 8% higher; in a fit over $0.004 \leq |t| \leq 0.035$ using (11.13) without the linear term ($A_1 = 0$) gave $\alpha = 0.02 \pm 0.05$. Given the low upper bound of the range, it is conceivable that the data were in the finite-size rounded regime, where analytic behavior is expected, and so might not be indicative of the critical behavior. Very recent work has also shown that trace Sn impurities can cause sizable shifts in T_c [11.87]. Possible explanations for the discrepant T_c's are recounted in [11.88].

An intriguing situation in which an Ising transition might occur is H/Ni{111} [11.89]. There H sits in both kinds of 3-fold sites with comparable probability [11.89,90], indicative of a small or negligible binding energy difference. On this honeycomb lattice, with nearest neighbor exclusions due to the short spacing, one expects both types of sites to be occupied equally at low θ (and high T). As θ is increased, we expect an Ising transition to a phase in which either one of the kinds of sites is preferentially occupied [11.91,92]. Since the ordered state is p(1x1), there are no new diffraction features. However, with ion channeling one can distinguish occupancy in the two kinds of sites [11.90,93]. Unfortunately, in a realistically large sample one would expect domains of both varieties. Since this experiment measures the equivalent of the magnetization rather than its magnitude, much cancellation will occur, making detection difficult. As an aside, we note that the well-known graphitic (2x2) phase on this honeycomb net [11.89] is predicted to disorder in the 4-state Potts class [11.15,91]. An unfortunate but presumably surmountable barrier to such a study with LEED is the weak scattering of electrons by H.

11.3.3 XY with Cubic Anisotropy

We have not spent much time on the XY with cubic anisotropy universality class. Since the exponents are non-universal, this class is not optimal for comparison with theory. But the model has a rich range of behavior that will prove fascinating once there are a few more studies at the level of the Munich group's investigation of O/Ru{0001} to demonstrate feasibility. The prototype of this model is a net of spins, with nearest-neighbor dot-product interaction which can point in any direction in a plane but with an anisotropy field favoring alignment along any of four mutually perpendicular directions (e.g., N, E, S, W or NE, SE, SW, NW) [11.94]. In the limit of infinitely strong field, the spin can only point along these four directions, and the model reduces to Ising behavior. In the other limit of vanishing anisotropy the model reduces to the Kosterlitz-Thouless model, mentioned above in conjuction with the transition between floating and (incommensurate) disordered phases. At this unusual transition, $\ln \xi \propto |T - T_{KT}|^{-1/2}$ (rather than $-\nu \ln|t|$) and $\beta = \infty$; we also note that there is no anomaly in the specific heat at the transition [11.95].

Two transitions in this class have been studied. The work of *Lyuksyutov* and *Fedorus* [11.96] on H/W{110} was mentioned in [11.3] in a "note added in proof". They looked only for the exponent β, using the convolution-with-Gaussian-smearing method mentioned above. They estimated T_c as that value which gave the most linear log-log plot of $0.05 < |t| < 0.20$. For saturation coverage $\theta = 1/2$, they quote $\beta = 0.13 \pm 0.04$ based on data over the range $-0.05 < -t < 0.20$. (Note that with this procedure they are including data above T_c!) They were particularly interested in the coverage dependence of T_c, and indicate that similar values of β were found at $\theta = 0.44$ and 0.51. Above $\theta = 0.60$, there is a (2x2) phase. At the saturation coverage $\theta = 0.75$, they estimate $\beta = 0.25 \pm 0.07$ with similar values at $\theta = 0.58$, 0.70, and 0.80. Given the rather inadequate nature of this early experiment, one must be cautious about placing too much weight on these numbers. Nonetheless, it is noteworthy that there is no indication of Fisher renormalization. This absence is consistent with the expected non-positive α for this model [11.97]. Then, in (11.15), the second, analytic term dominates near T_c and no renormalization occurs.

For W{100}, *Wendelken* and *Wang* [11.98] again followed the Russians' procedure. As in that study, only the long-range-order exponent β was measured. Wendelken and Wang simultaneously fit T_c, β, the amplitude of the assumed power law for the intensity, and the width of the Gaussian smearing function. The optimal set of values is quoted as 211 K, 0.144 ± 0.04, 1.76, and 3.26 K, respectively, for a flat surface. The primary goal of this brief report was to see how finite-size effects in the form of steps alter critical behavior. The domains on a flat surface are estimated to be of diameter 129 ± 20 Å, while on a stepped surface with terraces 30 Å wide; the lateral length of the domains is estimated as ~ 60 Å. For this stepped surface, the "best-fit" estimate of the four parameters is 217 K, 0.050 ± 0.01, 0.95, and 2.32 K, respectively. The strong dependence of β on T_c was noted, and the error bars on β were obtained by assuming ± 5 K error in T_c, with the comment that systematic errors should be in the same direction in both cases. In the flat case, pure power law behavior is observed only over the small range $0.03 < |t| < 0.08$; it would have been interesting to see how inclusion of a nonlinear term in the fitting form would have affected the results. It is not clear, however, that the data were robust enough to support an extra fitting parameter.

Another chemisoption system with a disordering transition in this class is the p(2x1)/W{110}. This system has been scrutinized by Lagally's group for nearly a decade [11.48]. Not only have equilibrium properties such as the phase diagram been thoroughly explored [11.99], even the kinetics of ordering have been investigated [11.100]. This system thus seems an excellent candidate unless it turns out that adsorption occurs in quasithreefold sites [11.60].

11.3.4 Se/Ni{100}: Realization of the Ashkin-Teller Model?

Based on symmetry arguments *Bak* et al. [11.68] suggested that the system Se/Ni{100} might be the first known realization of the Ashkin-Teller model [11.101]. Even though exponents have not yet been measured for this adsorption system, we include it here as an example of the exciting directions work in this field can take. The Ashkin-Teller model, which has been studied for over a quarter century for its intrinsic interest, consists of two distinct sets of Ising spins on a square lattice of sites, coupled by a quartic term. There are three phases: 1) at high T both Ising subsystems are disordered; 2) at low T both Ising subsystems are (ferromagnetically) ordered; 3) when the quartic inter- action parameter is larger than the Ising coupling parameter and when T is moderately high, there is a remarkable "polarized" phase in which each Ising subsystem is disordered but the product of spins of the two subsystems (at the same site) has a non-zero expectation value. In the Ashkin-Teller model, the transition between this polarized phase and either of the other phases, which bound it, is in the Ising universality class while the transition between the ordered and disordered phases at small quartic coupling) is in the XY with cubic anisotropy class. The point at which the three phase boundaries intersect is in the 4-state Potts class. In Se/Ni{100}, the three phases are disordered, p(2x2), and c(2x2). One then makes the connection that the intensities of the (1/2, 0) and (0, 1/2) spots correspond, respectively, to the squared expecta- tion value of each the two Ising subsystems, while the (1/2, 1/2) beam corresponds to the squared "polarization". Thus, the transition between the c(2x2) and either p(2x2) or disordered [(1x1)] is expected to be Ising-like while that between p(2x2) and disorder should be like XY with cubic anisotropy, consistent with Landau assignments (cf. Table 11.1). In addition, the multicritical point where the three lines meet, i.e., where the upper boundary of the c(2x2) phase drops to meet the top of the p(2x2) phase, should have 4-state Potts symmetry. The com- petition between the critical behavior associated with the different transitions leads to crossover effects (discussed generically in [11.3]), impeding extraction of exponents either in a numerical study (e.g. our transfer matrix scaling work) or in a diffraction measurement. Testing these predictions poses a formidable but intriguing challenge.

11.4 Conclusions

In assessing the current status of the field, we see that a few good systems have been well studied. It will be gratifying to find realizations of the various universality classes. The thermal range that is probed in present experiments is on the order of a single decade (an improvement over the half decade used for O/Ni{111} [11.4, 41]), going from reduced temperatures of order 10^{-2} to 10^{-1}. This range pales in comparison with 3D fluids (which, however, have much smaller critical fluctuations): ranges of $3 \cdot 10^{-6}$ - 10^{-4} for binary fluids [11.102] and 10^{-6} - 10^{-2} for liquid helium [11.103] have been attained. On the other hand, it is far superior to spin glasses [11.104], where the nonlinear susceptibility is measured over the range 0.15 - 0.3. The importance of large defect-free regions, to allow small $|t|_{min}$, should be obvious.

While the pure exponents of the various classes are firmly established, theoretical questions remain about which corrections to scaling are present in particular systems [11.23]. In order to contribute to this intriguing research, experiments will need to be improved by an order of magnitude, if not more. It now appears that in some cases, e.g., Pt{111} [11.49,105], defect-free regions as large as 1000-2000 Å have been achieved. (If better metallic surfaces cannot be obtained in other cases, attempts should be made to grow on high-grade graphite, mica, or MgO, although in most cases the metal will bead up; see for example [11.106].) This will allow the closer approach to T_c needed to investigate intriguing corrections to scaling. To take advantage of better surfaces, higher resolution will be needed. Several labs now have high-resolution LEED systems. MEMLEED promises even better resolution [11.85]. X-ray radiation from synchrotrons has already demonstrated it [11.107], but now must attain the sensitivity needed for measurements of submonolayers. Note that in this desired scenario, the critical scattering will be a much smaller fraction of the long-range order, heightening the need for improved sensitivity. In contrast with the integrated intensity approach, any surface science laboratory can look at critical properties in the course of characterizing a sample with a phase transition. But, as we have emphasized, it is important to have several exponents. Measurements below T_c, i.e., in the ordered state are more prone to systematic errors than those above; experimentalists should not just content themselves with extraction of the most obvious exponent, β.

At the same time, more systems should be studied with current methods so that as improved apparatus and sample preparation are possible, it will be clear where to apply them. Several possibilities have been mentioned in the text. *Unertl* [11.14] mentions as examples of systems ripe for measurement the following: Se/Ni{001}, S/Ni{111}, Te/Ni{111}, Bi/Cu{111}, H/W{001}, [H/]Mo{001}, and S/Mo{110}. It would be particularly interesting to find a good 3-state Potts transition; the disordering of ($\sqrt{3}$x$\sqrt{3}$) phases of I/Cu{111} [11.108] and S/Pt{111} [11.109] deserve further attention.

Two topics reviewed here in 1983 are especially suitable for investigation in chemisorption systems. Vigorous study of incommensuration [11.11] is worthwhile. This problem is being actively pursued in physisorption systems [11.107] and in crystal growth [11.50]. In the chemisorption literature, there are many reports of streaking or splitting of sharp spots as coverage is increased beyond the saturation value of an ordered monlayer, as well as of a region labeled "antiphase domains". It is important but difficult to distinguish between floating phases and incommensurate disordered phases. In "strong" chemisorption, the adatoms will always be in registry, so that domain walls will be sharp.

The systematics of finite-size [11.110] effects also deserve close attention. Once a high-quality surface has been attained and measured, L_s can be systematically reduced with the addition of point defects, either by the adsorption of very strongly bound "impurity" atoms [11.111] or by the creation of local defects using sputtering. Alternatively, as applied by *Wendelken* and *Wang* [11.98], one can consider vicinal surfaces, leading to asymmetric reduction in L_s.

The most exciting recent development in the theory of 2D phase transitions is the application of ideas from conformal invariance [11.112]. At T_c, when ξ is infinite, correlation functions are conformally covariant. (This local property is considerably more restrictive than the [global] covariance under scale change that underlies renormalization group procedures.) In 2D, this fact is particularly useful since all analytic functions are conformal transformations. The analysis only applies to systems with isotropic interactions (e.g., not to (3x1)/centered rectangular). This rather esoteric line of study has led to an explanation of why the critical exponents (including correction-to-scaling exponents [11.23]) are rather simple rational fractions. For purposes of comparison with experiment, one should consider the lineshape $S(k, T_c)$, (in principle also) including the effects of multiple scattering. *Kleban* et al. [11.88,113,114] have poineered this challenging research, presenting results for the Ising model, with explicit calculations in the simplifying approximation of s-wave scattering. Recall that the data around T_c were excluded from earlier scaling analyses as being contaminated by finite-size effects. Near T_c, one finds the scaling form [11.113,115]

$$S(k,T,L_s) \sim L_s^{4-\eta} X_\pm(k\xi,kL_s) \to L_s^{4-\eta} Y(kL_s) \; [T=T_c] \qquad (11.17)$$

where Y depends explicitly on the shape and boundary conditions of the scattering lattice and $\eta = 2 - \gamma/\nu \to 1/4$ for the Ising model. Since conformal invariance is a continuum theory, the simple form of (11.17) must break down when k gets so large that the lattice constant becomes significant; explicit tests [11.115] show that (11.17) holds for k less than about 1/4 of a reciprocal lattice spacing. Moreover, for geometries that are not very anisotropic, the shape dependence is relatively weak, so that one can approximate the domain as circular [11.88]. (In the limit of large anisotropy of the domain, such as vicinal surfaces, there may be notable anisotropic effects.) To a good approximation, the lineshape can be represented by a Gaussian plus a Lorentzian, with the ratio of their heights and their widths explicitly given. These predictions have been tested, with good agreement, on experimental data for the (1x2) reconstruction of Au{110} [11.88]. (The asymmetry of the microscopic interactions in the rectangular Au{110} can be trivially rescaled to attain an isotropic situation [11.116].) With the advent of better surfaces, more detailed tests should be possible.

With these new developments and possibilities mentioned earlier in this section, one would expect this exciting area to be booming; unfortunately, this is not the case. I offer some personal comments on the situation. The field is a demanding one for surface scientists in terms of both the background theory and the actual experiments. The rewards in terms of practical applications are less immediately apparent than in other pursuits in surface science. Much of the problem, however, is "sociological". The major seminal breakthroughs in theory came about a decade ago. Experiments were slow to follow, and many of the leading theorists moved on to other topics. Now that reasonable experiments are becoming feasible, the field is viewed by enough peer reviewers as passé to threaten funding in these competitive, trendy times. I find it tragic and perverse that the field is languishing while on the verge of great progress.

Acknowledgements. Most of our reseearch has been supported by the Department of Energy under grant DE-FG05-84ER45071. Computer facilities were supplied by the University of Maryland Computer Science Center. Partial funding presently was obtained from NSF grant DMR-85-04163 and NATO grant 86/0782. I also acknowledge the hospitality, as a guest worker of the Surface Science Division of the National Bureau of Standards, Gaithersburg, MD. The research described herein was done in extended collaboration with N.C. Bartelt and L.D. Roelofs, without whose energy and insight it would have been impossible. I thank the former also for helpful comments on the manuscript. Ongoing fruitful interactions with E.D. Williams and R.L. Park and their many talented students have been crucial to our program. Many colleagues elsewhere have contributed preprints and helpful conversations; at the risk of offending the omitted, I mention P.H. Kleban, W.N. Unertl, H. Pfnür, J.C. Campuzano, A.L. Stella, U. Glaus, S. Fishman, J.G. Amar.

References

11.1 For a review of reconstruction, see P.J. Estrup: In *Chemistry and Physics of Solid Surfaces V*, ed. by R. Vanselow, R. Howe (Springer, Berlin, Heidelberg 1984) p.205

11.2 For a review, see K. Binder: In *Phase Transitions and Critical Phenomena*, ed by C. Domb, J.L. Lebowitz, Vol.8 (Academic, London 1983), p.1

11.3 T.L. Einstein: In *Chemistry and Physics of Solid Surfaces IV*, ed. by R. Vanselow, R. Howe (Springer, Berlin, Heidelberg 1982) p.251

11.4 L.D. Roelofs, A.R. Kortan, T.L. Einstein, R.L. Park: Phys. Rev. Lett. 46, 1465 (1981)

11.5 N.C. Bartelt, T.L. Einstein, L.D. Roelofs: J. Vac. Sci. Tech. A 1, 1217 (1983)

11.6 N.C. Bartelt, T.L. Einstein, L.D. Roelofs: Phys. Rev. B 35, 1776 (1987)

11.7 N.C. Bartelt, T.L. Einstein, L.D. Roelofs: Phys. Rev. B 32, 2993 (1985)

11.8 N.C. Bartelt, T.L. Einstein, L.D. Roelofs: Phys. Rev. B 35, 4812 (1987)

11.9 N.C. Bartelt, T.L. Einstein, L.D. Roelofs: Phys. Rev. B 35, 6786 (1987)

11.10 An excellent general reference on critical phenomena is a) P. Pfeuty, G. Toulouse: *Introduction to the Renormalization Group and Critical Phenomena* (Wiley, New York 1977);
see also b) M.E. Fisher: In *Collective Properties of Physical Systems*, ed. by

B. Lundqvist, S. Lundqvist (Proc. 24th Nobel Symposium, Academic, New York 1973) p.16;
c) S.-K. Ma: *Modern Theory of Critical Phenomena* (Benjamin, Reading, Mass. 1976)

11.11 P. Bak: In *Chemistry and Physics of Solid Surfaces V*, ed. by R. Vanselow, R. Howe (Springer, Berlin, Heidelberg 1984) p.317

11.12 L. Passell, S.K. Satija, M. Sutton, J. Suzanne: In *Chemistry and Physics of Solid Surfaces VI*, ed. by R. Vanselow, R. Howe (Springer, Berlin, Heidelberg 1986) p.609

11.13 T.L. Einstein: In Proc. 10th Johns Hopkins Workshop. *Infinite Lie Algebras and Conformal Invariance in Condensed Matter and Particle Physics*, ed. by K. Dietz, V. Rittenberg (World Scientific, Singapore 1987) p.17

11.14 W.N. Unertl: Comments Cond. Mat. Phys. **12**, 289 (1986)

11.15 M. Schick: Prog. Surf. Sci. **11**, 245 (1981);
see also E. Domany, M. Schick, J.S. Walker, R.B. Griffiths: Phys. Rev. B **18**, 2209 (1978);
S. Alexander: Phys. Lett. **54A**, 353 (1075)

11.16 For a review of Potts models see F.Y. Wu: Rev. Mod. Phys. **54**, 235 (1982)

11.17 D.A. Huse, M.E. Fisher: Phys. Rev. Lett. **49**, 793 (1982); Phys. Rev. B **29**, 239 (1984)

11.18 M.P.M. den Nijs: J. Phys. A **17**, L295 (1984)

11.19 P.A. Bennett, M.B. Webb: Surf. Sci. **104**, 74 (1981);
E.G. McRae, R.A. Malic: Surf. Sci. **148**, 551 (1984)

11.20 M. Trenary, K.J. Uram, F. Bozso, J.T. Yates, Jr.: Surf. Sci. **146**, 269 (1984)

11.21 R.J. Phaneuf, M.B. Webb: Surf. Sci. **164**, 167 (1985)

11.22 A. Aharony, M.E. Fisher: Phys. Rev. B **27**, 4394 (1983)

11.23 H.W.J. Blöte, M.J.M. den Nijs: Phys. Rev. B **37**, 1766 (1987)

11.24 N.C. Bartelt, T.L. Einstein, L.D. Roelofs: Phys. Rev. Lett. **56**, 2881 (1986)

11.25 S. Dietrich, H. Wagner: Phys. Rev. Lett. **51**, 1469 (1983);
see also W. Moritz, M.G. Lagally: Phys. Rev. Lett. **56**, 865 (1986)

11.26 M. Fink, J. Ingram: At. Data **4**, 1 (1972);
M.B. Webb, M.G. Lagally: Solid State Phys. **28**, 301 (1973)

11.27 M.E. Fisher: Phys. Rev. **176**, 257 (1968)

11.28 R.L. Park, J.E. Houston, D.G. Schreiner: Rev. Sci. Instrum. **42**, 60 (1971)

11.29 M. Henzler: Surf. Sci. **152/153**, 963 (1985);
M.G. Lagally: Appl. Surf. Sci. **13**, 260 (1982);
M.G. Lagally, J.A. Martin: Rev. Sci. Instrum. **54**, 1273 (1983);
U. Scheithauer, G. Meyer, M. Henzler: Surf. Sci. **178**, 441 (1986)

11.30 T. Engel: In *Chemistry and Physics of Solid Surfaces V*, ed. by R. Vanselow, R. Howe (Springer, Berlin, Heidelberg 1984) p.257, and this volume
J.P. Toennies: private commun.

11.31 M. Schick: Phys. Rev. Lett. **51**, 1347 (1981)

11.32 J.F. Nicoll, J.K. Bhatarcharjee: Phys. Rev. B **23**, 389 (1981);
J.F. Nicoll, P.C. Albright: ibid. **31**, 4576 (1985)

11.33 K. Binder: J. Stat. Phys. **24**, 69 (1981);
K. Binder, D.P. Landau: Phys. Rev. Lett. **52**, 318 (1984);
see also M. Fähnle, J. Souletie: J. Phys. C **17**, L469 (1984); Phys. Rev. B **32**, 3328 (1985);
A.S. Arrott: ibid. **31**, 2851 (1985)

11.34 J.V. Sengers: private commun.

11.35 B. Nienhuis: J. Phys. A **15**, 199 (1982)

11.36 J.L. Cardy, M. Nauenberg, D.J. Scalapino: Phys. Rev. B **22**, 2560 (1980)

11.37 C.A. Tracy, B.M. McCoy: Phys. Rev. B **12**, 368 (1975);
E. Barouch, B.M. McCoy, T.T. Wu: Phys. Rev. Lett. **31**, 1409 (1973)

11.38 M.E. Fisher, J.S. Langer: Phys. Rev. Lett. **20**, 665 (1978);
see also M. Ferer, M.A. Moore, M. Wortis: ibid. **22**, 1382 (1969)

11.39 A.D. Bruce: J. Phys. C **7**, 2089 (1974)

11.40 N.C. Bartelt, T.L. Einstein, L.D. Roelofs: J. Vac. Sci. Tech. A **3**, 1568 (1985)

11.41 N.C. Bartelt, T.L. Einstein, L.D. Roelofs: In *The Structure of Surfaces*, ed. by M.A. van Hove, S.Y. Tong (Springer, Berlin, Heidelberg 1985) p.357

11.42 J. Henrion, G.E. Rhead: Surf. Sci. **29**, 20 (1972)

11.43 M. Kaufmann, D. Andelman: Phys. Rev. B **29**, 4010 (1984)

11.44 K.D. Miner, M.H.W. Chan, A.D. Migone: Phys. Rev. Lett. **51**, 1465 (1983)

11.45 J.H. Campbell, M. Bretz: Phys. Rev. B. **32**, 2861 (1985)

11.46 Y. Saito: Phys. Rev. B **24**, 6652 (1981);
K. Binder: J. Stat. Phys. **24**, 69 (1981);
W. Selke, J. Yeomans: Z. Phys. B **46**, 311 (1982)

11.47 An exception is K.K. Chan, M. Deutsch, B.M. Ocko, P.S. Pershan, L.B. Sorensen: Phys. Rev. Lett. **54**, 920 (1985)

11.48 M.G. Lagally, G.-C. Wang, T.-M. Lu: In *Chemistry and Physics of Solid Surfaces II*, ed. by R. Vanselow (CRC, Boca Raton 1979) p.153

11.49 K. Kern, P. Zeppenfeld, R. David, G. Comsa: In *The Structure of Surfaces-II*, ed. by J.F. van der Veen, M.A. Van Hove (Springer, Berlin, Heidelberg 1988)

11.50 J.H. van der Merwe: *Chemistry and Physics of Solid Surfaces V* ed. by R. Vanselow, R. Howe (Springer, Berlin, Heidelberg 1984) p.365;
J.M. Moison, C. Guille, M. Bensoussan: In *The Structure of Surfaces-II*, ed. by J.F. van der Veen, M.A. van Hove (Springer, Berlin, Heidelberg 1988)

11.51 P. Dutta, S.K. Sinha: Phys. Rev. Lett. **47**, 50 (1981)

11.52 D.A. Huse: J. Phys. A **16**, 4357 (1983)

11.53 R.J. Baxter: J. Phys. A **13**, L61 (1980); *Exactly Solved Models in Statistical Mechanics* (Academic, London 1982);
R.J. Baxter, P.A. Pearce: J. Phys. A **15**, 897 (1982)

11.54 A.L. Stella, X.-C. Xie, T.L. Einstein, N.C. Bartelt: Z. Phys. B **67**, 357 (1987), and references therein

11.55 M.E. Fisher: J. Stat. Phys. **34**, 667 (1984)

11.56 S.N. Coppersmith, D.S. Fisher, B.I. Halperin, P.A. Lee, W.F. Brinkman: Phys. Rev. B **25**, 349 (1982)

11.57 D. Nelson: In *Phase Transitions and Critical Phenomena*, Vol.7, ed. by C. Domb, J.L. Lebowitz (Academic, New York 1983)

11.58 R. Imbihl, R.J. Behm, K. Christman, G. Ertl, T. Matsushima: Surf. Sci. **117**, 257 (1982)

11.59 W. Kinzel, W. Selke, K. Binder: Surf. Sci. **121**, 13 (1982)

11.60 W. Moritz, R. Imbihl, R.J. Behm, G. Ertl, T. Matsushima: J. Chem. Phys. **83**, 1959 (1985)

11.61 L.D. Roelofs: In *Chemistry and Physics of Solid Surfaces IV*, ed. by R. Vanselow, R. Howe (Springer, Berlin, Heidelberg 1982), p.219

11.62 A.R. Kortan, R.L. Park: Phys. Rev. B **23**, 6340 (1981)

11.63 K. Binder, D.P. Landau: Phys. Rev. B **30**, 1477 (1984);
M.S.S. Challa, D.P. Landau, K. Binder: Phys. Rev. B **34**, 1841 (1986);
K. Binder, D. Stauffer: In *Application of the Monte Carlo Method in Statistical Physics*, ed. by K. Binder (Springer, Berlin, Heidelberg 1984) Chap.1;
K. Binder: Rep. Prog. Phys. **50**, 783 (1987), Ref.18.a

11.64 K. Binder: J. Stat. Phys. **24**, 69 (1981)

11.65 M.E. Fisher, A.N. Berker: Phys. Rev. B **26**, 2507 (1982)

11.66 M. Schick: Phys. Rev. Lett. **47**, 1347 (1981)

11.67 G. Grest, M. Widom: Phys. Rev. B **24**, 6508 (1981)

11.68 P. Bak, P.H. Kleban, W.N. Unertl, J. Ochab, G. Akinci, N.C. Bartelt, T.L. Einstein: Phys. Rev. Lett. **54**, 1539 (1985)

11.69 I.G. Enting: J. Phys. A **8**, 1681 (1975)

11.70 W. Witt, E. Bauer: Ber. Bunsenges. Phys. Chem. **90**, 248 (1986)

11.71 A recent review is: G. Grinstein: In *Fundamental Problems in Statistical Mechanics*, ed. by E.G.D. Cohen (Elsevier, Amsterdam 1985) Vol.6, p.147

11.72 W. Kinzel: Phys. Rev. B **27**, 5819 (1983);
L.D. Roelofs: Appl. Surf. Sci. **11/12**, 425 (1982);
R. Birgeneau, A.N. Berker: Phys. Rev. B. **26**, 3751 (1982)

11.73 H. Ohtani, C.-T. Kao, M.A. Van Hove, G.A. Somorjai: Prog. Surf. Sci. **23**, 155 (1987);
G.A. Somorjai: *Chemistry in Two Dimensions: Surfaces* (Cornell U. Press, Ithaca 1981);
G.A. Somorjai, F.Z. Szalkowski: J. Chem. Phys. **54**, 389 (1971)

11.74 Peder Estrup remarked to me recently, "There are no simple chemisorption systems!"

11.75 R.Q. Hwang; E.D. Williams, R.L. Park: Surf. Sci. Lett. **193**, L53 (1988)

11.76 P. Piercy, H. Pfnür: Phys. Rev. Lett. **59**, 1124 (1987);
P. Piercy, M. Maier, H. Pfnür: In *The Structure of Surfaces-II* , ed. by J.F. van der Veen, M.A. Van Hove (Springer, Berlin, Heidelberg 1988)

11.77 B. Nienhuis, E.K. Riedel,M. Schick: Phys. Rev. B **27**, 5625 (1983);
M.Schick: Surf. Sci. **125**, 94 (1983)

11.78 D.E. Taylor, R.L. Park: Surf. Sci. **125**, L73 (1983)

11.79 D.E. Taylor, E.D. Williams, R.L. Park, N.C. Bartelt, T.L. Einstein: Phys. Rev. B **32**, 4653 (1985)

11.80 G.-C. Wang, T.-M. Lu: Phys. Rev. B **31**, 5918 (1985)

11.81 P.M. Horn, R.J. Birgeneau, P. Heiney, E.M. Hammonds: Phys. Rev. Lett. **41**, 961 (1978)

11.82 M.E. Fisher: J. Chem. Soc., Faraday Trans. II **82**, 1569 (1986)

11.83 H.K. Kim, M.H.W. Chan: Phys. Rev. Lett. **53**, 170 (1984)

11.84 J.C. Campuzano, M.S. Foster, G. Jennings, R.F. Willis, W.N. Unertl: Phys. Rev. Lett. **54**, 2684 (1985);
J.C. Campuzano, G. Jennings, R.F. Willis: Surf. Sci. **162**, 484 (1985)

11.85 M.S. Foster, J.C. Campuzano, R.F. Willis, J.C. Dupuy: J. Microscopy **140**, 395 (1985)

11.86 D.E. Clark, W.N. Unertl, P.H. Kleban: Phys. Rev. B **34**, 4379 (1986)

11.87 E.G. McRae, T.M. Buck, R.A. Malic, G.H. Wheatley: Phys. Rev. B **36**, 2341 (1987)

11.88 P.H. Kleban, R. Hentschke, J.C. Campuzano: Phys. Rev. B **37**, 5788 (1988)

11.89 K. Christmann, R.J. Behm, G. Ertl, M.A. Van Hove, W.H. Weinberg: J. Chem. Phys. **70**, 4168 (1979)

11.90 F. Besenbacher, I. Stensgaard, K. Mortensen: In *The Structure of Surfaces-II*, ed. by J.F. van der Veen, M A. Van Hove (Springer, Berlin, Heidelberg 1988);
F. Besenbacher: private commun.

11.91 L.D. Roelofs, T.L. Einstein, N.C. Bartelt, J.D. Shore: Surf. Sci. **176**, 295 (1986)

11.92 W.M. Gibson: In *Chemistry and Physics of Solid Surfaces V*, ed. by R. Vanselow, R. Howe (Springer, Berlin, Heidelberg 1984) p.427;
T. Gustafsson: In *The Structure of Surfaces-II*, ed. by J.F. van der Veen, M.A. Von Hove (Springer, Berlin, Heidelberg 1988)

11.93 E. Domany, M. Schick: Phys. Rev. B **20**, 3828 (1979)

11.94 J.V. José, S. Kirkpatrick, L.P. Kadanoff, D.R. Nelson: Phys. Rev. B **16**, 1217 (1977)

11.95 J.M. Kosterlitz, D.J. Thouless: J. Phys. C **6**, 1181 (1973); Ref.38

11.96 I.F. Lyuksyutov, A.G. Fedorus: Sov. Phys. JETP **53**, 1317 (1981) [Zh. Eksp. Teor. Fiz. **80**, 2511 (1981)]

11.97   G.-Y. Hu, S.-C. Ying: [Physica 140 A, 585 (1987)] argue that ν increases monotonically from 1 to ∞ along the line of transitions, implying by hyperscaling that α decreases from 0 to $-\infty$.

11.98   J.F. Wendelken, G.-C. Wang: Phys. Rev. B 32, 7542 (1985)

11.99   G.-C. Wang, M.G. Lagally: Surf. Sci. 81, 69 (1979);
M.G. Lagally, T.-M. Lu, G.-C. Wang: In *Ordering in Two Dimensions*, ed. by S.K. Sinha (North-Holland, New York, Amsterdam 1980) p.113

11.100 M.C. Tringides, P.K. Wu, M.G. Lagally: Phys. Rev. Lett. 59, 315 (1987)

11.101 J. Ashkin, E. Teller: Phys. Rev. 64, 178 (1943)

11.102 R.F. Chang, H. Burstyn, J.V. Sengers: Phys. Rev. B 19, 866 (1979)

11.103 G. Ahlers: Rev. Mod. Phys. 52, 489 (1980)

11.104 P. Mazumdar, S.M. Bhagat: J. Mag. & Mag. Mat. 66, 263 (1987)

11.105 G. Comsa: private commun.

11.106 L. Holland: *Vacuum Deposition of Thin Films* (Wiley, New York 1961)

11.107 R.J. Birgeneau, G.S. Brown, P.M. Horn, D.E. Moncton, P.W. Stephens: J. Phys. C 14, L49 (1981);
P.W. Stephens, P.A. Heiney, R.J. Birgeneau, P. Horn, D.E. Moncton, G.S. Brown: Phys. Rev. B 29, 3512 (1984)

11.108 S.B. DiCenzo, G.K. Wertheim, D.N.E. Buchanan: Surf. Sci. 121, 411 (1982)

11.109 M. Auer, H. Leonhard, K. Hayek: Appl. Surf. Sci. 17, 70 (1983)

11.110 P.H. Kleban: In *Chemistry and Physics of Solid Surfaces V*, ed. by R. Vanselow, R. Howe (Springer, Berlin, Heidelberg 1984) p.339;
M.N. Barber: In *Phase Transitions and Critical Phenomena*, Vol.8, ed. by C. Domb, J.L. Lebowitz (Academic, London 1984) p.145

11.111 D.W. Goodman: In *Chemistry and Physics of Solid Surfaces VI*, ed. by R. Vanselow, R. Howe (Springer, Berlin, Heidelberg 1986) p.169

11.112 For an excellent introductory review, see J.L. Cardy: In *Phase Transitions and Critical Phenomena*, Vol.11, ed. by C. Domb, J.L. Lebowitz (Academic, London 1986)

11.113 P.H. Kleban, G. Akinci, R. Hentschke, K.R. Brownstein: J. Phys. A 19, 437 (1986); Surf. Sci. 166, 159 (1986)

11.114 R. Hentschke, P.H. Kleban: Surf. Sci., to be published

11.115 N.C. Bartelt, T.L. Einstein: J. Phys. A 19, 1429 (1986)

11.116 M.P. Nightingale, H. Blöte: J. Phys. A 16, 1657 (1983)

12. Surface Electronic Interactions of Slow Ions and Metastable Atoms

H.D. Hagstrum

30 Sweetbriar Road, Summit, NJ 07901, USA

The study of the surface interactions of slow ions and metastably excited atoms is a part, at the low energy end, of the field of particle-solid interactions. It has lead to a detailed understanding of a number of interesting and important phenomena. This, in turn, has made possible the development of two interrelated electron spectroscopies of surfaces based on electron ejection induced by ion neutralization or metastable atom de-excitation. These are among the most surface specific of the spectroscopies applied to surface studies because they involve the phenomenon of quantum mechanical electron tunneling. The purpose in this paper is to discuss, from an historical perspective, the nature of the atom-solid electronic interactions that underlie INS and MDS and give them their individual characteristics.

It was observed very early that electrons are ejected from solids when slowly moving ions are neutralized [12.1-3] or metastably excited atoms are de-excited [12.4-6] at the surface of a solid. These studies were undertaken, at least initially, in order to understand the means by which electrons are emitted from the cathode of a glow discharge. Later it was realized that the kinetic energy distributions of electrons ejected from surfaces by slow ions contained information on the electronic structure of the outermost atomic layers a solid and that this could be the basis for a surface-specific ion neutralization spectroscopy (INS) [12.7-9]. The de-excitation of a metastable atom as the basis of an electron specroscopy of surfaces was definitively established using adsorbate-covered metal surfaces for which the spectra could be compared with known spectral features from photoemission [12.10-13]. This spectroscopy has several names: metastable de-excitation spectroscopy (MDS), Penning ionization electron spectroscopy (PIES), and metastable quenching spectroscopy (MQS). The first of these will be employed here. However, since ions may turn to metastables or metastables to ions enroute to the surface, the spectroscopies in this paper will for the most part be identified by the operative ejection mechanism, Auger de-excitation, AD, or Auger neutralization, AN. When INS or MDS are used it will be understood that one is identifying the spectroscopy by the ejection mechanism, not by what particle is sent toward the surface. Ions and metastable atoms of the noble gases are used because they store large amounts of electronic energy and their parent gas does not adsorb on surfaces.

12.1 Electron Ejection by Auger Neutralization (AN) and Auger De-excitation (AD)

Electrons are ejected from surfaces by ions and metastables in Auger-type processes in which two electrons participate, one gaining energy and the other losing the same amout of energy in a fast radiationless process. The Auger neutralization (AN) of an incident ion is depicted schematically in Fig.12.1a. The so-called "down" electron tunnels into the well of the ion and then drops in energy to fill the vacant ground level. The "up" electron acquires the energy thus released and will escape the solid if it is directed away from the surface with sufficient momentum normal to the surface. Far from the solid the ejected electron has the kinetic energy E_k. The maximum value of E_k is equal to $E_i'-2\phi$, the effective ionization energy of the parent atom near the surface minus twice the work function, and is obtained when the two participating electrons lie initially at the Fermi level, E_F.

There are two basic types of Auger de-excitation (AD) of an incident metastable atom. In the tunneling or exchange process, whose transitions are shown as full lines in Fig.12.1b, an electron from the solid tunnels into the ground state of the incident atom with the simultaneous ejection of the metastable electron. In the nontunneling or nonexchange AD process, that depicted by the dashed lines in Fig.12.1b, no electron tunnels. This process requires relaxation of the metastability selection rule to allow the metastable electron to drop to the ground state with the energy thus released exciting an electron in

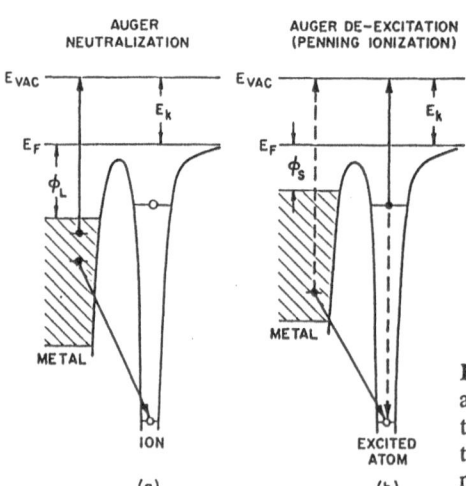

Fig.12.1 Electron-energy diagrams, appropriate to He, indicating the electronic transitions of the electron-ejection processes initiated by an ion and a metastable atom.

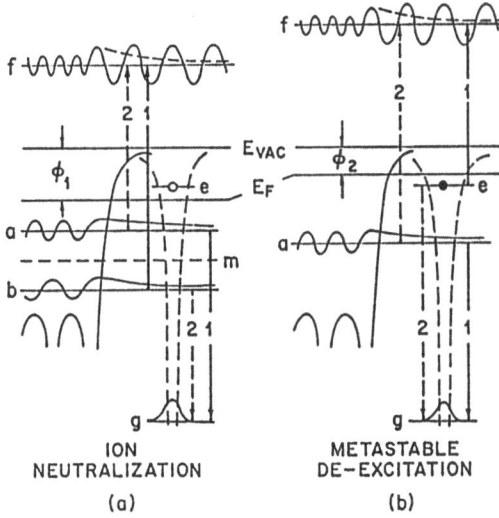

Fig.12.2 Wave functions involved in the electron-ejection processes depicted in Fig.12.1

the solid to a level above the vacuum level. For the helium metastables we expect this to be possible for the 2^1S state but not the 2^3S for which the relaxation of metastability would involve a magnetic spin flip.

In Fig.12.2 electron energy diagrams are presented in which wave functions are shown. Consider first the simpler case, namely Auger de-excitation or AD shown in (b). In the tunneling type (transitions 1) an electron from the solid at energy level a tunnels into the atomic well at an energy above the ground state g whose wave function has a narrow bell-shaped form. It is this coupling of the tunneling electron to the highly localized ground state that makes the ejection process surface specific. Transitions 2 indicate the nontunneling type. Even though it involves the transition of two electrons, AD is quasi one-electron since one electron moves equivalently between the discrete metastable and ground levels giving up a fixed amount of energy to the up electron independent of its original level a. Auger de-excitation thus resembles photoemission and the kinetic energy distribution (KED) of the excited electron yields directly the transition density of the process which is an inextricable mixture of initial state density and transition probability.

The process of ion neutralization shown in Fig.12.2a is seen to be a true two-electron Auger process since neither initial-state electron moves equivalently between two discrete levels. The energy given up by the down electron, whether by transition 1 or 2, is given to a second electron from the solid which may lie anywhere in the filled band of the solid. Furthermore, the probability that an electron will leave an initial state, a or b, is not equal to the probability that an electron will arrive at the final state f as is the case for AD or any one-electron process. In the elemental AN process the probability that an electron arrives at the final state is the product of the transition probabilities that each of the two electrons involved leave their initial states. Thus, we must think in terms of initial-state and final-state transition densities. The total probability that an electron will arrive at a specific final level, f in Fig.12.2a, the final-state transition density, is thus the integral of the product of the probabilities that electrons leave their initial states, a and b, the initial-state transition densities, for all pairs of initial states that are symmetrically disposed relative to that level, m, which lies halfway between the final state f and the gound state g. In terms of an energy variable, ζ, that runs positively downward into the filled band from zero at E_F we may write this expression as:

$$F_p(\zeta) = \int_{-\zeta}^{+\zeta} V(\zeta + \Delta) \, W(\zeta - \Delta) \, d\Delta = V \otimes W \,, \tag{12.1}$$

indicating that the up and down electron have, in general, different initial-state transition densities V and W, respectively.

In (12.1) $F_p(\zeta)$, the convolution product of V and W, is the function that determines the features of the measured KED. A viable electron spectroscopy must give us, however, as one-electron spectroscopies do, the initial-state transition density which is closely related to the local density of states where the electron transitions occur. Clearly, we cannot, from (12.1) determine V without knowing W or conversely. *Hagstrum* [12.9] suggested, however, that one can obtain from a measured KED of the AD process a function $U(\zeta)$ of the character of an initial-state transition density by relating the measured KED to F_s, the convolution square of $U(\zeta)$ thus:

$$F_s(\zeta) = \int_{-\zeta}^{+\zeta} U(\zeta + \Delta) \, U(\zeta - \Delta) d\Delta = U \otimes U \,. \tag{12.2}$$

Now it is possible to unfold the experimental function to obtain $U(\zeta)$. The next step, of course, is to interpret U in terms of V and W. Purely mathematically, it was demonstrated that U is a form of mean, dubbed the convolution mean, of V and W [12.14]. The physical reasonableness of this step was pointed out as follows. V and W are functions that give the local densities of states at two physical positions relative to the surface. Considering the screening of the Coulomb interaction potential, *Heine* [12.15] concluded that V for the up electron should be the local density of states at or near the surface. The initial-state transition density W for the down electron is localized at the position of the neutralizing ion by the narrow ground state wave function. *Appelbaum* and *Hamann* [12.16] agree with this conclusion on the basis of fitting cross folds of their calculated densities of states for semiconductor surfaces to experimental kinetic energy distributions. The functions V and W will each have peaks at the same energy levels corresponding to energy bands in the solid or resonances in adsorbed species on the surface. The peak intensities will differ both because W lies farther out on the decaying wave function tails outside the surface and because the rate of decay can differ for different kinds of wave functions, s or p versus d functions, for example. The U function will then show peaks and valleys undistortedly at the same energy levels as V and W but the intensities will be neither those of V nor W but a form of mean between them [12.14]. Thus, the distortions inherent in the U function resemble the distortions evident in other electron spectrosopies. Photoemission, for example, discriminates against the valence s electrons in alkali atoms adsorbed on metals whereas these are observed with high intensity by Auger de-excitation.

Clearly, the necessity to unfold the experimental kinetic energy distribution makes INS a more complicated spectroscopy than MDS and other one-electron spectroscopies, but understanding it and being capable of using it supplies us with another tunneling spectroscopy having its own unique characteristics that differ in several interesting ways from those of all other spectroscopies. AD is impossible at some surfaces, as we shall see, making the AN process the only possibility.

12.2 Resonance Tunneling Makes Two-Stage Processes Possible

Resonance tunneling can ionize a metastable atom or neutralize an ion to a metastable state. These processes are illustrated in Fig. 12.3. Resonance neutralization of an ion occurs at a surface with small work

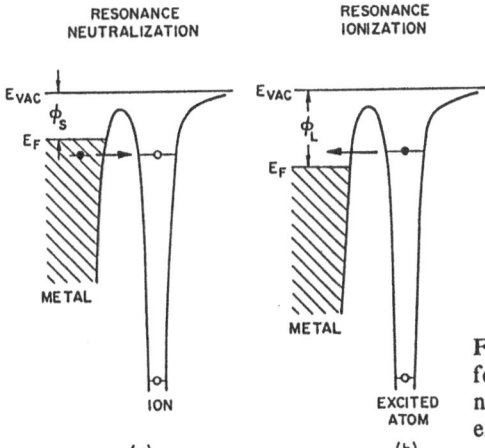

Fig. 12.3 Electron-energy diagrams for the two types of resonance-tunneling process involving an ion and an excited atom near a solid surface.

function (ϕ_S) and resonance ionization of a metastable atom at a surface with large work function (ϕ_L) as is illustrated in Fig.12.3a and b, respectively. Resonance processes can be inhibited by multilayered adsorption as is discussed in Sect.12.6. Because the wave function of the metastable level is spatially more extended than is that of the ground state function, the transition probabilities of the resonance processes become appreciable at larger atom-solid separation than do the Auger ejection processes. Thus, the resonance processes are candidates as precursors in two-stage processes of electron ejection from solids by either ions or metastable atoms.

Figure 12.4 indicates schematically the relationships between the two Auger ejection processes and the two resonance tunneling processes that make possible either one- or two-stage ejection. Three isoelectronic states are involved: the ion X^+ and n electrons X_M^- in the metal, the metastable atom X^* and (n-1) electrons in the metal, and the neutralized atom X with (n-2) electrons in the metal and a free or ejected electron e^-. One should note also that though the final states of AN and AD are written the same, the electronic configurations and the energies involved are quite different. Thus AN draws two electrons from the solid, AD only one. The kinetic energy distributions of the ejected electrons are also completely different.

Prior to 1930 it had been suggested only that ejection by ions could occur by virtue of the ion's electrostatic attraction for an electron in the solid overcoming the electron's attraction for its image [12.17], or by the ion being neutralized to an excited state after which the excited atom is de-excited and an electron is ejected in a "collision of the second kind" [12.18]. The first quantum theory by *Massey* [12.19] assumed the two-stage ejection process of RN + AD and yielded the experimentally unacceptable result that the AD process occurs farther from the surface than does the RN process. *Cobas* and *Lamb* [12.20] who later investigated the same problem, found an error in a matrix element in Massey's theory. When this was corrected, the theory gave the order of relative probabilities appropriate to the proper functioning of the two-stage ejection process. It is a curious fact that in neither of these investigations was the one-stage AN process considered as a possibility. The first quantum mechanical treatment of AN was published by *Shekhter* [12.21] seven years before the publication

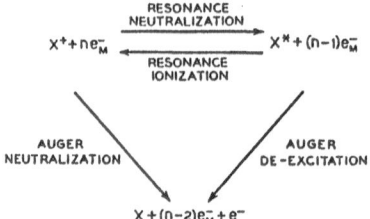

Fig.12.4 A "reaction diagram" indicating the interconnections of the Auger and resonance processes.

of *Cobas* and *Lamb*. That *Shekhter* wrote in Russian in all probability accounts for the fact that *Cobas* and *Lamb* appear to have been unaware of his contribution and made no mention of it. Shekhter also calculated the probabilities of the RN + AD process and found them to have the relative values appropriate to the two-stage process. In none of the above work was the possibility considered of the variation of energy levels in the atom near the solid surface (Sect. 12.4).

12.3 Matrix Elements, Transition Rates, and Related Probability Functions

Without attempting to calculate matrix elements, *Hagstrum* [12.7,8] has discussed the AN and AD processes in quantum mechanical terms based in part on the above theories in order to understand details of experimental results. For each type of electron ejection, the rate of an elemental process occurring at the atom-surface separation s is governed by the square of a matrix element multiplied by a density of final states. The matrix element for the AN process, H, may be written:

$$H = \iint u_f^*(1)u_b(1)(e^2/r_{12})u_g^*(2)u_a(2)d1\,d2 \ . \qquad (12.3)$$

In this expression, u_a and u_b are wave functions of electrons in initial states in the solid, u_g that for the incident atom's ground state, and u_f that of the electron which is excited to a level f where it may leave the solid as a free electron (Fig. 12.2a).

For the AD process we must write two matrix elements, one for the tunneling or exchange process, H'_t, the other for the nontunneling or nonexchange process, H'_n, already discussed. These are

$$H'_t = \iint u_f^*(1)u_e(1)(e^2/r_{12})u_g^*(2)u_a(2)d1\,d2 \ , \qquad (12.4)$$

$$H'_n = \iint u_f^*(2)u_a(2)(e^2/r_{12})u_g^*(1)u_e(1)d1\,d2 \ . \qquad (12.5)$$

In these expressions the new wave function, u_e, is that of the excited (metastable) state of the incident atom. We recognize that the interaction term, although written here in simple Coulomb form, should be screened appropriately to account for the presence of metal electrons [12.15].

In (12.3-5) the functions of the coordinates of one of the electrons, 1 or 2, are grouped together on one side of the interaction term, e^2/r_{12}, and the functions of the other electron on the other side. This enables us to see that each integral may be viewed as specifying a Coulomb interaction, screened as appropriate, between two "clouds" of electrons each specified by the product of two wave functions.

Let us look at the nature of each electron cloud for the three cases described by (12.3-5). For the tunneling processes of (12.3,4) the down-electron cloud is $u_g^* u_a$ and in (12.5) it is $u_g^* u_e$. These are all highly localized on the atom by virtue of the deep-lying, and hence narrow, component u_g^*. The up-electron cloud in (12.3,5) will lie closer to the metal surface since neither of its component functions is localized. In (12.4), however, the up-electron cloud, like the down-electron cloud, is localized on the atom by virtue of the metastable electron component u_e. Since u_e is a much broader function than u_g^*, $u_f^* u_e$ is not as compactly localized a function as $u_g^* u_a$. Each of the function properties we have just described is the basic reason for certain of the characteristics of the specific spectroscopy that the matrix element describes.

The transition rate for the elemental electron ejection process may be written:

$$r_t(s) = (2\pi/\hbar) \, |H|^2 \, N(E_k) d\Omega \ . \tag{12.6}$$

Here the atom is assumed at a separation s from the surface, electrons in specific initial states, and the ejected electron at energy E_k and with velocity vector in the element of solid angle $d\Omega$. The total transition rate, $R_t(s)$, for the atom at a distance s from the surface is obtained by integrating $r_t(s)$ over all energies and angles subject to the necessary energy conservation of the process. The early theories [12.20,21], as well as later ones [12.22,23], have yielded an exponential multiplied by a polynomial in s for the rates of the Auger processses. The polynomial causes the rate to fall below the exponential close to the surface. Despite this, it is convenient pedagogically to think in terms of an exponential rate for which it is easy to calculate related probability functions. The deviations from an exponential rate will shift functions somewhat relative to the surface and change the form of the functions to a certain extent but will not alter any of the conclusions we reach on the basis of an exponential rate. It is interesting to note that *Hentschke* et al. [12.23] find that the inclusion in their theory of an image perturbation compensates for the tendency of the rate to saturate near the surface extending the range of validity of the simple exponential. Using the exponential rate function

$$R_t(s) = A \, e^{-as} \ , \tag{12.7}$$

with the symbols R, A, a for AN; R', A', a' for AD, and R'', A'', a'' for RN and RI, we may proceed to determine related probability functions which illuminate some of the properties of the tunneling processes we are considering [12.20,7]. The probability that an incident ion, for example, approaching the surface from infinity with an assumed constant velocity, v_n, normal to the surface will reach the separation s without undergoing neutralization is

$$P_0(s,v_n) = \exp\left[-\int_s^\infty R_t(s)ds/v_n\right] . \tag{12.8}$$

The probability that neutralization will occur in ds at s is

$$P_t(s,v_n)ds = [R_t(s)\ ds/v_n]\ \exp\left[-\int_s^\infty R_t(s)ds/v_n\right] . \tag{12.9}$$

Insertion of (12.7) into (12.8,9) gives

$$P_0(s,v_n) = \exp[-(A/av_n)e^{-as}] \tag{12.10}$$

and

$$P_t(s,v_n) = (A/v_n)\ \exp[-(A/av_n)e^{-as} - as] . \tag{12.11}$$

The $P_t(s,v_n)$ function reaches its maximum at

$$s_m = (1/a)\ell n(A/av_n) , \tag{12.12}$$

at which

$$P_t(s_m) = a/e = 0.368a . \tag{12.13}$$

Its width at half maximum is

$$W(P_t) = 2.48/a . \tag{12.14}$$

Written in terms of s_m, Eqs.(12.8,9) become

$$P_o(s,v_n) = e^{-\exp[-a(s-s_m)]} \tag{12.15}$$

and

$$P_t(s,v_n) = a\ e^{-\exp[-a(s-s_m)] - a(s-s_m)} \tag{12.16}$$

The P_0 and P_t functions of expressions (12.15,16) are plotted relative to $a(s-s_m)$ in Fig.12.5. A reasonable figure for a is 3 Å^{-1} which makes one unit of $a(s-s_m)$ in Fig.12.5 equal 0.33 Å. The P_t function, for example, has interesting properties which should hold approximately for both resonance and Auger processes. Its form is independent of the atom's velocity but its position, via s_m, varies as the logarithm of its inverse velocity. The terms P_0, P_t refer to the AN process; P_0', P_t' to AD, and P_0'', P_t'' to RN and RI.

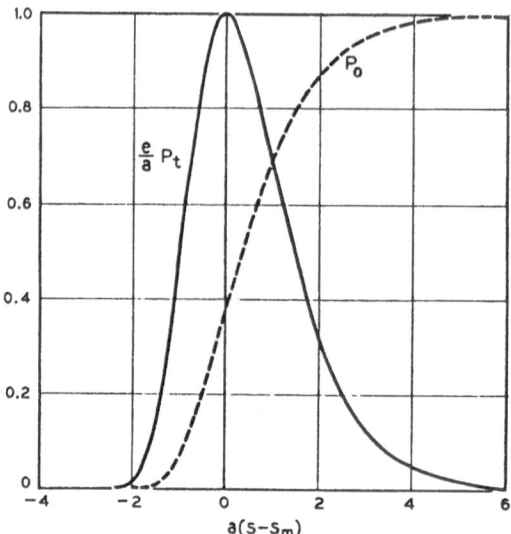

Fig.12.5 Plots of the probability functions: $P_0(s,v_n)$, (12.15), and $P_t(s,v_n)$, (12.16).

12.4 Variation of Atomic Energy Levels Near a Solid Surface

Energy level variation in an atom as a function of the atom-surface separation near a surface can be derived from the potential energies of interaction of the atom with the surface in its ground, excited, and ionized states [12.24,25,7]. Figure 12.6 shows such potential energy curves for He. The He atom interacts with the metal via the van der Waals force which is so small that its potential plot is a horizontal line on the energy scale of the drawing. He$^+$, on the other hand, has a substantial interaction, namely the image potential. This interaction, at least at larger s, is

$$E(He^+) = -e^2/4s = -3.6[eV·Å]/s[Å] . \qquad (12.17)$$

In this expression s is the separation measured from the image plane which lies in the range 0.6 to 0.8 Å outside the jellium edge [12.26]. The singlet and triplet metastable interactions include a van der Waals energy which is, however, still small on the scale of Fig.12.6. An electron energy level diagram, Fig.12.7, in which one plots the energies of electron levels below the vacuum level or ionization limit, can be derived from Fig.12.6.

The upward shift of the metastable levels in Fig.12.7 as the separation s decreases has an interesting consequence for the resonance tunneling processes [12.7]. Resonance neutralization (RN) can be written for atom X as the process

350

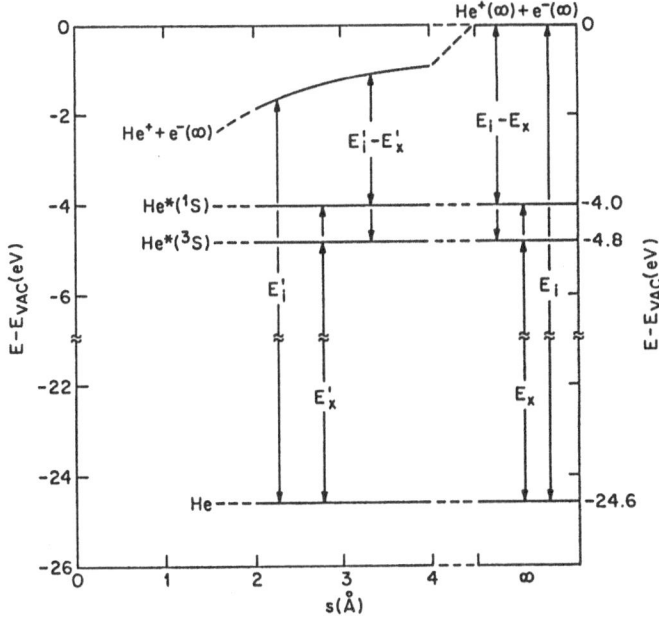

Fig.12.6 Potential-energy diagram of the interactions of four states of the He atom with a metal surface. The atom-surface separation, s, is measured from the image plane.

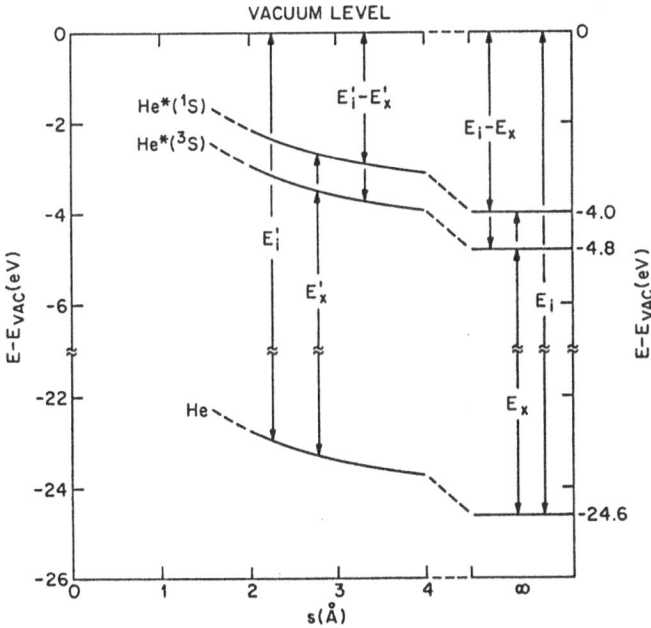

Fig.12.7 An electron-energy diagram, derived from **Fig.12.6,** showing the variation with s of the ground-state level and the two metastable levels of He near a metal surface.

$$X^+ + e_M^-(-\alpha) \rightarrow X^*, \ \alpha > \phi \tag{12.18}$$

in which α is the ionization energy of the electron in the solid that tunnels into the ion. Reversing the arrow and inequality in (12.18) gives the process of resonance ionization of a metastable for which α is then the energy below the vacuum level of the state into which the metastable electron tunnels. Balancing energies on the two sides of these equations yields in each case

$$\alpha = E_i^{*\prime} = E_i^* - 3.6[\text{eV}\cdot\text{Å}]/s[\text{Å}] \ . \tag{12.19}$$

Figure 12.7 shows us that α as a function of s decreases as s decreases. The special case when $\alpha = \phi$ defines a critical distance s_c separating the region in which only resonance neutralization (RN) of an ion is possible, $\alpha > \phi$, $s > s_c$, from that in which only resonance ionization (RI) of a metastable is possible $\alpha < \phi$, $s < s_c$. Thus, it was the inclusion of the image potential in the ion-solid interaction term that led to the concept of the critical distance, s_c [12.24,25,7]. It can play a role in the excitation conversion in an incident metastable He atom [12.27,28] (Sect. 12.6) and its magnitude is crucial in determining whether, as the work function is changed by alkali adsorption, one- or two-stage ejection processes occur for incident metasables [12.29] or ions [12.30]. (See also Fig. 12.12 discussed in Sect. 12.6.

The interpretation of an experiment involving the metastably excited ion $He^{+*}(2s)$ incident on the two surfaces Ni{100} and Ni{110}, requires in an interestingly confirmatory way the energy level variation due to the image potential [12.31]. Instead of being neutralized in an AN type process or de-excited in an AD process, this excited ion is first resonance neutralized to doubly excited states of the neutral He atom. The states in the requisite energy range are two sets of two states each: $He^{0**}(2s^2)^1S$ and a close $(2s,2p)^3P$ satellite that lie about two eV below $He^{0**}(2p^2)^1D$ and its close-lying $(2s,2p)^1P$ satellite. It was observed that when the surface was Ni{100} with $\phi = 5.1$ eV, only the lower energy pair are produced by resonance tunneling and that these states very quickly autoionize to $He^+(1s)$ emitting a single peak of electrons of 34.4 eV kinetic energy. Thus, only the $2s^2$ pair then lay below E_F with the $2p^2$ pair above. When the surface was Ni{110}, with $\phi = 4.7$ eV, two peaks of electrons appeared in the KED at 34.2 and 36.0 eV indicating that both state pairs were then below E_F. Each of these kinetic energies is 0.7 eV greater than those measured when the same doubly excited atoms autoionize in free space. This results from the fact that the final state of the He atom in the autoionization process is the ion He^+ so that the energy separation between it and the initial doubly excited state is greater near the surface by the image potential minus a much smaller van der Waals energy. This interpretation further requires that this same 0.7 eV must be the amount by which the two sets of excited levels rise near the surface above their free-space values. This is just the rise needed to put the states in the configurations relative to E_F described above for the two Ni surfaces and, thus, to provide a gratifying consistency to the interpretation.

12.5 Excitation Conversion of He*(^{1}S) to He*(^{3}S)

Several recent experiments have indicated that He*(^{1}S) converts to He*(^{3}S) near a metal surface whose work function lies in a specific range. *Roussel* [12.32] measured the scattering of incident helium metastables of each species as ions from a Ni{111} surface whose work function was reduced from the clean surface value by potassium atom adsorption. The ion signal is zero on the clean surface because ions formed by resonance tunneling of the metastable electron into an empty level above the Fermi level are bound to the surface by the image potential and are neutralized by the AN process. As K atoms are adsorbed as ions on the clean metal surface, the work function drops from the assumed value of 5.25 eV and there appear an increasing number of K$^+$ sites directly above each of which the interaction potential experienced by an ion is now repulsive over the separation range in which resonance tunneling processes occur. This makes it posssible for He$^+$ ions to leave the surface without neutralization if they are formed by resonance tunneling on or very near to the surface normal through a K atom. For each metastable species the intensity of scattered ions rises from zero for the clean surface and drops to zero again when the work function equals the ionization energy of the metastable, $E_i^{*\prime} = E_i^\prime - E_x^\prime$ because above this limit the metastable electron's energy level lies opposite filled levels in the solid into which it cannot tunnel. Thus, the current of reflected ions, N(He$^+$), for incident ^{3}S He* moves monotonically up and then down with decreasing ϕ until ϕ reaches the limiting value $\phi_3 = E_i^{*\prime}(^3S)$. See Fig. 12.8, taken from Roussel's paper, in which this critical value of ϕ is indicated by an arrow.

The ion current from the ^{1}S metastable, on the other hand, rises and falls in synchrony with the ^{3}S current and then rises again before dropping to zero at the low-ϕ end of its 1.3 eV range. Here ϕ has reached the value $\phi_1 = E_i^{*\prime}(^1S)$, again indicated by an arrow in Fig. 12.8. It is clear that the synchronous behavior of the ^{1}S and ^{3}S ion signals in the range of possible ionization of the ^{3}S is the result of tunneling of the ^{1}S electron into empty states above E_F followed immediately by resonance neutralization to the ^{3}S level. As the work function decreases below the ^{3}S limit at ϕ_3, ^{1}S atoms continue to be ionized at an increasing number of K$^+$ sites until the ultimate turn-

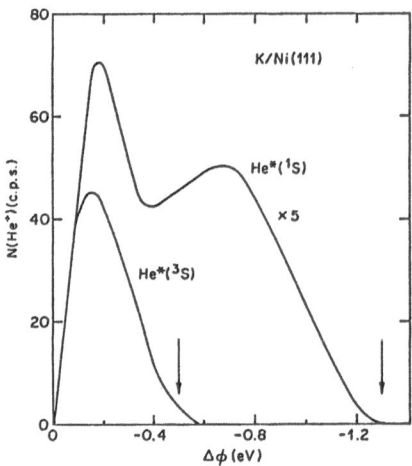

Fig.12.8 The relative numbers of He$^+$ ions produced by scattering the incident metastables, He*(^{3}S) and He*(^{1}S), from a metal surface as its work function is reduced from the clean-surface value by the adsorption of potassium atoms [12.32].

down toward the ^{1}S limit at ϕ_1 is required. *Makoshi* and *Newns* [12.33] have presented a theory of Roussel's experiment based on ionization and neutralization occurring at the intersection of potential curves and only for ions formed on the surface normal through the adsorbed potassium atoms. Their derived ionization probabilities reproduce the main features of the experimental data.

In an experiment in which the change in electronic structure of a potassium-covered, Ni{111} surface is observed as a function of K coverage using metastable de-excitation spectroscopy, *Lee* et al. [12.34] also conclude that excitation conversion from He*(^{1}S) to He*(^{3}S) occurs. De-excitation of the two metastables produces two narrow K(4s) electron peaks in the MDS kinetic energy spectrum that are separated by 0.8 eV, the difference between the ^{1}S and ^{3}S excitation energies. The relative peak heights determine the relative metastable abundance at the distance from the surface where de-excitation occurs. Conversion of ^{1}S to ^{3}S at greater distance from the surface was demonstrated by the observation that adding ^{1}S atoms to their incident beam by turning off their quench lamp produced a large increase in the K(4s) peak that is due to the de-excitation of ^{3}S atoms.

Lee et al. concluded that ^{1}S to ^{3}S conversion should occur because the ^{1}S level, having the smaller ionization energy, will rise above the Fermi level before the ^{3}S level does as the work function is reduced by increased K adsorption (Fig.12.7). Thus, in a certain range of K coverage one can achieve a situation in which the two metastable levels straddle the Fermi level as is shown in Fig.12.9a. In this configuration, the ^{1}S electron can be resonance ionized and the ^{3}S level filled by resonance neutralization. This mode of excitation conversion requires that the atom must reach a point between the critical separation s_c of the singlet level and that of the triplet level.

354

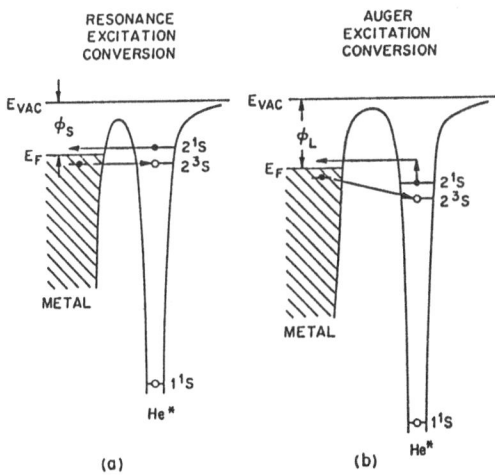

Fig.12.9 Electron-energy diagrams showing two types of excitation conversion of He*(^{1}S) to He*(^{3}S): a resonance type at (a), [12.24,25,7]; and an Auger type, after [12.35], at (b).

In discussing their results demonstrating excitation conversion at a Cs covered Cu{110} surface whose work function is 1.3 eV, *Woratschek* et al. [12.35] point out that the resonance type process of Fig.12.9a would have to occur at smaller surface-atom separation than that at which one expects the subsequent de-excitation process to occur. To avoid this impossible state of affairs, they propose a significant addition to the known set of atom-surface interactions. It is an Auger-type process of excitation conversion illustrated in Fig.12.9b, a reproduction of a figure in their paper. This has the important characteristic that it can occur at greater atom-surface separation than Auger excitation conversion and is thus an appropriate precursor to AD even for surfaces of low work function. This characteristic is rooted in the fact that the wave function of the metastable level to which the down electron transits extends much farther from the atom core than that of the deep-lying ground level to which the down electron transits in Auger de-excitation (compare Figs. 12.9b and 12.1b). In Fig.12.10, reproduced from [12.28], the two curves are the AD electron spectra for triplet (top) and singlet (bottom) metastables incident on a Cu{111} surface covered with 10 or more layers of Cs. Note that the AD spectrum for incident ^{1}S metastables displays a very small peak β at the kinetic energy where the Cs 6s electrons should appear if they are in fact ejected by AD of ^{1}S metastables. The spectrum of the great majority of ejected electrons α is identical to that ejected by ^{3}S metastables, demonstrating the almost total conversion of the incoming singlets to triplets before the AD process occurs.

At a K covered Ni{100} surface, incident ions convert to metastable atoms before the AD process occurs. The width of the ejected

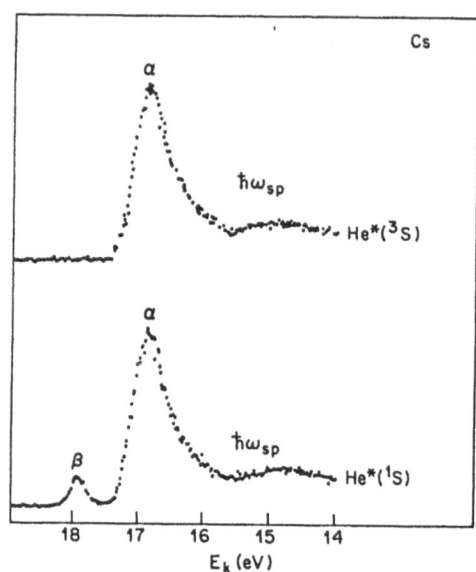

Fig.12.10 Kinetic energy distributions showing excitation conversion of $He^*(^1S)$ to $He^*(^3S)$ at a Cs-covered metal surface as explained in the text. After [12.28].

electron kinetic energy distribution is 18.2 eV which equals the excitation energy of the ^3S atom (19.8 eV) minus the surface work function (1.6 eV) [12.36]. Also, the difference in widths of the distribution for incident He^+ and Ne^+ is consistent only with the difference between the ^3S He^* and the Ne^* metastable excitation energies [12.37]. Each of these facts further demonstrates that even though both singlet and triplet metastables are formed by resonance neutralization of the incoming ion, essentially all of the singlets are converted to triplets before Auger de-excitation occurs.

12.6 What Determines the Ultimate Mode of Electron Ejection?

For a given type of surface, we ask whether the mode of electron ejection for an incident ion will be one-stage AN or two-stage RN + AD and for an incident metastable will be one-stage AD or two-stage RI + AN. For clean metal surfaces the answers to these questions depend upon where the critical distance s_c, defined in Sect.12.4, lies with respect to the distance region in which resonance and Auger processes occur. Since s_c is the distance at which the effective ionization energy of the metastable state, $E_i^{*\prime}$, equals the work function ϕ, its value is governed both by the type of surface and the nature of the metastable state of the noble gas atom involved. Here we consider surfaces of various ϕ but only incident He^+ and $He^*(^3S)$. Figure 12.11 depicts schematically the interrelation of $E_i^{*\prime}$ (taken from Fig.12.7), ϕ, and s_c. The magnitude of s_c is obtained from (12.19) with $\alpha = \phi$ as

$$s_c[\text{Å}] = 3.6\ [\text{eV·Å}]/(E_i^*[\text{eV}] - \phi[\text{eV}]) . \tag{12.20}$$

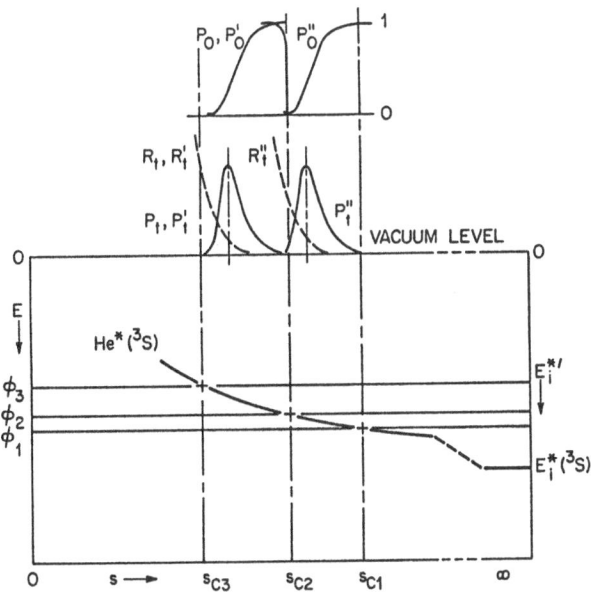

Fig.12.11 Electron-energy plot showing how the work function of a surface determines the position of the critical separation, s_c, relative to the functions P_0 and P_t for the resonance and Auger processes. s_c is the separation at which the work function equals the ionization energy of $He^*(^3S)$.

As one considers different solids or a metal with a variable amount of adsorbed alkali it is possible to vary ϕ so that s_c scans through the range of s in which the resonance tunneling and Auger processes occur. The occurrence of these processes can be displayed by either the P_0 or P_t function. Recall that, for a given process, $P_0(s,v_n)$ is the survival probability for the atom in its inital state and $P_t(s,v_n)ds$ is the probability that the process occurs in ds at s. [(12.15,16) of Sect.12.3]. These functions, shown at the top of Fig.12.11, are distinguished as in Sect.12.3 by writing P_0, P_t for AN, $P_0{}'$, $P_t{}'$ for AD, and $P_0{}''$, $P_t{}''$ for RN and RI. The positions on the s scale of each of these functions will depend on the atomic character (ion or metastable), the normal velocity v_n, and the parameters A and a of the transition rate function R_t $= Ae^{-as}$. P_t and $P_t{}'$ are superposed in the figure as are the two $P_t{}''$ for RI and RN. The $P_t{}'$ and $P_t{}''$ functions for incoming metastables will, of course, lie at a larger s than the P_t and $P_t{}''$ associated with the more rapidly moving incoming ions. Hence no numerical scale for s is given in the figure. The terms P_t and $P_t{}'$ for the Auger processes lie closer to the surface than the $P_t{}''$ of their corresponding resonance processes because resonance rates are higher than Auger rates at a given s.

There are two limiting critical distances s_c indicated in Fig.12.11. At s_{c1} both the resonance and Auger processes occur at $s < s_c$ where $E_i{}'$ is smaller than ϕ and thus RI is possible but RN is not. Thus the

357

electron ejection process is AN for both ions and metastables since ions undergo AN directly and metastables RI + AN. This condition prevails for clean metal surfaces whose ϕ exceeds ϕ_1. For $s_c = s_{c3}$, $E_i^{*\prime}$ is greater than ϕ, RN is possible but RI is not, and the ejection process for both incident particles is AD with metastables undergoing only AD and ions RN + AD. This condition prevails for surfaces whose ϕ is less than ϕ_3, examples being surfaces of the alkalis or of higher work function metals with a uniform alkali overlayer.

Matters become more complicated when s_c lies between s_1 and s_3. Consider as an example $s_c = s_{c2}$. The resonance process is now RN and incident ions will turn to metastables by the time the atoms reach s_{c2}. At this separation, however, the process that the P_0'' function governs changes from RN to RI. Since the rate function R_t'' of these resonance processes is now high, the newly formed metastables rapidly return to ions before the Auger ejection process occurs, as the rapidly rising P_0'' function at $s = s_{c2}$ indicates. This same situation applies to incident metastables, so for either incident particle, the ultimate ejection process is AN. *Sesselmann* et al. [12.38], who discuss the role of the resonance processes relative to the final ejection process for incident metastables, have concluded that the AD process occurs at separations between 1.4 and 4.4 Å measured relative to the image plane.

The usual method of varying ϕ continuously is to adsorb an alkali. The initial coverage is sparse, there being relatively large unoccupied spaces between the alkali atoms. Since the resonance and Auger processes are localized in the immediate vicinity of the incoming atom, it is necessary to think in terms of two local work functions, one in the bare spaces which is the same as that of the clean surface and the other, at the alkali sites, approximating that of the uniformly covered surface [12.39]. Thus, a principal complication in the work function range corresponding to the interval between s_{c1} and s_{c3} in Fig. 12.11 is due to this mixture of electron ejection from high and low work function patches on the partially covered surface.

Woratschek et al. [12.29] have plotted, on an energy scale that is the equivalent of energy above the Fermi level, the kinetic energy distributions of electrons ejected by incident He* as a Cu{110} surface covers with adsorbed potassium. Figure 12.12 presents a similar plot for He$^+$ ions incident on Cu{100} in which an identical sequence of distributions is found [12.30]. Curve 1 is for the clean surface where electron ejection occurs via AN yielding a folded spectrum. As the concentration of K atoms increases, the spectrum changes as shown in curves 2 through 9. In the concentration range of curves 4 and 5, the critical distance s_c passes through the s_{c1} to s_{c3} transition region depicted in Fig. 12.11 where for incident ions RN becomes possible, ions turn to metastables, and ejection occurs via RN + AD. In the case of

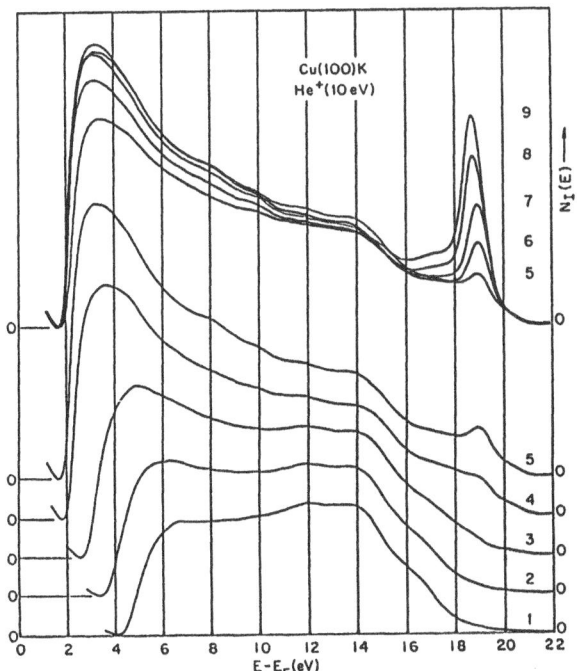

Fig.12.12 Kinetic energy distributions of electrons ejected by incident He⁺ ions from a Cu(100) surface which, from curve 1 through curve 9, has an increasing number of K atoms adsorbed upon it. Curve 1 is for the clean surface.

incident metastables [12.29] a converse situation prevails in which ejection occurs via RI + AN at the clean surface and, as K concentration increases, changes in the transition region to AD as RI is no longer possible.

Curve 1 of Fig.12.12 for the clean copper surface displays a shoulder at $E - E_F = 14$ eV that also appears at the same energy in curves 2 through 5 despite the large ϕ change as the concentration of adsorbed K increases. As *Woratschek* et al. [12.29] point out, this is direct experimental evidence that AN is a local probe contributing to the total distribution a component like curve 1 from clean patches on the partially K-covered surface. Similarly, the small peak at $E - E_F = 19$ eV in curves 3 through 5 consisting of electrons ejected from the filled portion of the K(4s) resonance below E_F is evidence that RN + AD is also a local probe of the electron energy distribution near K sites on the partially covered surface. In curves 5-9 the K(4s) peak increases in magnitude as the K concentration increases without appreciable change in ϕ. Note that the value of $E - E_F$ at the low energy, vacuum-level cutoff of each distribution is equal to the work function of the surface then prevailing.

Table 12.1. Electron Ejection Modes for Ions and Metastables Incident on Various Surfaces

Category	Surface	Ions	Metastables
1	high ϕ metal	AN	RI+AN
2	same, with adsorbed "non-metallic" monolayer ($n \leq 1$)	AN	RI+AN
3	same, with adsorbed or condensed multilayer ($n \geq 2$)	AN	AD
4	high ϕ metal, with adsorbed alkali	RN+AD	AD
5	same, with coadsorbed alkali and other atom or molecule	(RN+AD)	AD
6	high ϕ semiconductor, with wide forbidden gap	AN	AD
7	same, with narrow gap	AN	RI+AN

The surfaces we have considered above allow the resonance processes RI or RN to occur when the metastable electron level lies above or below E_F, respectively. This is so because these surfaces permit a strong overlap between the wave function of the metastable electron and the evanescent tail outside the surface of unfilled or filled bulk states for high or low ϕ, respectively. These are the surfaces in class 1 and 4 in Table 12.1. Another class of high ϕ surfaces that permit RI are the clean surfaces of the transition or noble metals with an adsorbed layer that is no more than a monolayer thick (Class 2, Table 12.1). Even though such a monatomic layer does not have a resonance in the energy range including the metastable atomic level the evanescent tail of the wave function of the unfilled substrate level does overlap that of the metastable electron sufficiently to permit resonance tunneling. Another important class of surface (Class 3, Table 12.1) comprises a clean high ϕ metal with a "nonmetallic" adsorbate or condensate of two or more atomic layers that is sufficiently thick to decouple the metastable wave function from the allowed and unfilled levels of the substrate thereby blocking RI and making the Auger ejection process AD for incident metastables and AN for incident ions. See Fig.12.13 which illustrates how an erectly adsorbed diatomic molecule (CO or NO) can do this. Clean high ϕ semiconductors with wide forbidden gaps do not permit RI but those with narrow forbidden gaps do (Classes 6 and 7, repectively, Table 12.1). Thus, several

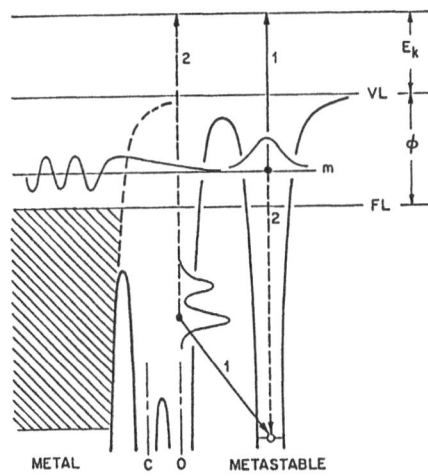

Fig.12.13 Electron-energy diagram showing how adsorbed molecules, two or more atomic layers in thickness, decouple the wave function of the metastable electron of He* from the evanescent tails of the wave functions of unfilled but allowed levels of the solid above the Fermi level, thus preventing resonance ionization, RI.

do (Classes 6 and 7, respectively, Table 12.1). Thus, several characteristics of a solid surface conspire to determine the nature of the ultimate mode of electron ejection for incident ions and metastable atoms.

12.7 A Selective Listing of Investigations with Some Comments

In this section, briefly annotated references, organized according to the categories of surface type in Table 12.1, will be made to those papers in the rather extended literature on INS and MDS that deal with the surface interactions that underlie these spectroscopies.

Category 1 - Incident Ions via AN

Theories of AN by *Heine* [12.15], *Appelbaum* and *Hamann* [12.16] *Horiguchi* et al. [12.22] and *Hentschke* et al. [12.23] have been referred to in earlier sections. The theoretical interpretation [12.15,16] of the kinetic energy distributions (KEDs) of AN as the cross fold $F = V \otimes W$ of the local density of states at the surface, V, and at the ion, W, is of particular importance. In the method of INS [12.9] the initial-state transition density is derived from the experimental KED, taken to be $F = V \otimes W$, as the function U, the deconvolution square root of F or $V \otimes W$ which may be written as $U = F^{1/2} = (V \otimes W)^{1/2}$. As indicated earlier, this deconvolution results in a function that is a convolution mean of V and W that displays peak or resonance features in the initial-state transition density at the correct energies but with some possible distortion of intensities depending on the nature of the wave function involved. In the work with incident ions, a step midpoint sequential unfolding procedure has been employed [12.9,14]. This is a sensitive method requiring very smooth data (obtained by averaging many runs) and a means of finding the correct zero of the function by the use of a Van Cittert type extrapolation of KEDs at two ion energies to a debroadened KED. Unfolding of the basic data is, of course, required for any AN process (see next section).

Because AN requires deconvolution, INS has two unique characteristics. The INS KED will be the true fold of the initial-state transition density only if the band of electronic states involved, after giving up the first electron, reconstitutes itself in a time short with respect to the lifetime of the AN process ($\simeq 10^{-15}$ s). This means that narrow bands such as surface states of semiconductors are, in general, not detectable by INS [12.16]. A second unique characteristic of INS relates to the possibility that a surface may present a nonfolding electron distribution to the incident ion ("Category 6 - Ions").

Energy broadening in the AN process for incident ions is certainly greater than that for incident metastables but clearly varies considerably from surface to surface. Its magnitude must certainly be related to the density of states immediately above the Fermi level into which band electrons are nonadiabatically excited by the action of the moving charge outside the surface [12.40]. This broadening is large for clean Ni{100}, for example, but much smaller for Cu as is seen in curve 1 of Fig.12.12. Comparing this curve with the corresponding curve obtained with AD for incident metastables [12.29] one concludes that this latter curve reaches zero at the energy $E-E_F$ = 18.5 eV in Fig.12.12. The broadened tail beyond this point in curve 1 of Fig.12.12 is small and is easily handled by the debroadening step of the INS data-handling procedure. It appears that the very large broadenings predicted by *Moyer* and *Orvek* [12.41] do not occur in INS.

E. Hood et al. [12.42] have discussed a phenomenological model of the AN ejection process demonstrating a fit to the KED of clean Ni{111} achieved by the appropriate choice of several parameters. A theory by *Modinos* et al. [12.43] produces an AN spectrum that is as intense as the experimental Ni{111} spectrum only in its higher energy range. These authors believe that this may be due to the presence of a large secondary electron component in the experimental spectrum. This appears to be contradicted by an earlier study of this compoment [12.44] and by the conclusion of *Hentschke* et al. [12.23] that "all energetically allowed transitions contribute significantly to the transition rate, and not just transitions from inital states lying energetically near E_F".

Category 1 - Incident Metastables via RI + AN

This process has been discussed in several papers [12.45,38] but deconvolution of the data has been done only by the Munich group [12.38] in whose paper the mathematical procedure using spline functions is presented and references are given to earlier papers reporting its development. Because of the smaller broadening accompanying slower ions, the method works without a debroadening procedure, thus, giving a somewhat broadened initial-state transition density. The paper by *Sesselmann* et al. [12.38] is a particularly good review of this category of electron ejection. *Onellion* et al. [12.46] have reported the perfection of a spin-polarized AN spectroscopy by means of which they can probe surface magnetism.

Category 2 - Incident Ions via AN
The work in this category involving principally the chalcogens O, S, Se, and Te adsorbed in ordered arrays on clean single crystal faces of nickel has been reviewed in [12.47].

Category 2 - Incident Metastables via RI +AN
Sesselmann et al. [12.38] provide in their Table III a listing of the adsorbate-covered surfaces at which RI + AN is observed experimentally.

Category 3 - Incident Ions via AN
No work has been done with ions incident on this category of surface for the very good reason that incident metastables eject electrons via AD not requiring deconvolution.

Category 3 - Incident Metastables via AD
It is both an important and convenient fact that adsorbate layers two or more atoms thick decouple the metastable level from states above the Fermi level at high ϕ surfaces making the AD process possible for this chemically very interesting category of surface. This possibility was first suggested by *Allison* et al. [12.48]. *Shibata* et al. [12.10], studying condensed anthracene, were the first to confirm that an AD spectrum reveals valence bands in the surface layer analogous to photoemission (UPS). *Munakata* et al. [12.49] first demonstrated that AD as a probe of the outermost layer of an adsorbate could be used to indicate the orientation of a complex molecule by comparing AD spectra with those from UPS and gas phase Penning ionization. *Harada* [12.50] has reviewed work on organic films showing that the characteristics of AD can reveal changes in the molecular orientation and electronic state of the outermost surface layer of such films during epitaxial growth and the amorphous to crystal transition. *Sesselmann* et al. [12.51] have studied for several polyatomic molecules the roles of surface concentration and molecular-surface geometry in determining when AD or RI + AN occur. *Bozso* et al. Studied tilting of adsorbed CO on Ni{100} [12.13] and the decomposition of NO on this surface precovered with O and N+O [12.52]. Auger de-excitation and thermal desorption have

been used to study multilayer formation of NH_3 and CO on Ni{111} [12.53]. Perhaps these are enough examples to show that AD is an important tool in the study of adsorbed and condensed multilayers.

Category 4 - Incident Ions via RN + AD

The work with incident ions on K covered Ni and Cu surfaces [12.36,37] and the data in Fig.12.12 have already been discussed.

Category 4 - Incident Metastables via AD

Here too the work in this category on K and Cs covered surfaces has already been discussed in connection with excitation conversion and the variation of ϕ to demonstrate change in the ultimate ejection mechanism, see [12.28,29,39,54].

Category 5 - Incident Ions via RN + AD

There is no work in this category.

Category 5 - Incident Metastables via AD

It is known that coadsorbed K in small amounts on transiton metals modifies substantially the properties of chemisorbed CO about which several papers using AD spectroscopy have been published [12.55-57].

Category 6 - Incident Ions via AN

Hagstrum and *Becker* studied the {111}, {110}, and {100} faces of Si and Ge, finding that the results using AN reflected the gross features of the bulk energy level structures but differed in detail both from the bulk and from each other [12.58]. However, one could not deconvolve the data for the Si {111} 7x7 surface without removing a high energy "tail" by the zero-finding procedure. Interpretation of the need for this unusual procedure led to the understanding that this surface exhibited an electronic inhomogeneity, producing a KED that is not a convolution square [12.59]. This illustrated another unique characteristic of the AN process.

Category 6 - Incident Metastables via AD

Nishigaki et al. [12.60] have shown that incident He* undergoes the AD process at the Si{111} 7x7 surface.

Category 7 - Incident Ions via AN

Hagstrum and *Becker* have shown that He* ions incident on multilayer ordered Te on Ni{100} undergo AN as expected since no electrons are available to neutralize them [12.61].

Category 7 - Incident Metastables via RI + AN

Harada et al. [12.62] have shown that He*(^{3}S) and Ne*(3P) deexcite at a Se surface via RI + AN since the metastable electron can in each case tunnel into allowed but unfilled levels above this semiconductor's narrow forbidden gap.

12.8 Summary

It is clearly evident from Table 12.1 that neither INS nor MDS alone can be used to study all possible surfaces. Thus a knowledge of the physics of both the AN and AD processes is essential. Wherever AD does occur, it is clearly the spectroscopy of choice because it is basically a one-electron process resembling UPS but with a much greater surface specificity and deconvolution is not necessary. Metastables are generally preferred to ions as incident particles because their lower velocity results in less broadening of the kinetic energy distributions of electrons ejected by either AN or AD. It is possible to claim that the interaction with surfaces of atoms that carry potential energy is now a quite well understood subfield of particle-solid interactions. Trying to understand AN and to devise a means of using it has been an interesting physico-mathematical endeavor. The use of AD in the study of surface chemistry is exciting, particularly because AD dominates category 3, high ϕ metals with larger molecular adsorbates. This is the area that will see the major effort in the future, certainly among chemists. The theory of the ejection processes is not easy, but one can hope for continued progress here as well.

References

12.1 F.M. Penning: Proc. R. Soc. Amsterdam 31, 14 (1928); 33, 841 (1930)
12.2 M.L.E. Oliphant: Proc. R. Soc. London A127, 373 (1930)
12.3 P.B. Moon: Proc. Camb. Phil. Soc. 27, 57 (1931)
12.4 M.L.E. Oliphant: Proc. R. Soc. London A124, 227 (1929)
12.5 W. Uyterhoeven, M.C. Harrington: Phys. Rev. 36, 709 (1930)
12.6 R. Dorrestein: Physica 9, 433 (1942)
12.7 H.D. Hagstrum: Phys. Rev. 96, 336 (1954)
12.8 H.D. Hagstrum: Phys. Rev. 122, 83 (1961)
12.9 H.D. Hagstrum: Phys. Rev. 150, 495 (1966)
12.10 T. Shibata, T. Hirooka, K. Kuchitsu: Chem. Phys. Lett. 30, 341 (1975)
12.11 H. Conrad, G. Ertl, J. Küppers, S.W. Wang, K. Gérard, H. Haberland: Phys. Rev. Lett. 42, 1082 (1979)
12.12 J. Roussel, C. Boiziau, R. Nuvolone, C. Reynaud: Surf. Sci. 110, L634 (1981)
12.13 F. Bozso, J.T. Yates, Jr., J. Arias, H. Metiu, R.M. Martin: J. Chem. Phys. 78, 4256 (1983)
12.14 H.D. Hagstrum, G.E. Becker: Phys. Rev. B 4, 4187 (1971)
12.15 V. Heine: Phys. Rev. 151, 561 (1966)
12.16 J.A. Appelbaum, D.R. Hamann: Phys. Rev. B 12, 5590 (1975)
12.17 G. Holst, E. Oosterhuis: Physica 1, 78 (1921)
12.18 M.L.E. Oliphant, P.B. Moon: Proc. R. Soc. London A 127, 388 (1930)
12.19 H.S.W. Massey: Proc. Camb. Phil. Soc. 26, 386 (1930); 27, 460 (1931)
12.20 A. Cobas, W.E. Lamb, Jr.: Phys. Rev. 65, 327 (1944)
12.21 S.S. Shekhter: J. Exp. Theor. Phys. USSR 7, 750 (1937)
12.22 S. Horiguchi, K. Koyama, Y.H. Ohtsuki: Phys. Stat. Sol. (b) 87, 757 (1978)
12.23 R. Hentschke, K.J. Snowdon, P. Hertel, W. Heiland: Surf. Sci. 173, 565 (1986)

12.24 H.D. Hagstrum: Phys. Rev. **91**, 543 (1953), Sect.V, Fig.10

12.25 L.J. Varnerin, Jr.: Phys. Rev. **91**, 859 (1953)

12.26 N.D. Lang, W. Kohn: Phys. Rev. **7**, 3541 (1973)

12.27 J. Lee, C. Hanrahan, J. Arias, F. Bozso, R.M. Martin, H. Metiu: Phys. Rev. Lett. **54**, 1440 (1985)

12.28 B. Woratschek, W. Sesselmann, J. Küppers, G. Ertl, H. Haberland: Surf. Sci. **180**, 187 (1887)

12.29 B. Woratschek, W. Sesselamnn, J. Küppers, G. Ertl, H. Haberland: Phys. Rev. Lett. **55**, 1231 (1985)

12.30 H.D. Hagstrum: to be published

12.31 H.D. Hagstrum, G.E. Becker: Phys. Rev. B **8**, 107 (1973)

12.32 J. Roussel: Physica Scripta T4, 96 (1983)

12.33 K. Makoshi, D.M. Newns: Surf. Sci. **159**, 149 (1985)

12.34 J. Lee, C. Hanrahan, J. Arias, F. Bozso, R.M. Martin, H. Metiu: Phys. Rev. Lett. **54**, 1440 (1985)

12.35 B. Woratschek, W. Sesselmann, J. Küppers, G. Ertl, H. Haberland: Phys. Rev. Lett. **55**, 611 (1985)

12.36 H.D. Hagstrum: Phys. Rev. Lett. **43**, 1050 (1979)

12.37 H.D. Hagstrum: J. Vac. Sci. Technol. **20**, 626 (1982)

12.38 W. Sesselmann, B. Woratschek, J. Küppers, G. Ertl, H. Haberland: Phys. Rev. B **35**, 1547 (1987)

12.39 H. Conrad, G. Ertl, J. Küppers, W. Sesselmann, H. Haberland: Surf. Sci. **100**, L461 (1980)

12.40 H.D. Hagstrum, Y. Takeishi, D.D. Pretzer: Phys. Rev. **139**, A526 (1965)

12.41 C.A. Moyer, K. Orvek: Surf. Sci. **114**, 295 (1982); **121**, 138 (1982)

12.42 E. Hood, F. Bozso, H. Metiu: Surf. Sci. **161**, 491 (1985)

12.43 A. Modinos, S.I. Easa: Surf. Sci. **185**, 569 (1987)

12.44 H.D. Hagstrum, Y. Takeishi: Phys. Rev. **137**, A304 (1965)

12.45 F. Bozso, J. Arias, C. Hanrahan, R.M. Martin, J.T. Yates, Jr., H. Metiu: Surf. Sci. **136**, 257 (1984)

12.46 M. Onellion, M.W. Hart, F.B. Dunning, G.K. Walters: Phys. Rev. Lett. **52**, 380 (1984)

12.47 H.D. Hagstrum: In *Electron and Ion Spectroscopy of Solids*, ed. by L. Fiermans, J. Vennik, W. Dekeyser (Plenum, New York 1978) p.273

12.48 W. Allison, F.B. Dunning, A.C.H. Smith: J. Phys. B **5**, 1175 (1972)

12.49 T. Munakata, K. Ohno, Y. Harada: J. Chem. Phys. **72**, 2880 (1980)

12.50 Y. Harada: Surf. Sci. **158**, 455 (1985)

12.51 W. Sesselmann, B. Woratschek, G. Ertl, J. Küppers, H. Haberland: Surf. Sci. **130**, 245 (1983); **146**, 17 (1984)

12.52 F. Bozso, J. Arias, C.P. Hanrahan, J.T. Yates, Jr., R.M. Martin, H. Metiu: Surf. Sci. **144**, 591 (1984)

12.53 H. Tochihara, G. Rocker, J.D. Redding, J.T. Yates, Jr., R.M. Martin, H. Metiu: Surf. Sci. **176**, 1 (1986)

12.54 J. Lee, C.P. Hanrahan, J. Arias, R.M. Martin, H. Metiu: Surf. Sci. **161**, L543 (1985)

12.55 J. Lee, C.P. Hanrahan, J. Arias, R.M. Martin: Phys. Rev. Lett. **51**, 1803 (1983)

12.56 J. Lee, J. Arias, C.P. Hanrahan, R.M. Martin: Phys. Rev. Lett. **51**, 1991 (1983)

12.57 J. Lee, J. Arias, C.P. Hanrahan, R.M. Martin: J. Chem. Phys. **82**, 485 (1985)

12.58 H.D. Hagstrum, G.E. Becker: Phys. Rev. B **8**, 1580 (1973)

12.59 H.D. Hagstrum, G.E. Becker: Phys. Rev. B **8**, 1592 (1973)

12.60 S. Nishigaki, K. Takao, T. Yamada, M. Arimoto, K. Kometsu: Surf. Sci. **158**, 473 (1985)

12.61 H.D. Hagstrum, G.E. Becker: In "The Physical Basis for Heterogeneous Catalysis", ed. by E.Drauglis, R.I. Jaffee (Plenum, New York 1975) p.173

12.62 Y. Harada, H.Ozaki, K. Ohno: Solid State Commun. **49**, 71 (1984)

13. Equilibrium Crystal Shapes and Interfacial Phase Transitions

Michael Wortis

Department of Physics, Simon Fraser University
Burnaby, BC V5A 1S6, Canada

This chapter reviews the present status of the theory of equilibrium crystal shapes, with particular emphasis on the connection between the crystal-shape problem and recent results in two-dimensional statistical mechanics. Interpretation of the Wulff construction as a Legendre transformation allows the crystal shape to be viewed as a free energy. Thus, edges on the crystal shape correspond to thermodynamic phase boundaries, and critical behavior is observed in the neighborhood of certain edges. The thermal evolution of the crystal shape is reinterpreted from this point of view, and results are compared with recent experiments.

13.1 Introduction

Most crystals which we see in nature or under conditions which are not specially controlled do not exhibit true equilibrium shapes. Examples are snowflakes, which are growth forms, and naturally occurring mineral crystals, which may or may not have been at equilibrium under ambient conditions (typically, high temperature and high pressure) when and where they were formed but are almost certainly not at equilibrium as observed in the laboratory. History independence is a characteristic of equilibrium. Thus, we know that snowflakes, for example, are not equilibrium forms, because many different snowflake shapes occur in the same snowstorm. Indeed, the shape of an individual snowflake carries a detailed record of the precise atmospheric conditions (principally, the degree of supersaturation) along the trajectory on which it fell.

In order to observe *equilibrium* crystal shapes, there are two requirements: stable coexistence and time. Figure 13.1 shows a generic material phase diagram. Equilibrium crystal shapes may only be seen when conditions of temperature T and pressure P are adjusted to lie along the solid/vapor or solid/liquid phase boundaries, $P_0(T)$. Under these conditions a chunk of solid material of fixed volume may coexist stably with the surrounding fluid; however, time is still necessary to achieve shape equilibrium. The physical processes which redistribute

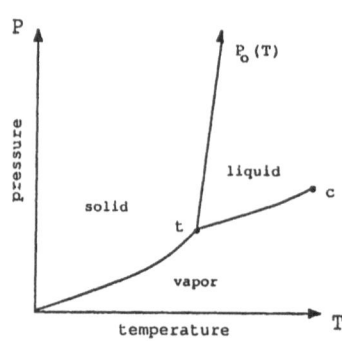

Fig. 13.1. Bulk phase diagram of generic material. Macroscopic crystals can coexist at equilibrium with a fluid phase only along the solid/vapor and solid/liquid phase boundaries.

material, thus allowing the solid chunk to reach its equilibrium shape, depend on ambient conditions but are typically fluid-phase transport and/or surface diffusion. These are hydrodynamic processes and may be very slow for crystals of macroscopic size. For example, micron-scale metal crystals (for instance, Au [13.1], Pb [13.2,3], and In [13.4,5]) in contact with a tenuous vapor may require times of several hours to equilibrate, even near the bulk melting temperature. Larger crystals cannot equilibrate on practical time scales. On the other hand, centimeter-scale crystals of solid ^{4}He in coexistence with the super-fluid [13.6] near 1.3 K equilibrate in a few seconds, because the coexisting superfluid phase provides excellent mass transport.

Precise experiments to measure equilibrium shapes are not easy, at least in part because of the above two requirements:

a) Times necessary for equilibration may depend sensitively on temperature, especially when surface diffusion is the dominant transport mechanism. For systems such as the metal crystals mentioned above, this may mean that there is only a narrow window of temperature over which experiments can be done at all. Above this window, the crystal melts; below it, equilibration takes impractically long.

b) Careful temperature control is important. If a fluctuation in temperature makes one region of the experimental cell momentarily cooler, then crystallites in that region will tend to grow at the expense of those in other regions, resulting in growth forms there and dissolution forms elsewhere.

c) The crystal must in principle be observed in situ, as equilibrated. Removing it from the growth chamber for observation changes the ambient conditions and will in general initiate alteration of the crystal shape. Experiments are often carried out at high temperatures (see a), but observations - typically by scanning electron microscopy - are made at a lower temperature. If the crystal is cooled sufficiently rapidly, then its overall shape cannot change significantly; however, the possibility of important local shape changes is more problematic.

d) Impurity effects are exceedingly important and not always easy to control. Crystal shape is determined, as we shall see below, by interface energetics. Impurity species may arrive at the interface from the interior of the crystal (surface segregation) or from the vapor phase (adsorption). Once there, they can migrate selectively to special surface sites such as steps and kinks. The corresponding energy modification can lead to qualitative changes in the crystal shape [13.7,8].

e) Finite-size effects. The theory in its classic form (see below) describes the outline of a large crystal, in the so-called *thermodynamic limit*, illustrated in Fig.13.2. The notion of a crystal shape is always somewhat fuzzy on a microscopic scale, both because of the dynamical motion of the atoms and in the statistical sense that two nominally identical samples differ due to thermal fluctuations. The thermodynamic limit of the crystal outline $R(\hat{h})$ is defined by considering a sequence of larger and larger crystals under the same ambient conditions. Each crystal is then photographed and all the images are expanded or reduced to some arbitrary common diameter. It is the limiting shape $r(\hat{h})$ of these images, in which the scale of atomic fuzziness shrinks to zero, that is described by the theory. A complete theory of the finite-size corrections to this thermodynamic limiting shape does not yet exist. Experiments necessarily treat finite samples, so the relationship of even the best experiment to the theoretical limit is not as clean as one would like. This is particularly so in experimental tests of the finer features of predicted shapes, such as the detailed behavior in the near vicinity of certain crystal edges.

Before discussing the theory, it is useful to have in mind what some typical equilibrium shapes look like, as seen in the laboratory. It is important to emphasize again that it is the macroscopic, thermodynamic limit, defined in e) above, in which we are interested. Interest-

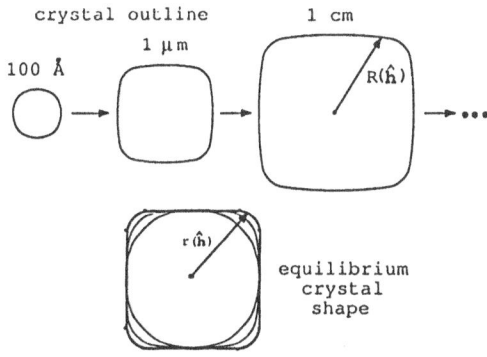

Fig.13.2. The thermodynamic limit. The equilibrium crystal shape is defined in the V(olume) $\rightarrow \infty$ limit, (13.1). The proportions of the crystal outline become scale invariant in this limit, under suitable conditions. The intrinsic atomic-scale fuzziness of the outline disappears when lengths are rescaled by $V^{1/3}$. Mathematically sharp features such as the strict planarity of facets, the sharpness of edges and corners, etc., are absent for finite V.

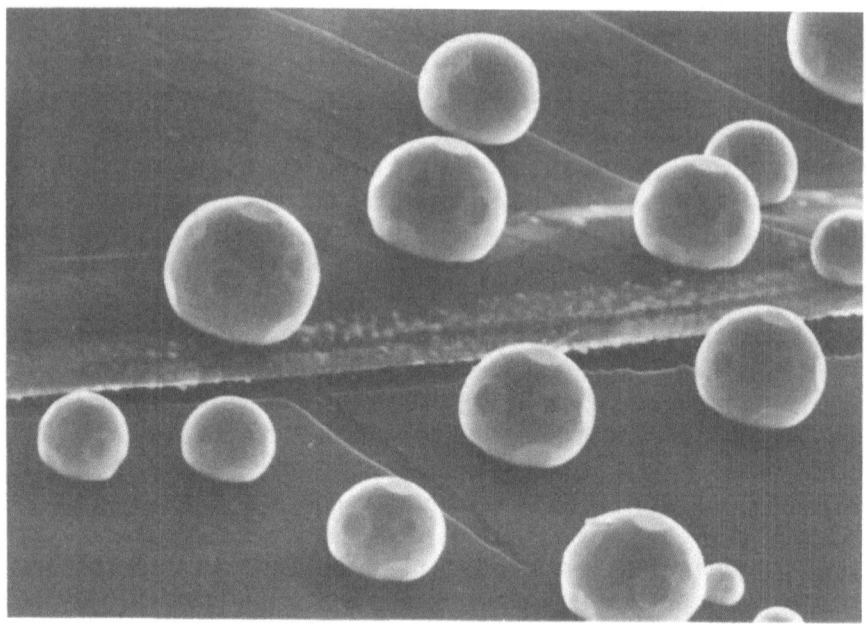

Fig. 13.3. Gold crystals on a graphite substrate at about 1000°C, prepared by Heyraud and Métois, as described in the text [13.1]. The larger crystals are about 5 μm in diameter. The presence of the substrate does not change the crystal shape in the thermodynamic limit, other than to truncate it. Crystallites of different sizes have essentially the same shape. Note the tendency of the crystals to orient on the substrate.

ing, mathematically singular, sharp shapes can only occur in this limit (just as finite-size samples cannot undergo sharp phase transitions). Laboratory crystals approach this limit but always exhibit some finite-size rounding of singular features.

Figure 13.3 shows crystals of gold on a graphite substrate prepared by *Heyraud* and *Métois* [13.1,9,10] at Marseille. This sample was prepared by vapor depositing 3000 Å of gold onto a graphite substrate at room temperature. At this temperature the gold atoms are not mobile but simply stick where they land, forming a continuous film. The sample was then heated to 1000°C in a sealed cell (to keep the system at the coexistence curve, since the vapor pressure of Au at this temperature is $\sim 5 \cdot 10^{-6}$ Torr [13.10]). This temperature is still well below the bulk melting point of gold ($T_t = 1063°C$), so atoms are ordered locally in a crystalline array; nevertheless, transport (predominantly surface diffusion) is sufficiently active so that the configuration of the bulk material can change, albeit rather slowly. What one observes over 10 to 50 hours is that the film is unstable, develops holes, and eventually breaks up into separate more or less spherical "droplets", much as a water film might break up on a waxed surface. This indicates that solid-phase gold does not "wet" graphite at these temperatures. Next, the individual droplets develop the flattened planar regions evident in the picture, making them look rather like truncated golf balls. Figure 13.3 represents the situation after about 70 hours, when no further detectable change is taking place. Of course, on an even longer time-scale, we might anticipate a coarsening or a ripening process in which the larger crystallites grow at the expense of smaller ones; however, the low vapor pressure effectively inhibits inter-crystallite transport, so we may in practice think of each crystallite as a closed system which has come sparately to equilibrium [13.11]. Note that the different crystallites vary in size but all have (at least as far as the eye can tell) the same shape. This is an example of the thermodynamic limiting behavior discussed above and provides further evidence that equilibrium has been achieved.

Figure 13.4 shows a single gold crystal at greater magnification [13.5]. The crystal has been oriented so that some of the flattened regions are in profile, and one can clearly see that these regions are planar "facets". There are two types of facets, distinguished by their

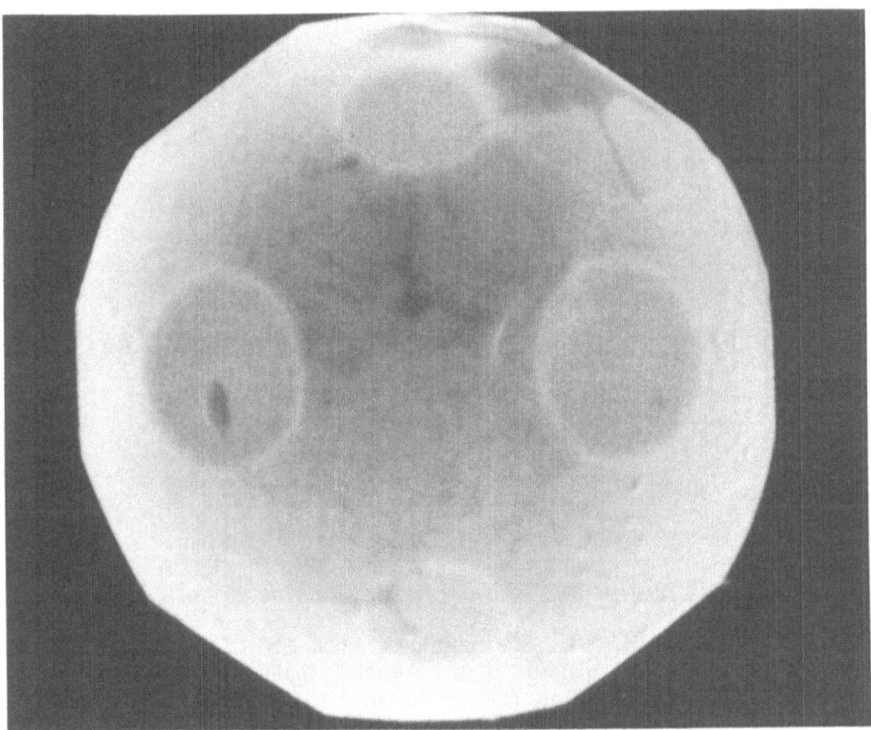

Fig. 13.4. A single gold crystal prepared as in Fig. 13.3, seen along a ⟨110⟩ direction. The overall symmetry is cuboctahedral, so in this view there are four visible facets, two {111} (larger) and two {100} (smaller). Four {111} and two {100} facets are visible in profile. The edges between the facets and the rounded regions appear to be sharp ("first order"), i.e., to involve a slope discontinuity.

diameters: the eight (larger) {111} facets and the six (smaller) {100} facets, corresponding to the cuboctohedral symmetry of the bulk fcc crystal structure. The rest of the crystal surface is smoothly rounded [13.12]. Separating the facets and the rounded regions are sharp edges, across which the tangent plane to the crystal surface changes discontinuously.

Figures 13.5,6 show lead crystals at about 300°C prepared in a similar manner [13.2, 3, 9, 10]. Lead melts at 327°C. It has a rather low vapor pressure (~10⁻⁹ Torr) at the temperature of the experiment, so the rate of evaporation is negligible. Thus, a sealed cell is not necessary and the sample is maintained under high vacuum. Otherwise, however, lead is qualitatively quite similar to gold. The principal difference is that here the edges separating the facets and the rounded regions are "smooth" in the sense that there is no evident discontinuity of the tangent plane (we shall return later to this point).

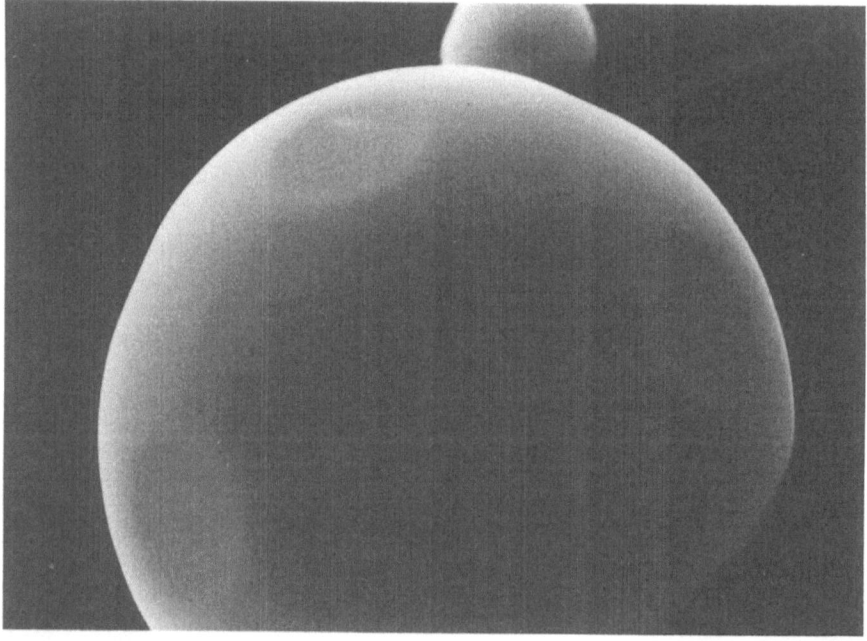

Fig. 13.5. Lead crystal on graphite substrate at about 300°C, prepared by Heyraud and Métois [13.2,3]. The diameter of the crystal is about 6 μm. The view is along ⟨100⟩. The three large facets are {111} (a fourth is truncated by the substrate). The {100} facets are present but smaller and hard to see.

Fig.13.6. Same as Fig.13.5, but viewed along ⟨110⟩, thus putting {111} facets in profile. The planarity of the {111} facets is evident. The edges between the facets and the rounded regions are smooth ("second order"), i.e., no slope discontinuity. The shape of the rounded region in the vicinity of the edge is well described by (13.2) with a critical index $\lambda = 1.60 \pm 0.15$ [13.3].

13.2 The Crystal Shape as a Free Energy

The equilibrium outline of a large crystal may be described mathematically by giving the length $R(\hat{h})$ of the radius vector from the center of the crystal to its boundary in each direction $\hat{h}$. The thermodynamic limiting process (Fig.13.2) involves rescaling $R(\hat{h})$ and taking the limit as the volume V of the crystal tends to infinity at a fixed temperature T. Thus, we shall define the "shape" of the crystal by

$$r(\hat{h}) \equiv \lim_{V \to \infty} \left[\frac{R(\hat{h})}{\alpha V^{1/3}} \right], \qquad (13.1)$$

where α is dimensionless and may be chosen arbitrarily. What is important is that for a macroscopic crystal the outline $R(\hat{h})$ is proportional to $V^{1/3} r(\hat{h})$ up to finite-size corrections. We shall often write $r(\hat{h}, T)$ to emphasize that the crystal shape is temperature dependent.

It is now useful to state one of the main points of this article: *The function $r(\hat{h}, T)$ is a free energy*, i.e., the crystal shape should be thought of as an ordinary free-energy surface! We shall discuss later why this is so. For the moment it provides a language and a set of in-

373

sights which are sufficiently useful so that the reader is asked to accept it as a working hypothesis, deferring to Sect. 13.3 a more formal development.

Thermodynamically speaking, there are several different free energies, depending on which variables are taken as fundamental. The free energy we have in mind here is one whose natural variables are "fields" as opposed to "densities". In the liquid-vapor problem, this would be the grand canonical free energy $g(\mu,T)$, where μ is the chemical potential; for a magnetic system, it would be the magnetic Gibbs free energy $g(h,T)$, where h is the external magnetic field. (Note in passing that this interpretation is at least not inconsistent with the fact that $r(\hat{h})$ is a continuous and convex function [13.13]!) We now draw some conclusions (Table 13.1):

a) Conventional phase boundaries are the loci of nonanalyticity of a free-energy surface, i.e., the set of values of its arguments at which the free energy is singular. The loci $\hat{h}(T)$, or $T(\hat{h})$, of nonanalyticity of the crystal-shape function $r(\hat{h},T)$ are the crystal edges. Thus, if the crystal shape is to be regarded as a free energy, then, correspondingly, the crystal edges should be regarded as phase boundaries.

b) Continuous regions of analytic behavior of a free energy are pure thermodynamic phases. Thus, the phases of the crystal-shape problem are the facets and the smoothly curved regions which we have seen in Figs. 13.3-6. Accordingly, we may refer to the (100) facet as the "(100) phase". The smoothly curved region of the crystal surface is called the "rough phase" for reasons which will become apparent below.

Table 13.1. Correspondences between the equilibrium crystal shape and ordinary thermodynamics.

Surfaces:	Phases:
(a) faceted region	(a) "frozen"
(b) rounded region	(b) "rough"
Edges:	Phase boundaries:
(a) sharp	(a) first-order
(b) smooth	(b) second-order
Crystal shape $r(\hat{h},T)$	Free energy $g(h,T)$ (or $g(\mu,T)$)
Interface free energy $f_i(\hat{m},T)$	Free energy $f(m,T)$ (or $f(n,T)$)
Variables:	Variables:
interface orientation $\hat{m}$	magnetic field h (or chemical potential μ)
crystal shape direction $\hat{h}$	magnetization m (or density n)
Geometrical relation $\hat{m}(\hat{h})$	Eq. of state $m(h,T)$ (or $n(\mu,T)$)

c) Thermodynamically, we distinguish between first-order phase boundaries, characterized by discontinuities in one or more first derivatives of the free energy, and second-order phase boundaries, characterized by continuous first derivatives. For crystal shapes, this distinction is between the sharp edges (as seen for the gold crystals) and the smooth edges (as seen for the lead crystals).

d) The behavior of a free energy in the near vicinity of a second-order phase transition is characterized by critical singularities, often of power-law type. Thus, we may expect that the shape of a crystal surface near a smooth edge is described by (Fig.13.7) behavior of the form

$$y \sim x^\lambda , \tag{13.2}$$

with a value of λ which is typically noninteger. In fact, the edge visible in Fig.13.6 between the facet and the rounded region has been analyzed [13.3] and gives $\lambda = 1.60 \pm 0.15$. Similarly, *Carmi* et al. [13.14] have analyzed vicinal surfaces of ^{4}He crystals and find $\lambda = 1.55 \pm 0.06$. We shall see in Sect.13.4.5 that there is reason to believe that λ is exactly equal to 3/2.

So far we have focused exclusively on the angular variable $\hat{h}$ in the crystal shape. We now turn to the dependence on temperature. Again, it is useful to review the phenomenology. There appear, crudely speaking, to be two types of crystal-shape thermal evolution [13.15]. Figure 13.8 shows schematically the kind of thermal evolution which we shall call "type-A". This type of thermal evolution is exhibited [13.6] by solid ^{4}He and (probably) by metal crystals such as lead [13.2,3] and indium [13.4,5] (but perhaps *not* gold, see Sect.13.5). At T=0 the crystal is polyhedral, i.e., it is bounded entirely by facets [13.16]. Rounded regions appear separating the facets at arbitrarily low

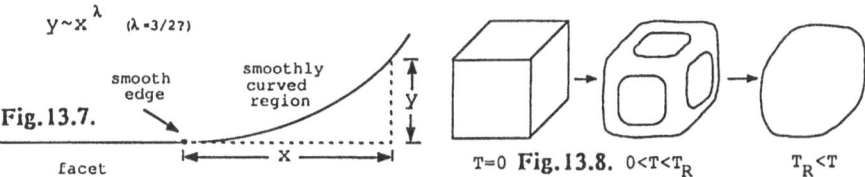

Fig.13.7.

Fig.13.8.

Fig.13.7. Critical behavior of the crystal shape near a smooth (second-order) edge. The shape of the smoothly curved region near the edge is described by a power law. Away from the edge there are corrections to this power-law behavior ("corrections to scaling"). For a crystal of any fixed, finite size, the sharp singularity at x = 0 is absent ("finite-size rounding").

Fig.13.8. Sketch of type-A thermal evolution of the crystal shape. The crystal is strictly polyhedral at T = 0, but at arbitrarily low nonzero T rounded regions appear, separating the facets. Each type of facet disappears at its own roughening temperature T_R (only (100) facets are shown in this example). Above T_R, facets are absent, and the crystal shape is smoothly rounded. In the laboratory this evolution may sometimes be cut off by bulk melting (T_t) below the roughening temperature(s).

nonzero temperature. As the temperature is increased, the facets shrink and the rounded regions grow. There is a sharp, well-defined temperature T_R, called the "faceting temperature" in the crystal-shape context or the "roughening temperature" in other contexts (Sect.13.4.1) at which the facets shrink to zero. Above T_R the crystal surface is everywhere smoothly rounded. There are two important modifications of this simplest picture. The first is that bulk melting may occur before the facets disappear, thus truncating the "full" thermal evolution: there is no special connection between bulk melting (which takes place at the triple point T_t) and roughening (which is entirely an interfacial phenomenon). The second is that the T=0 crystal shape may contain several distinct, symmetry-inequivalent facets [13.17] {hkl}. When this is so, the facets continue to be separated by rounded regions at any T > 0; however, each type of facet disappears at its own roughening temperature $T_{R\{hkl\}}$. Both these modifications are probably common.

The best-studied example of type-A thermal evolution is hcp ^{4}He in coexistence with the superfluid in the temperature range below 1.4 K (Chap.17). Figure 13.9 shows some pictures of ^{4}He crystals [13.18] at different temperatures, illustrating the successive appearance

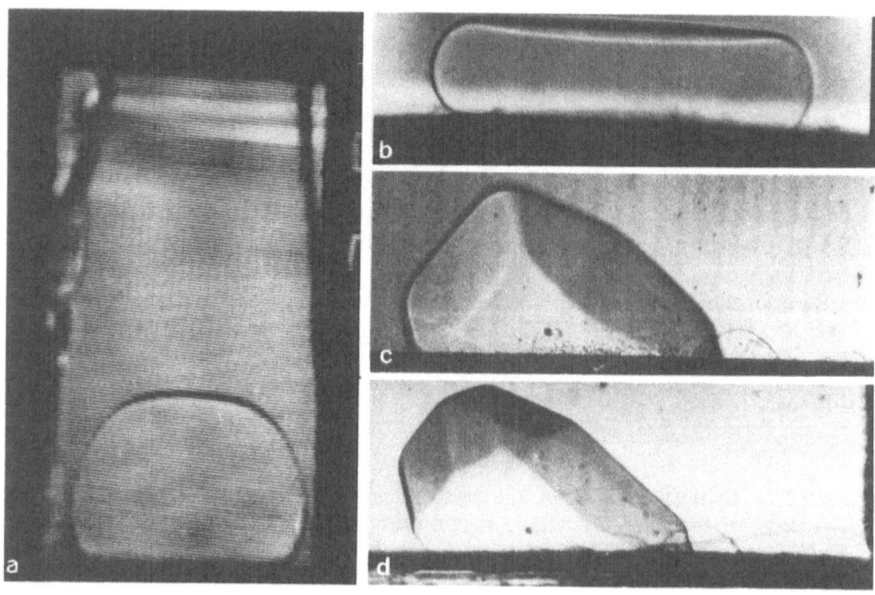

Fig.13.9. Crystals of hcp ^{4}He in rapid growth, showing the development of new facets as the temperature is lowered. Photos by Balibar et al. [13.18]. The temperatures are (a) T = 1.35 K (above all roughening transitions), (b) T = 1.1 K (the "c" facet is visible), (c) T = 0.5 K ("a" and "c" facets), and (d) T = 0.35 K ("a", "c", and "s" facets). Facets are accentuated by conditions of rapid growth, because rough orientations grow faster than facets, but also do appear at equilibrium.

of distinct facets as the temperature is lowered. These pictures were actually taken not at equilibrium but under conditions of rapid growth, where the facets show up more clearly; however, careful equilibrium observations [13.6] have revealed apparent type-A thermal evolution: above a highest roughening temperature T_{Rc} = 1.28 K the crystal shape is smoothly rounded. As the temperature is lowered, the {0001} or "c" facets appear at T_{Rc}, the {1$\bar{1}$00} or "a" facets appear at a second roughening temperature T_{Ra} ~ 1.0 K, and the {1$\bar{1}$01} or "s" facets appear at T_{Rs} ~ 0.35 K. No further transitions have been seen at temperatures down to 0.07 K. At all temperatures studied, the facets are separated from one another by rounded regions.

It is instructive to sketch the phase diagram $T(\hat{h})$ corresponding to the schematic type-A thermal evolution shown in Fig.13.8. This is done in Fig.13.10. In order to avoid three-dimensional complexity (due to the two angular variables in $\hat{h}$), Fig.13.10 treats only the equatorial plane of the full crystal shape. The figure is symmetric under rotations by $\pi/2$, and only a single period is shown. Note that at T=0 there is one singularity, located at $\theta = \pi/4$. At low but nonzero temperature, this singularity has split into two, symmetrically placed above and below $\pi/4$. As the temperature is further raised, the width of the rough phase increases until the {100} facets pinch off at T_R, leaving the crystal surface smoothly rounded. The phase boundaries in Fig.13.10 are second order, since the edges in Fig.13.8 lack slope discontinuity. It is therefore appropriate to ask what are the universality classes of the phase transitions represented. We shall return to this question in some detail in Sect.13.4.5. For the moment it will suffice to say that the cusp points θ=0, $T=T_R$ and $\theta=\pi/4$, T=0 are special, while the rest of the phase boundaries are in the so-called *Pokrovsky-Talapov* [13.19] universality class, characterized by the power-law behavior (13.2) with $\lambda = 3/2$.

The other pattern of thermal evolution, which we shall call "type-B", is illustrated schematically in Fig.13.11. Here, the crystal shape remains strictly polyhedral up to a nonzero temperature T_0, the "corner-rounding" temperature, above which smoothly rounded ("rough") regions appear at the corners. Between T_0 and some higher temperature T_1 the rounded regions grow but sharp (first-order) facet-facet edges still persist. At T_1 the sharp edges disappear, and above this the facets are separated by rounded regions. As the temperature is further increased, the remaining facets shrink and eventually disappear at a roughening temperature T_R, just as in type-A evolution. This scenario is at the moment based in considerable part on theoretical models [13.20-23]. It seems likely on this basis that the edges between facets and rounded regions are all of the Pokrovsky-Talapov form (13.2); however, it is important to state that solid experimental evidence for

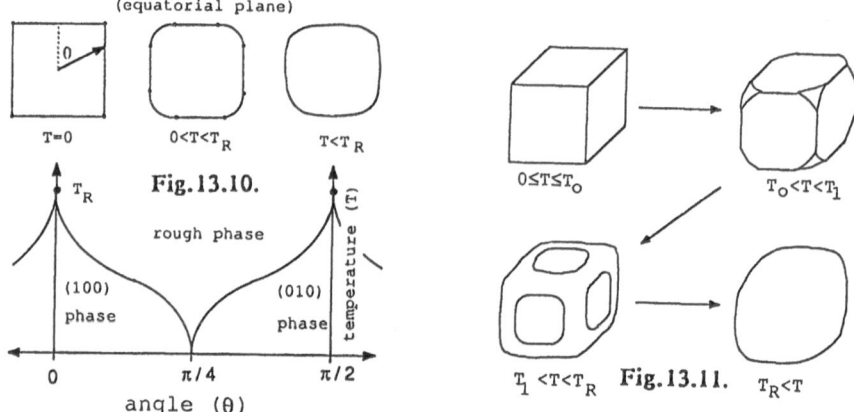

(equatorial plane)

T=0   0<T<T_R   T<T_R

Fig. 13.10.

rough phase

(100) phase   (010) phase

temperature (T)

0   π/4   π/2

angle (θ)

$0 \leq T \leq T_0$ $T_0 < T < T_1$

$T_1 < T < T_R$ **Fig. 13.11.** $T_R < T$

Fig. 13.10. Schematic phase diagram for type-A thermal evolution of the crystal shape. The full phase diagram is three-dimensional [13.22]. Shown here is the evolution of the equatorial plane only. Phase diagram plots the angular position θ of the crystal edges. For T > 0, phase boundaries are all second order. For crystals with short-range forces, the curved boundaries separating {100} and rough phases belong to the Pokrovsky-Talapov universality class (Sects. 13.4.5, 13.5), with $\lambda = 3/2$. Faceting at $T = T_R, \theta = 0$ is a Kosterlitz-Thouless transition (Sect. 13.4.5).

Fig. 13.11. Sketch of type-B thermal evolution of the crystal shape. The crystal remains polyhedral up to a nonzero temperature T_0, at which the cube corners begin to round. Segments of cube edges remain sharp (first order) between T_0 and a higher temperature T_1. Above T_1, the facets are separated by rounded regions, and evolution proceeds as for type-A crystals (Fig. 13.8). Edges between facets and rounded regions are always second order. In more complicated cases, when several symmetry-distinct facets are present at T = 0, there may be several different T_0's, T_1's, and T_R's.

this is still lacking. The "intermediate" shape $(T_0 < T < T_1)$ has not yet, as far as I am aware, been observed in an equilibrium experiment. When there are several crystallographically distinct facets at T = 0, more general type-B behavior, with several distinct corner-rounding transitions, can occur. Indeed, it appears quite possible on theoretical grounds [13.21, 22] that, for the same crystal, some T=0 edge-and-corner systems may behave in the type-B manner and others in the type-A manner. Physical examples of such mixed behavior are lacking. We shall discuss in Sect. 13.4.3 the microscopic, physical reason for the difference between type-A and type-B behavior.

The main experimental candidate [13.24] for type-B behavior is NaCl, pictures of which are shown as Figs. 13.12, 13. The vapor pressure of NaCl is high ($1.7 \cdot 10^{-2}$ Torr at 680°C) and the crystals are prepared in a sealed cell. The strictly cubical form at 620°C (Fig. 13.12a) persists up to a T_0 of about 650°C. By 710°C (Fig. 13.12b) the facets are entirely separated by rounded regions. It is not clear whether the "intermediate" shape $(T_0 < T < T_1)$ has been seen. The authors state in

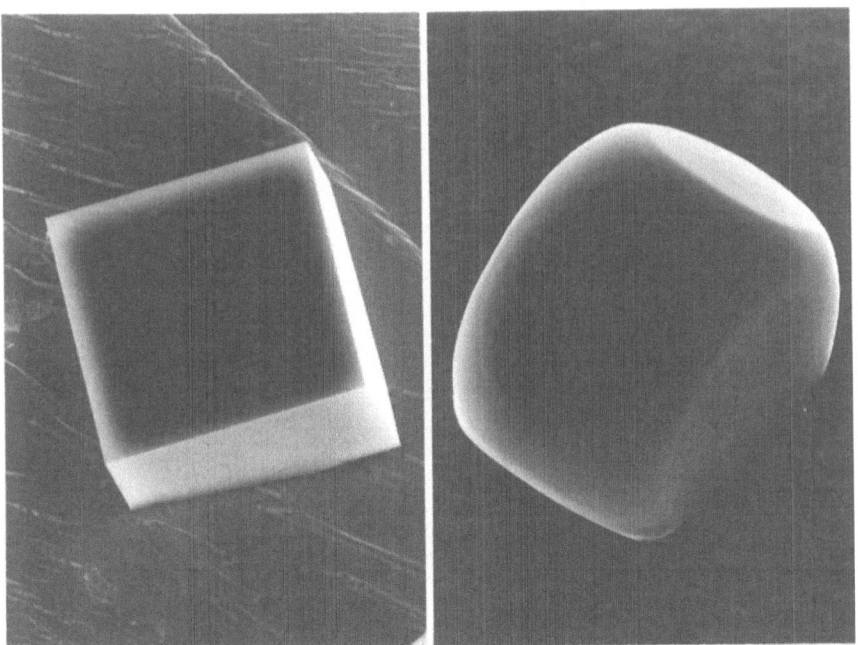

Fig.13.12. Equilibrium crystals of NaCl prepared by Heyraud and Métois [13.24] at temperatures of (a) 620°C and (b) 710°C. At the lower temperature the shape is cubic, with sharp edges between the facets and no evident rounded regions. At the higher temperature {100} facets remain but are separated by smoothly rounded (rough) regions. The corner-rounding temperature T_0 is near 650°C. The "intermediate" shape (expected theoretically for $T_0 < T < T_1$), with rounded corners but still some sharp edges, has not yet been seen at equilibrium. Crystal size is 5-10 μm.

their abstract [13.24], "For higher annealing temperatures (than 650°C), first the summits and later the ridges become strongly blunted." Figure 13.13 at 750°C does show some shapes of the "intermediate" type; however, it is clear that many different crystal morphologies are coexisting, a sure indication that the system is not at equilibrium. It appears likely, in retrospect, that Fig.13.13 represents growth (or dissolution) forms due to small thermal gradients in the experimental cell, since, when precautions were taken to reduce such gradients, the resulting shapes at this temperature were those of Fig.13.12b.

A schematic phase diagram for type-B behavior is shown in Fig.13.14. The dotted vertical line at $\theta = \pi/4$, representing the sharp, first-order edge between (100) and (010) facets, ends at T_1, when the rounded regions reach the equatorial plane. Behavior at this special point is probably bicritical [13.20,21,25]. Above T_1, the phase diagram is the same as Fig.13.11, with second-order, Pokrovsky-Talapov lines and a roughening transition at T_R.

This concludes our quick summary of crystal-shape phenomenology. It is perhaps appropriate to close this section with a formal ques-

Fig.13.13. NaCl crystals at 750°C observed in the laboratory of Heyraud and Métois [13.9]. Some of these crystals have shapes that are temptingly similar to the "intermediate" shapes expected at equilibrium (except that the edges between facets and rounded regions are sharp here). These crystals are not at full thermal equilibrium, and different crystallites exhibit quite different shapes. It seems likely that these are dissolution forms, since the rounded regions probably evaporate faster than the facets.

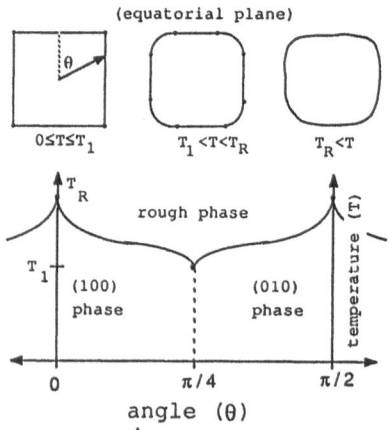

Fig.13.14. Schematic phase diagram for type-B thermal evolution of the crystal shape. Only the behavior of the equatorial plane is shown. Corner rounding reaches the equatorial plane at temperature T_1. The dotted (first-order) phase boundary between $T = 0$ and $T = T_1$ at $\theta = \pi/4$ represents the sharp edge between the {100} and {010} facets.

tion: If $r(\hat{h},T)$ is a free energy analogous to the grand canonical free energy $g(\mu,T)$, what is the analogue in the crystal-shape problem of the Legendre-conjugate canonical free energy $f(n,T)$, depending on the particle density n instead of the chemical potential μ? (In magnetic language, it would be the magnetic Helmholtz free energy $f(m,T)$, depending on the magnetization density m instead of the external field h.) The answer to this question will occupy the next section.

380

13.3 The Wulff Construction as Legendre Transformation

Crystals at equilibrium have interesting shapes because the interface free energy per unit area $f_i(\hat{m},T)$ depends on the orientation $\hat{m}$ of the interface relative to the crystallographic axes of the bulk material. The reason for this dependence at a microscopic level is illustrated in Fig.13.15: Crystalline solids are anisotropic, so the structure (and therefore the energetics) of an interface between a crystalline solid and a fluid depends on the orientation of the interface. All of the complicated microscopic considerations of solid-state physics and surface science enter into determining the details of this dependence. *Wulff* [13.26] in 1901 understood how, once given this dependence, the crystal shape follows rather simply [13.27].

Consider a large inert box (Fig.13.16) in a thermal bath at temperature T, attached to a reservoir of material via a feed tube. The box is initially empty and a metered quantity of material is allowed to flow through the feed tube. At first, the overall density in the box is very low and the material finds itself in the vapor state; however, as more material is added, there comes a point $[P = P_0(T)]$ at which more particles cannot fit in the vapor and a dense phase begins to condense. If the temperature is sufficiently low (below T_t), the condensed phase is solid. We shall assume for simplicity that gravity may be neglected [13.28] and that the properties of the wall are such that the crystal forms in the interior of the box, i.e., that the solid material does not

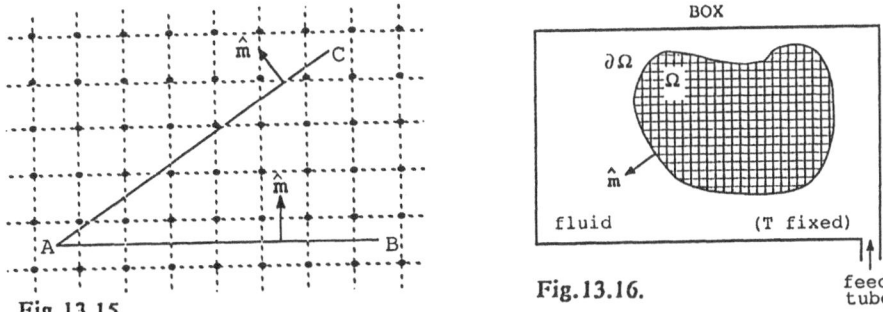

Fig.13.15.

Fig.13.16.

Fig.13.15. Anisotropy of the interfacial free energy per unit area $f_i(\hat{m})$. The anisotropy of the crystalline solid structure leads to a dependence of f_i on the interfacial orientation $\hat{m}$. To see this, imagine a "Kossel crystal", in which the atoms are bound by pair forces represented by the dotted lines. The interface free energy per unit area at T = 0 is then simply proportional to the number of bonds cut per unit interfacial area. The interface AB cuts vertical bonds only. The interface AC of the same area (length) cuts a different number of bonds per unit area. See also Fig..13.22 and (13.20-22).

Fig.13.16. A schematic experiment. The box has inert walls maintained at a fixed temperature T. Material is added to the interior via the feed tube until the solid phase condenses. The volume of the solid is fixed by the amount of material added; its shape adjusts in such a way as to minimize the total interfacial free energy $F_i(T)$.

wet the walls. Beyond this point (and until the solid completely fills the interior of the box), adding further material simply adjusts the relative volumes of the solid and vapor phases, without changing the pressure. Thus, for a fixed amount of added material, the bulk free-energy contributions of the two phases are fixed, and it is only the total free energy,

$$F_i(T) = \int_{\partial\Omega} dS \; f_i(\hat{m},T) = \int_{\partial\Omega} dS \cdot \hat{m} f_i(\hat{m},T) \; , \qquad (13.3)$$

associated with the interface $\partial\Omega$ of the region Ω occupied by the solid which can change. Here, dS is the element of interface area and $dS = \hat{m}dS$, where $\hat{m}$ is the unit outward normal to the interface. The volume,

$$V(\Omega) = \int_\Omega (dr) = \frac{1}{3} \int_{\partial\Omega} dS \cdot \mathbf{r} \; , \qquad (13.4)$$

of the region Ω occupied by the solid is fixed by the amount of material that has passed through the feed tube.

Now, $F_i(T)$ changes its value when $\partial\Omega$ changes its shape, and thermodynamic principles assert that the equilibrium outline of the crystal $R(\hat{h})$ is that which minimizes (13.3). Thus, determination of $R(\hat{h})$ requires minimization of (13.3) subject to the constraint given by (13.4). The constraint is conveniently incorporated by introducing a Lagrange multiplier 2λ and requiring stationarity $\delta\Omega = 0$ of the functional

$$\Phi \equiv F_i - 2\lambda V \; , \qquad (13.5)$$

where $\lambda = \lambda(V)$ sets the overall scale of the crystal outline. This constrained minimization problem has a simple and unique solution [13.27], which can be expressed algebraically in the form

$$R(\hat{h}) = \min_{\hat{m}} \left[\frac{f_i(\hat{m})/\lambda}{\hat{m} \cdot \hat{h}} \right] . \qquad (13.6)$$

Although (13.6) is a correct solution to the variational problem, the physical effects (such as edge and corner energies, curvature corrections to the interface free energy per unit area $f_i(\hat{m},T)$, and the like) which are responsible for finite-size corrections have already been thrown out in writing the total free energy in the form given by (13.3), so (13.6) is not strictly valid except in the limit (13.1) $V\to\infty$. In this limit [13.29], $\lambda \sim \alpha' V^{-1/3}$, so we can choose $\alpha\alpha' = 1$ and write simply

$$r(\hat{h}) = \min_{\hat{m}} \left[\frac{f_i(\hat{m})}{\hat{m} \cdot \hat{h}} \right] . \qquad (13.7)$$

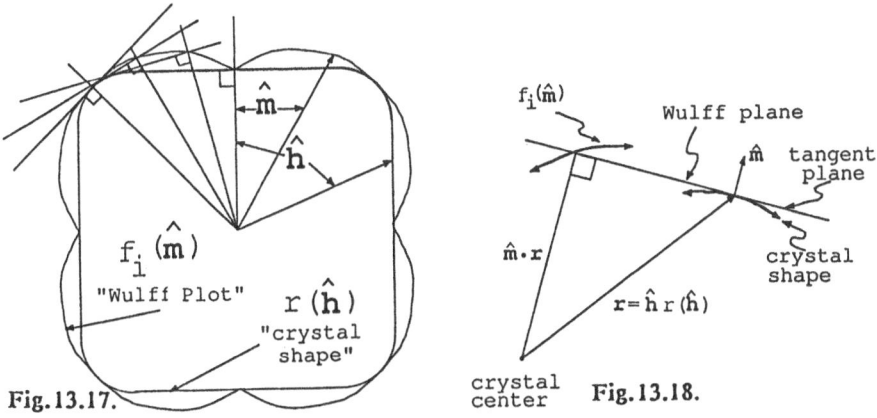

Fig. 13.17. The Wulff construction. The interfacial free energy per unit area $f_i(\hat{m})$ is plotted in polar form (the "Wulff plot"). Draw a radius vector in each direction $\hat{m}$ and construct a perpendicular plane where this vector hits the Wulff plot. The interior envelope of the family of "Wulff planes" thus formed, expressed algebraically in (13.7), is the crystal shape, up to an arbitrary overall scale factor which may be chosen equal to unity.

Fig. 13.18. Geometry of the Wulff construction. The vector $r(\hat{h})$ may be resolved into components along and perpendicular to $\hat{m}$. The component along $\hat{m}$ is just $f_i(\hat{m})$, (13.8a). The remainder is (13.8b). The fact that the Wulff plane associated with the direction $\hat{m}$ is tangent to the crystal shape at $\hat{h}$ provides a natural relation $\hat{h}(\hat{m})$ or $\hat{m}(\hat{h})$. We shall see below that this relation is just the equation of state.

Equation (13.7) has an elegant and celebrated geometric interpretation known as the Wulff construction, which is illustrated in Fig. 13.17. What the Wulff construction says is that, when the interface free energy per unit area $f_i(\hat{m})$ is drawn as a polar plot (the "Wulff plot", sometimes called "γ-plot"), the crystal shape is given as the interior envelope of the family of perpendicular planes passing through the ends of the radius vectors $\hat{m}f_i$. Figure 13.18 shows this geometry in more detail. It is convenient to resolve the vector $r(\hat{h}) \equiv \hat{h}r(\hat{h})$ into components along and perpendicular to $\hat{m}$, and thus to reexpress [13.30] the content of (13.7) as

$$r_{\hat{m}}^{\text{par}}(\hat{h}) = r \cdot \hat{m} = f_i(\hat{m}) \tag{13.8a}$$

and

$$r_{\hat{m}}^{\text{perp}}(\hat{h}) = r - \hat{m}r_{\hat{m}}^{\text{par}}(\hat{h}) = \nabla_{\hat{m}}f_i(\hat{m}) , \tag{13.8b}$$

where in (13.8b) the gradient refers to the two-dimensional plane perpendicular to $\hat{m}$. Notice that (13.7) [or (13.8)] provides a functional relationship $\hat{m}(\hat{h})$ [or $\hat{h}(\hat{m})$]. We shall argue that this is the equation of state of the crystal-shape problem. Finally, we remark that (13.7) has an inverse [13.27],

383

$$\frac{1}{f_i(\hat{m})} = \min_{\hat{h}} \left[\frac{1/r(\hat{h})}{\hat{m} \cdot \hat{h}} \right].$$ (13.9)

The geometrical interpretation [13.31] of (13.9) is that a Wulff construction on the inverse crystal shape function $1/r(\hat{h})$ gives the inverse free energy $1/f_i(\hat{m})$.

It was *Andreev* [13.32,33] in 1981 who first pointed out in this context that the Wulff construction is simply the geometrical version of a two-dimensional Legendre transformation between the free energy $f_i(\hat{m})$ and the crystal shape $r(\hat{h})$. From this observation it follows that the crystal shape is just as much a "free energy" as is the canonical free energy $f(n)$ in the liquid-gas problem (or $f(m)$ in the magnetic context). We propose to demonstrate this in two ways.

First, as an argument by analogy, we recall the mathematically well-known fact that an ordinary, one-dimensional Legendre transformation is entirely equivalent to a construction involving the interior envelope of a family of planes [13.34]. We choose the magnetic language and remind the reader that the Gibbs free energy, defined for lattice models in statistical mechanics by

$$g(h) \equiv \lim_{N \to \infty} \left[\frac{1}{N}(-k_B T) \ln \mathrm{Tr} \, \exp\!\left[-\beta(H = -h \sum_i s_i + ...)\right] \right],$$

(13.10)

is algebraically related to the corresponding Helmholtz free energy $f(m)$ by

$$g + hm = f .$$ (13.11)

From (13.10) we find that $\partial g/\partial h = -m(h)$, which gives the magnetic equation of state. It then follows from (13.11) that $\partial f/\partial m = h(m)$. Figure 13.19 illustrates the appearance of $g(h)$ and $f(m)$ for a symmetric magnet in its ferromagnetic phase. Note that $g(h)$ has a slope discontinuity at $h = 0$, corresponding to the jump in magnetization from $m = -m_0$ to $m = +m_0$ (m_0 is the spontaneous magnetization), as the external field h passes through zero. We shall see in a moment that this slope discontinuity is the analogue of a sharp edge in the crystal shape [13.35]. Figure 13.20 now shows a geometric interpretation of the transformation (13.11): the tangent to the curve $g(h)$ at the point h intersects the ordinate axis at a point y such that

$$\frac{y-g}{h} = -\frac{\partial g}{\partial h} = -m ,$$ (13.12)

which identifies the intercept as $y \equiv f$. Thus, $g(h)$ is the interior envelope of the family of lines defined by slope $-m$ and intercept $f(m)$, so we may write

384

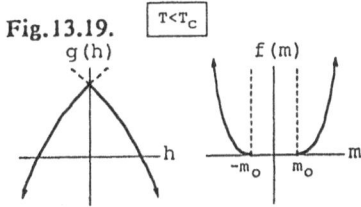

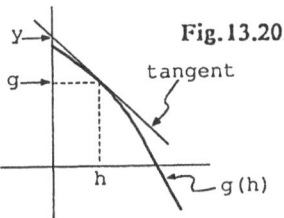

Fig.13.19. Schematic graphs of magnetic free energies for a ferromagnet below T_c. h is the external magnetic field. The slope of g(h) gives the magnetization m. The slope discontinuity of g(h) at h = 0 reflects the jump in magnetization from $-m_0$ to $+m_0$ as h goes from 0^- to 0^+. The free energy f(m) is not defined for values of m between $-m_0$ and $+m_0$, since no spatially uniform system with these magnetizations is stable. A straight-line interpolation between $-m_0$ and $+m_0$ follows the axis and is tangent to the curve f(m) at both ends. This "double-tangent" construction represents the average free energy of a mixed-phase system with the intermediate magnetizations. The cusp in g(h) is like a sharp edge on the crystal shape. The forbidden region is like the unstable region of a Wulff plot. The double-tangent construction is the analogue of the circle interpolation (13.23).

Fig.13.20. Geometric interpretation of the ordinary Legendre transformation. The tangent at field h to the free energy g(h) intercepts the vertical axis at a point y = f(m(h)). Thus, g(h) is the interior envelope of the family of lines with y intercept f(m) and slope -m.

$$g(h) = \frac{min}{m} \, [f(m) - hm] \, . \tag{13.13}$$

Crudely speaking, the Wulff construction (13.7) is just a two-dimensional version of this relation, referred to polar axes [13.36]. Correspondingly, the analogue of the magnetic equation of state is just the relation $\hat{m}(\hat{h})$ defined geometrically in Fig.13.17.

To make this more concrete, we express the Wulff problem in Cartesian form [13.33,20]: let $Z(X_1, X_2)$ describe the height Z of the crystal surface above the (X_1, X_2) plane [13.37]. Then,

$$F_i(T) = \int dX_1 dX_2 \tilde{f}_i(Z_1, Z_2)$$
and
$$V(\Omega) = \int dX_1 dX_2 Z(X_1, X_2) \, , \tag{13.14}$$

where $\tilde{f}_i \equiv f_i[(1+(\nabla Z)^2]^{1/2}$ and $Z_k \equiv \partial Z / \partial X_k$. Note that the first integrand depends only on the derivatives of Z. Minimizing F_i subject to the constraint of constant volume leads to the Euler-Lagrange equation,

$$\frac{\partial}{\partial X_k} \frac{\partial \tilde{f}_i(Z_1, Z_2)}{\partial Z_k} = -2\lambda \, , \tag{13.15}$$

where a sum on the repeated index k = 1,2 is understood. To extract the crystal shape, we must now take the V→∞ limit, as in passing between (13.6 and 7). This gives

$$\frac{\partial}{\partial x_k} \frac{\partial \tilde{f}_i(z_1,z_2)}{\partial z_k} = -2 , \tag{13.16}$$

where all lengths have been rescaled $x_k = X_k/\alpha V^{1/3}$ and the crystal shape is given by $z(x_1,x_2) = Z(X_1,X_2)/\alpha V^{1/3}$, so that $z_k = Z_k$.

Equation (13.16) is a highly nonlinear second-order partial differential equation for $z(x_1,x_2)$, subject to boundary conditions which we have not even specified. It is by no means clear at this point that it has the same content as (13.7). To see that it does, one verifies by direct substitution that (13.16) has a simple first integral,

$$z - x_\alpha z_\alpha = \tilde{f}_i(z_1,z_2) . \tag{13.17a}$$

Differentiation with respect to z_α leads to

$$x_\alpha = - \frac{\partial \tilde{f}_i}{\partial z_\alpha} . \tag{13.17b}$$

A little algebra now suffices to verify that (13.17a,b) are identical to (13.8a,b) and may, indeed, be written in the form

$$z(x_1,x_2) = \min_{z_k} (\tilde{f}_i + x_k z_k) . \tag{13.18}$$

The connection with the magnetic case should now be clear: (13.17a) and (13.18) correspond to (13.11) and (13.13), respectively. The crystal shape z is like the free energy g; $\tilde{f}_i$ is like f; the variables $\{x_k\}$ are like the magnetic field h; $\{z_k\}$ is like the magnetization m and its definition [13.38] corresponds to the magnetic equation of state $m = -\partial g/\partial h$; and, finally, (13.17b) corresponds to the inverse equation of state $h = \partial f/\partial m$. Of course, in the polar (Wulff) version of the problem, $\{z_k\}$ corresponds to $\hat{m}$ and $\{x_k\}$ to $\hat{h}$, as shown in Table 13.1.

The discussion in the last two paragraphs might be taken as a proof [13.39] of the Wulff construction (13.7); however, there remains one serious problem and that is the question of uniqueness: (13.16) is nonlinear. We have argued that (13.17a) is *a* solution; but it is by no means clear that it is the only solution, or (in case of multiple solutions) the solution which minimizes F_i. This question is not trivial and certainly hinges on the issue of boundary conditions, which we have so far ignored. I do not know how to prove uniqueness from this point of view and must rely on the work of previous authors [13.27,40]. It is perhaps worthwhile at this point to make a few remarks about the distinct but closely related problem of finding, for given free energy function $f_i(\hat{m})$, surfaces which minimize F_i subject to a fixed boundary. (The analogous isotropic problem is that of finding the shape of a soap film whose edges are constrained to lie on a fixed wire). This problem has attracted considerable interest recently [13.41]. It is possible to prove two nice results concerning the proper-

386

ties of these anisotropic minimizing surfaces: (a) Any tangent plane which they possess must exist as a plane tangent to the corresponding equilibrium crystal shape. (This is, of course, only interesting in situations where the equilibrium crystal shape lacks certain tangent planes, i.e., where it contains sharp edges, corners, etc., as in the low-temperature region of type-B evolution). (b) Any dihedral angle which they possess (i.e., any edge or "crease" where two different tangent planes intersect) must exist as a dihedral angle on the corresponding crystal shape or on its inverse (the shape of a fluid inclusion in the solid). One might guess at this point that the same kind of result would hold for corners. It turns out that this is not so: indeed, a specific counterexample has recently been constructed by *Taylor* and *Cahn* [13.41].

13.4 Applications

With this conceptual background, we are in a position to discuss a number of important applications. Some of these are standard, dating back to *Herring's* classic paper [13.27]; others are based on more recent work. Our aim is to focus on the physical understanding of certain generic features of crystal shapes and their thermal evolution.

It looks feasible at this stage of development to calculate T=0 crystal shapes from first principles. The necessary ingredient is the interface energy as a function of m̂. There is an extensive older literature [13.42] applicable to unrelaxed interfaces and atoms which may be regarded as interacting via known pair forces. More recently, density-functional techniques have been applied to treat situations where the full electron distribution plays an important role. These techniques have been shown capable even of describing reconstruction effects at semiconductor surfaces. So far, they have been applied mainly to the principal orientations m̂; however, extension to vicinal surfaces seems feasible. This T = 0 work is important in identifying the principal facets and in determining the type (A or B) of thermal evolution (Sect. 13.4.3).

Unfortunately, as we remarked in Sect. 13.1, kinetic constraints effectively prevent equilibration for most systems until close to the melting temperature (T_t), where thermal fluctuations can and do significantly influence the crystal shape. Although the conceptual framework is in place, microscopic treatment at T > 0 is in a much more primitive state than at T=0. We shall comment briefly on our own recent work in Sect. 13.4.6. For the most part, the physical parameters necessary to characterize statistical surface models are not yet available. Step, edge, and corner energies are crucial, and it would be nice to see a serious effort to measure and/or calculate these quantities for prototype systems. I hope that further motivation may be provided by the new generation of crystal-shape measurements now in progress and by the availability of relevant and tractable models requiring these quantities as input. However, this is mainly an agenda for the future.

Recent progress has focused on generic and "universal" behavior, which is independent of specific material parameters. Here, what one relies on is the connection between crystal shapes and thermodynamic phase diagrams, as sketched in Sects. 13.2 and 3. The interfacial region has nonzero width; but, at long length scales and in the thermodynamic limit, it may be regarded as strictly two-dimensional. There is an extensive literature on phase transitions and critical behavior in two dimensions [13.43], and major universality classes have been characterized in detail. Thus, by properly identifying features of the equilibrium crystal shape with their counterparts in d = 2 statistical mechanics, it is possible to make crystal-shape predictions. In particular, because of the universality of second-order critical behavior, certain features of the crystal shape in the vicinity of smooth edges can be predicted under rather broad hypotheses.

13.4.1 Facets, Cusps, and Roughening

Cusps in the Wulff plot, $f_i(\hat{m})$ are the origin of planar facets on the equilibrium crystal shape. How this comes about is illustrated in Fig. 13.21: suppose that there is a linear cusp at $\theta = 0$ (we consider a single angular variable, for simplicity),

$$f_i(\theta) = A + B|\theta| + \dots . \tag{13.19}$$

The Wulff plane associated with a small value $\theta>0$ intersects the (horizontal) $\theta=0$ Wulff plane at a distance $A \sin\theta + d(\theta)$ from the vertical axis. The crystal will have a horizontal facet if and only if $d(\theta)$ approaches a nonzero value as $\theta \rightarrow 0$. It is easy to see that for small θ, $\sin \theta \simeq B\theta/d(\theta)$, so that $d(0) = B$, i.e. the facet radius is given by the coefficient B of the linear cusp. For any weaker dependence on θ, for instance, $B|\theta|^\tau$ with $\tau > 1$, $d(0) = 0$, and there is no facet.

It remains to explain the physical origin of the cusp in (13.19). At T=0 and within the context of a nearest-neighbor Kossel-crystal model (for instance, Fig. 13.15) this is easy enough to see: Figure 13.22 shows an idealized crystal lattice (square, for simplicity only). If we imagine that the atoms are bound by nearest-neighbor forces, then the energy to make an interface by cleaving the crystal between the points marked by crosses is simply

$$E(X,Y) = v(|X| + |Y|) , \tag{13.20}$$

where v is the energy necessary to cut a single bond. The total interfacial "area" produced in this process is $2(X^2+Y^2)^{1/2}$ (the factor 2 reflects

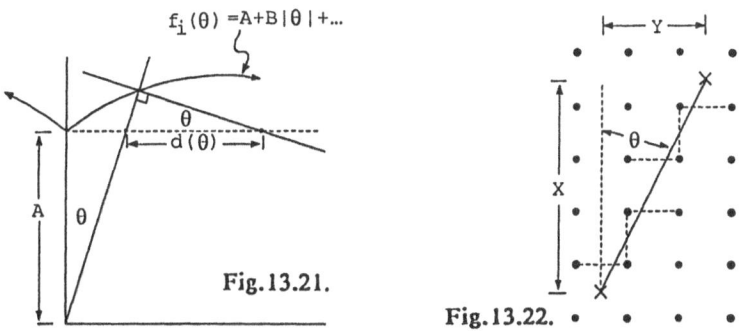

Fig.13.21.

Fig.13.22.

Fig.13.21. Wulff plot with a linear cusp at $\theta = 0$. If $d(\theta) \to 0$ as $\theta \to 0$, then there is no facet corresponding to $\theta = 0$, and the $\theta = 0$ Wulff plane (dotted line) is tangent to the crystal shape at one point only. Calculation shows that $d(0) = B$, so the presence of a cusp in the Wulff plot suffices to guarantee a facet of the corresponding orientation on the crystal shape.

Fig.13.22. Kossel crystal at $T = 0$. The energy to cleave the crystal along the slanted interface shown ($\tan\theta = Y/Z$) is $v(|X|+|Y|)$.

the fact that two surfaces, an upper and a lower, are produced in the cleaving process). There are no entropy effects at T=0, so the free energy per unit interface area is just

$$f_i(\theta) \equiv \frac{E(X,Y)}{\text{area}} = \frac{v}{2}\left(|\sin\theta| + |\cos\theta|\right), \qquad (13.21)$$

which exhibits a linear cusp at $\theta = 0$ (with A=B=v/2).

For T > 0 thermal fluctuations play a role and this simple argument does not suffice; nevertheless, it is not hard to explain what happens. Figure 13.23 shows a vicinal (small θ) surface at low but nonzero temperature. It consists of broad terraces of width 1 separated by occasional steps of height a and is tilted at an angle $\theta \simeq a/l$ relative to the horizontal. At T = 0, the terraces would be atomically flat

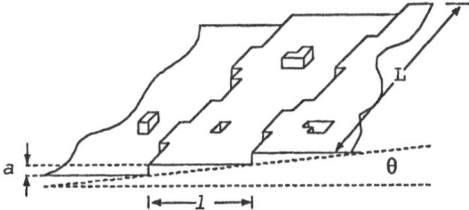

Fig.13.23. Sketch of a vicinal interface at low but nonzero temperature. The steps have height a and average separation depth l, with $\tan\theta = a/l$. For small θ, the steps are widely separated and the total free energy of the interface may be thought of as made up of the sum of two contributions, one proportional to the total horizontal terrace area and the other proportional to the total step length. The terrace free energy per unit area $f_t(T)$ and the step free energy per unit length $f_s(T)$ are both modified by the presence of thermal fluctuations at nonzero temperature.

389

(except for zero-point fluctuations) and the steps would be straight. At low but nonzero temperatures, the terraces have occasional thermally excited islands and holes and the steps wander about a little due to occasional kinks; however, it is reasonable to suppose that the total free energy of the interface is given by

$$F_i(T) = NL \, [lf_t(T) + f_s(T)] , \qquad (13.22)$$

where L is the width of the surface, N is the number of steps, $f_t(T)$ is the free energy per unit area of the terraces, and $f_s(T)$ is the free energy per unit length of the steps. In writing (13.22) we have assumed that the steps are far enough apart so that their interactions may be neglected. This is always justifiable in d = 3 as $\theta \to 0$, provided the elementary interactions are not long-ranged [13.44]. Dividing (13.24) by the macroscopic area of the surface, $L(l^2+a^2)^{1/2}$, gives an interface free energy per unit area of the form (13.19), with $A(T) = f_t(T)$ and $B(T) = f_s(T)/a$.

We have seen that the facet size is controlled by the step free energy per unit length $f_s(T)$. The characteristic behavior of $f_s(T)$ is shown in Fig.13.24. The function f_s contains energy and entropy contributions in the usual combination, $f_s = e_s - Ts_s$. As the temperature increases, kinks become more frequent and the step wanders, increasing its entropy and, thereby, decreasing $f_s(T)$. The function $f_s(T)$ goes to zero at a temperature denoted T_R and called the "faceting temperature", since the facet radius vanishes there. When the step free energy goes to zero, it becomes favorable for large islands (and islands upon islands, etc.) to form, and the surface loses its long-range flatness and becomes "rough" at long length scales [13.45]. For this reason T_R is also called the "roughening temperature". The roughening·transition, which takes place at T_R, has been extensively studied [13.46] (see also contribution by Van der Veen et al., this volume). All this technology may be applied to the behavior of the crystal shape at the faceting temperature (Sect.13.4.5). Each distinct T = 0 facet {hkl} disappears at its own roughening temperature $T_R(\{hkl\})$.

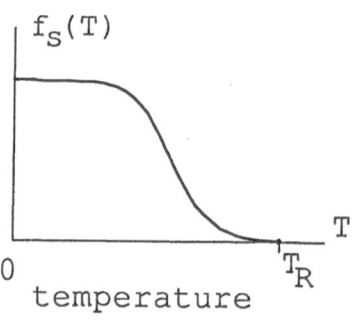

Fig.13.24. Typical step free energy f_s per unit length as a function of temperature. Thermal fluctuations increase the step entropy per unit length and thus decrease its free energy. f_s goes to zero at the roughening temperature T_R. Behavior near T_R is of the Kosterlitz–Thouless type, if atomic forces are short ranged. Facet radius is proportional to $f_s(T)$ and therefore vanishes at T_R.

13.4.2 Sharp Edges, Thermal Faceting, and Forbidden Regions of the Wulff Plot

When sharp edges and/or corners are present on the equilibrium crystal shape, certain tangent planes are absent from the Wulff construction interior envelope. How this comes about is illustrated schematically in Fig.13.25: if $f_i(\hat{m})$ is sufficiently elongated in some direction, there may be a set of orientations $\hat{m}$ whose Wulff planes lie outside of the envelope. When this happens, the plane tangent to the equilibrium shape jumps discontinuously across a set of "forbidden" orientations, forming a sharp edge. These missing orientations are the precise analogue of the forbidden magnetizations in a ferromagnet [13.35] below T_c (Fig.13.19) or of the forbidden densities at the liquid-vapor transition.

We have just seen that the angular coexistence which produces the dihedral at a sharp edge is the same as the coexistence of phases of different density at the liquid/vapor phase boundary. This correspondence goes further: the absence of a spatially uniform stable phase of density intermediate between the vapor density n_v and the liquid density n_l makes it impossible, even in principle, to define a canonical free energy f(n) for $n_v < n < n_l$ (except via the level rule or double-tangent construction [13.47]). Similarly, it is impossible to define a pure-phase interface free energy $f_i(\hat{m})$ for orientations $\hat{m}$ which do not appear on the equilibrium crystal shape. At first sight this seems strange, since it would appear possible (see Fig.13.26) to take a crystal, cleave it at angle $\hat{m}$, and simply measure the free energy of the resulting interface. The catch is that thermodynamic quantites such as $f_i(\hat{m})$ are only defined at *equilibrium*, so it is necessary to wait after cleaving

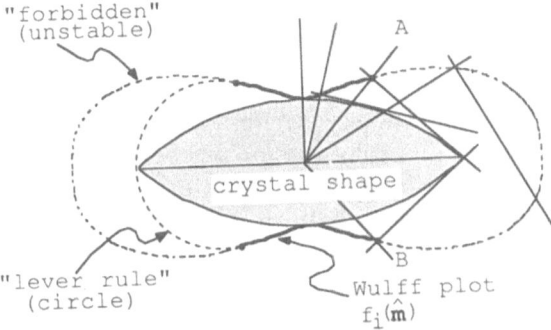

Fig.13.25. Wulff construction with forbidden regions. The outer curve is the nominal Wulff plot. The orientations $\hat{m}$ between A and B (dashed part of the outer curve) have Wulff planes which do not contribute to the interior envelope. Thus, the tangent to the crystal shape jumps discontinuously, leading to a sharp edge. These "forbidden" orientations are actually unstable and spontaneously decompose into macroscopic hill-and-valley formations (Fig.13.26) involving several coexisting stable orientations (full line). The average free energy per unit area of these macroscopically reconstructed surfaces follows the "lever-rule" curve (spherical in d=3) and is the analogue of the familiar double-tangent construction.

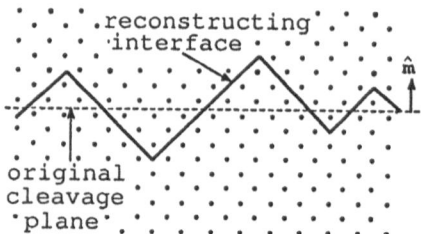

Fig. 13.26. Hill-and-valley rearrangement of an unstable interface. If the orientation $\hat{m}$ lies in a forbidden region, then the flat cleavage plane is unstable to the rearrangement shown. Initially, this rearrangement starts at a microscopic scale and can proceed quite rapidly, since it involves mass transport over short distances only. Given sufficient time, it will continue to coarsen until it reaches macroscopic scales and the original orientation has entirely disappeared; however, the later stages involve long-distance transport and are slow. In experiments, visible striations appear when the rearrangement has reached the scale of optical wavelengths.

before measuring. It is possible that what results is a relaxed interface which retains overall orientation $\hat{m}$; however, it is also possible that a lower overall free energy is achieved by rearranging the interface at a *macroscopic* level into a hill-and-valley formation using other nearby orientations. It is in fact not hard to prove [13.27] that the orientation $\hat{m}$ is stable against such a rearrangement if and only if it appears as a tangent plane of the equilibrium crystal shape.

The decomposition of an initially unstable orientation into a hill-and-valley formation is precisely what is seen in so-called "thermal faceting" experiments [13.7,48]. In these experiments, metal crystals such as Fe, Cu or Ag are cut to expose a high-Miller-index plane, polished, and then annealed at high temperature. Under appropriate conditions (i.e., when the cut plane corresponds to a forbidden orientation $\hat{m}$), optical striations eventually appear. These striations are just the optical manifestation of the hill-and-valley formation, when it has had time to grow in scale to optical wavelengths. In principle, this rearrangement continues to grow in scale until it reaches the size of the entire crystal, at which point the crystal is fully equilibrated and the unstable orientation $\hat{m}$ has disappeared. In practice, kinetic limitations normally make the later ("coarsening") stages of this process too slow, and it does not go to completion [13.49]. The parallel with spinodal decomposition is close.

Strictly speaking, the function $f_i(\hat{m})$ is not definable for "forbidden" orientations $\hat{m}$, so the dotted portion of the Wulff plot in Fig. 13.25 is really unphysical and should be dropped. Generally, there are gaps in the Wulff plot whenever there are sharp edges or corners on the equilibrium crystal shape. (Indeed, when the crystal shape is fully polyhedral, as is always the case [13.16] at T = 0, the Wulff plot reduces to a set of discrete points.) Nevertheless, in the literature it is common to draw $f_i(\hat{m})$ as continuous, even in the presence of sharp edges. Partly, I think, this is just carelessness; however, there are three quite legitimate senses in which it can be done: 1) as part of an approximate calculation of the mean-field type, wherein at first a van der Waals loop emerges [13.50], the central segment of which is then shown explicitly to be metastable or unstable; 2) at T=0, where there is no problem, in principle, in choosing any interface configuration (stable or unstable) and finding its energy; and 3) in the sense of the appropriately weighted average free energy of the coexisting stable facets used by the hill-and-valley rearrangement.

The last of these (#3) is the analogue of drawing in the lever-rule or double-tangent [13.47] line in the free energy f(m) (or f(n) in the liquid/vapor case). A brief calculation (Fig.13.27) shows that for the crystal-shape problem these lever-rule interpolations are actually spherical surfaces: suppose P and Q are points on the Wulff plot and the intermediate angles are unstable, so that there is an edge on the crystal shape at R. An interface at the intermediate orientation $\hat{m}$ rearranges into a hill-and-valley formation using the orientations P and Q, and it has, therefore, a macroscopically averaged free energy per unit area

$$[f_i(\hat{m})]_{av} = \frac{xf_i(P) + yf_i(Q)}{d} . \qquad (13.23)$$

It is left as an exercise for the reader to show that $(\hat{m},[f_i(\hat{m})]_{av})$ lies on the circle, so that the corresponding Wulff plane passes through the edge at R. We shall come back to this point in discussing T = 0 "degenerate" behavior in Sect.13.4.3.

13.4.3 T=0 Roughening and the Degeneracy of Corners and Edges

In Sect.13.2 we described at some length the distinction between type-A and type-B thermal evolution of the crystal shape. We are now in a position to understand the physical origin of this difference in beha-

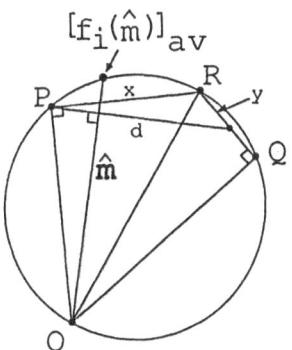

Fig.13.27. Crystal-shape analogue of the double-tangent construction. O is the center of the crystal. Points P and Q are on the (stable) Wulff plot but the region between them is unstable, so the crystal shape follows PRQ and has an edge at R. An interface at the intermediate orientation $\hat{m}$ breaks up into the orientations P and Q in the proportions x:y and, therefore, has an average free energy per unit area given by (13.23). It follows from (13.23) that $[f_i(\hat{m})]_{av}$ lies on the circle.

vior. The central issue is whether the T=0 facet edges, which are always sharp for interatomic forces of strictly finite range [13.16,17], are (type-B) or are not (type-A) stable to rounding (roughening) in the presence of arbitrarily weak thermal fluctuations. The physics is best approached by means of an example. Consider the nearest-neighbor Kossel crystal of Fig.13.22. The interface free energy at T=0 was calculated in (13.23) and gives a Wulff plot (Fig.13.28a) which consists of four intersecting circles (eight intersecting spheres in the full three-dimensional case), leading to a square T=0 equilibrium shape (cubic in d=3). Now, we have just seen in Sect.13.4.2 that angular coexistence leads to just this sort of circular (spherical) Wulff plot, in which a whole set of planes pass through the same edge (corner). What is special here is that the exact microscopic $T = 0$ free energy has this form. The crystal shape is "degenerate" in the sense that a whole class of real Wulff planes pass through one corner. At any nonzero temperature, thermal fluctuations break this degeneracy, each Wulff plane becomes separately tangent to the crystal shape, and the corners round at arbitrarily small T > 0, leading to type-A evolution.

One might think that this marginal behavior is accidental and should disappear when the nearest-neighbor model is modified in any way. This is partially true and partially false. If a second-neighbor interaction $v_2 = \sigma v$ is added to the Kossel-crystal model of Fig.13.22, the resulting $T = 0$ free energy is

$$f_i(\theta) = \frac{v}{2} (1+\sigma) (|\cos\theta| + |\sin\theta|) + \frac{v\sigma}{2} |(|\cos\theta| - |\sin\theta|)| . \qquad (13.24)$$

The corresponding Wulff plot consists of eight circular segments, as shown in Figs. 13.28a and c. When $\sigma < 0$ (second-neighbor-repulsive forces), the Wulff construction leads to a square (cubic in d=3) crystal shape in which the edges (corners) are no longer degenerate, so type-B thermal evolution is expected [13.51], since orientations other than

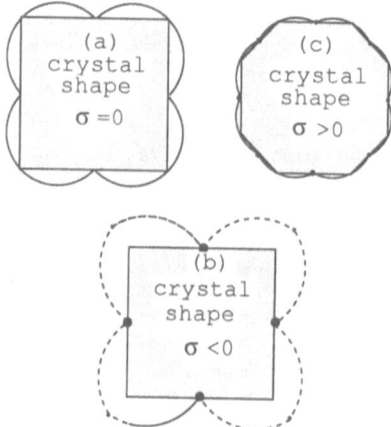

(a)
crystal
shape

$\sigma = 0$

(c)
crystal
shape

$\sigma > 0$

(b)
crystal
shape

$\sigma < 0$

Fig.13.28. $T = 0$ Wulff plots and crystal shapes with nearest- and next-nearest-neighbor interactions [13.22,27]. The second-neighbor interaction is σv. The $T = 0$ Wulff plots consist entirely of circular (spherical) segments. (a) When $\sigma = 0$, the crystal shape is square (cubic in d=3) with degenerate corners. (b) When $\sigma < 0$ (repulsive), it is square and nondegenerate. (c) When $\sigma < 0$ (attractive), it develops new facets and the new edges are degenerate.

{100} are disfavored by a nonzero energy at T = 0 and cannot be stabilized by arbitrarily weak thermal fluctuations. On the other hand, for $\sigma > 0$ (second-neighbor-attractive forces), the T = 0 crystal shape develops {11} facets (in the full d=3 case, {110} and {111} facets) but the new edges are degenerate, so the thermal evolution remains type-A.

It is not hard to show that the behavior in this example is, in fact, generic: when a crystal shape is degenerate (as all nearest-neighbor Kossel-crystal shapes are [13.27]), then appropriate small modifications of the interactions can lead either to a lifting of the degeneracy (and type-B evolution) or to the appearance of new facets and new degenerate edges (and type-A evolution). In situations more complicated than this example [13.20], some edges or corners may be degenerate at T = 0 and others not, so the classification, A or B, of the subsequent evolution must be made for each feature separately. Physically, the origin of all this degenerate behavior is the fact that, for forces of strictly finite range, r_0, steps separated by distances greater than r_0 do not interact at T = 0. An argument precisely equivalent to that surrounding (13.22) then shows that

$$f_i(\theta) = e_t |\cos\theta| + \frac{e_s}{a} |\sin\theta| , \tag{13.25}$$

for sufficiently small θ, where e_t and e_s are the T=0 terrace and step energies, respectively. (Eq.(13.21) is a special case.) The Wulff plot (13.25) is locally circular (spherical), so the facet edge is degenerate (Fig.13.27). At any nonzero temperature, all step configurations are included in the thermal ensemble, so that there is some probability that two adjacent steps approach closer than r_0, however large the average value of l. This leads at large l to a weakly repulsive interaction between steps, which changes the form of (13.25) in such a way as to round the corners (type-A). The presence of further short-range interactions may introduce new T = 0 facets but does not change this logic except when it favors configurations in which the steps clump into a spatially nonuniform state in which the steps are bound together at distances less than r_0. This happens for second-neighbor-repulsive interactions ($\sigma<0$ in the example above) and leads to a state which is stable to weak thermal perturbation (type-B).

13.4.4 The Statistical Mechanics of Crystal Shapes

So far in this section our considerations have been rather general. To develop an overall picture of generic phase diagrams and critical behavior, it is useful to have in mind some microscopic models, which can then in principle be treated by statistical mechanics. The Legendre-transform interpretation of the Wulff plot (Sect.13.3) provides an important technical simplification in this process.

Suppose one has a *microscopic* interface model which allows one to assign an energy $H[C]$ to each microscopic configuration C of the interface (imagine specifying C by giving microscopically $R(\hat{h})$ or $Z(X_1,X_2)$). How does one go about calculating from statistical mechanics the corresponding crystal shape? There are two routes. The traditional way is to calculate the interface free energy per unit area,

$$f_i(\hat{m},T) \equiv \lim_{S\to\infty} \left[\frac{1}{S} (-k_B T) \ln \text{Tr}_{\hat{m}} \exp(-\beta H[C]) \right], \qquad (13.26)$$

where S is the macroscopically defined area of the interface and the subscript $\hat{m}$ on the trace indicates that the allowed configurations C are restricted to have macroscopic arientation $\hat{m}$. Once this calculation is completed, we then use the Wulff construction to get the crystal shape. Alternatively [13.20], using (13.18), we may calculate the Legendre transformed free energy,

$$z(x_1,x_2) \equiv$$

$$\lim_{A\to\infty} \left(\frac{1}{A} (-k_B T) \ln \text{Tr} \exp\left[-\beta(H[C] + x_k \int_A dX_1 dX_2 Z_k) \right] \right), \quad (13.27)$$

where A is a macroscopic area in the (X_1,X_2) plane and the trace over configurations C is no longer restricted. The terms x_1 and x_2 are simply parameters which couple to the local slope of the surface $Z(X_1,X_2)$ in such a way that $\partial z/\partial x_k = \langle \partial Z/\partial X_k \rangle$. Equation (13.27) gives the crystal shape directly and has been used in a number of recent model calculations of crystal shape [13.20,21,23,52-54]. Note the precise parallel between (13.27) and (13.10). The variables x_k are like a two-dimensional magnetic field. They couple to a local slope and can be used to adjust its average to any accessible value in just the way that the chemical potential in a grand canonical ensemble is used to adjust the average particle density. The different routings (13.26 and 27) correspond simply to the canonical versus the grand canonical approach [13.55].

It may be useful to insert here a few remarks in justification of a description in terms of a Hamiltonian, $H[C]$. At first sight it may seem inconsistent (and, indeed, wrong!) to treat the interface as a strictly two-dimensional surface, since we know that as soon as we go beyond Kossel-crystal models the interfacial region is characterized by profiles of particle density, etc., which extend over distances of the same order of magnitude as the bulk coherence length(s). $H[C]$ is best thought of as a "mesoscale" description, in which short-wavelength degrees of freedom (which describe the important internal structure of the interface) have somehow been integrated out (in the renormalization-group sense [56]). Thus, the parameters of $H[C]$ should really be thought of as temperature dependent, except in the simplest models, and are not truly microscopic. These renormalization effects may be important in some systems. So far, they have not been considered seriously. For the kind of simple solid/vapor systems so far treated it seems likely that they are small. Furthermore, much of the information that we shall wish to infer from these models is "universal" (again, in the renormalization-group sense) and thus insensitive to this sort of short-distance effect.

Interface models used in studying the statistical mechanics of the crystal-shape problem have so far been exclusively of the so-called solid-on-solid (SOS) type. These arise naturally from the Kossel-crystal picture (a regular compact array of atoms truncated to form an interface). In SOS models, the X_1,X_2 (or X,Y) plane is made discrete, corresponding to the underlying crystal lattice, and the interface configuration C is specified by giving the height $Z(X_1,X_2)$ of the highest atom over the reference plane [13.37]. For most purposes, the problem is further simplified by restricting the height difference ΔZ between neighboring (X_1,X_2) lattice sites. The resulting restricted SOS (RSOS) models can in simple cases be mapped by ingenious transformations to two-dimensional statistical mechanics problems which have been solved - exactly or approximately - in other contexts. The reader is referred to the literature for details [13.20,21,23,52-54].

13.4.5 Critical Behavior of the Equilibrium Crystal Shape

Looking at the crystal shape in terms of phase diagrams motivates us to identify the phases and phase transitions which appear, just as we would in dealing with any other thermodynamic system. The interface is two-dimensional. There is an extensive literature on phases and phase transitions in two dimensions [13.57]. Correspondence with other two-dimensional models is in practice achieved directly via (13.27) and allows us to draw useful conclusions about the crystal-shape problem. It is not our purpose here to dwell on technical details, which are in any case spelled out in the literature. Nevertheless, a few examples may serve to demonstrate the power of the viewpoint which we have advocated in this article.

a) Phases. Two types of phases appear in the phase diagrams (Figs. 13.10,14), facets and rounded ("rough") regions. Both have been studied in great detail in the context of SOS models. Facets conform in most respects to the ideal picture of surface science: Height variations due to thermal fluctuations are atomic in scale, so the scattering of probes such as X-rays or atomic beams leads to ordinary (d=2) Bragg spots. The rough phase, on the other hand, behaves at long distances

[13.58] like a fluid/fluid interface. The vanishing of the step energy f_s per unit length allows long-wavelength capillary fluctuations, the presence of which gives height fluctuations that scale as $\ln(L/a)$ for a rough region of sice L [13.46]. This in turn means that sharp (δ-function) Bragg scattering peaks are replaced by diffuse scattering with a power-law peak at the Bragg position.

b) Pokrovsky-Talapov edges. The curved phase boundaries separating facets and rough regions in Figs. 13.10,14 describe the smooth (second-order) edges that we have seen in Pb, ^{4}He, and NaCl (710°C). Vicinal surfaces (on the rounded region but close to the edge) are characterized by an array of widely separated steps. At nonzero T, these steps are not straight but lower their free energy by wandering. Because the steps cannot overlap, this free-energy lowering is smaller when the steps are close together, and it creates an effective step-step repulsion of purely entropic origin. This effect has been studied in a variety of other statistical-mechanical contexts and leads to the $x^{3/2}$ behavior of (13.2) and Fig.13.7. One of the many ways of deriving this result is to treat the direction along the step as "time" and to represent the configuration of the step as the position of a "particle" moving in one dimension. The non-overlapping condition is incorporated by making the particle a fermion. The problem then reduces to that of noninteracting fermions in one dimension, which is easily soluble!

c) The roughening transition. The roughening transition which occurs at θ=0 as T passes through T_R has been extensively studied in the SOS representation. In particular, the step free energy per unit length is known to vanish as

$$f_s(T) \sim \exp(-\mathrm{const}/(T_R - T)^{1/2}) \,, \qquad (13.28)$$

as T approaches T_R from below. Because of the connection (Sect.13.4.1) between f_s and the cusp coefficient B in (13.19), the facet radius is predicted to vanish in the same manner. Despite some early discrepancies, recent experiments appear to confirm this reasoning [13.6,18,28,59]. An even more subtle and beautiful effect was first pointed out by *Jayaprakash* and coworkers [13.54] and is illustrated in Fig.13.29. Question: How does one expect the curvature at the center of a rough region to behave, as the roughening (faceting) temperature is approached from above? One's first guess is certainly that it should vanish smoothly. Wrong! What happens is that it decreases to a nonzero value and then jumps abruptly to zero. Moreover, in appropriate units (set by the step height) the value of the curvature at the jump is expected to be always exactly $2/\pi$ [13.60]. This prediction has been reasonably confirmed in recent experiments on ^{4}He crystals by *Balibar* and coworkers [13.18,61].

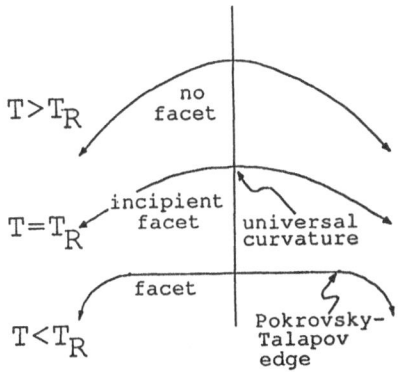

Fig. 13.29. Universal curvature at T_R. Above T_R the surface is rounded; below T_R it is faceted. When $T = T_R$, the curvature at the center of the surface, just where the new facet is about to appear, takes on a universal value.

d) Behavior near T_0 and T_1 for type-B crystals. A good deal of information is now available concerning the universal behavior of type-B crystals near the corner-rounding transition at T_0 and the sharp-edge-disappearing transition at T_1. The procedure is to consider Kossel crystals with the appropriate (repulsive) second-neighbor interactions. The corresponding RSOS models can then be mapped by clever transformations onto vertex models [13.20] or special plane antiferromagnetic models [13.23,52], whose universal properties are known. Two examples give the flavor of results [13.20]. In the "intermediate" type-B shape (Fig. 13.11), the sharp edges are predicted to meet the rounded regions at the cube corners with a slope discontinuity (this, despite the fact that the rounded regions meet the surrounding facets without slope discontinuity). Similarly, as T approaches T_1 from above, the curvature across the incipient sharp edge (Fig. 13.30) is predicted to diverge as $\text{const}/(T-T_1)$. Neither of these predictions has yet been tested experimentally.

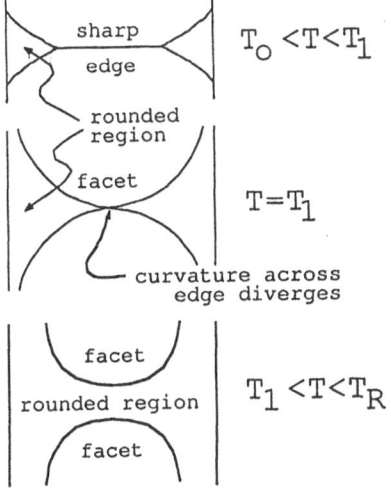

Fig. 13.30. Curvature near T_1. The type-B crystal of Fig. 13.11 is here viewed along a $\langle 110 \rangle$ direction. Below T_1 a segment of sharp interfacet edge remains; above T_1 it has disappeared. The curvature at the center of the rounded region in the direction perpendicular to the incipient edge diverges as $\text{const}/(T-T_1)$.

399

13.4.6 Calculating the Shapes of Real Crystals

The foregoing discussion of critical behavior may leave the impression that statistical mechanics are useful for the determination of "universal" behavior but cannot be used to calculate the shapes of real crystals at T>0. We have alluded at the beginning of this section to the practical difficulties of calculating surface properties from first principles. Nevertheless, in simple cases, I believe that progress can and will be made. As examples, let me briefly mention two recent calculations of our group. First [13.62], we considered what is certainly the simplest possible model of ^{4}He crystals: an ideal hcp Kossel crystal with nearest-neighbor interactions only. The T = 0 crystal shape of this model has been known for more than 50 years [13.42] and turns out to give correctly the identity of the principal facets seen in experiment (Fig. 13.9). In addition, this model predicts the observed type-A thermal evolution, including crude but reasonable estimates of the roughening temperatures of the three observed facets. Second [13.23], we considered NaCl and attempted to estimate the corner-rounding temperature T_0. The basic physics involved is simply an energy-entropy crossover: at T = 0 the {100} facets have the lowest energy and the crystal shape is cubic; however, fluctuations on the {100} facets cost a lot of energy. By contrast, the {111} interface is costly in energy but permits many low-energy fluctuations. Corner rounding occurs when the temperature is sufficiently high so the entropy advantage of the {111} orientation finally overcomes its energy disadvantage. A very simple calculation leads to a value of T_0 about a factor of two too high; however, we believe that including relaxation effects would improve this substantially.

13.5 Open Questions

The predictions of the statistical-mechanical approach to crystal shapes are only beginning to be tested. So far, things look good, and I think it is fair to say that there are no clear contradictions. On the other hand the evidence is certainly not all in, and what there is is not unequivocal. Let me comment briefly on the status of the (Pokrovsky-Talapov) $x^{3/2}$ law. Data [13.3,14] on both Pb and ^{4}He have been shown to be at least consistent with this form; however, because older mean-field-like theories [13.32,63] predict x^2, there has been some interest [13.5,10,63] in fitting extant data to both $\lambda=3/2$ and $\lambda = 2$ and then comparing the quality of the fit. In particular, *Saenz* and *Garcia* [13.63] have argued that they can obtain a superior fit to the Pb data [13.3] with $\lambda = 2$. The significance of this kind of comparison remains obscure, because the mean-field calculations are known to be incorrect in this instance (fluctuations, neglected in mean-field theory, always change the exponent from 2 to 3/2).

A deeper potential problem surrounds the nature of the forces which play a role in the crystal-shape problem. The models which have been treated in full calculational detail all assume forces of strictly finite range (nearest neighbors or nearest plus next-nearest neighbors). This is not a serious restriction, since universal behavior is expected to be insensitive to details of the potentials, provided that forces fall off sufficiently rapidly at long distance. *Jayaprakash* et al. [13.21], for example, have studied the Pokrovsky-Talapov transition in the fermion representation (Sect.13.4.5) and show that step-step interactions which fall off as $const/l^\tau$ with $\tau > 2$ (1 is the average step-step separation of Fig.13.23) do not modify the $x^{3/2}$ behavior. The difficulty is that both elastic [13.64] and dipolar forces lead to a step-step interaction with $\tau = 2$. Analysis of this limiting case shows that elastic interactions do not modify the $x^{3/2}$, while dipolar interactions of appropriate sign may drive the edge first order. On the other hand, recent work by *Rommelse* and *den Nijs* [13.65] seems to suggest that further-neighbor interactions, which serve to provide an energy difference between the step sequences up-up and up-down, can produce a new type of faceted phase which they call "disordered flat".

The behavior of Au crystals remains unexplained: Figure 13.4 shows a first-order edge between faceted and rounded regions. No edge of this type appears in the generic phase diagrams (Figs.13.10, 14), nor is such an edge present in any models so far solved exactly [13.66]. It is not clear whether this is an intrinsic shortcoming of the models or whether, on the other hand, it might be the result of a long-range force. It is also possible that the sharp edges observed on Au crystals are nonequilibrium effects, like those seen in Fig.13.13 for NaCl.

I hope that some of these problems will be clarified in the near future. I believe that the thermodynamic framework introduced above will remain valid, although it is entirely conceivable that long-range-force effects will require the study of some new models.

Detailed comparison with experiments, required for a really clear analysis of the exponent λ in x^λ or for verification of the universal curvature at T_R (Fig.13.29), requires an understanding of finite-size corrections and of kinetics. These are also interesting subjects in their own right. *Nozières* and coworkers [13.67] have recently made important progress in understanding dynamics near T_R. The fast (dendritic) growth of rough interfaces has received much recent attention. Much remains to be done, especially for faceted and vicinal interfaces.

Finally, there are many closely related interface problems which will probably receive attention in the future. At least two groups are studying the effects of adsorption on crystal shapes [13.8,68]. Similar methods and points of view have already proved useful in studying internal interfaces, such as grain boundaries [13.69].

Acknowledgement. I wish to acknowledge the help and inspiration of my former students, C. Jayaprakash, C. Rottman, A.-C. Shi, and M. Touzani, whose ideas and calculations have shaped my understanding of this field. I am grateful to F. Rosenberger for encouraging me to write up these ideas in what I hope is an accessible form. I thank S. Balibar, J.C. Heyraud, S. Lipson, and J.J. Métois for sharing their data and their enthusiasm and for taking the time to explain to a naive theorist what really goes on inside the box. I am indebted to S. Balibar, J.C. Heyraud, J.J. Métois, and F. Gallet for permission to use their beautiful photographs. This work was supported in part by the National Science Foundation under Grants No. DMR83-16981 and DMR84-15063.

References

13.1   J.C. Heyraud, J.J. Métois: J. Cryst. Growth 50, 571 (1980); Acta Metal. 28, 1789 (1980).
13.2   J.C. Heyraud, J.J. Métois: Surf. Sci. 128, 334 (1983).
13.3   C. Rottman, M. Wortis, J.C. Heyraud, J.J. Métois: Phys. Rev. Lett. 52, 1009 (1984).
13.4   T. Yanagihara: Jap. J. Appl. Phys. 21, 1554 (1982).
13.5   J.C. Heyraud, J.J. Métois: Surf. Sci. 177, 213 (1986) and J. Cryst. Growth 82, 269 (1987).
       J.J. Métois, J.C. Heyraud: Surf. Sci. 180, 647 (1987).
13.6   J. Landau, S.G. Lipson, L.M. Maattanen, L.S. Balfour, D.O. Edwards: Phys. Rev. Lett. 45, 31 (1980).
       S. Balibar, B. Castaing: J. Phys. Lett. 41, L329 (1980).
       K.O. Keshishev, A. Ya. Parshin, A.V. Babkin: Sov. Phys. JETP 53, 362 (1981).
       P.E. Wolf, S. Balibar, F. Gallet: Phys. Rev. Lett. 51, 1366 (1983). These works are representative only and contain many further references to this extensive literature.
13.7   M. Flytzani-Stephanopoulos, L.D. Schmidt: Prog. Surf. Sci. 9, 83 (1979); T. Wang, C. Lee, L.D. Schmidt: Surf. Sci. 163, 181 (1985). These articles will serve as an introduction to the literature.
13.8   I.N. Stranski: Bull. Soc. Fran. Min. Crist. 79, 359 (1956); A.-C. Shi: Phys. Rev. B 36, 9068 (1987).
13.9   J.C. Heyraud, J.J. Métois: private communication.
13.10 J.C. Heyraud: Thesis, Université de Droit, d'Economie et des Sciences d'Aix-Marseille, 1987 (unpublished).
13.11 One might worry that the equilibrium shape of the crystallites is influenced by the presence of the graphite substrate and does not represent the shape of a "free" gold crystal. Analysis shows that in the thermodynamic limit the only effect of the substrate is to truncate the free shape. Different substrates truncate it at different levels. See W.L. Winterbottom: Acta Metal. 15, 303 (1967).
13.12 One should not be mislead into thinking that the curved part of the crystal surface is strictly spherical. Although this is approximately true for gold at 1000°C, it is certainly not so in general.
13.13 A readable discussion of convexity and its significance in ordinary statistical mechanis is given by R.B. Griffiths in *Phase Transitions and Critical Phenomena*, ed. C. Domb, M.S. Green (Academic, London 1972) Vol.1, p.7ff
13.14 Y. Carmi, S.G. Lipson, E. Polturak: Phys. Rev. B 36, 1894 (1987).

13.15 This is a slight oversimplification. There is good reason to believe that mixtures of type-A and type-B thermal evolution will be seen. The designation should properly speaking attach to specific features of the T=0 crystal shape, according to whether they are (A) or are not (B) "degenerate" in a sense which will be made precise in Sect.13.4.3.

13.16 The polyhedral character of the T=0 crystal shape is believed to be entirely general. D.S. Fisher, J.D. Weeks: Phys. Rev. Lett. **50**, 1077 (1983); E. Fradkin: Phys. Rev. B **28**, 5338 (1983).

13.17 If the force law has the appropriate sign and is not strictly finite in range, then the number of distinct facets can be infinite. This may lead to fractal structure at T=0. Such structure does not persist to nonzero T and will be exceedingly difficult to observe because of kinetic constraints at low temperature. S.E. Burkov: J. Physique **46**, 317 (1985); J. Physique Lett. **46**, L805 (1985). M. den Nijs, E.K. Reidel, E.H. Conrad, T. Engel: Phys. Rev. Lett. **55**, 1689 (1985). See also L.A. Bol'shov, V.L. Pokrovsky, G.V. Uimin: J. Stat. Phys. **38**, 191 (1985).

13.18 S. Balibar, B. Castaing: Surf. Sci. Rep. **5**, 87 (1985). F. Gallet: Dissertation, Université de Paris (Paris 6), 1986 (unpublished). See also [13.60].

13.19 V.L. Pokrovsky, A.L. Talapov: Phys. Rev. Lett. **42**, 65 (1979). See also P.G. deGennes: J. Chem. Phys. **48**, 2257 (1968).

13.20 C. Jayaprakash, W.F. Saam: Phys. Rev. B **30**, 3916 (1984).

13.21 C. Jayaprakash, C. Rottman, W.F. Saam: Phys. Rev. B **30**, 6549 (1984).

13.22 C. Rottman, M. Wortis: Phys. Rep. **103**, 59 (1984); Phys. Rev. B **29**, 328 (1984).

13.23 A.-C. Shi, M. Wortis: Phys. Rev. B **37**, 7793 (1988)

13.24 J.C. Heyraud, J.J. Métois: J. Cryst. Growth **84**, 503 (1987) and private communication.

13.25 A tricritical behavior in this vicinity was found in [13.22]; however, it is likely that this was due to use of the mean-field approximation.

13.26 G. Wulff: Z. Kristallogr. Mineral. **34**, 449 (1901).

13.27 C. Herring: Phys. Rev. **82**, 87 (1951). See also C. Herring, in: *Structure and Properties of Solid Surfaces*, ed. by R. Gomer, C.S. Smith (Univ. Chicago Press, Chicago 1953), p.5ff.

13.28 This is so provided that the diameter of the solid is small on the scale of the appropriate capillary length, typically of the order of millimeters. For large soft crystals such as in the ^{4}He experiments, gravitational effects must be included in the analysis. See K.O. Keshishev, A. Ya. Parshin, A.I. Shal'nikov: "Surface phenomena in quantum crystals", in: *Physics Reviews* Vol.4, ed. I.M. Khalatnikov (Harwood Academic Publishers, Chur 1982), p.188ff.

13.29 In this limit, it can be shown that $2\lambda = P_s - P_v$, the pressure difference between the solid and the vapor. N. Cabrera: Surf. Sci. **2**, 320 (1964); J.W. Cahn, D.W. Hoffman: Acta Metal. **22**, 1205 (1974). For example, it is easy to verify for the isotropic case $f_i = \sigma$ (independent of $\hat{m}$) that $2\lambda = 2\sigma/R$.

13.30 In writing (13.8b) we have assumed that $f_i(\hat{m})$ is continuous and differentiable. When the crystal is fully faceted, this is not so, and some additional discussion is required (Sect.13.4.2).

13.31 This is sometimes called the Herring construction, if memory serves me. See also F.C. Frank, in: *Metal Surfaces*, (American Society for Metals, Ohio 1963) Chap.1; J.L. Meijering: Acta Metal. **11**, 847 (1963).

13.32 A.F. Andreev: Sov. Phys. JETP **53**, 1063 (1981). The result is also implicit in [13.33]. After going through the Legendre transformation, Andreev performs a mean-field calculation, leading to some *incorrect* conclusions, e.g., that $\lambda = 2$.

13.33 L.D. Landau, E.M. Lifshitz: *Statistical Physics* (Pergamon, Oxford 1980), 3rd ed. revised by E.M. Lifshitz, L.P. Pitaevskii, Part 1, p.520ff.

13.34 This elementary mathematical fact is notably missing from most thermodynamics texts. A happy exception is H.B. Callen: *Thermodynamics and an Introduction to Thermostatistics* (Wiley, New York 1985) p.137ff.

13.35 This magnetic analogue was pointed out by N. Garcia, J.J. Saenz, N. Cabrera: Physica 124b, 251 (1984), who exploited it in the context of a mean-field calculation.

13.36 In the sense that the intercept $f_i(\hat{m})$ is not on some fixed ordinate axis but rather on the radial line from the origin in the direction $\hat{m}$.

13.37 We shall treat $Z(X_1,X_2)$ as single-valued, thus excluding the possibility of local overhangs. This simplification is inessential.

13.38 There remains only a trivial difference in sign convention: it would be simple to redefine $z_k \equiv -\partial z/\partial x_k$.

13.39 Indeed, in [13.33] it *is* taken as a proof!

13.40 A. Dinghas: Z. Krist. 105, 304 (1944).

13.41 J.W. Cahn, J.E. Taylor: Scripta Metall. 18, 1117 (1984)
J.E. Taylor, J.W. Cahn: Acta Metal. 34, 1 (1986).

13.42 I.N. Stranski, R. Kaischew: Z. Kristallogr. 78, 373 (1931).
R. Lacmann: N. Jb. Miner. Abh. 122, 36 (1974).
M. Drechsler: In *Surface Mobilities on Solid Materials*, ed. by Vu Thien Bunh (Plenum, New York 1983) p.405ff.

13.43 *Phase Transitions and Critical Phenomena*, ed. C. Domb, M.S. Green, J.L. Lebowitz (Academic, New York 1972-1982) Vol.1-8.

13.44 In d=2 at T>0 the thermal wandering of the steps is sufficiently large so that the step-step interactions cannot be neglected. As a result, there are no facets and no edges at any T>0 and the d=2 crystal shape is always smoothly rounded.

13.45 Short length-scale singularities at T_R are very weak, so at an atomic scale not much happens to the interface at T_R.

13.46 See J.D. Weeks: In *Ordering in Strongly Fluctuating Condensed Matter Systems*, ed. by T. Riste (Plenum, New York 1980)

13.47 [Ref.13.34,p.243]

13.48 A.J.W. Moore, in: *Metal Surfaces: Structure, Energetics and Kinetics* (Am. Soc. Metals, Metals Park 1963) pp.155-198.

13.49 In practice, defect pinning may also play a role here.

13.50 Of course, this kind of an unstable region can never come out of a rigorous statistical mechanical calculation.

13.51 We speak here of three dimensions (even though we have been drawing pictures in d=2 for simplicity). In two bulk dimensions, the crystal surface is a one-dimensional system, and, so long as all forces are short-ranged, there can be no phase transition (i.e., no edges) of any kind for T>0.

13.52 H.W.J. Blote, H.J. Hilhorst: J. Phys. A 15, L631 (1982); B. Nienhuis, H.J. Hilhorst, H.W.J. Blote: J. Phys. A 17, 3559 (1984).

13.53 H.J. Schulz: J. Phys. 46, 257 (1985).

13.54 C. Jayaprakash, W.F. Saam, S. Teitel: Phys. Rev. Lett. 50, 2017 (1983)
W.F. Saam, C. Jayaprakash, S. Teitel, in: *Quantum Solids and Fluids*, ed. by E.D. Adams, G.G. Ihas (AIP, New York 1983) p.371.

13.55 Just as in the usual applications, certain hypotheses are needed to prove in a rigorous way the equivalence between ensembles. These hypotheses serve to control the scale of fluctuations about mean values such as $\langle \partial z/\partial x_k \rangle$. Hypotheses such as short-range pairwise forces normally suffice. We do not know of any work which addresses ensemble equivalence for the interface problem in a rigorous way. There are also important questions involved in the existence of an interface Hamiltonian such as $H[C]$ and in the proof directly from statistical mechanics of the Wulff construction. Some of these have been discussed in the review article of D.B. Abrahams, in: *Phase Transitions* and *Critical Phenomena*, ed. C. Domb, J.L. Lebowitz (Academic, New York 1986), Vol.10, p.2. There now exist a few rigorous results in two dimensions but much remains to be done. It is perhaps worth emphasizing, however, that for ordinary force laws everything in the literature suggests that the conventional viewpoint, which we take here, is correct.

13.56 This point of view has been discussed extensively in [13.43]. For a more elementary introduction, see P. Pfeuty, G. Toulouse: *Introduction to the Renormalization Group and to Critical Phenomena* (Wiley, New York 1977)

13.57 See [13.43] and also, e.g., *Ordering in Two Dimensions*, ed. by S.K. Sinha (North-Holland, New York 1980)

13.58 It is important to emphasize here that, although the height fluctuations become infinite in the thermodynamic limit, ln (L/a) grows so slowly with L that the "liquid-like" effects are difficult or impossible to observe in the laboratory. On short distance scales, rough interfaces and smooth (faceted) interfaces do not look very different.

13.59 T. Ohachi, I. Taniguchi: J. Cryst. Growth **65**, 84 (1983).

13.60 The behavior of the curvature is analogous to that of the superfluid density of two-dimensional ^{4}He at the Kosterlitz-Thouless transition.

13.61 P.E. Wolf, F. Gallet, S. Balibar, E. Rolley, P. Nozières: J. Physique **46**, 1987 (1985).

13.62 M. Touzani, M. Wortis: Phys. Rev. B **36**, 3598 (1987).

13.63 J.J. Saenz, N. Garcia: Surf. Sci. **155**, 24 (1985).

13.64 V.I. Marchenko, A. Ya. Parshin: Sov. Phys. JETP **52**, 129 (1981).

13.65 K. Rommelse, M. den Nijs: Phys. Rev. Lett. **59**, 2578 (1987)

13.66 Edges of this type did appear in [13.22] as the result of a mean-field calculation.

13.67 P. Nozières, F. Gallet: J. de Physique (March 1987); F. Gallet, S. Balibar, E. Rolley: J. de Physique (March 1987). See also F. Gallet, P. Nozières, S. Balibar, E. Rolley: Europhys. Lett. **2**, 701 (1986).

13.68 R. Kariotis, H. Suhl, B. Yang: Phys. Rev. B **32**, 4551 (1985).

13.69 Y. He, C. Jayaprakash, C. Rottman: Phys. Rev. B **32**, 12 (1985). See also C. Rottman: "Theory of phase transitions at internal interfaces"; J. Physique Colloq. (1988)

14. Experimental Aspects of Surface Roughening

Thomas Engel

Department of Chemistry, University of Washington
Seattle, WA 98195, USA

The roughening of surfaces has been an active area of research in recent years. Although the theoretical groundwork for the roughening transition was formulated by *Burton* et al. in 1951 [14.1], direct experimental evidence for such a transition has been available only since 1980. In this review, the experimental aspects of research in this area will be discussed. Reference will be made to theoretical work, but only where it is needed to place experimental studies in a proper framework. However, theoretical aspects of surface roughening and equilibrium crystal shapes are discussed in Chap.13 as well as elsewhere [14.2,3].

The configuration of a surface below and above the roughening transition temperature, T_R, is shown schematically in Fig.14.1. The

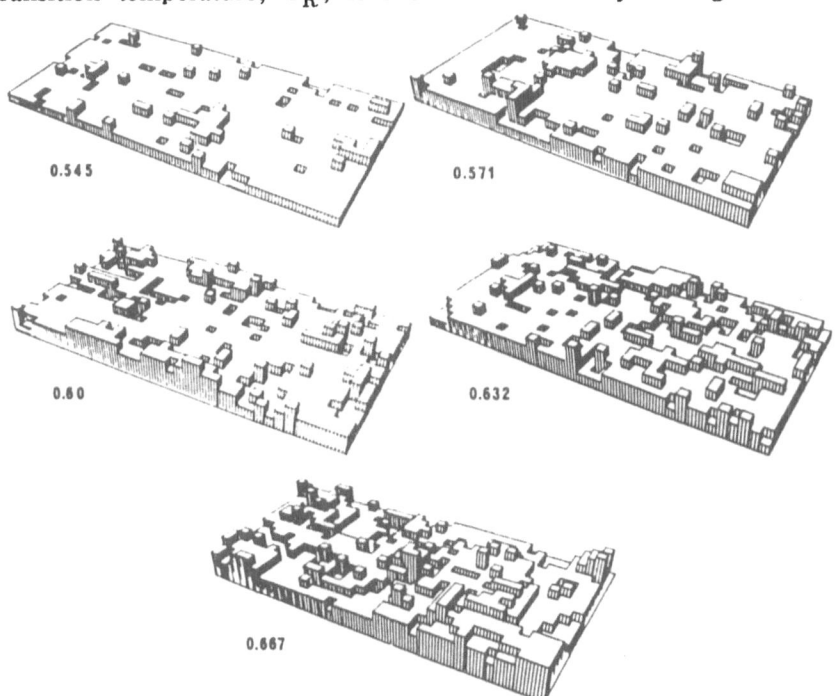

Fig.14.1. Surface configuration as a function of reduced temperature. The value of 0.57 corresponds to $T = T_R$ [14.3].

407

essential change in the surface configuration is that many atoms are promoted from lower lying layers in the solid to lattice positions in layers above the zero Kelvin surface plane. The driving force behind the phase transition is the gain in entropy produced by the reduction in long range order. At the transition temperature, the free energy required to form steps on the surface goes to zero and long range height fluctuations in the position of the topmost layer of atoms occur. The signature of the roughening transition is not that different atomic layers parallel to the surface are exposed, since this will be the case at temperatures well below T_R. The signature is that the height fluctuations of the surface become unbounded for an infinitely large crystal. For this reason, the surface plane is no longer well defined. Since real crystals have coherent domain sizes in the range 100-3000 Å depending on the material, height fluctuations above T_R will remain small even on an atomic scale, making the transition less dramatic. An important aspect of the surface configuration above T_R is that all atoms remain in lattice positions. Surface roughening can be distinguished from surface melting in this way. Roughening on an atomic scale as is shown in Fig. 14.1 can be probed using surface sensitive diffraction techniques as will be shown below.

An alternative way to diffraction methods in which surface roughening can be observed experimentally is by examining the equilibrium shapes of crystals as a function of temperature. This subject has been reviewed elsewhere [14.4,5] and is also the subject of Chap.13. In general, the shape of a crystal at equilibrium (at somewhat elevated temperature) will consist of flat facets separated by regions which show a rounded form. As the temperature is increased, the rounded portions of the crystal will increase in area at the expense of the flat facets. These rounded portions can be viewed as consisting of infinitesimal regions of high index or vicinal planes. As T approaches T_R for a given facet, the area of the facet will decrease and at $T = T_R$, it disappears into the neighboring rounded regions. This allows measurements of T_R to be made on a macroscopic scale rather than on an atomic scale as is the case for surface sensitive diffraction studies.

In the above discussion we have established a criterion for roughening in terms of temperature T_R such that for $T < T_R$, the surface is in the smooth phase and for $T > T_R$, it is in the rough phase. The roughening transitions should be viewed as one of a number of possible channels, which include surface or bulk melting. It is not clear a priori that surface roughening will occur below the bulk melting temperature. The roughening temperature is determined by the energy parameters characterizing the binding of atoms in the topmost layers of the surface and will vary with crystal face orientation for a given substance. As will be shown below, unambiguous evidence for roughening transitions has been found for solid helium and for vicinal metal surfaces. Some confusion exists in the published literature concerning the usage of "rough surface" since the term has been widely used to denote an uneven or disordered surface. The sense in which we will use it is in referring to an equilibrium surface configuration. In a roughening transition, the system can be made to proceed from the rough to the smooth phase and vice versa an arbitrary number of times. The other way in which the term has been used refers to a non-equilibrium case.

In this review, we will first discuss equilibrum crystal shapes in Sect.14.1. In Sect.14.2 we will discuss the detection of steps on surfaces. The energetics of step formation and surface diffusion will be considered in Sect.14.3. In Sect.14.4 we will discuss the use of diffraction methods to investigate the roughening transition. The dependence of the transition temperature on crystal face orientation will be discussed in Sect.14.5 and we will summarize our conclusions and mention some unresolved problems in Sect.14.6.

14.1 Equilibrium Crystal Shapes and Surface Roughening

Crystals show well-defined facets due to an anisotropy in the surface tension (or specific Gibbs surface free energy) of the solid. However, at elevated temperature, the anisotropy decreases and well-defined flat facets merge into their surrounding regions. This corresponds to roughening and the temperature at which the surface tension anisotropy vanishes and the facet disappears is the roughening temperature, T_R.

The first conclusive demonstrations of surface roughening were obtained by *Avron* et al. [14.6] and very shortly thereafter by *Balibar* et al. [14.7] and *Keshishev* et al. [14.8]. These studies were all carried out for helium crystals in equilibrium with superfluid helium. These studies showed that three different crystal faces roughen at temperatures which lie between 0.35 and 1.3 K. In addition, detailed studies of the variation of surface curvature with temperature [14.5] were able to show that the transition is of the Kosterlitz-Thouless universality class. Helium is ideally suited for for such studies because of the rapid kinetics of growth and the very high purity of the crystals which can be obtained.

Due to the importance of metals and semiconductors in technology, there has been an interest in determining whether roughening occurs for these materials. Since roughening leads to a dramatic increase in sites of high coordination number, substances may show a difference in chemical reactivity or electronic properties.

Equilibrium shapes of lead crystals have been studied in detail by *Heyraud* and *Métois* [14.9] under UHV conditions. Using the experi-

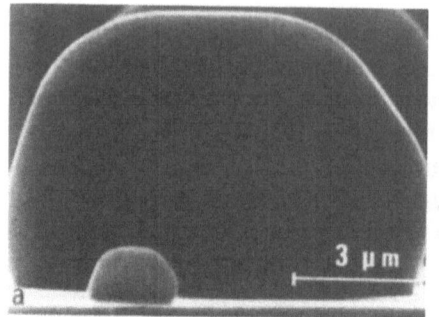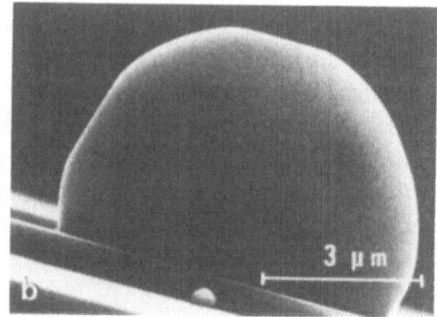

Fig.14.2. Equilibrium shape of a small lead crystallite in the ⟨110⟩ azimuth at a) 200°C and b) 300°C. The flat portions in a) proceeding from the left to right side of the figure have {111}, {111} and {100} orientations [14.9].

mentally observed shapes, these authors were able to determine the anisotropy of the specific Gibbs surface free energy as a function of temperature. Figure 2 shows the equilibrium shape of a lead crystal along the ⟨110⟩ azimuth at 200°C and 300°C. The rounded regions which appear near the top of the picture correspond to the ⟨110⟩ direction. Well developed facets are not seen in this direction in the temperature range over which the experiments were carried out. From this observation, it can be concluded that T_R for Pb{110} lies below 475 K.

Although these measurements give a direct visualization of the roughening process, kinetic limitations restrict the studies to elevated temperatures. Surface diffusion is too slow below 475 K to allow the equilibrium shape to be reached within a few days. For lead, this restricts the temperature range to $0.6T_m < T < T_m$ where T_m is the bulk melting temperature. In addition, it is well known that impurities can segregate to the surface of a metal at elevated temperatures and that the specific Gibbs surface free energy can be strongly altered. Therefore, it is necessary to ensure that impurity concentrations remain very low.

The studies of the roughening transition carried out using equilibrium crystal shapes have provided definitive proof that roughening can occur below the melting temperature. This method has the advantage that a number of crystal orientations can be investigated on a single sample. The disadvantage of the method is that no microscopic model of the surface structure can be obtained.

14.2 The Detection of Steps on Surfaces

The zero Kelvin stable state of a perfectly oriented single crystal surface is defect free. However, under equilibrium conditions at higher temperatures, there will be isolated atoms and islands lying above the

410

surface place as well as holes exposing lower atomic layers. Both islands and holes have boundaries consisting of straight steps combined with kinks. Near the roughening temperature, the fraction of atoms which lie at a step edge as opposed to on a flat terrace increases sharply with temperature. Before discussing roughening, we will first review what is known about step densities on surfaces under well-defined conditions.

The most powerful method of observing atomic steps is with a direct imaging technique such as scanning tunneling microscopy [14.10]. Figure 14.3 shows a scan over a Au{100} surface in which a step edge can be clearly seen. Note that it is made up of straight line segments rather than being composed of kinks and anti-kinks. In Fig.14.4, a gray scale image obtained from STM line scans is shown for a region containing a step on a Si{111}-(7x7) surface [14.11]. Here it can be clearly seen that the atoms along the step edge show no significant deviation from their ideal positions. Tunneling microscopy is still in its early stages, but it is clear that a great deal of information on atomic configurations on surfaces will become available using this technique.

Much has been learned about steps on surfaces using diffraction techniques and, in particular, by LEED studies [14.12,13]. Since steps will not usually have regular periodicity, they will not in general give rise to additional diffraction spots. An exception to this case is that of

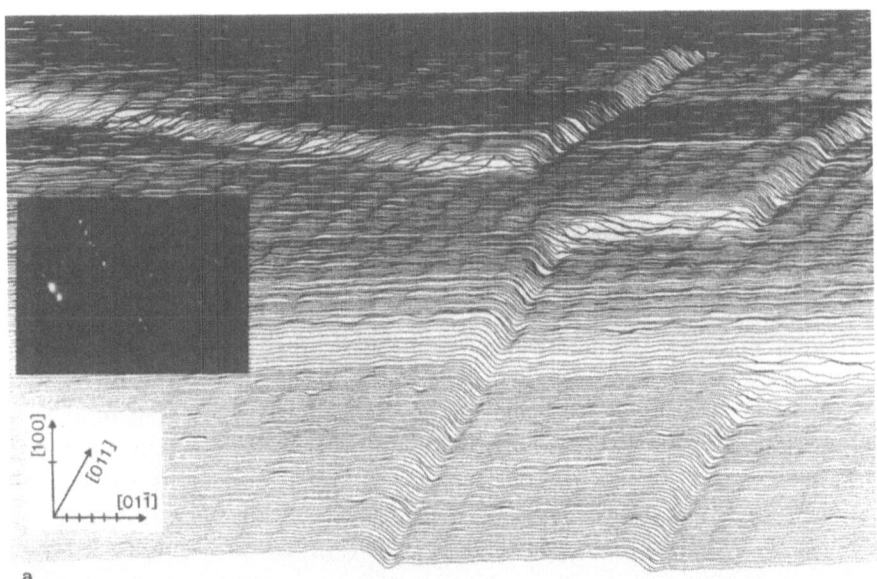

Fig.14.3. STM image of a clean reconstructed Au{100} surface. Divisions on the crystal axes are 5 Å with approximately 1.5 Å spacing between line scans. The inset shows a LEED image of the surface [14.10].

411

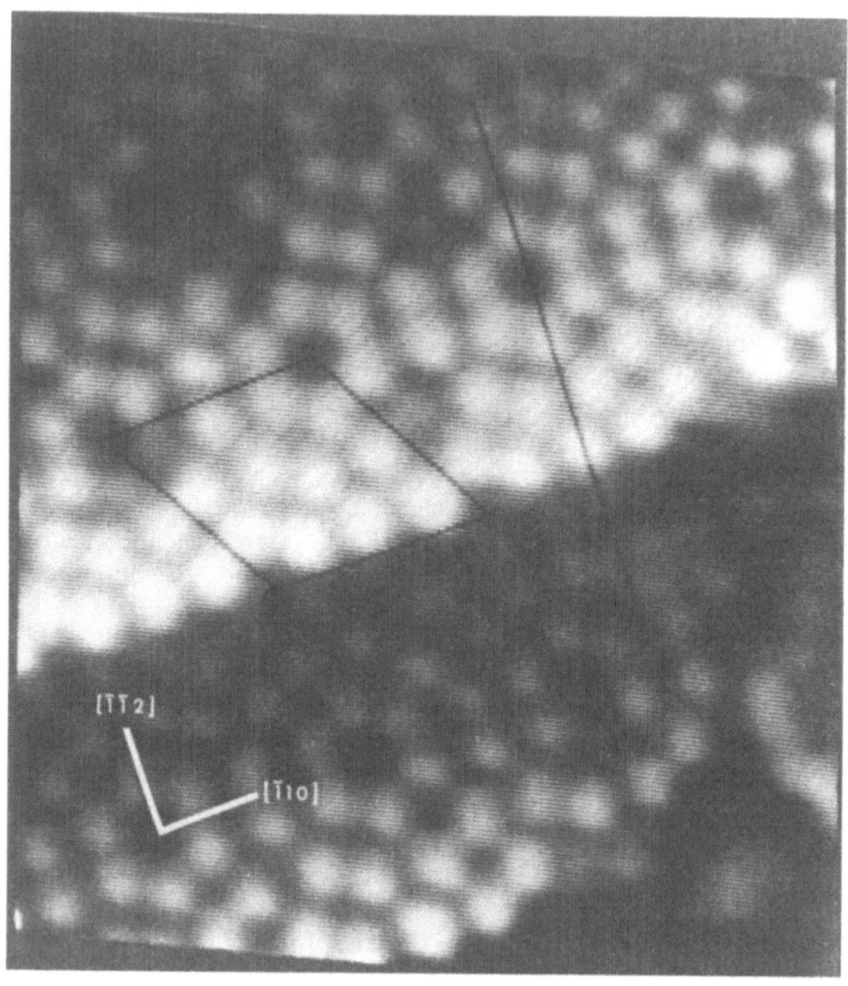

Fig.14.4. Gray scale image of a silicon (111) surface which shows the 7x7 reconstruction on terraces separated by atomic steps. White regions are elevated and dark ones depressed. The rhombic 7x7 unit cell is indicated, as is the orientation of the surface [14.11].

vicinial surfaces for which spot splitting is observed. However, steps which are randomly located on the surface will affect the diffraction line shape. The basic reason why this occurs is shown in Fig.14.5.

Consider the interference which occurs when a plane wave is scattered from two adjacent terraces separated in height by a single atomic step. For simplicity, only specular scattering will be considered. There is a characteristic set of angles for a given wavelength for which the waves scattered from the two terraces differ in phase by an integral multiple of 2π. Constructive interference will occcur and the surface will show a diffraction intensity and diffraction lineshape which is identical to that for a hypothetical step-free surface. There is

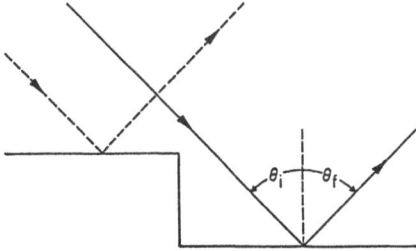

Fig.14.5. Schematic view of wave scattering from a surface region containing a surface step. Note that the path difference for scattering from the two terraces is different. This leads to in phase or out of phase interference depending on the angle of incidence.

a second set of incident angles for which the waves scattered from adjacent terraces differ in phase by an odd multiple of π. In this case destructive interference occurs. For a surface consisting of exactly equal areas of up and down terraces and for a perfect instrument, the intensity at the diffraction angle will be zero. However, convolution with the instrument response function and the presence of an unequal population of up and down terraces will lead to a reduced, but nonzero intensity at the diffraction angle when compared with the in-phase angles. Under these out of phase conditions, a diffraction technique can be very sensitive to the presence of steps on the surface. The intensity which is removed from the center of the diffraction peak is distributed into the wings of the peak leading to a peak broadening. This is illustrated in Fig.14.6 in which the half-width of the specular peak for helium scattered from Ni{115} is plotted as a function of the incident angle. The two minima correspond to in-phase conditions whereas the maximum half width, which is observed for a wavelength

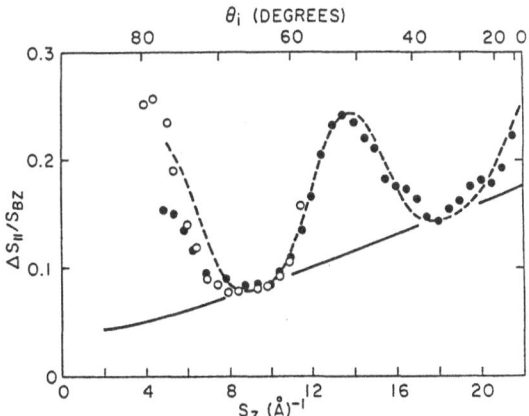

Fig.14.6. Peak width (FWHM) relative to ΔS at the zone boundary as a function of $S_{\perp}$ for in-plane helium scattering from Ni(115). The beam is incident in the $[\bar{5}\bar{5}2]$ azimuth which is perpendicular to the terrace edges. $(\cdot)k = 11.0$ Å^{-1}, (o) $k = 6.0$ Å^{-1}. The upper scale showing incident angles θ_i is for $k = 11.0$ Å^{-1} only [14.14].

413

of 0.57 Å near an incidence angle of 50°, corresponds to the out of phase condition.

Note that the dashed curve in Fig.14.6 follows a cosine behavior. This is to be expected if the steps on the surface are one atomic spacing in height. If a considerable fraction of the steps are a multiple of the minimum possible step height, the variation of the diffraction peak half width with angle will be more rapid that the cosine dependence shown in Fig.14.6 [14.14]. On metal surfaces, single step heights are generally observed.

Although steps can easily be observed with diffraction techniques through changes in the diffraction lineshape, quantification of this information is difficult. The reason for this is that the surface topography that can be deduced from experimentally determined lineshapes is not unique. Assumptions need to be made concerning the distributions in terrace widths and steps heights in order to extract quantities such as the average terrace width from diffraction peak shapes. Figure 14.7 shows the relationship obtained between diffraction peak half-width and the average terrace width for three different assumed surface configurations. Two of the curves assume a geometric distribution of terrace widths, one with only monatomic steps and the other with the probability of all possible step heights being equal. The geometric distribution is unrealistic in that the probability of a given terrace width increases monotonically with decreasing terrace width and is maximum at the minimum possible width. Since at small distances, step-step interactions are repulsive, a more realistic terrace width distribution would have the probability go to zero as the terraces became narrow. The resulting dependence of the half-width on average terrace width

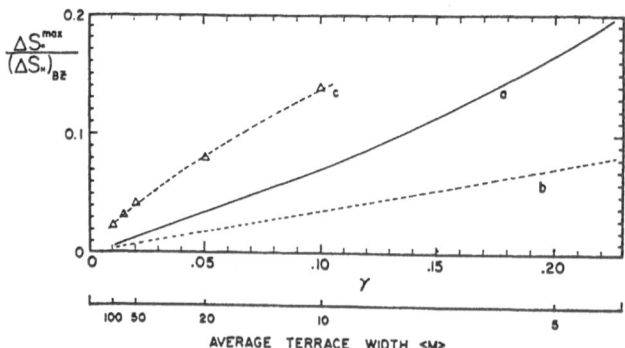

Fig.14.7. Maximum FWHM of any (hk) beam normalized to the Brillouin zone width as a function of step density in one direction. The following distributions are assumed: a) geometric distribution of terrace widths and monatomic steps, b) geometric distribution of terrace width and equal probability of all possible step heights, c) geometric distribution of terrace widths with a decrease as the width goes to its minimum possible value [14.15].

for a distribution of this type is shown as curve c in Fig.14.7 [14.15].
Since it is known that steps on metal surfaces are generally monatomic,
curves a and c are more realistic than curve b in comparing with data.
Applying these models to the data shown in Fig.14.6, we estimate the
Ni{115} surface to have a defect step after an average distance of
50-100 Å at low temperatures depending on which assumptions are
made regarding the distribution of terrace widths. Note that these
rather wide terraces still give rise to a step broadening which can
easily be measured, which shows the sensitivity of diffraction techniques.

Recently, LEED diffractometers have been disigned [14.16] which
allow the determination of large distances in direct space through a
very high resolution in reciprocal space. Figure 14.8 shows angular
profiles of the specular reflection from a silicon surface misoriented by
0.2° from the ⟨100⟩ direction; this has the nominal orientation
{1,1,400}. The satellite structure shows that distances corresponding to
400 lattice spacings are clearly resolved. Another diffractometer with
similarly high resolution [14.12] has been used to analyze the step
structure at an Si-SiO$_2$ interface. This has been done by etching away
the oxide layer and subsequently analyzing the interface using a diffraction lineshape analysis. The resulting distribution in terrace widths
is shown in Fig.14.9 . The analysis has assumed monatomic steps that
are uncorrelated.

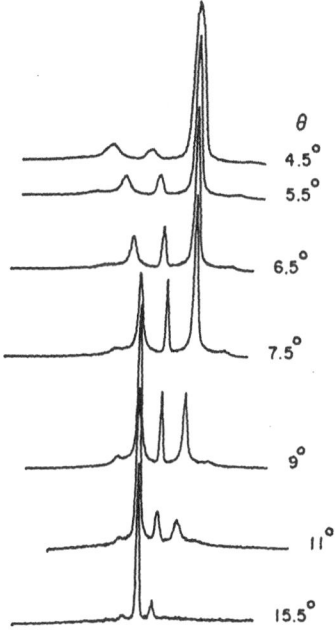

θ
4.5°
5.5°
6.5°
7.5°
9°
11°
15.5°

Fig.14.8. Angular profiles of the (00) beam
from a {1,1,400} silicon surface for different diffraction parameters. The direction of the detector
motion is normal to the nominal step edges. The
intensity in the direction along the step edges is
integrated by using a slit detector. E = 230 eV.
The abcissa is in arbitrary units of degrees. The
angular separation of the three satellites is less
than 0.5° [14.16].

415

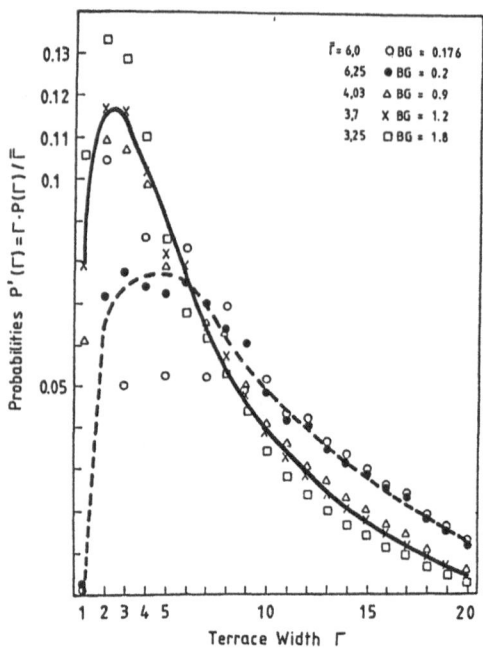

Fig.14.9. Probabilities that arbitrary surface atoms are found on terraces of width Γ for different background intensities (BG) at the zone boundary. Corresponding average terrace widths (Γ) are also shown. The dashed line is for BG = 0.176 after smoothing. The full line corresponds to BG = 0 and is assumed to be the best representation of the probability [14.12].

In summarizing this section, atomic steps on surfaces can be observed with both direct imaging (STM) and diffraction techniques. Although direct imaging studies will allow distributions of step heights and terrace widths to be measured directly, no systematic studies have been carried out to date. Clearly, a statistical analysis will play a vital role in such studies and many small area analyses must be combined to generate terrace width and step height distributions. In this respect, diffraction analyses have a decided advantage. Since the size of the spot is much larger than either the coherence width of the beam or the transfer width of the diffractometer in most cases, spot profiles are ensemble averages over $\simeq 10^6$ replica systems. Therefore, these results can be compared directly with equilibrium quantities derived from statistical mechanics. In addition, STM studies cannot easily be carried out at elevated temperatures at the present time due to thermal drifts. This limits studies to frozen in defects. However, until more is known about typical terrace width and step height distributions using direct imaging techniques, and analysis of surface topography using diffraction peak profiles cannot be truly quantitative. Although steps have been analyzed on many surfaces, the surfaces have not generally been in a well-defined equilibrium state. Therefore, associations of surface step density with surface tempertuare have not generally been made.

14.3 Energetics of Step Formation and Surface Diffusion

For a given material the roughening temperature will vary with crystal orientation. Whereas T_R is always defined as that temperature at which the free energy needed to form a surface step goes to zero, the bond strength at the surface dictates the magnitude of T_R. For this reason, closely packed planes which have more neighbors and shorter bond distances will have stronger bonds and, therefore, a higher T_R than more open planes. Experimental results on the variation of T_R with orientation will be presented in Sect.14.5.

Two models have been advanced to describe the energetics of surfaces with steps. These are the solid-on-solid (SOS) [14.17] and the terrace-step-kink (TSK) [14.4] models. In the SOS model the crystal is described by any array of vertical columns whose termination forms the surface. Two columns which differ in height by one atomic spacing differ in energy by the amount required to break one bond. The total energy of the surface configuration is given by

$$E = \sum_{j,\delta} V(|h_j - h_{j+\delta}|) \tag{14.1}$$

where the sum extends over neighboring columns of height j and j+δ. For single step heights as is typical on metal surfaces. $\delta = \pm 1$ or zero.

In the TSK model, the energy for a configuration of steps and kinks as is shown for a vicinal surface in Fig.14.10b is

$$E = \sum_{j,\delta} V(|\Delta_{ij} - \Delta_{ij+1}|) + \sum_{ij} U(\Delta_{ij}, \Delta_{kl}) , \tag{14.2}$$

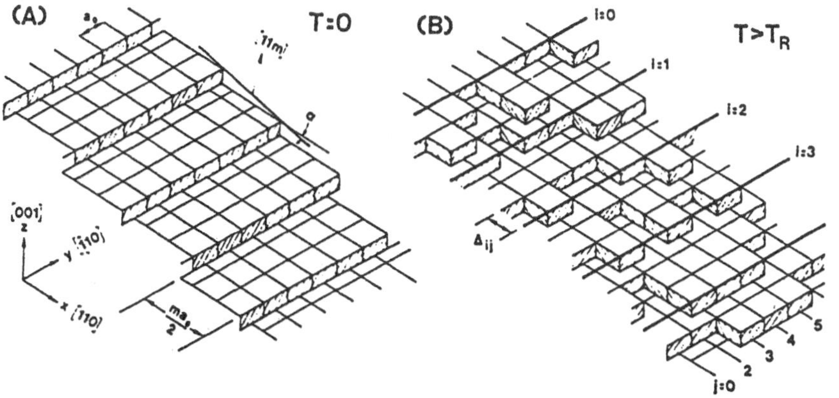

Fig.14.10. Schematic illustration of the structure of a {11m} surface in the fcc lattice. a) T = 0 equilibrium phase consisting of {001} terrace of width $ma_o/2$ separated by monatomic steps in the ⟨110⟩ direction. b) the rough phase $T > T_R$. Δ_{ij} is the displacement of the ij*th* terrace element from its T = 0 position which are indicated by the bold lines [14.18].

where Δ_{ij} is the local deviation of an edge atom from its zero Kelvin equilibrium position. The first term in this sum is the energy of forming a kink. Note that it is analogous to the SOS model if the columns are viewed as lying parallel to the terrace plane instead of perpendicular to it. The second term in (14.2) gives the interaction energy between edge elements on the same and different terraces. In this way, the interactions between surface elements can be brought in directly which is not done in the SOS model.

Both models are useful for studying phase transitions such as roughening from a theoretical point of view. Under certain simplifying assumptions about the nature and range of the potential parameters, predictions can be made about the universality class of the transition. For instance, it can be shown that the SOS models in the limit where height fluctuations about a reference plane are of long wavelength lead to a roughening transition which is of the Kosterlitz-Thouless class [14.17].

The models are less useful in looking at quantitative aspects of the roughening problem. For instance, the restriction of the interaction in both the SOS and TSK models to neighboring columns or step edge elements is unrealistic for metal surfaces in which the energy of the solid cannot be written as a sum of pairwise interaction energies. As will be shown in Sect.5, general statements can be made about the dependence of T_R on crystal orientation based on models such as those described above. However, they cannot be quantitative at our present level of understanding of the potential parameters of (14.1,2).

Both models have their utility and are directly applicable to vicinal surfaces which will be discussed in detail in Sect.14.4. In comparing the surface topography of vicinal surfaces such as are shown in Fig.14.10 for SOS and TSK models, a physically different picture emerges. In the limit of long wavelength fluctuations (which corresponds to single step heights in the SOS models or $\Delta = 1$ in the TSK model), the step edges will move in a more correlated fashion in the TSK than the SOS model due to the second term in (14.2). A more detailed description of these models can be found elsewhere [14.4,18]. Whether step edges move in a correlated or uncorrelated manner is important in choosing realistic potential parameters in (14.1,2) for modeling studies. Clearly, this will be a function of both the strength and range of the step-step interaction potential. At the present time, no quantitative information is available.

Whereas the energy parameters in (14.1,2) determine the magnitude of T_R, the activation energy for surface diffusion will determine how rapidly equilibrium is attained. For example, the data shown in Fig.14.6, although taken at 120 K are characteristic of disorder which is frozen in at higher temperatures. As a rough estimate, diffusion rates become negligible for vicinal nickel surfaces between 200 K and 300 K. Much higher values are found on semiconductor surfaces. For instance, diffraction peak widths on cleaved GaAs{110} samples were found to become smaller as the surface was heated to 750 K [14.19]. This indicates that the surface is becoming smoother and approaching equilibrum through a diffusion limited process rather than roughening which would lead to an increase in the diffraction peak width with increasing temperature.

Since roughening has been observed on vicinal nickel [14.18] and copper [14.20] surfaces, diffusion on such surfaces is of particular interest. Recently, surface self-diffusion on vicinal tungsten surfaces has been studied by *Gomer* et al. [14.21]. Field emission fluctuation techniques were used to sample regions whose linear dimensions are 50-100 Å. For different regions along the $\langle 111 \rangle$ zone connecting the {011} at the emitter apex and {112} planes, the following behavior was observed: At temperatures below 875 K, two-dimensional diffusion is seen. Above 875 K, additional one-dimensional diffusion which occurs on a much longer time scale is observed. Above 950 K, only two-dimensional diffusion is observed. These results are interpreted in terms of isolated atom diffusion over the vicinal surface for T < 875 K, and kink diffusion along the step edges between 875 and 950 K. This leads to large fluctuations of the step edge as the temperature increases until the one-dimensional character of the surface structure is lost. At that point, which should correspond to T >> T_R, diffusion of kinks along step edges would appear to be two dimensional.

The diffusion coefficient for the one dimensional motion was found to be a factor of 10^5 slower than two dimensional diffusion across the plane for T < T_R. This supports the idea that the diffusing species is larger than an individual atom. Estimates of the enthalpy of formation of an isolated atom on a terrace and of a kink on a step edge have also been extracted from the data. They are 7 and 19 kcal/mole respectively. It may be difficult to compare these values with macroscopic single crystals since thermally annealed field emission tips show a high level of disorder when imaged using the field ion microscope. Furthermore, the influence of the applied field on the diffusion behavior on the more open vicinal surfaces may be appreciable. However, to date these results are the only ones available to give some quantitative estimates on the relative diffusion rates and activation energies of isolated atoms on terraces and kinks along step edges.

14.4 Diffraction Studies of the Roughening Transition

As was discussed in Sect.14.2, steps on surfaces can be detected using diffraction techniques by an experiment which measures the lineshape.

LEED, RHEED, glancing angle x-ray and atom diffraction have all been used to study surface steps. Clearly, the greater the surface sensitivity, the more directly the experiment will reflect what is happening on the surface. Of the various diffraction techniques, atom diffraction has the highest surface sensitivity and to date is the only diffraction technique with which unambiguous information about surface roughening has been obtained.

In the following, it will be outlined how information on the roughening transition can be obtained from diffraction data. We will restrict our interest to the specific case of atom diffraction. Details which are not covered here can be found in the original literature [14.14,22,23]. The information concerning surface steps is found in the diffraction lineshape of each diffraction peak. This is to be compared with a surface structural determination which requires that the relative intensities of the diffraction peaks are known. This reduces the complexity of the experiment considered here since the shape of only one diffraction peak need be determined at an out of phase condition as a function of temperature. The specular peak is the obvious choice since it is not broadened by the velocity spread in the beam and, therefore, is the narrowest of the various diffraction peaks.

The diffraction lineshape from an infinite surface measured with a perfect instrument is the Fourier transform of the pair correlation function [14.24]. Since the observed lineshape is a convolution of the instrument response function with the lineshape for the finite domain surface, the instrument function and domain size must be properly taken into account. Since errors based on a deconvolution of data containing noise can be high, we have elected to calculate the expected lineshapes for different surface configurations rather than to determine the pair correlation function directly.

In proceeding to calculate the lineshape, we use the Eikonal approximation [14.25] to obtain the amplitude $A(Q)$ of the scattered wave as a function of the momentum transfer Q. This method would not be suitable for a surface structural determination since multiple scattering cannot be neglected. For the strongly corrugated vicinal surfaces [14.26] to which we will restrict ourselves. However, the lineshape should be unaffected by the multiple scattering.

The scattering intensity of the specular beam $I_0(Q)$ can be written as [14.18]

$$I_0(Q) = \left(\frac{k_{iz}}{k_{oz}}\right)^2 \frac{1}{Na^2} |F(Q)|^2$$

$$\cdot \sum_{ij} \langle \exp[iQ_\parallel \cdot (R_i - R_j) + iQ_z(z_i - z_j)] \rangle$$

(14.3)

where the vectors $\mathbf{Q}$ are defined with respect to the terrace plane rather than with respect to the vicinal surface. R_i is the distance of the ith terrace from the origin measured parallel to the reference and z_i is the perpendicular distance from the origin. k_{iz} and k_{oz} are the normal components of the incoming and diffracted beams and a is the lattice constant for these one-dimensionally corrugated surfaces. $F(\mathbf{Q})$ is the form factor which takes into account the scattering within the unit cell. The sum extends over all the terraces.

Convoluting the domain size and detector response function yields a single effective response function

$$j(\mathbf{Q}) = \frac{[W(Q_z)N]^2}{4\pi^2} \exp\left[-D^2 \frac{Q^2}{4\pi}\right] \tag{14.4}$$

where

$$D^2 = [W(Q_z)]^2 + \langle L \rangle^2 . \tag{14.5}$$

In these equations, the instrument response function has been assumed to be of Gaussian form with width $W(Q_z)$ and $\langle L \rangle = \langle Na \rangle$ is the mean domain size. After replacing the sum in (14.3) with an integral and using the convoluton theorem, we arrive at an expression for the specular intensity $P_0(\mathbf{Q})$ given by

$$P_0(\mathbf{Q}) = \iint d^2\mathbf{R} \ d^2\mathbf{R}' \ \exp\left[\pi \frac{(\mathbf{R} - \mathbf{R}')^2}{D^2}\right] \left(\exp[i\mathbf{Q}_\parallel \cdot (\mathbf{R} - \mathbf{R}')]\right) P(z_i - z_j) . \tag{14.6}$$

In this equation, $P(z_i - z_j)$ is the height-height correlation function given by

$$P(z_i - z_j) = \langle \exp[iQ_z(z_i - z_j)]\rangle . \tag{14.7}$$

To this point, the equation for the specular lineshape is not dependent on a particular form of the height-height correlation function. Below the roughening temperature, $P(z_i - z_j)$ decays rapidly with distance R_{ij} to a constant value. This will give rise to a Lorentzian lineshape.

Above T_R, the height-height correlation function diverges logarithmically with distance parallel to the surface. For large distance, R_{ij},

$$\left\langle \left[z_i - z_j - \frac{\sqrt{2}\chi_{ij}}{m}\right]^2 \right\rangle = \frac{2a_0^2 X_1}{\pi^2} \ln \frac{|R_{ij}|}{r_0} . \tag{14.8}$$

This expression has been written to include all {11m} planes of a substance which has the face centered cubic structure. These planes have terraces of {001} orientation with $\langle 110 \rangle$ oriented terrace edges and $\sqrt{2}/m = \tan\gamma$. Here γ is the tilt angle of the {11m} surface with respect to the {001} terraces. The third term of the left hand side of (14.8) takes into account the staircase structure of the surface when viewed relative to the {001} terraces.

Equation 14.8 is derived using statistical mechanics [14.27] and contains all the essential information needed to derive the diffraction lineshape for $T > T_R$. Since the height-height correlation function depends logarithmically on R_{ij}, the height fluctuations become unbounded for an infinite perfect single crystal. However, the logarithmic divergence is weak, so that the mean square height difference increases in magnitude by only a factor of 14 as the coherent domain size increases from 100 Å to 1 cm. In (14.8), X_1 is the roughness parameter and is a function of T and m. It takes on universal values at $T = T_R$ and will play a critical role, as will be discussed below, in determining whether a surface is in the smooth or rough state. The quantity a_0 is the nearest neighbor spacing in the terrace plane and r_0 is of the order of a_0.

Since the roughening transition belongs to the Kosterlitz-Thouless universality class, it can be shown that [14.27]

$$X_1^{(R)} = \begin{cases} \dfrac{2}{m^2} & m \text{ odd} \\[2mm] \dfrac{1}{2m^2} & m \text{ even} \\[2mm] \dfrac{1}{2} & m \to \infty \end{cases}$$ (14.9)

where the superscript R denotes the value at $T = T_R$.

Substituting (14.8) into (14.7) it can be shown that

$$\exp\left[iQ_z\left(z_i - z_j - \frac{2}{m}x_{ij}\right)\right] = \left[\left(\frac{y_{ij}}{2}\right)^2 + \left(\frac{x_{ij}}{N_m}\right)^2\right]^{-X_Q}$$ (14.10)

for $T > T_R$. The quantities X_Q and Q in (14.10) are given by

$$X_Q = N_m^2 X_1 f(Q), \quad Q \equiv \frac{S_1 \cdot b}{2\pi} \text{ modulo } 1$$ (14.11)

where the function $f(Q)$ describes the periodicity of the pair correlation function between the in-phase and out of phase scattering conditions. The quantity $N_m = m$ for m odd and $N_m = 2m$ when m is even. For the {001} surface $N_m = 2$. Although $f(Q)$ depends on the domain size for arbitrary values of Q, $f(Q) = 1/4$ at the out of phase conditions.

Using these relationships and (14.9, 11) it is found that $X_Q = 1/2$ at $T = T_R$ for all {11m} planes. This quantity is a physical observable, because it can be shown that the intensity of the specular peak as a

422

function of the momentum transfer parallel to the {11m} plane, $S_{\parallel}$, is given by (14.18)

$$I(S_{\parallel}) \propto (S_{\parallel} \cdot \mathbf{a})^{2X_Q - 2} . \qquad (14.12)$$

This equation holds in the limit $S_{\parallel} \gg \sigma$ where 2σ is the FWHM of the instrument response function.

Therefore, a log-log plot of the intensity against $(S_{\parallel} \cdot \mathbf{a})$ should be linear with a slope $2X_Q - 2$. The range over which the plot is linear is cut off at low momentum transfer by the restriction that $S_{\parallel} \gg \sigma$ and at high momentum transfer by the restriction that neighboring diffraction peaks should not make an appreciable contribution to the intensity. It should also be noted that we have assumed that the form factor remains constant over the range in $S_{\parallel}$ which we are considering. Clearly, this is more nearly valid the smaller the range chosen.

The experimental results for specular scattering of He from Ni{113}, plotted as described above are shown in Fig.14.11. As is seen there is a considerable region over which the linear dependence is observed. Since the slope of these lines which are taken for different temperatures is $2X_Q - 2$, X_Q and, therefore, X_1 can be determined directly. Since $X_Q = 0.5$ at $T = T_R$, a plot of X_Q vs T yields T_R. This yields $T_R = 750 \pm 50$ K for Ni{113} [14.28], as is shown in Fig.14.12.

From theoretical predictions, $X_Q = 0$ for $T < T_R$, $T_Q = 1/2$ at $T = T_R$, and X_Q should rise with T for $T > T_R$. The experimental

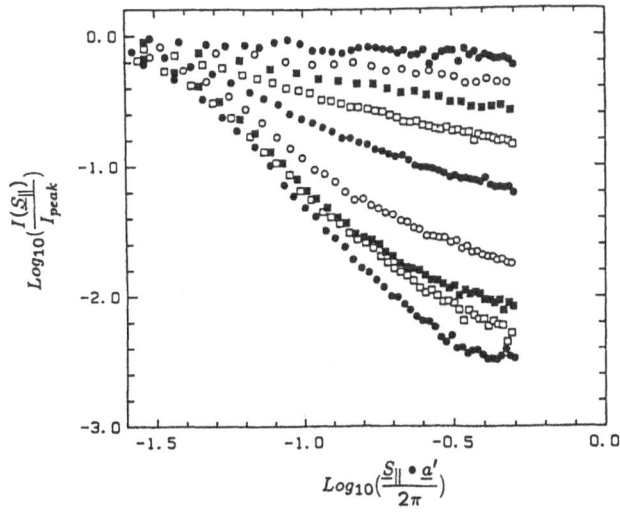

Fig.14.11. Log[$I(S_{\parallel})/I_{peak}$] versus log $(S_{\parallel} \cdot a'/2\pi)$ for helium scattering from Ni{113}. The beam is incident perpendicular to the steps, $k = 6.35$ Å^{-1} and $S_{\perp} = 8.88$ Å^{-1} corresponding to an out of phase angle. Temperatures are (top to bottom) 1400, 1250, 1050, 950, 750, 500, 400 and 200 K [14.28].

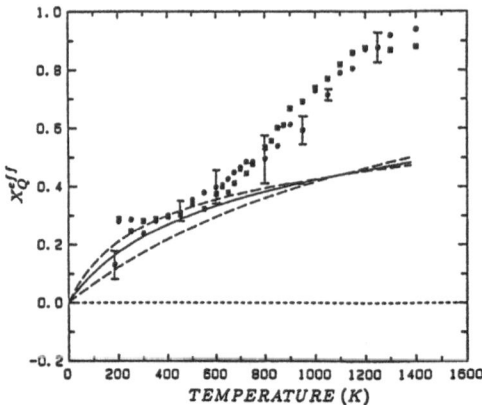

Fig.14.12. A plot of the measured effective X_Q values vs temperature. Filled circles are for data taken with the beam incident in the $[\bar{1}\bar{1}0]$ azimuth. Filled squares are data taken in the $[\bar{3}\bar{3}2]$ azimuth. The curves are the expected effective X_Q values due to inelastic scattering. The solid curve is calculated using an average value of the one phonon cross section. The dashed values are calculated using the upper, and lower bounds of the cross section as determined from experiment [14.28].

results indicate that $X_Q > 0$ for low temperature which is seemingly in contradiction with the above predictions. This difference is due to inelastic scattering. Since no velocity selection of the scattered molecules is carried out, the signal detected is composed of both elastic and inelastic components. This intensity in the wings of the peak can be written as

$$I(S_{\parallel}) = I_{el}(S_{\parallel}^{2X_Q-2}) + AI_{1p}(S_{\parallel}^{-1}) + BI_{mp}(S_{\parallel}^{0}) \qquad (14.13)$$

where the three terms refer to elastic, one phonon, and multiphonon contributions. The functional dependence of each of these components on $S_{\parallel}$ is also indicated in the above equation. Therefore, a plot of $\log I(S_{\parallel})$ vs $\log(S_{\parallel} \cdot a/2\pi)$ will give an effective X_Q which is the quantity plotted in Fig.14.12. By measuring lineshapes for the diffraction from Ni{001}, which does not roughen below 120 K [29], the quantities A and B can be estimated. This allows the values of X_Q to be corrected and a plot of X_Q vs T is shown in Fig.14.13. Within the error bars, the theoretically predicted behavior is observed. It can also be seen that the correction is only important for $T < T_R$. This can be easily understood. For $T = T_R$, the elastic intensity falls off as $S_{\parallel}^{-1}$ and for $T < T_R$ it falls off more rapidly with $S_{\parallel}$. Therefore, it can become masked by the one phonon inelastic term which falls off as $S_{\parallel}^{-1}$. However, for $T > T_R$, the elastic term falls off more slowly with $S_{\parallel}$ than the one phonon term and therefore, the correction will be small.

As we have shown above, the roughening temperature, T_R, can be determined from diffraction lineshapes since X_1 takes on a univer-

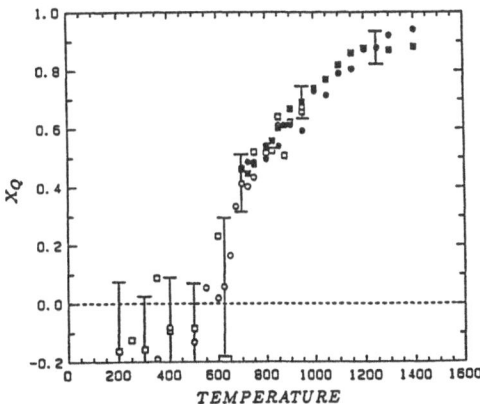

Fig. 14.13. The roughness exponent X_Q corrected for inelastic scattering vs temperature. Filled circles are the original data as in Fig. 14.12. Open symbols are the data calculated from log-log plots after inelastic scattering has been substracted [14.28].

sal value at T_R. Inelastic scattering becomes important near T_R such that corrections to X_Q become large for $T < T_R$. This masks the transition from the smooth to the rough state but has little influence on the determination of T_R.

14.5. The Dependence of T_R on Crystal Orientation

As was illustrated by Fig. 14.2 which shows the equilibrium crystal shape for lead, not all crystal planes roughen at the same temperature. This follows from the fact that T_R is determined by the surface bonding configuration which will differ with crystal face. Since the binding strength of an atom in the surface will increase with the number of nearest neighbors and decrease with the interatomic spacing in the surface plane, to first order, T_R should increase with the density of atoms in the surface plane.

The variation of T_R with orientation is of particular interest in considering the possible effect of roughening on chemical reaction rates. Industrial catalysts often consist of small single crystal metal hemispheres dispersed on an inert high area support. The active metal particle may achieve its equilibrium crystal shape at elevated temperatures and, therefore, may have regions which are in the smooth phase bounded by regions which are in the rough phase. If defects play a significant role in the reactivity, the rough regions may be particularly active sites.

Vicinal planes will be an important class of surfaces to consider since they bridge the regions between low index planes on spherical particles. To date, the only systematic measurements which have been made on closely related crystals are those on {11m} planes of Ni [14.14,28,29] and Cu [14.20].

These planes with m odd and m ≥ 3 differ in that the terrace width increases by one atomic row as m increases by 2. To be able to predict how T_R varies with m, one would need to know details of the

425

surface potential which are not currently available. However, it is clear that there will be two dominant terms which appear in the energy. The first of these is the energy to make a kink at a step edge which we will call W_0. The second of these is the step-step repulsion energy, measured per unit length. This repulsion energy may originate either through elastic interactions due to slight atomic relaxations at the step-edge or through a dipole-dipole repulsion due to the different fall-off lengths of positive and negative charge at the step-edge. A simplified model of the step-step repulsion can be formulated. In this model, the energy required to move a step away from its nearest neighboring step is w_m. If it is assumed that w_m takes on the values 0 and ∞ if the step-step spacing increases or decreases, respectively, the following equation can be derived [14.23].

$$\frac{w_m}{T_R} e^{T_R} = 2 .$$ (14.14)

If this model is applied to the {113} and {115} surfaces, w_m and W_0 can be determined from a measured value of T_R provided that it is known how w_m decays with the step-step spacing r. For an elastic interaction, w_m decays as r^{-3} [14.23]. The values shown in Table 14.1 have been derived from the currently available data. Clearly, the step-step interaction in this model is greatly simplified.

	T_R [K]	W_0 [K]	w_m [K]	Ref.
Ni{115}	450±50	1700±500	$10 < w_5 < 60$	14.14, 28
Ni{113}	750±50	1700±500	$100 < w_3 < 300$	14.28
Cu{115}	380	3000±500	$w_5 = 120$	14.20
Cu{113}	479			14.20
	Estimated from Debye-Waller measurements			14.20
Cu{117}	315			

Since the model is not very realistic, the cited values for W_0 and w_m should not be considered very accurate. However, T_R where determined by atom diffraction lineshape analysis is accurate. In some cases, T_R estimated from Debye-Waller measurements have also been included. A determination of roughening by this method is inaccurate [14.30] and these values are to be viewed only as estimates.

The trend is what would be expected in that T_R increases as the {11m} plane becomes more dense. There is a 300 K jump in T_R for nickel when m changes from 5 to 3. The {100} plane of nickel does not roughen below 1200 K which is within 30% of the bulk melting point. This indicates that roughening may not occur below the bulk melting point on surfaces in which the basic mechanism for roughening is by adatom-vacancy rather than kink-antikink creation.

14.6. Conclusions

The existence of a roughening transition, which was first suggested from theoretical consideration in 1951 [14.1] has been verified experimentally for solid helium in 1980 [14.5-8] and for vicinal metal surfaces of Ni [14.14,28] and Cu [14.20] in the past three years. For solid helium, the transition has been shown to belong to the Kosterlitz-Thouless universality class [14.5]. For metal surfaces, the data are consistent with this conclusion, but step-step interactions on vicinal surfaces may change the nature of the transition [14.31]. At present, there are not sufficiently accurate measurements near T_R to decide this question.

Although the existence of the roughening transition has been demonstrated, the energetics are not well understood. Diffraction methods are well suited to determine quantities such as the mean square height difference, but do not give the step density directly. Furthermore, little is known about the step-step interactions and correlations of kink-antikink excitations on different terraces. To date, measurements show that planes of low density undergo a roughening transition at rather low temperatures whereas nearly close packed planes may not roughen below the bulk melting point. It is clear that many more data are needed to understand the energetics of roughening and the dependence of T_R on crystal face orientation. However, substantial progress has been made since the first experimental observations in 1980.

Acknowledgement. The author gratefully acknowledges the assistance of many collaborators, especially E.H. Conrad. Part of the research described here was funded by the National Science Foundation under grant CHE-8109067.

References

14.1   W.K. Burton, N. Cabrera, F.C. Frank: Philos. Trans. R. Soc. London Sect. A243, 299 (1951).
14.2   D. Nenow: *Progress in Crystal Growth and Characterization* 9, 185 (Pergamon, London 1984)
14.3   J.D. Weeks, G.H. Gilmer: Adv. Chem. Phys. 40,147 (1979)
14.4   H. van Beijeren, I. Nolden: to be published.
14.5   P.E. Wolf, F. Gallet, S. Balibar, E. Rolley, P. Nozières: J. de Phys. 46, 1987 (1985).
14.6   J.F. Avron, L.S. Balfour, C.G. Kuper, J. Landau, S.G. Lipson, L.S. Schulmen: Phys. Rev. Lett. 45, 814 (1980).
14.7   S. Balibar, E. Castaing: J. Phys. Lett. 41 1329 (1980).
14.8   K.O. Keshishiv, A. Ya Parshin, A.V. Babkin: Sov. Phys. JETP 53, 362 (1981).

14.9   J.C. Heyraud and J.J. Métois: Surf. Sci. **128**, 334 (1983).

14.10   G.C. Binnig, H. Rohrer, Chr. Gerber, F. Stoll: Surf. Sci. **144**, 321 (1984); for an introduction see also R.J. Behm, W. Hösler: In *Chemistry and Physics of Solid Surfaces*, Vol.VI, ed. by R. Vanselow, R. Howe (Springer-Verlag, Berlin, Heidelberg 1986).

14.11   J. A. Golovchenko: Science **232**, 48 (1986).

14.12   H. Busch, M. Henzler: Surf. Sci. **167**, 534 (1986).

14.13   M.G. Lagally: *Methods of Experimental Physics*, Vol.22 (Academic, New York 1985), p237.

14.14   E.H. Conrad, R.M. Aten, D.S. Kaufman, L.R. Allen, T. Engel, M. den Nijs, E.K. Riedel: J. Chem. Phys. **84**, 1015 (1986). See also an Erratum, ibid. **85**, 4756 (1986).

14.15   T.-M. Lu, M.G. Lagally: Surf. Sci. **120**, 617 (1982)

14.16   D. Saloner, J.A. Martin, M.C.Tringides, D.E. Savage, C.E. Aumann, M.G. Lagally: J. Appl. Phys. **61**, 2884 (1987).

14.17   J.D. Weeks: *Ordering of Strongly Fluctuating Condensed Matter Systems.*, ed. by T. Riste (Plenum, New York 1980) p. 263.

14.18   E.H. Conrad, L.R. Allen, D.L. Blanchard, T. Engel: to be published.

14.19   D.G. Welkie, M.G. Lagally: J. Vac. Sci. Technol. **16**, 784 (1979).

14.20   F. Fabre, D. Gorse, B. Salanon, J. Lapujoulade: J. de Phys. **48**, 1017 (1987).

14.21   Y.M. Gong, R. Gomer: J. Chem. Phys. **88**, 1359 and 1370 (1988)

14.22   G. Blatter: Surf. Sci. **145**, 419 (1984).

14.23   J. Villain, D.R. Grempel, J. Lapujoulade: J. Phys. F**15**, 809 (1985).

14.24   M. Henzler: In *Electron Spectroscopy for Surface Analysis*, ed. by H. Ibach, Topics Current Phys., Vol.4 (Springer, Berlin, Heidelberg 1979) and references therein.

14.25   V. Garibaldi, A.C. Levi, R. Spandacini, G.E. Tommei: Surf. Sci. **48**, 649 (1975).

14.26   D.S. Kaufman, R.M. Aten, E.H. Conrad, L.R. Allen, T. Engel: J. Chem. Phys. **86**, 3682 (1987).

14.27   M. Den Nijs, E.K. Riedel, E.H. Conrad, T. Engel: Phys. Rev. Lett. **55**, 1689 (1985).

14.28   E.H. Conrad, L.R. Allen, D.L. Blanchard, T. Engel: Surf. Sci. in press.

14.29   E.H. Conrad, L.R. Allen, D.L. Blanchard, T. Engel: Surf. Sci. **184**, 227 (1987).

14.30   E.H. Conrad, D.S. Kaufman, L.R. Allen, R.M. Aten, T. Engel: J. Chem. Phys. **83**, 5286 (1985)

14.31   F.S. Rys: Surf. Sci. **178**, 419 (1986).

15. Relationship Between Anisotropy of Specific Surface Free Energy and Surface Reconstruction

H.P. Bonzel and K. Dückers

Institut für Grenzflächenforschung und Vakuumphysik
Kernforschungsanlage Jülich GmbH, D – 5170 Jülich, Fed. Rep. Germany

Some surfaces reconstruct by changing their morphology such that new surface orientations are being exposed. From general considerations it is clear that the surface free energy of such a reconstructed surface will be lower than that of the non-reconstructed form. It can be shown, however, that surface reconstruction will only be energetically favorable if the anisotropy of the surface free energy in the orientational range involving the non-reconstructed surface is high. Two different experiments carried out on Pt{110} support this idea. Measurements of surface self-diffusion on Pt{110} in the ⟨001⟩ direction are consistent with an anisotropy in surface free energy, $\gamma(\theta)$, of 8-10% at T > 1400 K. Spectroscopic data of surface atom core level shifts (SCLS) of clean reconstructed and non-reconstructed Pt surfaces also yield anisotropies of $\gamma(\theta)$ in excess of 10% along the [$1\bar{1}0$] zone. Reconstruction of the Pt{110} surface to the 1x2 lowers that anisotropy drastically. Measurements of SCLS of CO-covered Pt surfaces are also evaluated in terms of changes in $\gamma(\theta)$. It is found that CO adsorption moves the lowest energy cusp from {111} to {110}, and further that the non-reconstructed {110} is more stable than the reconstructed surface. This explains why CO lifts the 1x2 reconstruction of Pt {110}.

The analogy between microscopic morphological changes by reconstruction and surface faceting, as it is frequently observed in the presence of adsorbates or during surface reactions, is pointed out. Both can be understood in terms of the anisotropy in $\gamma(\theta)$ which can be severely altered by chemisorption. Several cases where adsorbates are known to cause surface reconstruction as well as surface faceting are being discussed. The resulting surface morphology is in general a function of adsorbate coverage and temperature which governs the kinetics of the morphological change.

15.1 Introduction

Reconstructed metal surfaces are - more than 20 years after their discovery - still a topic of great interest in surface science. Why do surfaces of a few materials reconstruct and others do not? What is the exact structure of these surfaces? What is the mechanism of converting a non-reconstructed into a reconstructed surface? Is reconstruction a clean surface phenomenon or are impurities participating? These and related questions have kept experimentalists and theorists busy for years, and even today we cannot say that surface reconstruction is completely understood.

In this chapter, we will examine specific surface free energies of materials and, in particular, their variation with crystallographic orientation θ (called "anisotropy" of specific surface free energy). Phenomenologically, it is clear that the total free energy of a crystal with a stable reconstructed surface has to be lower than that of the same crystal with a (metastable) non-reconstructed surface. Since reconstruction is taking place at the very surface involving not more than 2 or 3 atomic layers, we can probably say that the lowering in total free energy is identical with a lowering in the *surface* free energy. In fact, it is sometimes stated that the driving mechanism for reconstruction is a reduction in surface free energy. Therefore experimental and theoretical physicists have attempted to determine the specific surface free energies of non-reconstructed as well as reconstructed surfaces [15.1-4].

In the following, we will argue that surface reconstruction as a means of reducing the specific surface free energy, $\gamma(\theta)$, can only be successful if the anistropy of $\gamma(\theta)$ is large in that particular crystallographic zone which contains all surface orientations involved in reconstruction. Several experiments have produced data which are in support of this hypothesis [15.5,6]. These experimental results and their implications with respect to surface reconstruction will be discussed in detail.

Prior to presenting any specific data, we would like to offer a simple argument which will illustrate our statement made above. Figure 1a shows a section of the crystallographic dependence of $\gamma(\theta)$ with two particular orientations θ_1 and θ_2 for which the specific surface free energies are γ_1 and γ_2, respectively. These orientations may, for example, represent {111} and {110} of a fcc crystal. As $\gamma_2 > \gamma_1$, there is a possibility that a macroscopic surface of orientation θ_2, shown in Fig.15.1b, could reconstruct in order to lower its surface free energy. Two reconstruction modes may be envisaged: the epitaxial growth of a layer of orientation θ_1 on top of the crystal of orientation θ_2 (Fig.15.1c) or a microscopic faceting of the surface such that only segments of orientation θ_1 are exposed to the vacuum (Fig.15.1d). Let us now consider the surface free energy of each (fully relaxed) surface shown schematically in Fig.15.1b-d. It is assumed that the macroscopic surface area, O_2, remains the same in all cases to be discussed while the microscopic surface area, O_1, may change. In the first case of a

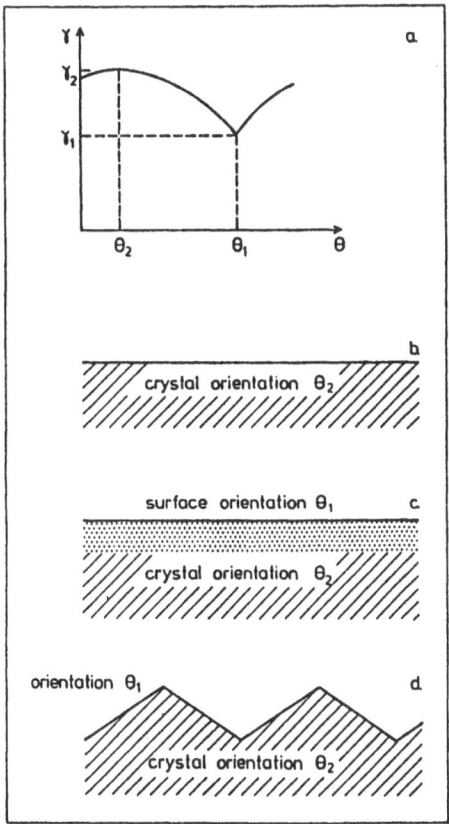

Fig.15.1 (a) Plot of specific surface free energy, γ, versus crystallographic orientation, θ. (b) Crystal with a macroscopic surface orientation θ_2. (c) Crystal with a reconstructed surface of orientation θ_1 epitaxed to the bulk orientation θ_2. (d) Crystal with a macroscopic (surface) orientation of θ_2 but with a reconstructed surface whose microscopic orientation is θ_1

reconstructed surface (Fig.15.1c) the total free energy contains in addition to the surface energy term, $O_2\gamma_1$, a second term, $O_2\Delta F_{inter}$, which accounts for the interface between the reconstructed layer of orientation θ_1 and the crystal of orientation θ_2. This term may also include an effect due to changes in interlayer spacing (relaxation) near the surface or near the interface. Thus with $O_1 = O_2$, we have

$$\gamma(\text{rec}) = \gamma_1 + \Delta F_{inter} \tag{15.1}$$

The difference between the reconstructed and the non-reconstructed surface is then:

$$\Delta\gamma = \gamma(\text{rec}) - \gamma_2 = \gamma_1 - \gamma_2 + \Delta F_{inter} \tag{15.2}$$

The difference $\gamma_1 - \gamma_2$ in (15.2) is negative while ΔF_{inter} is positive, i.e., $\Delta\gamma$ can in principle be positive, zero, or negative. For $\Delta\gamma < 0$ the reconstructed surface will be more stable, and this is only the case, when $|\gamma_1 - \gamma_2|$ is larger than ΔF_{inter}. In other words, reconstruction of a surface is favorable if the difference in specific free energies, $\gamma_1 - \gamma_2$, is sufficiently large, i.e., if the anisotropy of $\gamma(\theta)$ is sufficiently large.

In the second case (Fig.15.1d) the specific surface free energy contains at least three terms

431

$$\gamma'(rec) = \alpha\gamma_1 + \Delta F_{edge} + \Delta F_{relax} \qquad (15.3)$$

where α is the ratio of the microscopic surface area of the saw-tooth profile to the macroscopic surface area, the second term, ΔF_{edge}, is supposed to take care of the edges within the unit area of the profile, and the third term accounts for the changes in surface relaxation by going from orientation θ_1 to orientation θ_2. The difference

$$\Delta\gamma = \gamma'(rec) - \gamma_1 = \alpha\gamma_1 - \gamma_2 + \Delta F_{edge} + \Delta F_{relax} \qquad (15.4)$$

can also be positive, zero, or negative, just as with (15.2). It is likely to be negative only if the absolute difference between γ_1 and γ_2 is large. It follows that a *large anisotropy* in $\gamma(\theta)$ is a favorable condition for surface reconstruction to occur.

In view of these simple relationships, it is relatively safe to guess that only some surfaces of some materials may be able to lower their surface free energy by reconstructing. Those must be materials which exhibit unusually large anisotropies of $\gamma(\theta)$. Of course, the anisotropy of $\gamma(\theta)$ of a clean material may be changed by adsorbates [15.7,8]. The degree of anisotropy may be decreased or increased, such that surface reconstruction will become favorable for adsorbate covered surface orientations which in their clean state are non-reconstructed, or vice versa. Examples are known for both cases and will be discussed.

15.2. Anisotropies of Specific Surface Free Energy, $\gamma(\theta)$

15.2.1 Experiment

Measurements of absolute or relative specific surface free energies of solids for a variety of crystallographic orientations are cumbersome and difficult [15.9-11b]. They are usually carried out at high temperature (> 70% of the melting temperature) in order to assure thermodynamic equilibrium. As with all intrinsic surface properties, the cleanliness of the surface under investigation is of great importance. There are not many measurements of $\gamma(\theta)$ in the literature where the problem of surface purity has been adequately taken care of.

A collection of probably very reliable data on $\gamma(\theta)$ of clean metals [15.12-17] is presented in Table 15.1. They were taken at $T/T_m > 0.75$. All metals in this Table are fcc, and, therefore, three anisotropies of $\gamma(\theta)$, characteristic of pairs of the low-index orientations {111}, {100} and {110}, are given in %,

$$\frac{\Delta\gamma}{\gamma} = \frac{\gamma^\alpha - \gamma^\beta}{\gamma^\beta} \times 100 \qquad (15.5)$$

with α and β representing two orientations, and γ^β the smaller one of the two numbers. The specific surface free energy usually scales with the roughness of the surface [15.18], and, hence, for fcc crystals the

Table 15.1. Experimental anisotropies of specific surface free energy, $\gamma(\theta)$

	Ni	Cu	Pt	Au	Pb
T(K)	1300	1200	1773	1303	523
T/T$_m$	0.75	0.89	0.87	0.98	0.87
{100}-{111}	2.5	0.8	-	7.0[*]	2.0
{110}-{111}	3.0	1.4	7.8[**]	4.7	5.6
{110}-{100}	0.5	0.6	-	2.3	3.4
Reference	15.14	15.15	15.16	15.13	15.12

[*] additional value by *Heyraud* and *Métois* [15.17]: 1.9%
[**] maximum anisotropy, not necessarily between {110} and {111} orientations

difference in γ between {110} and {111} is expected to be large. The data in Table 15.1 show that this is true with the exception of Au [15.13]. The reason for this anomaly is not known. A more recent measurement of $\gamma(\theta)$ of Au reports 1,9% for {100} - {111} and 3.4% as the maximum anisotropy at 1273 K [15.17]. These data are more in line with the other sets although the maximum anisotropy is considerably lower.

When we inspect the maximum anisotropies $\Delta\gamma/\gamma$ between {110} and {111} in Table 15.1, we find numbers between 1.4% (Cu) and 7.8% (Pt). Taking into account that the anisotropy of $\gamma(\theta)$ decreases with increasing temperature [15.19], and that $\Delta\gamma/\gamma$ for Ni was measured at 0.75 of T$_m$, that for Au at 0.98 of T$_m$, we may conveniently divide the data into two categories: (a) $\Delta\gamma/\gamma < 3\%$, and (b) $\Delta\gamma/\gamma > 3\%$. The 3d-metals Cu and Ni fall into the first category and the 5d-metals Pb, Pt and Au fall into the second category. It is well known that clean Pt and Au low-index surfaces reconstruct [15.20,21] while Cu and Ni surfaces do not. Much less is known about the Pb low-index surfaces but thus far no reconstruction has been reported [15.22,23]. Our own study of Pb {110} has not revealed any evidence of recontruction at T > 200 K [15.24], not even in the presence of a low coverage of adsorbed potassium [15.25].

Based on the data of Table 15.1 and our knowledge of surface structure of these metals, we can cautiously suggest, that a correlation between high anisotropy of $\gamma(\theta)$ and surface reconstruction may exist. This concerns, in particular, the {110} orientation and much less the {100} orientation relative to the {111} which always has the lowest specific surface free energy, anyway. It is important to note at this point that the low-index surfaces {110} and {100} (and for Au perhaps also {111}) of Au and Pt, for which the specific surface free energy anisotropies were determined, were in their non-reconstructed (disordered) state because of the high temperatures. In that sense, the anisotropies in Table 15.1 are relevant to the problem of deciding whether reconstruction may yield a decrease in surface energy or not.

15.2.2 Theory

Whereas the experimental situation with specific surface free energy anisotropies is overall not brilliant, theoretical values are also scarce. Early theories have treated the specific surface free energy in terms of a pairwise interaction potential [15.18, 26-28]. In a qualitative sense the results of these calculations have been instructive, but the quantitative values of $\gamma(\theta)$ as well as the degrees of predicted anisotropy are generally not very useful.

There are a number of recent publications partly dealing with calculated specific surface free energies of metals. Two different approaches can be recognized: (1) the calculation of interaction energies of atoms by using a two-body pair potential have been extended by introducing an additional interaction term, in the form of a three-body potential [15.1], a volume dependent potential [15.29], a repulsive Born-Mayer interaction [15.2] or a volume dependent "glue" potential [15.3]. Also, the embedded-atom method of *Daw* and *Baskes* [15.30] has been combined with a repulsive pair-potential (accounting for core-core overlap) to calculate specific surface free energies of fcc metals [15.31]. In all these cases, surface and/or bulk properties are used to fit the parameters of the interaction potentials used in the calculation. (2) In the second approach true ab initio calculations of surface energies have been performed [15.4, 32, 33] where the sole input are electron orbital functions. Unfortunately, only a few specific surface free energies of low-index crystal faces were calculated in that way so that these results cannot be used for a detailed comparison with experimental data, in particular anisotropies of $\gamma(\theta)$.

Extensive calculations of $\gamma(\theta)$ for fcc metals were reported by *Machlin* [15.34] and more recently by *Wynblatt* [15.35]. *Machlin* used a complicated version of the Mie potential and arrived at values of γ which agree reasonably with (average) experimental data [15.36, 37]. *Wynblatt* also achieved good agreement with experiments on the basis of the potential proposed by *Baskes* and *Melius* [15.29]. His calculations were carried out for {100} surfaces of 6 fcc metals. The parameters of the potential were fitted to - among other properties - the stacking fault energy. *Wynblatt* suggested [15.35] that this may be partly responsible for the agreement between experiment and theory.

Another set of extensive theoretical $\gamma(\theta)$ data was recently reported by *Foiles* et al. [15.31] who used the embedded-atom-functions method. They calculated γ for all three low-index faces of 6 fcc metals so that the degree of anisotropy between {111}, {100} and {110} can be obtained and compared with the available data in Table 15.1. Their calculations are valid for T = 0 K, of course, so that a direct comparison of experiment and theory is difficult. In any case, the computed specific surface free energies of Foiles et al. are generally too low by at least 50%, and the anisotropies between {110} and {111} orientations range between 19 and 24% for all metals considered in their calculation [15.31]. Although this degree of $\gamma(\theta)$ anisotropy may not be out of reason at 0 K, it is not obvious that all metals should have about the same value. This result could be rationalized in the following way: we remember Herring's result that all theoretical γ-plots calculated on the basis of pairwise interaction potentials should consist of portions of spheres through the origin [15.26]. Now, if the whole [1$\bar{1}$0] zone between {111} and {110} is described by such a spherical section, then the maximum anisotropy between {111} and {110} is just equal to 22.4% by simple geometry. This value is close to those calculated [15.31] suggesting that there may be no more fundamental reason for these numbers than this one. Still, the very low γ values calculated by Foiles et al. render the used functions and pair potentials suspect.

Most recently, *Smith* and *Banerjea* [15.38] calculated the orientation dependent surface energies of several fcc and bcc metals. The approach is a modified total energy calculation involving perturbation theory. The results obtained for the close-packed surfaces agree quite well with experimental data but the anisotropies are very large, e.g., 35% for Cu at 0 K [15.38].

In context of the results reported lateron in this paper, it is of interest to point out various calculated surface energies of reconstructed and non-reconstructed surfaces, notably of Au{110} and Au{100}. *Ho* and *Bohnen* calculated the specific surface free energy for the Au{110} surface by a first principles approach and obtain 1380 erg/cm^2 for the 1x1 and 1310 erg/cm^2 for the 1x2 configuration, respectively [15.4]. Thus the reconstructed 1x2 surface is more stable by 70 erg/cm^2. The numbers compare quite favorably with the experimental average value of $\gamma(\theta)$ of Au extrapolated to 0 K, which is 1500 erg/cm^2 [15.36]. It is unfortunate that Ho and Bohnen do not report a γ value for Au {111} because it would be most interesting to see how their theory would predict the anisotropy of $\gamma(\theta)$ between {111} and {110} orientations.

Halicioglu et al. also computed surface energies for the reconstructed and non-reconstructed Au{110} surfaces which turned out to be 1515 and 1545 erg/cm^2, respectively [15.1]. Thus, the 1x2 configuration is stabilized by only 30 erg/cm^2. In their approach, the parameters of the interaction potential are matched to reproduce the specific surface free energy of the fully relaxed Au{111} surface, taken to be 1550 erg/cm^2 [15.38]. Thus, there is the somewhat disconcerting situation that the Au{110}-1x1 surface has a slightly lower surface energy than the Au{111}.

Tomanek computed the specific surface free energies of Pt{110} and Pt{100} surfaces in several configurations [15.2]. In this calculation he uses the bulk cohesive energy, the lattice parameter, and the bulk elastic modulus of Pt to determine the potential parameters. The specific surface free energies obtained are reported per surface atom (in eV) and no correction seems to have been made to account for different surface densities of 1x1 and reconstructed surfaces, for example. Although γ for the 1x2 missing row structure is with 1.77 eV/atom lower than 1.84 eV/atom for the 1x1 surface, the exact meaning of this difference remains unclear because the number of surface atoms are $1.38 \cdot 10^{15}$ and $0.9 \cdot 10^{15}$ cm^{-2}, respectively. The situation for Pt{100} is similar, with the added complication that the γ values reported for {100} are by a factor of about 0.5 smaller than those for {110} [15.2].

Ercolessi et al. compared calculated specific surface free energies of various Au{100} configurations. They have matched their calculational scheme to reproduce the specific surface free energy of a relaxed (but not reconstructed) Au{111} surface (value not quoted) [15.3]. The specific surface free energy of the ideal Au(100)-1x1 surface is reported as 1754 erg/cm^2 while that of the optimum reconstructed {100} surface comes out as 1637 erg/cm^2 for the 1x5 structure [15.3]. Thus, the reconstructed surface is stabilized by 117 erg/cm^2 for Au{111}, the stabilization energy amounts to 7.5% which is of the same order as the measured anisotropy in $\gamma(\theta)$ at high temperature [15.13]. A difference in γ between reconstructed and non-reconstructed surfaces is expected to be of the order of $\Delta\gamma/\gamma$, and, therefore, differences of 40-120 erg/cm^2 as obtained by the various groups [15.1-4] are quite reasonable.

15.3 Reconstruction of Pt{110} – Surface Self-Diffusion Measurements

The rate of mass transfer surface self-diffusion is governed by the anisotropy of the specific surface free energy, $\gamma(\theta)$. For example, morphological changes of a one-dimensional surface profile y(x) can be described by the following differential equation if surface self-diffusion is the dominant mode of transport [15.39,40]:

$$\frac{\partial y}{\partial t} = \frac{D_s \Omega^2}{kT} (1+y'^2)^{-1/2} \left[\frac{y'y''}{1+y'^2} \eta'(\theta, x) - \eta''(\theta, x) \right] \qquad (15.6)$$

where the prime denotes differentiation with respect to x and

$$\eta(\theta,x) = \left[\gamma(\theta) + \partial^2 \frac{\gamma(\theta)}{\partial\theta^2} \right] y''(1 + y'^2)^{-2/3} \qquad (15.7)$$

Here D_s is the mass transfer surface self-diffusion coefficient, Ω the atomic volume, k the Boltzmann constant, and T the absolute temperature. For slowly varying $\gamma(\theta)$, i.e., for a low anisotropy of $\gamma(\theta)$ and away from cusp orientations, the second derivative of $\gamma(\theta)$ with respect to θ in (15.7) can be neglected such that a much simpler equation results, which in many cases can be solved analytically [15.39]. Large anisotropies of $\gamma(\theta)$ and a cusp orientation can be accounted for by numerical solution of (15.6) whereby the cusp itself is approximated by a finite curvature [15.40]. The important point is that the rate of morphological change of a surface profile will be modified considerably by the $\gamma(\theta)$ function, or vice versa, that unusual decay characteristics observed during the decay of a surface profile may be traced to the anisotropy of $\gamma(\theta)$ for the orientational range that is covered by the profile shape.

In this context we assume that D_s is a constant for those crystallographic orientations present in the profile. This may not always be the case but for one-dimensional profiles y(x) where x runs parallel to a low-index crystallographic direction, the assumption seems justified. Those profiles consist generally of low-index terraces separated by steps [15.41].The surface self-diffusion coefficient, D_s, is then constant if diffusion across terraces is rate-determining and if steps (kink sites) serve as sources of diffusable species (e.g., adatoms).

During measurements of the directional anisotropy of surface diffusion on a Pt{110} surface it was observed that the rate of profile decay in ⟨110⟩ and ⟨001⟩ directions had very different characteristics [15.5,42]. Here, periodic profiles had been etched into the surface with the groove direction perpendicular to either the ⟨110⟩ or ⟨001⟩ direction in the surface. After a pre-anneal at high temperature the profile shape was nearly sinusoidal [15.42]. The amplitude of such profiles was then measured as a function of annealing time at a constant temperature. Figure 15.2 shows two examples of profile decay at 1650 K for the diffusion directions ⟨110⟩ and ⟨001⟩. The profile wavelength was 7.0 μm and the starting amplitude 0.3 - 0.4 μm so that the maximum slope was near 19°. One can see in Fig.15.2 that the decay is linear for the ⟨110⟩ direction but initially very non-linear for the ⟨001⟩ direction. Linear decay is expected for surface profiles with low or vanishing $\gamma(\theta)$ anisotropy; in those cases, the slope of the log A versus t plot is evaluated to yield $D_s(T)$ [15.39]. Non-linear decay as observed in Fig.15.2 for ⟨001⟩ can be interpreted in terms of (15.6) meaning that the $\gamma(\theta)$ in the [1$\bar{1}$0] zone is highly anisotropic.

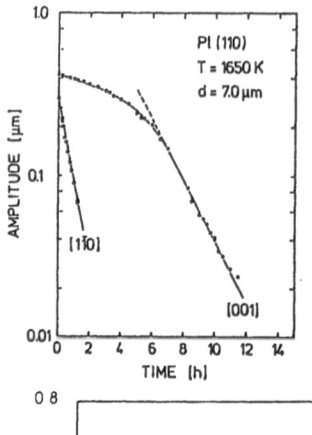

Fig.15.2 Semilog-plot of amplitude versus annealing time for periodic surface profiles on Pt{110}. The periodicity of the profiles was 7.0 μm. The diffusion direction was either <110> or <001>, perpendicular to the grooves of the profile. From [15.42]

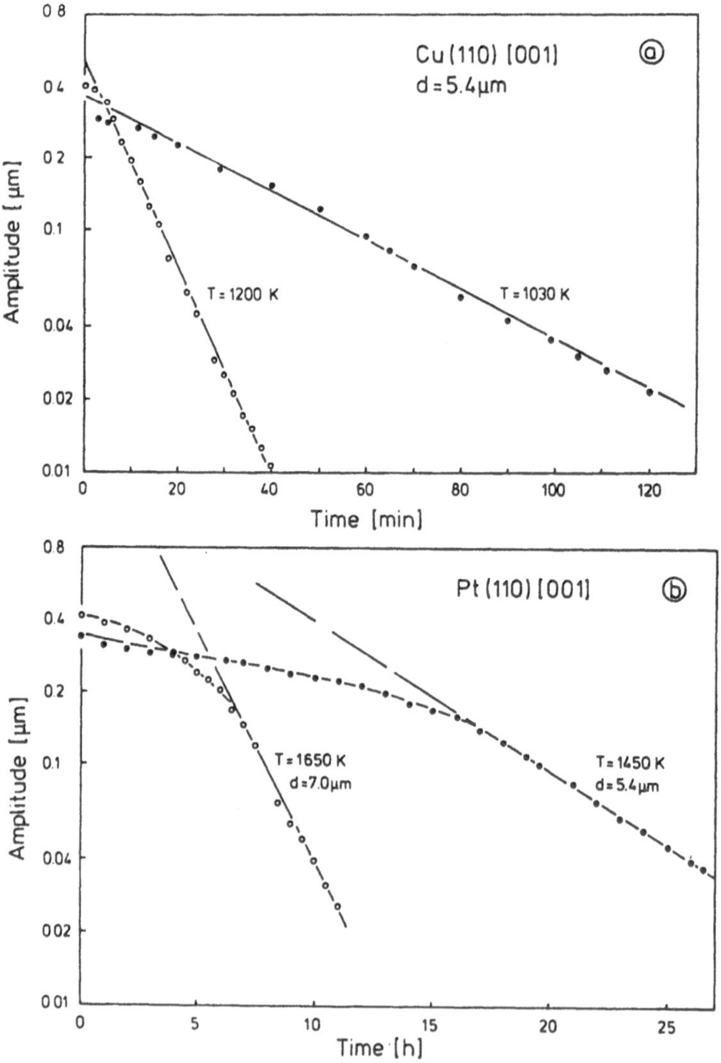

Fig.15.3 Comparison of periodic profile decay (log amplitude versus time) on Cu(110) and Pt(110) surfaces in <001> diffusion direction. From [15.5]

A comparison of D_s measurements for different fcc{110} metal surfaces shows that the ⟨001⟩ direction on Pt{110} is the only case so far where a pronounced non-linear behaviour in log A versus t for a large A/d ratios could be observed [15.5]. Figure 15.3 illustrates this point by comparing profile decay on Cu{110} and Pt{110} under rather similar A/d conditions. It was, therefore, conjectured that the specific surface free energy anisotropy of Pt along the [1$\bar{1}$0] zone on {110} was particularly large, such that the rate of profile decay was modified according to (15.6) by the orientation dependence of $\gamma(\theta)$, possibly also by a cusp at the {111} orientation.

Model calculations of profile decay were therefore carried out for anisotropic $\gamma(\theta)$ with various degrees of maximum anisotropy. Figure 15.4 shows the type of $\gamma(\theta)$ function used in these calculations [15.42]. There is a deep cusp at {111}, but negligible curvature $\partial^2\gamma/\partial\theta^2$ at {110}. The curvature in the cusp itself could be adjusted by a constant G (higher values of G indicate higher curvature). The result of the numerical profile decay based on (15.6) is shown in Fig.15.5 for several anisotropies $\Delta\gamma/\gamma$ between 0 and 10%. The starting profile was sinusoidal with the wavelength of 4 μm and an amplitude of 0.4 μm (maximum $\theta = 32°$). The calculation yields a linear decay for constant γ as expected. For large anisotropies of $\gamma(\theta)$, however, there is a slowly curved portion of the decay log A versus t at large amplitude where the extent of this curved region increases with increasing $\Delta\gamma/\gamma$. The curved region is followed by a linear portion with the slope parallel to that for γ = const. These decay curves for $\Delta\gamma/\gamma > 6\%$ are qualitatively

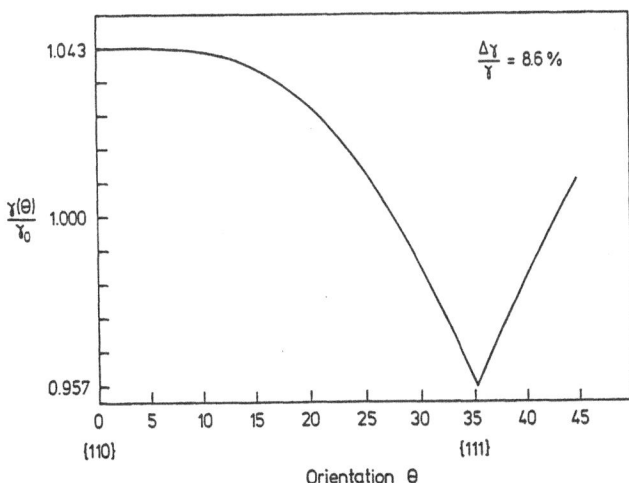

Fig.15.4 Model function of specific surface free energy versus orientation, Θ, between {110} and {111} orientations. This type of function exhibiting a cusp at {111} was used in calculating profile shape and decay kinetics. From [15.42]

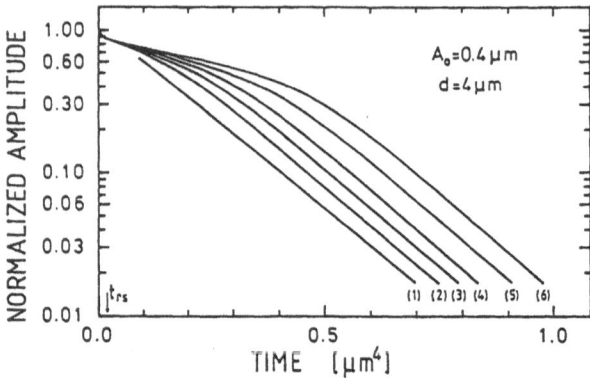

Fig.15.5 Calculated periodic profile decay for various anisotropies of specific surface free energy between {110} and {111}. The profile had a periodicity of 4.0 μm and a starting amplitude of 0.4μm. The anisotropies $\Delta\gamma/\gamma$ are: (1) zero; (2) 6%; (3) 7.6%; (4) 8.6%; (5) 9.6%; (6) 10%. From [15.42]

very similar to the experimental curves in Figs.15.2 and 15.3 for Pt{110} ⟨001⟩. From optimizing this comparison, we concluded that the $\gamma(\theta)$ anisotropy for Pt{110} along the zone [1$\bar{1}$0] should be about 8-10% [15.42]. On the other hand, the corresponding values for Cu or Ni, based on the straightness of profile decay [15.41,43], should be considerably less, i.e., smaller than 3%. This conclusion is in good agreement with the experimental data of Table 15.1.

The result of a high $\Delta\gamma/\gamma$ for Pt{110} in the [1$\bar{1}$0] zone is important in view of the fact that Pt{110} reconstructs to 1x2 in exactly this same zone [15.44]. This 1x2 reconstruction is stable at least up to 800 K and is now commonly attributed to a missing row structure [15.45,46] which consists of {111} oriented micro-facets, quite similar to the schematic shown in Fig.15.1d. Hence, we have here a good example for the simple energy consideration discussed in Sect. 15.2 which predicts surface reconstruction if the anisotropy $\Delta\gamma/\gamma$ is sufficiently large. Obviously this condition is fulfilled for Pt but not for Cu or Ni, for example.

Note that the estimate of $\Delta\gamma/\gamma$ obtained here for Pt {110} in the [1$\bar{1}$0] zone is valid at high temperatures, i.e., >1400 K. At this temperature it is likely (not yet proven) that the {110}-1x2 surface of Pt has undergone an order-disorder transition to a 1x1 structure analogous to that for Au{110}-1x2 [15.47] which occurs at 650 K. If we assume that T_c of this transition scales with the absolute melting points of these materials, T_c of Pt(110)-1x2 is expected at 1000 K. Investigations to this point are presently under way [15.48].

15.4 Surface-Atom Core Level Shifts of Pt

Another somewhat novel approach to getting information on anisotropies of surface energies is to measure electron core level binding energies of surface atoms and to relate differences of these to differences in surface energies [15.6]. The corresponding energy relationships are obtained by Born-Haber cycles [15.49]. For example, the difference in electron binding energies, E_B, of surface and bulk atoms with atomic number z, $\Delta E_{s,b}(z)$, is given by [15.49]

$$\Delta E_{s,b}(z) = E_B(\text{surf}) - E_B(\text{bulk}) = \epsilon_s(z+1) - \epsilon_s(z) + \epsilon_{imp}(z+1,z) \quad (15.8)$$

where $\epsilon_s(z+1)$ and $\epsilon_s(z)$ are surface energies (per atom) of elements with atomic number z+1 and z, respectively, and ϵ_{imp} is the energy of solution of z+1 impurities in a z matrix [15.49]. Considering two different surface orientations α and β, we obtain the following relationship:

$$\Delta E_{s,b}{}^{\alpha}(z) - \Delta E_{s,b}{}^{\beta}(z) = [\epsilon_s{}^{\alpha}(z+1) - \epsilon_s{}^{\beta}(z+1)] - [\epsilon_s{}^{\alpha}(z) - \epsilon_s{}^{\beta}(z)] \quad (15.9)$$

At this point it is important to realize that the surface energies ϵ_s in (15.8,9) are atomic quantities (in eV) characteristic only of those surface atoms that are actually represented by the surface core level shifts (SCLS). If, for example, a surface is heterogeneous in the sense that several kinds of surface atoms with different coordination number exist, then each kind of surface atom should exhibit its own binding energy peak and, hence, surface energy. Of course, sometimes these contributions cannot be resolved in an experimental spectrum. On the other hand, the main objective in this paper is to obtain an estimate of macroscopic specific surface free energies or their differences. These macroscopic specific surface free energies are an average energy per unit area (in erg/cm^2), with no information about the microscopic heterogeneities of that surface. In order to use a relationship such as (15.9), it has to be modified such that the atomic energies, ϵ_s, will be replaced by macroscopic specific surface free energies, γ. This can be done as follows. The macroscopic specific surface free energy, γ, is related to the surface energies per atom, ϵ_{si}, through summation

$$\gamma = f \sum_i N_i \, \epsilon_{si} \quad (15.10)$$

where N_i is the surface density (atoms/cm^2) of atoms with surface energy ϵ_{si}, and f is a conversion factor (= $1.6 \cdot 10^{-12}$ erg/eV). The surface energies contained in the equations resulting from a Born-Haber cycle, ϵ_{si}(eV/atom), are "partial" quantities. To obtain a relationship between the macroscopy γ and those partial surface energies, a correction factor K is introduced such that

$$K\gamma = f \, N_k \, \epsilon_{sk} \quad (15.11)$$

where here *only* those atoms (ϵ_{sk}) are included that are represented by the SCLS measurement. In order to estimate the correction factor K, the atomic surface energies ϵ_{si} may be approximated by a simple bond counting formula:

$$\epsilon_s = \frac{Z_b - Z_s}{Z_b} E_{coh} = \left(1 - \frac{Z_s}{Z_b}\right) E_{coh} \tag{15.12}$$

where Z_s and Z_b are surface and bulk coordination numbers, respectively, and E_{coh} is the (bulk) energy of cohesion. With this approximation, the correction factor K can be simply calculated according to

$$K = \frac{N_k(1 - Z_{sk}/Z_b)}{\sum_i N_i(1 - Z_{si}/Z_b)} \tag{15.13}$$

For the {110} surface, all surface atoms were taken into account, i.e., for the 1x1 surface the first and second layer, and for the 1x2 surface first, second and third layer atoms. Since only one surface peak was detected for the 1x2 surface, the first and second layer atoms were represented by an average coordination number of 8.33, while the third layer atoms have a coordination number of 11 (same as the second layer atoms for the 1x1 surface). The correction factors are calculated to be 0.833 and 0.917 for the 1x1 and 1x2 surface, respectively.

Using (15.11) as a relationship between ϵ_{sk} and macroscopic specific surface free energy γ, we can rewrite (15.9) as follows:

$$\Delta E_{s,b}{}^\alpha(z) - \Delta E_{s,b}{}^\beta(z) = \frac{1}{fN_{z+1}{}^\alpha} [K^\alpha \gamma^\alpha(z+1) - rK^\beta \gamma^\beta(z+1)]$$

$$- \frac{1}{fN_z{}^\alpha} [K^\alpha \gamma^\alpha(z) - rK^\beta \gamma^\beta(z)] \tag{15.14}$$

with $r = N_{z+1}{}^\alpha/N_{z+1}{}^\beta = N_z{}^\alpha/N_z{}^\beta$. Since we are interested in the anisotropy of $\gamma(\theta)$ of one element as a function of the same quantity of another element, we rearrange this equation accordingly. The final result is:

$$\left(\frac{\Delta\gamma}{\gamma}\right)_z = \frac{N_z{}^\alpha f}{\gamma^\beta(z)K^\alpha} (\Delta E_{s,b}{}^\alpha - \Delta E_{s,b}{}^\beta) + R^\alpha \frac{\gamma^\beta(z+1)}{\gamma^\beta(z)} \left(\frac{\Delta\gamma}{\gamma}\right)_{z+1}$$

$$- (1-rk)\left[1 - R^\alpha \frac{\gamma^\beta(z+1)}{\gamma^\beta(z)}\right] \tag{15.15}$$

where $R^\alpha = N_z^\alpha/N_{z+1}^\alpha$ and $k = K^\beta/K^\alpha$. This equation is then the basis for evaluating the anisotropy of $\gamma(\theta)$ of Pt between {111} and {110} orientations. The numerical input are measured (and in one case calculated) SCLS of Pt surfaces and published values of specific surface free energies of Pt and Au [15.34] which are taken to be characteristic of the {111} orientation, i.e., γ^β. Also, the measured anisotropy $\gamma(\theta)$ of Au was utilized in the evaluation.

We measured surface-atom core level shifts (SCLS) of several clean Pt surfaces by using either Y Mς radiation of a laboratory source at $h\omega = 132.3$ eV or monochromatized synchrotron radiation at $h\omega = 120$ eV (BESSY). Figure 15.6a shows the 4f levels for a Pt{111} surface with Y Mς radiation [15.50]. The experimental spectra were best-fitted by a convoluted Doniach-Šunjić and Gaussian line [15.50]. The SCLS in this case is -0.37 eV. A corresponding spectrum of clean Pt{110}-1x2 is presented in Fig.15.7a taken with synchrotron radiation.

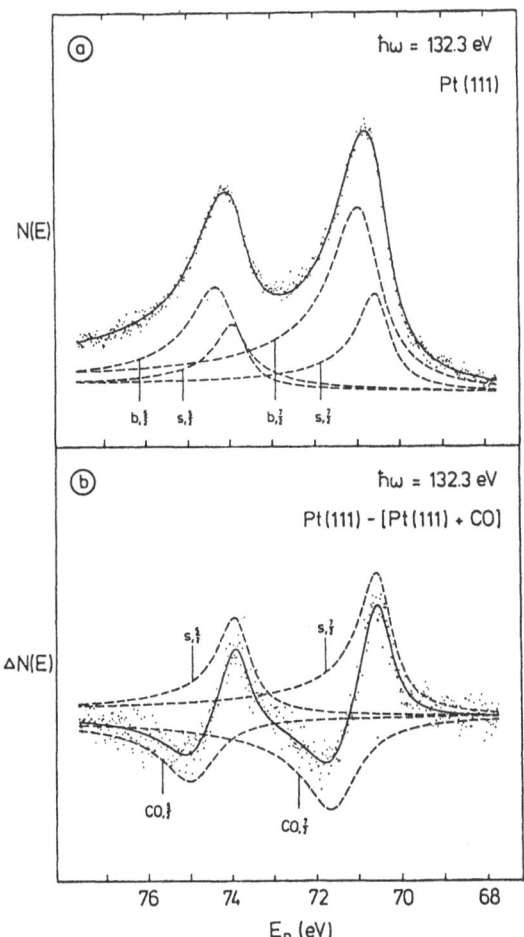

Fig.15.6 Photoemission spectra of Pt 4f levels taken with Y Mς radiation at $h\omega = 132.3$ eV. (a) Clean Pt (111) surface. (b) Difference spectrum between clean and CO-covered Pt{111} surface. From [15.50]

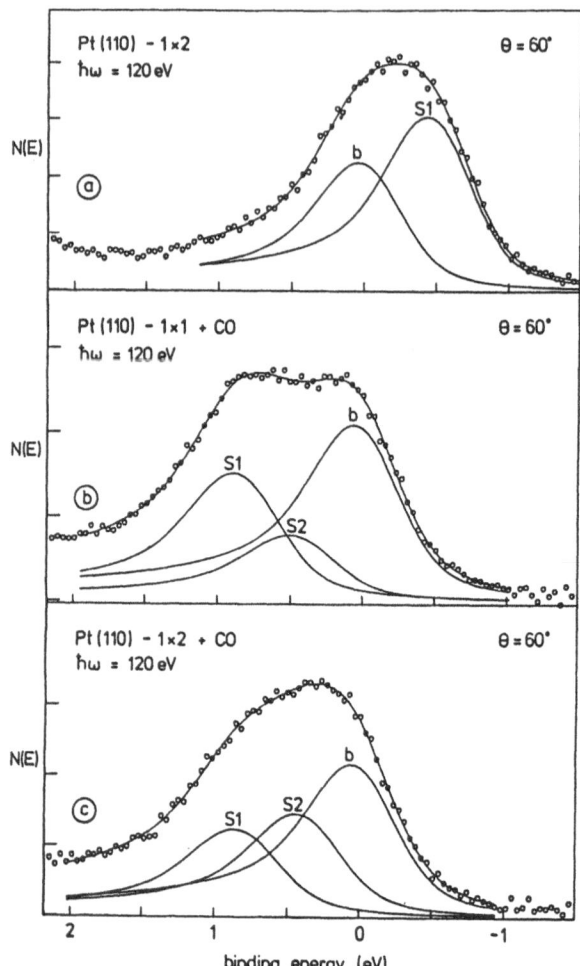

Fig.15.7 Photoemission spectra of Pt $4f_{7/2}$ level of Pt{110} taken with $\hbar\omega = 120$ eV. a) Clean surface. b) CO saturated surface in the 2x1 p1g1 configuration, i.e., on the non-reconstructed surface. c) After CO saturation at 110 K, i.e., on the reconstructed surface. From [15.51]

A best fit resulted in a SCLS of -0.46 eV [15.51]. These shifts are in reasonable agreement with previous investigations [15.52].

Unfortunately, the SCLS of the clean non-reconstructed Pt{110} surface could not be measured. Therefore, we estimated this quantity by using a formula from tight-binding theory [15.52]:

$$\Delta E_{s,b} = \text{const} \, (1 - \sqrt{Z_s/Z_b}) \qquad (15.16)$$

where Z_s and Z_b are the coordination numbers of surface and bulk atoms, respectively. The constant was evaluated from the data for the Pt{111} surface as -2.76 eV. With (15.16) $\Delta E_{s,b}$ of the non-reconstructed Pt{110} surface turned out to be -0.65 eV [15.51].

444

Using the available data of SCLS of Pt we calculated $(\Delta\gamma/\gamma)_{Pt}$ as a function of $(\Delta\gamma/\gamma)_{Au}$ according to (15.15). The atom densities N_z^α were $0.92 \cdot 10^{15}$ and $1.38 \cdot 10^{15}$ cm^{-2} for the 1x1 and 1x2 Pt{110} surface, respectively. The values of the specific surface free energies were $\gamma^\beta(\text{Pt}) = 2550$ erg/cm^2 and $\gamma^\beta(\text{Au}) = 1550$ erg/cm^2 [15.34], with $\beta = \{111\}$ and $\alpha = \{110\}$. Evaluating $(\Delta\gamma/\gamma)_{Pt}$ between {110} and {111} for the non-reconstructed as well as the 1x2 reconstructed surfaces yields the two straight lines shown in Fig.15.8. Without knowing $(\Delta\gamma/\gamma)_{Au}$, it is clear that the anisotropy of Pt is larger for the 1x1 than for the 1x2 surface relative to {111}. An experimental value of $(\Delta\gamma/\gamma)$ of Au for this crystallographic zone measured at 1303 K is 4.7% [15.13] (compare Table 15.1). At this temperature, Au{110} is in the non-reconstructed configuration [15.47]. Hence the anisotropy $(\Delta\gamma/\gamma)_{Pt}$ involving the non-reconstructed {110} would be 13.3%, i.e., high compared to the experimental value (15.16) and our previous value in Sect.15.3. However, all SCLS are measured near 300 K so that $\Delta\gamma/\gamma$ evaluated from Fig.15.8 cannot be compared immediately with those representative of high temperatures (1773 K in Table 15.1 or > 1400 K in D_s measurements [15.42]). An increase in $\Delta\gamma/\gamma$ with decreasing temperature is expected. In any case, $\Delta\gamma/\gamma$ for Pt is large for the non-reconstructed {110} surface relative to {111}, and secondly, it is greater than the value of $\Delta\gamma/\gamma$ expected for the reconstructed surface because if we estimate $\Delta\gamma/\gamma$ of Au to be about 2% for the reconstructed surface, $\Delta\gamma/\gamma$ of Pt should be < 10% (at 300 K). Thus, there is a considerable decrease in the degree of anisotropy of $\gamma(\theta)$ when the Pt{110} surface transforms from the non-reconstructed to the 1x2 configuration, and, hence, a decrease in the specific surface free energy of {110}.

Surface-atom core level shifts can also be measured for adsorbate covered surfaces. This opens up a way of estimating changes in surface

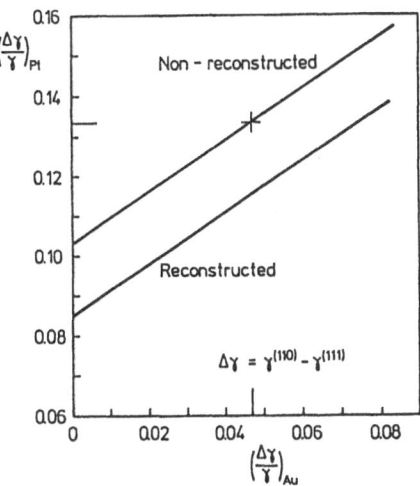

Fig.15.8 Plot of specific surface free energy anisotropy between {110} and {111} of Pt, $(\Delta\gamma/\gamma)_{Pt}$, versus the same quantitiy of Au

445

energies or their anisotropies as a consequence of adsorption. The relevant equation follows from (15.8):

$$\Delta E_{s,ad}(z) = \Delta E_{s,b}(z) - \Delta E_{s,b}{}^{ad}(z) = \Delta \epsilon_s{}^{ad}(z+1) - \Delta \epsilon_s{}^{ad}(z) \quad (15.17)$$

where $\epsilon_s{}^{ad}$ are surface energies per atom modified by the presence of an adsorbate. The relationship between the atomic surface energies and the macroscopic specific surface free energies (15.11) still holds such that for a given orientation

$$K \cdot \gamma^{ad} = f \cdot N_k{}^{ad} \epsilon_{sk}{}^{ad} \quad (15.18)$$

Assuming that K is equal for the clean and adsorbate covered surface and also for (z+1) and z material, we obtain:

$$\Delta E_{s,ad}(z) = \frac{K}{f} \left\{ \frac{\gamma(z+1)}{N_k(z+1)} - \frac{\gamma^{ad}(z+1)}{N_k{}^{ad}(z+1)} - \left[\frac{\gamma(z)}{N_k(z)} - \frac{\gamma^{ad}(z)}{N_k{}^{ad}(z)} \right] \right\} \quad (15.19)$$

As changes in γ due to the adsorbate are large (of the order of the adsorption energy) we will neglect differences in N_k and adopt an average surface atom density in (15.19) typical of that orientation. Rearranging the terms the final equation reads:

$$\Delta \gamma^{ad}(z) = \Delta \gamma^{ad}(z+1) - \frac{f N_k}{K} \Delta E_{s,ad}(z) \quad (15.20)$$

This equation holds for a particular orientation, where $\Delta \gamma_{ad} = \gamma$(clean) - γ(adsorbate). Pt4f spectra of CO covered {111} and {110} surfaces are shown in Figs.15.6b,7b,and 7c. The latter two represent CO on the non-reconstructed and on the reconstructed 1x2 Pt{110} surface [15.51]. The corresponding SCLS are -1.01 eV, -1.50 eV and -1.26 eV for Pt{111}, Pt{110}-2x1 p1g1, and Pt{110}-1x2, respectively.

In order to make use of these CO-induced SCLS for evaluating $\Delta \gamma_{ad}$(Pt), we have to estimate $\Delta \gamma_{ad}$(Au) first. This can be done via the adsorption energy of CO on Au which, unfortunately, is not known accurately. It has recently been estimated to be < 0.3 eV [15.54,55], i.e., CO is either weakly chemisorbed or physisorbed. Neglecting, therefore, any orientation dependence for CO on Au, we calculate $\Delta \gamma_{ad}$(Au) $\leq$ 400 erg/cm^2. In the framework of these approximations we calculate $\Delta \gamma_{ad}$(Pt) by means of (15.20) for the three surface orientations {111}, {110}-1x1 and {110}-1x2 of Pt. The ratio of $f N_k/K$, used for all orientations, was 1.6 · 10^3 erg/(cm^{2}eV). The resulting values of $\Delta \gamma^{ad}$(Pt) are listed in Table 15.2.

With the measured SCLS of Pt, the data of Fig.15.8, and with the absolute values of the "average" specific surface free energy of Pt [15.36] we have a sufficient data base to construct a section of the γ-plot (fix points at {111} and {110}) for clean and CO-covered Pt.

446

Table 15.2. Specific surface free energies of CO-covered Pt surfaces (no adsorption site correction)

	$\Delta\gamma_{ad}$ [erg/cm^2]	γ [erg/cm^2]
Pt{111}	2000	550
Pt{110}-1x2	2416	384
Pt{110}-1x1	2800	89

There are a number of steps to be taken in order to arrive at this goal. These will be described in the following. At first, we have chosen the absolute value of $\bar{\gamma}$ = 2550 erg/cm^2 [15.36] to correspond to the Pt {111} orientation because for well-annealed polycrystalline samples the dominant orientation is that of the lowest specific surface free enrgy, i.e., in this case {111}. The differences between Pt{111} and Pt{110} were then calculated by using (15.15) and the experimental anisotropies of Au: 0.047 for Au{110}-1x2 in the [1$\bar{1}$0] zone [15.13]. The results are summarized in Table 15.3 for both Pt{110} modifications. One can see that the macroscopic specific surface free energy of the reconstructed Pt{110} surface is stabilized by 89 erg/cm^2 relative to the 1x1. This is a rather small difference and corresponds to an energy of reconstruction of about 1.5 kcal/mole. The experimental difference $\Delta\gamma$ between the 1x1 and 1x2 modification of Pt{110} is of the same order of magnitude as the theoretical data for surface reconstruction {15.1-4}.

Next we calculate the changes in γ due to adsorbed CO. Most of what we need has already been accomplished in Sect. 15.4 and Table 15.2. With the data for $\Delta\gamma_{ad}$(Pt) in Table 15.2 and the absolute values of γ for the Pt surfaces in Table 15.3, we can determine the absolute values of γ of CO-covered Pt{111} and Pt{110} surfaces. These are also listed in Table 15.2.

We finally plot the various γ-values of Pt, summarized in Table 15.3, in a (Cartesian) γ versus θ diagram, Fig. 15.9. The γ values at

Table 15.3. Specific surface free energies of clean Pt surfaces, derived from SCLS data and on the basis of γ(Pt{111}) = 2550 erg/cm^2 [15.37]

	$\Delta\gamma$ [erg/cm^2]	γ [erg/cm^2]
Pt{110}-1x2	250	2800
Pt{110}-1x1	339	2889

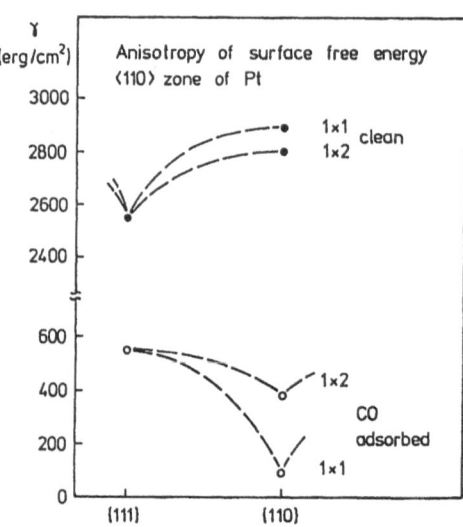

Fig.15.9 Partial γ-plot of Pt for clean and CO-covered {110} and {111} surfaces. Note different γ values for reconstructed and 1x1 surfaces of Pt{110}

{111} and at {110} have been connected by dashed lines in order to illustrate the expected functional behavior. Two important features become quite obvious from this plot: first, there is a deep cusp at the {111} orientation the depth of which is reduced considerably by the reconstruction of the {110} surface; secondly, adsorbed CO moves the cusp from {111} to {110} where now the Pt{110}-1x1 has the lower specific surface free energy. This means that adsorbed CO lifts the 1x2 reconstruction of the Pt{110} surface as long as the temperature is high enough for sufficient surface atom mobility [15.42,56]. Of course, the relative anisotropies, $\Delta\gamma/\gamma$, of CO covered Pt are very large compared to those of the clean surface, and as a consequence one would expect a CO induced reconstruction or faceting of Pt(111). Such an effect is not known to occur, perhaps because of the limited mobility of Pt surface atoms in the temperture range of CO adsorption.

15.5 Adsorbate–Induced Reconstruction

We have seen in Sects. 15.3,4 that a large specific surface free energy anisotropy of Pt for the [1$\bar{1}$0] zone may be responsible for the reconstruction of the clean Pt{110} surface to the 1x2 configuration. On the other hand, adsorbed CO changed the relative surface energies of {111} and {110} orientations, and inverted even those of the {110}-1x2 and 1x1 configurations, Fig.15.9. This explains the lifting of the 1x2 reconstruction by CO since the 1x1 with CO had the lower specific surface free energy.

In a more general sense, adsorbates can change the anisotropy of $\gamma(\theta)$ drastically, and, therefore, either lift or *induce* reconstruction. The latter is also a well known phenomenon in surface science. For

448

example, adsorbed oxygen induces a 2x1 surface reconstruction in Cu{110} and Ni{110} surfaces [15.57,58]. This reconstructon is again of the "missing row" type, but here every second row in <001> direction is removed. This would mean that {100} surfaces are preferentially exposed rather than {111} surfaces as it was the case with Pt{110}-1x2. Presumably, oxygen prefers {100}-type adsorption sites.

A complete study of $\gamma(\theta)$ of clean and oxygen-covered Cu has been carried out by *McLean* and *Hondros* [15.15]. Their results are partially reproduced in Fig.15.10 as lines of constant relative surface energy within the standard stereographic triangle. As pointed out in Table 15.1, the anisotropy of $\gamma(\theta)$ for clean Cu is small, with a maximum of 2.4%. The minimum of γ is located at {111}, in agreeement with theory [15.18]. After oxygen adsorption, on the other hand, the picture is completely changed, Fig.15.10b. Now the minimum surface energy is at {100}, and the maximum at {111}. The maximum anisotropy has increased to 3.25%. From this measurement it is clear that the preferred orientation in the presence of adsorbed oxygen is {100}, the same orientation that is preferentially exposed by the 2x1 reconstruction of Cu{110}-0 [15.57].

These γ-plots in Fig.15.10 were determined at 1173 K [15.15]. An interesting question in this context is whether the O-induced 2x1 reconstruction of Cu{110} is stable at this high temperature. If it were, the anisotropy measured at 1173 K would be typical for the reconstructed Cu system and presumably *lower* than for the non-reconstructed surface(s). The latter could be analyzed by SCLS measurements, for example.

A second illustrative example of adsorbate-induced changes in $\gamma(\theta)$ is the Fe/Si system studied in detail by *Mills* et al. [15.59]. No structural or surface compositional data are available here so that a connection between adsorption and reconstruction can only be inferred. Complete γ plots of 3% Si-Fe samples annealed in dry H_2 in the range of 1273 K to 1673 K were determined [15.59]. The maximum

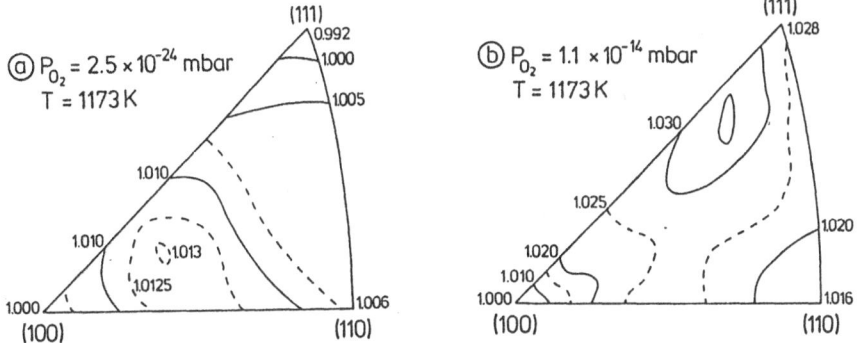

Fig.15.10 Complete γ-plot of clean and oxygen covered Cu surfaces at 1173 K. From [15.15]

449

anisotropies are in the range of 7-10% but the minimum value of γ appears at different orientations. For example, at 1273 K $\gamma\{111\}$ < $\gamma\{110\}$ < $\gamma\{100\}$, at 1473 K $\gamma\{110\}$ < $\gamma\{111\}$ < $\gamma\{100\}$, and at 1573 K $\gamma\{110\}$ < $\gamma\{100\}$ < $\gamma\{111\}$. These remarkable changes in $\gamma(\theta)$ are suggested to be due to the segregation of Si (or Si/0) to the Fe surface whereby the surface composition is temperature *and* orientation dependent [15.59]. At the highest temperature of 1673 K, however, the surface is proposed to be free of adsorbates because here the anisotropy $\gamma(\theta)$ exhibits the right sequence expected form theory for bcc metals [15.18]. It would be interesting to see whether adsorbed Si causes also a surface reconstruction of Fe.

Other adsorbate-induced reconstruction cases are alkali metals adsorbed on {110} surfaces of fcc metals, such as Ag [15.60], Cu [15.61], Ni [15.62] and Pd [15.63]. The alkali metals seem to cause a 1x2 reconstruction of the missing row type such that {111} orientations are again preferred. No corresponding specific surface free energy data are known. From the point of adsorption energies, it appears that the alkali metals must bond most strongly on {111} surfaces of fcc metals.

15.6 Reconstruction and Faceting - a Comparison

In a traditional sense, faceting is understood as the break-up of an initially flat surface into a hill and valley structure characterized by new surface orientations [15.26]. This process is driven by a decrease in the total surface free energy. In some cases surface reconstruction seems to follow the same rule on a microscopic scale. Good examples are the {110} surfaces of Ir, Pt and Au which all form a 1x2 missing row structure with {111} micro-facets. A striking difference is the apparent degree of order achieved in the processes of reconstruction and macroscopic faceting. Reconstructed surfaces (in the sense of Fig.15.1d) generally exhibit a high degree of order, i.e., there is a certain periodicity in the microfaceted profile giving rise to a LEED superstructure. Macroscopically faceted surfaces, on the other hand, have a more random structure. It is conceivable that this difference is due to the slow growth kinetics of facets prevalent for the conditions where reconstruction is initially observed. For example, it requires rather low temperatures to produce the 1x2 reconstruction of {110} surfaces of Ir, Pt or Au but larger microfacets such as present in 1xn (n > 2) structures are only occasionally reported [15.64].

Other reconstructions cannot really be called analogous to faceting because they do not exhibit a microscopic hill-and-valley structure. Instead, they show a surface structural change such as to expose a new surface orientation different from that of the macroscopic non-reconstructed surface. Examples are the {100} surfaces of Ir, Pt and Au

[15.65], the {111} surface of Au [15.66], or the {100} surface of W [15.67,68]. These examples are of the kind illustrated schematically in Fig.15.1c. However, there is also a decrease in surface free energy associated with that reconstruction, as outlined initially.

Further examples where reconstruction and faceting seem to be analogous have to do with adsorbate-induced changes in surface structure. There is a large number of faceting cases in the literature [15.69] although not all of them are equally well documented. Faceting of metal surfaces almost always occurs with oxygen adsorption or slight oxidation [15.69,70]. An illustration of a faceted Cu{110} surface that was heat-treated for several hours in 10^{-5} mbar of O_2 at 950 K [15.71] is shown in Fig.15.11. This surface has developed a lot of facets whose orientations are mostly {100}, in agreement with the oxygen-induced changes in $\gamma(\theta)$ [15.15]. As said before, this Cu {110} surface also reconstructs with adsorbed oxygen (low exposure at low temperature) to a 2x1. It would be interesting and important to study the transition from a reconstructed Cu{110}-O surface to a macroscopically faceted surface and to find out whether the same preferred orientations just grow or whether new ones develop with increasing oxidation.

Another elucidating case is K adsorption on Ni{110} which has recently been studied by *Behm* et al. [15.72]. These authors found that the Ni{110} surface reconstructs to a 1x2 with low coverages of adsorbed K. This is in line with other alkali-metal-induced 1x2

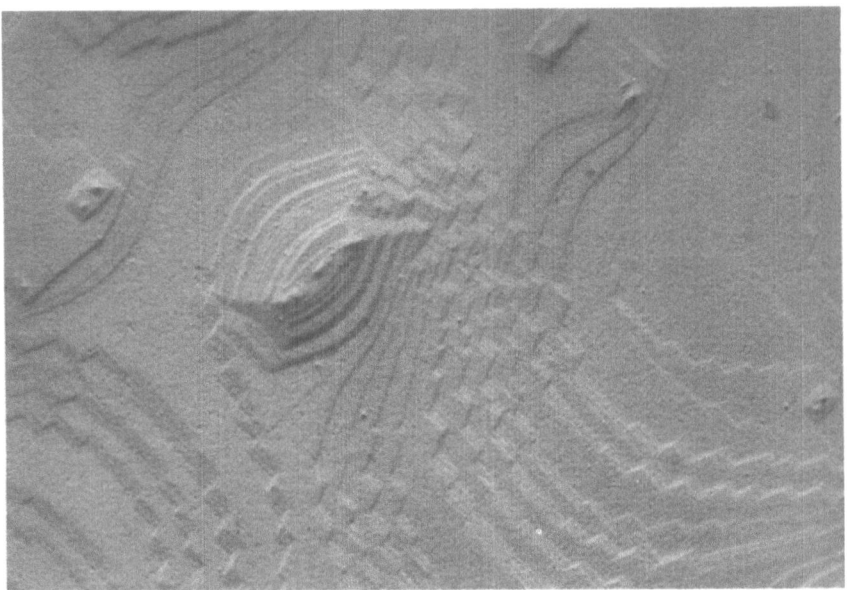

Fig.15.11 Electron micrograph (replica) of a Cu{110} single crystal surface annealed in O_2 ambient of about 10^{-5} mbar at 950 K for about 3 hours. Magnification is 20,900x [15.71]

451

reconstructions [15.60-63]. However, on heat treating this surface to 450 K for several minutes also 1x3 and 1x5 reconstruction was observed meaning that the size of the {111} facets grows as a function of time. It is not known how far this growth would go, but it is clearly a consequence of an increased anisotropy in $\gamma(\theta)$ due to the adsorbed alkali metal. It should be most interesting to continue this heat treatment of the Ni{110}+K surface in order to reach an "equilibrium" configuration.

If reconstruction and faceting are analogous processes (only different by their scale) a more complete knowledge about them should be of considerable practical importance in such technical areas as heterogeneous catalysis, metal corrosion, or semiconductor processing. Macroscopic faceting in context of heterogeneous reactions has been studied extensively and is a rather prevalent phenomenon [15.69]. The important question is whether it occurs also on very small metal particles of supported catalysts where microstructural changes would be extremely difficult to image. If we were allowed to generalize our ideas in this paper, it should occur and could have a profound influence on reaction kinetics and selectivity. Hence a detailed study of the transition from microscopic structural changes to macroscopic faceting - as they are induced by adsorbates or reaction intermediates - is of great importance. If these studies could be connected with reaction kinetic data, the understanding of structure sensitivity or insensitivity in heterogeneous catalysis could possibly be much improved.

Acknowledgement. We would like to thank Prof. Paul Wynblatt and Prof. W. Moritz for a critical reading of the manuscript and for valuable comments. Thanks are also due to Drs. P. Thiel and J. Behm as well as J. Smith for preprints of their papers prior to publication.

References

15.1 T. Halicioglu, T. Takai, W.A. Tiller: "The Structure and Surface Energy of Au(110) Studied by Monte Carlo Method", in *The Structure of Surfaces*, ed. by M.A. van Hove, S.Y. Tong, Springer Ser. Surf. Sci., Vol.2 (Springer, Berlin, Heidelberg 1985) p.231

15.2 D. Tomanek: Phys. Lett. A 113, 445 (1986)

15.3 F. Ercolessi, E. Tosatti, M. Parinello: Phys. Rev. Lett. 57, 719 (1986)

15.4 K.M. Ho, K.P. Bohnen: Phys. Rev. Lett. 59, 1833 (1987)

15.5 H.P. Bonzel, N. Freyer, E. Preuss: Phys. Rev. Lett. 57, 1024 (1986)

15.6 H.P. Bonzel, K. Dückers: Surf. Sci. 184, 425 (1987)

15.7 N.A. Gjostein: Acta Metall. 11, 957 and 969 (1963)

15.8 A.J.W. Moore: In *Metal Surfaces*, ed. by W.D. Robertson, N.A. Gjostein (Am. Soc. for Metals, Metals Park, Ohio 1963) p.155

15.9 E.D. Hondros: In *Techniques of Metals Research*, Vol.IV, Part 2, ed. by R.A. Rapp (Wiley, New York 1970) p.293

15.10 W.K. Winterbottom: In *Structure and Properties of Metals Surfaces*, Honda Memorial Series, No.1 (Maruzen Co., Tokyo 1973) p.36

15.11 J.M. Blakely: *Introduction to the Porperties of Crystal Surfaces* (Pergamon, New York 1973);
M. Drechsler: In *Surface Mobilities on Solid Materials*, ed. by Vu Thien Binh (Plenum, New York 1983) p.405
15.12 J.C. Heyraud, J.J. Métois: Surf. Sci. **128**, 334 (1983)
15.13 W.L. Winterbottom, N.A. Gjostein: Acta Metall. **14**, 1041 (1966)
15.14 T. Barsotti, J.M. Bermond, M. Drechsler: Surf. Sci. **146**, 467 (1984)
15.15 E.D. Hondros, M. McLean: Colloqu. Int'l. du CNRS, Nr.187, éditions du CNRS, Paris 1970
15.16 M. McLean, H. Mykura: Surf. Sci. **5**, 466 (1966)
15.17 J.C. Heyraud, J.J. Métois: Acta Metall. **28**, 1789 (1980)
15.18 J.K. MacKenzie, A.J.W. Moore, J.F. Nicholas: J. Phys. Chem. Solids **23**, 185 (1962)
15.19 E.E. Gruber, W.W. Mullins: J. Phys. Chem. Solids **28**, 875 (1967)
15.20 S. Hagstrom, H.B. Lyon, G.A. Somorjai: Phys. Rev. Lett. **15**, 491 (1965)
15.21 D.G. Fedak, N.A. Gjostein: Phys. Rev. Lett. **16**, 171 (1966)
15.22 R.M. Goodman, G.A. Somorjai: J. Chem. Phys. **52**, 6325 (1970)
15.23 J.W.M. Frenken, P.M.J. Marée, J.F. van der Veen: Phys. Rev. B **34**, 7506 (1986)
15.24 U. Breuer, K.C. Prince, H.P. Bonzel: Phys. Rev. Lett. **60**, 1146 (1988)
15.25 K.C. Prince: Surf. Sci. **193**, L24 (1988)
15.26 C. Herring: Phys. Rev. **82**, 87 (1951)
15.27 I.N. Stranski: Z. phys. Chem. **136**, 259 (1928); **11**, 421 (1931)
15.28 M. Drechsler, J.F. Nicholas: J. Phys. Chem. Solids **28**, 2609 (1967)
15.29 M.I. Baskes, C.F. Melius: Phys. Rev. B **20**, 3197 (1979)
15.30 M.S. Daw, M.I. Baskes: Phys. Rev. B **29**, 6443 (1984)
15.31 S.M: Foiles, M.I. Baskes, M.S. Daw: Phys. Rev. B **33**, 7983 (1986)
15.32 C.T. Chan, S.G. Louie: Phys. Rev. B **33**, 2861 (1986)
15.33 J.G. Gay, J.R. Smith, R. Richter, F.J. Arlinghaus, R.H. Wagoner: J. Vac. Sci. Technol. A **2**, 931 (1984)
15.34 E.S. Machlin: In *Interatomic Potentials and Crystalline Defects*, ed. by J.K. Lee (The Metall. Soc. of AIME, New York 1981) p.33
15.35 P. Wynblatt: Surf. Sci. **136**, L51 (1984); and private communication
15.36 W.R. Tyson, W.A. Miller: Surface Sci. **62**, 267 (1977)
15.37 A.R. Miedema: Z. Metallk. **69**, 287 (1978)
15.38 J.R. Smith, A. Banerjea: Phys. Rev. Lett. **59**, 2451 (1987)
15.39 W.W. Mullins: J. Appl. Phys. **28**, 333 (1957); dto. **30**, 77 (1959)
15.40 H.P. Bonzel, E. Preuss, B. Steffen: Appl. Phys. A **35**, 1 (1984); Surf. Sci. **145**, 20 (1984)
15.41 H.P. Bonzel, E.E. Latta: Surf. Sci. **76**, 275 (1978)
15.42 E. Preuss, N. Freyer, H.P. Bonzel: Appl. Phys. A **41**, 137 (1986)
15.43 N. Freyer: Ph.D. Thesis, Aachen 1986
15.44 H.P. Bonzel, R. Ku: J. Vac. Sci. Technol. **9**, 663 (1972)
15.45 H. Niehus: Surf. Sci. **145**, 407 (1984)
15.46 G.L. Kellogg: Phys. Rev. Lett. **55**, 2168 (1985)
15.47 J.C. Campuzano, M.S. Foster, G. Jennings, R.F. Willis, W. Unertl: Phys. Rev. Lett. **54**, 2684 (1985)
15.48 K. Dückers, H.P. Bonzel: submitted for publication
15.49 B. Johannsson, N. Mårtensson: Phys. Rev. B **21**, 4427 (1980)
15.50 K. Dückers, H.P. Bonzel, D. Wesner: Surf. Sci. **166**, 141 (1986)
15.51 K. Dückers, K.C. Prince, H.P. Bonzel, V. Cháb, K. Horn: Phys. Rev. B **36**, 6292 (1987)
15.52 R.C. Baetzold, G. Apai, E. Shustorovich, R. Jaeger: Phys. Rev. B **26**, 4022 (1982)

15.53 J.F. van der Veen, F.J. Himpsel, D.E. Eastman: Phys. Rev. Lett. **44**, 189 (1980)

15.54 D.A. Outka, R.J. Madix: Surf. Sci. **179**, 361 (1987)

15.55 P.R. Norton, R.L. Tapping, J.W. Goodale: Surface Sci. **72**, 33 (1978)

15.56 D.W. Bassett, P.R. Webber: Surf. Sci. **70**, 520 (1978)

15.57 H. Niehus, G. Comsa: Surf. Sci. **140**, 18 (1984)

15.58 A.M. Baró, G. Binnig, H. Rohrer, Ch. Gerber, E. Stoll, A. Baratoff, F. Salvan: Phys. Rev. Lett. **52**, 1304 (1984)

15.59 B. Mills, M. McLean, E.D. Hondros: Phil. Mag. **27**, 361 (1973)

15.60 B.E. Hayden, K.C. Prince, P.J. Davie, G. Paolucci, A.M. Bradshaw: Solid State Commun. **48**, 325 (1983)

15.61 C.J. Barnes, M.Q. Ding, M. Lindroos, R.D. Diehl, D.A. King: Surf. Sci. **162**, 59 (1985)

15.62 D.K. Flynn, K.D. Jamison, P.A. Thiel, G. Ertl, R.J. Behm: J. Vac. Sci. Technol. A **5**, 794 (1987)

15.63 M. Copel, W.R. Graham, T. Gustafsson, T. Yalisove: Solid State Commun. **54**, 695 (1985)

15.64 W. Moritz, D. Wolf: Surf. Sci. **88**, L29 (1979);
W. Moritz: private communication

15.65 M.A. van Hove, R.J. Koestner, P.C. Stair, J.P. Biberian, L.L. Kesmodel, I. Bartos, G.A. Somorjai: Surf. Sci. **103**, 189 (1981)

15.66 D.M. Zehner, J.F. Wendelken: Proc. of VII. Int'l Vac. Congr. and III. Int'l Conf. on Solid Surfaces, Vienna 1977, ed. by R. Dobrozemsky, (Berger and Söhne, Vienna 1977) p.517

15.67 M.K. Debe, D.A. King: Phys. Rev. Lett. **39**, 708 (1977)

15.68 R.A. Barker, P.J. Estrup: Solid State Commun. **25**, 375 (1978)

15.69 M. Ffytzani-Stephanopoulos, L.D. Schmidt: Progr. in Surf. Sci. **9**, 83 (1979)

15.70 A.T. Gwathmey, A.F. Benton: J. Chem. Phys. **8**, 431 (1940)

15.71 H.P. Bonzel: unpublished

15.72 R.J. Behm, D.K. Flynn, K.D. Jamison, P.A. Thiel, G. Ertl: to be published;
R.J. Behm: private communication

16. Surface Melting

J.F. van der Veen, B. Pluis, A.W. Denier van der Gon

FOM – Institute for Atomic and Molecular Physics
Kruislaan 407, NL – 1098 SJ Amsterdam, The Netherlands

In this chapter the role of the surface in initiating the melting of a solid is discussed. A thermodynamic model is presented which shows that, under certain conditions, a disordered (quasi-liquid) layer forms on the surface just below the triple point. Recently, a number of surface sensitive probes have been used to search for the effect. Surface melting is indeed observed on a variety of crystal faces. The experimental methods used are explained and the results are compared with the predictions of the thermodynamic model.

16.1 Melting and the Role of the Surface

Melting of a solid is one of the most commonly observed phase transitions in nature. The thermodynamic parameters involved are well-documented for almost all elements [16.1], yet the understanding of melting on a microscopic scale still forms an outstanding problem in condensed matter physics.

Bulk melting, being a first-order transition, is known as a heterogeneous process which involves the coexistence of two phases, with the liquid phase growing at the expense of the solid phase [16.2]. It is reasonable to assume that the melting nucleates at crystal defects such as vacancies, interstitials, dislocations and grain boundaries or at the surface [16.3]. Impurities could be obvious nucleation centers too, but henceforth they are assumed to be present in negligible quantities and not to play any role.

Of all possible sites where melting may start, the surface appears to be the most likely one. According to the Lindemann criterion of melting [16.4], which states that melting occurs when the root-mean-square (rms) thermal displacement of the atoms in the lattice reaches a critical fraction (typically 10%) of the interatomic distance, the top surface layer of a crystal should "melt" far below the bulk melting point. The loosely bound surface atoms, which lack a quarter to a half of their neighbors, have much higher rms displacements than atoms in the bulk. This has been confirmed by both experimental measurements and theoretical calculations [16.5]. Consequently, at the surface the

Lindemann criterion is fulfilled much earlier. This argument points to the important role that surfaces may play in the melting process and has inspired much of the earlier work on the subject [16.6-9].

The Lindemann criterion provides a too simple description of the melting process. The criterion represents a one-sided approach in that it only considers the solid phase. The latter, in fact, is a major short-coming of any lattice-instability theory of melting [16.10, 11]. Instead, we will use simple thermodynamics to explain the phenomenon of surface melting (Sect.16.2). We will derive a clear-cut criterion for the formation of a melted layer at the crystal-vapor interface and show that the melted layer thickness at the surface diverges as the triple point T_m is approached. The rate at which the melt penetrates into the bulk (logarithmically or with a power law) follows from a consideration of the interatomic forces in the system.

The term "surface melting" has caused some confusion in the literature. Occasionally, "surface melting" was taken to be synonymous to "surface roughening" [16.12]. There is, however, a definite distinction. Surface roughening implies the vanishing of the free energy associated with the formation of a step or kink on the crystal surface [16.12-14], while the atomic registry in the near-surface layers with respect to the underlying crystal is preserved (Chap.14). Surface melting, on the other hand, implies the actual dislodgement of atoms from lattice sites, resulting in a strong reduction of crystalline order and liquid-like properties of the surface region [16.15]. Since the atoms in a melted film of finite thickness "feel" the presence of the crystal underneath, there will always remain some crystalline order in it [16.16]. Hence, "surface melting" does *not* mean that the surface actually becomes a true liquid. Rather, the surface melt should be regarded as a *quasi-liquid* with properties intermediate between those of the solid and the bulk liquid. This issue too has been a source of confusion [16.17].

So, what precisely do we mean by "surface melting"? In our view it is the formation, just below T_m, of a quasi-liquid skin on the surface, which thickens as the temperature comes closer to T_m and finally converts the whole solid body into a melt at T_m. This phenomenon of surface-initiated melting, if it indeed commonly occurs in nature, offers an elegant explanation for the general absence of overheating effects in solids. The melted surface layer will act as a nucleation site for bulk melting at T_m, so that overheating is precluded. By analogy, a liquid cannot be undercooled upon passing the freezing point if it is in contact with the solid. It is interesting to note, that in cases where the influence of the surface on the melting process is suppressed, overheating of the solid can indeed be observed [16.18-22]. Thus the microscopic phenomenon of surface melting is seen to have far-reaching consequences on a macroscopic scale.

This chapter is organized as follows. In Sect.16.2 various thermodynamic aspects of surface melting are discussed. Recent observations of surface melting are reviewed in Sect.16.3, where a description is given of the experimental methods used. The chapter is concluded with an outlook in Sect.16.4.

The view presented here is that of the experimentalist. Therefore, relatively little attention has been given to more advanced theories of wetting and surface-induced disorder that have been developed very recently. This subject has been reviewed by *Dietrich* [16.23].

16.2. Thermodynamics of Surface Melting

16.2.1 Driving Force

First we briefly consider melting of bulk material. The thermodynamic parameter of interest here is the Gibbs free energy $G(P,T)$ at pressure P and temperature T. The T-dependence of G at constant P, is schematically drawn in Fig.16.1 for the bulk material in both the solid (s) and liquid (ℓ) state [16.2]. In thermodynamic equilibrium, the system is in the state of lowest free energy. Below T_m the state of lowest energy is the solid state, above T_m it is the liquid state. The melting point T_m is the temperature at which the free energy of the solid exactly equals that of the liquid. Note that the transition from solid to liquid is discontinuous (first-order). The latent heat of melting L is given by

$$L = T_m \left[\frac{\partial G_s}{\partial T} - \frac{\partial G_\ell}{\partial T} \right] . \tag{16.1}$$

Next we discuss the thermodynamics of *surface* melting. We define a specific Gibbs free energy γ for the surface, which represents the work involved in creating unit area of the surface at a given temperature and pressure [16.24].

In order to find out under which conditions surface melting is energetically favorable, we evaluate the specific surface free energy for a system in thermodynamic equilibrum, which lies on the solid-vapor coexistence line in the (P,T) phase diagram. Let us assume that the system is at a temperature close to the triple point T_m and that a quasi-liquid layer of thickness ℓ is present between solid and vapor (Fig.16.2). The free energy per unit area of the surface is given by

$$\gamma = \gamma_{sq} + \gamma_{qv} + L \cdot N \left[1 - \frac{T}{T_m} \right] \tag{16.2}$$

where γ_{sq} and γ_{qv} are the free energies per unit area of the solid/quasi-liquid and the quasi-liquid/vapor interfaces, respectively. L

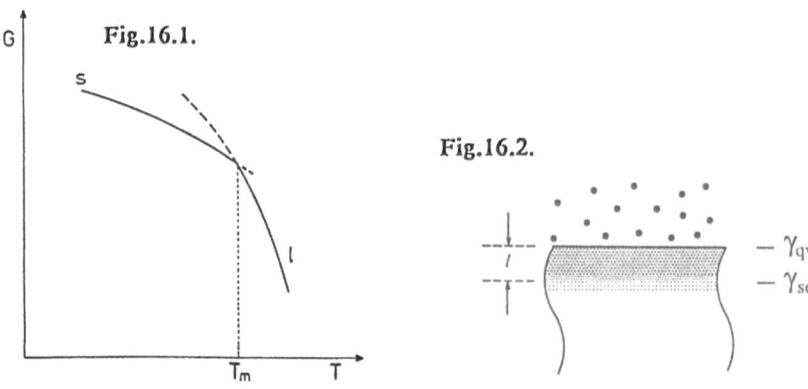

Fig.16.1. The Gibbs free energy versus the temperature for the solid (s) and liquid (ℓ) phases.

Fig.16.2. Wetting of a solid surface by its own melt. Symbols are defined in the text.

is the latent heat of melting per atom and N is the number of atoms per unit area in a quasi-liquid layer of thickness ℓ. N relates to the layer thickness ℓ through $N = n\ell$, where n is the atom concentration. The third term in (16.2) represents the specific free energy associated with undercooling the quasi-liquid layer. Here the approximation is made that the latent heat involved in the solid/quasi-liquid transition is equal to the latent heat involved in the solid/liquid transition. The specific free energy γ_{qv} for the quasi-liquid/vapor interface will lie between the specific free energy γ_{lv} of the liquid/vapor interface and the specific free energy γ_{sv} of the perfect solid/vapor interface (the surface without the quasi-liquid layer). The specific free energy γ_{sq} will be smaller than the solid/liquid specific interface energy γ_{sl}. So we write:

$$\gamma_{qv} = \gamma_{lv} + M(\gamma_{sv} - \gamma_{lv}) \tag{16.3a}$$

and

$$\gamma_{sq} = (1-M)\gamma_{sl}, \tag{16.3b}$$

where M is the effective crystalline order parameter, normalized such that it is unity for the perfect crystal and zero for the true liquid. For M one could take the Fourier component of the density having the periodicity of the crystal lattice. The order parameter in the quasi-liquid layer is assumed to be an exponentially decreasing function of the film thickness [16.25]:

$$M = e^{-\ell/\ell_0} = e^{-N/N_0} \tag{16.4}$$

where $N_0 = n\ell_0$ is a constant of microscopic dimensions, of the order of 10^{15} atoms/cm² (ℓ_0 is of the order of one interlayer spacing). Equa-

458

tion (16.4) is appropriate for a system governed by short-range forces [16.26]. It is readily checked that (16.3) and (16.4) satisfy the proper boundary conditions. For $\ell \to \infty$, the quasi-liquid becomes a true liquid. Therefore, $M \to 0$, $\gamma_{qv} \to \gamma_{lv}$ and $\gamma_{sq} \to \gamma_{sl}$. For $\ell \to 0$, the quasi-liquid film vanishes and so $M \to 1$, $\gamma_{qv} \to \gamma_{sv}$, and $\gamma_{sq} \to 0$. Substitution of (16.3) and (16.4) in (16.2) yields

$$\gamma = \gamma_{sl} + \gamma_{lv} + L \cdot N \left[1 - \frac{T}{T_m} \right] + (\gamma_{sv} - \gamma_{sl} - \gamma_{lv}) e^{-N/N_0} . \qquad (16.5)$$

The number of atoms in the quasi-liquid layer at equilibrium is obtained by minimizing γ with respect to N, resulting in

$$N_{eq}(T) = N_0 \ln \left[\frac{T_m \Delta \gamma}{(T_m - T) L N_0} \right] , \qquad (16.6)$$

where $\Delta \gamma \equiv \gamma_{sv} - \gamma_{sl} - \gamma_{lv}$ is the specific interfacial free energy at temperature T, which an ordered solid surface has in excess of a surface completely wetted with a liquid layer.

Obviously, surface melting occurs only if $\Delta \gamma > 0$, or

$$\gamma_{sv} > \gamma_{sl} + \gamma_{lv} . \qquad (16.7)$$

Equation (16.7) is equivalent to the well-known condition for wetting of a solid by its own melt [16.27]. Note that γ_{sv} in (16.7) is, in fact, not observable because it does not refer to an equilibrium quantity. However, γ_{sv} for the hypothetical dry solid-vapor interface can be estimated by extrapolating the equilibrium γ_{sv} value at lower temperature to the melting point T_m (Sect. 16.2.2).

Whereas the melting transition in the bulk is discontinuous, at the surface it proceeds continuously. As the temperature rises, the quasi-liquid film thickness grows critically as T_m is approached. The existence of a finite equilibrium thickness of the quasi-liquid at a temperature $T < T_m$ can be seen as a result of a balance between two opposing driving forces. According to (16.7), free energy is gained if a molten layer intervenes between vapor and solid. This gain, however, is to be balanced against the increase in free energy associated with undercooling the molten layer.

While the layer thickness has a logarithmic singularity at T_m for a system with short-range forces (16.6), it grows with a power law when long-range forces are present [16.28, 29]. In the latter case the surface order parameter decays slowly with thickness

$$M = \left[\frac{N_0}{N_0 + N} \right]^p \qquad (16.8)$$

where N_0 is a constant of microscopic dimension and the value of p depends on the type of long-range interaction (for non-retarded van der Waals interaction, $p = 2$ [16.26]). Modifying (16.5) accordingly and again minimizing γ with respect to N we obtain:

$$N_{eq}(T) = \left[\frac{T_m \Delta\gamma p N_0{}^p}{(T_m - T)L}\right]^{1/(p+1)} - N_0 \ . \tag{16.9}$$

Hence, $N_{eq}(T)$ grows asymptotically like $|T_m - T|^{-1/(p+1)}$ for such systems.

16.2.2 Which Surfaces Do Melt?

For a surface to melt below T_m, the wetting condition $\Delta\gamma > 0$ (16.7) must be fulfilled. Prediction of the sign of $\Delta\gamma$ requires knowledge of the specific interface energies γ_{sv}, γ_{lv} and γ_{sl}. For the determination of γ_{sl} several methods have been used [16.30]. The one most frequently employed is measuring the amount of undercooling before homogeneous nucleation of the solid occurs. This method is explained in [16.24]. For metals, the value of γ_{sl} is typically 10% to 20% of γ_{lv} [16.31]. The quantity γ_{lv} is known for most elements through measurements of the surface tension [16.32]. Obviously, the least accessible quantity is γ_{sv}. Tables of experimental values are available [16.33], but the scatter in the data is appreciable. Semi-empirical estimates (for $T=0$) have been given by *Miedema* for a large number of elements, along with a prescription to calculate their temperature dependence [16.33]. It was found that for metals, on average, the relationship $\gamma_{sv}(V_m{}^s)^{2/3} \simeq 1.13\gamma_{lv}(V_m{}^\ell)^{2/3}$ holds. Here, $V_m{}^s$ and $V_m{}^\ell$ are the molar volumes of the element in the solid and liquid state, which in general differ only by a few percent.

Estimates of γ_{sv}, γ_{sl}, γ_{lv} and $\Delta\gamma$ are given in Table 16.1 for a number of common elements near the melting point. They are based on *Miedema*'s data [16.31-33]. There are large uncertainties in the $\Delta\gamma$ values and also in their sign when they are close to zero. This is because $\Delta\gamma$ is the difference between nearly equal values of γ_{sv} and $\gamma_{lv}+\gamma_{sl}$ and especially the former one is not accurately known. Hence the table should only be used as a crude indicator of melting (+) or non-melting (-). It is concluded that, in fact, many elements are expected to exhibit surface melting. Table 16.1 also list values of $T_m\Delta\gamma/L$. The latter parameter forms the part in the right-hand side of (16.6) that is known from the other entries in Table 16.1. A large positive $T_m\Delta\gamma/L$ means that the element is likely to show a strong surface melting effect.

Table 16.1. Table of thermodynamic quantities related to surface melting for a number of common elements. The symbols are defined in the text. The values for γ_{sv} are from [16.33], for γ_{sl} from [16.31], and for γ_{lv} from [16.32]. $P(T=T_m)$ is the equilibrium vapor pressure at the melting point.

Z		T_m K	L 10^{-21}J/at	γ_{sv} mJ/m²	γ_{sl} mJ/m²	γ_{lv} mJ/m²	$\Delta\gamma$ mJ/m²	melting(+) non-melting(-)	$T_m\Delta\gamma/L$ 10^{18} K/m²	$P(T=T_m)$ Torr
11	Na	371	4.4	223	31	200	-8	(-)	-475	1.10^{-7}
12	Mg	923	15.3	679	115	570	-6	(-)	-362	3
13	Al	931.7	17.9	1032	154	865	13	(+)	677	3.10^{-9}
14	Si	1683	77.0	1038	416	800	-178	(-)	-3891	3.10^{-4}
23	V	2003	27.7	2280	375	1900	5	(+)	362	2.10^{-2}
24	Cr	2173	24.3	2031	381	1700	-50	(-)	-4471	3
25	Mn	1517	24.4	1297	183	1100	14	(+)	870	8.10^{-1}
26	Fe	1808	26.8	2206	326	1830	50	(+)	3373	2.10^{-2}
27	Co	1766	26.1	2197	345	1830	22	(+)	1489	2.10^{-3}
28	Ni	1728	29.3	2104	356	1750	-2	(-)	-118	3.10^{-3}
29	Cu	1356	21.7	1592	263	1310	19	(+)	1187	4.10^{-4}
30	Zn	692.7	47.8	895	119	770	6	(+)	87	1.10^{-1}
31	Ga	302.9	9.3	794	58	715	21	(+)	684	$< 1.10^{-11}$
32	Ge	1232	57.6	870	273	640	-43	(-)	-920	8.10^{-7}
41	Nb	2760	40.3	2314	399	1960	-45	(-)	-3082	8.10^{-4}
42	Mo	2883	40.3	2546	490	2130	-74	(-)	-5294	3.10^{-2}
44	Ru	2700	42.4	2591	443	2250	-102	(-)	-6495	2.10^{-2}
45	Rh	2240	36.2	2392	384	1970	38	(+)	2351	4.10^{-3}
46	Pd	1828	28.6	1808	302	1480	26	(+)	1662	3.10^{-2}
47	Ag	1234	20.1	1065	184	910	-29	(-)	-1780	3.10^{-3}
48	Cd	594.1	10.1	697	81	590	26	(+)	1529	1.10^{-1}
49	In	430	6.4	638	48	560	30	(+)	2016	$< 1.10^{-11}$
50	Sn	505.1	11.8	654	66	570	18	(+)	770	$< 1.10^{-11}$
73	Ta	3250	52.1	2595	477	2180	-62	(-)	-3868	5.10^{-3}
74	W	3650	58.5	2753	590	2340	-178	(-)	-11106	3.10^{-2}
75	Re	3440	54.9	3100	591	2650	-141	(-)	-8835	3.10^{-2}
76	Os	2970	44.5	3055	566	2500	-12	(-)	-801	2.10^{-2}
77	Ir	2727	45.9	2664	466	2250	-52	(-)	-3089	9.10^{-3}
78	Pt	2042.5	36.2	2223	334	1860	29	(+)	1636	2.10^{-4}
79	Au	1336	21.0	1363	200	1130	33	(+)	2099	1.10^{-5}
81	Tl	576.8	7.2	547	66	465	16	(+)	1282	2.10^{-8}
82	Pb	600.7	7.9	544	62	460	22	(+)	1673	4.10^{-9}
83	Bi	544.2	18.3	501	74	380	47	(+)	1398	2.10^{-10}

We are now in a position to calculate how close the melting point has to be approached in order to let the quasi–liquid layer grow to a thickness of at least a few monolayers. We take the element Pb as an example and require the layer thickness to be ~ 10 Å (i.e., $N_{eq} = 3.3 \cdot 10^{15}$ atoms/cm²). Substituting $T_m\Delta\gamma/L$ from Table 16.1 and

$N_0 \simeq 1.5 \cdot 10^{15}$ atoms/cm^2 in (16.6), we obtain $T_m - T \simeq 12$ K. This corresponds with $(T_m - T)/T_m = 2 \cdot 10^{-2}$. Because of the logarithmic growth law (16.6), these numbers will be of the same order for other metals. One could try to melt a near-macroscopic layer thickness of e.g. $\sim 10^2$ Å, but then the melting point has to be approached within $(T_m - T)/T_m \simeq 10^{-12}$! Experimentally, this is impossible to realize. Thus, for the detection of surface melting, a high surface sensitivity is desired. In general, sensitivity on the monolayer level is only available through the use of surface science equipment. In such equipment, the sample is placed in an ultrahigh vacuum (UHV) environment, so as to allow for unobstructed use of electron, photon or ion probes and to avoid surface contamination. This requirement on UHV compatibility of the experiment restricts our choice of sample material tremendously. Most metals have far too high vapor pressure at their melting point. Only elements such as Al, Ga, In, Sn, Pb and Bi have equilibrium vapor pressures low enough for ultrahigh vacuum conditions to be maintained (Table 16.1). We note, that under such vacuum conditions the surface cannot strictly be in equilibrium with its own vapor; the evaporating atoms will be pumped off or condense on the walls of the vacuum vessel. But the evaporation rates for the above elements are so small that this does not play any role in our melting model.

16.2.3 Crystal-Face Dependence of Surface Melting

The values of γ_{sv} upon which the above predictions have been based, are estimates for model surfaces having average packing density of atoms [16.33]. However, it is well known that the specific surface free energy of a crystal is anisotropic, i.e., it is dependent on the orientation of the crystal surface. In general, the most densely packed crystal face has the lowest specific surface free energy. For example, for the different low-index faces of an f.c.c. crystal the γ_{sv} values decrease in the order

$$\gamma_{sv}\{110\} > \gamma_{sv}\{100\} > \gamma_{sv}\{111\} \tag{16.10}$$

provided there is no reconstruction (see also Chap.15). For most metals the anisotropy is only a few percent at maximum [16.34-39]. Yet, such a small anisotropy can lead to a large percentage variation in $\Delta\gamma$ and thus to a dramatic change in melting behavior. Even a reversal in the sign of $\Delta\gamma$ may occur, resulting in melting and non-melting faces on one and the same crystal.

The anisotropy in γ_{sv} and its effect on the melting will now be discussed for other, higher-index, orientations $\{hkl\}$. We restrict ourselves to the very few crystals for which reliable data on the crystal-face dependence of γ_{sv} are available. Figure 16.3 shows for the f.c.c.

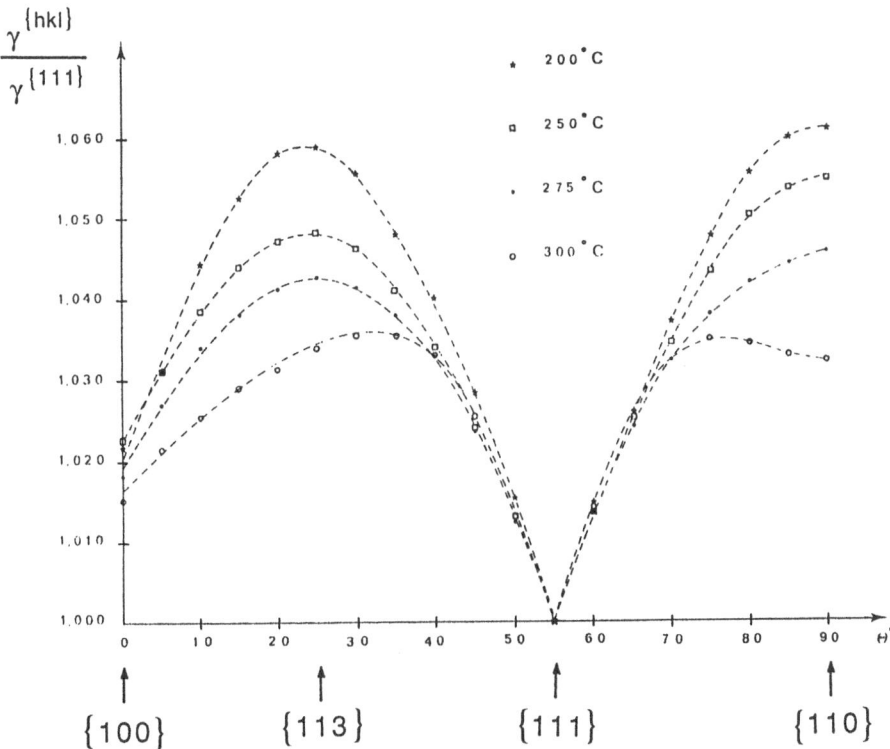

Fig.16.3. The normalized specific free energies $\gamma_{sv}\{hkl\}/\gamma_{sv}\{111\}$ of the Pb solid/vapor interface for a range of crystal-face orientations $\{hkl\}$ in the $\langle 110\rangle$ zone. From [16.38] with permission.

crystal of Pb a plot of the anisotropy $\gamma_{sv}\{hkl\}/\gamma_{sv}\{111\}$ at various temperatures as a function of orientation $\{hkl\}$ in the $\langle 110\rangle$ crystallographic zone. Note, that the specific free energies for the $\{110\}$, $\{100\}$ and $\{111\}$ orientations decrease indeed in the order given by (16.10). The data in Fig.16.3 were recently obtained by *Heyraud* and *Métois* [16.38] from measurements of the equilibrium shape of micron-sized lead crystallites and application of the *Wulff* theorem [16.40]. Similar data are available for In [16.39]. This method of determining the anisotropy in the surface energy by the Wulff construction is also discussed by M. Wortis in Chap.13. Absolute values of the surface energy cannot be obtained by this method; the data in Fig.16.3 are normalized to $\gamma_{sv}\{111\}$. The most striking features in Fig.16.3 are the cusped minima at the $\{111\}$ and $\{100\}$ orientations. The cusps are readily explained by the well-known terrace-ledge-kink (TLK) model in which an expression for the specific surface energy (i.e., the specific surface free energy at T = 0) is derived for orientations vicinal to a low-index orientation [16.12]. Consider a surface with terraces of width b and step ledges of height h (no kinks for simplicity), as shown in Fig.16.4.

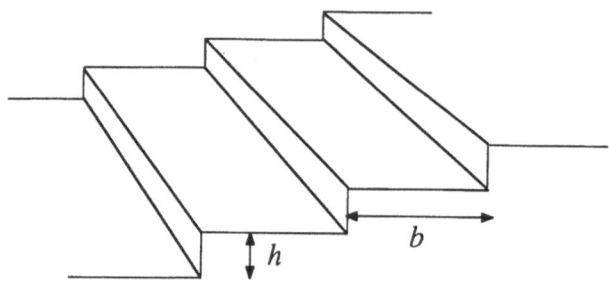

Fig.16.4. Schematic of stepped surface with terraces of width b and step ledges of height h.

The vicinal orientation angle θ is defined by $\tan\theta = h/b$. The surface energy per unit area is then given by

$$\gamma(\theta, T = 0) = \gamma_0 \cos|\theta| + \frac{\epsilon}{h}\sin|\theta| , \tag{16.11}$$

where γ_0 is the surface energy per unit area of the terrace and ϵ is the energy per unit length of the step ledge (step-step interaction can be neglected for large b). Equation (16.11) describes roughly the cusp forms seen in Fig.16.3, but a realistic theory should include the temperature as a parameter. As the temperature increases, the vicinal surface will roughen, i.e., the step ledges start to meander and the equilibrium concentration of kinks on the ledges will increase (Chap.14). This will raise the configurational entropy of the vicinal surface and thereby reduce the anisotropy around the cusp [16.41]. This is experimentally observed (Fig.16.3) and now also quantitatively explained by advanced theories of roughening (Chap.13). This temperature effect is also responsible for the absence of cusps around other, higher-index, orientations.

For an evaluation of the surface orientation dependence of $\Delta\gamma$ one is inclined to use γ_{sv} values as derived from the curve in Fig.16.3 measured at the temperature closest to $T_m(300°C)$. This, of course, is incorrect if at this temperature surface melting has already set in. In that case the γ_{sv} anistropies derived do in fact not correspond to those of a solid/vapor interface but rather to a quasi-liquid/vapor interface. From the γ-curve measured at a temperature of 200°C, we derive a maximum anisotropy of 6% for Pb. This corresponds to a variation of ~33 mJ/m² in the absolute value of γ_{sv}. This variation is larger than the tabulated value of $\Delta\gamma$ itself ($\Delta\gamma = 22$ mJ/m²). Hence it is very likely that $\Delta\gamma^{\{hkl\}}$ which is expected to be positive for most surface orientations, reverses its sign for orientations around the {111} plane and possibly also around the {100} plane, making these crystal faces non-melting. These expectations obviously make Pb an interesting object of experimental study, but similar orientation dependent effects may also occur on the crystal faces of other metals.

16.2.4 Molecular Dynamics Simulations and Phenomenological Theories

In molecular dynamics (MD) computer simulations, the classical equations of motion are solved for a system of N interacting particles. The interaction forces are usually derived from a model two-body potential, e.g., the Lennard-Jones (LJ) potential. MD simulations on crystal slabs generally show a disordering of the surface as the temperature increases [16.42,43], but the maximum temperature is often too far below T_m to permit the observation of a truly equilibrated quasi-liquid layer [16.44]. A very close approach to T_m is precluded by the temperature fluctuations which always occurs in a micro-canonical system of limited size. The fluctuations can be reduced by increasing the number of particles in the system but this costs an excessive amount of computation time. Despite these difficulties, extensive MD simulations on LJ crystal slabs have been performed to calculate the temperature dependence of the interface free energies and the quasi-liquid layer thickness [16.42].

On the theoretical side, substantial progress has recently been made with understanding the general characteristics of surface melting [16.16,23,45-49]. Most theories are phenomenological in that they assume the bulk and surface free energies to be known functions of the order parameter. Taking the Landau free-energy functional for a semi-infinite bulk as a starting point, *Lipowsky* and *Speth* [16.16] calculated the depth profile of the order parameter as a function of T_m-T. They showed, that, even though the bulk melting transition is first order, the surface may undergo a continuous order-disorder transition as T_m is approached. The disordered quasi-liquid surface layer wets the interface between the ordered phase and the vapor, and its thickness grows like $\ell \sim |\ln(T_m-T)|$ for the case of short-range forces. At the quasi-liquid/vapor interface the order parameter decays with temperature as $(1-T/T_m)^{\beta_1}$, with the value of the critical exponent β_1 being dependent on the degree of delocalization and roughness of the solid/quasi-liquid interface [16.16]. Very recently, a microscopic approach to the surface melting problem was developed, which is based on mean-field lattice theory [16.50]. When applied to a Lennard-Jones system with long-range interactions, the theory yields the growth law $N_{eq} \sim (T_m-T)^{-1/3}$ for the disordered layer thickness, with the thickness at a given temperature being larger on the open {110} face than on the {100} face.

16.3 Observations of Surface Melting

Since the beginning of the twentieth century [16.6] numerous attempts have been made to detect surface-initiated melting effects. However, many of the early experiments lacked the sensitivity to detect microscopic melt thicknesses and often there was no control on the presence of impurities at the surface. The latter is, of course, a major point of concern. Small amounts of impurities, segregated to the surface, may lower the melting point in a trivial way [16.51]. We also note, that some of the earlier studies, which were published as investigations of surface melting, were in fact concerned with the phenomenon of surface roughening [16.52].

It is only very recently, that the first direct observation of surface melting on a microscopic level was made on an atomically clean, well-characterized, surface [16.53,54]. For the detection of the positional lattice disorder, use was made of Rutherford backscattering of protons, in conjunction with shadowing and blocking [16.55]. Other techniques recently employed are calorimetry [16.56-61], ellipsometry [16.62,63], electron diffraction [16.64-66] and microscopy [16.67], neutron [16.68] and X-ray diffraction [16.69], quasi-elastic neutron [16.70] and He scattering [16.71,72] and last but not least, visual inspection with the naked eye [16.73,74]

In this section we shall discuss a number of these experiments along with descriptions of the methods upon which they are based. We do not attempt to place the subject in historical perspective, nor do we strive for completeness in our survey. References to the earlier work are given in reviews by *Nenow* [16.52] and *van der Veen* [16.15].

16.3.1 Calorimetry

This method is based on the measurement of anomalies in the specific heat near the melting point. In order to see a surface effect, one must use samples with a large surface to volume ratio. *Da-Ming Zhu* and *Dash* [16.56], in an investigation of the melting of argon surfaces, achieved this by adsorbing Ar films on exfoliated graphite foam placed in a calorimeter setup. Of course, the melting behaviour of a supported thin film is different from that of semi-infinite bulk, because of finite-size effects and interaction with the substrate, but it could be shown that this leads to rather small corrections for Ar film thicknesses larger than ~5 monolayers. Surface melting, if it indeed occurs, will introduce an anomaly in the heat capacity due to the conversion of solid into quasi-liquid. The heat absorbed in the conversion, per unit temperature interval and unit surface area, is given by

$$C(T) = L \frac{dN}{dT} \tag{16.12}$$

where L is the latent heat of melting per atom and N is the number of atoms per unit area in the quasi-liquid layer. Since the interatomic forces in the Ar crystal are of van der Waals type, N is given by (16.9), with p = 2. Substituting (16.9) in (16.12) one obtains

466

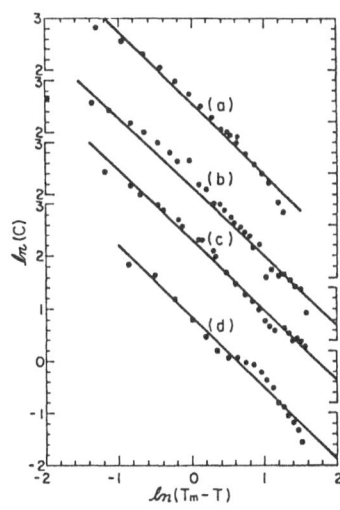

Fig.16.5. Heat capacities versus temperature for Ar/graphite films of increasing thickness in monolayers: (a) 10.5, (b) 8.9, (c) 8.0, and (d) 7.1. From [16.56] with permission.

$$C(T) = Q(T_m - T)^{-4/3} , \qquad\qquad (16.13)$$

where the constant Q is equal to $1/3(2T_m L^2 N_0{}^2 \Delta\gamma)^{1/3}$, but here of no further concern. Because C(T) diverges with a power law, it becomes the dominating term in the total heat capacity for temperatures close to T_m. A rise in C(T) with a power law has indeed been experimentally observed by *Da-Ming Zhu* and *Dash* [16.56]. In Fig. 16.5 ln(C) is plotted versus $\ln(T_m - T)$ for various film thicknesses. The exponent, which is given by the slope of the straight lines in the plot, was determined to be -1.36 ± 0.08. This is in excellent agreement with the theoretical exponent $-4/3$ (16.13). Very recently, the same group investigated edge melting of Ar monolayers. Again a power law was found, but with a different exponent due to the reduced dimensionality [16.57].

Anomalies in the specific heat were also detected on thin disks of the metals Ga and Na. These calorimetric measurements were performed by *Fritsch* et al. [16.58,59]. The anomaly for Ga was attributed to an intrinsic surface melting effect. In Na, dissolved impurities were found to play a role as well. Both surfaces were in contact with another material rather than their own vapor. This may have affected the results.

Willens et al. [16.60] performed differential scanning calorimetry on a stack of 20 Å Pb films sandwiched between layers of Ge. Anomalies below T_m were attributed to the melting of Pb-Pb grain boundaries in the polycrystalline films. The phenomenon of grain boundary melting is in nature quite similar to surface melting, and has received wide attention in the literature because of its implications for various high-temperature processes in metallurgy [16.28, 75-77].

16.3.2 Optical Measurements

The optical methods are relatively straightforward. They generally do not require ultrahigh vacuum conditions around the sample and are therefore ideal for studies of materials with high vapor pressure such as ice. The surface of ice has captured the interest of many researchers since Faraday's conjecture [16.78] that its slipperiness is caused by it being covered with a thin watery layer. Indeed, liquid-like layers have been detected below 0°C, with the use of ellipsometry [16.62,63]. This method involves the reflection of linearly polarized light from the surface and measuring the ellipticity coefficient of the reflected, elliptically polarized light. Because the refractive index of water is different from that of ice, the presence of a thin watery layer at the ice surface can be detected as a change in the reflective properties.

Furukawa et al. [16.63], using an ellipsometry setup in a cold room, investigated the melting of the {0001} and {10$\bar{1}$0} faces of a single crystal of ice and found the melted layer thickness to be smaller on the densely packed {0001} face than on the relatively open {10$\bar{1}$0} face (Fig.16.6). This finding is in qualitative agreement with the arguments in Sect.16.2.3. The refractive index of the quasi-liquid layer was found to be close to that of bulk water. A few questions, however, have remained unsolved. These concern the origin of the inflection of the {10$\bar{1}$0} surface melting curve in Fig.16.6 near -2°C (attributed by Furukawa et al. to a roughening of the solid/quasi-liquid interface) and the asymptotic dependence of layer thickness on T_m-T. A logarithmic growth with T_m-T was established by Fletcher [16.79], but Kuroda and Lacmann [16.80] predicted the growth law $\sim(T_m-T)^{-1/3}$, as expected for van der Waals bonding. A complicating factor is the strong polarity of the water molecules, which needs to be taken into account in the modeling of the intermolecular forces [16.81].

Ellipsometry has also been used in investigation of the melting of the {001} and {010} faces of a biphenyl crystal in contact with glass.

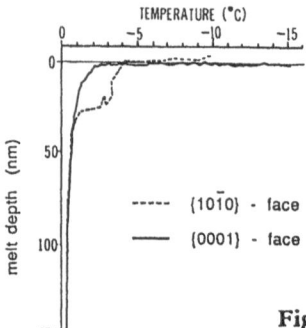

Fig.16.6. The melted layer thickness on the {10$\bar{1}$0} and {0001} faces of an ice crystal, as measured by ellipsometry. From [16.63] with permission.

468

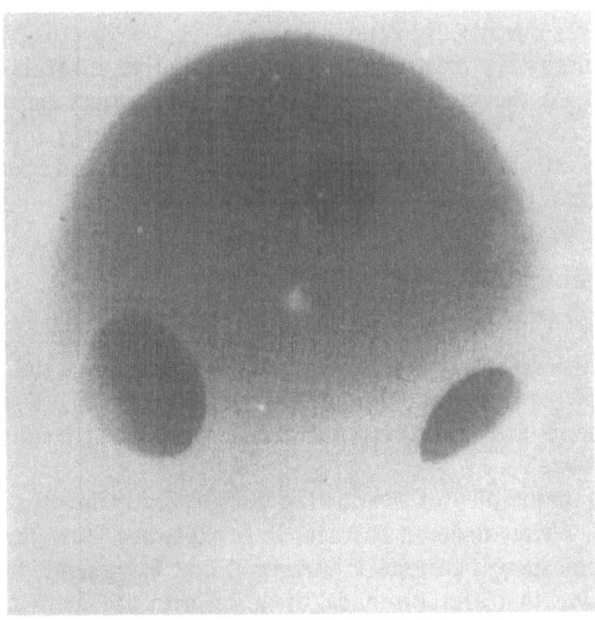

Fig.16.7. A melting sperical crystal of Cu, seen in its thermal light. The two black areas around the ⟨111⟩ poles and the smaller black area around the ⟨100⟩ pole are non-melting regions on the surface [16.73]

The initial growth was found to be logarithmic, but it changed to a power law close to T_m [16.82].

Surfaces of metals have received some attention as well. *Stock* and *Menzel* directly observed in a UHV setup the light emission from a spherical monocrystal of Cu [16.73, 74]. The solid and quasi-liquid phases were distinguished from one another by their difference in spectral emissitivity in the red and near-infrared wavelength range. Figure 16.7 shows a photograph of the glowing crystal, while it is placed in a small vertical temperature gradient around T_m (T increasing slightly from top to bottom). The gradual change from dark (solid) to bright (liquid) in the downward direction marks a transition region where the emissitivity is a superposition of emission from the solid bulk and emission from a surface melt of increasing thickness. However, well-defined circular regions around the ⟨111⟩ and ⟨100⟩ poles remain black, indicating the complete absence of a surface melt. The diameters of the non-melting region around {111} and {100} correspond to cone angles of 32° and 19°, respectively. The presence of these regions is related to cusps in the anisotropic specific free energy γ_{sv} of Cu around these directions, which cause the excess free energy $\Delta\gamma$ to change sign from positive to negative (see Sect.16.2.3). The observation of a larger region around {111} than around {100} is a consequence of the {111} cusp being deeper ($\gamma_{sv}\{111\} < \gamma_{sv}\{100\}$) [16.83]. From these optical emissitivity measurements no quantitative information could be obtained on the melt thickness and its temperature dependence. It was estimated by *Stock* [16.74] that a thickness of only 2 to 7 monolayers is sufficient to give a noticeable visual contrast and that this thickness range was reached at 1 to 2 K below T_m.

16.3.3 Rutherford Backscattering of Ions

This technique has monolayer sensitivity and allows for a quantitative determination of the disordered layer thickness [16.55]. To achieve this, use is made of shadowing and blocking effects in the crystal. The backscattering experiments are typically done with H^+ (or He^+) beams of medium energy (~100keV).

In Sect. 16.3.3a the principles of shadowing and blocking are discussed for the case of a well-ordered surface. It is explained how the backscattered yield from such a surface is measured and calculated. In Sect. 16.3.3b it is shown that the presence of a disordered, melted, surface layer leads to a substantially increased yield. Finally, some recent results of proton backscattering experiments on the melting of Pb surfaces are presented in Sects. 16.3.3c and d. The growth law of melted layer thickness is established and the predictions of Sect. 16.2.3 regarding the crystal-face dependence are tested.

a) Shadowing and Blocking

Surface sensitivity is obtained through the use of shadowing and blocking effects, which arise when a parallel beam of ions and a det-

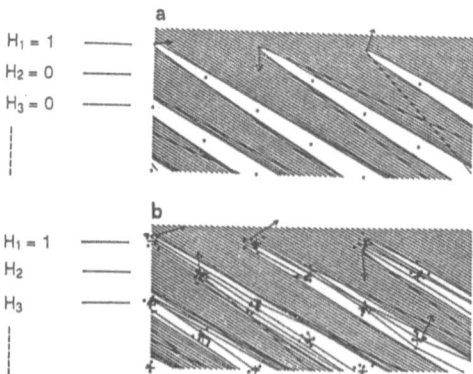

Fig. 16.8. Collection of computer-generated trajectories for an ion beam incident on a crystal along a low-index direction. Note the development of shadows behind the surface atoms and a few small-impact parameter collisions leading to backscattering. (a) Scattering from an ideal static lattice. (b) Scattering from a thermally vibrating lattice [16.55]

ector of backscattered ions are aligned with low-index directions in the crystal [16.55]. We first introduce the concepts of shadowing and blocking, then derive an expression for the backscattered yield from an ordered, non-melted, crystal surface.

To illustrate the shadowing effect, let us first consider an ideal static crystal with all atoms fixed on bulk lattice positions. The collection of ion trajectories in Fig.16.8a shows the development of the shadows behind the atom of the row with which the beam is aligned. The first atom is fully visible to the beam, i.e. it has a "hitting probability", H_1, of unity. The atoms further down the row have zero hitting probability $\{H_i=0, i\geq2\}$ since they are perfectly shadowed. In this hypothetical case only the top surface layer contributes to backscattering.

Now assume that the atoms are thermally vibrating but that the crystal is otherwise still ordered. Every new ion, as it travels along the row, then "sees" the atoms displaced from their equilibrium positions along the row. During the passage time of an ion, which is shorter than a typical vibration period by a factor of about $10^{-2}-10^{-3}$, the atoms remain essentially frozen in their thermally displaced positions. Figure 16.8b shows a computer-generated collection of superimposed "snaphots" of ion tracks in the surface region of the vibrating crystal. The shadows are still clearly present, but they are blurred. The hitting probabilities of the subsurface atoms have now become non-zero and are monotonically decreasing with layer number.

The shadow is formed by discrete small-angle deflections of the ions in the screened Coulomb field of the successive atoms along the row [16.84]. A sequence of possible scattering events is depicted schematically in Fig.16.9 for a single ion trajectory. A small-impact parameter backscattering event into the vacuum is assumed to take place from atom i along the row. It is further assumed that not only the incident beam direction e^1 but also the detection direction e^2 is aligned with an atomic row ("double alignment"). The alignment along the outgoing path causes a certain fraction of the ions backscattered from atom i to be blocked by the other atoms along the outgoing row. If we

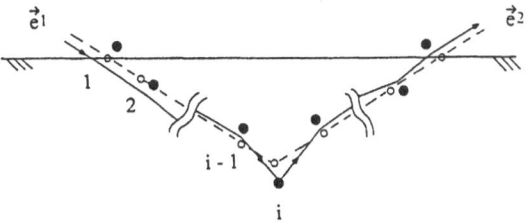

Fig.16.9. Ion trajectory in shadowing/blocking geometry. The black circles indicate the frozen thermally displaced positions of the atoms, the open circles the equilibrum positions. In- and outgoing directions of the ions are denoted by e^1 and e^2 [16.55]

define D_i as the probability for ions backscattered from ion i to be detected along e^2, then the joint probability of ions impinging along e^1 to backscatter from atom i and to be detected in the direction e^2, is to a very good approximation given by the product $L_i = H_i D_i$ [16.85]. The backscattering yield L is then given by

$$L = \sum_j L_j = \sum_j H_j D_j \qquad (16.14)$$

where the summation index j runs over all atoms j along the row for which the decreasing H_j and D_j are non-negligible. By definition, $H_1 = 1$ and $D_1 = 1$. The yield defined by (16.14) is in units of atoms per row. Alternatively, the backscattering yield may be expressed as an effective number of "visible" atoms per unit surface area by multiplying L with the areal density of atomic rows exposed to the beam.

The surface backscattering yield L is experimentally determined as follows [16.55]. Figure 16.10b shows schematically the energy spectrum of backscattered ions that is expected for the shadowing/blocking geometry of Fig.16.10a. The energy scale of the spectrum can be regarded as an inverted depth scale because of electronic stopping of the ions along their in- and outgoing paths [16.86]. The peak in the spectrum, the so-called surface peak, is composed of the decreasing backscattering contributions $L_i = H_i D_i$ from the successive layers. Its total area is equal to L. In an actual experiment the surface peak area is not directly measured in units of atoms per row (or atoms per unit area)

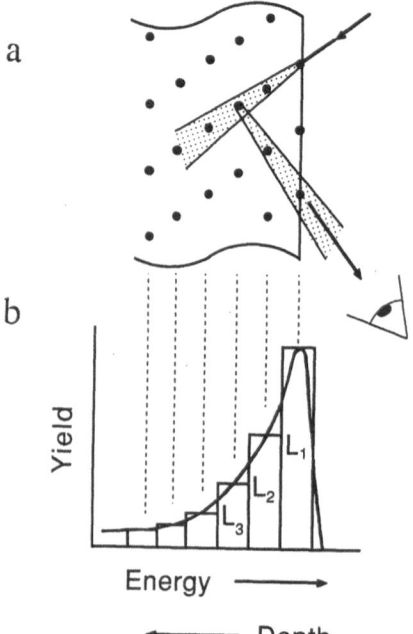

a

b

Yield

Energy ⟶

⟵ Depth

Fig.16.10. (a) Shadowing and blocking in the surface region of a well-ordered crystal. (b) The corresponding energy spectrum, composed of the different layer contributions L_i

but can be converted to these units by calibrating the measured peak area against a known Rutherford backscattering standard [16.55].

The hitting and detection probabilities H_i and D_i, and hence the total yield L, increase with temperature. This is caused by an increasing vibration amplitude of the atoms, rendering the shadowing and blocking less effective. It is of crucial importance to be able to calculate the rate at which L increases with temperature. A comparison with the measurements then allows us to distinguish a natural rise in L due to vibrations from an increase due to a melting-induced disordering of the surface.

The temperature dependence of H_i and D_i is calculated [16.87] by considering again the scattering process as sketched in Fig.16.9 for a single ion trajectory. Assume for each atom an isotropic and uncorrelated probability density of its thermal displacement of Gaussian form

$$G_j(\mathbf{x}_j) = \frac{1}{\sqrt{2\pi\sigma_j{}^3}} \exp\left[-\frac{|\mathbf{x}_j - \mathbf{x}_j{}^0|^2}{2\sigma_j{}^2}\right] , \tag{16.15}$$

where σ_j is the one-dimensional rms thermal displacement of atom j around its equilibrium position $\mathbf{x}_j{}^0$. The temperature dependence of σ_j is readily obtained from the Debye model. Suppose an ion enters the solid at some position $\mathbf{x}_0$. Before this ion reaches the backscattering atom i, it is deflected along the incoming path by atoms 1 to i-1, occupying the thermally displaced positions $\mathbf{x}_1$, $\mathbf{x}_2$,..., $\mathbf{x}_{i-1}$. The ion track approaches the equilibrium position of atom i closest at the passage point $\zeta_i \equiv \zeta_i(\mathbf{e}^1, \mathbf{x}_0, \mathbf{x}_1, \mathbf{x}_2,..., \mathbf{x}_{i-1})$. Atom i must be somewhere on the ion path if a backscattering collision is to take place. The probability density (PD) for this specific ion trajectory to lead to such a collision with atom i, is the product of the PD $G_1(\mathbf{x}_1) \cdot G_2(\mathbf{x}_2) \cdot ... G_{i-1}(\mathbf{x}_{i-1})$ for atoms 1 to i-1 to be at $\mathbf{x}_1$, $\mathbf{x}_2$,..., $\mathbf{x}_{i-1}$, with the (two-dimensional) PD $g_i(\zeta_i)$ for atom i to occupy a position somewhere on the ion path [16.87]. Integrating the PD for this trajectory over all possible ion impingement positions $\mathbf{x}_0$ in the surface plane and over all possible positions for atom 1 to i-1, we obtain the hitting probability of atom i:

$$H_i = \int..\int d^2\mathbf{x}_0 \left[\prod_{j=1}^{i-1} d^3\mathbf{x}_j \, G_j(\mathbf{x}_j)\right] g_i(\zeta_i) . \tag{16.16}$$

An analytical calculation of the integral in (16.16) is intractable, except for the case i = 2 [16.88-90]. The complication is largely hidden in the calculation of the passage position ζ_i, whose value is determined by the initial direction $\mathbf{e}^1$ and position $\mathbf{x}_0$ of the ion trajectory and by the sequence of reflections from the atoms positioned at $\mathbf{x}_1, \mathbf{x}_2,..., \mathbf{x}_{i-1}$. A natural way to evaluate the integral is by the Monte Carlo method. In this approach a large number of randomly chosen sets $\{\mathbf{x}_1, \mathbf{x}_2,..., \mathbf{x}_{i-1}\}$ is generated. The variable $\mathbf{x}_0$ runs over a regular mesh of sample points (or is randomly generated) in the surface plane and the random values $\mathbf{x}_1, \mathbf{x}_2,..., \mathbf{x}_{i-1}$ are drawn from the sampling distribution functions $G_i(\mathbf{x}_i), G_2(\mathbf{x}_2)...,$ and $G_{i-1}(\mathbf{x}_{i-1})$. For each set $\{\mathbf{x}_i, \mathbf{x}_2,..., \mathbf{x}_{i-1}\}$, the corresponding ion track and passage point ς_i, is evaluated, using for the calculation of the deflection angles the Molière scattering potential [16.84]. The hitting probability, H_i, is then obtained by summing the integrand $g_i(\varsigma_i)$ over a large number of random samples [16.91]

$$H_i = \frac{1}{\Phi} \sum_G g_i(\varsigma_i) , \qquad (16.17)$$

where Φ is the density of impingement positions $\mathbf{x}_0$ in a plane perpendicular to $\mathbf{e}^1$.

Expressions analogous to (16.16 and 17) can be derived for the detection probability, D_i, along the outgoing path. We note, that, because of time reversibility [16.92], the probability D_i of an ion backscattered from atom i to leave the crystal parallel to the direction $\mathbf{e}^2$ is equal to the probability H_i for the same atom to be hit by a beam incident along $-\mathbf{e}^2$.

Efficient computer codes have been developed for the evaluation of the $\{H_i\}$ and $\{D_i\}$, in which allowance is made for cross-overs of ions between neighboring atomic rows [16.87]. Input parameters in these calculations are the ion energy and ion type (H^+ or He^+), the thermal displacements $\{\sigma_j\}$, the equilibrium positions $\{\mathbf{x}_j^0\}$ and the in- and outgoing directions $\mathbf{e}^1$ and $\mathbf{e}^2$. The $\{\mathbf{x}_j^0\}$, which in Fig.16.9 were assumed to be lying exactly on a bulk crystal axis, may at the surface be displaced due to relaxation or reconstruction [16.55]. Moreover, the surface vibration amplitude σ_1 is, in general, substantially enhanced with respect to the bulk amplitude [16.5]. Both surface effects result in an enhanced value of L and should, therefore, be taken into account in any comparison with experimental data on surface melting.

b) Backscattering from a Melted Surface

As the surface region of the crystal becomes increasingly disordered with rising temperature, the shadowing and blocking effects will gradually vanish. Suppose the crystal is covered with a thin layer of completely disordered material, as is schematically indicated in Fig.16.11a. The atoms in this layer do no longer contribute to shadowing and

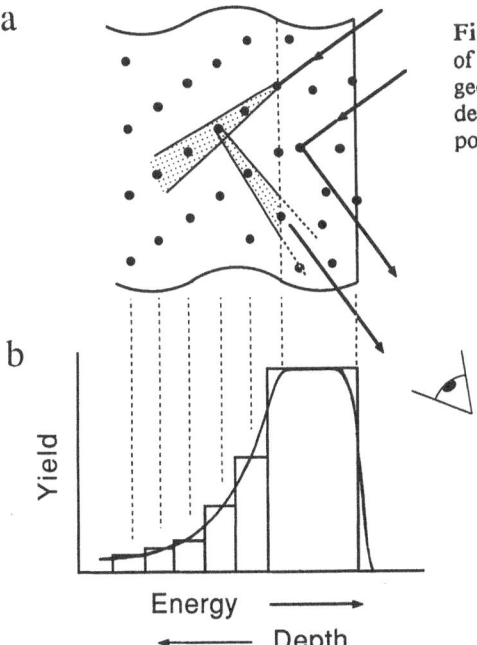

Fig.16.11. (a) Schematic representation of backscattering in shadowing/blocking geometry from a crystal with a disordered surface layer. (b) The corresponding energy spectrum

blocking and hence become fully visible to beam and detector. This causes the surface peak in the backscattering energy spectrum to broaden to lower energies as shown in Fig.16.11b. The corresponding increase in surface peak area directly yields the number of positionally disordered atoms per unit surface area. At still lower backscattering energies, the yield is significantly reduced by shadowing and blocking effects in the crystal underneath. From the shape of the surface peak, the depth distribution of the disorder is readily deduced, provided the energy (depth) resolution of the detector is high enough and the electronic stopping power is known.

We note that the shadowing/blocking technique is insensitive to roughening of the surface. On a roughened surface the atomic registry in the rows along the in- and outgoing directions is preserved. The only difference with respect to the smooth surface is that the atomic rows terminate at different heights. But this hardly influences the shadowing and blocking effects along the rows. It is only sideways displacements of the atoms away from the rows that leads to an increase in surface peak area. This insensitivity to roughening simplifies the interpretation of the data greatly.

The ion scattering measurements of surface melting were performed in the authors' laboratory using the equipment shown in Fig.16.12 [16.93,94]. An ion beam, collimated through the diaphragms, enters an ultrahigh vacuum chamber in which sample and detector are located. The sample is mounted on a high-precision goniometer [16.95]

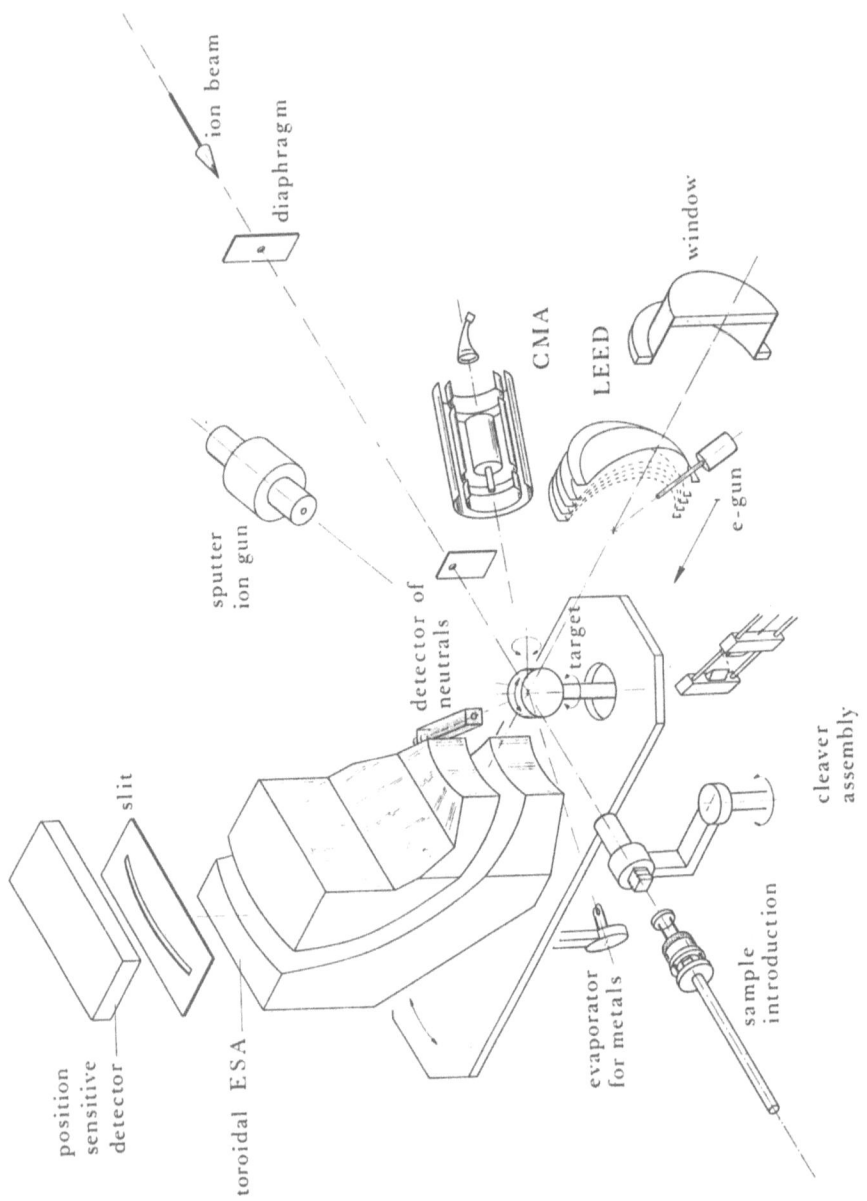

Fig.16.12. The experimental setup used for Rutherford backscattering studies of surface melting [16.55]

with three independent axes of rotation, allowing a crystal plane to be aligned with the horizontal scattering plane and a crystal axis in this plane to be aligned with the incident beam direction. A toroidal electrostatic analyzer [16.94] which is rotatable around the sample, enables

476

backscattered ions to be detected simultaneously in a 20° range of scattering angles centered around a selected blocking direction in the aligned crystal plane. Here, we will only consider the ions emerging from the crystal along a blocking direction. The backscattering setup also contains ancillary equipment for cleaning the surface (sputter-ion gun and sample heater) and characterizing its crystalline order (LEED) and cleanliness (cylindrical mirror analyzer for Auger electron spectroscopy). A special detector of neutrals serves to measure the neutralized fraction of backscattered projectiles. Knowledge of this fraction is required for calibration of the backscattered ion yield into the number of visible atoms per unit surface area [16.55].

c) Melting of Pb{110}

The element Pb is an obvious choice for surface melting studies (see Sect.16.2.3). The first microscopic observation of surface melting was made by Frenken et al. [16.53,54] on the {110} face of Pb. Surface peak areas were measured as a function of temperature using a beam of 98 keV protons in the shadowing/blocking geometry of Fig.16.13a. The number of visible monolayers, as determined from the surface peak area, first rises slowly with temperature (Fig.16.13b). This increase is due to an increasing thermal vibration amplitude of the atoms. Up to ~500 K, the measured number of visible monolayers is seen to be in perfect quantitative agreement with the number calculated for an ordered, thermally vibrating, lattice (solid line in

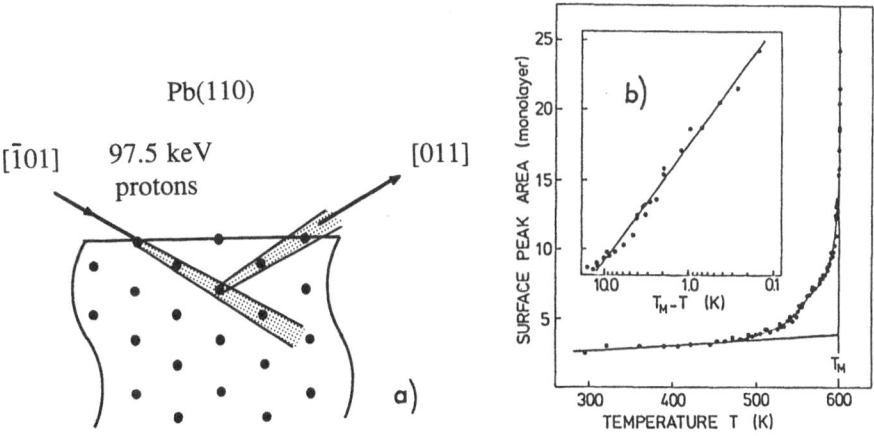

Fig.16.13. (a) Shadowing/blocking geometry used for the surface melting studies on Pb(110). (b) The surface peak area for Pb(110), expressed as the number of visible monolayers, as a function of temperature. The experimental conditions were as shown in Fig. 16.13a. The solid curve is the number of visible monolayers calculated by the Monte Carlo method for a well-ordered thermally vibrating crystal (see text). The inset is an expanded view of the highest 20 K interval, on a logarithmic scale [16.54]

Fig.16.13b). The calculations were performed using the Monte Carlo method discussed in Sect.16.3.3a. The bulk vibration amplitude σ_b in these calculations increases linearly from 0.18 Å at 300 K to 0.28 Å just below T_m [16.96] and the first-layer amplitude σ_1 is enhanced by 50% with respect to σ_b. The equilibrium positions of the atoms in the topmost layers are chosen to be in correspondence with previously determined relaxations of -15.9, +7.9, -6.8, and +0.7% of the first four interlayer spacings, respectively [16.97].

Above a temperature of 500 K, the measured yield starts to deviate significantly from the calculated yield and finally it diverges as T_m is approached. Close to T_m the number of disordered monolayers N increases logarithmically with T_m-T (inset of Fig.16.13b). This is precisely the growth law (16.6) expected for a system governed by short-range interactions. A good fit to the data is obtained with $N = N_0 \ln[T_0/(T_m-T)]$, where $T_0 = 55$ K and $N_0 = 3.6$ monolayers ($= 2.1 \cdot 10^{15}$ atoms/cm^2 or 6.3 Å). As expected, the parameter N_0 is of microscopic dimensions. The short-range nature of the interatomic interactions in a metal such as Pb is due to screening by the conduction electrons.

The depth resolution in the experiment was sufficiently high (7 Å) that the disordering of the surface region can be seen as a broadening of the surface peak. A collection of energy spectra taken at different temperatures is shown in Fig.16.14. Close to T_m the spectra assume the same shape as sketched in Fig.16.11b, giving direct evidence for the presence of a disordered film on top of a well-ordered substrate. Note, that the surface peak at the highest temperature (600.5 K, curve f)

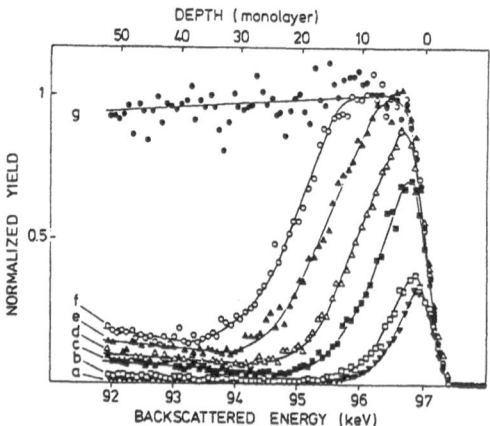

Fig.16.14. Energy spectra obtained with 97.5 keV protons in the scattering geometry of Fig. 16.13a, and calibrated with respect to the height measured at the surface of a bulk liquid (curve g): (a) 295 K, (b) 452 K, (c) 581 K, (d) 597 K, (e) 599.7 K, (f) 600.5 K, and (g) 600.8 K. The melting point is at $T_m = 600.7$ K. Solid curves serve to guide the eye [16.54]

attains a height close to that measured on a bulk liquid (curve g). These observations lead us to the following speculations regarding the microscopic structure of the disordered layer. Let us assume that the atoms in the disordered layer have Gaussian position distributions (16.15) around their original lattice positions but with much higher rms displacement, σ, than the thermally vibrating atoms in the ordered crystal underneath. A good fit to the area and shape of the surface-peak of spectrum f at 600.5 K is obtained if in the Monte Carlo calculation the atoms in the top 16 monolayers of the crystal are given an rms displacement of $\sigma = 1.0$ Å. This value is much higher than the bulk vibration amplitude of 0.28 Å at that temperature. In fact, it is so high, that the position distributions around neighboring lattice sites have substantial overlap (Fig.16.15). The probability density for an atom to be halfway between lattice sites is ~50% of that for an atom to be on lattice site. Hence, the atoms will diffuse readily within the layer, giving its properties intermediate between solid and liquid. This description in terms of a quasi-liquid is fully consistent with the ideas developed in Sect.16.2.1. We note, however, that other models for the position distribution may also give satisfactory fits to the data. In particular, the degree of disorder is likely to vary across the layer thickness [16.16] and a model featuring an increasing σ toward the surface would be reasonable. At the surface itself, the single-atom position distribution may be essentially flat, giving it strongly liquid-like properties.

While the logarithmic growth regime from 580 K to T_m is well understood, relatively little is known about the mechanism by which the surface begins to disorder. In the premelting regime from 450 K to 580 K the surface region disorders only gradually (Fig.16.13b). No specific growth law can be identified in this temperature interval. Above 580 K this premelting surface forms a partially disordered transition region between the well-ordered crystal and the strongly disordered film which builds up on top.

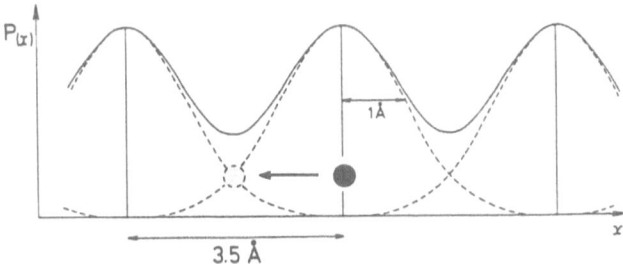

Fig.16.15. Gaussian position distribution along a (110) row of Pb with a rms displacement of 1.0 Å from lattice sites (dashed curves). The summed distribution (full curve) indicates the amount of order which may still be present in the quasi-liquid layer. An atom can easily diffuse from one site to the next [16.72]

479

d) Melting or Non–Melting of Pb{hkl}

Of all crystal faces of Pb the {111} face has the lowest specific surface free energy [16.38]. In Sect.16.2 it was predicted on the basis of thermodynamic arguments that this face does not melt. Here we discuss recent ion scattering experiments in which these predictions are tested [16.98,99]. In particular, we will discuss results obtained on a range of surface orientations along the ⟨110⟩ zone of the stereographic triangle, including the {111} face.

The thermal behavior of the Pb{111} surface was investigated using the backscattering geometry shown in the inset of Fig.16.16 [16.98]. The measured yield was found to closely follow the calculated yield for all temperatues up to the bulk melting point T_m. The Monte Carlo calculations were performed for an ordered unrelaxed surface having a 15% enhanced vibration amplitude. There is no sign of an intrinsic surface melting effect. The persistence of lattice order at the surface is also evident from the fact that, up to T_m, the measured surface peaks retain their shape and the angular distributions of backscattered protons exhibit a clear blocking minimum around the [001] exit direction.

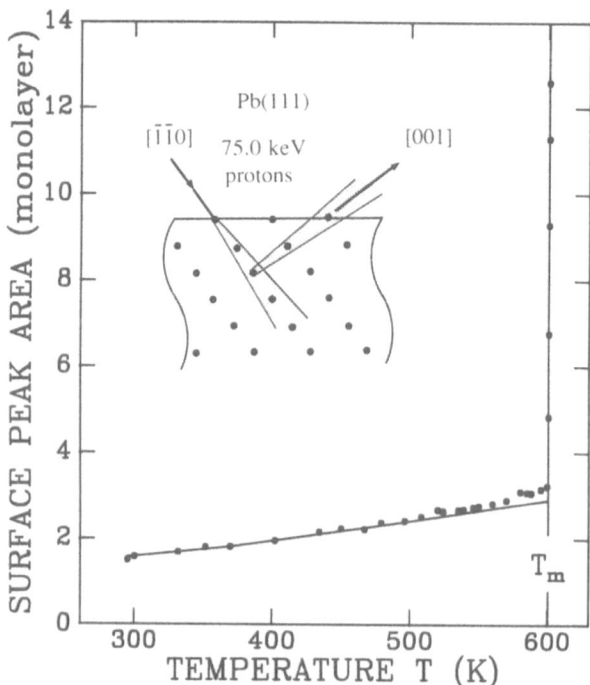

Fig.16.16. The surface peak area for Pb(111), expressed as the number of visible monolayers, as a function of temperature. The experimental conditions are given in the inset. The solid curve is the number of visible monolayers calculated by the Monte Carlo method for a well-ordered thermally vibrating crystal (see text) [16.98]

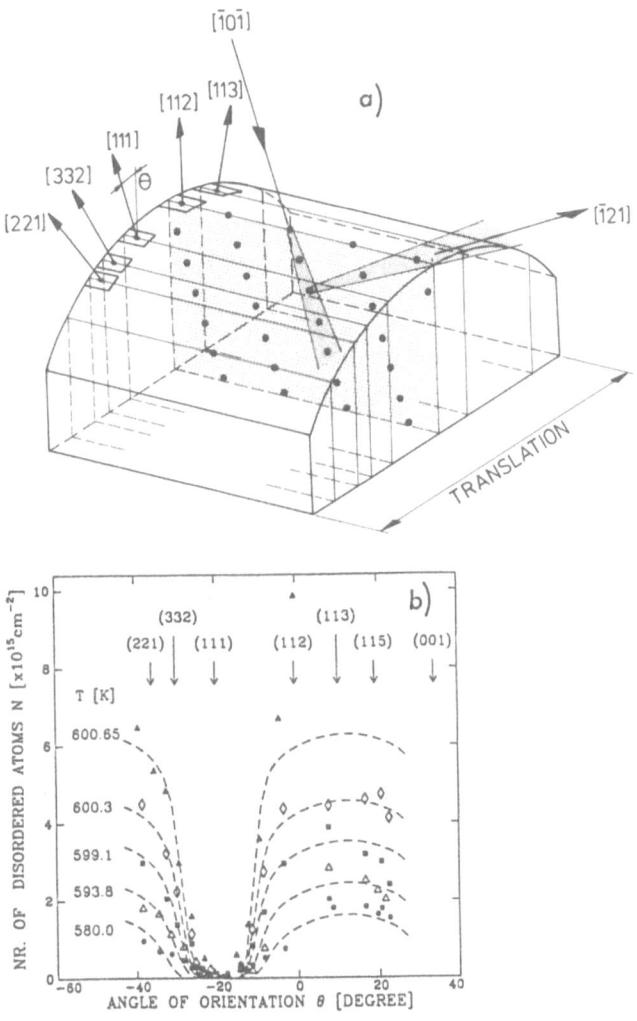

Fig.16.17. (a) Shadowing/blocking geometry in the $(11\bar{1})$ scattering plane of a cylindrically shaped Pb crystal which exposes a 73° range of surface orientations along the $[\bar{1}10]$ zone of the stereographic triangle [16.99]. (b) The number of disordered atoms per unit area N as a function of the surface orientation angle θ with respect to the [112] axis, measured at various temperatures starting from ~20 K below the bulk melting point of Pb ($T_m = 600.7$ K). The experimental conditions were as shown in Fig.16.17a. The dashed curves represent the optimal fit of (16.6) to the data (see text). From [16.99].

Further investigations of the crystal face dependence of surface melting were carried out with cylindrically shaped crystals [16.99, 100]. One such crystal is schematically shown in Fig.16.17a, along with the scattering geometry used in that experiment. The crystal exposes a 70° range of orientations within the $[\bar{1}10]$ zone of the stereographic triangle. The orientation of a particular face on the cylinder is defined by

481

the angle θ between its normal and the [112] direction. The dependence on surface orientation was investigated by translating the crystal at fixed T in discrete steps and measuring at each setting the number $L(T,\theta)$ of visible atoms per row for the corresponding orientation angle θ. In this procedure the alignement of beam and detector with respect to the [10$\bar{1}$] and [$\bar{1}$21] crystal axes is preserved. Hence, for a well-ordered surface, L essentially remains unchanged upon translation (apart from minor variations related to an orientation-dependent surface relaxation and vibration amplitude). For a melted surface, however, the measured $L(T,\theta)$ will be higher and also strongly dependent on θ and T. The number of disordered atoms per unit area $N(T,\theta)$ follows from $N(T,\theta) = N_s[L(T,\theta)-L_{ord}(T)]\cdot\cos\theta$. Here, N_s is the areal density of atomic rows terminating the (112) crystal face and $L_{ord}(T)$ is the number of atoms per row calculated by the Monte Carlo method for an ordered surface. In the calculation the surface was again assumed to be bulklike (no relaxation) and to have a 15% enhanced vibration amplitude.

For temperatures up to 480 K the measured $L(T,\theta)$ values are equal to the calculated $L_{ord}(T)$ values for all orientations on the cylinder surface, i.e., the surface remains well-ordered. Above 480 K the number of disordered atoms $N(T,\theta)$ becomes non-zero, except for a limited range of orientations centered around (111), which does not exhibit any disordering. The orientation dependence of $N(T,\theta)$ is plotted in Fig.16.17b for different temperatures ranging from 20 to 0.05 K below T_m. A pronounced minimum is observed around (111).

We now show, that the data of Fig. 16.17b are in excellent agreement with the predictions of (16.6). The parameter controlling the melting behavior is the anisotropic excess specific free energy $\Delta\gamma = \gamma_{sv}-\gamma_{sl}-\gamma_{lv}$. The term $\gamma_{sv}(\theta)$ in the excess specific free energy follows from the anisotropic ratio $\gamma_{sv}(\theta)/\gamma_{sv}^{\{111\}}$, as meausred by *Heyraud* and *Métois* (Fig.16.3) [16.38], and from the literature value for $\gamma_{sv}^{\{111\}}$ of 0.544 J/m^2 [16.33]. The solid/liquid interface energy $\gamma_{sl}(\theta)$ is obtained with the empirical rule $\gamma_{sl}(\theta) = 0.1\gamma_{sv}(\theta)$ [16.31]. The liquid/vapor interface energy is treated as a free parameter to be determined from fitting the expression for N_{eq} in (16.6) to the data in Fig.16.17b. Another parameter in the fit is N_0. The best fit to all $N(T,\theta)$ curves is obtained for $N_0 = 7.32\cdot10^{14}$ cm^{-2} and $\gamma_{lv} = 0.501$ J/m^2 (dashed lines). The latter value is close to the value of 0.46 J/m^2 known from the literature [16.32]. To allow for a comparison with the data, the best-fit $N(T,\theta)$ curves were convoluted with a Gaussian spread in θ, having a full-width-at-half-maximum of 3.6°. This corresponds to the area over the cylinder surface sampled by the proton beam.

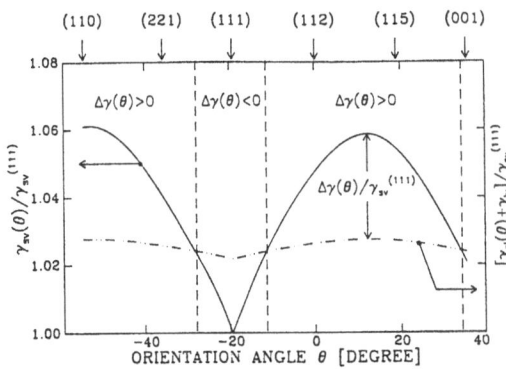

Fig.16.18. Graphical representation of the surface melting model for Pb. The solid curve represents the normalized specific free energy of the Pb solid/vapor interface $\gamma_{sv}(\theta)/\gamma_{sv}^{(111)}$, measured by *Heyraud* and *Métois* at T = 473 K (Fig.16.3 and [16.38]). The dash-dotted curve represents the normalized sum $[\gamma_{sl}(\theta)+\gamma_{lv}]/\gamma_{sv}^{(111)}$ [16.99]

A graphical representation of our melting model is given in Fig.16.18. It shows the curve for $\gamma_{sv}(\theta)$ measured by *Heyraud* and *Métois* at 470 K, and the sum $\gamma_{sl}(\theta) + \gamma_{lv}$, all normalized to the specific free energy of the (111) face. There is a well-defined zone around (111), for which $\Delta\gamma(\theta) < 0$. This zone does not exhibit surface melting. The graphical model predicts the (001) orientation to be on the verge of melting. Recent measurements on a cylindrical crystal containing a range of orientations in the [001] zone reveal indeed the existence of a narrow non-melting zone around (100) [16.100].

16.3.4 Diffraction

Whereas ion scattering monitors the short-range disordering of the lattice, diffraction techniques are sensitive to changes in long-range order. In this respect, the two methods are complementary. Surface sensitivity in diffraction is obtained through the use of low-energy electrons or beams of He atoms.

The first low-energy electron diffraction (LEED) experiment explicitly dealing with surface melting was performed in 1970 by *Goodman* and *Somorjai* [16.101] on the {111}, {100}, and {110} faces of Pb, the {110} face of Sn, and the {0001} and {01$\bar{1}$2} faces of Bi. Diffraction patterns were observed up to the bulk melting point and so it was concluded by these authors that the surfaces remained essentially ordered (a low concentration of disordered atoms could not be excluded). The negative finding for Pb{110} contradicts the ion scattering results by *Frenken* et al. [16.53,54]. Very recently, the controversy has been resolved by *Prince* et al. [16.66], who repeated the LEED experiment on the {110} and {111} surfaces of Pb and found the earlier LEED result for the {110} face to be in error. From ~50 K below T_m onwards the diffracted intensities from Pb{110} decreased much faster with temperature than is expected on the basis of the Debye-Waller factor, indicating substantial disordering. For Pb{111}, however, the intensities did follow the Debye-Waller temperature dependence up to T_m. These findings are fully consistent with the ion scattering results.

Similar LEED studies were performed by von Blanckenhagen et al. [16.65] on Al{110}. The diffracted intensities from this surface started to decrease anomalously at $\simeq 620$ K (for Al $T_m = 933$ K). Temperature-dependent He diffraction measurements by *Armand* et al. [16.102] on the {100} surface of Cu also showed an anomalous drop of the specular intensity. However, interpreting the latter data as evidence of surface melting would contradict the optical measurements by *Stock* and *Menzel* [16.73,74] (Fig.16.7). On the other hand, photoelectron diffraction measurements by *Trehan* and *Fadley* [16.103] on Cu{100} showed no sign of a disordering. In this experiment, however, the investigated temperature range was too limited to expect any significant disordering ($T/T_m < 0.74$).

As a general note we add that some of the above diffraction results may also be explained by anharmonic surface vibrations or surface roughening. A distinction between the different types of disorder, is, in principle, possible by analyzing the intensities and profiles of different (hk) reflections (see also Chap.14).

An unusual type of order-disorder transition has recently been observed on the Ge{111} surface [16.64]. The results indicated a loss of lateral crystalline order at 150 K below T_m, but, surprisingly, the disordering was found to remain restricted to the first few atomic layers only. This "blocked" melting behavior cannot be explained with the thermodynamic model of Sect.16.2, which even predicts the Ge surface to remain well-ordered up to T_m (Table 16.1). On the other hand, recent molecular dynamics simulations have indicated the possibility of blocked surface melting [16.104].

Finally, varous X-ray diffraction studies have been made of surface disordering phenomena. Using X-rays from a rotating anode source, Kouchi et al. [16.69] observed a clear melting effect at the surface of ice. This result corroborates earlier ellipsometric observations [16.63] (see also Sect.16.3.2). X-ray diffraction using synchrotron radiation [16.105] and electron diffraction [16.106] have also been applied to the study of various order-disorder transformations in thin films and overlayers, but the interpretation of these results in terms of surface melting is complicated by substrate interactions and finite-size effects.

16.3.5 Neutron and He Scattering

With the use of quasi-elastic scattering techniques, the diffusivity of the atoms in the disordered surface can be determined. Measurements of the atom mobility help us to decide whether the quasi-liquid film behaves like a true liquid or rather like a strongly disordered solid. The principle of the method is based on the fact that the elastic peak in the energy distribution of reflected neutrons or He atoms is somewhat broadened because of inelastic interactions with the moving atoms. In order to see an effect at all, a very monochromatic incident beam is required ($\leq 100~\mu eV$ width). In principle, by measuring the amount of broadening as a function of momentum transfer parallel to the surface, information about the pair correlation function of the diffusing species can be obtained.

Quasi-elastic neutron scattering measurements of surface melting were performed by *Bienfait* [16.70] on thin {100} oriented CH_4 films. By adsorbing these films on a powder of MgO{100}, a large surface to volume ratio was obtained. This maximizes the surface signal, which otherwise would be unobservable because of the very weak cross section for scattering of neutrons. The energy spectra revealed contributions from both quasi-liquid and solid, with the broad feature from the quasi-liquid growing at the expense of the sharp elastic peak from the solid as T_m is approached (Fig.16.19). The limited data base did not allow for a determination of the growth law, but from the quasielastic broadening the temperature-dependent translational diffusion coefficient could be derived. Close to T_m ($T_m = 90.7$ K) the atom mobility at the surface was found to exceed that in the bulk liquid by a factor of 2.5. The enhanced mobility reflects the reduced bonding of the surface molecules.

The first successful quasi-elastic He scattering experiment was recently reported by Frenken et al. [16.71,72]. A significant advantage of using He atoms instead of neutrons is that He atoms of thermal

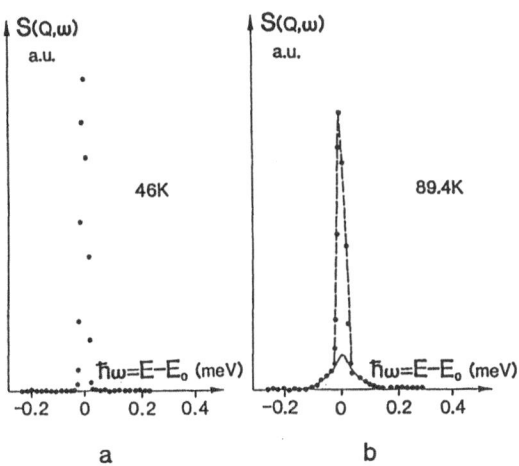

Fig.16.19. Measured energy distributions of neutrons scattered from a ~10 layer thick film of methane absorbed on MgO{100}, for a scattering vector $Q = 0.48$ Å^{-1}. a) Spectrum from the solid film at 46 K; b) Spectrum at 89.4 K (1.3 K below T_m), showing scattering contributions from 3.6 layers of quasi-liquid (broad solid curve) on top of the solid phase (sharp dashed curve). From [16.70] with permission.

energies interact exclusively with the atoms in the outermost atomic layer because of their high scattering cross section. The instrumental energy resolution could be improved to the point that the broadening of the reflected peak beomes measurable. Measurements on the Pb{110} surface in the temperature range 300<T<550 K revealed a liquid-like mobility at the surface at 50 K below T_m. The activation energy for diffusion was found to be intermediate between those for liquid and solid Pb, as expected.

16.4 Summary and Outlook

The existence of an intrinsic surface melting effect has now been established unambiguously by a variety of experimental techniques. The most direct evidence has come from ion scattering experiments on Pb crystals. In these experiments the temperature-dependent disordering of the Pb surfaces could be followed with monolayer sensitivity. The disordered layer thickness was found to diverge as $|\ln(T_m-T)|$, as predicted by a simple thermodynamic model. Remarkable is the strong crystal-face dependence of the disordering: whereas the open {110} face of Pb melts, the {100} and {111} faces do not. It was concluded that surface melting is driven by a positive orientation-dependent excess specific surface free energy $\Delta\gamma \equiv \gamma_{sv} - \gamma_{sl} - \gamma_{lv}$.

Of the elements listed in Table 16.1, for which surface melting is predicted to occur, only the metals Al, Cu and Pb have been investigated under well-defined experimental conditions. A disordering of the open crystal faces is indeed observed for these elements. It is not yet known which of the other elements listed in Table 16.1 exhibit melting on their open crystal faces. If the latter is generally the case, then we have identified a microscopic phenomenon of important macroscopic consequences. The surface of a solid of finite dimensions will then always contain regions where $\Delta\gamma > 0$. In all these cases, melting of the bulk solid will commence at the surface. Obviously, impurities or oxides, etc., which may be present on the surface under practical circumstances, modify the interfacial free energies and hence promote or suppress surface melting. The influence of such additives on the surface melting process needs to be further investigated.

A measurement of the critical behavior of the order parameter at the quasi-liquid surface will be of crucial importance for a proper understanding of the asymptotic surface melting regime. The most attractive experimental technique for that purpose appears to be two-dimensional X-ray diffraction. Obviously, a high X-ray intensity will be needed, but as more powerful surface diffraction equipment becomes available at synchrotron radiation sources, such experiments should become feasible. High-resolution LEED offers attractive possibilities too [16.107], but multiple scattering effects may complicate the interpretation of the results.

Still very little is known about the density profile $\rho(z)$ across the solid/quasi-liquid interface. Ion scattering measurements on Pb{110} have revealed the existence of a few monolayers wide transition region in which the crystalline order is gradually lost [16.54]. Measurements on other faces of Pb suggest that $\rho(z)$ depends on crystal orientation. This phenomenon is currently under investigation.

Acknowledgements. We have benefitted from discussions with J.W.M. Frenken, E.Tosatti, D. Frenkel and H.P. Bonzel. This work is part of the research program of the Stichting voor Fundamenteel Onderzoek der Materie (Foundation for Fundamental Research on Matter) and was made possible by financial support from the Nederlandse Organisatie voor Zuiver Wetenschappelijk Onderzoek (Netherlands Organization for Pure Scientific Research).

References

16.1   R.C. Weast (ed.): *Handbook of Chemistry and Physics* (CRC Press, Boca Raton, Fla 1981)

16.2   See, e.g., P.M. Morse: *Thermal Physics* (Benjamin-Cummings, Menlo Park, Calif.1964)

16.3   R.M. Cotterill, E.J. Jensen, W.D. Kristensen: In *Anharmonic Lattices, Structural Transitions and Melting*, ed. by T. Riste (Noordhoff, Leiden 1974) and references therein

16.4   F.A. Lindemann: Z. Phys. **14**, 609 (1910)

16.5   M.G. Lagally: In *Surface Physics of Materials*, Vol 2, ed. by J.M. Blakely (Academic, New York 1971) p.419

16.6   G. Tammann: Z. Phys. Chem. **68**, 205 (1910); Z.Phys. **11**, 609 (1910)

16.7   M. Volmer, O. Schmidt: Z. Phys. Chem. **B35**, 467 (1937)

16.8   I.N. Stranski: Z.Phys. **119**, 22 (1942)

16.9   I.N. Stranski: Naturwissenschaften **28**, 425 (1942)

16.10 L. Pietronero, E. Tosatti: Solid State Commun. **32**, 255 (1979)

16.11 C.S. Jayanthi, E. Tosatti, A. Fasolino, L. Pietronero: Surf. Sci. **152/153**, 155 (1985)

16.12 W. Burton, N. Cabrera, F.C. Frank: Philos. Trans. R. Soc. London **A143**, 199 (1951)

16.13 J.D. Weeks: In *Ordering in Strongly Fluctuating Condensed Matter Systems*, ed. by T. Riste (Plenum, New York 1980) p.293

16.14 F.S. Rys: Surf. Sci. **178**, 419 (1986)

16.15 J.F. van der Veen, J.W.M. Frenken: Surf. Sci. **178**, 382 (1986) and references therein

16.16 R. Lipowsky, W. Speth: Phys. Rev. **B28**, 3983 (1983)

16.17 P. Thiry, G. Jezequel, Y. Petroff: J. Vac. Sci. Technol. **A5**, 892 (1987)

16.18 S.E. Kaykin, N.P. Bene: (Dokl.) Acad. Sci. URSS **23**, 31 (1939)

16.19 N.G. Ainslie, J.D. Mackenzie, D. Turnbull: J. Chem. Phys. **65**, 1718 (1961)

16.20 R.L. Cormia, J.D. Mackenzie, D. Turnbull: J. Appl. Phys. **34**, 2239 (1963)

16.21 C.J. Rossouw, S.E. Donnelly: Phys. Rev. Lett. **55**, 2960 (1985)

16.22 J. Daeges, H. Gleiter, J.H. Perepezko: Phys. Lett. **A119**, 79 (1986)

16.23 S. Dietrich: In *Phase Transitions and Critical Phenomena*, Vol. 12 (Academic, London) to be published

16.24 D.P. Woodruff: *The Solid-Liquid Interface* (Cambridge Univ. Press, Cambridge 1973) and references therein

16.25   R. Lipowsky: Z. Phys. B55, 335, 345 (1984)
16.26   R. Lipowsky: Phys. Rev. Lett. 57, 2876 (1986)
16.27   See, e.g., P.G. de Gennes: Rev. Mod. Phys. 57, 827 (1985)
16.28   G.F. Bolling: Acta Metall. 16, 1147 (1968)
16.29   J.K. Kristensen, R.M.J. Cotterill: Philos. Mag. 36, 437 (1977)
16.30   D.R.H. Jones: J. Mater. Sci. 9, 1 (1974) and references therein
16.31   A.R. Miedema, F.J.A. den Broeder: Z. Metallkd. 70, 14 (1979) and refer-
         ences therein
16.32   A.R. Miedema, R. Boom: Z. Metallkd. 69, 183 (1978)
16.33   A.R. Miedema: Z. Metallkd. 69, 287 (1978) and references therein
16.34   M. McLean, H. Mykura: Surf. Sci. 5, 466 (1966)
16.35   E.D. Hondros, M. McLean: Colloq. Int. CNRS 187, 219 (1970)
16.36   M. McLean: Acta Metall. 19, 387 (1971)
16.37   T. Barsotti, J.M. Bermond,M. Drechsler: Surf. Sci. 146, 467 (1984)
16.38   J.C. Heyraud, J.J. Métois: Surf. Sci. 128, 334 (1983)
16.39   J.C. Heyraud, J.J. Métois: Surf. Sci. 177, 213 (1986)
16.40   G. Wulff: Z. Kristallogr. 34, 449 (1901)
16.41   E.E. Gruber, W.W. Mullins: J. Phys. Chem. Solids 28, 875 (1967)
16.42   J.Q. Broughton, G.H. Gilmer: Acta Metall. 31, 845 (1983)
16.43   J.Q. Broughton, G.H. Gilmer: J. Chem Phys. 79, 5095, 5105, 5119 (1983)
16.44   V. Pontikis, P. Sindzingre: Phys. Scr. 19 B, 375 (1987)
16.45   E. Tosatti,: In *The Structure of Surfaces II*, ed. by J.F. Van der Veen, M.A.
         Van Hove, Springer Ser. Surf. Sci., Vol. 11 (Springer, Berlin, Heidelberg
         1988) p.535 and references therein
16.46   A.C. Levi, E. Tosatti: Surf. Sci. 178, 425 (1986)
16.47   A.C. Levi, E. Tosatti: Surf. Sci. 189/190, 641 (1987)
16.48   A. Trayanov, D. Nenow: J. Cryst. Growth 74, 375 (1986)
16.49   R.Ts. Lyzhva, A.Ya. Mitus, A.Z. Patashinskil: Sov. Phys.-JEPT 54, 1168
         (1981)
16.50   A. Trayanov, E. Tosatti: Phys. Rev. Lett. 59, 2207 (1987)
16.51   See, e.g., M. Hansen, K. Anderko, R.E. Elliot: *Constitution of Binary Alloys,
         First Supplement*; F.A. Shunk: *Second Supplement* (McGraw-Hill, New York
         1985)
16.52   D. Nenow: Prog. Cryst. Growth Charact. 9, 1893 (1984) and references
         therein
16.53   J.W.M. Frenken, J.F. van der Veen: Phys. Rev. Lett. 54, 134 (1985)
16.54   J.W.M. Frenken, P.M.J. Marée, J.F. van der Veen: Phys. Rev. B34, 7506
         (1986)
16.55   J.F. van der Veen: Surf. Sci. Rep. 5, 199 (1985)
16.56   Da-Ming Zhu, J.G. Dash: Phys. Rev. Lett 57, 2959 (1986)
16.57   Da-Ming Zhu, J.G. Dash: Phys. Rev. B 37, 5586 (1988)
16.58   G. Fritsch, R. Lachner, H. Diletti, E. Lüscher: Philos. Mag. A46, 829 (1982)
16.59   G. Fritsch, H. Diletti, E. Lüscher: Philos. Mag. A50, 545 (1984)
16.60   R.H. Willens, A. Kornblit, L.R. Testardi, S. Nakahara: Phys. Rev. B25, 290
         (1982)
16.61   J.L. Tallon, W.H. Robinson, S.I. Smedley: J. Phys. Chem. 82, 1277 (1978)
16.62   D. Beaglehole, D. Nason: Surf. Sci. 96, 357 (1980)
16.63   Y. Furukawa, M. Yamato, T. Kuroda: J. Cryst. Growth 82, 665 (1987)
16.64   E.G. McRae, R.A. Malic: Phys. Rev. Lett. 58, 1437 (1987)
16.65   P. von Blanckenhagen, W. Schommers, V. Voegele: J. Vac. Sci. Technol. A5,
         649 (1987)
16.66   K.C. Prince, U. Breuer, H.P. Bonzel: Phys. Rev. Lett. 60, 1146 (1988)
16.67   G. Devaud, R.H. Willens: Phys. Rev. Lett. 57, 2683 (1986)
16.68   J. Krim, J.P. Coulomb, J. Bouzidi: Phys. Rev. Lett. 58, 583 (1987)

16.69 A. Kouchi, Y. Furukawa, T. Kuroda: J. de Phys., C1, Suppl. 3, **48**, 675 (1987)

16.70 M. Bienfait,: Europhys. Lett. **4**, 79 (1987); M. Bienfait, J.P. Palmari: In *The Structure of Surfaces II*, ed. by J.F. van der Veen, M.A. Van Hove, Springer Ser. Surf. Sci., Vol 11 (Springer, Berlin, Heidelberg 1988) p.559

16.71 J.W.M. Frenken, J.P. Toennies, Ch. Wöll: Phys. Rev. Lett. **60**, 1727 (1988)

16.72 J.W.M. Frenken, J.P. Toennies, Ch. Wöll, B. Pluis, A.W. Denier van der Gon, J.F. van der Veen: In *The Structure of Surfaces II*, ed. by J.F. van der Veen, M.A. Van Hove, Springer Ser. Surf. Sci, Vol. 11 (Springer, Berlin, Heidelberg 1988) p.545

16.73 K.D. Stock, E. Menzel: Surf. Sci. **61**, 272 (1976)

16.74 K.D. Stock: Surf. Sci. **91**, 655 (1980)

16.75 R. Kikuchi, J.W. Cahn: Phys. Rev. **B21**, 1893 (1980)

16.76 J.Q. Broughton, G.H. Gilmer: Phys. Rev. Lett. **56**, 2692 (1986)

16.77 T. Nguyen, P.S. Ho, T. Kwok, C. Nitta, S. Yip: Phys. Rev. Lett. **57**, 1919 (1986)

16.78 M. Faraday: Proc. R. Soc. London **10**, 440 (1860)

16.79 N.H. Fletcher: Surf. Sci. **115**, L103 (1982)

16.80 T. Kuroda, R. Lacmann: J. Cryst. Growth **56**, 189 (1982)

16.81 D. Nason: J. Chem. Phys. **64**, 3930 (1976)

16.82 A.A. Chernov, V.A. Yakovlev: JETP Lett. **45**, 160 (1987)

16.83 M. McLean: Acta Metall. **19**, 387 (1971)

16.84 G. Molière: Z. Naturforschung **2a**, 133 (1947)

16.85 R.M. Tromp, J.F. van der Veen: Surf. Sci. **133**, 159 (1983)

16.86 H.H. Andersen, J.F. Ziegler: *The stopping and Ranges of Ions in Matter*, Vol. 3 (Pergamon, New York 1977)

16.87 J.W.M. Frenken, R.M. Tromp, J.F. van der Veen: Nucl. Instrum. Methods, Phys. Res. Sect. **B17**, 334 (1986)

16.88 O.S. Oen: Phys. Lett. **19**, 358 (1965)

16.89 O.S. Oen: Surf. Sci. **131**, L407 (1983)

16.90 J.F. van der Veen, J.B. Sanders, F.W. Saris: Surf. Sci. **77**, 337 (1978)

16.91 J.H. Barrett: Phys. Rev. **B3**, 1527 (1971)

16.92 J. Lindhard: K. Dan. Vidensk. Selsk., Mat.-Fys. Medd. **34**, No.14 (1965)

16.93 W.C. Turkenburg, W. Soszka, F.W. Saris, H.H. Kersten, B.G. Colenbrander: Nucl. Instrum. Methods **132**, 587 (1976)

16.94 R.G. Smeenk, R.M. Tromp, H.H. Kersten, A.J.H. Boerboom, F.W. Saris: Nucl. Instrum. Methods **195**, 581 (1982)

16.95 C.W. Turkenburg, E. de Haas, A.F. Neutenboom, J. Ladru, H.H. Kersten: Nucl. Instrum. Methods **126**, 241 (1975)

16.96 E.V. Zarochentsev, S.P. Kravchuk, T.M. Tarusina: Fiz. Tverd. Tela **18**, 43 (1976) [Soviet. Phys.-Solid State **18**, 239 (1976)] and references therein

16.97 J.W.M. Frenken, J.F. van der Veen, R.N. Barnett, U. Landman, C.L. Cleveland: Surf. Sci. **172**, 319 (1986)

16.98 B. Pluis, J.W.M. Frenken, J.F. van der Veen: Phys. Scr. T **19** B, 382 (1987)

16.99 B. Pluis, A.W. Denier van der Gon, J.W.F. Frenken, J.F. van der Veen: Phys. Rev. Lett. **59**, 2678 (1987)

16.100 B. Pluis, A.W. Denier van der Gon, J.F. van der Veen: to be published

16.101 R.M. Goodman, G.A. Somorjai: J. Chem. Phys. **52**, 6325 (1970)

16.102 G. Armand, D. Gorse, J. Lapujoulade, J.R. Manson: Europhys. Lett. **3**, 1113 (1987)

16.103 R. Trehan, C.S. Fadley: Phys. Rev. **B34**, 6784 (1986)

16.104 P. Carnevali, F. Ercolessi, E. Tosatti: Surf. Sci. **189**, 645 (1987)

16.105 S. Brennan, P.H. Fuoss, P. Eisenberger: In *The Structure of Surfaces*, ed. by M.A. Van Hove, S.Y. Tong, Springer Ser. Surf. Sci., Vol.2 (Springer, Berlin, Heidelberg 1985) p.421

16.106 J. Henrion, G.E. Rhead: Surf. Sci. **29**, 20 (1972)
16.107 E.G. McRae, R.A. Malic: Surf. Sci. **148**, 551 (1984)

17. The Surface of Solid Helium

Humphrey J. Maris

Department of Physics, Brown University, Providence, RI 02912, USA

In this chapter we will review recent work on the surface of solid helium, mainly ^{4}He. No attempt will be made to make this an exhaustive survey of the field; the aim is to present those aspects of the subject which are of most general interest to workers in the field of surface science, and to describe some of the features which are unique to the helium surface.

17.1 Review of the Properties of Helium-4

We begin with a very brief review of the properties of liquid and solid helium (for a more extensive survey of helium properties see *Wilks* [17.1]). Helium is the only element which does not solidify when cooled to low temperatures. To produce solid helium it is therefore necessary to apply pressure to force the system into the lower-molar volume crystalline state. The phase diagram of helium-4 is shown in Fig.17.1. Notice that there is no triple point, and no point on the phase diagram where gas and solid phases coexist. Consequently, one can never have a surface of solid helium with gas above. Thus, when one discusses the surface of solid helium, one is always referring to the interface between the solid and liquid phases.

The peculiar phase diagram of helium occurs because of the large effect of quantum mechanical zero-point motion. The interaction potential between helium atoms is very weak (Lennard-Jones energy of $\simeq 10$ K) and the mass is low. Thus, the zero-point energy, E_0, per atom (which is of order $\hbar\omega_D$, ω_D: Debye frequency) is of the same order of magnitude as the binding enrgy, E_B. This binding energy is clearly less (i.e., more negative) in the solid phase than in the liquid, but E_0 is much larger in the solid. Hence, in the absence of an applied pressure the system prefers to be in the liquid phase.

Solid helium-4 exists in three crystallographic phases. The low temperature phase is hcp. There is a very small region of bcc, which lies close to the melting curve, and is present only for temperatures in the range 1.5 to 1.8 K. At high pressures (P>1100 bars) and high temperatures (T>15 K) the solid is fcc. (This is not shown in Fig.17.1). A

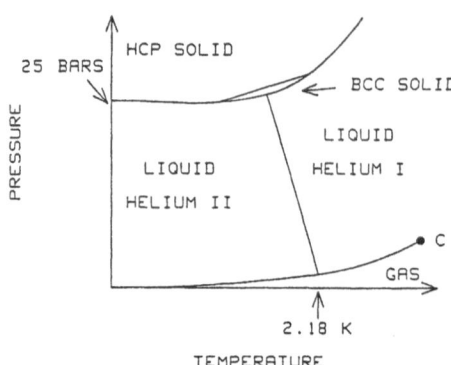

Fig.17.1. Phase diagram of helium-4 (not to scale). The fcc solid phase is not included

remarkable feature of helium is the existence of two distinct liquid phases, helium I and helium II. Helium I which is the high temperature phase, behaves as a fairly normal liquid. The boundary in the P-T plane between helium I and II is called the λ-line. Helium II is a superfluid and has many unusual properties. The viscosity is zero and currents which are set up by stirring the liquid do not damp out (this is analogous to persistent currents in superconductivity). The unique properties of superfluid helium are believed to be due to Bose condensation of atoms into the zero-momentum state. Thus, there is no transition in the same temperature range for the ^{3}He isotope which is a fermion.

When superfluid helium is cooled below the λ-line the entropy naturally decreases. It is found that as the temperature T is lowered towards zero, the entropy, S, of the liquid tends to zero. In one sense, this is as expected according to the third law of thermodynamics. However, it is a remarkable property for a liquid to possess, since one is so used to the idea that an essential feature of liquids is disorder, both in real and momentum space. The point is that over a certain range of pressures the lowest energy quantum mechanical state of the system corresponds to a "liquid" configuration (i.e., short range order, but no periodic translational order). Thus, as $T \rightarrow 0$ the system is in this single state and has zero entropy.

Why do these properties make helium an interesting system from the point of view of surface science? The main reasons are as follows:

1) Since helium is the only element that is liquid at low temperatures, all other elements freeze out on the walls of the container. Thus, the interface and the bulk phases are completely free of impurities.

2) The large zero-point motion in the solid means that point defects and dislocations have a high mobility even at T = 0 K. Thus, defects produced during crystal growth tend to anneal out quickly.

3) When a crystal grows, the latent heat of fusion is liberated at the surface. For a conventional material, the rate of crystal growth is often determined by the rate at which latent heat can be removed

from the interface (by conduction or convection), rather than by the microscopic kinetic processes occurring at the liquid-solid interface. Thus, growth kinetics are hard to study. For helium at low T the entropies of the solid and liquid phases are vey small, and, therefore, the latent heat of fusion also tends to zero as T → 0 (this has the consequence that the slope dP/dT of the melting curve tends to zero when T → 0, as shown in Fig.17.1). In addition heat flow is very rapid in both liquid and solid helium at low temperatures. Thus, the latent heat liberated in the growth process is unimportant for helium, and it is easy to study the kinetics of growth.

17.2 Growth of Solid Helium-4

The rate v_g at which a crystal grows is a function of the difference $\Delta\mu$ (=$\mu_l-\mu_s$) in the chemical potentials μ_l and μ_s of the liquid and solid. The growth coefficient K is defined by the relation

$$v_g = K\Delta\mu . \tag{17.1}$$

The growth coefficient K depends on the orientation of the surface and on whether the surface is atomically smooth or rough [17.2]. For atomically smooth surfaces (facets), growth proceeds either by nucleation of entire new atomic layers, or is associated with screw dislocations which intersect the surface. For atomically rough surfaces, there are many places where atoms can be added to the surface, and so K is much larger. Let us consider (following *Andreev* and *Parshin* [17.3]) the surface shown in Fig. 17.2. This is rough and contains steps and kinks. It is convenient to describe the addition of atoms to the solid which occurs during the growth process in terms of the motion of the kinks. A motion of the kinks in the direction of the arrows corresponds to addition of the atoms to the solid phase. (It is, of course, possible for growth to occur in other ways, for example, by the addition of atoms at points on the steps which are far removed from the location of the kinks. This is topologically equivalent to the production of kink-anti-kink pairs). For an ordinary material the addition of atoms to the solid is a random process; thus, the kinks move at random times with changes of position of one interatomic spacing. *Andreev* and *Parshin* [17.3] proposed that kinks on the surface of solid helium would move in a different way. The energy of a kink (when it is far from other kinks) is a periodic function of its position along a step. Thus, the kink can behave as an elementary excitation, i.e., it can propagate freely along the step. The energy E_k of the kink will depend on its momentum k, and the E_k vs k relation will have the usual form for an excitation in a periodic structure (Fig.17.3). The key point is that a kink with a non-zero momentum corresponds to a stationary state of

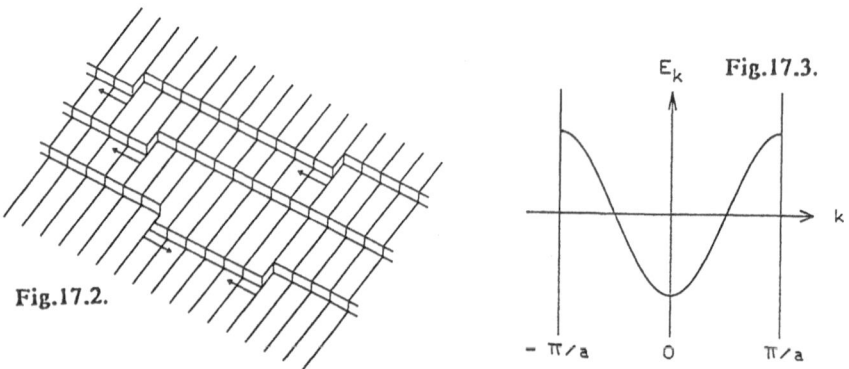

Fig.17.2. Schematic drawing of a crystal surface showing steps and kinks. Motions of the kinks in the directions shown transfers atoms from the liquid to the solid

Fig.17.3. Energy as a function of momentum for a kink. The kink bandwidth is Δ_{kink}

the system in which atoms are being transferred from the liquid to the solid at a constant rate. Thus, the solid can have a steady growth rate in the absence of a chemical potential difference. This means from (17.1) that the growth coefficient K is infinite.

The discussion just given assumes implicitly that the temperature is zero. For infinite T there will be thermal excitations (phonons and rotons) in the liquid and solid phases. These will reduce K to a finite value. The excitations will be reflected from the moving interface and will absorb energy (as a result of a Doppler shift) [17.3-5]. This energy has to come from the work done by a difference in chemical potential, and so $\Delta\mu$ must now be non-zero. A second mechanism is a possible direct interaction [17.6] between phonons and rotons and kinks. Since K is inversely proportional to the rate at which energy is lost from the moving interface, one might expect that

$$\frac{1}{K} \simeq AT^4 + Be^{-\Delta/T} \tag{17.2}$$

where A and B are approximately constant and Δ is the roton energy gap ($\simeq 7$ K). The idea is that the energy density of phonons in both the liquid and the solid goes as T^4, while the number of rotons thermally excited is proportional to $\exp(-\Delta/T)$. We will discuss some difficulties with this simple picture in Sect.17.4.

The Andreev-Parshin picture thus predicts that the growth of solid helium will have a character quite different from ordinary solids. For an ordinary solid one considers that an increase in temperature gives a higher mobility for atoms in the liquid, and hence increases the rate at which atoms can find energetically-favorable sites on the solid surface. Thus K should increase with increasing T, which is the opposite of the behavior predicted for helium. One can consider that the

494

different behavior occurs because the parameters describing kinks in helium have values which are very different from ordinary solids. The weakness of the interatomic potential, together with the low atomic mass, has the consequence that in helium the kink bandwidth is large and the kink velocity is high (this velocity is of order Δ_{kink} a/h; a = lattice parameter). For ordinary solids this velocity will be much smaller. For helium the kink damping vanishes as T → 0; for ordinary solids the liquid-solid interface only exists above the triple point where the number of thermal excitations is large and kinks are highly damped. Thus, except in the case of helium, kinks are highly-damped and it is not useful to think of them as propagating excitations.

In the next sections we describe experiments which have been performed to study the growth of helium under various conditions. We will discuss just the properties of atomically rough surfaces, and defer consideration of facets to Sect.17.3.

17.2.1 Melting–Freezing Waves

At first sight it would appear that one could measure the growth of solid helium simply by applying an excess pressure ΔP above the equilibrium melting pressure $P_m(T)$, and then measuring the growth rate. Experimentally this is impossible to do (an exception occurs if the crystal is completely faceted, as we discuss later). The crystal grows so fast that one cannot maintain a measurable ΔP, at least by ordinary methods. The first measurement of K(T) was performed in the following way. Let us suppose that by some means it is possible to produce a liquid-solid interface whose profile is as shown in Fig.17.4a. In this configuration both the gravitational and the surface energies are larger than for a flat horizontal interface. Consequently, melting of the peaks and freezing of liquid in the valleys occurs. For an ordinary material, this happens slowly, and the interface simply relaxes monotonically back to a horizontal configuration. For helium, the change in shape occurs very rapidly. At the instant when the interface first becomes flat (Fig.17.4b) there are still currents flowing in the liquid (these currents arise from the material in the solid mountains which has melted and flows to freeze in the valleys). The inertia of these currents causes the system to overshoot and continue melting and freezing until the configuration (Fig.17.4c) is reached. The process then reverses, giving

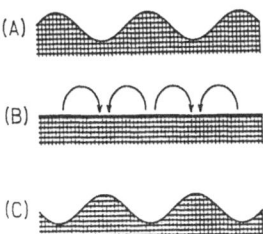

Fig.17.4. Melting-freezing waves. a) Initial configuration of the surface. b) The arrows show the flow in the liquid which exists at the instant that the surface has become flat. c) Shape of the surface after one half cycle of oscillation

a standing wave oscillation, referred to as a melting-freezing wave. It is clear that a travelling wave can also be produced. The dispersion relation for these waves is [17.3]

$$\omega^2 = \frac{\rho_1}{(\rho_s - \rho_1)^2} \left[\tilde{\alpha}k^3 + (\rho_s - \rho_1)gk - \frac{i\omega k\rho_s}{K} \right],$$ (17.3)

where ρ_1 and ρ_s are the densities of the liquid and solid, and

$$\tilde{\alpha} = \alpha + \frac{\partial^2 \alpha}{\partial\phi^2}.$$ (17.4)

The quantity α is the liquid-solid surface energy, and ϕ is the angle between the vertical (the z-axis) and the normal to the displaced surface. The term $\tilde{\alpha}$ (rather than α) enters into the dispersion relation because, when the surface is distorted as in Fig. 17.4a, there are two contributions to the increase in energy. The total area of the surface has increased, giving an energy change proportional to α. In addition, the orientation of the surface changes. The term in $\partial\alpha/\partial\phi$ gives zero when integrated over the surface, but a contribution to the energy remains which is proportional to $\partial^2\alpha/\partial\phi^2$.

Melting freezing waves were first observed by *Keshishev et al.* [17.7,8]. They measured the dispersion relation for frequency in the range $\simeq$ 50 to 4000 Hz, and were able to determine $\tilde{\alpha}$ in this way. The waves were generated by applying an oscillating electric field to the part of the surface near to where it made contact with the wall of the experimental cell. The amplitude of the waves was 0.01 to 0.5 mm. From the attenuation β of the waves with distance it was possible to measure the growth coefficient K. For large K one can show from (17.3) that

$$\beta = \frac{\rho_1^{1/3} \rho_s \omega^{1/3}}{3K(\rho_s - \rho_1)^{2/3} \tilde{\alpha}^{2/3}}.$$ (17.5)

This assumes that k is sufficiently large that the surface energy term in (17.3) is much larger than the gravity term. Results for K obtained in this way are shown in Fig. 17.5. These data clearly confirm the general idea of *Andreev* and *Parshin* [17.3] that the growth coefficient should increase as the temperature is lowered. The temperature-dependence is fairly well described by the sum of phonon and roton contributions as predicted by (17.2), together with a constant term of unknown origin.

It is important to distinguish between melting-freezing waves and Rayleigh surface waves. Rayleigh waves are elastic waves which propagate on the free surface of a solid, or at the boundary between a liquid and a solid. The restoring force is elasticity and, except at wavelengths of the order of the interatomic spacing, surface energy has

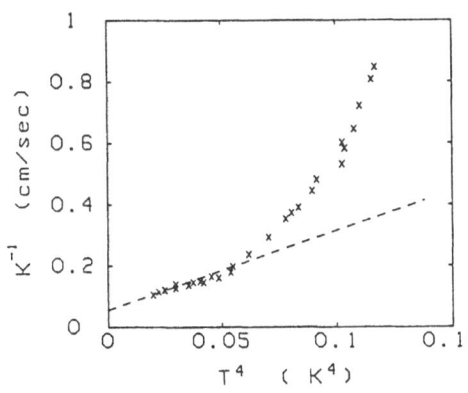

Fig.17.5. The reciprocal of the growth coefficient K as measured in [17.8] plotted versus T^4. The frequency was 232 Hz. The dashed line is a fit to the data in the low temperature range where phonons dominate the damping of the interface

a negligible effect. The atoms in the solid move in ellipses in the plane containing the wave vector and the normal to the surface. For melting-freezing waves, on the other hand, elasticity plays a minor role. In the derivation of the dispersion relation (17.3) it was assumed that the liquid and the solid are incompressible, i.e., effects of elasticity are ignored. One can show [17.9] that this is a good approximation for wavelengths much larger than the lattice spacing. Under these conditions, which are well satisfied in the experiments of *Keshishev* et al [17.7,8], the atoms in the solid do not move except when they melt and enter the liquid. Thus, the two types of wave motions have very different character. For very short wavelengths ($k>10^6$ cm^{-1}) there is a mixing of these two types of waves, and the dispersion relations are modified [17.9,10]. In addition, two other effects may be important at large k. A melting-freezing wave can spontaneously decay into two lower energy waves, thereby giving a finite damping (proportional to k^5) even at T = 0 K. Secondly, there should be corrections to the dispersion relation which come from the microscopic structure of the surface; these have been ignored in the above discussion. On a vicinal surface the spacing ℓ of steps will be large, and this should modify the melting-freezing wave dispersion relation when $k\ell > 1$ [17.11]. No experimental study of these short wavelength waves has yet been made.

17.2.2 Boundary Conditions at a Moving Interface

We have just described how a measurement of the damping of melting-freezing waves can be used to determine the growth coefficient K(T). It was realised by *Castaing* and *Nozieres* [17.12] that in addition to the growth coefficient, other parameters are needed to describe the kinetics at the interface. Let us first consider the naive view of transport processes at a moving interface. The growth velocity is determined from $\Delta\mu$ and K via (17.1). Latent heat is liberated at the interface and is conducted away into the bulk phases according to

Fourier's law. One can ask how much of the latent heat is deposited onto either side of the interface. For an ordinary material this question is unimportant, however. Since the interface is considered, at this level at least, to be just a mathematical surface it is immaterial on which side of the surface the latent heat appears. For helium, however, there is a large thermal boundary resistance (referred to as Kapitza resistance) that occurs at the interface. Thus, a heat current J_E across the interface from the solid to the liquid produces a discontinuity in temperature, ΔT ($\equiv T_1 - T_s$). If there is no mass flow, J_E and ΔT are related by

$$\Delta T = R_K J_E \tag{17.6}$$

where R_K is the Kapitza resistance [17.13]. This resistance arises because phonons (or rotons) incident on the interface have some probability of being reflected, i.e., the discontinuities in physical properties, such as density and sound velocity, that occur at the interface impede the heat flow. The existence of this boundary resistance has the consequence that the question about which side of the boundary the latent heat appears now becomes physically significant. *Castaing* and *Nozieres* [17.12] show that in this case the correct boundary conditions are

$$\frac{\Delta \mu}{T} = a J_M + b J_E \; ; \tag{17.7}$$

$$\frac{\Delta T}{T^2} = b' J_M + c J_E \; , \tag{17.8}$$

where J_M is the mass current across the interface from solid to liquid. By comparison with (17.1,6)

$$a = \frac{1}{K \rho_s T} \; ; \tag{17.9}$$

$$c = \frac{R_K}{T^2} \; . \tag{17.10}$$

It can be shown [17.12] by an Onsager relation that $b = b'$. One can consider that the terms involving b in (17.7 and 8) represent a Peltier effect for helium atoms crossing the interface.

For an ordinary material, the Kapitza resistance at room temperature is very small, and is hard to measure because it always appears in series with the thermal resistance of the bulk phases. In helium at low temperatures R_K is much larger and the thermal resistance of the bulk phases is very small. Hence, R_K is fairly easy to measure.

In the next two sections we describe measurements that have been made of the Kapitza resistance and the growth coefficient, together with a theoretical discussion of these quantities.

498

17.2.3 Kapitza Resistance

At the microscopic level, heat flow across an interface between two insulators must arise from the transmission of phonons. There will be a flux of energy, $\dot{Q}_1(T_1)$, onto the boundary from the liquid side, and $\dot{Q}_s(T_s)$ from the solid side. If the average transmission probabilities from the two sides are $\langle\langle t_1 \rangle\rangle$ and $\langle\langle t_s \rangle\rangle$, the net heat flow is

$$\dot{Q}_{ls} = \dot{Q}_1(T_1) \, \langle\langle t_{ls} \rangle\rangle - \dot{Q}_s(T_s) \, \langle\langle t_{sl} \rangle\rangle \, . \tag{17.11}$$

When T_1 and T_s are equal, the heat flows must vanish. Hence, for small ΔT we have

$$\dot{Q}_{ls} = \frac{d}{dT_1} \left[\dot{Q}_1(T_1) \, \langle\langle t_{ls} \rangle\rangle \right] \Delta T \, . \tag{17.12}$$

Thus, the Kapitza resistance is given by

$$\frac{1}{R_K} = \frac{d}{dT_1} \left[\dot{Q}_1(T_1) \, \langle\langle t_{ls} \rangle\rangle \right] \, . \tag{17.13}$$

It is straightforward to calculate $\dot{Q}_1(T_1)$. If the only heat carriers are phonons we have (by analogy with the result for black-body radiation of photons)

$$\dot{Q}_1 = \frac{E_1(T_1)c_1}{4} \, , \tag{17.14}$$

where E_1 is the phonon energy density at temperature T_1 and c_1 is the sound velocity in the liquid. The problem thus reduces to a calculation of $\langle\langle t_{ls} \rangle\rangle$, and this raises interesting questions about the physics of the interface. For an ordinary interface the calculation of $\langle\langle t_{ls} \rangle\rangle$ is a straightforward, but rather complicated problem, in acoustics and elasticity. A phonon incident from the liquid will be partly reflected, and partly transmitted as longitudinal and transverse waves in the solid. The boundary conditions at the surface are continuity of normal displacement, and continuity of the normal components of the stress. The result of this calculation [17.14], allowing for phonons of all possible angles of incidence is that $\langle\langle t_{ls} \rangle\rangle$ has the value of 0.16 independent of temperature. E_1 is proportional to T_1^4, and the final result for R_K is

$$R_K = 0.154 T^{-3} \, KW^{-1} \, cm^2 \, . \tag{17.15}$$

Now let us consider whether this calculation is correct when applied to helium. The boundary conditions that have been used assume implicitly that there is no melting or freezing at the interface. This is clearly a questionable assumption because of the high value of the growth coefficient K. On the other hand, one might argue that the frequency of a thermal phonon ($\simeq 10^{10}$ Hz) is so large that the interface could not possibly respond. As a start, consider the transmission coefficient for a wave of frequency Ω at normal incidence to the boundary. In each phase we have

$$\delta\mu = \frac{\delta P}{\rho} - s\delta T \tag{17.16}$$

where δP and δT are the pressure and temperature variation, ρ is the density, and s is the entropy per unit mass. We then have to solve the boundary condition problem, i.e., satisfy the conditions (17.7 and 8). The solution of this problem depends on the equations governing heat propagation (diffuse, ballistic phonons, or second sound) in the two media [17.12], and the final expressions are complicated. One can obtain some understanding of the problem if the thermal effects are ignored, i.e., we set b = c = 0, and ignore the latent heat of fusion. One then has a purely mechanical problem and the fraction of the energy which is transmitted is

$$t_{ls} = \frac{4\rho_1 c_1 \rho_s c_s}{[\rho_1 c_1 + \rho_s c_s + (\rho_s - \rho_1)^2 c_1 c_s \, K/\rho_1]^2} \, , \tag{17.17}$$

where c_s is the sound velocity in the solid (note the somewhat surprising result that t_{ls} does not depend on the frequency Ω). For small K, t_{ls} reduces to the ordinary acoustic mismatch result

$$t_{ls} = \frac{4Z_1 Z_s}{(Z_s + Z_1)^2} \, , \tag{17.18}$$

where $Z_1 = \rho_1 c_1$, $Z_s = \rho_s c_s$ are the acoustic impedances. As $K \rightarrow \infty$, $t_{ls} \rightarrow 0$. In this limit, the solid can melt and freeze so rapidly that the pressure at the instantaneous position of the interface is always the equilibrium pressure at the liquid-solid phase boundary. Thus, there is no oscillating pressure at the surface that can launch a wave into the interior of the solid, and so the transmission must be zero. From (17.17) it is clear that t_{ls} begins to decrease when K becomes greater than a critical value K_c given by

$$K_c = \frac{\rho_1(\rho_1 c_1 + \rho_s c_s)}{(\rho_s - \rho_1)^2 c_1 c_s} \simeq 0.005 \text{ s·cm}^{-1} \, . \tag{17.19}$$

One can see from the results for K obtained from the damping of melting-freezing waves that for T < 0.6 K, K is much larger than K_c, and so the transmission should be very small and the Kapitza resistance very large.

500

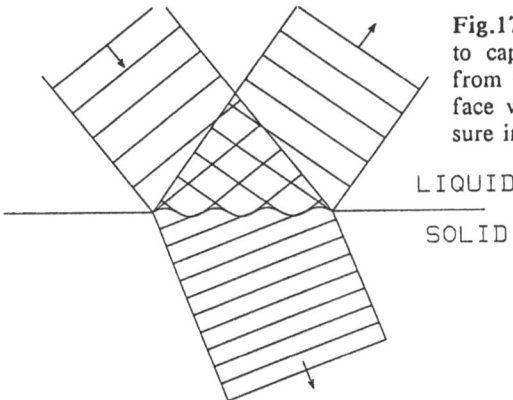

Fig.17.6. Transmission of phonons due to capillary effects. An incident wave from the liquid creates a curved interface which leads to an oscillating pressure in the solid

LIQUID

SOLID

It was realised by *Marchenko* and *Parshin* [17.15] that capillary effects can have a large effect on the transmission of phonons at oblique angles. Their idea is shown in Fig.17.6. A phonon is incident at an angle θ from the liquid side of the interface. This wave, and the reflected wave, distort the interface as shown. For a point on the interface where the principal radii of curvature are R_1 and R_2 the boundary conditions are now

$$-\sigma_{zz}{}^s = P^l + \alpha \left(\frac{1}{R_1} + \frac{1}{R_2} \right) ; \tag{17.20}$$

$$\sigma_{zz}{}^s = \sigma_{yz}{}^s = 0 \quad , \tag{17.21}$$

where z is the direction normal to the surface, $\sigma_{ij}{}^s$ is the stress tensor in the solid, P^l is the pressure in the liquid, and we have neglected the dependence of the surface energy, α, on the orientation of the surface. If the growth coefficient is infinite the chemical potentials at the boundary must be equal. In the absence of thermal effects $\delta\mu = \delta P/\rho$, and so

$$(-\sigma_{zz}^s - P_m)/\rho_s = (P^l - P_m)/\rho_l , \tag{17.22}$$

where P_m is the equilibrium melting pressure ($-\sigma_{zz}{}^s$ is equivalent to the pressure in the solid). The presence of the surface energy term in (17.20) means that the incident wave produces an oscillating stress inside the solid. From (17.20,22) one has

$$\delta\sigma_{zz}^s = - \frac{\rho_s}{\rho_s - \rho_l} \alpha \left(\frac{1}{R_1} + \frac{1}{R_2} \right) . \tag{17.23}$$

Since $\delta\sigma_{zz}{}^s$ is proportional to the curvature of the interface, this mechanism gives a transmission which is larger for short wavelength phonons. Results of a calculation [17.14] of the transmission (averaged over angles of incidence) as a function of Ω are shown in Fig.17.7. For

501

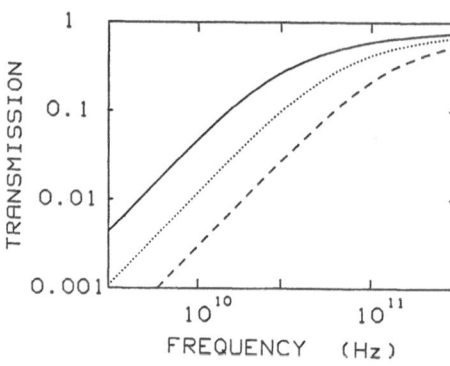

Fig.17.7. Transmission of phonons from liquid to solid helium as a function of frequency. It is assumed that the growth coefficient K is infinite. The surface energy is assumed to be (---) 0.1 erg·cm⁻², (...) 0.2 erg·cm⁻², and (-) 0.4 erg cm⁻²

low Ω, t_{ls} is proportional to Ω^2, and at high Ω, t_{ls} saturates at a value close to 1. Since the average frequency of a thermal phonon is of the order of $kT/\hbar$, the average transmission, $\langle\langle t_{ls}\rangle\rangle$, goes as T^2 at low T, and consequently from (17.13)

$$R_K \propto T^{-5} \tag{17.24}$$

This contrasts with $R_K \propto T^{-3}$ if there is no melting and freezing. Hence, a measurement of R_K can be used as a probe of the mobility of the interface at high frequencies. Measurements of R_K have been made by several groups [17.16-18]. These measurements show clearly that R_K does vary as T^{-5} below about 0.2 K, and thus imply the remarkable result that helium can melt and freeze at frequencies at least as high as 10^{10} Hz.

A more detailed look at the comparison of theory and experiment raises some fundamental questions about the equations governing the motion of a surface. The calculations just described [17.14] predict that in the low temperature regime

$$R_K = 3.6 \cdot 10^{-3} \alpha^{-2} T^{-5} \; [\text{K} \cdot \text{W}^{-1} \text{cm}^2] \; . \tag{17.25}$$

The best experimental data [17.18] give the result

$$R_K = 2.5 \cdot 10^{-2} T^{-5} \; [\text{K} \cdot \text{W}^{-1} \text{cm}^2] \; . \tag{17.26}$$

The data thus agree with theory if a value of 0.38 erg cm⁻² is assumed for α. This value is implausible since the melting-freezing experiments gave a value of α which was around 0.2 erg cm⁻² or less.

This discrepancy has led to interesting discussions about the boundary conditions that hold at the surface. For static equilibrium the conditions were derived by *Gibbs* [17.19], who clearly had no reason to envisage the possibility of an interface moving at high frequency. In a dynamic situation there is the general possibility that time derivatives of various quantities might be added to the boundary conditions (17.20-22), and this has been discussed in several papers [17.11,20,21]. *Puech* and *Castaing* [17.20] point out that while atoms are making the transition from liquid to solid there must be an excess kinetic energy, E_K, as the rearrangement takes place. This energy will be proportional to J_M^2, and can be written as

$$E_K = \frac{\sigma_K J_M^2}{2\rho_s \rho_l} ,$$ (17.27)

thereby making σ_K a mass per unit area. To accelerate the interface, this energy must be supplied by the difference in the chemical potentials of the liquid and solid. Thus, the equality of μ which holds in static equilibrium is replaced by

$$\mu_s = \mu_l + \frac{\sigma_K}{\rho_s \rho_l} \frac{\partial J_M}{\partial t} .$$ (17.28)

(This is for the case that the growth rate is infinite). *Kosevich and Kosevich* [17.11] have made a microscopic calculation of σ_K considering the kinetic energy of the flowing liquid in the vicinity of a moving step. The introduction of σ_K has the consequence that phonons at normal incidence to the interface now have a finite transmission coefficient which, like the transmission due to capillary effects, varies as Ω^2. Consequently, this can reduce the magnitude of the Kapitza resistance and bring it into agreement with experiment. The required value of σ_K is estimated to be $2 \cdot 10^{-10}$ g cm^{-2}, which corresponds to a layer of thickness 0.1 $\hat{A}$o with density equal to that of the solid.

In the Gibbs theory of capillarity, a choice has to be made regarding the location of the dividing surface S. This can be chosen so that the surface density of one particular quantity (e.g., mass) vanishes. It is clear that if the surface mass density does not vanish there will be an extra term in the boundary condition (17.20), which for a static situation expresses the mechanical balance of normal forces at the interface. This extra term is just the mass density, σ_M, times the acceleration of the interface. It would seem at first sight that since σ_M is arbitrary, being dependent on the choice of the dividing surface, the results for physical quantities such as the sound transmission would also be dependent on the choice of S. However, one can show [17.21] that a change in the dividing surface also changes σ_K, and the net effect leaves all physical quantities invariant.

The value of σ_K (when, for example, S is chosen so that $\sigma_M = 0$) can be determined accurately from a measurement of the transmission of phonons of a definite frequency with wave vector at normal incidence to the interface. This has not yet been done, but appears to be feasible.

The discussion given so far in this section concerns R_K for an atomically rough surface. For a facet, the growth coefficient K is much smaller (Sect.17.2.4). Then the Kapitza resistance should vary as T^{-3}, (17.15), and this has been confirmed experimentally [17.17].

17.2.4 Growth Coefficient

We have already described the results obtained for K by the study of the damping of melting-freezing waves [17.7,8]. These data and more recent results of *Bodensohn* et al. [17.22] can be fit reasonably well by a phonon term, T^4, and a roton term, $\exp(-\Delta/T)$, as in (17.2). (There is in addition some information about the growth from measurements of sound transmission across the interface [17.23-25]).

It turns out that when the theory is looked at in more detail, several difficulties arise. The phonon contribution to K has been calculated quantitatively by *Bowley* and *Edwards* [17.5]. Their result is in reasonably good agreement with the experimental results of *Keshishev* et al. [17.7,8]. However, *Andreev* and *Knizhik* [17.4] point out that in the upper part of the temperature range studied by Keshishev et al. ($\simeq 0.6$ K), the simple model of ballistic phonons scattering off the surface requires modification. The phonon mean-free-path, Λ, in the solid has decreased so that $k\Lambda < 1$ (k: wave number of the melting-freezing waves). Under these conditions, the phonon gas has to be treated hydrodynamically, rather than as a gas of independent excitations. Nevertheless, the fit to the data is still reasonably good (see note added in [17.5]).

A more serious problem concerns the roton contribution. Because of the peculiar form of the dispersion relation for rotons [17.1], a roton of momentum **p** incident on the surface can be reflected with two different possible momenta even if the momentum parallel to the surface is conserved. The momentum transfer will either be large (of the order of **p**) or very small ("anomalous" scattering). If one assumes that the fraction ξ of large momentum transfers is 1/2 and considers the rotons to propagate ballistically, one obtains a magnitude and temperature dependence of K which is in good agreement with experiment. However, the roton mean free path is actually very short [17.26-28] and so one should treat the roton gas hydrodynamically. A second difficulty is that there are arguments [17.4] that ξ is very small. Then most of the scattering is "anomalous", and the momentum transferred is reduced by a large factor. Both of these considerations would make K much smaller than the experimental value.

Two proposals have been made to resolve this dilemma. In a broad sense, the physics of these ideas is the same. For the ballistic approach to be correct, one needs the length scale, ℓ, of the perturbation acting on the roton gas to be less than the roton mean free path Λ_R. Secondly, the argument [17.4] that ξ is $<< 1$ assumes that the rotons are scattered from a flat surface. *Mukherjee* and *Edwards* [17.6] have pointed out that there should be a direct interaction between rotons and moving kinks. Since kinks are small objects (size of order 1 interatomic spacing) ℓ will be less than Λ_R (so the ballistic roton theory should apply), and there is no reason why ξ should be $\ll 1$. *Castaing* [17.29] has argued that the periodic potential of the solid produces static density oscillations in the liquid, and that the scattering of rotons from these oscillations gives a ξ which is not small. As with the Muk-herjee-Edwards approach, the fairly short range of the density oscilla-tions has the consequence that one should use ballistic roton theory. One might be able to test these ideas by means of an experiment in which a roton beam is directed at the solid surface, and the momen-tum distribution of the scattered rotons measured.

There is only one measurement of the Peltier effect [17.18], i.e., the b coefficient in the boundary conditions (17.7) and (17.8). The result is consistent with the calculation of *Bowley* and *Edwards* [17.5] allowing for the uncertainty in both theory and experiment.

17.2.5 High Velocity Growth

The Kapitza resistance experiments show that the interface can melt and freeze at frequencies as high as 10^{10} Hz. It is also interesting to study the maximum velocity that the interface can have. In the melting freezing wave experiments, for example, a wave of frequency 4000 Hz and amplitude 0.01 mm has a maximum growth velocity of 25 cm·s^{-1}. This is a very high growth rate for an ordinary crystal. There must be some limit to the growth velocity and so one expects that for large v_g, K must decrease. Measurements of K as a function of v_g have been made by *Graf* and *Maris* [17.30]. In these experiments the transmission coefficient t_{ls} for a sound wave normally incident on the interface was measured. The growth coefficient K was determined from t_{ls} using (17.17). The temperature was $\simeq 0.1$ K so that the effect of phonons and rotons on the growth rate was negligible. By changing the amplitude of the sound wave, the velocity of the interface could be varied and the dependence of K on velocity studied. Some typical results are shown in Fig.17.8. For small v_g, K is very large, in fact too large to measure by this method. When v_g increases above about 500 cm·s^{-1}, K decreases

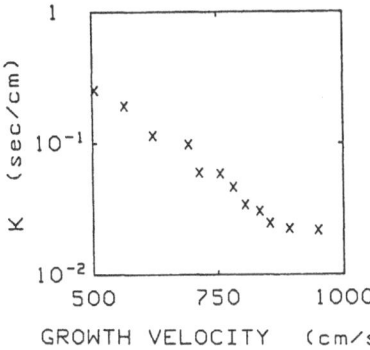

Fig.17.8. Growth coefficient K as a function of growth velocity, v_g. Results are from [17.30]

505

rapidly. Nevertheless, it was possible in these experiments to melt and freeze helium at velocities as large as 1000 cm·s⁻¹.

A finite value of K means that there is a difference in chemical potential between the liquid and solid. The work done by this potential difference must be balanced by dissipation occurring at the interface. Thus, to understand the variation of K with v_g, one has to develop at theory for this dissipation. At the most fundamental level, one can understand the dissipation in the following way. At low temperatures, the entropies of both the liquid and solid are very small. Thus, to a good approximation, we may consider the liquid and the solid to be in their respective ground states. When the solid grows slowly, we may consider that the ground state of the liquid is adiabatically distorted into the ground state of the solid. This process will be reversible and free of dissipation. At high growth rates, however, the change from liquid to solid will occur too rapidly, and the process will be nonadiabatic. The solid will then be formed in an excited state, which will eventually relax (by emitting phonons or rotons) and dissipate energy. This general picture leads to a growth coefficient [17.30] of the form

$$K = \frac{A \exp(v_c/v_g)}{v_g} , \tag{17.29}$$

where A and v_c are constants. This gives a reasonable fit to the data. Values of the critical velocity v_c found in this way range from 1600 to 5100 cm·s⁻¹. The variation may be related with the orientation of the crystal surface but this has not yet been established.

A microscopic understanding of the dissipation is lacking. One possibility is that it is related to the maximum velocity of kinks. Since the density of kinks is much less when the surface orientation is close to a high symmetry direction (such as the c-axis), for a given macroscopic growth rate v_g the kinks have to move faster for these orientations. Hence v_c should be much smaller, and this idea can be tested by experiments on surfaces of known orientation.

17.3 Facets and Facetting Transitions

So far, we have mostly discussed the properties of atomically rough surfaces. For ordinary substances one expects that at sufficiently low temperatures, the surface of a crystal in thermodynamic equilibrium will contain atomically flat regions (facets). The properties of facets and the transition between the facetted and unfacetted state (roughening transition) are difficult to study because of the long time required to each thermodynamic equilibrium. This time can be made shorter by using very small crystals, as in the experiments of *Heyraud* et al. [17.31], for example. Helium has the great advantage that the equilibrium time is short because of the high growth rate and the very small latent heat.

17.3.1 Roughening Temperatures

At high temperatures (e.g., 1.5 K) a crystal of solid helium is observed to have a smoothly rounded surface, indicating that on the atomic scale the surface is rough. The appearance of facets as the temperature is lowered was first studied by *Balibar* and *Castaing* [17.32] and by *Avron* et al. [17.33,34]. At T_c = 1.28 K facets appear normal to the c-axis ({0001} faces). When the temperature is lowered these facets become larger, and at $T_a \simeq 1.0$ K additional facets (a-facets) appear on the six {1100} faces. The third type of facet (s-facet) which has so far been discovered appears at 0.36 K [17.35], and is on the set of twelve faces of the {1101} type.

Experimentally, no further roughening transitions have been observed down to 70 mK. It is also clear that, at this temperature, a significant part of the surface is still rough. It is conceivable that extra facets have in fact been formed but are too small to be easily detected. One possibility is to study larger crystals (linear dimensions greater than the capillary length) for which the effects of gravity are important. If the orientation of the crystal is such that a facet is close to horizontal, the size of the facet will be greatly enhanced.

17.3.2 Theoretical Situation

Initially, it had been conjectured by *Andreev* and *Parshin* [17.3] that the surface of helium might remain rough all the way down to T = 0 K. The idea was that the quantum mechanical zero-point motion might be so large as to destroy the facets. Since facets are certainly observed, the question becomes: What, if any, influence does quantum mechanics have on the temperatures at which roughening transitions occur? A renormalization group calculation by *Fisher* and *Weeks* [17.36] gives the result that, regardless of the strength of quantum fluctuations, the roughening temperature T_R of any crystal face is finite. Quantum effects may reduce the surface energy (by making the interface thicker), but at low enough temperatures, the periodic potential of the crystal in the direction normal to the interface will always pin the interface and produce facetting. Thus, apart from the need to use a renormalized surface energy, the general theoretical predictions for roughening transitions should apply to helium.

These results include [17.37,38]:

1) The value of α at the roughening temperature should be given by

$$\alpha(T_R) = \frac{\pi K T_R}{2a^2} ,$$ (17.30)

where a is the spacing of crystal planes parallel to the facet plane.

2) At the boundary between a facet and rough surface, there should be no sudden change in the orientation of the surface. For a facet which lies in the plane $z = z_0$, the equation giving z as a function of distance r from the facet edge is

$$z = z_0 - Ar^{3/2} ,$$ (17.31)

where A is a temperature-dependent coefficient.

3) For $T < T_R$ the length L of a facet tends to zero according to the equation

$$L \propto \exp[-B/(T_R - T)^{1/2}] ,$$ (17.32)

where B is a constant.

17.3.3 Experimental Results Near Roughening Transitions

Measurements of crystal shapes near to the roughening transitions have been made by *Babkin* et al. [17.39,40] and by *Wolf* et al. [17.41]. Even given the advantages of helium, these experiments are very difficult. The problem is that close to T_R the facets are very small, and it is hard to distinguish the edge of the facet from the beginning of the curved surface because there is no change in slope at the edge (17.32). Given these uncertainties, one can say that the data confirm the value (17.30) of α at T_R to within experimental error. Predictions 2) and 3) have not been quantitatively tested, except that it has been verified that there is no discontinuous change in surface orientation at the edge of the facet.

The best test of critical behavior obtained so far probably comes from measurements of the growth rate as a function of T near to T_R. For a facet, two distinct growth mechanisms are possible. Growth can occur by the spiral motion of a step around a screw dislocation that intersects the surface. The other possible mechanism is the nucleation of a new layer of solid atoms. The temperature dependence of the growth rate of an a-facet as measured by *Wolf* et al. [17.41] is shown in Fig. 17.9. For low T, the growth velocity increases with decreasing T. In this regime it is believed that the main contribution comes from spiral growth. The velocity of the step around a dislocation is limited by scattering of phonons and rotons off the step, and this velocity, therefore, increases as T goes down. At high T, as the roughening transition is approached, the growth rate increases with increasing T. In this regime growth occurs by the nucleation of new layers. To produce a new layer of radius R, requires an amount of energy

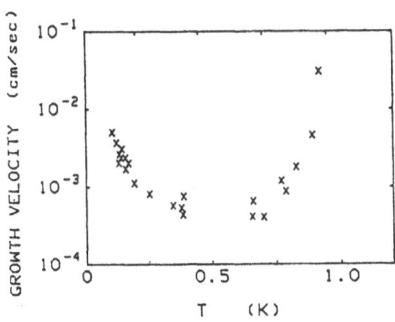

Fig.17.9. Growth velocity of an *a*-facet as a function of temperature [17.41]. The chemical potential difference driving the growth is slightly different for different data points, and is approximately 40 cgs units

$$\Delta E = 2\pi R E_{s} - \pi R^2 a \rho_s \Delta\mu \ . \tag{17.33}$$

The first term is the energy of the step which runs around the circumference of the new layer (E_s: step energy per unit length); the second term is the decrease in energy due to the volume of solid formed. ΔE has a maximum at $R = E_s/(a\rho_s \Delta\mu)$, and so the nucleation barrier is

$$\Delta E_B = \frac{\pi E_s^2}{a\rho_s \Delta\mu} \ . \tag{17.34}$$

Thus, the nucleation rate Γ is proportional to

$$\exp\left[-\frac{\pi E_s^2}{a\rho_s \Delta\mu kT}\right] , \tag{17.35}$$

and the growth rate can be related to this Γ. Thus, from the measurement of K one can determine the step energy E_s and its temperature dependence. The growth rate should depend on $\Delta\mu$, and this therefore provides a check on the interpretation. The final results [17.42] for the step energy have a temperature dependence near to T_R which is in good agreement with the theoretically predicted [17.37] critical behavior

$$E_s \propto \exp\left[-\frac{B}{(T_R - T)^{1/2}}\right] . \tag{17.36}$$

[Note that E_s has the same critical behavior as the facet size (17.32)].

17.4 Helium-3

Helium-3 has a phase diagram which is similar to that of helium-4, but with some significant differences. The low-temperature low-pressure solid phase has the bcc structure. The liquid phase does not become superfluid until 2.7 mK, and since it is a Fermi liquid the entropy at higher T varies linearly with T. The solid has a large entropy ($\simeq R\ell n2$) due to spin disorder in the entire temperature range above 1mK. Thus, in contrast to ^{4}He, there is a significant latent heat ℓ_{ls} of melting. At $\simeq 0.3$ K, however, ℓ_{ls} does pass through zero and so this is an especially favorable temperature for studies of crystal growth.

Because of the existence of Fermi excitations in the liquid, it was anticipated [17.3] that the growth coefficient for ^{3}He would be much smaller than for ^{4}He. An estimate gave [17.3]

$$K = 1/(v_F) \tag{17.37}$$

where v_F is the Fermi velocity. This assumes implicitly that $T \ll T_F$, and is thus not strictly valid at 0.3 K. An experiment by *Puech* et al. [17.43] did not succeed in measuring K, but did show that it was at least 200 times less than the estimate of (17.37). It was argued that (17.37) should contain a correction factor of order

$$\frac{\rho_s \rho_l}{(\rho_s - \rho_l)^2} . \tag{17.38}$$

At lower temperatures (<80 mK), measurements were made of the Kapitza resistance. The Kapitza resistance, R_K, was found to be significantly greater than expected for an immobile interface, and this also supports the conclusion that K is fairly large.

At temperatures substantially below 1 mK, the entropies of both phases become very small and it should be possible to propagate melting-freezing waves. These waves might have interesting coupling to spins in the liquid and solid.

For ^{3}He, only one facetting transition has so far been discovered [17.44]. This is at 80 mK and the facets are {110} faces.

Finally, there is the possibility of studies of solutions of ^{3}He in ^{4}He. There is evidence [17.33,34] that the addition of ^{3}He enhances the size of facets on small crystals and eliminates the rough parts of the interface. However, Kapitza resistance measurements on large crystals, whose shape is influenced by gravity, clearly shows that the surface is still rough [17.45]. One can measure in these experiments two distinct Kapitza resistances. These control the flow of heat between the solid and the phonons in the liquid, and the flow between the solid and the ^{3}He.

17.5 Summary

In this chapter we have given a short review of the properties of the surface of solid helium. For each topic discussed, we have attempted to list some of the significant unresolved problems. It is clear that because of the unique properties of the liquid helium-solid helium interface, research in this area will play a special role in surface science for some time to come.

Acknowledgement. This work was supported in part by the National Science Foundation through grant DMR 8612207.

References

17.1 J. Wilks: *Liquid and Solid Helium* (Oxford. London 1967)
17.2 See, for example, D.P. Woodruff: *The Solid-Liquid Interface* (Cambridge Univ. Press, London 1973)
17.3 A.F. Andreev, A.Y. Parshin: Sov. Phys. JETP **48**, 763 (1979)
17.4 A.F. Andreev, V.G. Knizhnik: Sov. Phys. JETP **56**, 226 (1982)
17.5 R.M. Bowley, D.O. Edwards: J. Phys. **44**, 723 (1983)
17.6 S. Mukherjee, D.O. Edwards: Jpn. J. Appl. Phys. **26**, 395 (1987)
17.7 K.O. Keshishev, A.Y. Parshin, A.V. Babkin: JETP Lett. **30**, 57 (1979)
17.8 K.O. Keshishev, A.Y. Parshin, A.V. Babkin: Sov. Phys. JETP **53**, 362 (1981)
17.9 M. Uwaha, G. Baym: Phys. Rev. **B26**, 4928 (1982)
17.10 Y.A. Kosevich: Sov. Phys. JETP **63**, 278 (1986)
17.11 A.M. Kosevich, Y.A. Kosevich: Sov. J. Low Temp. Phys. **7**, 394 (1982)
17.12 B. Castaing, P. Nozières: J. Phys. **41**, 701 (1980)
17.13 P.L. Kapitza: J. Phys. Moscow **4**, 181 (1941)
17.14 H.J. Maris, T.E. Huber: J. Low Temp. Phys. **48**, 99 (1982)
17.15 V.I. Marchenko, A.Y. Parshin: JETP Lett. **31**, 724 (1980)
17.16 T.E. Huber, H.J. Maris: Phys. Rev. Lett. **47**, 1907 (1980); J. Low Temp. Phys. **48**, 463 (1982)
17.17 L. Puech, B. Hebral, D. Thoulouze, B. Castaing: J. Phys. Lett. **43**, 809 (1982)
17.18 P.E. Wolf, D.O. Edwards, S. Balibar: J. Low Temp. Phys. **51**, 489 (1983)
17.19 J.W. Gibbs: *Collected Works* (Longmans, New York 1928)
17.20 L. Puech, B. Castaing: J. Phys. Lett. **43**, L601 (1982)
17.21 M.Y. Kagan: Sov. Phys. JETP **63**, 288 (1986)
17.22 J. Bodensohn, K. Nicolai, P. Leiderer: Z. Phys. B **64**, 55 (1986)
17.23 B. Cataing, S. Balibar, C. Laroche: J. Phys. **41**, 897 (1980)
17.24 M.B. Manning, M.J. Moelter, C. Elbaum: J. Low Temp. Phys. **61**, 447 (1985)
17.25 M.J. Moelter, M.B. Manning, C. Elbaum: Phys. Rev. **B34**, 4924 (1986)
17.26 B. Castaing, A. Libchaber: J. Low Temp. Phys. **31**, 887 (1978)
17.27 H.J. Maris, R.W. Cline: Phys. Rev. **B23**, 3308 (1981)
17.28 I.M. Khalatnikov: *An Introduction to the Theory of Superfluidity* (Benjamin, New York 1965)
17.29 B. Castaing: J. Phys. Lett. **45**, L233 (1984)
17.30 M.J. Graf, H.J. Maris: Phys. Rev. **B35**, 3142 (1987)
17.31 J.C. Heyraud, J.J. Métois: Surf. Sci. **128**, 334 (1983)
17.32 S. Balibar, B. Castaing: J. Phys. Lett. **41**, L329 (1980)
17.33 J.E. Avron, L.S. Balfour, C.G. Kuper, J. Landau, S.G. Lipson, L.S. Schulman: Phys. Rev. Lett. **45**, 814 (1980)
17.34 J. Landau, S.G. Lipson, L.M. Määtänen, L.S. Balfour, D.O. Edwards: Phys. Rev. Lett. **45**, 31 (1980)
17.35 P.E. Wolf, S. Balibar, F. Gallet: Phys. Rev. Lett. **51**, 1366 (1983)
17.36 D.S. Fisher, J.D. Weeks: Phys. Rev. Lett. **50**, 1077 (1983)
17.37 C. Jayaprakash, W.F. Saam, S. Teitel: Phys. Rev. Lett. **50**, 2017 (1983)
17.38 C. Rottman, M. Wortis: Phys. Rev. **B29**, 328 (1984)

17.39 A.V. Babkin, K.O. Keshishev, D.B. Kopeliovich, A.Y. Parshin: JETP Lett. **39**, 633 (1984)

17.40 A.V. Babkin, D.B. Kopeliovich, A.Y. Parshin: Sov. Phys. JETP 62, 1322 (1985)

17.41 P.E. Wolf, F. Gallet, S. Balibar, E. Rolley, P. Nozières: J. Phys. **46**, 1987 (1985)

17.42 F. Gallet, P. Nozières, S. Balibar, E. Rolley: Europhys. Lett. 2, 701 (1986)

17.43 L. Puech, G. Bonfait, B. Castaing: J. Low Temp. Phys. **62**, 315 (1986)

17.44 E. Rolley, S. Balibar, F. Gallet: Europhys. Lett. 2, 247 (1986)

17.45 M.J. Graf, R.M. Bowley, H.J. Maris: Phys. Rev. Lett. **53**, 1176 (1986); J. Low Temp. Phys. **58**, 209 (1985)

18. Solving Complex and Disordered Surface Structures with Electron Diffraction

M.A. Van Hove

Materials and Chemical Sciences Division, Lawrence Berkeley Laboratory, Berkeley, CA 94720, USA

The history of surface structure determination with low-energy electron diffraction (LEED) will be briefly reviewed, setting the stage for a discussion of recent and future developments. The aim of these developments is to solve complex and disordered surface structures. Some efficient solutions to the theoretical and experimental problems will be presented. Since the theoretical problems dominate, the emphasis will be on theoretical approaches to the calculation of the multiple scattering of electrons through complex and disordered surfaces.

18.1 Introduction

18.1.1 The Past

Since its inception as a surface structural tool in the early 1970s [18.1-5], low-energy electron diffraction (LEED) has been used to determine at least 200 structures. These range from simple clean metal surfaces to more complicated metal or semiconductor reconstructions, as well as molecular adsorbate layers. The vast majority of these structures is of the ordered type, i.e. the structure is based on diffraction spot intensities that correspond to ordered parts of the surface. One structure, that of oxygen on W{100}, consists of a disordered overlayer and its structure was obtained via the diffuse intensity that exists between the diffraction spots [18.6].

Another 50 or so structures have been determined relatively completely by other techniques of surface structure analysis. Foremost among these are medium and high energy ion scattering (MEIS and HEIS), surface extended x-ray absorption fine structure (SEXAFS), angle-resolved photoelectron emission fine structure (ARPEFS) and other variants of photoelectron diffraction. The structures determined with these techniques include simple and ordered as well as complex and disordered cases. The more complex structures have often been studied incompletely, e.g. with near-edge x-ray absorption fine structure (NEXAFS). But there has always been a preference for simple ordered structures with these techniques as well, because of a higher chance of success and fewer unknown parameters to be determined.

Recently, a database of solved surface structures has been compiled. It comes in the form of a handbook [18.7] containing a "catalog" of those structures, as well as in the form of an electronic database with computer graphics program [18.8]. Over 250 structures are included: they demonstrate the considerable achievements of surface structural analysis over some 15 years. However, they also clearly point towards new types of surface structure that were so far considered too complicated to solve.

The most complex and "complete" structural determinations to date contain four to five structural parameters. Most of these structures were analyzed by LEED. Examples are:

    - the metal surface reconstructions of Au [18.9], Ir [18.10] and Pt{110}-(1x2) [18.11], illustrated in Fig. 18.1;

    - the structure of S adsorbed on reconstructed Ir{110}-(1x2) [18.12], shown in Fig. 18.2;

fcc (110) − (1 × 2) missing-row model

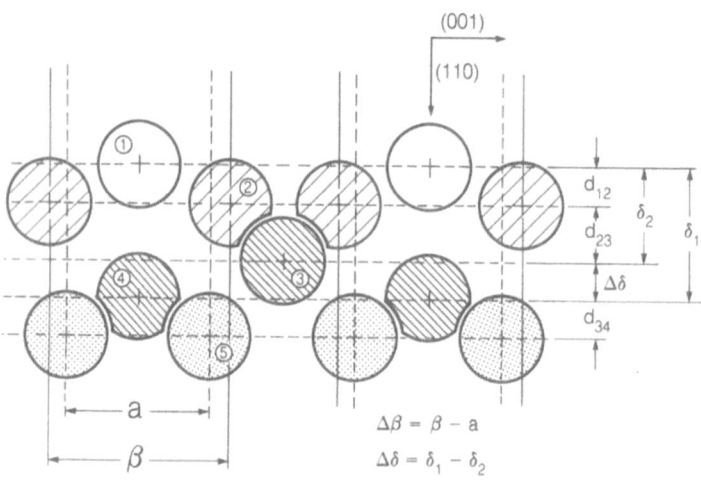

Fig.18.1. Structural diagram of reconstructed fcc(110)-(1x2), representative of the corresponding Au, Ir and Pt surface structures

Ir (110) − (2 × 2) − 2S

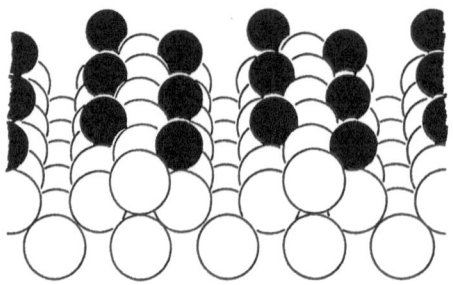

Fig.18.2. Perspective view of sulfur adatoms adsorbed on reconstructed Ir{110}-(1x2)

- the reconstruction of Ir{100}-(1x5) [18.13];
- the semiconductor reconstructions of Si{100}-(2x1) [18.14], Si{111}-(2x1) [18.15] and GaAs{111}-(2x2) [18.16];
- molecular overlayer structures on metal substrates, such as Rh [18.17] and Pd{111}-(3x3)-C_6H_6+2CO [18.18], illustrated in Figs.18.3 and 4, respectively.

This is not to say that other techniques did not contribute to the results. In fact, LEED alone could not have solved many of these structures. Outside input has been essential in eliminating many *a priori* possible models. However, one may say that LEED often produces the most accurate final result for relatively complex structures.

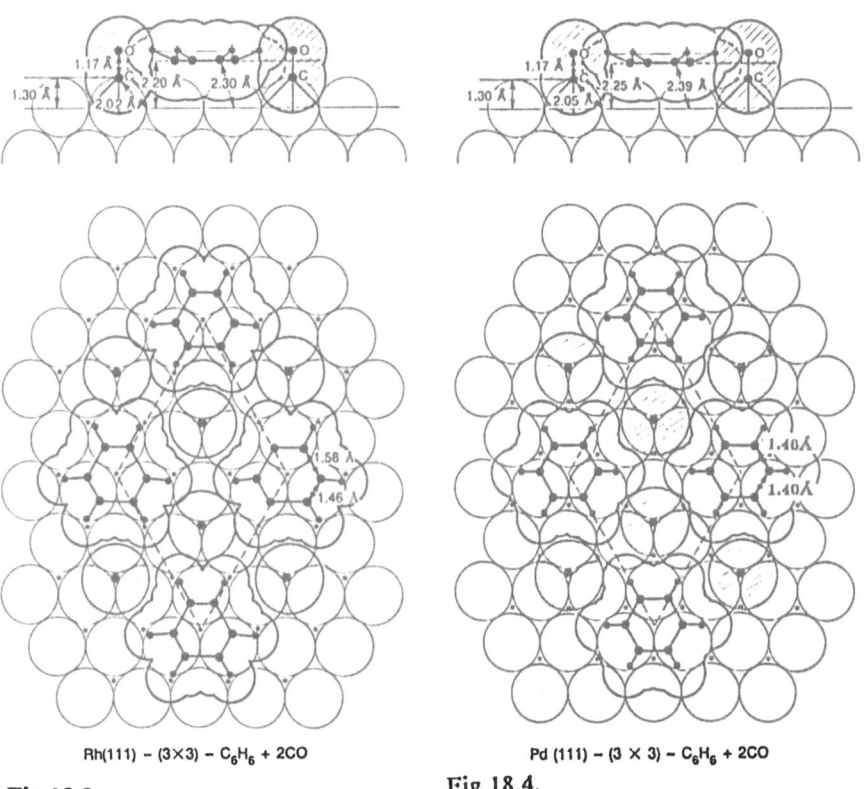

Rh(111) – (3×3) – C_6H_6 + 2CO   Pd (111) – (3 × 3) – C_6H_6 + 2CO

Fig.18.3.   Fig.18.4.

Fig.18.3. Side view (top panel) and top view (main panel) of Rh{111}-(3x3)-C_6H_6+2CO. Van der Waals contours are given for the molecules. The hydrogen positions are assumed. A (3x3) unit is outlined. Selected distances are shown, indicating a large C_6 ring distortion relative to the gas-phase C_6H_6 molecule (for which the C-C distances are all 1.397 Å)

Fig.18.4. Similar to Fig.18.3, but for Pd as a substrate rather than Rh. Note the minor C_6 ring distortion in this case relative to the Rh case. Also, the molecules in this case have adopted different hollow sites than on Rh

18.1.2 The Objective

Clearly, one would like to solve many more complex structures. For instance: stepped surfaces in the clean state or with adsorbates, large-unit cell semiconductor reconstructions and the effect of adsorbates thereupon, overlayers of large molecules, disordered overlayers, defect structures in clean or adsorbate-covered surfaces, and incommensurate overlayers.

To be more specific, here is a list of directions in which it is desirable to develop the LEED technique, as well as other surface-structure techniques:
- structures with many structural parameters;
- structures with large unit cells;
- disordered structures, with lattice-gas disorder, or with other defects;
- stepped surfaces, clean and adsorbate-covered;
- structures with incommensurate overlayers.

We shall look at the recently proposed approaches for dealing with these various forms of more complex and disordered surface structures. We shall limit ourselves to LEED and some of its closely-related electron diffraction techniques.

18.2 Towards Diffraction from Complex and Disordered Surfaces

18.2.1 The Problem

The central difficulty in extending LEED to more complex and disordered structures lies in the theory of multiple scattering. Multiple scattering leads to many possible scattering paths and to selfconsistency requirements. The complications are the same as in solid-state band-structure and molecular-orbital problems. The issue is then to find suitable approximations.

Conventional LEED theory [18.1,4,5] has identified the necessary ingredients of a successful description of multiple scattering in relatively simple ordered structures. The most fundamental ingredients are the muffin-tin model and the description of electron scattering by atoms with the help of partial (i.e., spherical) waves. A partial wave expansion of the scattered wave has to be truncated at a finite value ℓ_{max} of the angular momentum ℓ. This yields $(\ell_{max}+1)^2$ partial waves at each atom, due to the allowed values of the magnetic quantum number m. The (complex) amplitudes of all these partial waves at all inequivalent surface atoms have to be determined in the process of simulating LEED intensities on a computer. Thus the number of atoms in the two-dimensional surface unit cell enters in the case of ordered surfaces; this unit cell extends as deep into the surface as the electrons can penetrate. For disordered surfaces, the unit cell effectively becomes infinitely large and so does the number of inequivalent atoms N. However, many forms of disorder are in some way repetitive if the short-range order is considered (e.g. identical adsorption sites arranged without long-range order), so that the number of inequivalent atoms N need not be large. We shall later discuss more thoroughly why and how the long-range and short-range order can be separated for this purpose.

In Table 18.1 the cost of the conventional LEED computation is characterized in terms of its dependence on ℓ_{max} and N: the high power with which this cost rises clearly indicates where the problem lies. But the problem crops up in different ways as well. We may use the popular plane-wave representation between atomic layers. Then, with ordered structures, the total number of diffracted beams comes into play. This number is equal to the number of plane waves used, and both are simply proportional to the two-dimensional unit cell area A. Table 18.1 shows how the computational cost scales with A: depending on the calculational scheme used, a power law with a power between 2 and 3 is found. This again creates serious problems as the surface structure becomes complex, e.g. with large unit cell reconstructions or with large adsorbates.

With the plane-wave representation, the number of plane waves is also affected by the internuclear spacing between atomic layers: small spacings d require relatively many plane waves. As a result, the computational cost can rise, as Table 18.1 shows, with a strongly negative power of d. This is unfortunate, because the plane-wave representation would otherwise have been an attractive alternative to the spherical-wave representation.

Table 18.1. Dependence of the cost of conventional LEED computations on basic parameters [ℓ_{max}: cutoff angular momentum, N: no. atoms in unit cell, A: unit cell area ($\propto$ no. beams), d: interlayer spacing (small for steps), and E: kinetic energy]

In spherical-wave space	In plane-wave space
$\ell_{max}^{4 \text{ to } 6}$	
$N^{4 \text{ to } 6}$	
	$A^{2 \text{ to } 3}$
	$d^{-4 \text{ to } -6}$
$E^{2 \text{ to } 3}$	$E^{2 \text{ to } 3}$

Finally, we must consider the energy dependence of the computational cost of conventional LEED. With increasing energy E, the value of ℓ_{max} increases slowly but surely. More seriously, the number of required plane waves increases rapidly. These two effects are combined in Tab.18.1 to show a square-to-cubic dependence on E.

A different kind of difficulty with LEED for complex and disordered surfaces is the issue of finding an efficient procedure to determine the many unknown structural parameters. This problem is not unique to LEED, however. It exists equally well, for example, in standard x-ray crystallography, despite its relatively simple diffraction theory.

On the experimental side, two demands arise when dealing with complex or disordered surfaces. First, complexity requires an adequate data base to determine the additional unknown parameters. Thus, rapid and automated data acquisition is desirable. Second, for disordered structures the diffuse LEED intensity needs to be measured. This leads to new approaches, which are already becoming available. We shall treat these experimental issues in Sect.18.6.

18.2.2 Basic LEED Methods

Before entering into descriptions of the newly introduced theoretical techniques, it is necessary to state which basic ingredients these will use. We shall not get into the mathematical formalisms, but stay with a physical description of the processes that are included in the formalisms. More detailed treatments are available in the literature [18.1,4,5].

The basic ingredients to be introduced next are common to virtually all the LEED techniques, whether old or new, and should remain in force for the foreseeable future.

As mentioned earlier, the muffin-tin model is utilized to represent the scattering potential of the surface atomic lattice. It consists of spherically-symmetrical ion-core potentials, surrounded by a constant muffin-tin level. However, this muffin-tin level may change from one atomic layer to another. This happens of course in particular at the interface between lattice and vacuum.

A layered structure is often adopted in LEED to describe a surface. Atomic layers parallel to the surface are defined, in a way appropriate for each particular theoretical method. In a number of cases, a combined-space representation is chosen, which depends intimately on this layer approach. Namely, the wavefield is expanded in spherical waves within those layers, while it is expanded in plane waves in the gaps between those layers.

In the spherical-wave representation it is most common to use free-space Green functions to describe wave propagation from one atom to another. In this representation there are various ways of obtaining self-consistent solutions to the multiple-scattering problem. A "giant-matrix" inversion as proposed by Beeby does the job in a closed form. Then there are perturbation expansions, such as reverse scattering perturbation, which can converge to the same result when multiple scattering is not too strong.

In the plane-wave representation many methods are available to treat multiple scattering. One is the Bloch-wave method with wave matching across the surface. Another is layer doubling. Both of these have a closed form, with the second requiring sufficient damping of the wavefield into the bulk. A popular perturbation expansion of this problem is offered by renormalized forward scattering, in which the predominance of forward scattering is used to achieve rapid convergence (as long as multiple scattering is not too strong and damping not too weak).

We have mentioned damping: this is the effect of inelastic energy losses, which reduce the elastically surviving flux of electrons. The most common way of taking this into account is through a small imaginary part of the scattering potential. It is usually assumed homogeneous and isotropic. Another equivalent approach is to introduce a mean free path length, that also exponentially dampens waves.

Finally, all LEED theories include Debye-Waller factors: these represent the effect of thermal vibration, namely the elastic or quasi-inelastic removal of electron flux into diffuse directions of diffraction. At each scattering in a multiple-scattering chain a Debye-Waller factor is allowed to remove some electron flux.

The new methods introduced for complex and disordered surfaces use the same physical ingredients and many of the same calculational techniques described in this section. The difference often lies in a new packaging of multiple-scattering paths into units that are better adapted to the new tasks. Or some multiple-scattering paths are neglected. The end result either is equivalent to that of the old way, or is a suitable approximation to it.

18.2.3 Theoretical Solutions

We shall now describe general strategies that have been proposed to overcome the computational problems of LEED discussed in Sect. 18.2.1. We shall also briefly introduce specific methods. Some of these will be discussed individually in more detail in Sect. 18.4,5.

a) Cluster Methods

With complex and disordered surfaces, plane waves lose some of their attractive features: their number increases rapidly with large unit cells, while their suitability becomes less obvious in the disordered case, since beams become diffuse. In addition, the long-range periodicity becomes a secondary aspect of the structure, in the sense that the immediate environment of each atom shows no sign of long-range order.

Thus, cluster methods have been proposed: in these the spherical-wave representation is used within clusters representing suitable neighborhoods of surface atoms. "Suitable" means: if one chooses a particular surface atom, the cluster should be large enough to include all significant multiple-scattering paths which pass through that particular surface atom. The finite electronic mean free path ensures that the cluster radius can be made finite, for instance as small as perhaps 10 Å. As a consequence, one may have to choose a different cluster to represent the neighborhood of each inequivalent surface atom. This is schematically illustrated in Fig.18.5. In general, the clusters based on different atoms overlap each other.

The problem is hereby reduced to solving the multiple scattering within each cluster and then somehow combining the individual results. In this way, the computing time is made proportional to the number of translationally inequivalent surface atoms, a great improvement over its second or third power. But it does remain proportional to a high power of the number of atoms within each cluster.

Early cluster methods include those proposed by *Duke* et al. [18.19] and by *Moritz* et al. [18.20] for the calculation of diffuse intensities due to surface disorder. These methods were rather restricted in their applicability due to high computational cost. For example, the approach of *Moritz* et al. could only handle very small clusters to rep-

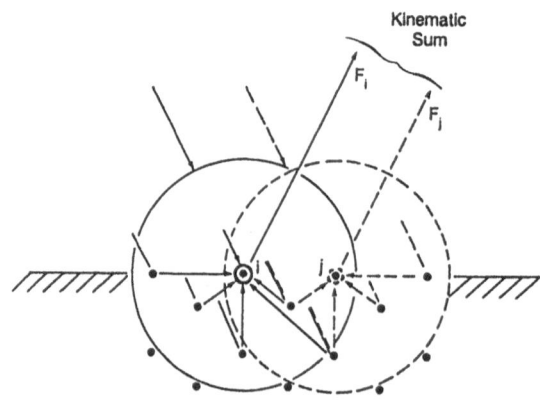

Fig.18.5. Schematic representation of two clusters built around two different surface atoms. The two spherical clusters overlap in the sense that some scattering paths are common to both clusters

resent the multiple-scattering effect of each atom's neighborhood, while the number of inequivalent clusters must remain small. But these approaches paved the way for future developments. They already included the concept of "kinematic cluster addition" (KCA), in which the total scattered wave is composed of kinematically added waves leaving inequivalent surface atoms after multiple scattering within each cluster.

A later development was the one-center expansion [18.21], in which the spherical-wave scattering properties of an entire cluster were obtained selfconsistently and then were available for calculating plane-wave diffraction in many situations. The approach was made efficient by breaking down the clusters into concentric shells and separating the problem of the scattering by individual shells from that of the scattering by the assembly of shells [18.22,23]. A variant of this method was developed for steps, or, more generally, for linear defects [18.24a]. Here, cylindrical waves were used to describe the multiple scattering within individual chains of atoms, which are chosen parallel to the linear defect. These chains were then combined into three-dimensional surfaces (this method has close analogies with the earlier "chain method" developed for medium and high energy electron diffraction [18.25]). Another cluster method was proposed by *Marcus* et al. [18.26]. These methods exhibit the common feature that the computing time scales in simple proportion to the number of inequivalent surface atoms.

b) Reducing Multiple-Scattering Paths

Another strategy starts from the observation that the computational cost is directly due to the large number of multiple-scattering paths that need to be calculated. Perhaps one could identify and ignore large classes of such paths that contribute only weakly to the diffracted intensities. In dense metal surfaces, for which the conventional LEED theory was developed, it is difficult to find classes of weak paths.

But in many other materials, multiple scattering is much less important. For example, aluminum and silicon are more "kinematical" materials as far as LEED is concerned. And so are many other open-lattice semiconductors, as well as many low-density materials and overlayers, especially molecular overlayers.

Two factors reduce multiple scattering: a loose packing of the material and a light nuclear weight. A loose packing means larger interatomic distances on the average, which reduces the number of possible multiple-scattering paths, given a constant mean free path. A light nuclear weight reduces the amplitude of any scattered wave, especially of a multiple-scattered wave.

Weak multiple-scattering paths can be eliminated in several ways. First, one may simply shrink the size of the above-mentioned clusters, within which one allows multiple scattering around each atom. In this manner, the individual cluster calculation is made much faster. This approach already had to be followed by *Moritz* et al. [18.20]. It stands the best chance of being reasonably accurate in loosely packed materials, where the average distance between atoms and their various shells of neighbors is relatively large. Thus, semiconductors and molecular materials are good candidates for this treatment.

It is also possible to prevent long distances between successive scatterings in a multiple-scattering path: far fewer paths are then allowed. Thus, in the method called "near-neighbor multiple scattering" (NNMS) [18.27], a wave is only allowed to travel unscattered from a given atom to its nearest neighbors. In other words, a multiple-scattering path is only allowed to consist of short hops between nearest neighbors (but a string of short hops may lead far away). There may be as many hops as the mean free path allows. This approach does not reduce the scaling power of the computing time dependence on N, the number of atoms in a unit cell or cluster, but it does reduce its prefactor considerably.

c) Kinematic Sublayer Addition

A simple but very effective variant is "kinematic sublayer addition" (KSLA) [18.28]. It applies to the case where the clusters are disjoint, i.e. no multiple-scattering path can hop from one cluster to another. An example is molecular layers, where the hop from an atom in one molecule to an atom in another molecule is often too large to provide a significant contribution to intensities. In that case the scattering properties of the separate molecules can be calculated independently. They can then be kinematically combined and recombined for many different relative positions of the different molecules.

d) Forward Focusing

One may also exploit the well-known predominance of forward scattering of electrons by atoms. The idea is to neglect paths that include many sharp bends. An old implementation of this approach is the "quasi-dynamical" theory [18.29,5], in which intralayer multiple scattering is calculated kinematically, while interlayer multiple scattering is calculated more exactly.

This kind of approach is especially useful at higher energies, above about 200 eV, where forward scattering is so pronounced as to become forward focusing. A cluster-oriented method that specifically exploits forward focusing is called "near-field expansion in clusters" (NFEC) [18.30]. Here the wave field scattered by an atom is expanded about a nearby atom center located most often in the forward direction, using a "Taylor series, magnetic quantum number expansion" (TS-

MQNE) [18.31]. This replaces the conventional accurate representation of the scattered wavefield in all directions around the scattering atom. The resulting computational gain is considerable. The cost becomes proportional to the square root of the energy and directly proportional to the number of inequivalent atoms. And at higher energies more experimental data can be gathered, as a larger part of reciprocal space is accessible.

e) Beam Set Neglect

Next we consider a method that is plane-wave oriented. Thus, the intention is to identify and eliminate sets of plane waves that do not contribute significantly. In this case, one holds on to the plane-wave representation as much as possible, but simplifies it to bring its escalating cost under control. Such methods often combine well with cluster approaches, cumulating the benefits.

The "beam set neglect" (BSN) method [18.32] applies to overlayers or reconstructed layers with a superlattice or a disordered lattice on a perfect substrate. It recognizes the fact that only a very limited set of plane waves (beams) contribute significantly to the detected intensities: many sets of beams can be neglected in the calculation. As a result, the dependence of the computation cost on the unit-cell area A and the energy E fall from the second or third power to no dependence. In other words, the computation cost can be made independent of the energy, the unit-cell area or the presence of disorder. This approach works also for incommensurate overlayers. BSN can be very effectively combined with cluster methods, in particular the KSLA method.

f) Tensor LEED

A third basic approach is not go give up multiple-scattering paths, but to approximate their contributions as being linear expansions from a nearby surface geometry that was treated exactly. In other words, one would compute exactly the LEED intensities for a given reference structure, preferably a simple one: for example, a highly symmetrical or undistorted structure. Then, the intensities for a somewhat deviating structure would be computed, using a linear expansion in terms of the structural parameters. This yields excellent computation times, especially if many deviating structures around the reference structure are explored, because the linear expansion itself is a very simple operation. This is the philosophy of tensor LEED [18.33], which has already been applied to periodic surface reconstructions and to disordered overlayers that induce substrate distortions.

g) Combinations

It should be pointed out that many of the techniques described above can be combined to cumulate their advantages. This is especially so

523

because in LEED one often decomposes a surface into layers parallel to the surface. Each layer can be defined and treated with a method that best suits its structure: often this will be a spherical-wave and/or cluster method. And between layers one may apply plane waves and choose approximations accordingly. This gives great latitude for finding efficient, although case-dependent, methods to solve complex and disordered surface structures.

If one uses approximate methods as proposed above, one stands the obvious risk of obtaining incorrect structural results. For that reason, a sequential approach can be used: one may start a structural search with a broad analysis of many possible structures using an approximate and efficient computational method. In this way many unpromising structures can be rapidly eliminated from consideration. The next step is to concentrate on the relatively few remaining promising structures and to test them with a more accurate LEED theory. The process can be repeated with increasing accuracy, yielding not only a (hopefully) unique structural solution but at the same time a refined structure.

18.3 Order vs Disorder and Diffraction

It will be very helpful for our further discussions to analyze more clearly how order and disorder affect the diffraction of electrons by a surface. This is rather well understood for kinematic diffraction (e.g. in x-ray diffraction), but often leads to some confusion in LEED due to multiple scattering.

Figure 18.6 summarizes a fundamental distinction that has to be made here. It illustrates the case of an ordered surface. Such a surface is characterized by a "lattice" and a "basis". The lattice describes the

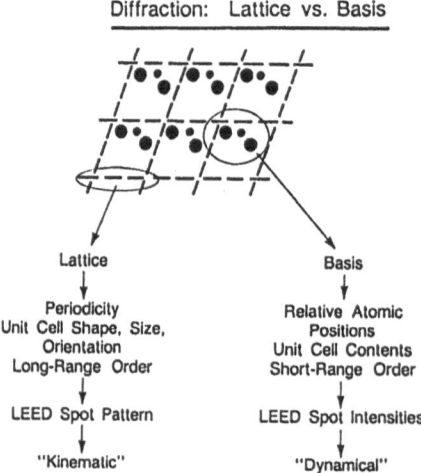

Diffraction: Lattice vs. Basis

Lattice	Basis
Periodicity Unit Cell Shape, Size, Orientation Long-Range Order	Relative Atomic Positions Unit Cell Contents Short-Range Order
LEED Spot Pattern	LEED Spot Intensities
"Kinematic"	"Dynamical"

Fig.18.6. The distinction between lattice and basis is emphasized, together with the different implications for electron diffraction

524

periodicity of the surface, i.e. the shape, size and orientation of the unit cell, but not its contents. Thus, the lattice describes only the long-range structure. It is responsible for the presence of sharp LEED beams, which correspond to the reciprocal lattice of the surface lattice, as given by the Bragg diffraction conditions. It is important to realize that although these are purely kinematic considerations, they are not affected by the presence of multiple scattering. Indeed, the LEED spot pattern can be very accurately predicted with the kinematic Bragg conditions. We shall call this lattice-induced contribution the "structure factor", in close analogy with x-ray terminology.

The situation is different with the "basis". The basis is the set of atoms that is contained within any unit cell, together with their individual positions and scattering properties. The basis therefore relates primarily to the short-range order. The intensities of the LEED beams (rather than their existence and direction) is governed by the basis. These intensities are strongly affected by multiple scattering. They depend in particular on the relative positions of the basis atoms, not only through the kinematic phase factors, but also very much through the multiple scattering. Therefore, intensities are often said to be "dynamical" and this is of course the origin of the label "dynamical LEED". We shall call this basis-induced contribution the "form factor".

Thus, the diffraction pattern is primarily determined by long-range order, while the diffraction intensity is primarily determined by short-range order. In fact, one may view the observed LEED to be the product of the smoothly varying basis-induced form factor and the sharply spiked lattice-induced structure factor.

This realization is central to understanding diffuse LEED due to disordered surfaces. Let us take an ordered surface and gradually disorder it in such a way that the short-range order is not changed appreciably. This is most readily visualized with a low-coverage overlayer that becomes disordered without changing the individual adsorbate sites (so-called lattice-gas disorder). Then the multiple scattering is essentially unchanged, because the neighborhood of each atom is largely unchanged. What matters here is the neighborhood within, say, one mean free path length. Thus the smooth "basis-induced" form factor is almost unchanged from the ordered case. However, the loss of long-range order has a profound effect on the "lattice-induced" kinematic part of the problem: there are no Bragg conditions anymore, but instead diffraction may occur in any direction. The structure factor and therefore the LEED pattern become diffuse (however, there may still be sharp spots, due solely to the still-ordered substrate).

Now again, the final pattern can be viewed as the product of a kinematic structure factor reflecting the long-range disorder and a dynamical form factor reflecting the local short-range order. However, it is no longer easy to distinguish these two contributions. The structure factor of the disordered part of the surface no longer stands out as sharp spots. But an I-V curve, which shows the intensity measured as a function of electron energy for constant momentum transfer parallel to the surface, will look very much like that of a sharp spot: most of the structure of such an I-V curve is due to the short-range order, whether the surface is ordered or disordered. This is because a constant momentum transfer parallel to the surface keeps the structure factor constant in this measurement (at least if there is purely two-dimensional disorder only). A more mathematical treatment of these questions is available elsewhere [18.34].

18.4 Cluster-Oriented Approaches

We shall now discuss in some more detail a few modern LEED methods that are based on clusters. They involve primarily the spherical-wave representation. Here again, we shall refrain from showing detailed formalisms and concentrate on their physical meaning.

18.4.1 Kinematic Cluster Addition

Let us consider the electrons that are detected after the diffraction process from a surface. They are represented by a wave that has left the surface. It is easy to imagine this wave to be composed of individual waves that have left each atom of the surface. Let us choose one such wave having left a particular atom. This wave includes all possible ways for electrons to travel from the source (electron gun) into the surface, via any number of scatterings, until final scattering by that particular atom. In other words, we have classified all possible multiple-scattering paths according to which is the last scattering atom in these paths. Figure 18.5 illustrates the situation: waves with amplitudes F_i and F_j leave atoms i and j, respectively. They are built up of all possible scattering paths that end up in atoms i and j, respectively.

Since we have decomposed the final wave into a superposition of waves coming from different surface atoms, the task is now reduced to calculating the waves coming from each surface atom separately. Then, their superposition is a kinematic addition of wave amplitudes, i.e. an addition of amplitudes that takes into account the phases of the different waves due to the relative positions of the last scattering atoms. Such an addition of amplitudes is the same as that practiced in kinematic x-ray diffraction, for example. The difference is that in x-ray diffraction the wave amplitudes are simply given by the atomic scattering amplitudes, while in LEED they are much more complicated due to multiple scattering before arrival at each surface atom.

The significance of this approach is that the computational problem has been made proportional to the number of inequivalent atoms at the surface, rather than scaling with a high power of that number. (We say "inequivalent" atoms, because if two atoms have identical surroundings, the waves leaving those two atoms towards the detector will be identical, except for the simple kinematic phase factor.)

We are, however, left with the significant problem of calculating the individual waves leaving the inequivalent surface atoms. This is where the cluster concept enters. We choose a cluster of finite size around such a surface atom, with a radius comparable to the mean free path. We then allow the incident electron to travel in any way that it wishes through that cluster up to the chosen surface atom. The result is the desired wave amplitude leaving that surface atom.

Practical methods of solving this multiple-scattering problem in a cluster have been slow in coming. Perhaps the most attractive at present is that of *Pendry* and coworkers [18.22, 23]. They propose that a cluster be decomposed into concentric shells of atoms around the surface atom of interest. First the scattering properties for each shell taken separately are calculated. This is done in the spherical-wave representation to take advantage of the spherical geometry. The incoming electron wave is also decomposed into spherical waves centered on the surface atom in question. The spherical waves are then allowed to propagate inward from shell to shell (including any number of back-reflections between shells) until they arrive at the central atom. The result of the last scattering by the central atom is the desired wave amplitude. This approach is analogous to the layer-by-layer approach of layer doubling or renormaiized forward scattering in the plane-wave representation: the role of the layers is now played by the shells.

Note that the method has made no assumption about long-range order: it applies equally well to ordered and disordered surfaces. In fact, one may treat many kinds of surfaces with this approach. Any form of point defect can be handled, such as an adatom, a vacancy, a substitution, an interstitial, including local distortions caused by such defects. Ordered defects represent ordered overlayers, reconstructions, alloys, etc. Disordered defects represent random adsorbates, impurities, distortions, etc.

The method has been applied to the solution of the adsorption structure of disordered oxygen on W{100} [18.6]. Not only was the oxygen position determined from the diffuse intensities, but more recently a significant oxygen-induced distortion of the surrounding W lattice was found [18.33]: the oxygen atoms occupy fourfold hollow sites of the metal substrate, and they pull the four surrounding W atoms inward, reducing their mutual W-W distances.

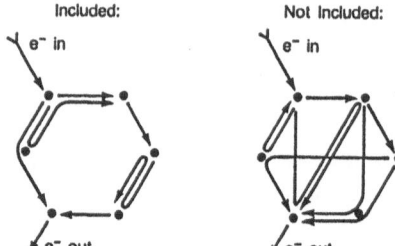

Near-Neighbor Multiple Scattering (NNMS)

Included:

e⁻ in

e⁻ out

Not Included:

e⁻ in

e⁻ out

Fig.18.7. The principle of near-neighbor multiple scattering is exhibited by examples of included paths (left diagram) and examples of excluded paths (right diagram). The scattering cluster is taken as a hexagonal ring of atoms. An electron is allowed to scatter only between nearest neighbors in the NNMS method

An approximation to the cluster method is achieved by restricting hops in a multiple-scattering path to only nearest neighbors. Figure 18.7 illustrates what kind of path is accepted and what kind is not. This "near-neighbor multiple scattering" scheme [18.27] severely reduces the number of possible paths and thereby reduces the computational effort. As mentioned in Sect.18.2.3, this approach reduces the prefactor of the computing time rather than the power law of its dependence on the number of atoms in the unit cell or in the cluster. It therefore does not in itself make a very big difference, but does help other methods perform somewhat faster.

18.4.2 Near-Field Expansion in Clusters

We describe here a method [18.30] that utilizes the fact that forward scattering of electrons by atoms dominates over scattering in other directions. This "near-field expansion in clusters" (NFEC) stems from a method developed for angle-resolved photoelectron diffraction at surfaces: ARPEFS ("angle-resolved photoelectron fine structure"). It was designed for slightly higher energies than is customary in LEED, but can be used down to LEED energies.

Forward scattering has long been known to be important in LEED, especially towards higher energies (>100 eV). In fact, it was the basis for the renormalized forward scattering method. However, all LEED methods so far have used an exact description of the scattered wave not only in the forward direction, but all around the scattering atom. This is a waste of effort if most scatterings occur in the forward direction. The NFEC method therefore performs an expansion of the scattered wave about the next atom that will receive it "downstream" (the procedure is repeated if there are other atoms downstream of a given scatterer). The expansion is done in two dimensions: radially and azimuthally. By choosing suitable local coordinates for each scattering, the expansion has few terms, resulting in very efficient computation.

This is a cluster method, because the scatterings from atom to atom are followed up individually in chains of multiple scatterings that are restricted to a cluster. The shape and size of the cluster is governed by the mean free path and by the cone formed by the forward scattering. The algorithm continuously adjusts that shape and size as it proceeds from scattering to scattering.

Being a cluster method, the computation cost is simply proportional to the number of inequivalent atoms in the surface. In terms of energy dependence, NFEC scales as the square root of the energy. Being a spherical-wave method, its performance does not depend on the unit cell size or on the distances between atomic layers. And it scales only linearly with the number of partial waves used, ℓ_{max}.

NFEC has been applied to reproduce and explain in more physical detail than possible so far the LEED intensities from simple surfaces. And it has behaved as efficiently as expected for larger energies, providing a very attractive way to calculate medium-energy electron diffraction intensities. It can now also be used to attack complex and disordered surfaces.

A variant of NFEC has been developed [18.35,36] to study the emissions of electrons from surfaces by Auger and other inelastic processes at the higher energies, say 500 to 1000 eV. This is directly analogous to angle-resolved photoelectron emission: in some manner an electron of a given energy is created within the surface; it then elastically travels out, undergoing multiple scattering in the process; the resulting emitted electron distribution reflects the surface structure and can be used to obtain it. Relative to the LEED problem, there are two main differences here. First, the electrons of interest are deemed to start their useful life at some atomic location within the surface rather than at an electron gun outside the surface. Second, the electron "creation" is an incoherent event, unrelated in its phase to any other electron creation elsewhere.

Corresponding full multiple-scattering computations have been performed for the first time and match experimental data well [18.36]. For instance, Auger electron emission data were analyzed for monolayers of Cu embedded within a Ni{100} surface: the depth at which the Cu layers were embedded could be determined. Also, inelastic "Kikuchi-like" electron emission was reproduced, which allowed the site determination of oxygen adsorbed on Mg{0001}: the oxygen atoms were found to be preferentially buried interstitially in octahedral sites below the first and second metal layers.

An interesting feature of this electron emission is that forward-focusing peaks are observed experimentally: along crystallographic chains of atoms one systematically finds an enhancement in electron emission, within a cone of perhaps 10° half width. The reason for this is that a spherical wave leaving one atom is focussed behind each neighboring atom as if the latter acted like a converging lens. This phenomenon was first observed and explained by *Egelhoff* in the case of photoemission and later Auger electrons [18.37].

However, the NFEC calculations [18.36] clearly show that other peaks not related to interatomic directions exist as well. These are due to interference effects between different paths leaving the surface. Thus one has to be careful not to interpret every peak as an interatomic direction: a full calculation is normally required to extricate these two kinds of peaks due to focusing vs. interference.

18.5 Beam-Oriented Approaches

18.5.1 Plane Waves Despite Large Unit Cells and Disorder

Despite the advantages of spherical waves for treating diffraction by complex and disordered surfaces, plane waves retain a great attraction. They are mathematically always far easier to treat and lead to very efficient computation whenever they can be used. This has been recognized a long time ago in the case of electron diffraction at surfaces. For example, photoelectron emission has been treated with plane waves for over a decade, even though there are no naturally obvious plane waves in the problem. As mentioned in the last section, such electrons start out as spherical waves emitted by some atom and then propagate via multiple collisions out from the surface.

However, the detector by its very presence and narrow angular aperture defines a clear direction: from surface to detector. This direction in turn singles out a plane-wave component of the emitted spherical wave, while the other plane-wave components travel undetected past the detector. Thus a well-defined plane wave can be chosen. It can then be traced back toward the surface and, in reverse time, its diffraction by surface layers can be followed. If a periodical surface layer is encountered, the diffraction leads to a complete set of plane waves that correspond to the two-dimensional reciprocal lattice of that layer. This is directly analogous to the plane-wave diffraction process in standard LEED, where a periodic surface generates a set of well-defined diffracted beams. In the time-reversed thinking process, one can follow up the waves back to their source. This might be an electron gun, in the LEED case. It could also be a photoemission or Auger electron or other electron emission process.

In the particular case of LEED from disordered surfaces, there is a diffuse diffraction process to be included. Incoming electrons may diffract as plane waves from perfectly periodical layers, but as soon as a defect is encountered, any plane wave scatters into spherical waves. However, we are again only interested in those electrons that will travel in the one direction in which the detector is located. Thus, again plane waves are defined and the calculation can focus on those and ignore other scattered waves.

The general picture we get is the following. If there is an electron gun in the problem, it defines an incident plane wave, together with scattered plane waves related to it by the surface reciprocal lattice. And if there is an angle-resolving detector as well, it defines an emitted plane wave, together with other plane waves related to it by the same surface reciprocal lattice. Thus we obtain two sets of important plane waves. These can be selected out, while any other plane waves can be neglected. One does pay a price for this in the case of large-unit-cell or disordered overlayers. Some plane waves with non-vanishing amplitude are neglected in this process. We shall discuss below for individual techniques whether the results are adversely affected.

18.5.2 Beam Set Neglect

As discussed earlier, there is great benefit in reducing the number of plane waves that a calculation needs. This is because all the plane waves consistent with the periodicity of the surface are interconnected by the multiple scattering: as a result, all the plane-wave amplitudes must in principle be computed simultaneously in a selfconsistent manner, and thus large calculations result when there are many beams.

In the layer treatment of surfaces, we would compute the diffraction properties of the overlayer and of the substrate taken separately. These diffraction properties would be obtained for all the beams (plane waves) consistent with the periodicity: in the ordered case, the beams are all the superlattice-implied beams; in the disordered-overlayer case, one would have to include the infinite set of all possible plane waves, an impractical proposition. Then, with layer-stacking techniques such as renormalized forward scattering or layer doubling, the overlayer would be added onto the substrate, yielding the final beam intensities.

a) Two Important Beam Sets

The beam-set-neglect method simplifies this problem as follows. Let us focus our attention on a particular detected plane wave and ask for its intensity at the detector. According to Sect. 18.5.1, two sets of plane waves are then particularly important. We shall define those two sets more accurately now for our present purposes [18.32]:

1) The set of plane waves $\mathbf{0}+\{\mathbf{G}\}$, where $\mathbf{0}$ represents the incident plane wave and $\{\mathbf{G}\}$ represents all the two-dimensional reciprocal-lattice vectors of the substrate; this defines the "integral-order beams" and we shall call them the "incident set". It is illustrated in Fig. 18.8 as the beams $\mathbf{k}_{in}+\mathbf{g}_{1x1}$, to stress the role of the substrate periodicity.

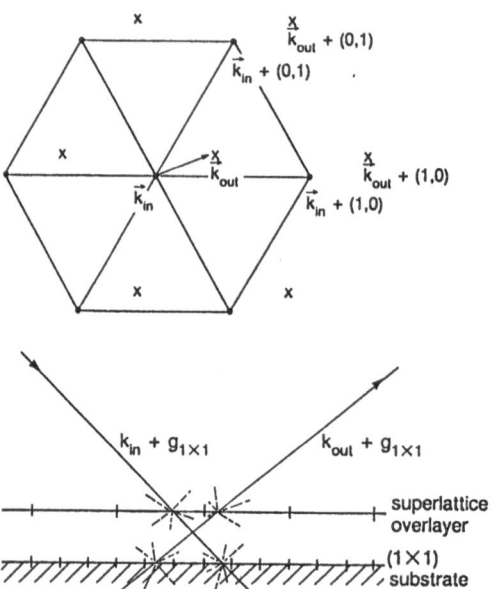

Beam Set Neglect

Fig.18.8. The two beam sets used in the beam-set-neglect method are shown. In the top diagram, the dots represent beams (plane waves) obtained from the incident beam by diffraction from the substrate lattice. The crosses represent a similar set of beams obtained from the outgoing detected beam. These two sets are schematically shown in the bottom diagram, emphasizing scattering at the overlayer and separately at the substrate

2) The set of plane waves $k_f+\{G\}$, where k_f represents the final detected plane wave and $\{G\}$ again represents all the two-dimensional reciprocal-lattice vectors of the substrate; this defines a shifted copy of the "integral-order beams", such that one of them goes straight to the detector: this set can be called the "detected set". Figure 18.8 shows this set as the beams $k_{out}+g_{1x1}$, again stressing the role of the substrate periodicity.

The reason for the choice of the substrate vectors $\{G\}$ is this: the substrate lattice can only diffract a beam by adding a vector G to its parallel component. This means that the substrate cannot diffract a beam out of set 1 into set 2 or any other such shifted set of beams. Only the overlayer, whether ordered or disordered, can diffract out of these two beam sets. As we shall see below, the result is that only higher-order multiple-scattering terms are excluded by restricting ourselves to these two beam sets.

If we are dealing with an ordered overlayer, the detected set corresponds to choosing a particular "fractional-order beam" $k_f=g$, and including all the beams obtained from it by the substrate reciprocal lattice: $g+\{G\}$. Thus, if there are n "integral-order beams" included in the calculation, the calculation would use 2n plane waves. Let us compare this with the conventional LEED calculation. If the overlayer has a unit cell with an area F times larger than the substrate unit cell, then the conventional approach would include Fn rather than 2n plane waves. When F is large, this makes a big difference, since it is the second or third power of the number of beams that comes into play: we obtain a reduction in computational effort by a factor of about $(2/F)^2$ or better, which can easily be one or more orders of magnitude. For example, with a (3x3) overlayer, the reduction factor is $(2/9)^2 \simeq$ 0.05 or better.

However, the 2n-beam calculation only computes the intensities for those 2n beams, since all the others have been excluded. In many cases, this subset of 2n beams may already be sufficient to do a structural determination: then the beam-set-neglect approach has delivered a result at a computational cost which does not depend on F, the unit cell size.

For more accuracy, and especially when the structure is complex and involves many unknown parameters, one may wish to use more than 2n beams in the comparison with experiment. This can be handled quite simply by adding more computations similar to the one just described: one chooses a new detected beam that was not already computed and makes a new computation with the new pair of beam sets based on the incident beam and this new detected beam. This produces the intensities of 2n beams, n of which are duplicates of the previously calculated integral-order beams. That leaves n new intensities, giving 3n overall so far. This can be repeated for other emitted directions, such that each additional computation adds n more intensities. One can even go all the way to exhaust all F sets of beams, requiring a total of F-1 computations. Thus, even in this worst case, the total computational cost is only directly proportional to the area of the unit cell, rather than to its second or third power.

Other combinations of beam sets can be used to enhance the accuracy of the computation, but at a cost in computational effort. For example, there is no need to neglect all but two beam sets. One might include more than two such beam sets at once in the same computation. This reintroduces a number of multiple-scattering pathways and thereby enhances the accuracy. In the limit of inclusion of all beam sets in a single calculation, one simply recovers the full conventional LEED formalism. Thus the accuracy of the beam-set-neglect approach can be tailored to one's needs.

This is illustrated by Fig.18.9, in which different levels of accuracy are applied to the calculation of I-V curves for an overlayer of benzene molecules on a Rh{111} substrate, with a periodicity of (3x3). The curves labeled (3x3) are calculated using all beams, i.e. without beam set neglect (but without multiple scattering within the molecular layer, which is not relevant to the comparisons made here). The integral-order beams (10) and (01) also show curves labeled (1x1). When the detected direction coincides with one of the integral-order beams due to the incident beam, then the two sets 1 and 2 defined above are identical to each other. Therefore, the two beam sets collapse into just the set 0+{G}, i.e. the set of waves that the clean surface would produce. However, their intensities are much different from the clean-surface intensities, also shown in Fig.18.9. It can be seen how well beam set neglect reproduces the "exact" (3x3) curves of the (10) and (01) beams.

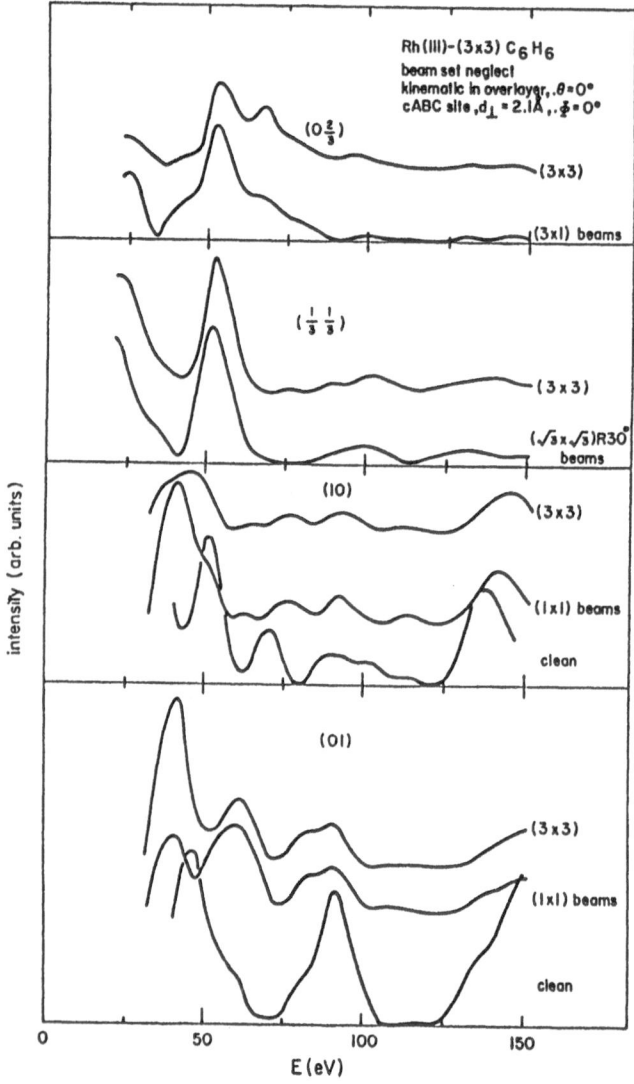

Fig.18.9. Calculated I-V curves for a layer of benzene on Rh{111} with a (3x3) periodic superlattice. The curves marked "clean" correspond to the adsorbate-free substrate. The curves marked "(3x3)" are due to an all-beam calculation: they can be viewed as "exact" (except for the constant neglect of multiple scattering in the overlayer in all cases shown here). The other curves are due to various levels of beam set neglect, as described in the text

534

Two "fractional-order" beams are shown in Fig. 18.9. The intensities of one of them has been approximated by using the "(3x1) beams". This is a particular group of three sets of beams, rather than the two sets of the standard beam-set-neglect method. Again one sees very good agreement in the structure of the I-V curves, especially in the positions of peaks and valleys. These are the most important features of I-V curves for structural determination. The other fractional-order beam is also treated with a group of three (other) sets of beams, and the same comments apply to it.

b) Accuracy of Beam Set Neglect

We must next discuss in more detail the question of the accuracy of beam set neglect relative to the "exact" conventional calculation, which does not neglect plane waves. To that end, we can ask which multiple-scattering paths are ignored due to beam set neglect. The answer is simple: only scattering paths with three or more scatterings are ignored (none of these scatterings being in the forward direction). And, since three or more scatterings (other than purely in the forward direction) give weak contributions, we are only ignoring weak corrections to the final intensity.

That we are only ignoring paths with three or more scatterings can be easily seen as follows. The neglect of a certain beam g' not included in the two sets defined above prevents multiple-scattering paths of the kind: scattering from the incident beam 0 to beam g' by the overlayer, then reflection from g' to g' by the substrate, and finally scattering from g' to g by the overlayer, where g is the final detected beam. Thus, three scatterings are needed in this path. More generally, at least three scatterings will be found in any path that includes an intermediate beam of the excluded type g'.

Thus, whenever multiple scattering is not too strong, beam set neglect should work well. This favors systems where the scatterers in either the overlayer or the substrate are not too strong, e.g. organic overlayers on metallic substrates. Figure 18.10 illustrates the accuracy of beam set neglect in a practical situation: a layer of benzene molecules lying on a Rh{111} surface with a $c(2\sqrt{3}x4)$rect periodicity [18.32]. The figure quantifies the agreement between theory and experiment for two different hollow adsorption sites, and as a function of the spacing between the top Rh nuclear plane and the C_6 nuclear plane. The comparison with experiment is made both without and with beam set neglect. The differences between the two cases are seen to be very small, and both clearly favor the same geometry: the "bABC hollow" site with indistinguishable layer spacings. (The multiple scattering within the molecules was in this case calculated using the near-neighbor multiple scattering method).

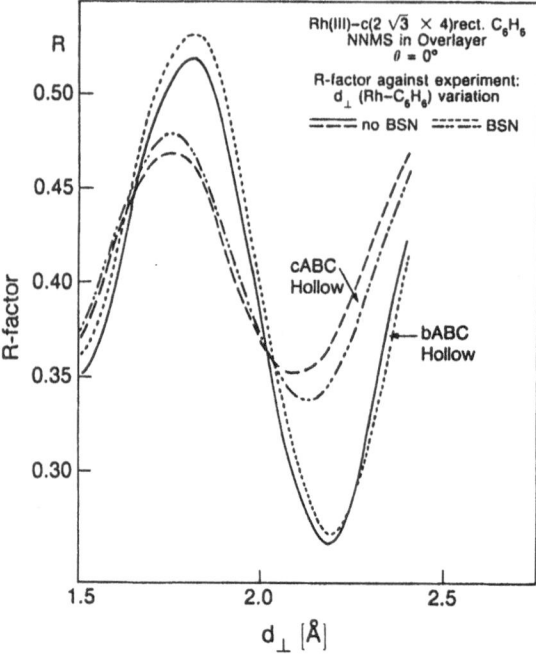

Fig.18.10. I-V curves calculated with and without beam set neglect are compared with experiment for benzene on Rh{111}, by means of an R-factor. The interlayer spacing between substrate and molecules is varied to show the effect of BSN on structural determination. Two adsorption sites (cABC hollow and bABC hollow) are tested

c) Surface Reconstruction

The beam-set-neglect method can be applied to surface reconstructions as well as overlayers. One then merely considers the reconstructed layer to constitute one overlayer. More generally, beam set neglect assumes that the surface consists of a substrate that supports an overlayer which has a superlattice or disorder. The method exploits the contrast between the substrate lattice and the overlayer lattice. This, however, implies that the method has no advantage in the case of a very thick overlayer or reconstruction, since then the electrons do not reach the substrate and the method cannot benefit from its different lattice.

d) Disorder

We have not much differentiated between ordered and disordered surfaces in the discussion so far about beam set neglect. Indeed, there is little difference in the theoretical justification of the method for these two cases. For diffuse intensity calculations, one has the option of obtaining an I-V curve for the emitted direction, or a two-dimensional distribution of intensity across the screen.

In the former case, our reasoning [18.34] and theoretical tests indicate that one gets the same I-V curve appearance as if the surface were ordered. In other words, to measure the I-V curve of a spot due to an ordered surface or of the disorder-induced diffuse intensity in the same position as the spot gives the same result. This confirms that it is the local short-range order that gives rise to the peak-and-valley structure of I-V curves. However, one must pay attention to the fact that traditional spot I-V curves are measured by following the spot as it moves across the screen with energy. This keeps the parallel momentum transfer constant. The corresponding I-V measurement of diffuse intensities must also maintain a constant parallel momentum transfer, i.e. track the position where a spot would be if the surface were ordered.

A similar word of caution applies to the measurement of the two-dimensional distribution of intensity at a given energy. A typical distribution which we have calculated with beam set neglect is presented in Fig.18.11. One often will need the logarithmic derivative of the in-

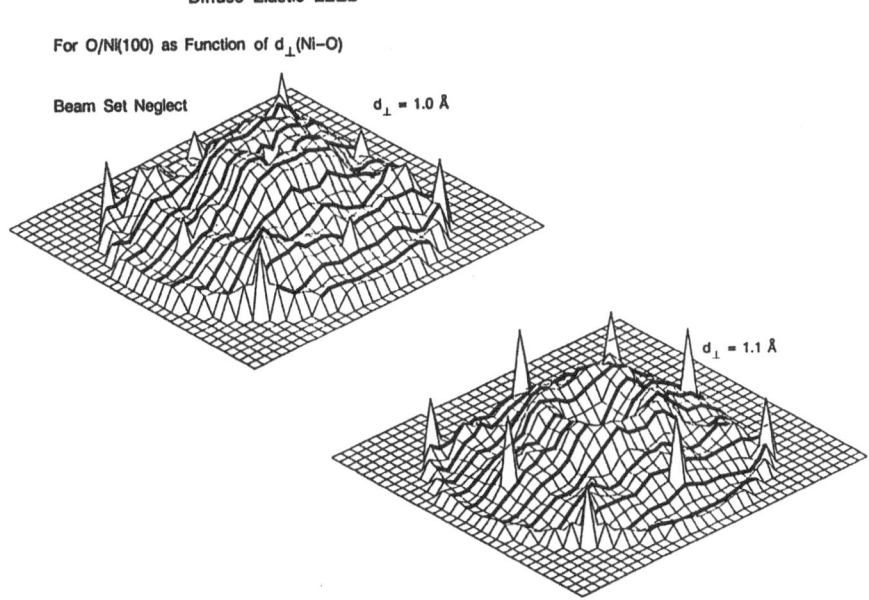

Diffuse Elastic LEED

For O/Ni(100) as Function of $d_\perp$(Ni–O)

Beam Set Neglect $d_\perp = 1.0$ Å

$d_\perp = 1.1$ Å

Fig.18.11. Diffuse LEED intensities calculated for disordered oxygen atoms in hollow sites on Ni{100}, for two different Ni-O interlayer spacings. The spikes represent sharp integral-order substrate-induced beams (not to scale in either width or height). The grazing-emergence condition gives rise to the intensity cutoff forming a circle around the specular reflection direction. Normal incidence is assumed, giving fourfold rotational symmetry and mirror planes

tensity with respect to the energy in order to eliminate the effect of
the long-range structure factor (Sect.18.6). This is based on the independence of the structure factor from the parallel momentum transfer
(at least for two-dimensional disorder). The implication is that the logarithmic derivative of the intensity with respect to energy should be
measured at constant parallel momentum transfer, i.e. as if one were
tracking a spot position [18.34,38].

e) Incommensurate Overlayers

Beam set neglect has provided a better solution to the problem of calculating diffracted intensities from incommensurate overlayers [18.39].
By incommensurate we mean that the overlayer has a two-dimensional
lattice which is independent of that of the substrate. This situation is
common with overlayers that are strongly cohesive and can ignore the
periodicity of the substrate on which they lie: for instance, graphite
and oxides or other strong compounds form overlayers with their own
bulk lattice, which in general does not match that of the substrate.

By using exactly the same arguments as above for ordered and
disordered overlayers, one can easily show that acceptably accurate calculations can be performed with just the two sets of beams defined
earlier. Again, the effect is to ignore weak third- and higher-order
multiple-scattering paths. This approach has been applied to the structure determination of a graphite layer grown from hydrocarbon decomposition on a Pt{111} substrate [18.39]. The result is illustrated in
Fig.18.12. It was found that a single graphite layer rests on an incomplete layer of individual carbon atoms chemisorbed in hollow sites on
the Pt substrate. The interlayer distances found are very reasonable
when compared with known Pt-C bonds and with Van der Waals
dimensions for graphite.

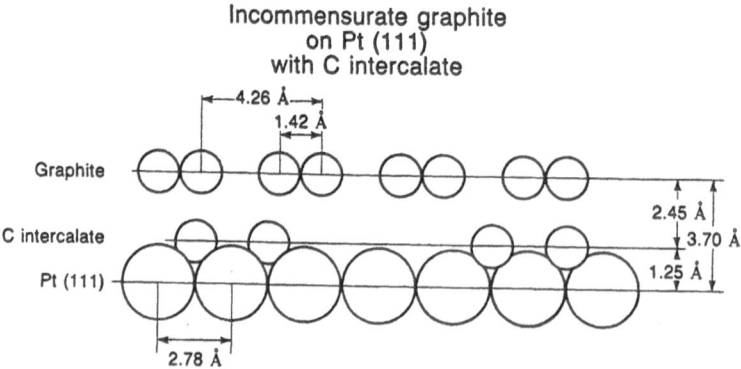

Fig.18.12. Structural diagram of a graphite layer adsorbed on Pt{111}, with an intercalated chemisorbed carbon layer. The graphite layer is incommensurate with the
substrate, while the individual carbon atoms are bonded in hollow sites of the substrate

f) Combination with Other Methods

Finally, it must be pointed out that beam set neglect can be very efficiently combined with the other approaches described in this text. This is because beam set neglect does not prescribe how the scattering of the overlayer or of the substrate are to be calculated. These can be obtained with any other suitable method. For instance, beam set neglect can be combined with cluster methods, in particular with kinematic sublayer addition (Sect.18.5.3). This combination was in fact used by us to analyze the structures of benzene and coadsorbed CO deposited on Rh [18.17,28], Pt [18.40] and Pd{111} [18.18]. In each case the computation was efficient enough to allow 1000 to 2000 structures to be investigated.

g) Advantages and Disadvantages of BSN

Let us summarize here the advantages and disadvantages of beam set neglect (BSN):

- BSN is applicable to any superlattice, including low-density adsorbate structures, large adsorbates, many adsorbates per unit cell (e.g. regular out-of-phase domains) and complex surface reconstructions;

- BSN allows higher energies, since fewer beams are used, where more experimental data are accessible;

- BSN allows calculating intensities for only a small number of beams, permitting efficient elimination of many incorrect structural models;

- Many structural parameters can be optimized in a preliminary fashion with BSN at low cost;

- BSN is easily programmed as a natural extension of the existing combined space method [18.2,5];

- BSN applies to disordered and to incommensurate overlayers.

Disadvantages of BSN are:

- BSN assumes that the overlayer is thin enough to let the electrons reach the substrate, which in addition must have a different two-dimensional lattice from that of the overlayer;

- Since approximations are involved in BSN, structural results are less accurate than with a full calculation (in our experience, the inaccuracies are often smaller than those already inherent in the LEED method);

- To benefit from linearity of the computation time with the number of inequivalent atoms, it is necessary that multiple scattering not take place over large distances, i.e. the inelastic mean free path should be small compared to, say, the unit cell dimensions.

18.5.3 Kinematic Sublayer Addition

We consider here a particular cluster method that simplifies into a very efficient plane-wave method. Kinematic sublayer addition [18.28] is concerned with clusters that are well separated. The most obvious example is a molecular layer, where each molecule defines a cluster. Within each cluster multiple scattering can be calculated with any desired accuracy. But we neglect the direct scattering of an electron from one molecule to any other molecule. The greatest advantage of this approach comes when one deals with more than one molecule per unit cell in the ordered case, and with two or more inequivalent adsorbates in a disordered overlayer.

The specification "*direct* scattering" refers to the following assumption, which in turn is based on the use of plane waves. We treat the layer of clusters (e.g., molecules) as a separate overlayer, whose plane-wave diffraction properties are to be calculated in the absence of the substrate. Then we combine the overlayer with the substrate by using any of the plane-wave layer-stacking techniques, e.g. renormalized forward scattering or layer doubling. This layer-stacking step introduces multiple scattering between the overlayer and the substrate. As a result, it remains possible for electrons to first scatter from an adsorbed molecule, then be reflected from the substrate and finally scatter from another adsorbed molecule. Thus *indirect* scattering between clusters is allowed via the substrate, even if *direct* scattering is not.

The plane waves that are to be used in this layer-stacking process depend on the application. If one deals with an ordered overlayer, one may use the complete set of plane waves defined by the overlayer periodicity. However, it is also possible in the same case to use beam set neglect, i.e. to limit the plane waves to the "incident" and "detected" sets: the incident plane wave together with the substrate-diffracted plane waves, and the final detected plane wave of interest together with those plane waves which are scattered into it by the substrate. This process is to be partially repeated to treat other emission directions.

The advantages of this approach are clear. First, the multiple scattering within the individual clusters can be precomputed, stored and then used in many geometrical positions of the clusters relative to each other and relative to the substrate. This is because the positioning of the overlayer and its components is very efficiently done in the plane-wave representation, and thus can be repeated many times at little extra cost.

Second, it is almost no extra effort to include a higher density of the same molecules in a molecular overlayer: the additional molecules occur as further terms in a simple kinematic sum over the molecules contained in the layer. And the extra molecules need not occupy the same adsorption sites as the first ones (but they should have the same orientation, or else additional calculations are needed, which are proportional to the number of orientations).

Third, the method can be equally well used for ordered and disordered overlayers.

Fourth, one may just as well mix molecules with individual adatoms, rather than with other molecules. For that matter, it is also possible to mix different adatoms in this way.

The question is, however, whether one obtains a reasonable approximation by ignoring direct scattering between molecules. The answer depends simply on how closely the molecules are packed together in an overlayer. If the molecules are very tightly packed, there is a bigger chance of an electron scattering from one molecule to a neighboring molecule. Figure 18.13 shows the geometry. Around each adsorbate a sphere is drawn, within which multiple scattering is assumed to be important. When one sphere includes another adsorbate ("overlapping spheres"), scattering from one adsorbate to the other can be important and affect the result.

We have made test calculations for one of the worst cases [18.27b]: closest packing of CO in an overlayer. This is the case of a (2x2) structure of CO formed on Rh{111}, which has a coverage of 0.75 molecules per surface metal atom. The result was that the optimized structural parameters, such as bond lengths, for this system were less than 0.1 Å off when kinematic sublayer addition was used, relative to the case of full multiple-scattering calculations. In this particular worst case, one would need a better approximation for fine tuning of the structural parameters.

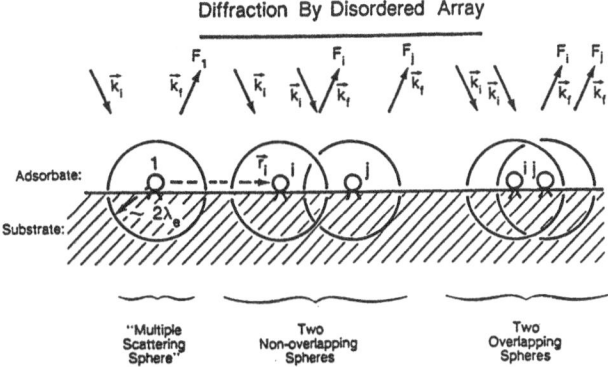

Fig.18.13. Diffuse diffraction by a disordered overlayer of molecules. The large circles represent the clusters within which multiple scattering is important for the molecule at its center. When the molecules are tightly packed, giving overlapping spheres, electrons can scatter from one molecule to the other, damaging the kinematic–sublayer–addition method

On the other hand, we have performed the same tests for overlayers involving hydrocarbons. Since hydrogen is a very weak scatterer, the hydrogens form "cushions" between the different molecules. The space occupied by the hydrogens produces no detectable diffraction and thus effectively separates the scatterers in the different molecules by 5 Å or more. Multiple scattering is quite negligible between such light atoms as carbon, nitrogen and oxygen over these distances. Thus one obtains very good structural results in such cases. We have tested this in particular for benzene overlayers coadsorbed with CO.

We have extensively applied kinematic sublayer addition to the structural determination of a number of molecular coadsorption structures on metal substrates. These include benzene coadsorbed with CO, two examples of which are illustrated in Figs.18.2,3 [18.17,18,28,40], ethylidyne coadsorbed with CO or NO [18.41], and CO coadsorbed with Na [18.42].

18.6 Experimental Requirements

On the whole, the experimental techniques have needed relatively less evolution than did theory to attack complex and disordered surfaces. For complex ordered structures with many diffraction spots, it is mostly a matter of more conveniently acquiring the larger amount of data. This favors automated systems, for instance video LEED [18.43-45].

With disordered surfaces, the experimental challenge is to measure the diffuse intensity between any sharp spots due to the substrate. Such measurements have long been performed with Faraday cups, often as a byproduct of measuring spot profiles. However, here again one welcomes automation to acquire larger amounts of experimental data. Video LEED has already been successfully applied to this task [18.6,33,38].

A more novel tool ideally suited to the measurement of diffuse LEED intensities is the position-sensitive detector [18.46,47]. This detector replaces the traditional LEED display screen with its requirement for conversion of an optical image into a digital set of data. Instead, one uses an electronic counting device which feeds its output directly in digital form into a computer: hence the name "digital LEED". To keep the counting rate down to a manageable value for the electronic counters, one uses nano-ampere to pico-ampere incident beam currents, and amplifies single electrons into detectable bursts through microchannel plates. The position-sensitive detector may be a resistive anode or a strip-and-wedge plate. In either case, the detector outputs three or four analog currents which are electronically combined to produce the position of the detected burst representing a single electron.

produce the position of the detected burst representing a single electron.

After recording of the experimental data by either video or digital LEED, they are manipulated to produce a suitable database for comparison with theory. In the case of ordered overlayers, a computer must identify the spots, make background corrections, integrate the spot intensities over their cross sections and string together the data into I-V curves.

The diffuse LEED intensities require some more treatment. Several new considerations arise, compared to spot intensities [18.6,33,34,38]. First, because the diffuse intensities are typically weak, there is a danger that they are affected by inelastic intensity contributions. The energy filtering that is normally applied in LEED does not exclude inelastic electrons which have suffered phonon losses or other energy losses of about 0.25 eV or less. This issue has been studied by *Ibach* and *Lehwald* [18.48]: they found that on the whole inelastic losses do not affect the measured diffuse intensities adversely. However, there are particular combinations of angles and energies where the elastic intensity is very low (due to destructive interference) while the inelastic losses are larger. One might apply a narrower energy-acceptance window, but this is not easy because of the small energy differences involved and the nonuniformities of the filtering grids. An other alternative, suggested by Ibach and Lehwald, is to subtract the diffuse intensities of the clean surface from those of the overlayer-covered surface. At least those phonon losses common to both surfaces will then approximately cancel out.

Another concern with diffuse intensities goes back to our discussion in Sect.18.3 about the structure factor and the form factor. The diffuse intensities can be viewed as the product of an intensity due to the long-range order (the structure factor) and an intensity due to the short-range order (the form factor). If we are interested in the short-range order, i.e. the local bonding geometry, then we should eliminate the perturbing influence of the structure factor, because it is not normally included in the LEED theory: it could only be included if one knew the long-range order in the first place. This is relatively easily accomplished in the case of purely two-dimensional disorder (thus excluding step disorder, for example). One can then use the logarithmic derivative of the diffuse intensity with respect to energy, keeping the parallel momentum transfer constant [18.6, 33, 34, 38]. Since the structure factor remains constant for constant parallel momentum transfer, it disappears. But one must then also in the theory produce the logarithmic derivative of the calculated diffuse intensity: that is straightforward.

18.7 Conclusions

Our main conclusion is that the computational barrier to the structural determination of complex and disordered structures has been largely broken down. The prohibitive power laws governing the computational cost as a function of complexity have been reduced to direct proportionality or even more favorable behaviors.

The only major remaining barrier to structural determination is the issue of how to search through the high-dimensional structural parameter space for the correct solution. This is the same issue facing x-ray crystallography. There are no universal mathematical methods that are effective at finding the correct structural solution in a unique and reliable manner in a high-dimensional parameter space.

Thus, LEED crystallography has progressed to the point where it is faced with the same problems that x-ray crystallography faces. However, x-ray crystallography benefits from the existence of "direct methods", which help in certain not-too-complex situations. These methods are based on the kinematic nature of x-ray diffraction and it is not yet clear whether similar methods might be viable in the case of LEED.

LEED theory is now capable of treating not only surfaces with any large unit cell, but also disordered surfaces. In addition, incommensurate overlayers can now be treated effectively.

The next stage, already partly realized, is that of vacancies [18.49] and substitutional or interstitial impurities in surfaces. Also, steps at surfaces [18.24] and adsorbates attached to those steps are becoming accessible.

Other types of defects, such as local distortions, are already covered by present theoretical techniques.

On the experimental side, automated data acquisition systems are already available to handle the demand for larger data bases to solve more complex structures. Diffuse intensities due to disordered surfaces can also be measured, either with a video camera or, better, with a position-sensitive detector.

A word of caution is needed, however, with respect to disordered and defected surfaces. Their experimental preparation is more difficult than with ordered surfaces, because the sharp LEED pattern is no longer available to judge the reproducibility of the surface. In other words, it is more difficult to determine the state of a disordered or defected surface. It is therefore harder to reproduce desired surface conditions in such cases.

Acknowledgments. This work was supported in part by the Director, Office of Energy Research, Office of Basic Energy Sciences, Materials Sciences Division of the U.S. Department of Energy under contract No. DE-AC03-76SF00098. Supercomputer time was also made available by the Office of Energy Research of the U.S. Department of Energy. The NFEC development was in part funded by the Army Research Office. Many colleagues have contributed significantly to our work described in this text: J.J. Barton, G.S. Blackman, C.-M. Chan, Z.P. Hu, C.-T. Kao, R.J. Koestner, R.F. Lin, D.F. Ogletree, H. Ohtani, J.B. Pendry, P.J. Rous, D.K. Saldin, G.A. Somorjai, E. Sowa, and M.L. Xu.

References

18.1 J.B. Pendry: *Low-Energy Electron Diffraction* (Academic, London 1974)

18.2 M.A. Van Hove, S.Y. Tong: *Surface Crystallography by LEED*, Springer Ser. Chem. Phys., Vol.2 (Springer, Berlin, Heidelberg 1979)

18.3 M.A. Van Hove, G.A. Somorjai: *Adsorbed Monolayers on Solid Surfaces*, Structure and Bonding, Vol.38 (Springer, Berlin, Heidelberg 1979)

18.4 L.J. Clarke: *Surface Crystallography: An Introduction to LEED* (Wiley, London 1985)

18.5 M.A. Van Hove, W.H. Weinberg, C.-M. Chan: *LEED: Experiment, Theory and Structural Determination*, Springer Ser. Surf. Sci., Vol.6 (Springer, Berlin, Heidelberg 1986)

18.6 K. Heinz, D.K. Saldin, J.B. Pendry: Phys. Rev. Lett. **55**, 2312 (1985)

18.7 J.M. McLaren, J.B. Pendry, P.J. Rous, D.K. Saldin, G.A. Somorjai, M.A. Van Hove, D.D. Vvedensky: *Surface Crystallographic Information Service: A Handbook of Surface Structures* (Reidel, Dordrecht 1987)

18.8 J.B. Pendry: *Surface Crystallographic Information Service: Database and Graphics Programs* (Reidel, Dordrecht 1987)

18.9 W. Moritz, D. Wolf: Surf. Sci. **163**, L655 (1985)

18.10 C.-M. Chan, M.A. Van Hove: Surf. Sci. **171**, 226 (1986)

18.11 E. Sowa, M.A. Van Hove, D.L. Adams: to be published

18.12 C.-M. Chan, M.A. Van Hove: Surf. Sci. **183**, 303 (1987)

18.13 E. Lang, K. Müller, K. Heinz, M.A. Van Hove, R.J. Koestner, G.A. Somorjai: Surf. Sci. **127**, 347 (1983)

18.14 W.S. Yang, F. Jona, P.M. Marcus: Phys. Rev. **B28**, 2049 (1983)

18.15 F.J. Himpsel, P.M. Marcus, R. Tromp, I.P. Batra, M.R. Cook, F. Jona, H. Liu: Phys. Rev. **B30**, 2257 (1984)

18.16 S.Y. Tong, W.N. Mei, G. Xu: J. Vac. Sci. Technol. **B2**, 393 (1984)

18.17 R.F. Lin, G.S. Blackman, M.A. Van Hove, G.A. Somorjai: Acta Crys. B **43**, 368 (1987)

18.18 H. Ohtani, M.A. Van Hove, G.A. Somorjai: to be published

18.19 C.B. Duke, G.E. Laramore: Phys. Rev. **B2**, 4765 (1970); Phys. Rev. **B2**, 4783 (1970)

18.20 W. Moritz, H. Jagodzinski, D. Wolf: Surf. Sci. **77**, 233 and 249 (1978)

18.21 J.B. Pendry: In *Determination of Surface Structure by LEED*, ed. by P.M. Marcus, F. Jona (Plenum, New York, 1984)

18.22 J.B. Pendry, D.K. Saldin: Surf. Sci. **145**, 33 (1984)

18.23 D.K. Saldin, D.D. Vvedensky, J.B. Pendry: In *The Structure of Surfaces*, ed. by M.A. Van Hove, S.Y. Tong, Springer Ser. Surf. Sci., Vol.2 (Springer, Berlin, Heidelberg 1985) p.131

18.24 P.J. Rous, J.B. Pendry: Surf. Sci. **173**, 1 (1986)
D.K. Saldin, J.B. Pendry: Surf. Sci. **162**, 941 (1985)

18.25 N. Masud and J.B. Pendry, J. Phys. **C9**, 1833 (1976)

18.26 F. Jona, J.A. Strozier Jr., P.M. Marcus: In *The Structure of Surfaces*, ed. by M.A. Van Hove and S.Y. Tong, Springer Ser. Surf. Sci., Vol.2 (Springer, Berlin, Heidelberg 1985) p.92

18.27 M.A. Van Hove and G.A. Somorjai, Surf. Sci. **114**, 171 (1982)
M.A. Van Hove, R.J. Koestner, J.C. Frost, G.A. Somorjai, Surf. Sci. **129**, 482 (1983)

18.28 M.A. Van Hove, R.F. Lin, G.A. Somorjai: J. Am. Chem. Soc. 108, 2532 (1986)

18.29 S.Y. Tong, M.A. Van Hove, B.J. Mrstik: In Proc. 7th Int'l. Vacuum Congress and 3rd Int'l. Conf. Solid Surfaces (Vienna 1977) p.2407

18.30 J.J. Barton, M.A. Van Hove: Bull. Am. Phys. Soc. 31, 425 (1986); and to be published

18.31 J.J. Barton, D.A. Shirley: Phys. Rev. B32, 1906 (1985)

18.32 M.A. Van Hove, R.F. Lin, G.A. Somorjai: Phys. Rev. Lett. 51, 778 (1983)

18.33 P.J. Rous, J.B. Pendry, K. Heinz, K. Müller, N. Bickel: Phys. Rev. Lett. 57, 2951 (1986)

18.34 D.K. Saldin, J.B. Pendry, M.A. Van Hove, G.A. Somorjai: Phys. Rev. B31, 1216 (1985)

18.35 J.J. Barton, M.L. Xu, M.A. Van Hove: to be published

18.36 M.L. Xu, J.J. Barton, M.A. Van Hove: to be published

18.37 W.F. Egelhoff Jr.: Phys. Rev. B30, 1052 (1984)
W.F. Egelhoff Jr.: J. Vac. Sci. Technol. A3, 1511 (1985)
R.A. Armstrong, W.F. Egelhoff Jr.: Surf. Sci. 154, L225 (1985)

18.38 K. Heinz, K. Müller, W. Popp, H. Lindner: Surf. Sci. 173, 366 (1986)

18.39 Z.P. Hu, D.F. Ogletree, M.A. Van Hove, G.A. Somorjai: Surf. Sci. 180, 433 (1987)

18.40 D.F. Ogletree, M.A. Van Hove, G.A. Somorjai: Surf. Sci. 183, 1 (1987)

18.41 C.-T. Kao, G.S. Blackman, C.M. Mate, B.E. Bent, M.A. Van Hove, G.A. Somorjai: to be published

18.42 G.S. Blackman, C.M. Mate, M.A. Van Hove, G.A. Somorjai: to be published

18.43 P. Heilmann, E. Lang, K. Heinz, K. Müller: Appl. Phys. 19, 247 (1976)

18.44 E. Lang, P. Heilmann, G. Hanke, K. Heinz, K. Müller: Appl. Phys. 19, 287 (1979)

18.45 D.F. Ogletree, G.A. Somorjai, J.E. Katz: Rev. Sci. Instr. 57, 3012 (1986)

18.46 P.C. Stair: Rev. Sci. Instr. 51, 132 (1980)

18.47 D.F. Ogletree, G.A. Somorjai, J.E. Katz: to be published

18.48 H. Ibach, S. Lehwald: Surf. Sci. 176, 629 (1986)

18.49 P.J. Rous, J.B. Pendry: Surf. Sci. 155, 241 (1985)

19. Recent Developments in Scanning Tunneling Microscopy and Related Techniques

R.M. Tromp

IBM Thomas J. Watson Research Center, Yorktown Heights, NY 10598, USA

This chapter discusses some of the recent developments in scanning tunneling microscopy and related techniques. The aim is to give a flavor of some of the current activities and to indicate new directions in which these exciting techniques are developing.

19.1 Scanning Tunneling Microscopy

The technique of scanning tunneling microscopy (STM) was invented in 1982 by *Binnig, Rohrer* and coworkers at the IBM Research Center in Switzerland [19.1]. When a sharp metal tip is brought close enough to a sample, such that the wave functions of sample and tip overlap, a small tunneling current can be established by applying an external bias voltage across the junction. The magnitude of the tunneling current depends exponentially on the distance between sample and tip and changes roughly by a factor 100 per nm. Therefore, the magnitude of this tunneling current can be used very sensitively to probe the sample-to-tip distance. When the tip is scanned along the surface by a three axes piezo-electric transducer the sample-tip distance can be kept constant by keeping the tunneling current constant. A feedback circuit applies a correction voltage to the Z-drive normal to the surface in order to achieve this. The magnitude of this correction voltage can be used to obtain a map of the surface corrugations as the tip scans along the surface.

19.1.1 Studies of Surface Atomic and Electronic Structure

One of the amazing feats of the STM is that single atoms can be resolved with relative ease in a variety of environments: vacuum, air, oils, acids, etc. [19.2]. Figure 19.1 shows an STM topograph of the Si{111}-(7x7) surface obtained in ultra high vacuum. One unit cell is outlined. White is high, black is low and the black-to-white range is 0.2 nm. Each white fuzzy dot is a single Si atom adsorbed on top of the underlying Si crystal [19.3,4]. The sides of the unit cell are 7 times longer in the surface than in the bulk because of the tendency of the

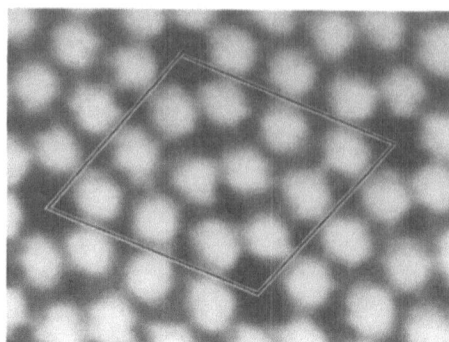

Fig.19.1. STM image of the Si{111}-(7x7) surface, tunneling into the emtpy states of the sample (bias +2V). A single unit cell is outlined. Black-to-white range is 0.2 nm

surface atoms to rearrange themselves and thereby minimize the number of broken bonds. This "surface reconstruction" is very common on semiconductor surfaces and it is one of the great powers of the STM that it can study such reconstruction phenomena with atomic resolution.

Since the tunneling current results from an overlap of wavefunctions on the tip with wavefunctions on the sample, the tunneling current depends not only on distance but also on the precise nature of those wavefunctions. This fact, together with the high spatial resolution has allowed STM to study the electronic structure of surfaces with atomic resolution. On the Si{111}-(7x7) surface, for instance, the surface states long known from photoemission and inverse photoemission experiments could be observed with STM [19.5,6]. Their locations inside the (7x7) unit cell made it possible to derive the origin of these surface states by a direct correlation with the underlying atomic structure. Figure 19.2 shows real space images of surface states between the Fermi level E_F and 0.35 eV below E_F (A) and between 0.6 and 1.0 eV below E_F(B). The outline of a single unit cell is shown. Comparison with Fig.19.1 shows that the surface states closest to (and actually straddling) E_F are localized on the adatoms visible in Fig.19.1, while the second surface state (B) is located at 7 different positions between the adatoms where atoms in the first layer of the underlying crystal expose broken bond oribtals to the vacuum ("restatoms").

Counting the number of bright "dots" in Figs.19.2A,B, we conclude that there are 19 broken bonds per unit cell. This and the location of these broken bond orbitals is in good agreement with the structural model proposed by *Takayanagi* et al. [19.7] and with calculations for subunits of this model by *Northrup* [19.8]. Figs.19.2A,B represent the first real space, energy resolved observations of surface electronic states. Similar studies have now been made on the Si{001}-(2x1) [19.9], Si{111}-(2x1) [19.10], GaAs{110}-(1x1) [19.11] and several adsorbate covered semiconductor surfaces.

Although the ability to study surface electronic structure in such detail presents unique opportunities for surface structure studies, it is

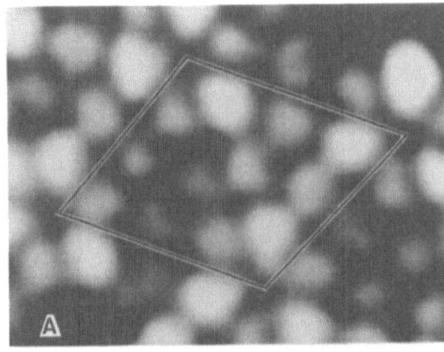

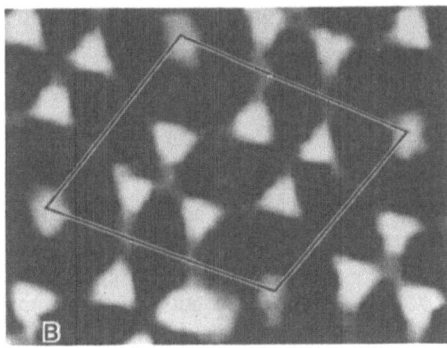

Fig.19.2. Surface states of the Si{111}-(7x7) surface.
a) States between E_F and 0.35eV below E_F, located on the adatoms.
b) States between 0.6 and 1.0 eV below E_F, located on the restatoms

not always obvious how to interpret experimental results. One example may be the Si{111}-($\sqrt{3}$x$\sqrt{3}$) Ag surface. On this surface hexagonal arrays of maxima are observed [19.12,13] (Fig.19.3). While we know that the surface contains both Si and Ag it is not at all obvious whether the bumps in Fig.19.3 are due to Ag or Si (more precisely: whether the bumps correspond to wave functions localized on Ag or Si atoms). In fact, different groups have argued for both interpretations and the question remains controversial [19.12-14]. (Unfortunately, the Ag d-electrons cannot be observed with tunneling spectroscopy [19.12], apparently because the d-wavefunctions decay rapidly and their local density of states at the tip position is negligible).

The tunneling process does not distinguish wavefunctions on the sample from wavefunctions on the tip. Thus, the experiment determines not simply the electronic structure of the sample, but the convolution of the wavefunctions of the sample with the wavefunctions of the tip. In many cases it appears that the electronic structure of the tip is relatively unimportant and experimental results can be interpreted as being characteristic of the sample only.

Experimentally it was recently observed that sometimes atomic resolution is obtained for one polarity (say, probing the empty states of the sample), while at the *same time* very poor resolution is obtained for the opposite polarity (probing filled states of the sample) [19.15].

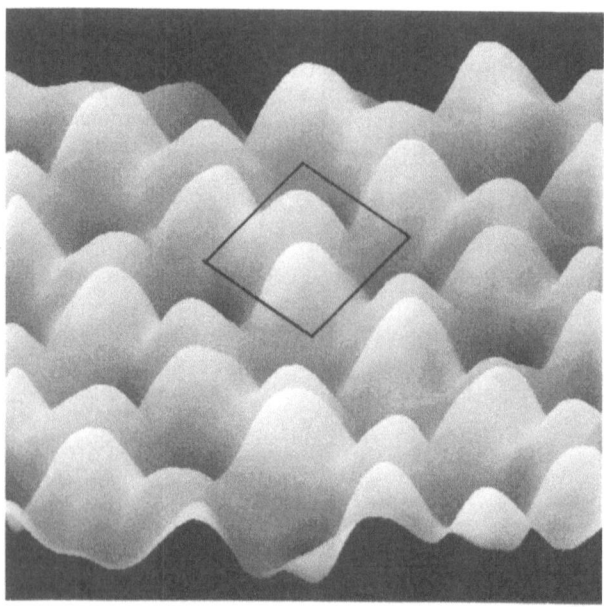

Fig.19.3. STM image of the Si(111)-(3x 3) Ag surface, tunneling from the filled states of the sample into the tip (bias: 1 V). A single unit cell is outlined. The amplitude of the corrugation is 0.15 nm

This situation can be explained on the basis of the detailed theoretical work by Lang [19.16], who treated sample and tip electronic structure on an equal footing. Suppose that the probe tip were blunt. It is well known that under such conditions atomic resolution is not achievable for either filled or empty states. Now suppose that we adsorb a single strongly electronegative atom at the apex of this tip. This atom adds to the local density of filled states in the vacuum region, but not to the local density of empty states (Fig.19.4). This tip can atomically resolve empty states of the sample (electrons tunneling from the atomically sharp peak in the filled density of states on the tip to the sample), but it cannot resolve filled states (when electrons tunnel from the sample to the "blunt" empty density states of the tip). Detailed calculations for a S atom adsorbed on a (jellium) tip scanning over a Na atom on a (jellium) sample showed that the S tip can atomically resolve the

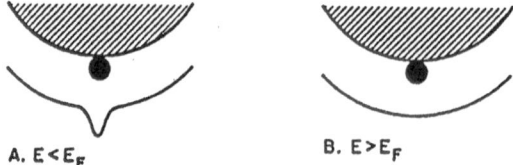

A. $E < E_F$ B. $E > E_F$

Fig.19.4. Schematic representation of a tip with a strongly electronegative atom adsorbed at its apex. Shown are contours at a constant charge density in the vacuum region for filled states (A) and empty states (B)

550

empty states of the Na atom, but it does not "see" the Na atoms when electrons tunnel from sample to tip [19.15, 16]. These experimental and theoretical results stress the importance of the electronic structure of the tip.

19.1.2 Other Applications of STM

In the previous section we have discussed the STM as an analytical tool used in ultrahigh vacuum on well-defined surfaces. However, its applicability is much wider. Instruments working in liquid nitrogen have been used to image charge density waves in layered compounds [19.17]. Atomic resolution images of graphite in air have been obtained routinely by many different groups. Even more remarkable, atomic resolution images of graphite have been obtained while the surface was immersed in highly acetic solutions used for electroplating [19.18]. Imaging under salty solutions is also possible which is of interest for further developments of STM for biological applications [19.19]. Several successful attempts have been made to image biological material such as DNA, although not yet with atomic resolution [19.20, 21].

STM's may be used not only to look at surfaces but also to change them. Since the tip is very small and the tunneling current flows in a very restricted volume, one may think of the STM as the ultimate electron beam source and use it to perform e-beam lithography. One example of such an application is the recording of bits in the surface of a metallic glass [19.22]. In order to write one bit the tunneling current is increased to the point where the surface melts and a Taylor cone is formed in the electric field between tip and sample. When the tunneling current is reduced again this cone freezes and is left as a permanent mark on the surface. When the tip is scanned in the high current mode one can draw a line of the surface. The use of STM as a lithographic tool is still in its infancy but the example given here will convey some of the promise the technique has in this technologically important area.

19.2 The Atomic Force Microscope

One of the limitations of STM is that one needs to establish a tunneling current, and this restricts the materials and substrates one can study to reasonably good conductors. Unfortunately, many of the materials which one would like to study are insulators (oxides, ceramics, biological materials, polymers, etc.) The atomic force microscope, introduced by *Binnig* and co-workers [19.23], resolves this problem in an elegant fashion.

Again a very sharp tip is brought very close to the sample. When the tip approaches the sample it will first start to undergo an attractive force, pulling it towards the sample. At very close distance this changes in a repulsive force, when the tip actually touches the sample. When the tip is mounted on a lever these forces will give rise to a very small deflection of the lever and to a shift in the resonance frequency of the lever. The deflection is proportional to the magnitude (and sign) of the force, the shift in resonance frequency is proportional to the gradient of the force with distance. When the sample is scanned along the tip the tip-to-sample distance can again be controlled by a feedback circuit maintaining either constant deflection of constant resonance frequency. The lever deflection or resonance frequency can be measured with a second tip tunneling to the lever (which gives an exponential dependence of tunneling current with respect to lever deflection [19.24] or - more conveniently - with a laser interferometer (deflections of 0.01 nm can be measured easily [19.24, 25]). Several groups have now obtained atomic resolution images of graphite using the AFM in the repulsive regime. As mentioned, however, much of the interest in AFM arises from the desire to study materials of technological interest that cannot easily be studied otherwise.

Figure 19.5 shows two micrographs of narrow grooves etched into a Si wafer [19.25]. Figure 19.5A was obtained with a conventional scanning electron microscope (SEM). Due to charging of the sample the resolution is not very good and the exact shape of the grooves cannot be deduced from the micrograph. Figure 19.5B shows an AFM image (attractive regime) of the same sample. The much sharper definition and superior resolution are evident. Studies at higher magnification indicated a lateral resolution of better than 10 nm in this particular study.

Figure 19.6 shows an image obtained with a magnetic tip scanning over an optical recording medium (TbFe) in which bits have been written thermomagnetically using a focussed laser beam [19.26]. These bits are recognized as black dots where the recording medium was magnitized when (due to laser heating) the disk was locally brought above the Curie point while kept in an external magnetic field. The magnetic bits exert a force on the tip and change the resonance frequency of the lever. Thus one can obtain high resolution, high contrast images of localized magnetic regions.

19.3 Related Microscopies

In addition to STM and AFM, a number of related microscopic have been developed over the last two or three years. Particularly noteworthy are the thermal profiler [19.27] and the scanning potentiometer [19.28]. In the thermal profiler an extremely fine thermocouple is brought close to the sample. The thermocouple signal will depend

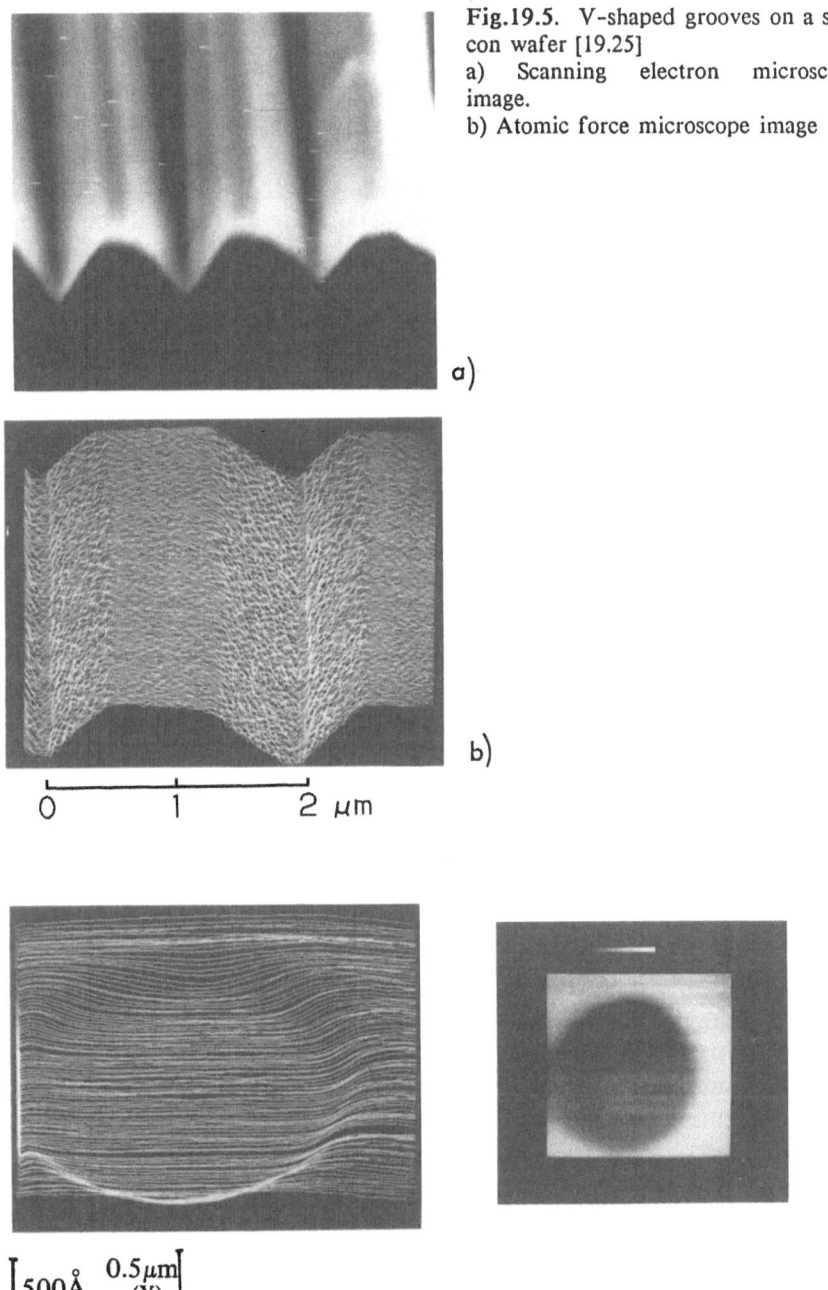

Fig.19.5. V-shaped grooves on a silicon wafer [19.25]
a) Scanning electron microscope image.
b) Atomic force microscope image

0   1   2 μm

500Å
(Z)   0.5μm
(Y)   0.5μm(X)

Fig.19.6. High magnification image of laser written magnetic domain in TbFe film. This image was obtained with an atomic force microscope using a magnetic probe tip [19.26]

553

strongly on the local temperature of the sample and the distance between sample and thermocouple and can be used to drive a feedback circuit in the familiar way. Microscopes with lateral resolutions in the range of 10 nm have been successfully operated on semiconductors, metals and biological material (red blood cells).

The scanning potentiometer is basically an STM used to measure voltage distributions along the surface. When one crosses a diode junction, for instance, there will be a gradual change in the position of conduction and valence bands relative to the Fermi level. This has been measured in cleaved GaAs/GaAlAs laser junctions [19.29]. The applicability of this technique in the study of superlattices and cleaved interfaces in general is obvious, in particular if one realizes that atomic resolution is available.

In addition to the techniques discussed here, a number of other microscopies have been and are being developed, all based on the idea of having some probe close to a sample with the sample-to-tip distance regulated by a feedback circuit that responds to some physical characteristic of the junction (tunneling current, lever deflection, heat flow, voltage drop, ...). Thus, scanning tunneling microscopy has been the start of a new generic type of scanning tip microscopies (STIM), with many different properties being probed by a large variety of probe tips. It is clear that we are currently at the beginning of a new era in high resolution surface microscopy. The feasibility of inelastic tunneling spectroscopy has recently been demonstrated [19.30, 31] and holds the promise of real space, atomically resolved imaging of surface chemical processes by utilizing molecular vibrations as the measured quantity. Tunneling with a spin polarized tip would allow direct access to surface magnetic phenomena with atomic resolution. Scanning tunneling lithography is in its infancy, but offers ultimate resolution and control. With microscopes becoming simpler and more reliable and less sensitive to environmental noise (vibrations!) one may expect the proliferation of these techniques outside highly specialized research laboratories, in particular in development and production areas where non-atomic resolution microscopy is needed in some non-vacuum environment with lateral resolution beyond the wavelength of visible light.

Acknowledgements. I am grateful to Joe Demuth, Bob Hamers and Evert van Loenen for a most stimulating and rewarding collaboration in studying the surfaces of silicon. I thank S. Chiang, Y. Martin, A. McDonald, H. Wickramasinghe and C. Williams for stimulating discussions on many of the subjects covered in this paper.

References

19.1 G. Binnig, H. Rohrer: Helv. Acta 55, 726 (1982)
G. Binnig, H. Rohrer, CH. Gerber, E. Weibel: Phys. Rev. Lett. 49, 57 (1982)
19.2 Several review papers have appeared recently: G. Binnig, H. Rohrer: IBM J. Res. Develop. 30, 335 (1986)
J. Golovchenko: Science 232, 48 (1986)
R.M. Tromp, R.J. Hamers, J.E. Demuth: Science 234, 304 (1986)
P.K. Hansma, J. Tersoff: J. Appl. Phys. 61, R1 (1987)
19.3 G. Binnig, H. Rohrer, Ch. Gerber, E. Weibel: Phys. Rev. Lett. 50, 120 (1983)
19.4 R.M. Tromp, R.J. Hamers, J.E. Demuth: Phys. Rev. B34, 1388 (1986)
19.5 R.J. Hamers, R.M. Tromp, J.E. Demuth: Phys. Rev. Lett. 56, 1972 (1986)
19.6 R.M. Tromp, R.J. Hamers, J.E. Demuth: Science 234, 304 (1986)
19.7 K. Takayanagi, T. Tanishiro, M. Takahashi, S. Takahashi: J. Vac. Sci. Technol. A3, 1502 (1985)
19.8 J.E. Northrup: Phys. Rev. Lett. 57, 154 (1986)
19.9 R.J. Hamers, R.M. Tromp, J.E. Demuth: Surf. Sci. 181, 346 (1987)
19.10 R.M. Feenstra, J.A. Stroscio, A.P. Fein: Surf. Sci. 181, 295 (1987)
19.11 R.M. Feenstra, J.A. Stroscio, J. Tersoff, A.P. Fein: Phys. Rev. Lett. 58, 1192 (1987)
19.12 E.J. van Loenen, J.E. Demuth, R.M. Tromp, R.J. Hamers: Phys. Rev. Lett. 58, 373 (1987)
19.13 R.J. Wilson, S. Chiang; Phys. Rev. Lett. 58, 369 (1987)
19.14 T. Yokotsuka, S. Kono, S. Suzuki, T. Sagawa: Surf. Sci. 127, 35 (1983)
19.15 R.M. Tromp, E.J. van Loenen, J.E. Demuth, N.D. Lang: Phys. Rev. Lett. B in press
19.16 N.D. Lang: Phys. Rev. B34, 5947 (1986); N.D. Lang: Phys. Rev. Lett. 58, 45 (1987) and references therein
19.17 R.V. Coleman, W.W. McNairy, C.G. Slough, P.K. Hansma, B. Drake: Surf. Sci. 181, 112 (1987) and references therein
19.18 R. Sonnenfeld, B. Schardt: Appl. Phys. Lett. 49, 1172 (1986)
19.19 B. Drake, R. Sonnenfeld, J. Schneir, P.K. Hansma: Surf. Sci. 181, 92 (1987) and references therein
19.20 A.M. Baro, R. Miranda, J. Alaman, N. Garcia, G. Binnig, H. Rohrer, Ch. Gerber, J.L. Carrascosa: Nature 315, 253 (1985)
19.21 G. Travaglini, H. Rohrer, M. Amrein, H. Gross: Surf. Sci. 181, 380 (1987)
19.22 U. Staufer, R. Wiesendanger, L. Eng, L. Rosenthaler, H.R. Hidber, H.-J. Guntherodt, N. Garcia: Appl. Phys. Lett. 51, 244 (1987)
19.23 G. Binnig, C.F. Quate, Ch. Gerber: Phys. Rev. Lett. 56, 930 (1986)
19.24 G.M. McLelland, R. Erlandsson, S. Chiang: Review of Progress in Quantitative Non-Destructive Evaluation, Vol.6 (Plenum, New York) to be published
19.25 Y. Martin, C.C. Williams, H.K. Wickramasinghe: J. Appl. Phys. 61, 4723 (1987)
19.26 Y. Martin, D. Ruga, H.K. Wickramasinghe: Appl. Phys. Lett. in press
19.27 C.C. Williams, H.K. Wickramasinghe: Appl. Phys. Lett. 49, 1587 (1986)
19.28 P. Muralt, D.W. Pohl: Appl. Phys. Lett. 48, 514 (1986); P. Muralt, D.W. Pohl, W. Denk: IBM J. Res. Develop. 30, 443 (1986)
19.29 P. Muralt: Surf. Sci. 181, 324 (1987)
19.30 D.P.E. Smith, G. Binnig, C.V. Quate: Appl. Phys. Lett. 49, 1641 (1986)
19.31 D.P.E. Smith, M.D. Kirk, C.V. Quate: J. Chem. Phys. 86, 6034 (1987)

20. Growth Kinetics of Silicon Molecular Beam Epitaxy

E. Kasper and H. Jorke

AEG Research Center, Sedanstr. 10, D – 7900 Ulm, Fed. Rep. Germany

Modern microelectronics is based on semiconductor materials, which contain differently doped (n- or p-type) regions. An often crucial step for the preparation of the active semiconductor regions is epitaxy, the growth of a thin, oriented single crystalline film on a substrate. Due to its unique combination of properties, silicon is mainly used as semiconductor material for the fabrication of integrated circuits (microchips). The epitaxy technique for today's mass production utilizes chemical vapour deposition (CVD). But for future device application the method of molecular beam epitaxy (MBE) is emerging, making possible sophisticated device structures with precise submicron structures and heterojunctions. The material properties of the semiconductor itself can be artificially influenced by periodic nanometer structures, so-called superlattices. The integration of this novel class of man-made semiconductors with conventional circuits will improve the performance of microchips and open a wide variety of new applications.

In the MBE process the epitaxial layer is formed by condensation of molecular beams consisting of the matrix materials and the dopant elements. The single crystallinity of the film is obtained by proper adjustment of the growth conditions (flux densities, temperatures, clean environment). Figure 20.1 shows the scheme of a silicon molecular beam epitaxy (Si-MBE) apparatus used for the preparation of microwave device structures. A detailed description of principles, methods and applications of Si-MBE is given in [20.1]

The primary aim of epitaxy is the growth of a thin layer of bulk material with high crystal perfection. But, indeed, this generation of a piece of bulk material is performed by a two step process, the first step being the absorption of atoms at the surface (Fig.20.2). Therefore, the surface properties are essential for understanding the epitaxy process. As shown later, the motion of adatoms and the step configuration are of primary influence. On the other hand, epitaxy can give valuable information about surface properties. In the following sections we will specially address the connection between surface properties and material growth by Si-MBE.

Silicon Molecular Beam Epitaxy (Si -MBE)

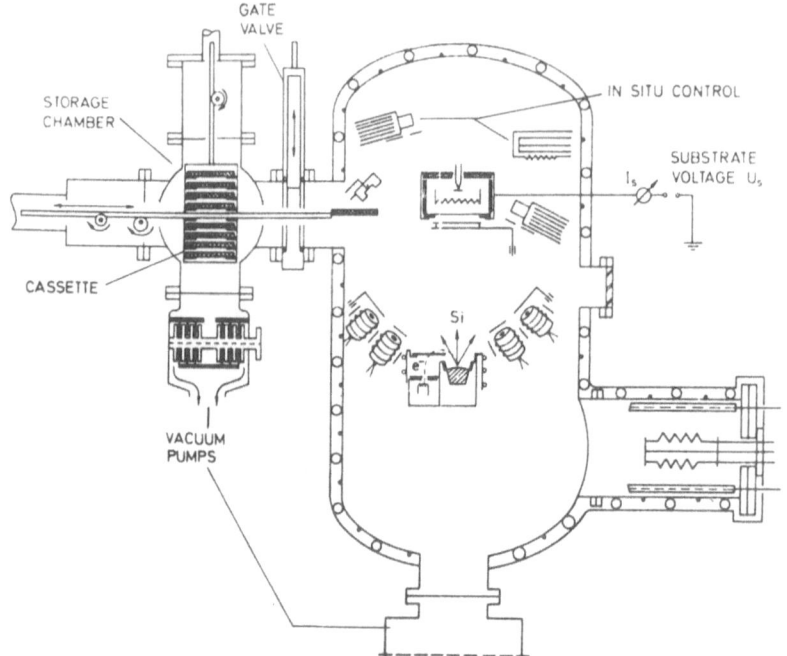

Scheme of apparatus

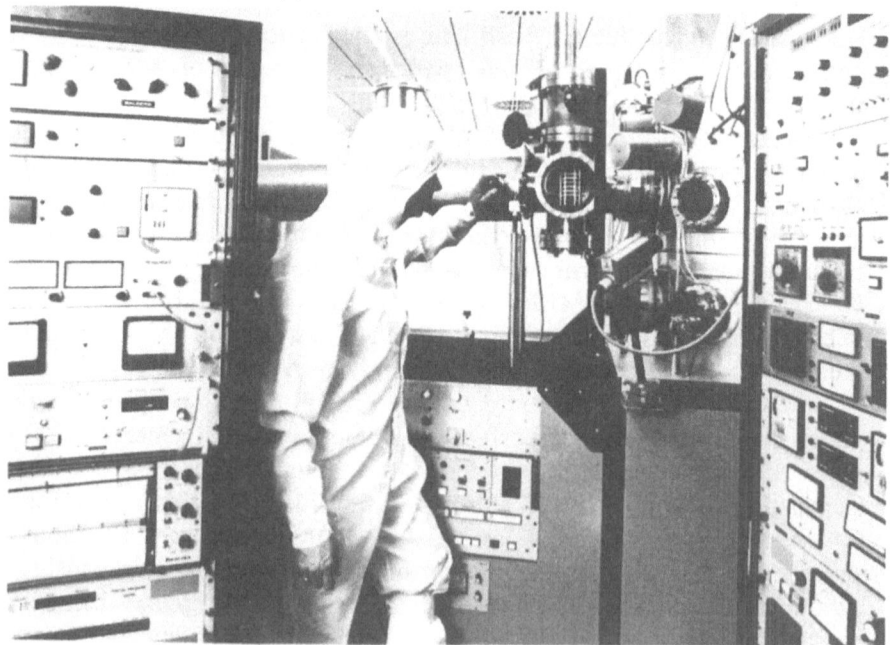

Industrial equipment at AEG, Ulm

Fig.20.1 Scheme of Si-MBE apparatus used for device fabrication [20.1]

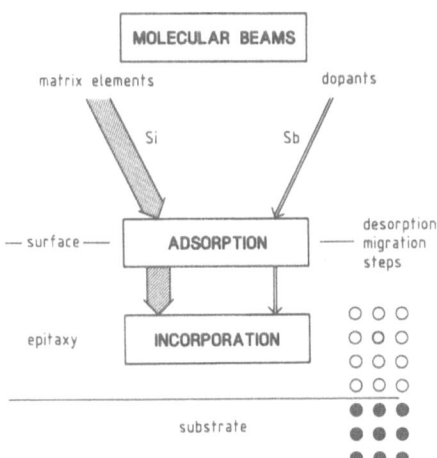

Fig.20.2 Epitaxy as a two step process with surface adsorption as the first step

20.1 Two Dimensional Growth of the Matrix

For modelling the silicon growth, we have to consider the special growth conditions typical for MBE. Especially important for understanding the growth behaviour are the following features:

(i)   Growth at temperatures far below the melting point.

(ii)   Clean surfaces without adsorption layers from residual and carrier gases, or chemical reaction products apart from the intentional dopant material.

(iii) Irradiation of the surface by electrons, ions and electromagnetic radiation originating from the electron gun evaporator.

(IV) Complete condensation of the Si-beam. Negligible reflection or desorption of Si-atoms (condensation coefficient $\eta = 1$).

Standard chemical vapour deposition (CVD) of silicon takes place 200 - 350 K below its melting point (1690 K), whereas the MBE process for doped layers is usually performed at far lower temperatures (850 - 1050 K) near half the melting point, but well above the Debye temperature   (660 K). At present, no consensus exists about absolute values of the lowest epitaxial temperatures. This is probably because of the variety of growth conditions used by various investigators (cleanliness of environment, irradiation conditions) and their different investigative methods (LEED, RHEED, Rutherford backscattering). The most rigorous investigation was carried out by *de Jong* [20.2,3] using LEED for the characterization of the ordering of MBE grown Si layers. The general results of his investigation are as follows:

(i)   Epitaxial growth with ordered, reconstructed surfaces is obtained above a certain temperature T_{epi} which is called the epitaxial temperature. Table 20.1 lists values of T_{epi} found by different authors [20.3-6].

559

Table 20.1. Epitaxial temperature T_{epi} above which ordered bulk-like growth is obtained. Values obtained by different authors [20.3-6] on {100}, {111} and 4° vicinal {111} surfaces.

Surface	T_{epi} [K]	de Jong [20.3]	Shiraki [20.4]	Grossmann [20.5]	Gronwald [20.6]	Herzog [20.8]
{100}	T_{epi}	470	440	570	–	510
{111}	T_{epi}	870	–	790	970	760
vicinal {111}	T_{epi}	770	–	–	–	–

(ii) The epitaxial temperature, T_{epi}, depends on the surface orientations. T_{epi} for the {100} surface is much lower than T_{epi} for the {111} surface. Vicinal {111} surfaces, tilted a few degrees, exhibit a lower T_{epi} than {111} surfaces, which were specified to deviate less than 0.5° from the corresponding crystallographic plane.

(iii) Below the epitaxial temperature, T_{epi}, disordered growth occurs indicated by a LEED pattern with enhanced background and gradually decreasing fractional-order spots. With very thin layers some sort of ordering seems to be present. This is in accordance with the earlier observations of *Jona* [20.7].

Gossmann et al. [20.5] used the He backscattering to investigate the order/disorder transition. Their results (Table 20.1) differ considerably from the results obtained by LEED measurements only. Additionally, they discovered an interesting property of interface reordering at room temperature deposition. The surface of a Si {100} 2x1 substrate is re-ordered to bulk positions if an amorphous layer is deposited. The Si {111} 7x7 is not reordered after deposition of the amorphous layer. The different reordering behaviour of {100} and {111} surfaces is in agreement with the observed tendency to lower epitaxial temperatures of the {100} surface.

In our laboratory [20.8] the following experiment was performed to investigate the absolute value of epitaxial temperature, T_{epi}. The basic idea was to start growth at a temperature high enough to ensure epitaxial ordering. Thereupon the substrate temperature was gradually decreased to room temperature but maintaining a constant Si molecular beam flux of about $1 \cdot 10^{15}$ atoms/cm²s. The samples were grown in high-throughput equipment normally used for our device work [20.9]. Examination of the grown layer by electrical profiling and Rutherford backscattering yields the ordering at each depth corresponding to the accompanying substrate temperature (Fig.20.3).

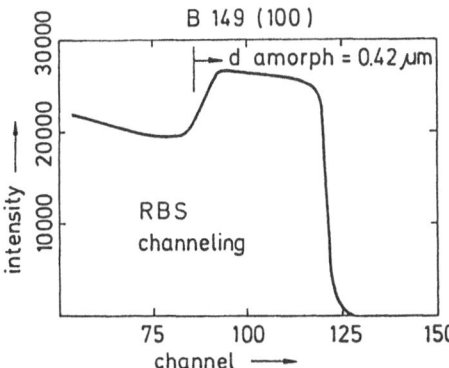

Fig.20.3 Growth experiment with varying substrate temperatures. After a thermal anneal at 1175 K, growth was started at 1025 K and decreased to 300 K with increasing thickness of the (100) layer. Rutherford backscattering result for this sample showing the amorphous overlayer grown with temperatures $T < T_{epi}$

The low growth temperature regime of Si-MBE does not overlap with the high temperature growth regime (1200-1500 K) of conventional chemical vapour deposition (CVD) of epitaxial silicon layers. We have performed some MBE experiments with enhanced growth temperature up to 1225 K [20.10] to elucidate such general properties as condensation behaviour of silicon epitaxy. The condensation coefficient, η, measures the portion of the incoming Si flux which condenses to create the epitaxial film. With CVD, the growth rate around 1200 K decreases with decreasing temperature [20.11]. This is partly ascribed to the desorption of Si atoms indicating a slow motion of Si adatoms to capture sites. *R. Farrow* [20.12] deduced from mass spectrometric measurements of silane pyrolysis at 1200 K, a condensation coefficient $\eta = 0.15$ and an activation energy for surface diffusion of 1.6 eV (CVD). However, Si MBE reveals a condensation coefficient near unity ($\eta = 1$) which indicates mobile adatoms with low activation energy for surface diffusion [20.2,10,13]. The different condensation behaviour of MBE and CVD grown films may be attributed to the adsorption of carrier gas and chemical reaction products on the growing surface.

The surface of a solid is in equilibrium with its vapour if the flux of atoms adsorbed from the vapour equals the flux desorbed from the surface. The equilibrium flux density, F_0, of incident atoms is connected to the equilibrium vapour pressure, P_0, by

$$F_0[m^{-2}s^{-1}] = \sqrt{\frac{N_A}{2\pi MkT}}\ P_0 = 8.33\cdot 10^{22}\ \frac{P_0[Pa]}{(MT)^{1/2}} \tag{20.1}$$

where N_A is Avogadro's number, M the molecular weight, k Boltzmann's constant, and T is the temperature.

561

Table 20.2. Equilibrium flux densities F_0 of Si atoms [20.15] as a function of temperature T. Vapour pressures P_0 have been extrapolated from *Honig* et al. [20.16]. All values are given in SI units.

T [K]	723	823	923	1023	1123	1223
P_0 [P_a]	$3.6 \cdot 10^{-20}$	$2.5 \cdot 10^{-16}$	$2.7 \cdot 10^{-13}$	$7.1 \cdot 10^{-11}$	$6.9 \cdot 10^{-9}$	$3.2 \cdot 10^{-7}$
F_0 [$m^{-2}s^{-1}$]	$6.6 \cdot 10^{2}$	$4.4 \cdot 10^{6}$	$4.3 \cdot 10^{9}$	$1.1 \cdot 10^{12}$	$1 \cdot 10^{14}$	$4.6 \cdot 10^{15}$

Table 20.2 gives numerical values of the equilibrium flux for silicon. This table shows the drastic decrease of desorption events on decreasing the temperature from the upper MBE temperature regime (1200 K) to the lower regime (750 K). In general, only a part of the incident atoms or molecules are adsorbed on the surface, the remainder being reflected. The adsorbed part is characterized by the sticking coefficient. The sticking coefficient for silicon atoms on the silicon surface is unity, at least in the MBE temperature regime. A schematic picture of the processes ocurring on an equilibrated surface is shown in Fig. 20.4. A direct evaporation process would mean removal of an atom from a kink position at a surface step to the vapour. This direct evaporation process [20.14] requires a relatively high silicon binding energy, W, of 4.55 eV. Therefore, the following step process is more likely. As a first step an atom at a kink position jumps to a neighbouring free site on the surface (adsorbed atom). The activation energy for this jump is W_s. The equilibrium density, n_{sequ}, of adsorbed atoms is given by

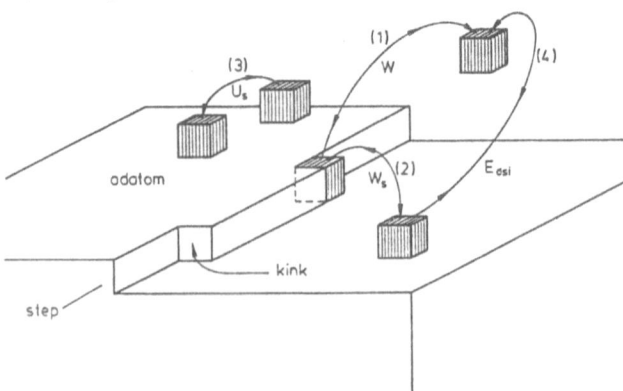

Fig.20.4 Model of a solid surface in equilibrium with its vapour. The crystal grows by repeated addition of an atom from the vapour to a kink position on a step (1). The adjoining energy gain is the binding energy W of the crystal. The most frequent path from the vapour to a step kink is via adsorption on the surface (2) with energy gain E_{dSi}, migration (3) of the adatom and capture by a step (4) with energy gain W_s for the jump from an adatom position to the kink position

$$n_{sequ} = N_{os} \exp(-W_s/kT) \tag{20.2}$$

with a density N_{os} of the surface sites. The adsorbed atoms migrate rather easily in a random walk on the surface. The energy barrier for the migration process is given by the diffusion barrier U_s. The surface diffusivity, D_s, is given by

$$D_s = a'^2\nu \exp(-U_s/kT) \tag{20.3}$$

with a' the jump distance between surface sites and ν the vibration frequency. The adsorbed atom desorbs after a mean time τ_0. The desorption energy barrier E_{dSi}, is added to W_s to give the binding energy W

$$W = W_s + E_{dSi} . \tag{20.4}$$

The mean time τ_0 for desorption is given by

$$1/\tau_0 = \nu'' \exp(-E_{dSi}/kT) \tag{20.5}$$

where the pre-exponential factor ν'' is a vibration frequency similar to the one in (20.3). For the following considerations we make the simplifying assumption that $\nu = \nu''$. The diffusion length λ_s which gives the mean migration distance before desorption is given by

$$\lambda_s = (D_s\tau_0)^{1/2} = a' \exp\left[\frac{E_{dSi} - U_s}{2kT}\right] . \tag{20.6}$$

Note the increasing diffusion length, λ_s, with decreasing temperature because generally $U_s < E_{dSi}$ [20.15]. The desorbing flux, which equals F_0 at equilibrium, is given by

$$F_0 = n_{sequ}/\tau_0 = N_{os}\nu'' \exp(-W/kT) . \tag{20.7}$$

From vapour pressure data [20.16] we calculate the energy $W = 4.55$ eV and for pre-exponential term $N_{os}\nu'' = 2.5 \cdot 10^{34}\,\mathrm{m^{-2}s^{-1}}$ using (20.1) and (20.7).

Etching or growth of the solid proceeds from unsaturated or supersaturated vapors, respectively. A measure of the supersaturation, σ, under MBE conditions is the ratio

$$\sigma = (F_{Si}/F_0) - 1 \tag{20.8}$$

where F_{Si} is the actual Si flux density, and F_0 the equilibrium flux density calculated from (20.1).

Table 20.3 gives some numerical values of the supersaturation for typical MBE growth rates [20.15]. Often, crystal growers prefer low supersaturation for growth of high quality material. As Table 20.3 shows, Si-MBE experiments are always performed under extremely high supersaturation, maybe the highest ever reported for high quality growth of single crystals.

Table 20.3. The supersaturation σ (20.8) as a function [20.15] of the growth temperature T for a growth rate of 0.4 nm/s obtained with a Si flux density $F_{si} = 2 \cdot 20^{19}$ $m^{-2}s^{-1}$

T [K]	732	823	923	1023	1123	1223
σ	$3 \cdot 10^{16}$	$4.5 \cdot 10^{12}$	$4.7 \cdot 10^{9}$	$1.8 \cdot 10^{7}$	$2 \cdot 10^{5}$	$4.4 \cdot 10^{3}$

20.1.1 Vertical Growth by Lateral Motion of Surface Steps

A real surface contains surface steps at thermal equilibrium between vapour and solid. The origins of these surface steps at equilibrium are

(i)   dislocations,
(ii) misorientation of the surface relative to the low index plane,
and under supersaturation the mechanism of
(iii) two-dimensional nucleation which generates additional surface steps.

Introduction of a dislocation into a solid creates a surface step. The step runs along the intersection of the slip plane with the surface. Often, as an example, a screw dislocation penetrating the surface is sketched [20.14]. During growth the steps originating from dislocations wind up in spirals creating growth pyramids (Fig. 20.5) as was elegantly demonstrated by liquid phase epitaxy (LPE) experiments [20.17]. We believe that steps from dislocations play a minor role in MBE because substrates are used with low dislocation densities.

Commercially available substrates are misoriented relative to the nominal low index plane by less than a degree. But this, nevertheless, would imply low index terraces separated by misorientation steps. In an idealized situation (Fig. 20.6), a periodic array of straight steps is given with step separation L_0

$$L_0 = h/\sin i \tag{20.9}$$

where i is the angle of misorientation and h the step height.

Even from small misorientations, a high density of surface steps results, e.g., $1/l_0 = 10^7 \, m^{-1}$ for $i = 0.25°$, $h = 0.384$ nm.

Adatoms migrating on the surface may join to form a two-dimensional nucleus. Under supersaturation, a nucleus above the critical size is stable and can grow further, but a nucleus below the critical size is unstable. The critical size is a function of supersaturation. The higher the supersaturation the lower is the critical size. Steps from nuclei annihilate each other after growth of a monolayer. Then a nucleation process restarts. In general, two-dimensional nucleation should be of

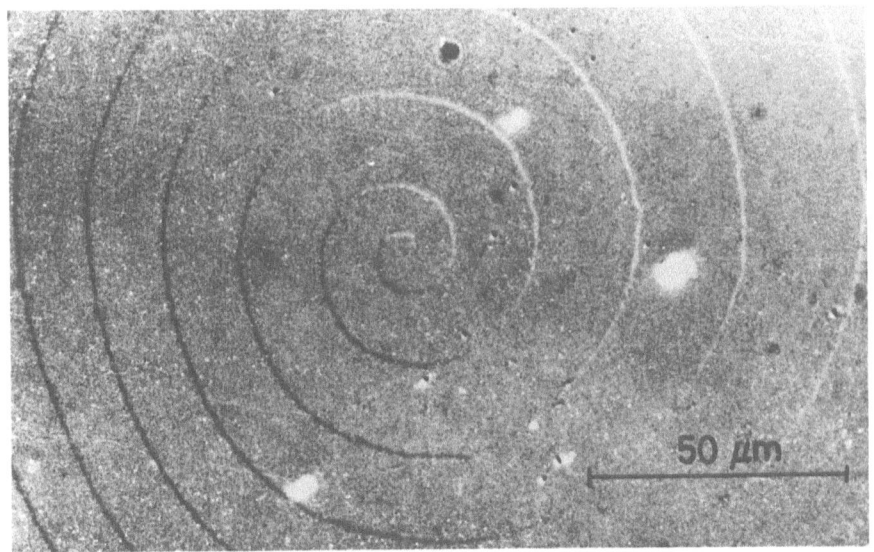

Fig.20.5 Optical micrograph [20.17] (Normarski differential interference contrast) of a growth spiral with atomic steps orginating from a dislocation intersecting the surface

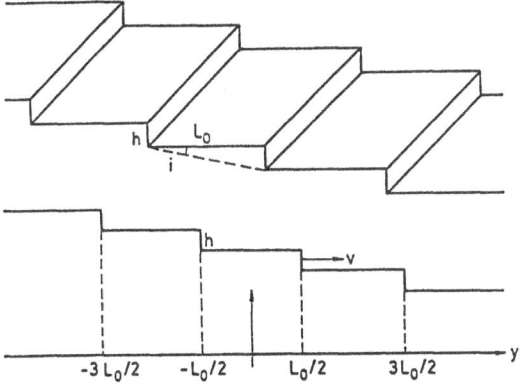

Fig.20.6 (a) Scheme of a misoriented surface (angle i) with low–index planes separated by misorientation steps of height h (step separation distance L_0)

minor importance on stepped surfaces (vicinal surfaces) whereas it must be the dominant source of steps on well prepared surfaces with very small misorientations.

The lateral motion of microscopic surface steps causes a macroscopic vertical growth rate, R, which is step flux times step height. Trains of steps progress along the surface. The step flux at a fixed position is given by $(N_{st} v)$ independently of the origin of the steps. The surface step density, N_{st}, is defined as the length of steps per unit area, with h the step height and v the velocity:

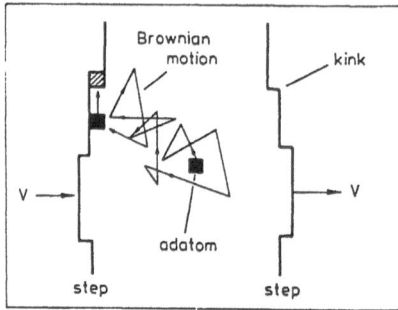

Fig.20.7 Lateral motion of steps caused by the capture of adatoms resulting from surface diffusion due to Brownian motion

$$R \equiv (N_{st} v) h \ . \tag{20.10}$$

The capture of adatoms by steps causes lateral step motion (Fig. 20.7). A clear treatment of the capture of adatoms may be given within the framework of adatom diffusion theory as described below.

20.1.2 Burton-Cabrera-Frank (BCF) Theory

Burton, *Cabrera*, and *Frank* considered the diffusion of adatoms towards steps driven by the supersaturation, σ. The step itself was considered as a perfect sink of adatoms, that means the adatom density n_s drops at the step to the equilibrium value n_{sequ}. The net flux density, j_v, of silicon atoms going from vapour to the surface is given by the difference between the incident Si flux, F_{Si}, and the desorbing flux, n_s/τ_0,

$$j_v = F_{Si} - n_s/\tau_0 \ . \tag{20.11}$$

The current density, j_s, of Si atoms on the surface is given by the diffusion equation

$$j_s = -D_s \ \text{grad} \ n_s \ . \tag{20.12}$$

For stationary solutions the continuity equation holds

$$\text{div} \ j_s = j_v \ . \tag{20.13}$$

Combining (20.6,8,11-13) yields

$$\lambda_s^2 \Delta n_s + n_s = n_{sequ}(\sigma+1) \tag{20.14}$$

with the boundary condition $n_s = n_{sequ}$ at step positions.

Burton, *Cabrera*, and *Frank* solved (20.14) for some step configurations. They neglected in their solution the movement of the step boundary, which is not allowed at high supersaturations. For a straight step at $y = 0$ they found

$$n_s = n_{sequ}\{1 + \sigma [1 - \exp(-y/\lambda_s)]\} \tag{20.15}$$

566

with distance y > 0, normal to the step. For a parallel set of straight steps they found

$$n_s/n_{sequ} = 1 + \sigma \left[1 - \frac{\cosh (y/\lambda_s)}{\cosh (L_0/2\lambda_s)} \right] \qquad (20.16)$$

with the origin of the coordinate system halfway between two steps as given in Fig. 20.6 ($L_0/2 > y > -L_0/2$).

Under the influence of supersaturated vapour, the adatom density, n_s, increases above the equilibrium value n_{sequ}. In an analogous manner to that for the vapour supersaturation we can define a local surface adatom supersaturation, σ_s,

$$\sigma_s = (n_s/n_{sequ}) - 1 \qquad (20.17)$$

The surface supersaturation, σ_s, is always lower than the vapour supersaturation, σ. The smaller the step distance L_0 the lower is σ_s at constant supersaturation σ.

From known solutions of $n_s(y)$, all other interesting properties such as desorption, velocity, v, of the steps, condensation coefficient, η, and growth rate, R, may be determined as was shown for special step configurations [20.10,14,15].

Mullins [20.18] proved that the mathematical solution of the BCF theory becomes inaccurate at extremely high supersaturations. For a single step, the accurate solution for the velocity, v, has a singularity when a parameter, b, equals unity

$$b = \sigma \, n_{sequ}/N_{os} \qquad (20.18)$$

Recently, *Voigtländer* [20.15] and *Fuenzalida* [20.19] extended the solutions to a parallel sequence of steps using a movable coordinate system. For b > 0.5 large deviations from the classical BCF theory can take place for the adatom distribution, the step velocity and the condensation coefficient. The asymmetry of the adatom distribution is caused by the unidirectional movement of the step train.

20.1.3 LEED/RHEED Oscillations

A milestone in understanding MBE growth was reached with the discovery of LEED/RHEED oscillations from growing surfaces by the group of *Henzler* [20.20] (LEED) and *Joyce* [20.21] (RHEED). The intensity of electron diffraction (ED) spots from the growing surface oscillates both in low energy (LEED) and reflection high energy (RHEED) experiments. The period of oscillations usually coincides with the growth of a monolayer. It is believed that the oscillations are a proof of (i) the two-dimensional growth mode and, (ii) the occurrence of two-dimensional nucleation processes (Fig.20.8). Neither three-dimensional (3D) growth mode with rough surfaces nor two-

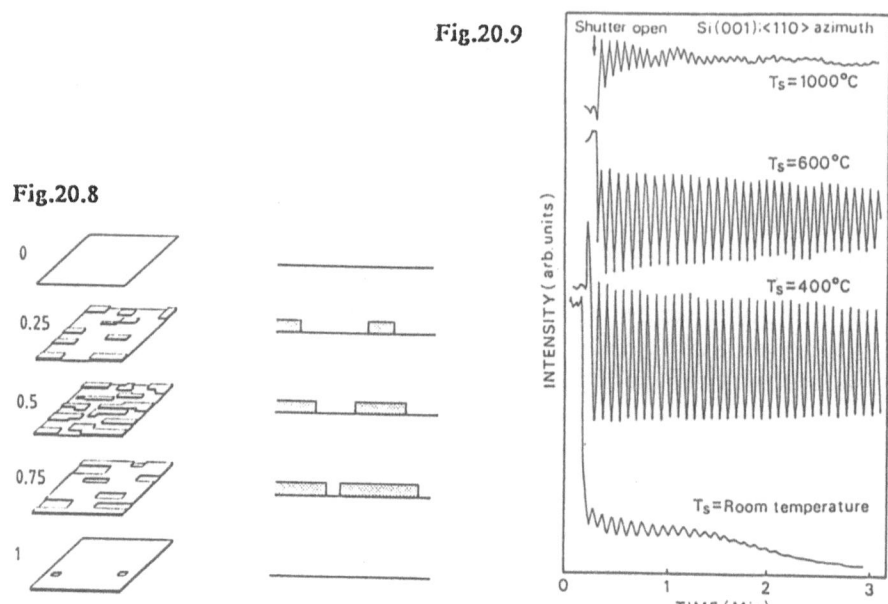

Fig.20.9

Fig.20.8

Fig.20.8 Two-dimensional growth on a low index plane without misorientation steps. Growth proceeds by nucleation and lateral expansion of the nucleus via capture of adatoms at the steps. After completion of a monolayer, the process is repeated periodically leading to oscillations in the LEED and RHEED intensities [20.21]

Fig.20.9 Dependence of the amplitude of RHEED oscillations on growth temperature T_s. Within the epitaxial growth range ($T_s > T_{epi}$) the amplitude increases with decreasing temperature [20.22]

dimensional (2D) growth mode via dislocation steps or misorientation steps would generate such intensity oscillations. Note that 3D growth would create a rough surface, 2D growth via misorientation steps would create the smoothest surface available at a given misorientation. It is not easy to observe electron diffraction intensities with high accuracy in a Si-MBE apparatus because of electrons from the electron gun evaporator. *Henzler* et al. used a special growth apparatus without an electron gun evaporator. Also they polished the substrate surfaces for accurate orientation whereas commercially available substrates are slightly misoriented ($i \approx 5 \cdot 10^{-3}$). Recently, *Sakamoto* [20.22] observed clear RHEED oscillations in a Si-MBE apparatus. Two striking features of such oscillations (Fig.20.9) are:

(i) the oscillations fade out with time, i.e. with layer thickness;
(ii) the strength of the oscillations is strongly dependent on growth temperatures.

These features are often interpreted in terms of a competing 3D growth mode which roughens the surface with time. 2D growth via misorientation steps could also be used to interpret these features, at least partially. Remember, one expects a basic misorientation step density of 10 nm on commercial substrates. With decreasing temperature the supersaturation is increasing and the diffusivity of adatoms is decreasing which favours two-dimensional nucleation at lower temperatures. It should be mentioned that Sakamoto specially treated the surface (1000°C anneal of very well oriented surfaces) to obtain rather large facets without misorientation steps. At low temperatures (500°C) the periodic process of nucleation dominates leading to strong RHEED oscillations. At higher temperatures, the faster diffusion of surface adatoms favours the continuous flow of misorientation steps leading to weaker RHEED oscillations. As a rough rule for many material systems, *Ishizaka* [20.23] claimed that growth occurs via two dimensional nucleation at temperatures $T_m/4 < T < T_m/2$ ($T_m \equiv$ melting point) and via misorientation step flow at $T_m/2 < T < 3T_m/4$. Considering the foregoing sections, we can add that the exact transition temperature ($T_m/2$ in Ishizaka's rule) should be influenced by the particular surface properties, such as low index plane, misorientation, step configuration.

20.2 Impurity Incorporation into Si–MBE Layers

Though process temperatures of Si-MBE are far below temperatures usually needed for activation of solid state diffusion, fabrication of arbitrarily fine doping structures is not a simple matter. Initial experiments using coevaporated antimony for n-type doping have shown pronounced profile smearing [20.24]. In addition, low sticking coefficients, i.e., ratios of dopant in the film to dopant evaporated onto the surface have been established [20.25]. Except for boron [20.26], such or a similar behaviour was found for many of the common dopants such as As [20.27], Sb [20.24-29], Al [20.30], Ga [20.31], and In [20.32]. Fig.20.10 illustrates schematically this behaviour. A dopant surface

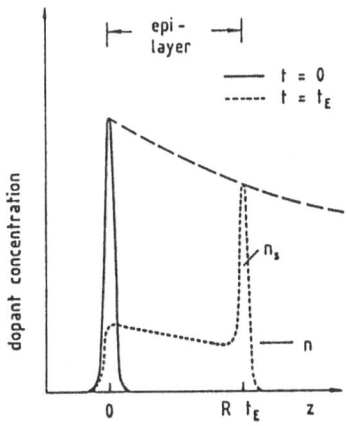

Fig.20.10 Incorporation behaviour of common dopants in Si-MBE (schematically). A dopant surface layer (< one monolayer) deposited onto the substrate, remains at the surface, leaving behind only a weakly doped MBE film ($t_E \equiv$ epitaxy duration, $R \equiv$ growth rate)

layer, deposited prior to MBE growth, remains almost totally in front of the growing surface, leaving behind only a weakly doped MBE film.

A first model to describe this unexpected impurity incorporation behaviour was suggested by *Iyer* et al. [20.31]. They assumed that a dopant atom entrapped on the Si surface, either desorbs or becomes incorporated after a mean residence time. Thus, an accumulated dopant adlayer, n_s, is present at the surface which is assumed to obey a first order kinetic equation

$$\frac{dn_s}{dt} = j - n_s/\tau_D - n_s/\tau_I . \qquad (20.19)$$

τ_D and τ_I are the time constants for desorption and incorporation, respectively, j is the incident dopant flux. A similar first order kinetic equation was suggested also by *M. Tabe* et al. [20.28]. Since desorption and incorporation are considered to be thermally activated processes, τ_D and τ_I are written as

$$\tau_D = \tau_{D,0} \exp(E_D/kT) \qquad (20.20a)$$

$$\tau_I = \tau_{I,0} \exp(E_I/kT) . \qquad (20.20b)$$

Table 20.4 includes values of τ_D and τ_I for the system Sb on Si {111} for various growth temperatures [20.29]. Both, τ_D and τ_I, are considerably larger than the time typically needed for one monolayer MBE growth ($t_m \simeq 1$ s). Particularly at higher temperatures, τ_I is significantly larger than τ_D and, accordingly, the sticking coefficient, S, given by the ratio $S = \tau_D/\tau_I$, is very low at these temperatures. The flux density of dopant atoms into the growing MBE film corresponds to the term n_s/τ_I of (20.19) and so the bulk concentration n immediately below the surface is given by

$$n R = n_s/\tau_I \qquad (20.21)$$

where R is the growth rate. *Tabe* et al., who assume an equilibrium reaction between Sb (surface) and Sb (bulk) propose a similar relation at low coverages as (20.21) [20.28]. The difference to (20.21) is that n

Table 20.4. Time constants for desorption (τ_D) and incorporation (τ_I), sticking coefficients S, and profile broadening Δ, for Sb on Si {111} at various growth temperatures T_s (growth rate $R \simeq 1$ Å/s), data are from [20.29]

T_s [C]	τ_D [s]	τ_I [s]	S	Δ [cm]
600	$1.0 \cdot 10^5$	$3.1 \cdot 10^6$	$3.2 \cdot 10^{-2}$	$1.0 \cdot 10^{-3}$
700	$3.5 \cdot 10^3$	$1.8 \cdot 10^6$	$1.9 \cdot 10^{-3}$	$3.5 \cdot 10^{-5}$
800	$2.4 \cdot 10^2$	$1.1 \cdot 10^6$	$2.2 \cdot 10^{-4}$	$2.4 \cdot 10^{-6}$
900	$2.5 \cdot 10$	$7.8 \cdot 10^5$	$3.2 \cdot 10^{-5}$	$2.5 \cdot 10^{-7}$

tends to infinity when n_s approaches the total number of adsorption sites. The set of equations (20.19-21) enables simulation of doping profiles for different doping procedures (coevaporation, pre-deposition and flash-off by use of the appropriate initial conditions. With coevaporation the profile smearing or broadening, Δ, is found to be

$$\Delta = R/(\tau_D^{-1} + \tau_I^{-1}) \ . \tag{20.22}$$

In Table 20.4 values of Δ are shown for the system Si{111}: Sb for some temperatures. The term Δ strongly decreases with increasing temperature. Unfortunately, the sticking coefficient simultaneously decreases drastically. By Sb adlayer pre-adjusting, however, while temporarily arresting silicon growth, arbitrarily sharp profiles can be grown even at lower temperatures. Since within the framework of the model, τ_I is still large, only low doping levels can be achieved in both cases.

The adlayer model discussed so far describes dopant accumulation in a purely phenomenological way, assuming thermally activated transition processes into vapour and bulk (20.20). *Barnett* and *Greene* discussed surface segregation as the underlying mechanism for this accumulation phenomenon [20.33]. Driven by a free energy of segregation, ΔG^s dopant atoms localized in the bulk, segregate to the growing surface at a drift velocity given by [20.34]

$$\nu = \frac{D}{kT} \frac{d\Delta G^s}{dz} \tag{20.23}$$

where D is the dopant diffusivity which is assumed to be markedly enhanced in the near surface region [20.35]. At sufficiently high growth temperatures, the drift velocity exceeds the growth rate. In this temperature range the equilibrium surface concentration, n_s, is related to the bulk concentration, n, by [20.34]

$$n_s/n \simeq \exp{(-\Delta G_0^s/kT)} \ . \tag{20.24}$$

In this equilibrium regime this so-called segregation ratio $r = n_s/n$ [20.33] (which is normalized to an atomic length unit in (20.24)) corresponds to the time constant τ_I for the incorporation used in the adlayer-model (20.21):

$$r = R\tau_I \ . \tag{20.25}$$

By comparing (20.20b) with (20.24), the free energy of segregation ΔG_0^s can be identified with the activation energy for incorporation (except for the sign). When the process temperature is lowered, however, the drift velocity (20.23) strongly decreases and, finally, subsides below the growth rate. Thus the segregation model predicts the existence of a transition from equilibrium to kinetically-limited segregation. In early experiments some evidence of this transition was found for Sb on Si {111} at temperatures below 650°C where the sticking coefficient is significantly higher than the values predicted by the adlayer-model (Fig.20.11). In recent experiments, where dopant incorporation was studied in a relatively low temperature regime, the transition was clearly observed for Sb on Si {100} (Fig.20.12 [20.36]). These experiments revealed a well-defined transition temperature, $T^* = 560°C$ at $R = 3$ Å/s growth rate. In addition, the transition temperature was found to shift to lower values when the growth rate is lowered [20.36].

In a recent paper, one of the authors suggested a microscopic model of impurity incorporation which assumes that surface segregation observed in MBE consists of an exchange process solely between two states [20.36]. Model studies on group V {111} surfaces have shown almost doubly occupied (unoccupied) dangling hybrids of these impurities, forming surface states below (above) the Fermi energy [20.37]. When substitutionally incorporated, however, these impurities form

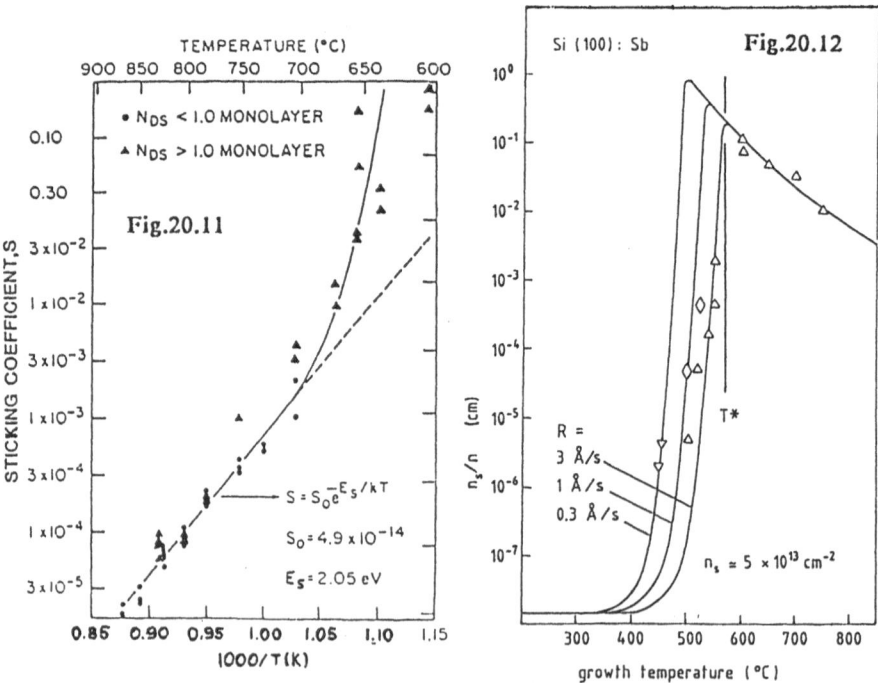

Fig.20.11 Sticking coefficient of Sb on Si vs growth temperature. At $T \leq 650°C$ the sticking coefficient significantly exceeds the adlayer-model prediction (data are from [20.29])

Fig.20.12 Segregation coefficient, $r = n_s/n$ versus growth temperature in the vicinity of the transition temperature, T^* (data are from [20.36])

572

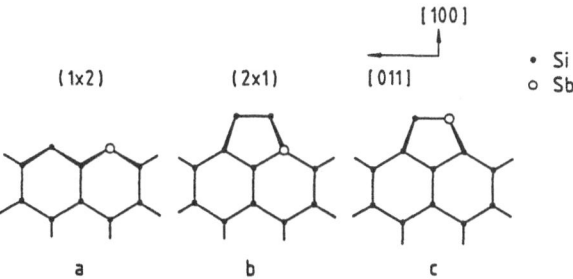

Fig.20.13 Bond geometries of Sb on Si {100} which may be related to the surface (a and c) and subsurface state (b)

states above (below) the Fermi energy. Thus, migration onto the surface is energetically favoured [20.37]. Figure 20.13 shows bond geometries of a donor atom at the Si{100} surface during MBE growth. Initially, the atom is assumed to be in the threefold coordinated site at the (1x2) reconstructed surface. Immediately after monolayer deposition, the atom is in a fourfold coordinated subsurface site. By electronic driving forces the surface state may be energetically favoured and so Sb migrates to the (2x1) reconstructed surface, again into a threefold coordinated surface site. A schematic potential energy diagram, as shown Fig.20.14, can be used to describe this exchange process in a general way. By iteratively solving a coupled set of rate equations for the impurity densities in the surface and subsurface states, the following approximate equations for the surface and bulk concentrations n and n_s, respectively, are obtained [20.36]

$$n_s = nRt_m \left[\exp(-E_I/kT) + \exp(\lambda t_m) \right]^{-1} \tag{20.26}$$

$$\frac{dn_s}{dt} = -n_s \nu \exp(-E_D/kT) - n_s \left[\frac{\exp(\lambda t_m)}{t_m} \right] \tag{20.27}$$

where $\lambda = -\nu \exp(-E_A/kT)$, t_m is the time needed for one monolayer MBE growth. Equations (20.26) and (20.27) approach the adlayer model equations (20.19,21) for $-\lambda t_m \gg 1$, i.e., for sufficiently high temperatures. The segregation coefficient, r, and the profile broadening, Δ, follow from (20.26) and (20.27)

$$r = \frac{a_0/4}{\exp(-E_I/kT_a) + \exp(\lambda t_m)} \tag{20.28}$$

$$\Delta = \frac{a_0/4}{t_m \nu \exp(-E_D/kT) + \exp(\lambda t_m)} \tag{20.29}$$

where a_0 is the lattice constant of Si. Choosing the pre-exponential factor ν to be $\nu = 2 \cdot 10^{12} s^{-1}$ [20.36] and the potential energy parameters to be $E_I = 1.2$ eV, and $E_A = 1.78$ eV (Fig.20.14), the segregation

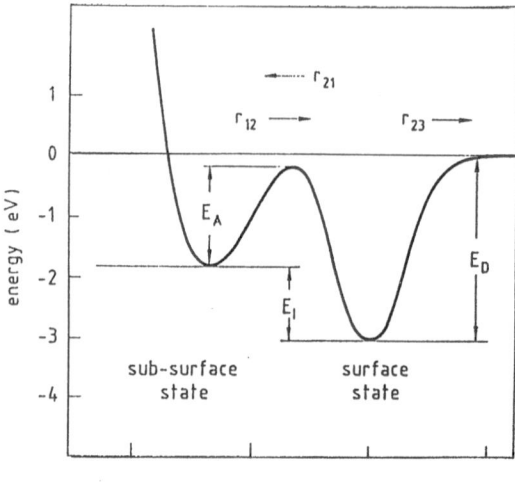

reaction coordinate

Fig.20.14 Potential energy diagram to describe exchange processes between surface and subsurface states. Desorption is also accounted for

coefficient found experimentally can be matched in the whole temperature range, as shown in Fig.20.12 (full line). For $T > T^*$, the segregation coefficient slowly decreases due to an increasing occupation of the higher energy subsurface state by Boltzmann statistics ($n_s/n = (a_0/4)$ exp (E_I/kT)). The profile broadening, Δ, on the other hand, rapidly decreases due to the onset of additional desorption processes. At $T \simeq T^*$, r and Δ become identical, showing a strong decrease for $T < T^*$. In this low temperature range, equilibrium occupation cannot be reached between surface and subsurface states. This increasingly favours the occupation of subsurface states with decreasing temperature (Figs. 20.13, 14).

20.2.1 Secondary Implantation

The extremely high surface segregation, particularly of antimony on silicon, induces high surface coverages at simultaneously low bulk concentrations. For instance, an adlayer density of $n_s = 1 \cdot 10^{14}$ Sb atoms/cm^2 yields a bulk concentration of below 1×10^{16} cm^{-3} at $T_s = 650°C$ growth temperature (Fig.20.12). This offers new feasibilities to study secondary implantation effects, which are similar to well established processes of recoil implantation and collisional mixing ([20.38] and references therein), at growing surfaces. Exciting features of this novel application of these techniques are the use of submonolayer dopant coverages and, accordingly, the use of very low energy projectiles. Thus, multiple collisions between dopant atoms, as in ordinary recoil implantation experiments [20.38], usually need to be considered here - dopant incorporation results from a single collision between the primary projectile and the dopant adatom.

Secondary implantation of Sb into Si MBE layers has been demonstrated in a recent experiment [20.39]. With this experiment, low energy Si^+ ions (200 eV), which were generated by partially ionizing the Si beam, were used as primary projectiles. The energy of the Si^+ ions was adjusted by applying a proper voltage between substrate and grounded Si source. After growing a buffer layer, Sb was evaporated onto the growing surface. The MBE process was performed at $T_s = 650°C$ where extreme surface segregation (Fig.20.12) restricts the corresponding bulk concentration to $n \simeq 10^{16}$ cm^{-3} (Fig.20.15). When the growing film was irradiated by low energy Si^+ ions, however, the bulk concentration increased by three orders of magnitude up to 10^{19} cm^{-3}. Subsequently, the adlayer was depleted by incorporation, causing an exponential doping decay. The amount of Sb incorporated is almost identical to the amount deposited initially. This and similar experiments [20.40, 41] have shown that

(i) sputtering of Sb atoms is neglibible at least up to 1500 eV Si^+ ion energy.

(ii) Sb atoms are completely substitutionally incorporated.

(iii) Electron mobilities are comparable to bulk mobilities even at high doping levels.

To determine recoil yields, thin MBE films (about 0.1 μm) were grown after pre-deposition of a submonolayer of Sb onto the substrate surface. Results obtained from these experiments are summarized in Fig.20.16. The doping level increases proportionally both to the adlayer density n_s and to the Si^+ ion flux density, j_I, according to

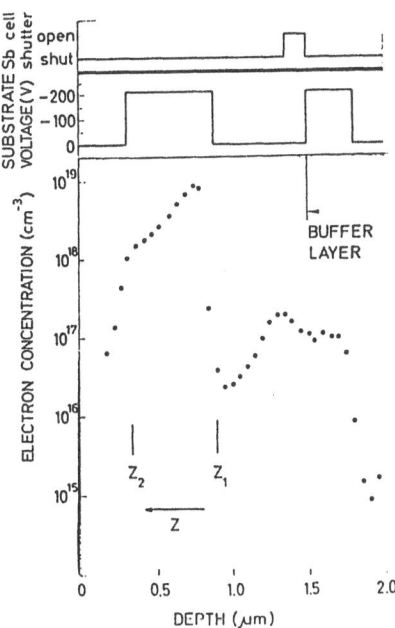

Fig.20.15 Carrier concentration vs depths evaluated by speading resistance analysis. The program of the Sb cell shutter and the applied substrate voltage is shown in the upper part (Doping by secondary implantation - DSI)

575

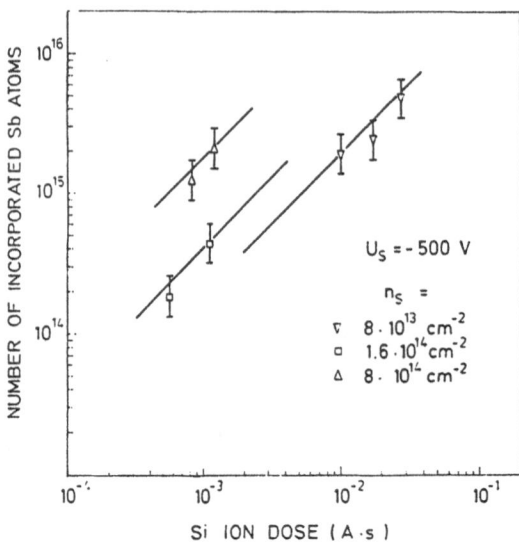

Fig.20.16 Number of incorporated Sb atoms (integrated over the whole wafer area) vs Si⁺ ion dose at different predeposited Sb adatom densities

$$n = \sigma_I j_I n_s / R \qquad\qquad (20.30)$$

where R is the growth rate and σ_I the cross section for incorporation. The cross section σ_I is estimated from these experiments to be

$$\sigma_I = 5 \cdot 10^{-16} \ cm^2 \ .$$

This value approximately agrees with the size of an Sb atom $(\pi\rho^2 = 5.81 \times 10^{-16} \, cm^2)$, using a covalent radius of $\rho = 1.36$ Å [20.42]. Consequently, the recoil yield, i.e., the number of Sb atoms incorporated per Si⁺ ion, amounts to about 0.4 at a monolayer Sb coverage.

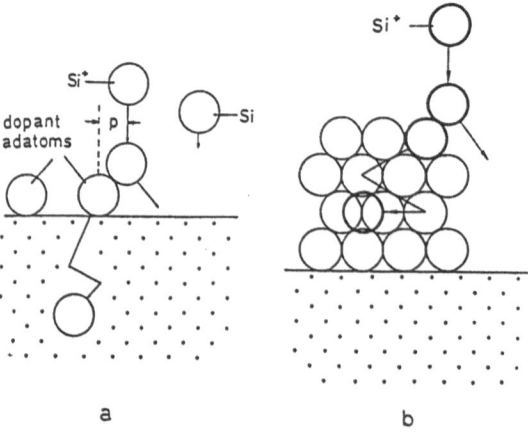

Fig.20.17 Incorporation of dopant adatoms induced by interaction with incident ions. Recoil yields are reduced in the case of cluster formation

Similar experiments to those discussed so far, using also other dopants than Sb, were described by *Kubiak* et al. [20.43,44]. They found a response to an applied substrate potential also for As. P-type doping using Ga, however, showed negligible response to the substrate potential. Though the situation concerning Ga is not yet clear, clustering is likely the reason for this different behaviour (Fig.20.17). Island formation of Ga on Si{111} has also been suggested to explain Auger electron spectrometry measurements [20.31].

20.3 Limits of the Two-Dimensional (2D) Growth Mode

Under ideal conditions, as described in the foregoing sections, the two-dimensional growth mode governs Si-MBE: The lateral movement of steps is driven by diffusion of adatoms and their capture by steps. For commercially available substrates with unintentional misorientation of a few tenths of degree, the steps originate from misorientation (dominant at higher temperatures) and from 2D nucleation (dominant at lower temperatures). Under real conditions the 2D-growth mode can be disturbed by contamination of the surface or by heteroepitaxy (growth of a film different in chemical composition from the substrate). The classification scheme of growth modes is given in Fig.20.18. The extremes are: 2D-growth mode (Frank/V.d.Merwe) and three-dimensional (3D) growth mode (Volmer/Weber). In between is a growth mode which starts with 2D and then switches to 3D (Stranski/Krastanov).

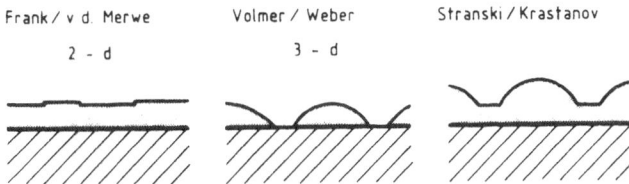

Fig.20.18 Scheme of growth modes

20.3.1 Surface Contamination

The UHV environment of the MBE process is clean enough to avoid disturbance of the 2D-growth mode. But, often, the substrate does not meet the severe requirements of contamination free surfaces. This was proven by laser light scattering experiments [20.45] showing a peak typical for 3D growth of the inital layer. Qualitatively, the breakdown of 2D growth via lateral motion of steps can be understood if one assumes a partial pinning of growth steps by surface contaminants [20.46,47].

20.3.2 Heteroepitaxy

Heteroepitaxy means growth of a film of different material B on a substrate A. The driving forces for changing the 2D-growth mode are the chemistry of the system A/B (differences in surface and interface energies) and the strain energy connected with lattice mismatch between A and B. A qualitative insight into the influence of chemistry is given by the well-known droplet model, which predicts 3D growth if the specific surface free energy of the film, ϕ_B, is higher than the difference of the substrate's specific surface free energy, σ_A, and specific interfacial free energy, σ_i.

Lattice mismatch favours 3D growth, because strain can be relaxed better by an island than by a uniform film. The equilibrium shape of the islands was calculated by *Stoop* [20.48] neglecting chemistry ($\sigma_A = \sigma_B$). As a general result, the 2D-growth switches at a certain mismatch η_0 to 3D growth with cap shaped islands inclinded to the surface by an angle ϕ. The mismatch η_0 at the switching point depends on the size of the nucleus (Fig.20.19). The combined influence of chemistry and mismatch is shown in a simplified model [20.49], summing Lennard-Jones type atom interactions.

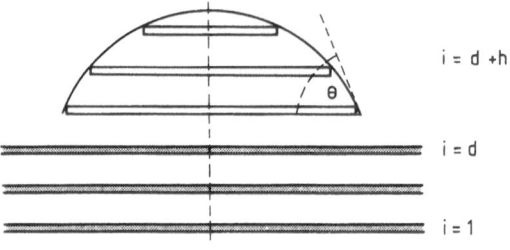

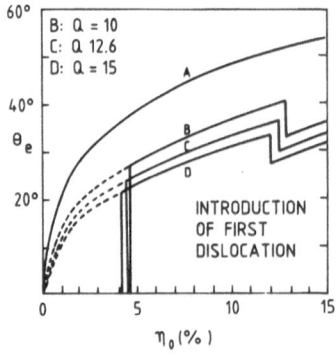

Fig.20.19 3D-growth mode for lattice mismatched heteroepitaxy [20.48]. Angle ϕ as a function of mismatch, η_0, for different nuclei sizes (curves B, C, D). Curve A: Simplified model

20.3.3 The SiGe/Si System

The GaAs/Si and the SiGe/Si systems are the most important hetero-epitaxial couples on Si substrates. The SiGe/Si system can be considered as a model system for studying the pure influence of lattice mismatch (η_0 = 4.2% for Ge/Si) because of complete miscibility and of chemical similarity of Si and Ge. The interface structure is described elsewhere [20.50]. The surface morphology depends strongly on growth temperature [20.51] resulting in rather smooth {100} surfaces at temperatures below 550°C, whereas at 750°C [20.52], 3D growth with large islands dominates for lattice mismatch $\eta_0 > 1$% (Fig.20.20). The reason for this strong temperature dependence is not well understood. Our own explanation suggests that the distance between nuclei and, therefore, the island size, decreases strongly with decreasing temperature. Smaller nuclei will switch to 3D growth at higher mismatch in accordance with predictions of *Stoop* [20.48]. Recently, *Maree* [20.53] investigated the surface morphology of {111} oriented Ge/Si films. He observed the Stranski-Krastanov growth mode for Ge on Si and the Volmer-Weber growth mode for Si on Ge.

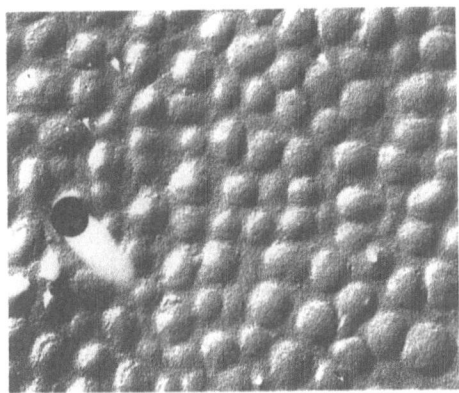

Fig.20.20   SiGe/Si epitaxy. 3D growth at 750°C growth temperature. TEM image of surface replica

20.4 Conclusion

It has been shown that the surface plays a key role for understanding growth mode, dopant incorporation and heteroepitaxy using Si-MBE. From the view point of Si MBE, a better quantitative description of adatom properties, and experimental verification of models and new analytical techniques applicable to in situ process monitoring are most desirable. Examples of research in these three directions are: i) calculations of adatom diffusion [20.54,55], which predict activation energies U_d for adatom motion ranging from 0.16 eV for {100} surfaces to 1.36 eV for {111} surfaces, (ii) the direct observation of surface steps by in situ micro-RHEED [20.56], (iii) optical monitoring of the surface during the MBE process using spectroscopic ellipsometry [20.57] and second harmonic generation [20.58].

Acknowledgement. The valuable help of S. Lindenmaier and H.J. Herzog in conducting electron microsocpy of Si-MBE layers is gratefully acknowledged. Part of the work (doping techniques) was sponsored by the ESPRIT program of the European Community.

References

20.1 E. Kasper, J.C. Bean: *Silicon Molecular Beam Epitaxy* (CRC Press, Boca Raton, USA 1987)

20.2 T. de Jong: Thesis, University of Amsterdam

20.3 T. de Jong, W.A.S. Douma, L. Smit, V.V. Korablev, F.W. Saris: J. Vac. Sci. Technol. B 1, 808 (1983)

20.4 Y. Shiraki, Y. Katayama, K. Kobayashi, K.F. Komatsubara: J. Crystal Growth 45, 287 (1978)

20.5 H.J. Grossmann, L.C. Feldman: Appl. Phys. A 38, 171 (1985)

20.6 K.D. Gronwald, M. Henzler: Surf. Sci. 117, 180 (1982)

20.7 F. Jona: Appl. Phys. Lett. 9, 235 (1966)

20.8 H.-J. Herzog, E. Kasper, P. Eichinger, H. Kibbel: unpublished

20.9 E. Kasper, K. Wörner: J. Electrochem. Soc. 123, 2481 (1985)

20.10 E. Kasper: Appl. Phys. A 28, 129 (1982)

20.11 J. Bloem: J. Crystal Growth 50, 581 (1980)

20.12 R.F.C. Farrow: J. Electrochem. Soc. 121, 899 (1974)

20.13 H.C. Abbink, R.M. Broudy, G.P. McCarthy: J. Appl. Phys. 39, 4673 (1968)

20.14 W.K. Burton, N. Cabrera, F.C. Frank: Philos. Trans. Roy. Soc. (London) 243 A, 299 (1951)

20.15 K. Voigtländer, H. Risken, E. Kasper: Appl. Phys. A 39, 31 (1986)

20.16 R.E. Honig, D.A. Kramer: RCA Rev. 30, 285 (1969)

20.17 D. Kass, M. Warth, W. Appel. H.P. Strunk, E. Bauser: Proc. 1st Int'l Symp. Si-MBE, Toronto, ed. by J.C. Bean, Vol.85-7 (The Electrochem. Soc., Pennington, NJ, USA 1985) p.250

20.18 W.W. Mullins, H.P. Hirth: J. Phys. Chem. Solids 24, 1391 (1963)

20.19 V. Fuenzalida, I. Eisele: J. Crystal Growth 74, 597 (1986)

20.20 M. Henzler: Appl. Surf. Sci. 11, 450 (1982) or Appl. Phys. A 34, 205 (1984)

20.21 B.A. Joyce, H.J. Neave, P.J. Dobson, P.K. Larsen: Phys. Rev. B 29, 814 (1984)

20.22 T. Sakamoto, N.J. Kawai, T. Nakagawa, K. Ohta, T. Kojima, G. Hashiguchi: Coll. Pap. MSS-2, Kyoto (1985) p.282

20.23 A. Ishizaka: Int'l Workshop on Electron Devices - Superlattice Devices, Tokyo (Feb. 1987)

20.24 J.C. Bean: Appl. Phys. Lett. 33, 654f (1978)

20.25 U. König, H. Kibbel, E. Kasper: J. Vac. Sci. Technol. 16, 985 (1979)

20.26 R.A.A. Kubiak, W.Y. Leong, E.H.C. Parker: Appl. Phys. Lett. 44, 878 (1984)

20.27 Y. Ota: J. Electrochem. Soc. 126, 1761 (1979)

20.28 M. Tabe, K. Kajiyama: Japan. J. Appl. Phys. 22, 423 (1983)

20.29 R.A. Metzger, F.G. Allen: J. Appl. Phys. 55, 423 (1984)

20.30 G.E. Becker, J.C. Bean: J. Appl. Phys. 48, 3395 (1977)

20.31 S.S. Iyer, R.A. Metzger, F.G. Allen: J. Appl. Phys. 55, 831 (1981)

20.32 J. Knall, J.-E. Sundgren, S.E. Greene, A. Rockett, S.A. Barnett: Appl. Phys. Lett. 45, 689 (1984)

20.33 S.A. Barnett, J.E. Greene: Surf. Sci. 151, 67 (1985)

20.34 A. Rockett, T.S. Drummond. S.E. Greene, H. Morcoc: J. Appl. Phys. 53, 7085 (1982)

20.35 G. Bajor, J.E. Greene: J. Appl. Phys. 54, 1579 (1983)

20.36 H. Jorke: Surf. Sci. 193, 569 (1988)

20.37 C. Menendez, J.A. Verges: Surf. Sci. 112, 359 (1981)

20.38 J.J. Grob, N. Mesli, A. Grob, P. Siffert: Appl. Phys. A 35, 161 (1984)

20.39 H. Jorke, H.-J. Herzog, H. Kibbel: Appl. Phys. Lett. 47, 511 (1985)

20.40 H. Jorke, H. Kibbel: Proc. 1st Int'l Symp. on Si MBE, ed. by J.C. Bean, Vol.PV85-7, (Electrochem. Soc., Pennington, N.J. 1985)

20.41 H. Jorke, H. Kibbel: J. Electrochem. Soc. 133, 774 (1986)

20.42 Ch. Kittel: *Einführung in die Festkörperphysik* (R. Oldenborg Verlag, München, Wien 1976)

20.43 R.A.A. Kubiak, W.Y. Leong, E.H.C. Parker: Appl. Phys. Lett. 46, 565 (1985)

20.44 R.A.A. Kubiak, W.Y. Leong, E.H.C. Parker: J. Electrochem. Soc. 132, 2738 (1985)

20.45 D.J. Robbins, A.J. Pidduck, A.G. Cullis, N.G. Chew, R.W. Hardeman, D.B. Gasson, C. Pickering, A.C. Daw, M. Johnson, R. Jones: J. Cryst. Growth 81, 421 (1987)

20.46 N. Cabrera, D.A. Vermilya: In *Growth and Perfection of Crystals*, ed. by R.H. Doremus, B.W. Roberts, D. Turnbull (Wiley, New York 1958) p.393

20.47 E. Kasper: Wiss. Ber. AEG-Telefunken 53, 170 (1980)

20.48 L.C.A. Stoop: Thin Solid Films 24, 243 (1974)

20.49 T. Halicioglu: J. Cryst. Growth 29, 40 (1975)

20.50 E. Kasper: Surf. Sci. 174, 630 (1986)

20.51 J.C. Bean, T.T. Sheng, L.C. Feldman, A.T. Fiory, R.T. Lynch: Appl. Phys. Lett. 44, 102 (1984)

20.52 E. Kasper, H.-J. Herzog, H. Kibbel: Appl. Phys. 8, 199 (1975)

20.53 P. Maree: "Silicon Heteroepitxy"; Thesis, University of Utrecht (1987)

20.54 I. Noor Batcha, L.M. Raff, D.C. Thomson: J. Chem. Phys. 81, 3715 (1984)

20.55 E.M. Pearson, T. Halicioglu, W.A. Tiller: J. Cryst. Growth 83, 499 (1987)

20.56 M. Ichikawa, T. Doi, K. Hayakawa: Surf. Sci. 159, 133 (1985)

20.57 J.P. Delrue, S. Andrieu, A. d'Avitaya: private communication

20.58 S. Iyer, T.F. Heinz, M.M.T. Loy: J. Vac. Sci. Technol. B 5, 709 (1987)

Subject Index

Gibbs free energy 374,384
Glacing angle XRD 420
Gottlieb functions 159
Grand canonical free energy 374, 380
Green's function
- matching method 156
- spectral density calculations 155
Growth coefficients, He 493,504

Hausdorff measure 293
Heisenbeg model 327
He scattering 67,68,76,466
- cross-section 75
- diffraction 69,76
- diffuse elastic 75,76
- inelastic 76,99
- inert gases on Ag{111} 174
- Pb{110} 485
- specular 76
- surface phonon dispersion
  curves 166
- time of flight 77,99
Helmholtz free energy 380,384
Henry's law 298
Heteroepitaxy 578
- lattice mismatch 578
- SiGe/Si 579
High energy electron diffraction 420
- oscillations in MBE 567
High resolution He scattering
spectrometer 76,77
HREELS, see electron energy loss spectroscopy

IEELS, see inelastic electron
energy loss spectroscopy
Image potential 350
Incommensurate adatoms 174
Incommensurate disordered phases 321,333
Incommensurate overlayers 539
Incommensurate phases
- hexagonal 84,85
- hexagonal rotated 85

- striped 82,84,85
Inelastic electron energy loss
spectroscopy 201
- acetonitrile 208
- benzene 212
- 2-butene 213
- CO 205
- ethane 208,213
- ethylene 205
- hexane 212
- hydrogen cyanide 205
- methanol 205
- methyl formate 212,213
Infrared reflection absorption
spectroscopy 225
Infrared spectra
- hydrogen on Ge{100} 111,130, 135,136,138
- hydrogen on Ge{111} 144
- hydrogen on Si{100} 110,117, 120,121,130,135-142
- hydrogen on Si{111} 130, 141-143,145
- SiO_2 174
- water on Si{100} 111,140,141
Infrared spectrometer 126,146
Infrared spectroscopy 109,110, 131,135,136,140,141,144,146, 147
Inhomogeneous broadening 139
INS, see ion neutralization
spectroscopy
Integrated circuits 557
Interfacial free energy 459
Interfacial phase transitions 367
Interferometer 126,127,141,146
Internal energy distribution 35
Internal reflection, multiple 110, 121-129,143,146,147
Ion channelling 330
Ion neutralization spectroscopy 341
IRAS, see infrared reflection
absorbtion spectroscopy
Ising model 307,310,311,318,321, 328,329

Contents of **Chemistry and Physics of Solid Surfaces IV**
(Springer Series in Chemical Physics, Vol. 20)

597

Contents of **Chemistry and Physics of Solid Surfaces V**
(Springer Series in Chemical Physics, Vol. 35)

Contents of **Chemistry and Physics of Solid Surfaces VI**

Springer Series in Surface Sciences, Vol. 5)